PERRY'S STANDARD
TABLES AND
FORMULAS FOR
CHEMICAL ENGINEERS

PERRY'S STANDARD TABLES AND FORMULAS FOR CHEMICAL ENGINEERS

James G. Speight

McGRAW-HILL

New York   Chicago   San Francisco   Lisbon   London   Madrid
Mexico City   Milan   New Delhi   San Juan   Seoul
Singapore   Sydney   Toronto

The McGraw·Hill Companies

Library of Congress Cataloging-in-Publication Data
Speight, J. G.
    Perry's standard tables and formulas for chemical engineers / James G.
Speight.
        p.   cm.
    Includes index.
    ISBN 0-07-138777-3
    1. Chemistry—Tables.   I. Title.
    QD65 .S584   2002
    540′.2′1—dc21   2002035990

1 2 3 4 5 6 7 8 9 0 DOC/DOC 0 9 8 7 6 5 4 3

ISBN 0-07-138777-3

*The sponsoring editor for this book was Kenneth P. McCombs and the pro-
duction supervisor was Sherri Souffrance. It was set in Times Roman by SNP
Best-set Typesetter Ltd., Hong Kong.*

Printed and bound by RR Donnelley.

McGraw-Hill books are available at special quantity discounts to use as pre-
miums and sales promotions, or for use in corporate training programs. For
more information, please write to the Director of Special Sales, Professional
Publishing, McGraw-Hill, Two Penn Plaza. New York, NY 10121-2298. Or
contact your local bookstore.

CONTENTS

PREFACE

Industrial chemistry and engineering are exhaustively treated in a whole series of encyclopedic works, but it is not always simple to rapidly grasp information from the various sources. There is, thus, a growing demand for a book such as this *Standard Tables and Formulas* that provides the necessary information in a quick and concise manner for the various chemical properties and chemical processes of industrial operations.

This book of *Standard Tables and Formulas* is single comprehensive and authoritative source book containing tables on formulas that will provide chemical engineers with a quick, ready reference. The tables of data and formulas are designed for easy reference and use for application to chemical processing of raw materials into usable products and also to provide information for chemical process operations and the design of the equipment.

Dr. James G. Speight
Laramie, Wyoming, USA
December 2002

PERRY'S STANDARD TABLES AND FORMULAS FOR CHEMICAL ENGINEERS

ACID–BASE INDICATORS

An acid–base indicator is a conjugate acid–base pair of which the acid form and the base form are of different colors. These indicators are used to show the relative acidity or alkalinity of the test material.

Acid–base indicators are dyes that are themselves weak acids and bases. The conjugate acid–base forms of the dye are of different colors. An indicator does not change color from pure acid to pure alkaline at specific hydrogen ion concentration, but, rather, color change occurs over a range of hydrogen ion concentrations. This range is termed the *color change interval* and is expressed as a pH range. The chemical structures of the dyes are often complex but can be represented chemically by the symbol HIn. The acid–base indicator reaction is represented as:

$$HIn + H_2O \quad H_3O^+ + In \tag{1}$$

ACID–BASE INDICATORS — TABLE 1. Acid–Base Indicators

Indicator	pH range		Color	
	Minimum	Maximum	Acid	Alkaline
Brilliant cresyl blue	0.0	1.0	red-orange	blue
Methyl violet	0.0	1.6	yellow	blue
Crystal violet	0.0	1.8	yellow	blue
Ethyl violet	0.0	2.4	yellow	blue
Methyl Violet 6B	0.1	1.5	yellow	blue
Cresyl red	0.2	1.8	red	yellow
2-(p-Dimethylaminophenylazo) pyridine	0.2	1.8	yellow	blue
Malachite green	0.2	1.8	yellow	blue-green
Methyl green	0.2	1.8	yellow	blue
Cresol red (o-Cresolsulfonephthalein)	1.0	2.0	red	yellow
Quinaldine red	1.0	2.2	colorless	red
p-Methyl red	1.0	3.0	red	yellow
Metanil yellow	1.2	2.3	red	yellow
Pentamethoxy red	1.2	2.3	red-violet	colorless
Metanil yellow	1.2	2.4	red	yellow
p-Phenylazodiphenylamine	1.2	2.6	red	yellow
Thymol blue (Thymolsulfonephthalein)	1.2	2.8	red	yellow
m-Cresol purple	1.2	2.8	red	yellow
p-Xylenol blue	1.2	2.8	red	yellow
Benzopurpurin 4B	1.2	3.8	violet	red
Tropeolin OO	1.3	3.2	red	yellow
Orange IV	1.4	2.8	red	yellow
4-o-Tolylazo-o-toluidine	1.4	2.8	orange	yellow
Methyl violet 6B	1.5	3.2	blue	violet
Phloxine B	2.1	4.1	colorless	pink
Erythrosine, disodium salt	2.2	3.6	orange	red
Benzopupurine 4B	2.2	4.2	violet	red
N,N-dimethyl-p-(m-tolylazo) aniline	2.6	4.8	red	yellow
2,4-Dinitrophenol	2.8	4.0	colorless	yellow
N,N-Dimethyl-p-phenylazoaniline	2.8	4.4	red	yellow

ACID–BASE INDICATORS—TABLE 1. Acid–Base Indicators (*Continued*)

Indicator	pH range		Color	
	Minimum	Maximum	Acid	Alkaline
Methyl yellow	2.9	4.0	red	yellow
Bromophenol blue	3.0	4.6	yellow	blue-violet
Tetrabromophenol blue	3.0	4.6	yellow	blue
Direct purple	3.0	4.6	blue-purple	red
Congo red	3.1	4.9	blue	red
Methyl orange	3.1	4.4	red	yellow
Bromochlorophenol blue	3.2	4.8	yellow	blue
Ethyl orange	3.4	4.8	red	yellow
p-Ethoxychrysoidine	3.5	5.5	red	yellow
Alizarin sodium sulfonate	3.7	5.2	yellow	violet
α-Naphthyl red	3.7	5.7	red	yellow
Bromocresol green	3.8	5.4	yellow	blue
Resazurin	3.8	6.4	orange	violet
Bromophenol green	4.0	5.6	yellow	blue
2,5-Dinitrophenil	4.0	5.8	colorless	yellow
Methyl red	4.2	6.2	red	yellow
2-(p-Dimethylaminophenylazo) pyridine	4.4	5.6	red	yellow
Lacmoid	4.4	6.2	red	blue
Azolitmin	4.5	8.3	red	blue
Litmus	4.5	8.3	red	blue
Alizarin red S	4.6	6.0	yellow	red
Chlorophenol red	4.8	6.4	yellow	red
Cochineal	4.8	6.2	red	violet
Propyl red	4.8	6.6	red	yellow
Hematoxylin	5.0	6.0	red	blue
Bromocresol purple	5.2	6.8	yellow	violet
Bromophenol red	5.2	7.0	yellow	red
Chlorophenol red	5.4	6.8	yellow	red
p-Nitrophenol	5.6	6.6	colorless	yellow
Alizarin	5.6	7.2	yellow	red
Bromothymol blue	6.0	7.6	yellow	blue
Indo-oxine	6.0	8.0	red	blue
Bromophenol blue	6.2	7.6	yellow	blue
m-Dinitrobenzoylene urea	6.4	8.0	colorless	yellow
Phenol red (Phenolsulfonephthalein)	6.4	8.0	yellow	red
Rosolic acid	6.4	8.0	yellow	red
Brilliant yellow	6.6	7.9	yellow	orange
Quinoline blue	6.6	8.6	colorless	blue
Neutral red	6.8	8.0	red	orange
Phenol red	6.8	8.4	yellow	yellow
m-Nitrophenol	6.8	8.6	colorless	yellow
Cresol red (o-Cresolsulfonephthalein)	7.0	8.8	yellow	red
α-Naphtholphthalein	7.3	8.8	yellow	blue
Curcumin	7.4	8.6	yellow	red
m-Cresol purple (m-Cresolsulfonephthalein)	7.4	9.0	yellow	violet
Tropeolin OOO	7.6	8.9	yellow	rose-red

ACID–BASE INDICATORS—TABLE 1. Acid–Base Indicators (*Continued*)

Indicator	pH range		Color	
	Minimum	Maximum	Acid	Alkaline
2,6-Divanillydenecyclohexanone	7.8	9.4	yellow	red
Thymol blue (Thymolsulfonephthalein)	8.0	9.6	yellow	purple
p-Xylenol blue	8.0	9.6	yellow	blue
Turmeric	8.0	10.0	yellow	orange
Phenolphthalein	8.0	10.0	colorless	red
o-Cresolphthalein	8.2	9.8	colorless	red
p-Naphtholphthalein	8.2	10.0	colorless	pink
Ethyl bis(2,4-dimethylphenyl acetate)	8.4	9.6	colorless	blue
Ethyl bis(2,4-dinitrophenyl acetate)	8.4	9.6	colorless	blue
α-Naphtholbenzein	8.5	9.8	yellow	green
Thymolphthalein	9.4	10.6	colorless	blue
Nile blue A	10.0	11.0	blue	purple
Alizarin yellow CG	10.0	12.0	yellow	lilac
Alizarin yellow R	10.2	12.0	yellow	orange red
Salicyl yellow	10.0	12.0	yellow	orange-brown
Diazo violet	10.1	12.0	yellow	violet
Nile blue	10.1	11.1	blue	red
Curcumin	10.2	11.8	yellow	red
Malachite green hydrochloride	10.2	12.5	green-blue	colorless
Methyl blue	10.6	13.4	blue	pale violet
Brilliant cresyl blue	10.8	12.0	blue	yellow
Alizarin	11.0	12.4	red	purple
Nitramine	11.0	13.0	colorless	orange brown
Poirier's blue	11.0	13.0	blue	violet-pink
Tropeolin O	11.0	13.0	yellow	orange
Indigo carmine	11.4	13.0	blue	yellow
Sodium indigosulfonate	11.4	13.0	blue	yellow
Orange G	11.5	14.0	yellow	pink
2,4,6-Trinitrotoluene	11.7	12.8	colorless	orange
1,3,5-Trinitrobenzene	12.0	14.0	colorless	orange
2,4,6-Trinitrobenzoic acid	12.0	13.4	blue	violet-pink
Clayton yellow	12.2	13.2	yellow	amber

ACIDITY (IONIZATION) CONSTANTS

The acidity constant, K, of an acid–base equilibrium $A \leftrightarrow B + H^+$ is the limiting value at infinite dilution of $[B][H^+]/[A]$.

For a strong acid:

$$HA + H_2O \rightarrow H_3O^+ + A^-$$

(2)

For a weak acid:

$$HA + H_2O \leftrightarrow H_3O^+ + A^-$$ (3)

For the weak acid, the acidity constant (K_a) can be written as:

$$K_a = a(H_3O^+)a(A^-)/a(HA \text{ or as}$$ (4)

$$K_a = a(H^+)a(A^-)/a(HA \text{ or as}$$ (5)

$$K_a = [H^+][A^-]/[HA]$$ (6)

pK is $-\log_{10} K_a$ when K_a is measured in moles per kilogram at 25°C (77°F).

ACIDITY (IONIZATION) CONSTANTS — TABLE 1.
Ionization Constants of Hydrocarbons and Heteroatom
Compounds

	Acidity constant	pK_a
Tricyanomethane	1.0 E + 05	−5.0
Dinitromethane	2.5 E − 04	3.6
5,5-Dimethyl-1,3-cyclohexandione	1.6 E − 05	4.8
Malonaldehyde	1.0 E − 05	5.0
Meldrum's acid	8.0 E − 06	5.1
2,4-Pentanedione	2.6 E − 09	8.9
Nitromethane	6.3 E − 11	10.2
Malononitrile	1.0 E − 11	11.0
Ethyl acetoacetate	1.0 E − 11	11.0
Bis(methylsulfonyl)methane	2.0 E − 13	12.7
Diethylmalonate	3.2 E − 14	13.5
Cyclopentadiene	1.0 E − 16	16.0
Phenylacetylene	1.0 E − 19	19.0
Indene	1.0 E − 20	20.0
Tri(phenylthio)methane	1.0 E − 23	23.0
Acetylene	1.0 E − 25	25.0
1,3-Dithiane	1.0 E − 31	31.0
Triphenylmethane	3.2 E − 32	31.5
Diphenylmethane	3.2 E − 34	33.5
Toluene	1.0 E − 40	40.0
Benzene	1.0 E − 43	43.0
Ethylene	1.0 E − 44	44.0
Cyclopropane	1.0 E − 46	46.0
Ethane	1.0 E − 48	48.0

ACIDITY (IONIZATION) CONSTANTS—TABLE 2.

Ionization Constants of Inorganic Acids in Aqueous Solution

Acid	Acidity constant*	pK
Perchloric acid	1.0 E + 10	−10.0
Hydrogen iodide	1.0 E + 07	−7.0
Sulfuric acid	K1: 2.4 E + 06	−6.6
	K2: 1.0 E − 02	2.0
Hydrogen bromide	1.0 E + 06	−6.0
Hydrogen chloride	1.0 E + 03	−3.0
Thiocyanic acid	70.0	−1.9
Nitric acid	30.0	−1.5
Chloric acid	10.0	−1.0
Chromic acid	K1: 3.6 E + 0	−0.6
	K2: 3.4 E − 07	6.5
Iodic acid	0.2	0.8
Pyrophosphorous acid	K1: 3.0 E − 02	1.5
	K2: 4.4 E − 03	2.4
	K3: 2.5 E − 07	6.6
	K4: 5.6 E − 10	9.3
Sulfurous acid	K1: 1.7 E − 02	1.8
	K2: 6.0 E − 03	7.2
Phosphorous acid	K1: 1.6 E − 02	1.8
	K2: 6.3 E − 07	6.2
Hypophosphorous acid	1.0 E − 02	2.0
Phosphoric acid	K1: 7.1 E − 03	2.2
	K2: 6.2 E − 08	7.2
	K3: 4.6 E − 13	12.3
Arsenic (V) acid		2.3
Hydrogen fluoride	6.6 E − 04	3.2
Nitrous acid	5.0 E − 04	3.3
Cyanic acid	3.5 E − 04	3.5
Hydrazoic acid	2.4 E − 05	4.6
Carbonic acid	K1: 4.4 E − 07	6.4
	K2: 4.7 E − 11	10.3
Hydrogen sulfide	K1: 9.0 E − 08	7.0
	K1: 1.0 E − 15	15.0
Hypochlorous acid	3.0 E − 08	7.5
Hypobromous acid	5.8 E − 10	8.7
Boric acid	K1: 7.2 E − 10	9.1
	K2: 1.8 E − 13	12.7
	K3: 1.6 E − 14	13.8
Arsenic (III) acid		9.2
Boric Acid		9.2
Hydrocyanic acid	5.8 E − 10	9.3
Silicic acid		9.9
		11.9
Hypoiodous acid	1.0 E − 10	10.0
Hydrogen peroxide	2.5 E − 12	11.6
Water	1.8 E − 16	15.7
Deuterium oxide	1.8 E − 17	16.7
Ammonia	1.0 E − 34	34.0

* Example: $2.4\ E + 06 \equiv 2.4 \times 10^6$

ACIDITY (IONIZATION) CONSTANTS—TABLE 3.

Ionization Constants of Onium Acids

	Ka	pKa
Aliphatic nitro compounds	1.0 E + 12	−12.0
Aryl nitro compounds	1.0 E + 11	−11.0
Nitriles	1.0 E + 10	−10.0
Aliphatic aldehydes	1.0 E + 08	−8.0
Aryl aldehydes	1.0 E + 07	−7.0
Aliphatic ketones	1.0 E + 07	−7.0
Esters	3.0 E + 06	−6.5
Aryl alkyl ethers	3.0 E + 06	−6.5
Phenols	3.0 E + 06	−6.5
Aryl ketones	1.0 E + 06	−6.0
Carboxylic acids	1.0 E + 06	−6.0
Sulfides	1.0 E + 06	−6.0
Triphenylamine	1.0 E + 05	−5.0
Diethyl ether	3.0 E + 03	−3.5
Dioxan	1.0 E + 03	−3.0
Pyran	3.0 E + 02	−2.5
Tetrahydrofuran	1.0 E + 02	−2.0
Alcohols	1.0 E + 02	−2.0
Aryl amides	1.0 E + 02	−2.0
Indole	1.0 E + 02	−2.0
Water	55.0	−1.7
Aliphatic amides	3.2	−0.5
Dimethyl sulfoxide	1.0	0.0
Pyrrole	1.0	0.0
Urea	0.8	0.1
Diphenylamine	0.2	0.8
Pyridine N-oxide	0.1	1.0
p-Nitroaniline	0.1	1.0
Aniline	2.5 E − 05	4.6
Trimethylamine N-oxide	2.5 E − 05	4.6
N,N-dimethylaniline	1.0 E − 03	5.1
Pyridine	6.3 E − 06	5.2
Hydroxylamine	1.3 E − 06	5.9
2,6-Dimethylpyridine	2.0 E − 07	6.7
Imidazole	1.0 E − 07	7.0
Hydrazine	1.0 E − 08	8.0
Alkyl phosphines	1.0 E − 08	8.0
Aziridine	1.0 E − 08	8.0
2,2,2-Trifluoroethylamine	5.0 E − 09	8.3
Morpholine	5.0 E − 09	8.3
4-Dimethylaminopyridine	2.0 E − 20	9.7
Ethylamine	2.0 E − 11	10.7
Triethylamine	1.8 E − 11	10.8
Diethylamine	1.0 E − 11	11.0
Piperidine	1.0 E − 11	11.0
Pyrrolidine	6.3 E − 12	11.2
Guanidine	2.0 E − 14	13.6

ACIDITY (IONIZATION) CONSTANTS—TABLE 4. Ionization Constants of Organic Acids and Bases in Aqueous Solution

	Acidity constant	pK
Trifluoromethanesulfonic acid	1 E + 13	−13.00
Benzene sulfonic acid	1 E + 03	−2.50
Methanesulfonic acid	3 E + 02	−2.00
Acetamide	2.34 E − 01	−1.10
2-Nitroaniline		−0.28
Trifluoroacetic acid	1.00	0.00
Urea	7.94 E − 01	0.18
Picric acid (2,4,6-Trinitrophenol)	0.50	0.30
N-Acetoaniline		0.61
Trichloroacetic acid	0.23	0.77
Diphenylamine	1.62 E − 01	0.80
4-Nitroaniline		0.98
Oxalic acid	K1: 6.5 E − 02	1.20
Squaric acid	K1: 0.33	1.50
Dichloroacetic acid	5.5 E − 02	1.25
2-Nitrobenzoic acid	6.93 E − 03	2.17
Glycine	4.46 E − 03	2.35
3-Nitroaniline		2.45
Cyanoacetic acid	3.65 E − 03	2.47
2-Chloroaniline		2.56
Fluoroacetic acid	2.5 E − 03	2.60
2-Chloropropionic acid	1.47 E − 03	2.83
2-Chlorobutyric acid	1.39 E − 03	2.84
Malonic acid	1.40 E − 03	2.86
Chloroacetic acid	1.36 E − 03	2.88
Bromoacetic acid	1.38 E − 03	2.90
2-Chlorobenzoic acid	1.20 E − 03	2.94
Phthalic acid	K1: 1.3 E − 03	2.95
o-Hydroxybenzoic (salicylic) acid	K1: 1.06 E − 03	2.99
Tartaric acid	K1	3.03
Citric acid	K1: 7.4 E − 04	3.13
Iodoacetic acid	7.5 E − 04	3.18
4-Nitrobenzoic acid	3.94 E − 04	3.43
3-Nitrobenzoic acid	3.4 E − 04	3.45
3-Chloroaniline		3.46
Squaric acid	K2: 3.3 E − 04	3.50
Methoxyacetic acid		3.53
Acetoacetic acid	2.62 E − 04	3.58
Formic acid	1.77 E − 04	3.75
Hydroxyacetic acid (Glycolic acid)	1.52 E − 04	3.83
3-Chlorobenzoic acid	1.51 E − 04	3.83
Lactic acid	1.39 E − 04	3.86
2-Methylbenzoic acid	1.22 E − 04	3.91
1-Naphthylamine	1.2 E − 04	3.92
4-Chloroaniline		3.93
4-Chlorobenzoic acid	1.04 E − 04	3.99
2,4-Dinitrophenol	1.1 E − 04	4.01
3-Chlorobutyric acid	8.9 E − 05	4.06

ACIDITY (IONIZATION) CONSTANTS—TABLE 4. Ionization Constants of Organic Acids and Bases in Aqueous Solution (*Continued*)

	Acidity constant	pK
3-Hydroxybenzoic acid	8.7 E − 05	4.08
3-Chloroproprionic acid	1.3 E − 04	4.10
2-Naphthylamine	6.92 E − 05	4.11
Ascorbic acid	K1: 6.7 E − 05	4.17
3-Hydroxyaniline		4.17
Succinic acid	K1: 6.63 E − 05	4.20
Oxalic acid	K2: 6.1 E − 05	4.20
Benzoic acid	6.3 E − 05	4.20
3-Methylbenzoic acid	5.32 E − 05	4.24
Vinylic acid		4.26
Benzylic acid		4.31
4-Methylbenzoic acid	4.33 E − 05	4.34
Tartaric acid	K2	4.37
2-Methylaniline	3.63 E − 05	4.38
N,N-dimethylaniline		4.38
N-Methylaniline		4.40
Adipic acid	3.72 E − 05	4.43
4-Chlorobutyric acid	3.0 E − 05	4.52
4-Hydroxybenzoic acid	3.3 E − 05	4.58
Isophthalic acid (m-Phthalic acid)	2.5 E − 05	4.60
Aniline	2.34 E − 05	4.62
3-Methylaniline	1.86 E − 05	4.67
2-Hydroxyaniline		4.72
Acetic acid	1.55 E − 05	4.75
Citric acid	K2: 1.5 E − 05	4.76
3-Aminobenzoic acid	1.67 E − 05	4.78
n-Butyric acid	1.54 E − 05	4.82
Terephthalic acid	1.5 E − 05	4.82
Isobutyric acid	1.50 E − 05	4.85
Pentanoic acid		4.86
n-Propionic acid	1.34 E − 05	4.87
Quinoline	1.25 E − 05	4.91
4-Aminobenzoic acid	1.2 E − 05	4.92
4-Methylaniline	8.32 E − 06	5.00
Trimethylethanoic acid		5.05
o-Phthalic acid	K2: 3.9 E − 06	5.41
4-Hydroxyaniline		5.47
Succinic acid	K2: 2.8 E − 06	5.64
Citric acid	K3: 4.0 E − 07	6.40
Thiophenol	2.5 E − 07	6.60
Tropolone	2.0 E − 07	6.70
2-Aminobenzoic acid	1.07 E − 05	6.97
p-Nitrophenol	5.7 E − 08	7.15
o-Nitrophenol	6.8 E − 08	7.21
2,4,6-Trichlorophenol	1 E − 06	7.60
Nicotine	9.55 E − 09	8.02
Peracetic acid	5.7 E − 09	8.20
Morphine	6.16 E − 09	8.21

ACIDITY (IONIZATION) CONSTANTS—TABLE 4. Ionization Constants of Organic Acids and Bases in Aqueous Solution (*Continued*)

	Acidity constant	pK
Morpholine	4.67 E − 09	8.33
m-Nitrophenol	5.3 E − 09	8.40
2-Chlorophenol		8.48
3-Chlorophenol		9.02
Benzylamine		9.37
4-Chlorophenol		9.38
Succinimide	2.5 E − 10	9.60
Triethylamine	9.77 E − 12	9.80
Resorcinol (1,3-Dihydroxybenzene)	1.55 E − 10	9.81
Piperazine	1.48 E − 10	9.83
Catechol (1,2-Dihydroxybenzene)	1.4 E − 10	9.85
Glycine	1.68 E − 10	9.87
Phenol	1 E − 10	10.00
Chloral hydrate ·	1 E − 10	10.00
m-Cresol	9.8 E − 11	10.09
Benzenesulfonamide	8 E − 11	10.10
p-Cresol	0.7 E − 11	10.26
o-Cresol	6.3 E − 11	10.29
Hydroquinone (1,4-Dihydroxybenzene)	4.5 E − 11	10.35
n-Butylamine	1.69 E − 11	10.61
Methylamine	2.7 E − 11	10.64
Cyclohexylamine	2.19 E − 11	10.64
Triethylamine	9.77 E − 12	10.64
Dimethylamine	1.85 E − 11	10.72
Ethylamine	1.56 E − 11	10.73
n-Butylamine	1.48 E − 11	10.83
Propylamine	1.96 E − 11	10.84
Diethylamine	3.24 E − 11	10.93
Ascorbic acid	K2: 2.5 E − 12	11.60
o-Hydroxybenzoic (salicylic) acid	K2: 3.6 E − 14	2.99
Urea	1.5 E − 14	13.82
Thiourea	1.1 E − 15	14.96
Methyl alcohol		15.50
Ethyl alcohol		16.00

ACTIVATION ENERGIES

The activation energy of a reaction is the amount of energy needed to initiate the reaction and is the minimum energy needed to form an activated complex during a collision between reactants. In slow reactions the fraction of molecules in the system that are moving fast enough to form an activated complex when a collision occurs is low so that most collisions do not produce a reaction. However, in a fast reaction the fraction is high so that most collisions produce a reaction.

For a given reaction the rate constant, K, is related to the temperature of the system by what is known as the Arrhenius equation:

$$K = Ae^{(-Ea/RT)} \tag{7}$$

where R is the ideal gas constant [8.314 J/(mol-°K)]; T is the temperature in degrees Kelvin; Ea is the activation energy in joules per mol; and A is a constant called the frequency factor, which is related to the fraction of collisions between reactants having the proper orientation to form an activated complex.

ACTIVATION ENERGIES—TABLE 1. Activation Energy of Selected Reactions

Reaction	Catalyst	Activation energy
$H_2 + I_2 = 2HI$	None	157
$2HI = H_2 + I_2$	None	183
	Gold	105
	Platinum	58
$H_2 + Cl_2 = 2HCl$	None	25
$2NH_3 = N_2 + 3H_2$	None	330
	Tungsten	163
$2N_2O = 2N_2 + O_2$	None	245
	Gold	121
	Platinum	136
$2NO_2 = 2NO + O_2$	None	112
$2NOCl = 2NO + Cl_2$	None	99
$2NOBr = 2NO + Br_2$	None	58
$2H_2O_2 = 2H_2O + O_2$	None	75
	Platinum	49
	Enzyme	23
$CH_3CHO = CH_4 + CO$	None	190
	Iodine	136
$C_2H_5OC_2H_5$	None	224
	Iodine	143

AMINO ACIDS

An amino acid is an organic compound containing an amine group ($—NH_2$) and a carboxylic acid group ($—CO_2H$) in the same molecule. Although there are many forms of amino acids, all the important amino acids found in living organisms are alpha-amino acids. The carboxylic acid group and the amino group of alpha-amino acids are attached to the same carbon atom.

The simplest amino acid is glycine (H_2NCH_2COOH), which contains no asymmetric carbon atoms (tetrahedral carbon atoms with four different groups attached). All the other amino acids contain an asymmetric carbon atom and are therefore optically active. Under physiological aqueous conditions a proton transfer from the acid to the base occurs, forming a dipolar ion, or zwitterion, because the carboxylic acid is a much stronger acid than is the ammonium ion. The structure of glycine in solution, for example, is $^+H_3NCH_2COO^-$ at pH 7 rather than H_2NCH_2COOH. At very low pH the acid group can be protonated, and at very high pH the ammonium group can be deprotonated; however, the forms of amino acids relevant to living organisms are the zwitterions.

AMINO ACIDS—TABLE 1. Sructural Formulas of Amino Acids

Name	Abbreviation	Linear structural formula
Alanine	ala	$CH_3—CH(NH_2)—COOH$
Arginine	arg	$HN\!\!=\!\!C(NH_2)—NH—(CH_2)_3—CH(NH_2)—COOH$
Asparagine	asn	$H_2N—CO—CH_2—CH(NH_2)—COOH$
Aspartic acid	asp	$HOOC—CH_2—CH(NH_2)—COOH$
Cysteine	cys	$HS—CH_2—CH(NH_2)—COOH$
Glutamine	gln	$H_2N—CO—(CH_2)_2—CH(NH_2)—COOH$
Glutamic acid	glu	$HOOC—(CH_2)_2—CH(NH_2)—COOH$
Glycine	gly	$NH_2—CH_2—COOH$
Histidine	his	$NH—CH\!\!=\!\!N—CH\!\!=\!\!C—CH_2—CH(NH_2)—COOH$
Isoleucine	ile	$CH_3—CH_2—CH(CH_3)—CH(NH_2)—COOH$
Leucine	leu	$(CH_3)_2—CH—CH_2—CH(NH_2)—COOH$
Lyine	lys	$H_2N—(CH_2)_4—CH(NH_2)—COOH$
Methionine	met	$CH_3—S—(CH_2)_2—CH(NH_2)—COOH$
Phenylalanine	phe	$C_6H_5—CH_2—CH(NH_2)—COOH$
Proline	pro	$NH—(CH_2)_3—CH—COOH$
Serine	ser	$HO—CH_2—CH(NH_2)—COOH$
Threonine	thr	$CH_3—CH(OH)—CH(NH_2)—COOH$
Tryptophan	trp	$C_6H_4—NH—CH\!\!=\!\!C—CH_2—CH(NH_2)—COOH$
Tyrosine	tyr	$HO—p—C_6H_4—CH_2—CH(NH_2)—COOH$
Valine	val	$(CH_3)_2—CH—CH(NH_2)—COOH$

AMINO ACIDS—TABLE 2. Molecular Weight and Density of Amino Acids

Name	Symbol	Molecular weight	Density
Alanine	Ala	71	1.401
Arginine	Arg	156	1.1
Asparagine	Asn	114	1.66
Aspartic acid	Asp	15	1.54
Cysteine	Cys	103	
Glutamic acid	Glu	129	1.46
Glutamine	Gln	128	
Glycine	Gly	57	1.607
Histidine	His	137	
Isoleucine	Ile	113	
Leucine	Leu	113	1.191
Lysine	Lys	128	
Methionine	Met	131	1.34
Phenylalanine	Phe	147	
Proline	Pro	97	
Serine	Ser	87	1.537
Threonine	Thr	101	
Tryptophan	Trp	186	
Tyrosine	Tyr	163	1.456
Valine	Val	99	1.23

API GRAVITY

Density is defined as the mass of a unit volume of material at a specified temperature and has the dimensions of grams per cubic centimeter (a close approximation to grams per milliliter). Density is measured at a standard temperature, most often 15.6 or 20°C (60 or 68°F), and is often written in the form of d_4^{20} to indicate that it was measured at 20°C and in a pycnometer calibrated with water at 4°C (39°F; i.e., with a medium of density equal to 1.0000).

Specific gravity is the ratio of the mass of a volume of the substance to the mass of the same volume of water and is dependent on two temperatures, those at which the masses of the sample and the water are measured. When the water temperature is 4°C (39°F), the specific gravity is equal to the density in the centimeter-gram-second (cgs) system, because the volume of 1 g of water at the temperature is, by definition, 1 ml. Thus the density of water, for example, varies with temperature, and its specific gravity at equal temperatures is always unity. The standard temperatures for a specific gravity in the petroleum industry in North America are 60/60°F (15.6/15.6°C).

In the petroleum industry, the use of density or specific gravity has largely been replaced by API (American Petroleum Institute) gravity as the preferred property. It is one criterion that is used in setting prices for petroleum. In the United States, API gravity has been used for the classification of a reservoir and the accompanying tax and royalty consequences. This property was derived from the Baumé scale:

$$\text{Degrees Baumé} = 140/\text{sp gr} @60/60\,°F - 130 \qquad (8)$$

However, a considerable number of hydrometers calibrated according to the Baumé scale were soon found to be in error by a consistent amount, which led to the adoption of the equation:

$$\text{Degrees API} = 141.5/\text{sp gr} @60/60\,°F - 131.5 \qquad (9)$$

ATMOSPHERE

The Earth is surrounded by a blanket of air, the atmosphere that extends more than 350 miles (560 kilometers) from the surface of the Earth. This envelope of gas that surrounds the Earth is divided into four distinct layers, the differences of which are based on thermal characteristics (temperature changes), chemical composition, movement, and density.

Troposphere

The troposphere starts at the Earth's surface and extends 5 to 9 miles (8 to 14.5 kilometers) high. The tropopause separates the troposphere from the next layer, the *stratosphere*. The tropopause and the troposphere are known as the *lower atmosphere*.

Stratosphere

The stratosphere starts just above the troposphere and extends to 31 miles (50 kilometers) high. Compared to the troposphere, this part of the atmosphere is dry and less dense. The ozone layer, which absorbs and scatters the solar ultraviolet radiation, is in this layer. Ninety-nine percent of air is located in the troposphere and stratosphere. The stratopause separates the stratosphere from the next layer, the *mesosphere*.

Mesosphere

The mesosphere starts just above the stratosphere and extends to 53 miles (85 kilometers) high. The chemicals in the *mesosphere* are in an excited state, as they absorb energy from the Sun. The mesopause separates the stratosphere from the next layer, the *thermosphere*.

Thermosphere

The thermosphere starts just above the mesosphere and extends to 372 miles (600 kilometers) high. Chemical reactions occur much faster here than on the surface of the Earth. This layer is known as the *upper atmosphere*.

Compositon of the Atmosphere

The atmosphere is primarily composed of nitrogen (N_2, 78%), oxygen (O_2, 20%), argon(Ar, 0.9%), and a variety of other components that include water (H_2O, 0.1%), with trace amounts of the rare gases (helium, neon, krypton, xenon), as well as trace amounts of hydrogen, ozone, carbon dioxide, carbon monoxide, nitrogen oxides, sulfur dioxide, and chlorofluorocarbons. Several of these gases originated in the atmosphere (permanent), and others are a result of the effect of anthropogenic activities (variable).

ATMOSPHERE—TABLE 1. Composition of the Atmosphere

Component	Abundance, ppm by volume	Comment
Nitrogen	7.81 E + 08	Permanent
Oxygen	2.01 E + 08	Permanent
Argon	9.34 E + 06	Permanent
Water	1.00 E + 06	Permanent
Carbon dioxide	3.50 E + 05	Variable
Neon	1.80 E + 04	Permanent
Helium	5.20 E + 03	Permanent
Methane	1.60 E + 03	Variable
Krypton	1.00 E + 03	Permanent
Hydrogen	5.00 E + 02	Variable
Nitrous oxide	3.00 E + 02	Variable
Carbon monoxide	1.00 E + 02	Variable
Sulfur dioxide	0.75 E + 02	Variable
Ozone	0.75 E + 02	Variable
Xenon	1.00 E + 01	Permanent
Nitrogen oxides (other than nitrous oxide)	0.50 E + 01	Variable
Chlorofluorocarbons	1.00 E – 01	Variable

ATOMIC DATA

An *atom* is the smallest part of an element with no net electrical charge that can enter into chemical combinations. The *isotopic abundance (atom abundance)* is the relative number of atoms of a particular isotope in a mixture of the isotopes of a chemical element and is expressed as a fraction of all the atoms of the element.

The *natural isotopic abundance* of a specified isotope of an element is the isotopic abundance in the element as found in nature. The *mass abundance* is the relative mass of atoms of a specified isotope in a mixture of the isotopes of an element and is expressed as a fraction of all the mass of the atoms of the element. The *isotopic composition* of an element is the set of isotopic abundances characterizing the atomic composition of an element.

ATOMIC DATA—TABLE 1.   Isotope Abundances of the Elements

Hydrogen	^{1}H	99.985	Phosphorus	^{31}P	100
	^{2}H	0.015	Sulfur	^{32}S	95.02
Helium	^{3}He	0.000137		^{33}S	0.75
	^{4}He	99.999863		^{34}S	4.21
Lithium	^{6}Li	7.5		^{36}S	0.02
	^{7}Li	92.5	Chlorine	^{35}Cl	75.77
Beryllium	^{9}Be	100		^{37}Cl	24.23
Boron	^{10}B	19.9	Argon	^{36}Ar	0.337
	^{11}B	80.1		^{38}Ar	0.063
Carbon	^{12}C	98.90		^{40}Ar	99.600
	^{13}C	1.10	Potassium	^{39}K	93.2581
Nitrogen	^{14}N	99.634		^{40}K	0.0117
	^{15}N	0.366		^{41}K	6.7302
Oxygen	^{16}O	99.762	Calcium	^{40}Ca	96.941
	^{17}O	0.038		^{42}Ca	0.647
	^{18}O	0.200		^{43}Ca	0.135
Fluorine	^{19}F	100		^{44}Ca	2.086
Neon	^{20}Ne	90.48		^{46}Ca	0.004
	^{21}Ne	0.27		^{48}Ca	0.187
	^{22}Ne	9.25	Scandium	^{45}Sc	100
Sodium	^{23}Na	100	Titanium	^{46}Ti	8.0
Magnesium	^{24}Mg	78.99		^{47}Ti	7.3
	^{25}Mg	10.00		^{48}Ti	73.8
	^{26}Mg	11.01		^{49}Ti	5.5
Aluminum	^{27}Al	100		^{50}Ti	5.4
Silicon	^{28}Si	92.23	Vanadium	^{50}V	0.250
	^{29}Si	4.67		^{51}V	99.750
	^{30}Si	3.10			

ATOMIC DATA—TABLE 1. Isotope Abundances of the Elements (*Continued*)

Chromium	^{50}Cr	4.345		Strontium	^{84}Sr	0.56
	^{52}Cr	83.789			^{86}Sr	9.86
	^{53}Cr	9.501			^{87}Sr	7.00
	^{54}Cr	2.365			^{88}Sr	82.58
Iron	^{54}Fe	5.8		Rubidium	^{85}Rb	72.165
	^{56}Fe	91.72			^{87}Rb	27.835
	^{57}Fe	2.2		Yttrium	^{89}Y	100
	^{58}Fe	0.28				
Manganese	^{55}Mn	100		Zirconium	^{90}Zr	51.45
					^{91}Zr	11.22
Nickel	^{58}Ni	68.077			^{92}Zr	17.15
	^{60}Ni	26.223			^{94}Zr	17.38
	^{61}Ni	1.140			^{96}Zr	2.80
	^{62}Ni	3.634		Molybdenum	^{92}Mo	14.84
	^{64}Ni	0.926			^{94}Mo	9.25
Cobalt	^{59}Co	100			^{95}Mo	15.92
					^{96}Mo	16.88
Copper	^{63}Cu	69.17			^{97}Mo	9.55
	^{65}Cu	30.83			^{98}Mo	24.13
Zinc	^{64}Zn	48.6			^{100}Mo	9.63
	^{66}Zn	27.9		Niobium	^{93}Nb	100
	^{67}Zn	4.1				
	^{68}Zn	18.8		Ruthenium	^{96}Ru	5.52
	^{70}Zn	0.6			^{98}Ru	1.88
					^{99}Ru	12.7
Gallium	^{68}Ga	60.108			^{100}Ru	12.6
	^{71}Ga	39.892			^{101}Ru	17.0
Germanium	^{70}Ge	21.23			^{102}Ru	31.6
	^{72}Ge	27.66			^{104}Ru	18.7
	^{73}Ge	7.73		Palladium	^{102}Pd	1.02
	^{74}Ge	35.94			^{104}Pd	11.14
	^{76}Ge	7.44			^{105}Pd	22.33
Selenium	^{74}Se	0.89			^{106}Pd	27.33
	^{75}Se	9.36			^{108}Pd	26.46
	^{77}Se	7.63			^{110}Pd	11.72
	^{78}Se	23.76		Rhenium	^{103}Rh	100
	^{80}Se	49.61				
	^{82}Se	8.73		Cadmium	^{106}Cd	1.25
Astatine	^{75}As	100			^{108}Cd	0.89
					^{110}Cd	12.49
Krypton	^{78}Kr	0.35			^{111}Cd	12.80
	^{80}Kr	2.25			^{112}Cd	24.13
	^{82}Kr	11.6			^{113}Cd	12.22
	^{83}Kr	11.5			^{114}Cd	28.73
	^{84}Kr	57.0			^{116}Cd	7.49
	^{86}Kr	17.3				
Bromine	^{79}Br	50.69				
	^{81}Br	49.31				

ATOMIC DATA—TABLE 1.   Isotope Abundances of the Elements (*Continued*)

Silver	^{107}Ag	51.839	Lanthanum	^{138}La	0.0902
	^{109}Ag	48.161		^{139}La	99.9098
Tin	^{112}Sn	0.97	Praseodymium	^{141}Pr	100
	^{114}Sn	0.65	Neodymium	^{142}Nd	27.13
	^{115}Sn	0.34		^{143}Nd	12.18
	^{116}Sn	14.53		^{144}Nd	23.80
	^{117}Sn	7.68		^{144}Nd	8.30
	^{118}Sn	24.23		^{146}Nd	17.19
	^{119}Sn	8.59		^{148}Nd	5.76
	^{120}Sn	32.59		^{150}Nd	5.64
	^{122}Sn	4.63			
	^{124}Sn	5.79	Samarium	^{144}Sm	3.1
				^{147}Sm	15.0
Indium	^{113}In	4.3		^{148}Sm	11.3
	^{115}In	95.7		^{149}Sm	13.8
Tellurium	^{120}Te	0.096		^{150}Sm	7.4
	^{122}Te	2.603		^{152}Sm	26.7
	^{123}Te	0.908		^{154}Sm	22.7
	^{124}Te	4.816	Europium	^{151}Eu	47.8
	^{125}Te	7.139		^{153}Eu	52.2
	^{126}Te	18.95	Gadolinium	^{152}Gd	0.20
	^{128}Te	31.69		^{154}Gd	2.18
	^{130}Te	33.80		^{155}Gd	14.80
Antimony	^{121}Sb	57.36		^{156}Gd	20.47
	^{123}Sb	42.64		^{157}Gd	15.65
Xenon	^{124}Xe	0.10		^{158}Gd	24.84
	^{126}Xe	0.09		^{160}Gd	21.86
	^{128}Xe	1.91	Dysprosium	^{156}Dy	0.06
	^{129}Xe	26.4		^{158}Dy	0.10
	^{130}Xe	4.1		^{160}Dy	2.34
	^{131}Xe	21.2		^{161}Dy	18.9
	^{132}Xe	26.9		^{162}Dy	25.5
	^{134}Xe	10.4		^{163}Dy	24.9
	^{136}Xe	8.9		^{164}Dy	28.2
Iodine	^{127}I	100	Terbium	^{159}Tb	100
Barium	^{130}Ba	0.106	Erbium	^{162}Er	0.14
	^{132}Ba	0.101		^{164}Er	1.61
	^{134}Ba	2.417		^{166}Er	33.6
	^{135}Ba	6.592		^{167}Er	22.95
	^{136}Ba	7.854		^{168}Er	26.8
	^{137}Ba	11.23		^{170}Er	14.9
	^{138}Ba	71.70			
Cesium	^{133}Cs	100	Holmium	^{165}Ho	100
Cerium	^{135}Ce	0.19			
	^{138}Ce	0.25			
	^{140}Ce	88.4			
	^{142}Ce	11.08			

ATOMIC DATA—TABLE 1. Isotope Abundances of the Elements (*Continued*)

Ytterbium	^{166}Yb	0.13	Platinum	^{190}Pt	0.01
	^{170}Yb	3.05		^{192}Pt	0.79
	^{171}Yb	14.3		^{194}Pt	32.9
	^{172}Yb	21.9		^{195}Pt	33.8
	^{173}Yb	16.12		^{196}Pt	25.3
	^{174}Yb	31.8		^{198}Pt	7.2
	^{176}Yb	12.7	Iridium	^{191}Ir	37.3
Thulium	^{169}Tm	100		^{193}Ir	62.7
Hafnium	^{174}Hf	0.162	Mercury	^{196}Hg	0.15
	^{175}Hf	5.206		^{198}Hg	9.97
	^{177}Hf	18.606		^{199}Hg	16.87
	^{178}Hf	27.297		^{200}Hg	23.10
	^{179}Hf	13.629		^{201}Hg	13.18
	^{180}Hf	35.100		^{202}Hg	29.86
Lutetium	^{175}Lu	97.41		^{204}Hg	6.87
	^{176}Lu	2.59	Gold	^{197}Au	100
Tantalum	^{180}Ta	0.012	Thallium	^{203}Tl	29.524
	^{181}Ta	99.988		^{205}Tl	70.476
Tungsten	^{180}W	0.13	Lead	^{204}Pb	1.4
	^{182}W	26.3		^{206}Pb	24.1
	^{183}W	14.3		^{207}Pb	22.1
	^{184}W	30.67		^{208}Pb	52.4
	^{186}W	28.6	Bismuth	^{209}Bi	100
Osmium	^{184}Os	0.02	Thorium	^{232}Th	100
	^{186}Os	1.58	Uranium	^{234}U	0.0055
	^{187}Os	1.6		^{235}U	0.7200
	^{188}Os	13.3		^{238}U	99.2745
	^{189}Os	16.1			
	^{190}Os	26.4			
	^{192}Os	41.0			
Rhenium	^{185}Re	37.40			
	^{187}Re	62.60			

ATOMIC DATA—TABLE 2.   Atomic Numbers, Periods, and Groups of the Elements (The Periodic Table)

Group / Period	1	2	3	4	5	6	7	8	9	10	11	12	13	14	15	16	17	18
1	1 H																	2 He
2	3 Li	4 Be											5 B	6 C	7 N	8 O	9 F	10 Ne
3	11 Na	12 Mg											13 Al	14 Si	15 P	16 S	17 Cl	18 Ar
4	19 K	20 Ca	21 Sc	22 Ti	23 V	24 Cr	25 Mn	26 Fe	27 Co	28 Ni	29 Cu	30 Zn	31 Ga	32 Ge	33 As	34 Se	35 Br	36 Kr
5	37 Rb	38 Sr	39 Y	40 Zr	41 Nb	42 Mo	43 Tc	44 Ru	45 Rh	46 Pd	47 Ag	48 Cd	49 In	50 Sn	51 Sb	52 Te	53 I	54 Xe
6	55 Cs	56 Ba	* 71 Lu	72 Hf	73 Ta	74 W	75 Re	76 Os	77 Ir	78 Pt	79 Au	80 Hg	81 Tl	82 Pb	83 Bi	84 Po	85 At	86 Rn
7	87 Fr	88 Ra	** 103 Lr	104 Unq	105 Unp	106 Unh	107 Uns	108 Uno	109 Mt	110 Uun	111 Uuu	112 Uub	113 Uut	114 Uuq	115 Uup	116 Uuh	117 Uus	118 Uuo

*Lanthanides	57 La	58 Ce	59 Pr	60 Nd	61 Pm	62 Sm	63 Eu	64 Gd	65 Tb	66 Dy	67 Ho	68 Er	69 Tm	70 Yb
†Actinides	89 Ac	90 Th	91 Pa	92 U	93 Np	94 Pu	95 Am	96 Cm	97 Bk	98 Cf	99 Es	100 Fm	101 Md	102 No

ATOMIC DATA—TABLE 3. Atomic Weights of the Elements in Order of Name

Name	Atomic number	Symbol	Atomic weight
Actinium	89	Ac	[227]
Aluminium	13	Al	26.981538
Americium	95	Am	[243]
Antimony	51	Sb	121.76
Argon	18	Ar	39.948
Arsenic	33	As	74.9216
Astatine	85	At	[210]
Barium	56	Ba	137.327
Berkelium	97	Bk	[247]
Beryllium	4	Be	9.012182
Bismuth	83	Bi	08.98038
Bohrium	107	Bh	[264]
Boron	5	B	10.811
Bromine	35	Br	79.904
Cadmium	48	Cd	112.411
Caesium	55	Cs	132.90545
Calcium	20	Ca	40.078
Californium	98	Cf	[251]
Carbon	6	C	12.0107
Cerium	58	Ce	140.116
Chlorine	17	Cl	35.4527
Chromium	24	Cr	51.9961
Cobalt	27	Co	8.9332
Copper	29	Cu	63.546
Curium	96	Cm	[247]
Dubnium	105	Db	[262]
Dysprosium	66	Dy	162.5
Einsteinium	99	Es	[252]
Erbium	68	Er	167.26
Europium	63	Eu	151.964
Fermium	100	Fm	[257]
Fluorine	9	F	18.9984032
Francium	87	Fr	[223]
Gadolinium	64	Gd	157.25
Gallium	31	Ga	69.723
Germanium	32	Ge	72.61
Gold	79	Au	196.96655
Hafnium	72	Hf	178.49
Hassium	108	Hs	[265]
Helium	2	He	4.002602
Holmium	67	Ho	164.93032
Hydrogen	1	H	1.00794
Indium	49	In	114.818
Iodine	53	I	126.90447
Iridium	77	Ir	192.217
Iron	26	Fe	55.845
Krypton	36	Kr	83.8
Lanthanum	57	La	138.9055
Lawrencium	103	Lr	[262]

ATOMIC DATA—TABLE 3. Atomic Weights of the Elements in Order
of Name (*Continued*)

Name	Atomic number	Symbol	Atomic weight
Lead	82	Pb	207.2
Lithium	3	Li	6.941
Lutetium	71	Lu	174.967
Magnesium	12	Mg	24.305
Manganese	25	Mn	54.938049
Meitnerium	109	Mt	[268]
Mendelevium	101	Md	[258]
Mercury	80	Hg	200.59
Molybdenum	42	Mo	95.94
Neodymium	60	Nd	144.24
Neon	10	Ne	20.1797
Neptunium	93	Np	[237]
Nickel	28	Ni	58.6934
Niobium	41	Nb	92.90638
Nitrogen	7	N	14.00674
Nobelium	102	No	[259]
Osmium	76	Os	190.23
Oxygen	8	O	15.9994
Palladium	46	Pd	106.42
Phosphorus	15	P	30.973761
Platinum	78	Pt	195.078
Plutonium	94	Pu	[244]
Polonium	84	Po	[209]
Potassium	19	K	39.0983
Praseodymium	59	Pr	140.90765
Promethium	61	Pm	[145]
Protactinium	91	Pa	231.03588
Radium	88	Ra	[226]
Radon	86	Rn	[222]
Rhenium	75	Re	186.207
Rhodium	45	Rh	102.9055
Rubidium	37	Rb	85.4678
Ruthenium	44	Ru	101.07
Rutherfordium	104	Rf	[261]
Samarium	62	Sm	150.36
Scandium	21	Sc	44.95591
Seaborgium	106	Sg	[263]
Selenium	34	Se	78.96
Silicon	14	Si	28.0855
Silver	47	Ag	107.8682
Sodium	11	Na	22.98977
Strontium	38	Sr	87.62
Sulfur	16	S	32.066(6)
Tantalum	73	Ta	180.9479
Technetium	43	Tc	[98]
Tellurium	52	Te	127.6
Terbium	65	Tb	158.92534
Thallium	81	Tl	204.3833
Thorium	90	Th	232.0381

ATOMIC DATA—TABLE 3. Atomic Weights of the Elements in Order of Name (*Continued*)

Name	Atomic number	Symbol	Atomic weight
Thulium	69	Tm	168.93421
Tin	50	Sn	118.71
Titanium	22	Ti	47.867
Tungsten	74	W	183.84
Ununbium	112	Uub	[277]
Ununnilium	110	Uun	[269]
Unununium	111	Uuu	[272]
Uranium	92	U	238.0289
Vanadium	23	V	50.9415
Xenon	54	Xe	131.29
Ytterbium	70	Yb	173.04
Yttrium	39	Y	88.90585
Zinc	30	Zn	65.39
Zirconium	40	Zr	91.224

ATOMIC DATA—TABLE 4. Atomic Weights of the Elements in Order of Atomic Number

Atomic number	Symbol	Name	Atomic weight
1	H	Hydrogen	1.00794
2	He	Helium	4.002602
3	Li	Lithium	6.941
4	Be	Beryllium	9.012182
5	B	Boron	10.811
6	C	Carbon	12.0107
7	N	Nitrogen	14.00674
8	O	Oxygen	15.9994
9	F	Fluorine	18.9984032
10	Ne	Neon	20.1797
11	Na	Sodium	22.98977
12	Mg	Magnesium	24.305
13	Al	Aluminum	26.981538
14	Si	Silicon	28.0855
15	P	Phosphorus	30.973761
16	S	Sulfur	32.066
17	Cl	Chlorine	35.4527
18	Ar	Argon	39.948
19	K	Potassium	39.0983
20	Ca	Calcium	40.078
21	Sc	Scandium	44.95591
22	Ti	Titanium	47.867
23	V	Vanadium	50.9415
24	Cr	Chromium	51.9961

ATOMIC DATA—TABLE 4. Atomic Weights of the Elements in Order of Atomic Number (*Continued*)

Atomic number	Symbol	Name	Atomic weight
25	Mn	Manganese	54.938049
26	Fe	Iron	55.845
27	Co	Cobalt	58.9332
28	Ni	Nickel	58.6934
29	Cu	Copper	63.546
30	Zn	Zinc	65.39
31	Ga	Gallium	69.723
32	Ge	Germanium	72.61
33	As	Arsenic	74.9216
34	Se	Selenium	78.96
35	Br	Bromine	79.904
36	Kr	Krypton	83.8
37	Rb	Rubidium	85.4678
38	Sr	Strontium	87.62
39	Y	Yttrium	88.90585
40	Zr	Zirconium	91.224
41	Nb	Niobium	92.90638
42	Mo	Molybdenum	95.94
43	Tc	Technetium	[98]
44	Ru	Ruthenium	101.07
45	Rh	Rhodium	102.9055
46	Pd	Palladium	106.42
47	Ag	Silver	107.8682
48	Cd	Cadmium	112.411
49	In	Indium	114.818
50	Sn	Tin	118.71
51	Sb	Antimony	121.76
52	Te	Tellurium	127.6
53	I	Iodine	126.90447
54	Xe	Xenon	131.29
55	Cs	Cesium	132.90545
56	Ba	Barium	137.327
57	La	Lanthanum	138.9055
58	Ce	Cerium	140.166
59	Pr	Praseodymium	140.90765
60	Nd	Neodymium	144.24
61	Pm	Promethium	[145]
62	Sm	Samarium	150.36
63	Eu	Europium	151.964
64	Gd	Gadolinium	157.25
65	Tb	Terbium	158.92534
66	Dy	Dysprosium	162.5
67	Ho	Holmium	164.93032
68	Er	Erbium	167.26
69	Tm	Thulium	168.93421
70	Yb	Ytterbium	173.04
71	Lu	Lutetium	174.967
72	Hf	Hafnium	178.49
73	Ta	Tantalum	180.9479

ATOMIC DATA—TABLE 4. Atomic Weights of the Elements in Order
of Atomic Number (*Continued*)

Atomic number	Symbol	Name	Atomic weight
74	W	Tungsten	183.84
75	Re	Rhenium	186.207
76	Os	Osmium	190.23
77	Ir	Iridium	192.217
78	Pt	Platinum	195.078
79	Au	Gold	196.96655
80	Hg	Mercury	200.59
81	Tl	Thallium	204.3833
82	Pb	Lead	207.2
83	Bi	Bismuth	208.98038
84	Po	Polonium	[209]
85	At	Astatine	[210]
86	Rn	Radon	[222]
87	Fr	Francium	[223]
88	Ra	Radium	[226]
89	Ac	Actinium	[227]
90	Th	Thorium	232.0381
91	Pa	Protactinium	231.03588
92	U	Uranium	238.0289
93	Np	Neptunium	[237]
94	Pu	Plutonium	[244]
95	Am	Americium	[243]
96	Cm	Curium	[247]
97	Bk	Berkelium	[247]
98	Cf	Californium	[251]
99	Es	Einsteinium	[252]
100	Fm	Fermium	[257]
101	Md	Mendelevium	[258]
102	No	Nobelium	[259]
103	Lr	Lawrencium	[262]
104	Rf	Rutherfordium	[261]
105	Db	Dubnium	[262]
106	Sg	Seaborgium	[263]
107	Bh	Bohrium	[264]
108	Hs	Hassium	[265]
109	Mt	Meitnerium	[268]
110	Uun	Ununnilium	[269]
111	Uuu	Unununium	[272]
112	Uub	Ununbium	[277]

BAUMÉ GRAVITY

Density is defined as the mass of a unit volume of material at a specified temperature and has the measurement of grams per cubic centimeter (a close approximation to grams per milliliter). Density is measured at a standard temperature, most

often 15.6 or 20°C (60 or 68°F), and is often written in the form of d_4^{20} to indicate that it was measured at 20°C and in a pycnometer calibrated with water at 4°C (39°F; i.e., with a medium of density equal to 1.0000).

Specific gravity is the ratio of the mass of a volume of the substance to the mass of the same volume of water and is dependent on two temperatures, those at which the masses of the sample and the water are measured. When the water temperature is 4°C (39°F), the specific gravity is equal to the density in the centimeter-gram-second (cgs) system, because the volume of 1 g of water at that temperature is, by definition, 1 ml. Thus the density of water, for example, varies with temperature, and its specific gravity at equal temperatures is always unity. The standard temperatures for a specific gravity in the petroleum industry in North America are 60/60°F (15.6/15.6°C).

In the petroleum industry, the use of density or specific gravity has largely been replaced by API (American Petroleum Institute) gravity as the preferred property. It is one criterion that is used in setting prices for petroleum. In the United States, API gravity has been used for the classification of a reservoir and the accompanying tax and royalty consequences. This property was derived from the Baumé scale:

$$\text{Degrees Baumé} = 140/\text{sp gr @ }60/60\text{F}° - 130 \qquad (10)$$

However, a considerable number of hydrometers calibrated according to the Baumé scale were soon found to be in error by a consistent amount, which led to the adoption of the equation:

$$\text{Degrees API} = 141.5/\text{sp gr @ }60/60\text{F}° - 131.5 \qquad (11)$$

BOILING POINT

The normal boiling point (boiling temperature) of a substance is the temperature at which the vapor pressure of the substance is equal to atmospheric pressure. At the boiling point, a substance changes its state from liquid to gas. A stricter definition of boiling point is the temperature at which the liquid and the vapor (gas) phases of a substance can exist in equilibrium. When heat is applied to a liquid, the temperature of the liquid rises until the *vapor pressure* of the liquid equals the pressure of the surrounding atmosphere (gases). At this point there is no further rise in temperature, and the additional heat energy supplied is absorbed at *latent heat* of vaporization to transform the liquid into gas. This transformation occurs not only at the surface of the liquid (as in the case of *evaporation*) but also throughout the volume of the liquid, where bubbles of gas are formed. The boiling point of a liquid is lowered if the pressure of the surrounding atmosphere (gases) is decreased. Conversely, if the pressure of the surrounding atmosphere (gases) is increased, the boiling point is raised. For this reason, it is customary when the boiling point of a substance is given to include the pressure at which it is observed, if that pressure is other than standard (i.e., 760 mm of mercury or 1 atmosphere [STP, Standard Temperature and Pressure]). The boiling point of a solution is usually higher than that of the pure solvent; this boiling-point elevation is one of the colligative properties common to all solutions.

BOILING POINT—TABLE 1. Boiling Points of Common Organic Compounds at Selected Pressures

Temperature,°C

Compound	1 mm	5 mm	10 mm	20 mm	40 mm	60 mm	100 mm	200 mm	400 mm	760 mm
Acenaphthene	s.	114.8	131.2	148.7	168.2	181.2	197.5	222.1	250.0	277.5
Acetaldehyde	-81.5	-65.1	-56.8	-47.8	-37.8	-31.4	-22.6	-10.0	4.9	-20.2
Acetamide	s.*	92.0	105.0	120.0	135.8	145.8	158.0	178.3	200.0	222.0
Acetic acid	s.	s.	17.5	29.9	43.0	51.7	63.0	80.0	99.0	118.1
Acetic anhydride	1.7	24.8	36.0	48.3	62.1	70.8	82.2	100.0	119.8	139.6
Acetone	-59.4	-40.5	-31.1	-20.8	-9.4	-2.0	+7.7	22.7	39.5	56.5
Acetonitrile	s.	-26.6	-16.3	-5.0	+7.7	15.9	27.0	43.7	62.5	81.8
Acetophenone	37.1	64.0	78.0	92.4	109.4	119.8	133.6	154.2	178.0	202.4
Adipic acid	159.5	191.0	205.5	222.0	240.5	251.0	265.0	287.8	312.5	337.5
Allyl alcohol	-20.0	+0.2	10.5	21.7	33.4	40.3	50.0	64.5	80.2	96.6
n-Amyl alcohol	+13.6	34.7	44.0	55.8	68.0	75.5	85.8	102.0	119.8	137.8
iso-Amyl alcohol	+10.0	30.9	40.8	51.7	63.4	71.0	80.7	95.8	113.7	130.6
sec-Amyl alcohol	1.5	22.1	32.2	42.6	54.1	61.5	70.7	85.7	102.3	119.7
Aniline	34.8	57.9	69.4	82.0	96.7	106.0	119.9	140.1	161.9	184.4
Anisole	5.4	30.0	42.2	55.8	70.7	80.1	93.0	112.3	133.8	155.5
Anthracene	s.	s.	s.	s.	s.	231.8	250.0	279.0	310.2	342.0
Benzaldehyde	26.2.	50.1	62.0	75.0	90.1	99.6	112.5	131.7	154.1	179.0
Benzene	s.	s.	s.	s.	7.6	15.4	26.1	42.2	60.6	80.1
Benzoic acid	s.	s.	131.1	146.7	162.6	172.8	186.2	205.8	227.0	249.2
Benzoic anhydride	143.8	180.0	198.0	218.0	239.8	252.7	270.4	299.1	328.8	360.0
Benzoin	135.6	170.2	188.0	207.0	227.6	241.1	258.0	284.4	313.5	343.0
Benzonitrile	28.2	55.3	69.2	83.4	99.6	109.8	123.5	144.1	166.7	190.6
Benzophenone	108.2	141.7	157.6	175.8	195.7	208.2	224.4	249.8	276.8	305.4
Benzoyl chloride	32.1	59.1	73.0	87.6	103.8	114.7	128.0	149.5	172.8	197.2
Benzyl alcohol	58.0	80.8	92.6	105.8	119.8	129.3	141.7	160.0	183.0	204.7
Benzyl chloride	22.0	47.8	60.8	75.0	90.7	100.5	114.2	134.0	156.8	179.4
Benzyl cinnamate	173.8	206.3	221.5	239.3	255.8	267.0	281.5	303.8	326.7	350.0

BOILING POINT—TABLE 1. Boiling Points of Common Organic Compounds at Selected Pressures (*Continued*)

Compound	Temperature,°C									
	1 mm	5 mm	10 mm	20 mm	40 mm	60 mm	100 mm	200 mm	400 mm	760 mm
α-Bromonaphthalene	84.2	117.5	133.6	150.2	170.2	183.5	198.8	224.2	252.0	281.1
Bromobenzene	2.9	27.8	40.0	53.8	68.6	78.1	90.8	110.1	132.3	156.2
Bromoform	s.	22.0	34.0	48.0	63.6	73.4	85.9	106.1	127.9	150.5
n-Butane	-101.5	-85.7	-77.8	-68.9	-59.1	-52.8	-44.2	-31.2	-16.3	-0.5
n-Butyl alcohol	-1.2	+20.0	30.2	41.5	53.4	60.3	70.1	84.2	100.8	117.5
iso-Butyl alcohol	-9.0	+11.0	21.7	32.4	44.1	51.7	61.5	75.9	91.4	108.0
tert-Butyl alcohol	s.	s.	s.	s.	s.	31.0	39.8	52.7	68.0	82.9
n-Butyric acid	25.5	49.8	61.5	74.0	88.0	96.5	108.0	125.5	144.5	163.5
iso-Butyric acid	14.7	39.3	51.2	64.0	77.8	86.3	98.0	115.5	134.5	154.5
Carbon disulphide	-73.8	-54.3	-44.7	-34.3	-22.5	-15.3	-5.1	+10.4	28.0	46.5
Carbon tetrabromide	s.	s.	s.	s.	96.3	106.3	119.7	139.7	163.5	189.5
Carbon tetrachloride	s.	s.	-19.6	-8.2	+4.3	12.3	23.0	38.3	57.8	76.7
Chloracetic acid	s.	68.3	81.0	94.2	109.2	118.3	130.7	149.0	169.0	189.5
Chlorobenzene	-13.0	10.6	22.2	35.3	49.7	58.3	70.7	89.4	110.0	132.2
Chloroform	-58.0	-39.1	-29.7	-19.0	-7.1	+0.5	10.4	25.9	42.7	61.3
Cinnamyl alcohol	72.6	102.5	117.8	133.7	151.0	162.0	177.8	199.8	224.6	250.0
o-Cresol	38.2	64.0	76.7	90.5	105.8	115.5	127.4	146.7	168.4	190.8
m-Cresol	52.0	76.0	87.8	101.4	116.0	125.8	138.0	157.3	179.0	202.8
p-Cresol	53.0	76.5	88.6	102.3	117.7	127.0	140.0	157.7	179.4	201.8
Cumene	2.9	26.8	38.3	51.5	66.1	75.4	88.1	107.3	129.2	152.4
Cyclohexane	s.	s.	s.	s.	+6.7	14.7	25.5	42.0	60.8	80.7
Cyclohexanol	s.	44.0	56.0	68.8	83.0	91.8	103.7	121.7	141.4	161.0
Cyclohexanone	1.4	26.4	38.7	52.5	67.8	77.5	90.4	110.3	132.5	155.6
p-Cymene	19.0	44.6	57.6	71.5	87.0	96.8	110.1	130.0	151.8	175.0
cis-Decalin	22.5	50.1	64.2	79.8	97.2	108.0	123.2	145.4	169.9	194.6
trans-Decalin	-0.8	+30.6	47.2	65.3	85.7	98.4	114.6	136.2	160.1	186.7
Dibenzyl	86.8	119.8	136.0	153.7	173.7	186.0	202.8	227.8	255.0	284.0

n-Dibutyl phthalate	148.2	182.1	198.2	216.2	235.8	247.8	263.7	287.0	313.5	340.0
Diethylamine	s.	s.	-33.0	-22.6	-11.3	-4.0	+6.0	21.0	38.0	55.5
Difluorodichloro-methane	-118.5	-104.6	-97.8	-90.1	-81.6	-76.1	-68.6	-57.0	-43.9	-29.8
Difluoromono-chloromethane	-122.8	-110.2	-103.7	-96.5	-88.6	-83.4	-76.4	-65.8	-53.6	-40.8
Dimethylamine	-87.7	-72.7	-64.6	-56.0	-46.7	-40.7	-32.6	-20.4	-7.1	+7.4
Dimethylaniline	29.5	56.3	70.0	84.8	101.6	111.9	125.8	146.5	169.2	193.1
Epichlorohydrin	-16.5	+5.6	16.6	29.0	42.0	50.6	62.0	79.3	98.0	117.9
Ethane	-159.5	-148.5	-142.9	-136.7	-129.8	-125.4	-119.3	-110.2	-99.7	-88.6
Ether (diethyl)	-74.3	-56.9	-48.1	-38.5	-27.7	-21.8	-11.5	+2.2	17.9	34.6
Ethyl acetate	-43.4	-23.5	-13.5	-3.0	+9.1	16.6	27.0	42.0	59.2	77.1
Ethyl acetoacetate	28.5	54.0	67.3	81.1	96.2	106.0	118.5	138.0	158.2	180.8
Ethyl alcohol	-31.3	-12.0	-2.3	+8.0	19.0	26.0	34.0	48.4	63.5	78.4
Ethylamine	s.	-66.4	-58.3	-48.6	-39.8	-33.4	-25.1	-12.3	+2.0	16.6
Ethylbenzene	-9.8	+13.9	25.9	38.6	52.8	61.8	74.1	92.7	113.8	136.2
Ethly benzoate	44.0	72.0	86.0	101.4	118.2	129.0	143.2	164.8	188.4	213.4
Ethyl bromide	-74.3	-56.4	-47.5	-37.8	-26.7	-19.5	-10.0	+4.5	21.0	38.4
Ethyl chloride	-89.8	-73.9	-65.8	-56.8	-47.0	-40.6	-32.0	-18.6	-3.9	12.3
Ethylene	-168.3	-158.3	-153.2	-147.6	-141.3	-137.3	-131.8	-18.6	-3.9	-112.3
Ethylenediamine	s.	+10.5	21.5	33.0	45.8	53.8	62.5	81.0	99.0	117.2
Ethylene dibromide	s.	s.	18.6	32.7	48.0	57.9	70.4	89.8	110.1	131.5
Ethylene dichloride	s.	-24.0	-13.6	-2.4	+10.0	18.1	29.4	45.7	64.0	82.4
Ethylene glycol	53.0	79.7	92.1	105.8	120.0	129.5	141.8	158.5	178.5	197.3
Ethylene oxide	-89.7	-73.8	-65.7	-56.6	-46.9	-40.7	-32.1	-19.5	-4.9	+10.7
Ethyl formate	-60.5	-42.2	-33.0	-22.7	-11.3	-4.3	+3.4	20.0	37.1	54.3
Ethyl iodide	-54.4	-34.3	-24.3	-13.1	-0.9	+7.2	18.0	34.1	52.3	72.4
Ethyl mercaptan	-76.7	-59.1	-50.2	-40.7	-29.8	-22.4	-13.0	+1.5	17.7	35.0
Ethyl oxalate	47.4	71.8	83.8	96.8	110.6	119.7	130.8	147.9	166.2	185.7
Ethyl salicylate	61.2	90.0	104.2	119.3	136.7	147.6	161.5	183.7	207.0	231.5
Ethyl sulphate	47.0	74.0	87.7	102.1	118.0	128.6	142.5	162.5	185.5	209.5d.*
Eugenol	78.4	108.1	123.0	138.7	155.8	167.3	182.2	204.7	228.3	253.5

BOILING POINT – TABLE 1. Boiling Points of Common Organic Compounds at Selected Pressures *(Continued)*

Compound	1 mm	5 mm	10 mm	20 mm	40 mm	60 mm	100 mm	200 mm	400 mm	760 mm
Fluorobenzene	s.	−22.8	−12.4	−1.2	+11.5	19.6	30.4	47.2	65.7	84.7
Formaldehyde	s.	s.	−88.0	−79.6	−70.6	−65.0	−57.3	−46.0	−33.0	−19.5
Formic acid	s.	s.	s.	10.3	24.0	32.4	43.8	61.4	80.3	100.6
Furfuryl alcohol	31.8	56.0	68.0	81.0	95.7	104.0	115.9	133.1	151.8	170.0
Glycerol	125.5	153.8	167.2	182.2	198.0	208.0	220.1	240.0	263.0	290.0
n-Heptane	−34.0	−12.7	−2.1	+9.5	22.3	30.6	41.8	58.7	78.0	98.4
n-Hexane	−53.9	−34.5	−25.0	−14.1	−2.3	+5.4	15.8	31.6	49.6	68.7
Hydroquinone	s.	s.	s.	174.6	192.0	203.0	216.5	238.0	262.5	286.2
Indene	16.4	44.3	58.5	73.9	90.7	100.8	114.7	135.6	157.8	181.6
Isoprene	−79.8	−62.3	−53.3	−43.5	−32.6	−25.4	−16.0	−1.2	15.4	32.6
Isooctane	−36.5	−15.0	−4.3	7.5	20.7	29.1	40.7	58.1	78.0	99.2
Isoquinoline	63.5	92.7	107.8	123.7	141.6	152.0	167.6	190.0	214.5	240.5
Maleic anhydride	s.	63.4	78.7	95.0	111.8	122.0	135.8	155.9	179.5	202.0
Mesitylene	9.6	34.7	47.4	61.0	76.1	85.8	98.9	118.6	141.0	164.7
Methane	s.	s.	s.	s.	s.	s.	−181.4	−175.5	−168.8	−161.5
Methyl acetate	−57.2	−38.6	−29.3	−19.1	−7.9	−0.5	+9.4	24.0	40.0	57.8
Methyl alcohol	−44.0	−25.3	−16.2	−6.0	+5.0	12.1	21.2	34.8	49.9	64.7
Methylamine	s.	−81.3	−73.8	−65.9	−56.9	−51.3	−43.7	−32.4	−19.7	−6.3
Methylaniline	36.0	62.8	76.2	90.5	106.0	115.8	129.8	149.3	172.0	195.5
Methyl benzoate	39.0	64.4	77.3	91.8	107.8	117.4	130.8	151.4	174.7	199.5
Methyl bromide	s.	−80.6	−72.8	−64.0	−54.2	−48.0	−39.4	−26.5	−11.9	+3.6
Methyl chloride	s.	s.	−92.4	−84.8	−76.0	−70.4	−63.0	−51.2	−38.0	−24.0
Methylene bromide	−35.1	−13.2	−2.4	+9.7	23.3	31.6	42.3	58.5	79.0	98.6
Methylene chloride	−70.0	−52.1	−43.3	−33.4	−22.3	−15.7	−6.3	+8.0	24.1	40.7
Methyl ethylketone	−48.3	−28.0	−17.7	−6.5	+6.0	14.0	25.0	41.6	60.0	79.6

Methyl formate .	−74.2	−57.0	−48.6	−39.2	−28.7	−21.9	−12.9	+0.8	16.0	32.0
Methyl iodide .	s.	−55.0	−45.8	−35.6	−24.2	−16.9	−7.0	+8.0	25.3	42.4
Methyl salicylate	54.0	81.6	95.3	110.0	126.2	136.7	150.0	172.6	197.5	223.2
Monofluorotri-chloromethane	−84.3	−67.6	−59.0	−49.7	−39.0	−32.3	−23.0	−9.1	+6.8	23.7
Naphthalene .	s.	s.	85.8	101.7	119.3	130.2	145.5	167.7	193.2	217.9
α-Naphthol . .	s.	125.5	142.0	158.0	177.8	190.0	206.0	229.6	255.8	282.5
β-Naphthol . .	s.	128.6	143.5	161.8	181.7	193.7	209.8	234.0	260.6	288.0
α-Naphthylamine	104.3	137.7	153.8	171.6	191.5	203.8	220.0	244.9	272.2	300.8
β-Naphthylamine	108.0	141.6	157.6	175.8	195.7	208.1	224.3	249.7	277.4	306.1
Nicotine . .	61.8	91.8	107.2	123.7	142.1	154.7	169.5	193.8	219.8	247.3
Nitrobenzene .	44.4	71.6	84.9	99.3	115.4	125.8	139.9	161.2	185.8	210.6
Nitroethane .	−21.0	+1.5	12.5	24.8	38.0	46.5	57.8	74.8	94.0	114.0
Nitromethane .	−29.0	−7.9	+2.8	14.1	27.5	35.5	46.6	63.5	82.0	101.2
1-Nitropropane .	−9.6	+13.5	25.3	37.9	51.8	60.5	72.3	90.2	110.6	131.6
2-Nitropropane .	−18.8	+4.1	15.8	28.2	41.8	50.3	62.0	80.0	99.8	120.3
n-Octane . .	−14.0	+8.3	19.2	31.5	45.1	53.8	65.7	83.6	104.0	125.6
n-Octyl alcohol .	54.0	76.5	88.3	101.0	115.2	123.8	135.2	152.0	173.8	195.2
Oleic acid . .	176.5	208.5	223.0	240.0	257.2	269.8	286.0	309.8	334.7	360.0 d.
Palmitic acid .	153.6	188.1	205.8	223.8	244.4	256.0	271.5	298.7	326.0	353.8
n-Pentane . .	−76.6	−62.5	−50.1	−40.2	−29.2	−22.2	−12.6	+1.9	18.5	36.1
Phenanthrene .	118.2	154.3	173.0	193.7	215.8	229.9	249.0	277.1	308.0	340.2
Phenol . .	s.	62.5	73.8	86.0	100.1	108.4	121.4	139.0	160.0	181.9
Phthalic anhydride	s.	s.	134.0	151.7	172.0	185.3	202.3	228.0	256.8	284.5
Piperidine . .	s.	−7.0	+3.9	15.8	29.2	37.7	49.0	66.2	85.7	106.0
Propane . .	−128.9	−115.4	−108.5	−100.9	−92.4	−87.0	−79.6	−68.4	−55.6	−42.1
n-propyl acetate .	−26.7	−5.4	+5.0	16.0	28.8	37.0	47.8	64.0	82.0	101.8
n-Propyl alcohol	−15.0	+5.0	14.7	25.3	36.4	43.5	52.8	66.8	82.0	97.8
iso-Propyl alcohol	−26.1	−7.0	+2.4	12.7	23.8	30.5	39.5	53.0	67.8	82.5
Propylene . .	−131.9	−120.7	−112.1	−104.7	−96.5	−91.3	−84.1	−73.3	−60.9	−47.7
Pyridine . .	−18.9	+2.5	13.2	24.8	38.0	46.8	57.8	75.0	95.6	115.4

BOILING POINT—TABLE 1. Boiling Points of Common Organic Compounds at Selected Pressures (*Continued*)

Compound	1 mm	5 mm	10 mm	20 mm	40 mm	60 mm	100 mm	200 mm	400 mm	760 mm
					Temperature, °C					
Pyrocatechol	s.	104.0	118.3	134.0	150.6	161.7	176.0	197.7	221.5	245.5
Pyrogallol	s.	151.7	167.7	185.3	204.2	216.3	232.0	255.3	281.5	309.0 d.
Resorcinol	s.	138.0	152.1	168.0	185.3	195.8	209.8	230.8	253.4	276.5
Salicylic acid	s.	s.	s.	s.	172.2	182.0	193.4	210.0	230.5	256.0
Stearic acid	173.7	209.0	225.0	243.4	263.3	275.5	201.0	316.5	343.0	370.0 d.
Styrene	-7.0	+18.0	30.8	44.6	59.8	69.5	82.0	101.3	122.5	145.2
Succinic anhydride	s.	s.	128.2	145.3	163.0	174.0	189.0	212.0	237.0	261.0
Tetralin	38.0	65.3	79.0	93.8	110.4	121.3	135.3	157.2	181.8	207.2
Thiophene	s.	-20.8	-10.9	0.0	+12.5	20.1	30.5	46.5	64.7	84.4
Toluene	-26.7	-4.4	+6.4	18.4	31.8	40.3	51.9	69.5	89.5	110.6
o-Toluidine	44.0	69.3	81.4	95.1	110.0	119.8	133.0	153.0	176.2	199.7
m-Toluidine	41.0	68.0	82.0	96.7	113.5	123.8	136.7	157.6	180.6	203.3
p-Toluidine	42.0	68.2	81.8	95.8	111.5	121.5	133.7	154.0	176.9	200.4
Trichloroethylene	-43.8	-22.8	-12.4	-1.0	+11.9	20.0	31.4	48.0	67.0	86.7
Trimethylamine	-97.1	-81.7	-73.8	-65.0	-55.2	-48.8	-40.3	-27.0	-12.5	+2.9
n-Valeric acid	42.2	67.7	79.8	93.1	107.8	116.6	128.3	146.0	165.0	184.4
iso-Valeric acid	34.5	59.6	71.3	84.0	98.0	107.3	118.9	136.2	155.2	175.1
o-Xylene	-3.8	+20.2	32.1	45.1	59.5	68.8	81.3	100.2	121.7	144.4
m-Xylene	-6.9	+16.8	28.3	41.1	55.3	64.4	76.8	95.5	116.7	139.1
p-Xylene	-8.1	+15.5	27.3	40.1	54.4	63.5	75.9	94.6	115.9	138.3

* s: sublimes; d: decomposes.

BOILING POINT—TABLE 2. Boiling Points of n-Paraffins

Carbon number	Boiling point, °C	Boiling point, °F
5	36	97
6	69	156
7	98	209
8	126	258
9	151	303
10	174	345
11	196	385
12	216	421
13	235	456
14	253	488
15	271	519
16	287	548
17	302	576
18	317	602
19	331	627
20	344	651
21	356	674
22	369	696
23	380	716
24	391	736
25	402	755
26	412	774
27	422	792
28	432	809
29	441	825
30	450	841
31	459	858
32	468	874
33	476	889
34	483	901
35	491	916
36	498	928
37	505	941
38	512	958
39	518	964
40	525	977
41	531	988
42	537	999
43	543	1009
44	548	1018

BOND LENGTHS AND BOND ENERGIES

Distances between centers of bonded atoms are called *bond lengths*, or *bond distances*. Bond lengths vary depending on many factors, but in general they are consistent. The bond orders affect bond length, but bond lengths of the same order for the same pair of atoms in various molecules are consistent.

The *bond order* is the number of electron pairs shared between two atoms in the formation of the bond. Bond order for C=C and O=O is 2. The amount of energy required to break a bond is called *bond dissociation energy*, or simply *bond energy*. Because bond lengths are consistent, bond energies of similar bonds are also consistent.

Bonds between the same type of atom are *covalent bonds*, and bonds between atoms with slightly differing electronegativity are also predominant covalent in character. Theoretically, even ionic bonds have some covalent character. Thus, the boundary between ionic and covalent bonds is not a clear line of demarcation.

For covalent bonds, bond energies and bond lengths depend on many factors: electron afinities, sizes of atoms involved in the bond, differences in their electronegativity, and the overall structure of the molecule. There is a general trend in that *the shorter the bond length, the higher the bond energy*, but there is no formula to show this relationship because of the widespread variation in bond character.

BOND LENGTHS AND BOND ENERGIES—TABLE 1.   Covalent Bond Lengths

Bond	nm	Bond	nm
H—H	0.074	C—H	0.109
C—C	0.154	Si—H	0.146
C=C	0.134	N—H	0.101
C≡C	0.120	P—H	0.142
C⋯C (in benzene)	0.139	O—H	0.096
Si—Si	0.235	S—H	0.135
N—N	0.146	F—H	0.092
N=N	0.120	Cl—H	0.128
N≡N	0.110	Br—H	0.141
P—P (P_4)	0.221	I—H	0.160
O—O	0.148		
O=O	0.121	C—O	0.143
S—S (S_8)	0.207	C=O	0.122
S=S	0.188	C⋯O (in phenol)	0.136
F—F	0.142	C—N	0.147
Cl—Cl	0.199	C=N	0.127
Br—Br	0.228	C≡N	0.116
I—I	0.267	C⋯N (in phenylamine)	0.135
		C—F	0.138
		C—Cl	0.177
		C⋯Cl (in cholorobenzene)	0.169
		C—Br	0.193
		C—I	0.214
		Si—O	0.150

BOND LENGTHS AND BOND ENERGIES—TABLE 2. Bond Lengths in Organic Compounds

Compound	Bond	Interatomic distance, Å	Compound	Bond	Interatomic distance, Å
CBr_4	C—Br	1.942	$CHCl_3$	C—H	1.073
CH_2CHBr	C—Br	1.891	CH_2CH_2	C—H	1.085
C_6H_5Br	C—Br	1.86	C_6H_6	C—H	1.084
CHCBr	C—Br	1.795	C_2H_2	C—H	1.059
CCl_4	C—Cl	1.766	CH_3I	C—I	2.139
$CHCl_3$	C—Cl	1.762	CH_3CHI	C—I	2.092
CH_2CHCl	C—Cl	1.719	C_6H_5I	C—I	2.05
C_6H_5Cl	C—Cl	1.70	CH_3CCI	C—I	1.99
CHCCl	C—Cl	1.632	CH_3NH_2	C—N	1.474
CH_3F	C—F	1.385	$C_2H_5NO_2$	C—N	1.479
CH_2CHF	C—F	1.344	HCN	C—N	1.1554
C_6H_5F	C—F	1.332	CH_3OH	C—O	1.428
$Fe(CO)_5$	C—Fe	1.84	CH_2CH_2O (Epoxide)	C—O	1.435
CH_4	C—H	1.094	HCO_2H	C—O	1.312
CD_4	C—D	1.092	CO	C—O	1.128
CH_3Br	C—H	1.095	CO_2	C—O	1.1618
CH_2Cl_2	C—H	1.068	$(CH_3)_3P$	C—P	1.841

BOND LENGTHS AND BOND ENERGIES—TABLE 3. Bond Dissociation Energies, kcal per mole

Single bonds	ΔH	Single bonds	ΔH	Multiple bonds	ΔH
H̃—H	104.2	B̃—F	154	C=C	146
C̃—C	83	B̃—O	123	N=N	109
Ñ—N	38.4	C̃—N	73	O=O	119
Õ—O	35	Ñ—CO	86	C=N	147
F̃—F	36.6	C̃—O	85.5	C=O (CO_2)	192
S̃i—Si	42	ÕC—O	110	C=O (aldehyde)	177
P̃—P	51	C̃—S	65	C=O (ketone)	178
S̃—S	64	C̃—F	116	C=O (ester)	179
C̃l —Cl	58	C̃—Cl	81	C=O (amide)	179
B̃r—Br	46	C̃—Br	68	C=O (halide)	177
Ĩ—I	36	C̃—I	51	C=S (CS_2)	138
H̃—C	99	C̃—B	94	N=O ($HONO_2$)	143
H̃—N	93	C̃—Si	83	P=O ($POCl_3$)	109
H̃—O	111	C̃—P	73	P=S ($PSCl_3$)	70
H̃—F	135	Ñ—O	55	S=O (SO_2)	128
H̃—Cl	103	S̃—O	87	S=O (DMSO)	93
H̃—Br	87.5	S̃i—F	132	C≡C	200
H̃—I	71	S̃i—Cl	86	N≡N	226
H̃—B	90	S̃i—O	110	C≡N	213
H̃—S	81	P̃—Cl	79		
H̃—Si	90	P̃—Br	65		
H̃—P	77	P̃—O	98		

TABLE 4. Bond Dissociation Energies of Organic Compounds, kilocalories per mole

Atom or group	Methyl	Ethyl	Iso-propyl	t-Butyl	Phenyl	Benzyl	Allyl	Acetyl	Vinyl
H	103	98	95	93	110	85	88	87	112
F	110	110	109		124	94		119	
Cl	85	82	81	80	95	68	70	82	90
Br	71	70	69	66	79	55	56	68	80
I	57	54	54	51	64	40	42	51	
OH	93	94	92	91	111	79	82	107	
NH$_2$	87	87	86	85	104	72	75	95	
CN	116	114	112		128	100			128
CH$_3$	88	85	84	81	101	73	75	81	98
C$_2$H$_5$	85	82	81	78	99	71	72	78	95
(CH$_3$)$_2$CH	84	81	79	74	97	70	71	76	93
(CH$_3$)$_3$C	81	78	74	68	94	67	67		89
C$_6$H$_5$	101	99	97	94	110	83	87	93	108
C$_6$H$_5$CH$_2$	73	71	70	67	83	59	59	63	81

CHANGE OF STATE

All substances can exist in one of three forms (also called *states* or *phases*) that basically depend on the temperature of the substance. These states or phases are (1) solid, (2) liquid, and (3) gas.

The solid-to-liquid transition is a melting process, and the heat required is the heat of melting. The liquid-to-solid transition is the reverse process, and the heat liberated is the heat of freezing. The solid-to-gas transition is a sublimation process, and the heat required is the heat of sublimation. The liquid-to-gas transition is a vaporization process, and the heat required is the heat of vaporization (heat of boiling). Both the gas-to-solid and the gas-to-liquid processes are condensation processes and have an associated heat of condensation.

Each change of state is accompanied by a change in the energy of the system. Wherever the change involves the disruption of intermolecular forces, energy must be supplied. The disruption of intermolecular forces accompanies the state going toward a less ordered state. As the strengths of the intermolecular forces increase, greater amounts of energy are required to overcome them during a change in state. The melting process for a solid is also referred to as fusion, and the enthalpy change associated with melting a solid is often called the heat of fusion (ΔH_{fus}). The heat needed for the vaporization of a liquid is called the heat of vaporization (ΔH_{vap}).

COAL

Coal is a combustible organic rock composed primarily of carbon, hydrogen, and oxygen. Coal is burned to produce energy and is used to manufacture steel. It is also an important source of chemicals used to make medicine, fertilizers, pesticides, and other products. Coal comes from ancient plants buried over millions of years in Earth's crust. Coal, petroleum, natural gas, and oil shale are all known as fossil fuels because they come from the remains of ancient life buried deep in the Earth.

Coal is rich in hydrocarbonaceous compounds made up of the elements carbon, hydrogen, nitrogen, oxygen, and sulfur and was formed from ancient plants that died, decomposed, and were buried under layers of sediment during the carboniferous period, about 360 million to 290 million years ago. As more and more layers of sediment formed over this decomposed plant material, the overburden exerted increasing heat and pressure on the organic matter. Over millions of years, these physical conditions caused coal to form from the carbon, hydrogen, oxygen, nitrogen, sulfur, and inorganic mineral compounds in the plant matter. The coal is formed in layers known as seams.

COAL—TABLE 1. Classification by Rank (ASTM)

Class/group	Fixed carbon limits (dry, mineral-matter-free basis), %		Volatile matter limits (dry, mineral-matter-free basis), %		Gross calorific value limits (moist, mineral-matter-free basis)*				Agglomerating character
					MJ/kg		Btu/lb		
	Equal or greater than	Less than	Greater than	Equal or less than	Equal or greater than	Less than	Equal or greater than	Less than	
Anthracitic:									
Meta-anthracite	98	—	—	2	—	—	—	—	Nonagglomerating
Anthracite	92	98	2	8	—	—	—	—	
Semianthracite†	86	92	8	14	—	—	—	—	
Bituminous:									
Low-volatile bituminous coal	78	86	14	22	—	—	—	—	
Medium-volatile bituminous coal	69	78	22	31	—	—	—	—	
High-volatile A bituminous coal	—	69	31	—	32.6	—	14,000‡	—	Commonly agglomerating§
High-volatile B bituminous coal	—	—	—	—	30.2	32.6	13,000‡	14,000	
High-volatile C bituminous coal	—	—	—	—	26.7	30.2	11,500	13,000	Agglomerating
					24.4	26.7	10,500	11,500	
Subbituminous:									
Subbituminous A coal	—	—	—	—	24.4	26.7	10,500	11,500	Nonagglomerating
Subbituminous B coal	—	—	—	—	22.1	24.4	9,500	10,500	
Subbituminous C coal	—	—	—	—	19.3	22.1	8,300	9,500	
Lignitic:									
Lignite A	—	—	—	—	14.7	19.3	6,300	8,300	
Lignite B	—	—	—	—	—	14.7	—	6,300	

* *Moist* refers to coal containing its natural inherent moisture but not including visible water on the surface of the coal.
† If agglomerating, classify in low-volatile group of the bituminous class.
‡ Coals having 69% or more fixed carbon on the dry, mineral-matter-free basis shall be classified according to fixed carbon, regardless of gross calorific value.
§ It is recognized that there may be nonagglomerating varieties in these groups of the bituminous class and that there are notable exceptions in the high-volatile C bituminous group.

COAL—TABLE 2. Coal Analysis by Rank

Classification by rank	State	County	Bed	Proximate, %				Ultimate, %					Calorific Value, Btu/lb*
				Moisture	Volatile matter	Fixed carbon	Ash†	Sulfur	Hydrogen	Carbon	Nitrogen	Oxygen	
Meta-anthracite	Rhode Island	Newport	Middle	13.2	2.6	65.3	18.9	0.3	1.9	64.2	0.2	14.5	9,310
Anthracite	Pennsylvania	Lackawanna	Clark	4.3	5.1	81.0	9.6	0.8	2.9	79.7	0.9	6.1	12,880
Semianthracite	Arkansas	Johnson	Lower Hartshorne	2.6	10.6	79.3	7.5	1.7	3.8	81.4	1.6	4.0	13,880
Low-volatile bituminous coal	West Virginia	Wyoming	Pocahontas no. 3	2.9	17.7	74.0	5.4	0.8	4.6	83.2	1.3	4.7	14,400
Medium-volatile bituminous coal	Pennsylvania	Clearfield	Upper Kittanning	2.1	24.4	67.4	6.1	1.0	5.0	81.6	1.4	4.9	14,310
High-volatile A bituminous coal	West Virginia	Marion	Pittsburgh	2.3	36.5	56.0	5.2	0.8	5.5	78.4	1.6	8.5	14,040
High-volatile B bituminous coal	Kentucky, western field	Muhlenburg	No. 9	8.5	36.4	44.3	10.8	2.8	5.4	65.1	1.3	14.6	11,680
High-volatile C bituminous coal	Illinois	Sangamon	No. 5	14.4	35.4	40.6	9.6	3.8	5.8	59.7	1.0	20.1	10,810
Subbituminous A coal	Wyoming	Sweetwater	No. 3	16.9	34.8	44.7	3.6	1.4	6.0	60.4	1.2	27.4	10,650
Subbituminous B coal	Wyoming	Sheridan	Monarch	22.2	33.2	40.3	4.3	0.5	6.9	53.9	1.0	33.4	9,610
Subbituminous C coal	Colorado	El Paso	Fox Hill	25.1	30.4	37.7	6.8	0.3	6.2	50.5	0.7	35.5	8,560
Lignite	North Dakota	McLean	Unnamed	36.8	27.8	29.5	5.9	0.9	6.9	40.6	0.6	45.1	7,000

* Btu/lb × 2.325 = J/g; Btu/lb × 0.5556 = g-cal/g.
† Ash is part of both the proximate and ultimate analyses.

COAL—TABLE 3. Conversion Factors for Components Other Than Hydrogen and Oxygen (ASTM D3180)*

Given	As determined (ad)	As received (ar)	Dry (d)	Dry ash-free (daf)
As determined (ad)	—	$\dfrac{100 - M_{ar}}{100 - M_{ad}}$	$\dfrac{100}{100 - M_{ar}}$	$\dfrac{100}{100 - M_{ad} - A_{ad}}$
As received (ar)	$\dfrac{100 - M_{ad}}{100 - M_{ar}}$	—	$\dfrac{100}{100 - M_{ar}}$	$\dfrac{100}{100 - M_{ar} - A_{ar}}$
Dry (d)	$\dfrac{100 - M_{ad}}{100}$	$\dfrac{100 - M_{ar}}{100}$	—	$\dfrac{100}{100 - A_{d}}$
Dry, ash-free (daf)	$\dfrac{100 - M_{ad} - A_{ad}}{100}$	$\dfrac{100 - M_{ar} - A_{ar}}{100}$	$\dfrac{100 - A_{d}}{100}$	—

M = percent moisture by weight; A = percent ash by weight.
* For example, given ad, to find ar use the formula
$$ar = ad \times \frac{100 - M_{ar}}{100 - M_{ad}}$$

COAL—TABLE 4. Dielectric Constants of Petroleum, Coal, and Their Products

Compound	Dielectric constant	Loss tangent $\tan \Delta \times 10^{4}$
Asphalt (75°F)	2.6	
Bunker C oil	2.6	
Coal tar	2.0–3.0	
Coal, powder, fine	2.0–4.0	
Coke	1.1–2.2	
Gasoline (70°F)	2.0	
Heavy oil	3.0	
Kerosene (70°F)	1.8	
Liquefied petroleum gas (LPG)	1.6–1.9	
Lubricating oil (68°F)	2.1–2.6	
Paraffin oil	2.19	1.0
Paraffin wax	2.1–2.5	2
Petroleum (68°F)	2.1	
Vaseline	2.08	5

COAL—TABLE 5. Elements in Coal

Constituent	Range
Arsenic	0.50–93.00 ppm
Boron	5.00–224.00 ppm
Beryllium	0.20–4.00 ppm
Bromine	4.00–52.00 ppm
Cadmium	0.10–65.00 ppm
Cobalt	1.00–43.00 ppm
Chromium	4.00–54.00 ppm
Copper	5.00–61.00 ppm
Fluorine	25.00–143.00 ppm
Gallium	1.10–7.50 ppm
Germanium	1.00–43.00 ppm
Mercury	0.02–1.60 ppm
Manganese	6.00–181.00 ppm
Molybdenum	1.00–30.00 ppm
Nickel	3.00–80.00 ppm
Phosphorus	5.00–400.00 ppm
Lead	4.00–218.00 ppm
Antimony	0.20–8.90 ppm
Selenium	0.45–7.70 ppm
Tin	1.00–51.00 ppm
Vanadium	11.00–78.00 ppm
Zinc	6.00–5350.00 ppm
Zirconium	8.00–133.00 ppm
Aluminum	0.43–3.04%
Calcium	0.05–2.67%
Chorine	0.01–0.54%
Iron	0.34–4.32%
Potassium	0.02–0.43%
Magnesium	0.01–0.25%
Sodium	0.00–0.20%
Silicon	0.58–6.09%
Titanium	0.02–0.15%
Organic sulfur	0.31–3.09%
Pyritic sulfur	0.06–3.78%
Sulfate sulfur	0.01–1.06%
Total sulfur	0.42–6.47%
Sulfur by x-ray fluorescence	0.54–5.40%

COAL—TABLE 6. Fischer Assay

ASTM classification by rank		Coke, wt %	Tar, gal/ton	Light oil, gal/ton	Gas, ft³/ton	Water, wt %
Class	Group					
Bituminous	1. Low-volatile bituminous	90	8.6	1.0	1760	3
	2. Medium-volatile bituminous	83	18.9	1.7	1940	4
	3. High-volatile A bituminous	76	30.9	2.3	1970	6
	4. High-volatile B bituminous	70	30.3	2.2	2010	11
	5. High-volatile C bituminous	67	27.0	1.9	1800	16
Subbituminous	1. Subbituminous A	59	20.5	1.7	2660	23
	2. Subbituminous B	58	15.4	1.3	2260	28
Lignite	1. Lignite A	37	15.2	1.2	2100	44

Note: To convert gallons per ton to liters per kilogram, multiply by 0.004; to convert cubic feet per ton to cubic meters per kilogram, multiply by 3.1×10^{-5}.

COAL—TABLE 7. Coal Gasification: Chemical Reactions

Reaction	Reaction heat, kJ/(kg·mol)	Process
Solid-gas reactions		
$C + O_2 \rightarrow CO_2$	+393,790	Combustion
$C + 2H_2 \rightarrow CH_4$	+74,900	Hydrogasification
$C + H_2O \rightarrow CO + H_2$	−175,440	Steam-carbon
$C + CO_2 \rightarrow 2CO$	−172,580	Boudouard
Gas-phase reaction		
$CO + H_2O \rightarrow H_2 + CO_2$	+2,853	Water-gas shift
$CO + 3H_2 \rightarrow CH_2 + H_2O$	+250,340	Methanation
Pyrolysis and hydropyrolysis		
CH_x	$\left(1 - \dfrac{X}{4}\right)C + \left(\dfrac{X}{4}\right)CH_4$	Pyrolysis
$CH_x + m\,H_2$	$\left[1 - \left(\dfrac{X+2m}{4}\right)\right]C + \left(\dfrac{X+2m}{4}\right)CH_4$	Hydropyrolysis

COAL—TABLE 8. Coal Gas Properties

	Coke-oven gas	Producer gas	Water gas	Carbureted water gas	Synthetic coal gas
Reactant system	Pyrolysis	Air + steam	Steam	Steam + oil	Oxygen plus
Analysis, volume %*			(cyclic-air)	(cyclic-air)	steam at pressure
Carbon monoxide, CO	6.8	27.0	42.8	33.4	15.8
Hydrogen, H_2	47.3	14.0	49.9	34.6	40.6
Methane, CH_4	33.9	3.0	0.5	10.4	10.9
Carbon dioxide, CO_2	2.2	4.5	3.0	3.9	31.3
Nitrogen, N_2	6.0	50.9	3.3	7.9	
Other[†]	3.8	0.5	0.5	9.8	2.4
Fuel value, MJ/m^3	22.0	5.6	11.5	20.0	10.8
Btu/ft³	(590)	(150)	(308)	(536)	290
Uses	Fuel, chemicals	Fuel	Fuel, chemicals	Fuel	Fuel, chemicals

* Analyses and fuel values vary with the type of coal and operating conditions.
† Other contents include hydrocarbon gases other than methane, hydrogen sulfide, and small amounts of other impurities.

COAL—TABLE 9. Minerals Commonly Associated with Coal

Group	Species	Formula
Shale	Muscovite	$(K, Na, H_2O, Ca)_2(Al, Mg, Fe, Ti)_4$
	Hydromuscovite	$(Al, Si)_8O_{20}(OH, F)_4$ (general formula)
	Illite	$(HO)_4K_2(Si_6 \cdot Al_2)Al_4O_{20}$
	Montmorillonite	$Na_2(Al\ Mg)Si_4O_{10}(OH)_2$
Kaolin	Kaolinite	$Al_2(Si_2O_5)(OH)_4$
	Livesite	$Al_2(Si_2O_5)(OH)_4$
	Metahalloysite	$Al_2(Si_2O_5)(OH)_4$
Sulfide	Pyrite	FeS_2
	Marcasite	FeS_2
Carbonate	Ankerite	$CaCO_3 \cdot (Mg, Fe, Mn)CO_3$
	Calcite	$CaCO_3$
	Dolomite	$CaCO_3 \cdot MgCO_3$
	Siderite	$FeCO_3$
Chloride	Sylvite	KCl
	Halite	$NaCl$
Accessory minerals	Quartz	SiO_2
	Feldspar	$(K, Na)_2O \cdot Al_2O_3 \cdot 6\ SiO_2$
	Garnet	$3\ CaO \cdot Al_2O_3 \cdot 3\ SiO_2$
	Hornblende	$CaO \cdot 3\ FeO \cdot 4\ SiO_2$
	Gypsum	$CaSO_4 \cdot 2\ H_2O$
	Apatite	$9\ CaO \cdot 3\ P_2O_5 \cdot CaF_2$
	Zircon	$ZrSiO_4$
	Epidote	$4\ CaO \cdot 3\ Al_2O_3 \cdot 6\ SiO_2 \cdot H_2O$
	Biotite	$K_2O \cdot MgO \cdot Al_2O_3 \cdot 3\ SiO_2 \cdot H_2O$
	Augite	$CaO \cdot MgO \cdot 2\ SiO_2$
	Prochlorite	$2\ FeO \cdot 2\ MgO \cdot Al_2O_3 \cdot 2\ SiO_2 \cdot 2\ H_2O$

COAL—TABLE 9. Minerals Commonly Associated with Coal (*Continued*)

Group	Species	Formula
Accessory	Diaspore	$Al_2O_3 \cdot H_2O$
minerals	Lepidocrocite	$Fe_2O_3 \cdot H_2O$
(*Continued*)	Magnetite	Fe_3O_4
	Kyanite	$Al_2O_3 \cdot SiO_2$
	Staurolite	$2 \ FeO \cdot 5 \ Al_2O_3 \cdot 4 \ SiO_2 \cdot H_2O$
	Topaz	$2 \ AlFO \cdot SiO_2$
	Tourmaline	$3 \ Al_2O_3 \cdot 4 \ BO(OH) \cdot 8 \ SiO_2 \cdot 9 \ H_2O$
	Hematite	Fe_2O_3
	Penninite	$5 \ MgO \cdot Al_2O_3 \cdot 3 \ SiO_2 \cdot 2 \ H_2O$
	Sphalerite	ZnS
	Chlorite	$10(Mg, Fe)O \cdot 2 \ Al_2O_3 \cdot 6 \ SiO_2 \cdot 8 \ H_2O$
	Barite	$BaSO_4$
	Pyrophillite	$Al_2O_3 \cdot 4 \ SiO_2 \cdot H_2O$

COAL—TABLE 10. Parr Formulas for Calorific Value and Fixed Carbon

The Parr formulas are used for classifying coal according to rank using calculations for calorific value and fixed carbon:

$$F' = \frac{100(F - 0.15S)}{100 - (M + 1.08A + 0.55S)}$$

$$F' = \frac{100F}{100 - (M + 1.1A + 0.1S)}$$

$$V' = 100 - F'$$

$$Q' = \frac{100(Q - 50S)}{100 - (M + 1.08A + 0.55S)}$$

$$Q' = \frac{100Q}{100 - (1.1A + 0.1S)}$$

$M, F, A,$ and S are the weight percentages (moist basis) of moisture, fixed carbon ash, and sulfur, respectively, F' and V' are the percentages (dry basis) of fixed carbon and volatile matter, respectively. Q (moist basis) and Q' (moist mineral matter free basis) are calorific values (Btu/lb, 1 Btu/lb – 2326 J/kg).

CONVERSION FORMULAS

CONVERSION FORMULAS—TABLE 1. Alphabetical Listing of Common Conversions

To convert	Into	Multiply by
A		
acres	sq chains (Gunter's)	10
acres	sq rods	160
acres	square links (Gunter's)	1×10^5
acres	hectares or sq hectometers	0.4047
acres	sq ft	43,560.0
acres	sq meters	4,047
acres	sq miles	1.562×10^{-3}
acres	sq yd	4,840
acre-feet	cu ft	43,560.0
acre-feet	gallons	3.259×10^5
amperes/sq cm	amps/sq in	6.452
amperes/sq cm	amps/sq meter	10^4
amperes/sq in	amps/sq cm	0.1550
amperes/sq in	amps/sq meter	1,550.0
amperes/sq meter	amps/sq cm	10^{-4}
amperes/sq meter	amps/sq in	6.452×10^{-4}
ampere-hours	coulombs	3,600.0
ampere-hours	faradays	0.03731
angstrom unit	inches	$3,937 \times 10^{-9}$
angstrom unit	meters	1×10^{-10}
angstrom unit	microns or (μ)	1×10^{-4}
astronomical units	kilometers	1.495×10^8
atmospheres	ton/sq in	0.007348
atmospheres	cm of mercury	76.0
atmospheres	mm of mercury	760.0
atmospheres	torrs	760.0
atmospheres	ft of water (at 4°C)	33.90
atmospheres	in of mercury (at 0°C)	29.92
atmospheres	kg/sq cm	1.0333
atmospheres	kg/sq meter	10,332
atmospheres	lb/sq in	14.70
atmospheres	tons/sq ft	1.058
B		
barrels (US, dry)	cu in	7,056
barrels (US, dry)	quarts (dry)	105.0
barrels (US, liq)	gallons	31.5
barrels (oil)	gallons (oil)	42.0
bars	atmospheres	0.9869
bars	dynes/sq cm	10^6
bars	kg/sq meter	1.020×10^4
bars	lb/sq ft	2,089
bars	lb/sq in	14.50
baryes	dynes/sq cm	1.000

CONVERSION FORMULAS—TABLE 1. Alphabetical Listing of Common Conversions (*Continued*)

To convert	Into	Multiply by
Btu	liter—atmosphere	10.409
Btu	ergs	1.0550×10^{10}
Btu	foot-lb	778.3
Btu	gram-calories	252.0
Btu	horsepower-hr	3.931×10^{-4}
Btu	joules	1,054.8
Btu	kilogram-calories	0.2520
Btu	kilogram-meters	107.5
Btu	kilowatt-hr	2.928×10^{-4}
Btu/hr	foot-lb/sec	0.2162
Btu/hr	gram-cal/sec	0.0700
Btu/hr	horsepower-hr	3.929×10^{-4}
Btu/hr	watts	0.2931
Btu/min	foot/lb/sec	12.96
Btu/min	horsepower	0.02356
Btu/min	kilowatts	0.01757
Btu/min	watts	17.57
Btu/sq ft/min	watts/sq in	0.1221
bucket (UK dry)	cubic cm	1.818×10^4
bushels	cu ft	1.2445
bushels	cu in	2,150.42
bushels	cu meters	0.03524
bushels	liters	35.24
bushels	pecks	4.0
bushels	pints (dry)	64.0
bushels	quarts (dry)	32.0
C		
calories, gram (mean)	Btu (mean)	3.9685×10^{-3}
candle/sq cm	lamberts	3.142
candle/sq inch	lamberts	0.4870
centares (centiares)	sq meters	1.0
centigrade	Fahrenheit	$1.8°C + 32$
centigrams	grams	0.01
centiliters	ounces (fl)	0.3382
centiliters	cu in	0.6103
centiliters	drams	2.705
centiliters	liters	0.01
centimeters	feet	3.281×10^{-2}
centimeters	inches	0.3937
centimeters	kilometers	10^{-5}
centimeters	meters	0.01
centimeters	miles	6.214×10^{-6}
centimeters	millimeters	10.0
centimeters	mils	393.7
centimeters	yards	1.094×10^{-2}
centimeter-dynes	cm-grams	1.020×10^{-3}
centimeter-dynes	meter-kg	1.020×10^{-8}
centimeter-dynes	lb-ft	7.376×10^{-8}
centimeter-grams	cm-dynes	980.7

CONVERSION FORMULAS—TABLE 1. Alphabetical Listing of Common
Conversions (*Continued*)

To convert	Into	Multiply by
centimeter-grams	meter-kg	10^{-5}
centimeter-grams	lb-ft	7.233×10^{-5}
centimeters of mercury	atmospheres	0.01316
centimeters of mercury	ft of water	0.4461
centimeters of mercury	kg/sq meter	136.0
centimeters of mercury	lb/sq ft	27.85
centimeters of mercury	lb/sq in	0.1934
centimeters/sec	ft/min	1.1969
centimeters/sec	ft/sec	0.03281
centimeters/sec	kilometers/hr	0.036
centimeters/sec	knots	0.1943
centimeters/sec	meters/min	0.6
centimeters/sec	miles/hr	0.02237
centimeters/sec	miles/min	3.728×10^{-4}
centimeters/sec/sec	ft/sec/sec	0.03281
centimeters/sec/sec	km/hr/sec	0.036
centimeters/sec/sec	meters/sec/sec	0.01
centimeters/sec/sec	miles/hr/sec	0.02237
chain	inches	792.00
chain	meters	20.12
circumference	radians	6.283
coulombs	faradays	1.036×10^{-5}
coulombs/sq cm	coulombs/sq in	64.52
coulombs/sq cm	coulombs/sq meter	10^{4}
coulombs/sq in	coulombs/sq cm	0.1550
coulombs/sq in	coulombs/sq meter	1,550
coulombs/sq meter	coulombs/sq cm	10^{-4}
coulombs/sq meter	coulombs/sq in	6.452×10^{-4}
cubic centimeters	cu ft	3.531×10^{-5}
cubic centimeters	cu in	0.06102
cubic centimeters	cu meters	10^{-6}
cubic centimeters	cu yards	1.308×10^{-4}
cubic centimeters	drams (fl)	0.2705
cubic centimeters	gallons (US)	2.642×10^{-4}
cubic centimeters	gills	8.454×10^{-3}
cubic centimeters	liters	0.001
cubic centimeters	milliliters	1.0
cubic centimeters	minims	16.231
cubic centimeters	ounces (fl)	0.0338
cubic centimeters	pints (liq)	2.113×10^{-3}
cubic centimeters	quarts (liq)	1.057×10^{-3}
cubic feet	bushels	0.8036
cubic feet	cu cm	28,316.85
cubic feet	cu in	1,728.0
cubic feet	cu meters	0.02832
cubic feet	cu yards	0.03704
cubic feet	drams (fl)	7,660.05
cubic feet	gallons (US)	7.48052
cubic feet	gallons (UK)	6.229

CONVERSION FORMULAS—TABLE 1. Alphabetical Listing of Common Conversions (*Continued*)

To convert	Into	Multiply by
cubic feet	gills	239.38
cubic feet	liters	28.32
cubic feet	milliliters	28,316.85
cubic feet	minims	459,603.1
cubic feet	ounces (fl)	957.51
cubic feet	pecks	3.2143
cubic feet	pints (dry)	51.428
cubic feet	pints (liq)	59.84
cubic feet	quarts (dry)	25.714
cubic feet	quarts (liq)	29.92
cubic feet/min	cu cm/sec	472.0
cubic feet/min	gallons/sec	0.1247
cubic feet/min	liters/sec	0.4720
cubic feet/min	lb of water/min	62.43
cubic feet/sec	million gals/day	0.646317
cubic feet/sec	gallons/min	448.831
cubic inches	bushels	4.650×10^{-4}
cubic inches	cu cm	16.39
cubic inches	cu ft	5.787×10^{-4}
cubic inches	cu meters	1.639×10^{-5}
cubic inches	cu yards	2.143×10^{-5}
cubic inches	drams (fl)	4.4329
cubic inches	gallons (US)	4.329×10^{-3}
cubic inches	gallons (UK)	3.605×10^{-3}
cubic inches	gills	0.1385
cubic inches	liters	0.01639
cubic inches	milliliters	16.39
cubic inches	mil-ft	1.061×10^{5}
cubic inches	minims	265.974
cubic inches	ounces (fl)	0.5541
cubic inches	pecks	1.860×10^{-3}
cubic inches	pints (dry)	0.0298
cubic inches	pints (liq)	0.03463
cubic inches	quarts (dry)	0.0149
cubic inches	quarts (liq)	0.01732
cubic meters	bushels	28.38
cubic meters	cu cm	10^{6}
cubic meters	cu ft	35.31
cubic meters	cu in	61,023.74
cubic meters	cu yards	1.308
cubic meters	gallons (US)	264.2
cubic meters	gallons (UK)	220.0
cubic meters	liters	1,000.0
cubic meters	milliliters	10^{6}
cubic meters	pecks	113.51
cubic meters	pints (dry)	1,816.166
cubic meters	pints (liq)	2,113.0
cubic meters	quarts (dry)	908.083
cubic meters	quarts (liq)	1,057

CONVERSION FORMULAS—TABLE 1. Alphabetical Listing of Common Conversions (*Continued*)

To convert	Into	Multiply by
cubic yards	cu cm	7.646×10^5
cubic yards	cu ft	27.0
cubic yards	cu in	46,656.0
cubic yards	cu meters	0.7646
cubic yards	gallons (US)	202.0
cubic yards	gallons (UK)	168.2
cubic yards	liters	764.6
cubic yards	milliliters	7.646×10^5
cubic yards	pints (liq)	1,615.9
cubic yards	quarts (liq)	807.9
cubic yards/min	cubic ft/sec	0.45
cubic yards/min	gallons/sec	3.367
cubic yards/min	liters/sec	12.74

D

Dalton	gram	1.650×10^{-24}
days	seconds	86,400.0
decigrams	grams	0.1
deciliters	liters	0.1
decimeters	meters	0.1
degrees (angle)	quadrants	0.01111
degrees (angle)	radians	0.01745
degrees (angle)	seconds	3,600.0
degrees/sec	radians/sec	0.01745
degrees/sec	revolutions/min	0.1667
degrees/sec	revolutions/sec	2.778×10^{-3}
dekagrams	grams	10.0
dekaliters	liters	10.0
dekameters	meters	10.0
drams (avdp)	drams (apoth)	0.4557
drams (avdp)	grains	27.3437
drams (avdp)	grams	1.7718
drams (avdp)	kilograms	1.7718×10^{-3}
drams (avdp)	milligrams	1,771.85
drams (avdp)	ounces (apoth or troy)	0.0570
drams (avdp)	ounces (avdp)	0.0625
drams (avdp)	pennyweights	1.139
drams (avdp)	pounds (apoth or troy)	4.747×10^{-3}
drams (avdp)	pounds (avdp)	3.906×10^{-3}
drams (avdp)	scruples	1.367
drams (apoth)	drams (avdp)	2.1943
drams (apoth)	grains	60.0
drams (apoth)	grams	3.8879
drams (apoth)	kilograms	3.888×10^{-3}
drams (apoth)	milligrams	3,887.93
drams (apoth)	ounces (apoth or troy)	0.125
drams (apoth)	ounces (avdp)	0.1371429
drams (apoth)	pennyweights	2.5
drams (apoth)	pounds (apoth or troy)	0.0104
drams (apoth)	pounds (avdp)	8.571×10^{-3}

CONVERSION FORMULAS—TABLE 1. Alphabetical Listing of Common Conversions (*Continued*)

To convert	Into	Multiply by
drams (apoth)	scruples	3.0
drams (fl)	cu cm	3.6967
drams (fl)	cu ft	1.3055×10^{-4}
drams (fl)	cu in	0.2256
drams (fl)	gallons (US)	9.7656×10^{-4}
drams (fl)	gills	0.03125
drams (fl)	liters	3.6967×10^{-3}
drams (fl)	milliliters	3.6967
drams (fl)	minims	60.0
drams (fl)	ounces (fl)	0.125
drams (fl)	pints (liq)	7.8125×10^{-3}
drams (fl)	quarts (liq)	3.9063×10^{-3}
dynes	grams	1.020×10^{-3}
dynes	joules/cm	10^{-7}
dynes	joules/meter (newtons)	10^{-5}
dynes	kilograms	1.020×10^{-6}
dynes	poundals	7.233×10^{-5}
dynes	pounds	2.248×10^{-6}
dynes/cm	ergs/sq mm	0.01
dynes/sq cm	atmospheres	9.869×10^{-7}
dynes/sq cm	bars	10^{-6}
dynes/sq cm	in of mercury at 0°C	2.953×10^{-5}
dynes/sq cm	in of water at 4°C	4.015×10^{-4}
	E	
ergs	Btu	9.480×10^{-11}
ergs	dyne-centimeters	1.0
ergs	foot-lb	7.367×10^{-8}
ergs	gram-calories	0.2389×10^{-7}
ergs	gram-cm	1.020×10^{-3}
ergs	horsepower-hr	3.7250×10^{-14}
ergs	joules	10^{-7}
ergs	kg-calories	2.389×10^{-11}
ergs	kg-meters	1.020×10^{-8}
ergs	kilowatt-hr	0.2778×10^{-13}
ergs	watt-hr	0.2778×10^{-10}
ergs/sec	Btu/min	5.668×10^{-6}
ergs/sec	dyne-cm/sec	1.000
ergs/sec	ft-lb/min	4.427×10^{-6}
ergs/sec	ft-lb/sec	7.3756×10^{-8}
ergs/sec	horsepower	1.341×10^{-10}
ergs/sec	kg-calories/min	1.433×10^{-9}
ergs/sec	kilowatts	10^{-10}
	F	
Fahrenheit	centigrade/celcius	$0.556 F° - 17.8$
farads	microfarads	10^6
faradays/sec	ampere (abs)	9.6500×10^4
faradays	ampere-hr	26.80

CONVERSION FORMULAS—TABLE 1. Alphabetical Listing of Common Conversions (*Continued*)

To convert	Into	Multiply by
faradays	coulombs	9.649×10^4
fathoms	meters	1.828804
fathoms	feet	6.0
feet	centimeters	30.48
feet	inches	12.0
feet	kilometers	3.048×10^{-4}
feet	meters	0.3048
feet	miles (nautical)	1.645×10^{-4}
feet	miles (statute)	1.894×10^{-4}
feet	millimeters	304.8
feet	mils	1.2×10^4
feet	yards	0.333
feet of water	atmospheres	0.02950
feet of water	in of mercury	0.8826
feet of water	kg/sq cm	0.03048
feet of water	kg/sq meter	304.8
feet of water	lb/sq ft	62.43
feet of water	lb/sq in	0.4335
feet/min	cm/sec	0.5080
feet/min	feet/sec	0.01667
feet/min	km/hr	0.01829
feet/min	meters/min	0.3048
feet/min	miles/hr	0.01136
feet/sec	cm/sec	30.48
feet/sec	km/hr	1.097
feet/sec	knots	0.5921
feet/sec	meters/min	18.29
feet/sec	miles/hr	0.6818
feet/sec	miles/min	0.01136
feet/sec/sec	cm/sec/sec	30.48
feet/sec/sec	km/hr/sec	1.097
feet/sec/sec	meters/sec/sec	0.3048
feet/sec/sec	miles/hr/sec	0.6818
foot-candles	lumens/sq meter	10.764
foot-pounds	Btu	1.286×10^{-3}
foot-pounds	ergs	1.356×10^7
foot-pounds	gram-calories	0.3238
foot-pounds	hp-hr	5.050×10^{-7}
foot-pounds	joules	1.356
foot-pounds	kg-calories	3.24×10^{-4}
foot-pounds	kg-meters	0.1383
foot-pounds	kilowatt-hr	3.766×10^{-7}
foot-pounds/min	Btu/min	1.286×10^{-3}
foot-pounds/min	foot-lb/sec	0.01667
foot-pounds/min	horsepower	3.030×10^{-5}
foot-pounds/min	kg-calories/min	3.24×10^{-4}
foot-pounds/min	kilowatts	2.260×10^{-5}
foot-pounds/sec	Btu/hr	4.6263
foot-pounds/sec	Btu/min	0.07717

CONVERSION FORMULAS—TABLE 1. Alphabetical Listing of Common Conversions (*Continued*)

To convert	Into	Multiply by
foot-pounds/sec	horsepower	1.818×10^{-3}
foot-pounds/sec	kg-calories/min	0.01945
foot-pounds/sec	kilowatts	1.356×10^{-3}
furlongs	miles (US)	0.125
furlongs	rods	40.0
furlongs	feet	660.0
G		
gallons (US)	cu cm	3,785.0
gallons (US)	cu ft	0.1337
gallons (US)	cu in	231.0
gallons (US)	cu meters	3.785×10^{-3}
gallons (US)	cu yards	4.951×10^{-3}
gallons (US)	drams (fl)	1,024.0
gallons (US)	gallons (UK)	0.83267
gallons (US)	gills	32.0
gallons (US)	liters	3.785
gallons (US)	milliliters	3,785.0
gallons (US)	minims	61,440.0
gallons (US)	ounces (fl)	128.0
gallons (US)	pints (liq)	8.0
gallons (US)	quarts (liq)	4.0
gallons (UK)	cu ft	0.1605
gallons (UK)	cu in	277.4
gallons (UK)	cu meters	4.546×10^{-3}
gallons (UK)	cu yards	5.946×10^{-3}
gallons (UK)	gallons (US)	1.20095
gallons (UK)	liters	4.546
gallons of water	lb of water	8.3453
gallons/min	cu ft/sec	2.228×10^{-3}
gallons/min	liters/sec	0.06308
gallons/min	cu ft/hr	8.0208
gauss	lines/sq in	6.452
gauss	webers/sq cm	10^{-8}
gauss	webers/sq in	6.452×10^{-8}
gauss	webers/sq meter	10^{-4}
gilberts	ampere-turns	0.7958
gilberts/cm	amp-turns/cm	0.7958
gilberts/cm	amp-turns/in	2.021
gilberts/cm	amp-turns/meter	79.58
grams	drams (apoth)	0.2572
grams	drams (avdp)	0.5644
grams	dynes	980.7
grams	grains	15.43
grams	joules/cm	9.807×10^{-5}
grams	joules/meter (newtons)	9.807×10^{-3}
grams	kilograms	0.001
grams	milligrams	1,000.0
grams	ounces (apoth or troy)	0.03215
grams	ounces (avdp)	0.03527

CONVERSION FORMULAS—TABLE 1. Alphabetical Listing of Common Conversions (*Continued*)

To convert	Into	Multiply by
grams	pennyweights	0.643
grams	pounds (apoth or troy)	2.679×10^{-3}
grams	pounds (avdp)	2.205×10^{-3}
grams	poundals	0.07093
grams	scruples	0.7716
grams	slugs	6.852×10^{-5}
grams	tons (short)	1.102×10^{-6}
grams/cm	lb/in	5.600×10^{-3}
grams/cu cm	lb/cu ft	62.43
grams/cu cm	lb/cu in	0.03613
grams/cu cm	lb/mil-ft	3.405×10^{-7}
grams/liter	grains/gal	58.417
grams/liter	lb/1,000 gal	8.345
grams/liter	lb/cu ft	0.062427
grams/liter	parts/million	1,000.0
grams/sq cm	lb/sq ft	2.0481
gram-calories	Btu	3.9693×10^{-3}
gram-calories	ergs	4.1868×10^{7}
gram-calories	ft/lb	3.0880
gram-calories	horsepower-hr	1.5596×10^{-6}
gram-calories	kilowatt-hr	1.1630×10^{-6}
gram-calories	watt-hr	1.1630×10^{-3}
gram-calories/sec	Btu/hr	14.286
gram-centimeters	Btu	9.297×10^{-8}
gram-centimeters	ergs	980.7
gram-centimeters	joules	9.807×10^{-5}
gram-centimeters	kg-cal	2.343×10^{-8}
gram-centimeters	kg-meters	10^{-5}
H		
horsepower	Btu/min	42.44
horsepower	ft-lb/min	33,000
horsepower	ft-lb/sec	550.0
horsepower	horsepower (metric)	1.014
horsepower	kg-calories/min	10.68
horsepower	kilowatts	0.7457
horsepower	watts	745.7
horsepower (boiler)	Btu/hr	33,479
horsepower (boiler)	kilowatts	9.803
horsepower (metric)	horsepower	0.9863
horsepower (metric)	ft-lb/sec	542.5
horsepower-hours	Btu	2,547
horsepower-hours	ergs	2.6845×10^{13}
horsepower-hours	ft-lb	1.98×10^{6}
horsepower-hours	gm-cal	641,190
horsepower-hours	joules	2.684×10^{6}
horsepower-hours	kg-calories	641.1
horsepower-hours	kg-meters	2.737×10^{5}
horsepower-hours	kilowatt-hr	0.7457
hours	days	4.167×10^{-2}

CONVERSION FORMULAS—TABLE 1. Alphabetical Listing of Common Conversions (*Continued*)

To convert	Into	Multiply by
hours	weeks	5.952×10^{-3}
hundredweights (long)	cwt (short)	1.12
hundredweights (long)	kilograms	50.802
hundredweights (long)	ounces	1,792.0
hundredweights (long)	pounds	112
hundredweights (long)	slugs	3.4811
hundredweights (long)	tons (long)	0.05
hundredweights (long)	tons (short)	0.056
hundredweights (short)	cwt (short)	0.8929
hundredweights (short)	kilograms	45.359
hundredweights (short)	ounces	1,600.0
hundredweights (short)	pounds	100.0
hundredweights (short)	slugs	3.1081
hundredweights (short)	tons (metric)	0.0453592
hundredweights (short)	tons (long)	0.0446429
hundredweights (short)	tons (short)	0.05

I

To convert	Into	Multiply by
inches	centimeters	2.540
inches	feet	0.0833
inches	meters	2.540×10^{-2}
inches	miles	1.578×10^{-5}
inches	miles (naut)	1.3715×10^{-5}
inches	millimeters	25.40
inches	mils	1,000.0
inches	yards	2.778×10^{-2}
inches of mercury	atmospheres	0.03342
inches of mercury	ft of water	1.133
inches of mercury	kg/sq cm	0.03453
inches of mercury	kg/sq meter	345.3
inches of mercury	lb/sq ft	70.73
inches of mercury	lb/sq in	0.4912
inches of water (at 4°C)	atmospheres	2.458×10^{-3}
inches of water (at 4°C)	in of mercury	0.07355
inches of water (at 4°C)	kg/sq cm	2.540×10^{-3}
inches of water (at 4°C)	oz/sq in	0.5781
inches of water (at 4°C)	lb/sq ft	5.204
inches of water (at 4°C)	lb/sq in	0.03613
international ampere	ampere (abs)	0.9998
international volt	volts (abs)	1.0003
international volt	joules	9.654×10^{4}

J

To convert	Into	Multiply by
joules	Btu	9.480×10^{-4}
joules	ergs	10^{7}
joules	ft-lb	0.7376
joules	kg-calories	2.389×10^{-4}
joules	kg-meters	0.1020
joules	watt-hr	2.778×10^{-4}

CONVERSION FORMULAS—TABLE 1. Alphabetical Listing of Common Conversions (*Continued*)

To convert	Into	Multiply by
joules/cm	grams	1.020×10^4
joules/cm	dynes	10^7
joules/cm	joules/meter (newtons)	100.0
joules/cm	poundals	723.3
joules/cm	pounds	22.48
L		
liters	milliliters	1,000
liters	minims	16,230.73
liters	ounces (fl)	33.814
liters	pecks	0.1135
liters	pints (dry)	1.8162
liters	pints (liq)	2.113
liters	quarts (dry)	0.9081
liters	quarts (liq)	1.057
liters/min	cu ft/sec	5.886×10^{-4}
liters/min	gal/sec	4.403×10^{-3}
lumens/sq ft	foot-candles	1.0
lumen	spherical candle power	0.07958
lumen	watt	0.001496
lumen/sq ft	lumen/sq meter	10.76
lux	foot-candles	0.0929
M		
meters	centimeters	100.0
meters	feet	3.281
meters	inches	39.37
meters	kilometers	0.001
meters	miles (naut)	5.396×10^{-4}
meters	miles (stat)	6.214×10^{-4}
meters	millimeters	1,000.0
meters	rods	0.1988
meters	yards	1.094
meters	varas	1.179
meters/min	cm/sec	1.667
meters/min	ft/min	3.281
meters/min	ft/sec	0.05468
meters/min	km/hr	0.06
meters/min	knots	0.03238
meters/min	miles/hr	0.03728
meters/sec	feet/min	196.8
meters/sec	feet/sec	3.281
meters/sec	kilometers/hr	3.6
meters/sec	kilometers/min	0.06
meters/sec	miles/hr	2.237
meters/sec	miles/min	0.03728
meters/sec/sec	cm/sec/sec	100.0
meters/sec/sec	ft/sec/sec	3.281
meters/sec/sec	km/hr/sec	3.6
meters/sec/sec	miles/hr/sec	2.237

CONVERSION FORMULAS—TABLE 1.   Alphabetical Listing of Common
Conversions (*Continued*)

To convert	Into	Multiply by
meter-kilograms	cm-dynes	9.807×10^7
meter-kilograms	cm-gm	10^5
meter-kilograms	lb-ft	7.233
microns	meters	1×10^{-6}
miles (naut)	feet	6,080.27
miles (naut)	inches	7.2913×10^4
miles (naut)	kilometers	1.853
miles (naut)	meters	1,853
miles (naut)	miles (statute)	1.1516
miles (naut)	yards	2,027
miles (statute)	centimeters	1.609×10^5
miles (statute)	feet	5,280
miles (statute)	inches	6.336×10^4
miles (statute)	kilometers	1.609
miles (statute)	meters	1.609
miles (statute)	miles (naut)	0.8684
miles (statute)	rods	320
miles (statute)	yards	1,760
miles/hr	cm/sec	44.70
miles/hr	ft/min	88
miles/hr	ft/sec	1.467
miles/hr	km/hr	1.609
miles/hr	km/min	0.02682
miles/hr	knots	0.8684
miles/hr	meters/min	26.82
miles/hr	miles/min	0.1667
miles/hr/sec	cm/sec/sec	44.70
miles/hr/sec	ft/sec/sec	1.467
miles/hr/sec	km/hr/sec	1.609
miles/hr/sec	meters/sec/sec	0.4470
miles/min	cm/sec	2,682
miles/min	ft/sec	88
miles/min	km/min	1.609
miles/min	knots/min	0.8684
miles/min	miles/hr	60.0
mil-feet	cu in	9.425×10^{-6}
milliers	kilograms	1,000
milligrams	drams (apoth)	2.572×10^{-4}
milligrams	drams (avdp)	5.644×10^{-4}
milligrams	grains	0.01543236
milligrams	grams	0.001^3
milligrams	kilograms	10^{-6}
milligrams	ounces (apoth or troy)	3.215×10^{-5}
milligrams	ounces (avdp)	3.527×10^{-5}
milligrams	pennyweights	6.43×10^{-4}
milligrams	pounds (apoth or troy)	2.679×10^{-6}
milligrams	pounds (avdp)	2.2046×10^{-6}
milligrams	scruples	7.7162×10^{-4}
milligrams/liter	parts/million	1.0

CONVERSION FORMULAS—TABLE 1. Alphabetical Listing of Common
Conversions (*Continued*)

To convert	Into	Multiply by
millihenries	henries	0.001
milliliters	cu cm	1.0
milliliters	cu ft	3.531×10^{-5}
milliliters	cu in	0.06102
milliliters	drams (fl)	0.2705
milliliters	gallons (US)	2.642×10^{-4}
milliliters	gills	8.454×10^{-3}
milliliters	liters	0.001
milliliters	minims	16.231
milliliters	ounces (fl)	0.0338
milliliters	pints (liq)	2.113×10^{-3}
milliliters	quarts (liq)	1.057×10^{-3}
millimeters	centimeters	0.1
millimeters	feet	3.281×10^{-3}
millimeters	inches	0.03937
millimeters	kilometers	10^{-6}
millimeters	meters	0.001
millimeters	miles	6.214×10^{-7}
millimeters	mils	39.37
millimeters	yards	1.094×10^{-3}
millimeters of Hg	atmospheres	1.316×10^{-3}
millimeters of Hg	torr	1.0
millimicrons	meters	1×10^{-9}
million gals/day	cu ft/sec	1.54723
mils	centimeters	2.540×10^{-3}
mils	feet	8.333×10^{-5}
mils	inches	0.001
mils	kilometers	2.540×10^{-8}
mils	yards	2.778×10^{-5}
miner's inches	cu ft/min	1.5
minims (UK)	cu cm	0.059192
minims	cu cm	0.061612
minims	cu ft	2.176×10^{-6}
minims	cu in	3.7598×10^{-3}
minims	drams (fl)	0.0167
minims	gallons (US)	1.628×10^{-5}
minims	gills	5.208×10^{-4}
minims	liters	6.161×10^{-5}
minims	milliliters	0.061612
minims	ounces (fl)	0.0021
minims	pints (liq)	1.302×10^{-4}
minims (liq)	quarts (liq)	6.51×10^{-5}
minutes (angles)	degrees	0.01667
minutes (angles)	quadrants	1.852×10^{-4}
minutes (angles)	radians	2.909×10^{-4}
minutes (angles)	seconds	60.0
N		
newtons	dynes	1×10^{5}

CONVERSION FORMULAS—TABLE 1.   Alphabetical Listing of Common
Conversions (*Continued*)

To convert	Into	Multiply by
	O	
ohm (Int)	ohm (abs)	1.0005
ohms	megohms	10^{-6}
ohms	microhms	10^{6}
ounces (avdp)	cwt (long)	5.5804×10^{-4}
ounces (avdp)	cwt (short)	6.25×10^{-4}
ounces (avdp)	drams (apoth)	7.292
ounces (avdp)	drams (avdp)	16.0
ounces (avdp)	grains	437.5
ounces (avdp)	grams	28.349527
ounces (avdp)	kilograms	0.0283
ounces (avdp)	milligrams	28,349.5
ounces (avdp)	ounces (apoth or troy)	0.9115
ounces (avdp)	pennyweights	18.23
ounces (avdp)	pounds (apoth or troy)	0.0759
ounces (avdp)	pounds (avdp)	0.0625
ounces (avdp)	scruples	21.875
ounces (avdp)	slugs	1.9426×10^{-3}
ounces (avdp)	tons (long)	2.790×10^{-5}
ounces (avdp)	tons (metric)	2.835×10^{-5}
ounces (avdp)	tons (short)	3.125×10^{-5}
ounces (apoth or troy)	drams (apoth)	8.0
ounces (apoth or troy)	drams (avdp)	17.554
ounces (apoth or troy)	grains	480.0
ounces (apoth or troy)	grams	31.103481
ounces (apoth or troy)	kilograms	0.0311
ounces (apoth or troy)	milligrams	31,103.48
ounces (apoth or troy)	ounces (avdp)	1.09714
ounces (apoth or troy)	pennyweights	20.0
ounces (apoth or troy)	pounds (apoth or troy)	0.08333
ounces (apoth or troy)	pounds (avdp)	0.0686
ounces (apoth or troy)	scruples	24.0
ounces (fl)	cu cm	29.5735
ounces (fl)	cu ft	1.0444×10^{-3}
ounces (fl)	cu in	1.805
ounces (fl)	drams (fl)	8.0
ounces (fl)	gallons (US)	7.8125×10^{-3}
ounces (fl)	gills	0.25
ounces (fl)	liters	0.02957
ounces (fl)	milliliters	29.5735
ounces (fl)	minims	480.0
ounces (fl)	pints (liq)	0.0625
ounces (fl)	quarts (liq)	0.03125
ounces/sq in	dynes/sq cm	4309
ounces/sq in	pounds/sq in	0.0625
	P	
parsecs	miles	19×10^{12}
parsecs	kilometers	3.084×10^{13}
parts/million	grains/US gal	0.0584

CONVERSION FORMULAS—TABLE 1. Alphabetical Listing of Common
Conversions (*Continued*)

To convert	Into	Multiply by
parts/million	grains/UK gal	0.07016
parts/million	lb/million gal	8.345
pints (dry)	bushels	0.0156
pints (dry)	cu ft	0.0194
pints (dry)	cu in	33.60
pints (dry)	cu meters	5.506×10^{-4}
pints (dry)	liters	0.5506
pints (dry)	pecks	0.0625
pints (dry)	quarts (dry)	0.5
pints (liq)	cu cm	473.2
pints (liq)	cu ft	0.01672
pints (liq)	cu in	28.875
pints (liq)	cu meters	4.732×10^{-4}
pints (liq)	cu yards	6.189×10^{-4}
pints (liq)	drams (fl)	128.0
pints (liq)	gallons (US)	0.125
pints (liq)	gills	4.0
pints (liq)	liters	0.4732
pints (liq)	milliliters	473.2
pints (liq)	minims	7,680.0
pints (liq)	ounces (fl)	16.0
pints (liq)	quarts (liq)	0.5
Planck's quantum	erg-sec	6.624×10^{-27}
poise	gm/cm sec	1.00
poundals	dynes	13,826
poundals	grams	14.10
poundals	joules/cm	1.383×10^{-3}
poundals	joules/meter (newtons)	0.1383
poundals	kilograms	0.01410
poundals	pounds	0.03108
pounds (avdp)	cwt (long)	8.929×10^{-3}
pounds (avdp)	cwt (short)	0.01
pounds (avdp)	drams (apoth)	116.67
pounds (avdp)	drams (avdp)	256.0
pounds (avdp)	dynes	44.4823×10^{4}
pounds (avdp)	grains	7,000.0
pounds (avdp)	grams	453.5924
pounds (avdp)	joules/cm	0.04448
pounds (avdp)	joules/meter (newtons)	4.448
pounds (avdp)	kilograms	0.4536
pounds (avdp)	milligrams	453,592.37
pounds (avdp)	ounces (apoth or troy)	14.5833
pounds (avdp)	ounces (avdp)	16.0
pounds (avdp)	pennyweights	291.667
pounds (avdp)	poundals	32.17
pounds (avdp)	pounds (apoth or troy)	1.21528
pounds (avdp)	scruples	350.0
pounds (avdp)	slugs	3.108×10^{-2}
pounds (avdp)	tons (long)	4.464×10^{-4}

CONVERSION FORMULAS—TABLE 1. Alphabetical Listing of Common
Conversions (*Continued*)

To convert	Into	Multiply by
pounds (avdp)	tons (metric)	4.536×10^{-4}
pounds (avdp)	tons (short)	5.0×10^{-4}
pounds (apoth or troy)	drams (apoth)	96.0
pounds (apoth or troy)	drams (avdp)	210.65
pounds (apoth or troy)	grains	5,760.0
pounds (apoth or troy)	grams	373.2417
pounds (apoth or troy)	kilograms	0.3732
pounds (apoth or troy)	milligrams	373,241.72
pounds (apoth or troy)	ounces (apoth or troy)	12.0
pounds (apoth or troy)	ounces (avdp)	13.1657
pounds (apoth or troy)	pennyweights	240.0
pounds (apoth or troy)	pounds (avdp)	0.822857
pounds (apoth or troy)	scruples	288.0
pounds (apoth or troy)	tons (long)	3.6753×10^{-4}
pounds (apoth or troy)	tons (metric)	3.7324×10^{-4}
pounds (apoth or troy)	tons (short)	4.1143×10^{-4}
pounds of water	cu ft	0.01602
pounds of water	cu in	27.68
pounds of water	gallons	0.1198
pounds of water/min	cu ft/sec	2.670×10^{-4}
pound-feet	cm-dynes	1.356×10^{7}
pound-feet	cm-gm	13,825
pound-feet	meter-kg	0.1383
pounds/cu ft	gm/cu cm	0.01602
pounds/cu ft	kg/cu meter	16.02
pounds/cu ft	lb/cu in	5.787×10^{-4}
pounds/cu ft	lb/mil-foot	5.456×10^{-9}
pounds/cu in	gm/cu cm	27.68
pounds/cu in	kg/cu meter	2.768×10^{4}
pounds/cu in	lb/cu ft	1,728
pounds/cu in	lb/mil-foot	9.425×10^{-6}
pounds/ft	kg/meter	1.488
pounds/in	gm/cm	178.6
pounds/mil-foot	gm/cu cm	2.306×10^{6}
pounds/sq ft	atmospheres	4.725×10^{-4}
pounds/sq ft	ft of water	0.01602
pounds/sq ft	in of mercury	0.01414
pounds/sq ft	kg/sq meter	4.882
pounds/sq ft	lb/sq in	6.944×10^{-3}
pounds/sq in	atmospheres	0.06804
pounds/sq in	ft of water	2.307
pounds/sq in	in of mercury	2.036
pounds/sq in	kg/sq meter	703.1
pounds/sq in	lb/sq ft	144.0
Q		
quadrants (angle)	degrees	90.0
quadrants (angle)	minutes	5,400.0
quadrants (angle)	radians	1.571
quadrants (angle)	seconds	3.24×10^{5}

CONVERSION FORMULAS—TABLE 1. Alphabetical Listing of Common Conversions (*Continued*)

To convert	Into	Multiply by
quarts (dry)	bushels	0.0313
quarts (dry)	cu ft	0.0389
quarts (dry)	cu in	67.20
quarts (dry)	cu meters	1.101×10^{-3}
quarts (dry)	liters	1.1012
quarts (dry)	pecks	0.125
quarts (dry)	pints (dry)	2.0
quarts (liq)	cu cm	946.4
quarts (liq)	cu ft	0.03342
quarts (liq)	cu in	57.75
quarts (liq)	cu meters	9.464×10^{-4}
quarts (liq)	cu yards	1.238×10^{-3}
quarts (liq)	drams (fl)	256.0
quarts (liq)	gallons (US)	0.25
quarts (liq)	gills	8.0
quarts (liq)	liters	0.9464
quarts (liq)	milliliters	946.4
quarts (liq)	minims	15,360.0
quarts (liq)	ounces (fl)	32.0
quarts (liq)	pints (liq)	2.0
R		
radians	degrees	57.30
radians	minutes	3,438
radians	quadrants	0.6366
radians	seconds	2.063×10^5
radians/sec	degrees/sec	57.30
radians/sec	rev/min	9.549
radians/sec	rev/sec	0.1592
radians/sec/sec	rev/min/min	573.0
radians/sec/sec	rev/min/sec	9.549
radians/sec/sec	rev/sec/sec	0.1592
revolutions	degrees	360.0
revolutions	quadrants	4.0
revolutions	radians	6.283
revolutions/min	degrees/sec	6.0
revolutions/min	radians/sec	0.1047
revolutions/min	rev/sec	0.01667
revolutions/min/min	radians/sec/sec	1.745×10^{-3}
revolutions/min/min	rev/min/sec	0.01667
revolutions/min/min	rev/sec/sec	2.778×10^{-4}
revolutions/sec	degrees/sec	360.0
revolutions/sec	radians/sec	6.283
revolutions/sec	rev/min	60.0
revolutions/sec/sec	radians/sec/sec	6.283
revolutions/sec/sec	rev/min/min	3,600.0
revolutions/sec/sec	rev/min/sec	60.0
S		
seconds (angle)	degrees	2.778×10^{-4}
seconds (angle)	minutes	0.01667

CONVERSION FORMULAS—TABLE 1. Alphabetical Listing of Common Conversions (*Continued*)

To convert	Into	Multiply by
seconds (angle)	quadrants	3.087×10^{-6}
seconds (angle)	radians	4.848×10^{-6}
sphere	steradians	12.57
square centimeters	circular mils	1.973×10^{5}
square centimeters	sq ft	1.076×10^{-3}
square centimeters	sq in	0.1550
square centimeters	sq meters	0.0001
square centimeters	sq miles	3.861×10^{-11}
square centimeters	sq mm	100.0
square centimeters	sq yd	1.196×10^{-4}
square feet	acres	2.296×10^{-5}
square feet	circular mils	1.833×10^{8}
square feet	sq cm	929.0
square feet	sq in	144.0
square feet	sq meters	0.09290
square feet	sq miles	3.587×10^{-8}
square feet	sq mm	9.290×10^{4}
square feet	sq yd	0.1111
square inches	circular mils	1.273×10^{6}
square inches	sq cm	6.452
square inches	sq ft	6.944×10^{-3}
square inches	sq mm	645.2
square inches	sq mils	10^{6}
square inches	sq yd	7.716×10^{-4}
square kilometers	acres	247.1
square kilometers	sq cm	10^{10}
square kilometers	sq ft	1.076×10^{7}
square kilometers	sq in	1.550×10^{9}
square kilometers	sq meters	10^{6}
square kilometers	sq miles	0.3861
square kilometers	sq yd	1.196×10^{6}
square meters	acres	2.471×10^{-4}
square meters	sq cm	10^{4}
square meters	sq ft	10.76
square meters	sq in	1,550
square meters	sq miles	3.861×10^{-7}
square meters	sq mm	10^{6}
square meters	sq yd	1.196
square miles	acres	640.0
square miles	sq ft	2.788×10^{7}
square miles	sq km	2.590
square miles	sq meters	2.590×10^{6}
square miles	sq yd	3.098×10^{6}
square millimeters	circular mils	1,973
square millimeters	sq cm	0.01
square millimeters	sq ft	1.076×10^{-5}
square millimeters	sq in	1.550×10^{-3}
square mils	circular mils	1.273
square mils	sq cm	6.452×10^{-6}

CONVERSION FORMULAS—TABLE 1. Alphabetical Listing of Common
Conversions (*Continued*)

To convert	Into	Multiply by
square mils	sq in	10^{-6}
square yards	acres	2.066×10^{-4}
square yards	sq cm	8,361
square yards	sq ft	9.0
square yards	sq in	1,296
square yards	sq meters	0.8361
square yards	sq miles	3.228×10^{-7}
square yards	sq mm	8.361×10^{5}
	T	
temperature (°C) + 273	absolute temperature (°C)	1.0
temperature (°C) + 17.78	temperature (°F)	1.8
temperature (°F) + 460	absolute temperature (°F)	1.0
temperature (°F) − 32	temperature (°C)	5/9
tons (long)	cwt (long)	20
tons (long)	cwt (short)	22.4
tons (long)	kilograms	1,016
tons (long)	ounces (avdp)	35,840.0
tons (long)	pounds (avdp)	2,240.0
tons (long)	slugs	69.621
tons (long)	tons (metric)	1.0160
tons (long)	tons (short)	1.120
tons (metric)	cwt (short)	22.046
tons (metric)	kilograms	1,000.0
tons (metric)	ounces (avdp)	35,273.96
tons (metric)	pounds (avdp)	2,205
tons (metric)	tons (long)	0.9842
tons (metric)	tons (short)	1.1023
tons (short)	cwt (long)	17.857
tons (short)	cwt (short)	20.0
tons (short)	grams	9.072×10^{5}
tons (short)	kilograms	907.1847
tons (short)	ounces (apoth or troy)	29,166.66
tons (short)	ounces (avdp)	32,000.0
tons (short)	pounds (apoth or troy)	2,430.56
tons (short)	pounds (avdp)	2,000.0
tons (short)	slugs	62.16
tons (short)	tons (long)	0.89286
tons (short)	tons (metric)	0.9072
tons (short)/sq ft	kg/sq meter	9,765
tons (short)/sq ft	lb/sq in	2,000
tons of water/24 hr	lb of water/hr	83.333
tons of water/24 hr	gal/min	0.16643
tons of water/24 hr	cu ft/hr	1.3349
torr	mm of mercury	1.0
torr	atmospheres	1.316×10^{-3}
	V	
volt/inch	volt/cm	0.39370
volt (abs)	statvolts	0.003336

CONVERSION FORMULAS—TABLE 1. Alphabetical Listing of Common Conversions (*Continued*)

To convert	Into	Multiply by
W		
watts	Btu/hr	3.4129
watts	Btu/min	0.05688
watts	ergs/sec	10^7
watts	ft-lb/min	44.27
watts	ft-lb/sec	0.7378
watts	horsepower	1.341×10^{-3}
watts	horsepower (metric)	1.360×10^{-3}
watts	kg-calories/min	0.01433
watts	kilowatts	0.001
watts (abs)	Btu (mean)/min	0.056884
watts (abs)	joules/sec	1
watt-hours	Btu	3.413
watt-hours	ergs	3.60×10^{10}
watt-hours	ft-lb	2,656
watt-hours	gm-cal	859.85
watt-hours	horsepower-hr	1.341×10^{-3}
watt-hours	kg-cal	0.8605
watt-hours	kg-meters	367.2
watt-hours	kilowatt-hr	0.001
watt (int)	watt (abs)	1.0002
webers	maxwells	10^8
webers	kilolines	10^5
webers/sq in	gauss	1.550×10^7
webers/sq in	lines/sq in	10^8
webers/sq in	webers/sq cm	0.1550
webers/sq in	webers/sq meter	1,550
webers/sq meter	gauss	10^4
webers/sq meter	lines/sq in	6.452×10^4
webers/sq meter	webers/sq cm	10^{-4}
webers/sq meter	webers/sq in	6.452×10^{-4}
Y		
yards	centimeters	91.44
yards	feet	3.0
yards	inches	36.0
yards	kilometers	9.144×10^{-4}
yards	meters	0.9144
yards	miles (naut)	4.934×10^{-4}
yards	miles (stat)	5.682×10^{-4}
yards	millimeters	914.4

CONVERSION FORMULAS—TABLE 2. Kinematic-Viscosity Conversion Formulas

Viscosity scale	Range of t, sec	Kinematic viscosity, stokes
Saybolt Universal	$32 < t < 100$	$0.00226t–1.95/t$
	$t > 100$	$0.00220t–1.35/t$
Saybolt Furol	$25 < t < 40$	$0.0224t–1.84/t$
	$t > 40$	$0.0216t–0.60/t$
Redwood No. 1	$34 < t < 100$	$0.00260t–1.79/t$
	$t > 100$	$0.00247t–0.50/t$
Redwood Admiralty		$0.027t–20/t$
Engler		$0.00147t–3.74/t$

CONVERSION FORMULAS—TABLE 3. Values of the Gas-Law Constant

Temperature scale	Pressure units	Volume units	Weight units	Energy units	R
Kelvin			g-moles	calories	1.9872
			g-moles	joules (abs)	8.3144
			g-moles	joules (int)	8.3130
	atm	cm^3	g-moles	atm cm^3	82.057
	atm	liters	g-moles	atm liters	0.08205
	mm Hg	liters	g-moles	mm Hg-liters	62.361
	bar	liters	g-moles	bar-liters	0.08314
	kg/cm^2	liters	g-moles	$kg/(cm^2)(liters)$	0.08478
	atm	ft^3	lb-moles	atm-ft^3	1.314
	mm Hg	ft^3	lb-moles	mm Hg-ft^3	998.9
			lb-moles	chu or pcu	1.9872
Rankine			lb-moles	Btu	1.9872
			lb-moles	hp-hr	0.0007805
			lb-moles	kw-hr	0.0005819
	atm	ft^3	lb-moles	atm-ft^3	0.7302
	in Hg	ft^3	lb-moles	in Hg-ft^3	21.85
	mm Hg	ft^3	lb-moles	mm Hg-ft^3	555.0
	lb/in^2abs	ft^3	lb-moles	$(lb)(ft^3)/in^2$	10.73
	lb/ft^2abs	ft^3	lb-moles	ft-lb	1,545.0

CRUDE OIL

Crude oil (*petroleum*) is the term used to describe myriad hydrocarbon-rich fluids that have accumulated in subterranean reservoirs. Crude oil varies dramatically in color, odor, and flow properties that reflect the diversity of its origin.

Crude oil may be called *light* or *heavy* depending on the amount of its low-boiling constituents and its relative density (specific gravity). Likewise, odor is used to distinguish between *sweet* (low sulfur) and *sour* (high sulfur) crude oil. Viscosity indicates the ease of (or, more correctly, the resistance to) flow.

Although not directly derived from oil's composition, the terms *light* or *heavy* or *sweet* and *sour* are convenient for use in descriptions. For example, *light crude oil* (often referred to as *conventional crude oil*) is usually rich in low-boiling constituents and waxy molecules, whereas *heavy crude oil* contains greater proportions of higher-boiling, more aromatic, and heteroatom-containing (N-, O-, S-, and metal-containing) constituents. Heavy crude oil is more viscous than conventional crude oil and requires enhanced methods for recovery. *Bitumen* is *near solid* or *solid* and cannot be recovered through enhanced oil recovery methods.

Conventional (light) crude oil comprises hydrocarbons and smaller amounts of organic compounds of nitrogen, oxygen, and sulfur as well as still smaller amounts of compounds containing metallic constituents, particularly vanadium, nickel, iron, and copper. The processes by which crude oil was formed dictate that petroleum composition varies and is *site specific*, thus leading to a wide variety of compositional differences. The use of the term *site specific* it is intended to convey that petroleum composition is dependent on the proportion of the various precursors that went into the formation of the *protopetroleum*, as well as variations in temperature and pressure to which the precursors were subjected.

DIELECTRIC CONSTANT

The *dielectric constant* (also referred to as the *relative permittivity, K*) is the ratio of the permittivity of the material to the permittivity of free space and is the property of a material that determines the relative speed that an electrical signal will travel through that material.

$$K = \varepsilon_T/\varepsilon_0 \tag{12}$$

Signal speed is roughly inversely proportional to the square root of the dielectric constant. A low dielectric constant results in a high signal propagation speed, and a high dielectric constant results in a much slower signal propagation speed.

The *dielectric loss factor* is the tangent of the loss angle, and the *loss tangent* (tan Δ) is defined by the relationship:

$$\tan \Delta = 2\sigma/\varepsilon\upsilon \tag{13}$$

where σ is the electrical conductivity, ε is the dielectric constant, and υ is the frequency. The loss tangent is roughly wavelength independent.

DIELECTRIC CONSTANT—TABLE 1. Dielectric Constants of Hydrocarbons and Petroleum Products

Material	Temperature °C	°F	Dielectric constant
n-Hexane	0	32	1.918
	20	68	1.890
	60	140	1.817
n-Heptane	0	32	1.958
	20	68	1.930
	60	140	1.873
Benzene	10	50	2.296
	20	68	2.283
	60	140	2.204
Cyclohexane	20	68	2.055
Petroleum products			
Gasoline	20	68	1.8–2.0
Kerosene	20	68	2.0–2.2
Lubricating oil	20	68	2.1–2.6

DIELECTRIC CONSTANT—TABLE 2. Dielectric Constants of Inorganic Compounds

Compound	Dielectric constant	Loss tangent $\tan \Delta \times 10^4$
Air	1.0	
Air (dry) (68°F)	1.00	
Alumina	4.5	
Aluminum bromide (212°F)	3.4	
Aluminum fluoride	2.2	
Aluminum hydroxide	2.2	
Aluminum oleate (68°F)	2.4	
Aluminum phosphate	6.0	
Ammonia (−74°F)	25.0	
Ammonia (−30°F)	22.0	
Ammonia (40°F)	18.9	
Ammonia (69°F)	16.5	
Ammonium bromide	7.2	
Ammonium chloride	7.0	
Antimony trichloride	5.3	
Antimony pentachloride (68°F)	3.2	
Antimony tribromide (212°F)	20.9	
Antimony trichloride (166°F)	33.0	

DIELECTRIC CONSTANT—TABLE 2. Dielectric Constants of Inorganic Compounds (*Continued*)

Compound	Dielectric constant	Loss tangent tan $\Delta \times 10^4$
Antimony trichloride	5.3	
Antimony tri-iodide (347°F)	13.9	
Argon (−376°F)	1.5	
Argon (68°F)	1.00	
Arsenic tribromide (98°F)	9.0	
Arsenic trichloride (150°F)	7.0	
Arsenic trichloride (70°F)	12.4	
Arsenic tri-iodide (302°F)	7.0	
Arsine (−148°F)	2.5	
Asbestos	3.0–4.8	
Ash (Fly)	1.7–2.6	
Barium chloride	9.4	
Barium chloride (anhydrous)	11.0	
Barium chloride ($2H_2O$)	9.4	
Barium nitrate	5.8	
Barium sulfate (60°F)	11.4	
Bromine (68°F)	3.1	
Bromine (32°F)	1.01	
Carbon dioxide (68°F)	1.00	
Calcium fluoride	7.4	
Calcite	8.0	
Calcium	3.0	
Calcium carbonate	6.1–9.1	
Calcium fluoride	7.4	
Calcium oxide, granule	11.8	
Calcium sulfate	5.6	
Calcium sulfate (H_2O)	5.6	
Calcium superphosphate	14–15	
Carbon black	2.5–3.0	
Carbon dioxide (32°F)	1.6	
Carbon dioxide, liquid	1.6	
Cassiterite	23.4	
Cement	1.5–2.1	
Cement, Portland	2.5–2.6	
Cesium iodide	5.6	
Charcoal	1.81	
Chlorine (−50°F)	2.1	
Chlorine (32°F)	2.0	
Chlorine (142°F)	1.5	
Chlorine, liquid	2.0	
Chromite	4.0–4.2	
Chromyl chloride (68°F)	2.6	
Clay	1.8–2.8	
Copper oleate (68°F)	2.8	
Copper oxide	18.1	
Cupric oleate	2.8	

DIELECTRIC CONSTANT—TABLE 2. Dielectric Constants of Inorganic
Compounds (*Continued*)

Compound	Dielectric constant	Loss tangent $\tan \Delta \times 10^4$
Cupric oxide (60°F)	18.1	
Cupric sulfate	10.3	
Cupric sulfate (Anhydrous)	10.3	
Cupric sulfate ($5H_2O$)	7.8	
Deuterium (68°F)	1.3	
Deuterium oxide (77°F)	78.3	
Diamond	5.68	
Dinitrogen oxide (32°F)	1.6	
Dinitrogen tetroxide (58°F)	2.5	
Dolomite	6.8–8.0	
Ferric oleate (68°F)	2.6	
Ferrochromium	1.5–1.8	
Ferromanganese	5.0–5.2	
Ferrous oxide (60°F)	14.2	
Ferrous sulfate (58°F)	14.2	
Fluorine (−332°F)	1.5	
Fluorspar	6.8	
Fly ash	1.7–2.6	
Fuller's earth	1.8–2.2	
Germanium tetrachloride (77°F)	2.4	
Glass	3.7–10	
Glass (Silica)	3.8	
Graphite	12–15	
Gypsum (68°F)	6.3	
Helium-3 (58°F)	1.065	
Helium, liquid	1.05	
Hydrazine (68°F)	52.0	
Hydrochloric acid (68°F)	4.6	
Hydrocyanic acid (70°F)	2.3	
Hydrocyanic acid (32°F)	158.0	
Hydrogen (440°F)	1.23	
Hydrogen (212°F)	1.00	
Hydrogen iodide (72°F)	2.9	
Hydrogen bromide (24°F)	3.8	
Hydrogen bromide (−120°F)	7.0	
Hydrogen chloride (82°F)	4.6	
Hydrogen chloride (−188°F)	12.0	
Hydrogen cyanide (70°F)	95.4	
Hydrogen fluoride (32°F)	84.2	
Hydrogen fluoride (−100°F)	17.0	
Hydrogen iodide (72°F)	2.9	
Hydrogen peroxide (32°F)	84.2	
Hydrogen peroxide 100%	70.7	
Hydrogen peroxide 35%	121.0	
Hydrogen sulfide (−84°F)	9.3	

DIELECTRIC CONSTANT—TABLE 2.   Dielectric Constants of Inorganic Compounds (*Continued*)

Compound	Dielectric constant	Loss tangent tan $\Delta \times 10^4$
Hydrogen sulfide (48°F)	5.8	
Hydrofluoric acid (32°F)	83.6	
Ilmenite	6.0–7.0	
Iodine (107°F)	118.0	
Iodine (250°F)	118.0	
Iodine (granular)	4.0	
Iron oxide	14.2	
Lead oxide	25.9	
Lead acetate	2.5	
Lead carbonate (60°F)	18.1	
Lead chloride	4.2	
Lead monoxide (60°F)	25.9	
Lead nitrate	37.7	
Lead oleate (64°F)	3.2	
Lead oxide	25.9	
Lead sulfate	14.3	
Lead sulfite	17.9	
Lead tetrachloride (68°F)	2.8	
Lime	2.2–2.5	
Liquefied air	1.5	
Liquefied hydrogen	1.2	
Lithium chloride	11.1	
Manganese dioxide	5.2	
Magnesium oxide	9.7	
Magnesium sulfate	8.2	
Malachite	7.2	
Mercuric chloride	3.2	
Mercurous chloride	9.4	
Mercury (298°F)	1.00	
Mica	6.9–9.2	2
Neon (68°F)	1.00	
Nitric acid (14°F)	50.0	
Nitrogen (336°F)	1.45	
Nitrogen (68°F)	1.00	
Nitrosyl bromide (4°F)	13.0	
Nitrosyl chloride (10°F)	18.0	
Nitrous oxide (32°F)	1.6	
Oxygen (−315°F)	1.51	
Oxygen (68°F)	1.00	
Phosgene (32°F)	4.7	
Phosphine (−76°F)	2.5	
Phosphorus (93°F)	4.1	

DIELECTRIC CONSTANT—TABLE 2. Dielectric Constants of Inorganic Compounds (*Continued*)

Compound	Dielectric constant	Loss tangent tan $\Delta \times 10^4$
Phosphorus oxychloride (72°F)	14.0	
Phosphorus pentachloride (320°F)	2.8	
Phosphorus tribromide (68°F)	3.9	
Phosphorus trichloride (77°F)	3.4	
Phosphorus, red	4.1	
Phosphorus, yellow	3.6	
Phosphoryl chloride (70°F)	13.0	
Potassium aluminum Sulfate	3.8	
Potassium carbonate (60°F)	5.6	
Potassium chlorate	5.1	
Potassium chloride	4.6	
Potassium iodide	5.6	
Potassium nitrate	5.0	
Potassium sulfate	5.9	
Potassium chloride	5.0	
Quartz (68°F)	4.49	
Rutile	6.7	
Salt	3.0–15.0	
Sand (dry)	2.5–5.0	
Selenium (68°F)	6.1	
Selenium (482°F)	5.4	
Selenium (249°F)	5.4	
Silica aluminate	2.0	
Silicon	11.0–12.0	
Silicon dioxide	4.5	
Silicon tetrachloride (60°F)	2.4	
Silver bromide	12.2	
Silver chloride	11.2	
Silver cyanide	5.6	
Slaked lime	2.0–3.5	
Sodium carbonate	5.3–8.4	
Sodium carbonate (anhydrous)	8.4	
Sodium carbonate ($10H_2O$)	5.3	
Sodium chloride	5.9	
Sodium cyanide	7.55	
Sodium dichromate	2.9	
Sodium nitrate	5.2	
Sodium oleate (68°F)	2.7	
Sodium perchlorate	5.4	
Sodium phosphate	1.6–1.9	
Sodium perchlorate	5.4	
Sodium sulfide	5.0	
Stannic chloride (72°F)	3.2	
Sulfur	1.6–1.7	7–14
Sulfur dioxide (−4°F)	17.6	
Sulfur dioxide (32°F)	15.0	

DIELECTRIC CONSTANT—TABLE 2.   Dielectric Constants of Inorganic
Compounds (*Continued*)

Compound	Dielectric constant	Loss tangent tan $\Delta \times 10^4$
Sulfur monochloride (58°F)	4.8	
Sulfur trioxide (64°F)	3.1	
Sulfurous oxychloride (72°F)	9.1	
Sulfuryl chloride (72°F)	10.0	
Sulfur (244°F)	3.5	
Sulfur (450°F)	3.5	
Sulfur dioxide (32°F)	15.6	
Sulfur trioxide (70°F)	3.6	
Sulfur, liquid	3.5	
Sulfur, powder	3.6	
Sulfuric acid (68°F)	84.0	
Sulfuric acid (25°C)	100.0	
Sulfuric oxychloride (72°F)	9.2	
Tantalum oxide	11.6	
Thallium chloride	46.9	
Thionyl bromide (68°F)	9.1	
Thionyl chloride (68°F)	9.3	
Thiophosphoryl chloride (70°F)	5.8	
Thorium oxide	10.6	
Tin tetrachloride (68°F)	2.9	
Titanium tetrachloride (68°F)	2.8	
Titanium dioxide	110.0	
Titanium oxide	40–50	
Titanium tetrachloride (68°F)	2.8	
Tourmaline	6.3	
Vanadium oxybromide (78°F)	3.6	
Vanadium oxychloride (78°F)	3.4	
Vanadium sulfide	3.1	
Vanadium tetrachloride (78°F)	3.0	
Water (32°F)	88.0	
Water (68°F)	80.10	
Water (80°F)	80.0	
Water (212°F)	55.3	
Water (390°F)	34.5	
Water (steam)	1.01	
Zinc oxide	1.7–2.5	
Zinc sulfide	8.2	
Zircon	12.0	
Zirconium oxide	12.5	
Zirconium silicate	5.0	

DIELECTRIC CONSTANT—TABLE 3. Dielectric Constants of Organic Compounds

Compound	Dielectric constant	Loss tangent $\tan \Delta \times 10^4$
Acenaphthene (70°F)	3.0	
Acetal (70°F)	3.6	
Acetal Bromide	16.5	
Acetal dioxime (68°F)	3.4	
Acetaldehyde (41°F)	21.8	
Acetamide (68°F)	41.0	
Acetamide (180°F)	59.0	
Acetanilide (71°F)	2.9	
Acetic acid (68°F)	6.2	
Acetic acid (36°F)	4.1	
Acetic anhydride (66°F)	21.0	
Acetone (77°F)	20.7	
Acetone (127°F)	17.7	
Acetone (32°F)	1.0159	
Acetonitrile (68°F)	37.5	
Acetonitrile (70°F)	37.5	
Acetophenone (75°F)	17.3	
Acetoxime (24°F)	3.0	
Acetyl acetone (68°F)	23.1	
Acetyl bromide (68°F)	16.5	
Acetyl chloride (68°F)	15.8	
Acetyl acetone (68°F)	25.0	
Acetylene (32°F)	1.0217	
Acetylmethyl hexyl ketone (66°F)	27.9	
Allyl alcohol (58°F)	22.0	
Allyl bromide (66°F)	7.0	
Allyl chloride (68°F)	8.2	
Allyl iodide (66°F)	6.1	
Allyl isothiocyanate (64°F)	17.2	
Amyl acetate (68°F)	5.1	
iso-Amyl acetate (68°F)	5.6	
iso-Amyl alcohol (74°F)	15.3	
Amyl alcohol (−180°F)	35.5	
Amyl alcohol (68°F)	15.8	
Amyl alcohol (140°F)	11.2	
Amylamine (72°F)	4.6	
Amyl benzoate (68°F)	5.1	
Amyl bromide (50°F)	6.3	
iso-Amyl bromide (76°F)	6.1	
iso-Amyl butyrate (68°F)	3.9	
Amyl chloride (52°F)	6.6	
iso-Amyl chloride (64°F)	6.4	
iso-Amyl chloroacetate (68°F)	7.8	
iso-Amyl chloroformate (68°F)	7.8	
Amyl ether (60°F)	3.1	
Amyl formate (66°F)	5.7	
Amyl iodide (62°F)	6.9	
iso-Amyl Iodide (65°F)	5.6	

DIELECTRIC CONSTANT—TABLE 3.   Dielectric Constants of Organic Compounds (*Continued*)

Compound	Dielectric constant	Loss tangent $\tan \Delta \times 10^4$
Amyl nitrate (62°F)	9.1	
Amyl thiocyanate (68°F)	17.4	
iso-Amyl propionate (68°F)	4.2	
iso-Amyl salicylate (68°F)	5.4	
iso-Amyl valerate (19°F)	3.6	
Aniline (32°F)	7.8	
Aniline (68°F)	7.21	
Aniline (212°F)	5.5	
Anisaldehyde (68°F)	15.8	
Anisole (68°F)	4.3	
Azoxyanisole (122°F)	2.3	
Azoxybenzene (104°F)	5.1	
Azoxyphenitole (302°F)	6.8	
Benzal chloride (68°F)	6.9	
Benzaldehyde (68°F)	17.8	
Benzaldoxime (68°F)	3.8	
Benzene (50°F)	2.29	
Benzene (68°F)	2.28	
Benzene (140°F)	2.20	
Benzil (202°F)	13.0	
Benzonitrile (68°F)	26.0	
Benzophenone (122°F)	11.4	
Benzophenone (68°F)	13.0	
Benzotrichloride (68°F)	7.4	
Benzoyl chloride (70°F)	22.1	
Benzoyl chloride (32°F)	23.0	
Benzoylacetone (68°F)	29.0	
Benzyl acetate (70°F)	5.0	
Benzyl alcohol (68°F)	13.0	
Benzyl benzoate (68°F)	4.8	
Benzyl chloride (68°F)	6.4	
Benzyl cyanide (68°F)	18.3	
Benzyl cyanide (155°F)	6.0	
Benzyl salicylate (68°F)	4.1	
Benzylamine (68°F)	4.6	
Benzylethylamine (68°F)	4.3	
Benzylmethylamine (67°F)	4.4	
Biphenyl	20.0	
Bornyl acetate (70°F)	4.6	
Bromal (70°F)	7.6	
Bromoacetyl bromide (68°F)	12.6	
Bromoaniline (68°F)	13.0	
m-Bromoaniline (66°F)	13.0	
Bromoanisole (86°F)	7.1	
Bromobenzene (68°F)	5.4	
Bromobutylene (68°F)	5.8	
Bromobutyric acid (68°F)	7.2	

DIELECTRIC CONSTANT—TABLE 3. Dielectric Constants of Organic
Compounds (*Continued*)

Compound	Dielectric constant	Loss tangent $\tan\Delta \times 10^4$
Bromo-octadecane	3.5	
Bromodecane (76°F)	4.4	
Bromododecane (76°F)	4.1	
Bromododocosane (130°F)	3.1	
Bromododecane (75°F)	4.07	
Bromoform (68°F)	4.4	
Bromoheptane (76°F)	5.3	
Bromohexadecane (76°F)	3.7	
Bromohexane (76°F)	5.8	
Bromoisovaleric acid (68°F)	6.5	
Bromomethane (32°F)	9.8	
Bromonaphthalene (66°F)	5.1	
Bromooctadecane (86°F)	3.5	
Bromopentadecane (68°F)	3.9	
o-Bromotoluene (68°F)	5.1	
o-Bromotoluene (137°F)	4.3	
m-Bromotoluene (137°F)	5.4	
p-Bromotoluene	5.5	
Bromotridecane (50°F)	4.2	
Bromoundecane (15°F)	4.7	
Butane (30°F)	1.4	
Butanol (1) (68°F)	17.8	
Butanone (68°F)	18.5	
Butyric anhydride (20°F)	12.0	
n-Butyl acetate (68°F)	5.01	
n-Butyl acetate (19°F)	5.1	
n-Butyl Alcohol (77°F)	17.51	
iso-Butyl acetate (68°F)	5.6	
iso-Butyl alcohol (−112°F)	31.7	
iso-Butyl alcohol (68°F)	16.68	
iso-Butyl alcohol (32°F)	20.5	
iso-Butyl alcohol (68°F)	18.7	
iso-Butylamine (70°F)	4.5	
iso-Butylbenzene (62°F)	2.3	
iso-Butylbenzoate (68°F)	5.9	
iso-Butyl bromide (20°F)	4.0	
n-Butyl bromide (68°F)	6.6	
n-Butyl iodide (77°F)	6.1	
iso-Butyl iodide (68°F)	5.8	
iso-Butyl nitrate (66°F)	11.9	
Butylamine (70°F)	5.4	
iso-Butylamine (70°F)	4.5	
iso-Butyl butyrate (68°F)	4.0	
Butyl chloral (64°F)	10.0	
n-Butyl chloride (68°F)	7.39	
iso-Butyl chloride (68°F)	7.1	
iso-Butyl chloroformate (68°F)	9.2	
iso-Butyl cyanide (74°F)	13.3	

DIELECTRIC CONSTANT—TABLE 3. Dielectric Constants of Organic Compounds (*Continued*)

Compound	Dielectric constant	Loss tangent $\tan \Delta \times 10^4$
iso-Butylene bromide (68°F)	4.0	
n-Butyl formate (−317°F)	2.4	
iso-Butyl formate (66°F)	6.5	
iso-Butyl iodide (68°F)	5.8	
iso-Butyl nitrate (66°F)	11.9	
Butyl oleate (77°F)	4.0	
Butyl stearate (80°F)	3.1	
Butyraldehyde (79°F)	13.4	
n-Butyric acid (68°F)	2.9	
iso-Butyric acid (68°F)	2.7	
Butyric Anhydride (68°F)	12.0	
Butyronitrile (70°F)	20.7	
iso-Butyric acid (68°F)	2.6	
iso-Butyric acid (122°F)	2.7	
iso-Butyric anhydride (68°F)	13.9	
iso-Butyyronitrile (77°F)	20.8	
iso Butyronitrile	20.8	
Camphanedione (398°F)	16.0	
Camphene (68°F)	2.7	
Camphene (104°F)	2.3	
Camphor, Crystal	10–11	
Caproic acid (160°F)	2.6	
Caprolactam	1.7	
iso-Capronitrile (68°F)	15.7	
Carbon disulfide, liquid	2.6	
Carbon disulphide (68°F)	2.62	
Carbon disulphide (180°F)	2.2	
Carbon tetrachloride (68°F)	2.24	
Carvenone (68°F)	18.4	
Carvol (64°F)	11.2	
Carvone (71°F)	11.0	
Cellulose	3.2–7.5	
Cellulose acetate	3.2–7.0	300–400
Cellulose nitrate (proxylin)	6.4	
Cetyl iodide (68°F)	3.3	
Chloroacetic acid (140°F)	12.3	
Chloral (68°F)	4.9	
Chlorhexanone oxime	3.0	
Chloroacetic acid (68°F)	21.0	
Chloroacetone (68°F)	29.8	
m-Chloroanaline (66°F)	13.4	
Chlorobenzene (77°F)	5.62	
Chlorobenzene (68°F)	5.6	
Chlorobenzene (100°F)	4.7	
Chlorobenzene (230°F)	4.1	
Chlorocyclohexane (76°F)	7.6	
Chloroform (32°F)	5.5	

DIELECTRIC CONSTANT—TABLE 3. Dielectric Constants of Organic Compounds (*Continued*)

Compound	Dielectric constant	Loss tangent $\tan \Delta \times 10^4$
Chloroform (68°F)	4.81	
Chloroform (212°F)	3.7	
Chloroheptane (71°F)	5.5	
Chlorohexanone oxime (192°F)	3.0	
Chlorohydrate (68°F)	3.3	
Chloromethane (−4°F)	12.6	
Chloronaphthalene (76°F)	5.0	
Chlorooctane (76°F)	5.1	
Chloroheptane	5.4	
o-Chlorophenol (66°F)	8.2	
p-Chlorophenol (130°F)	9.5	
o-Chlorotoluene (68°F)	4.5	
m-Chlorotoluene (68°F)	5.6	
p-Chlorotoluene (68°F)	6.1	
Cholesterol (80°F)	2.9	
Chorine (170°F)	1.7	
Cinnamaldehyde (75°F)	16.9	
Citraconic anhydride (68°F)	40.3	
Citraconic nitrile	27.0	
Cocaine (68°F)	3.1	
o-Cresol (77°F)	11.5	
m-Cresol (75°F)	5.0	
p-Cresol (24°F)	5.0	
p-Cresol (70°F)	5.6	
p-Cresol (137°F)	9.9	
Cumene (68°F)	2.4	
Cyanoacetic acid (40°F)	33.0	
Cyanoethyl acetate (68°F)	19.3	
Cyanogen (73°F)	2.6	
Cyclohexane (68°F)	2.02	
Cyclohexanecarboxylic acid (88°F)	2.6	
Cyclohexanemethanol (140°F)	9.7	
Cyclohexanone (68°F)	18.2	
Cyclohexanol (77°F)	15.0	
Cyclohexanone (68°F)	18.2	
Cyclohexanone oxime (192°F)	3.0	
Cyclohexene (68°F)	18.3	
Cyclohexylamine (−5°F)	5.3	
Cyclohexylphenol (130°F)	4.0	
Cyclohexyltrifluoromethane (68°F)	11.0	
Cyclopentane (68°F)	1.97	
p-Cymene (63°F)	2.3	
Decahydronaphthalene (68°F)	2.2	
Decanal	8.1	
Decane (68°F)	2.0	
Decanol (68°F)	8.1	
Diacetoxybutane (76°F)	6.64	

DIELECTRIC CONSTANT—TABLE 3. Dielectric Constants of Organic
Compounds (*Continued*)

Compound	Dielectric constant	Loss tangent $\tan \Delta \times 10^4$
Diallyl sulfide (68°F)	4.9	
Dibenzofuran (212°F)	3.0	
Dibenzyl sebacate (68°F)	4.6	
Dibenzylamine (68°F)	3.6	
Dibromoheptane (24°F)	5.08	
p-Dibromobenzene (190°F)	4.5	
Dibromobutane (68°F)	5.7	
Dibromoethylene (Cis-1, 2) (32°F)	7.7	
Dibromoheptane (76°F)	5.1	
Dibromohexane (76°F)	5.0	
Dibromomethane (50°F)	7.8	
Dibromopropane (68°F)	4.3	
Dibromopropyl alcohol (70°F)	9.1	
Dibutyl phthalate (86°F)	6.4	
Dibutyl sebacate (86°F)	4.5	
Dibutyl tartrate (109°F)	9.4	
Dichloroacetic acid (20°F)	10.7	
Dichloroacetic acid (72°F)	8.2	
Dichloroacetone (68°F)	14.0	
o-Dichlorobenzene (68°F)	9.93	
o-Dichlorobenzene (77°F)	7.5	
m-Dichlorobenzene (77°F)	5.0	
p-Dichlorobenzene (68°F)	2.86	
p-Dichlorobenzene (120°F)	2.4	
1,1-Dichloroethane	10.7	
1,2-Dichloroethane (68°F)	10.66	
1,2-Dichloroethane (77°F)	10.7	
Dichloroethylene (62°F)	4.6	
Dichloromethane (68°F)	8.93	
Dichlorostyrene (76°F)	2.6	
Dichlorotoluene (68°F)	6.9	
Dicyclohexyl adipate (95°F)	4.8	
Diebenzylamine (68°F)	3.6	
1,2-Diethoxyethane (75°F)	3.8	
Diethyl disulfide (66°F)	15.9	
Diethyl glutarate (86°F)	6.7	
Diethyl ketone (58°F)	17.3	
Diethyl malonate (70°F)	7.9	
Diethyl oxalate (70°F)	8.2	
Diethyl sebacate (86°F)	5.0	
Diethyl succinate (86°F)	6.6	
Diethyl sulfide (68°F)	7.2	
Diethyl sulfite (68°F)	15.9	
Diethyl tartrate (68°F)	4.5	
Diethyl zinc (68°F)	2.6	
Diethylamine (68°F)	3.7	
Diethylaniline (66°F)	5.5	

DIELECTRIC CONSTANT—TABLE 3. Dielectric Constants of Organic Compounds (*Continued*)

Compound	Dielectric constant	Loss tangent $\tan \Delta \times 10^4$
Dimethylamine (64°F)	2.5	
Di-isoamylene (62°F)	2.4	
Di-iodoethylene 1 (80°F)	4.0	
Di-iodomethane (77°F)	5.3	
Di-isoamyl (62°F)	2.0	
Di-isoamylene	2.4	
Di-isobutylamine (71°F)	2.7	
Dimethoxybenzene (73°F)	4.5	
Dimethylacetamide (77°F)	37.78	
3-Dimethyl-2-butanone (293°F)	13.1	
Dimethyl ethyl (68°F)	11.7	
Dimethyl ethyl Carbinol (68°F)	11.7	
N,N-Dimethylformamide (77°F)	36.71	
Dimethyl malonate (68°F)	10.4	
Dimethyl oxalate (68°F)	3.0	
Dimethyl pentane (20°F)	1.9	
Dimethyl phthalate (75°F)	8.5	
Dimethyl sulfate (68°F)	55.0	
Dimethyl sulfoxide (68°F)	46.68	
Dimethyl sulfide (68°F)	6.3	
Dimethyl-1-hydroxybenzene (62°F)	4.8	
Dimethyl-2-hexane (68°F)	2.4	
Dimethylamine (32°F)	6.3	
Dimethylaniline (68°F)	4.4	
Dimethylbromoethylene (68°F)	6.7	
Dimethylpentane (68°F)	1.9	
Dimethylquinoxaline (76°F)	2.3	
Dimethyltoluidine (68°F)	3.3	
m-Dinitrobenzene (68°F)	2.8	
Dioctyl phthalate (76°F)	5.1	
1,4-Dioxane (68°F)	2.25	
1,4-Dioxane (77°F)	2.20	
Dipalmitin (161°F)	3.5	
Dipentene (68°F)	2.3	
Dipenylamine (125°F)	3.3	
Diphenylmethane (230°F)	2.4	
Diphenylmethane (62°F)	12.6	
	2.5	
Diphenyl ether (82°F)	3.9	
Diphenylamine (124°F)	3.3	
Diphenylethane (110°F)	2.38	
Diphenylmethane (62°F)	2.6	
Dipropyl ketone (62°F)	12.6	
Dipropylamine (70°F)	2.9	
Distearin (172°F)	3.3	
Docosane (122°F)	2.0	
Dodecane (68°F)	2.0	
Dodecanol (76°F)	6.5	
Dodecyne (76°F)	2.2	

DIELECTRIC CONSTANT—TABLE 3. Dielectric Constants of Organic Compounds (*Continued*)

Compound	Dielectric constant	Loss tangent $\tan \Delta \times 10^4$
Epichlorohydrin (68°F)	22.9	
Ethanediamine (68°F)	14.2	
Ethanethiol (58°F)	6.9	
Ethanethiolic acid (68°F)	13.0	
Ethanol (77°F)	24.3	
Ethoxy-3-methylbutane (68°F)	4.0	
Ethoxybenzene (68°F)	4.2	
Ethoxyethyl acetate (86°F)	7.6	
Ethoxypentane (73°F)	3.6	
Ethoxytoluene (68°F)	3.9	
Ethyl acetate (68°F)	6.4	
Ethyl acetate (77°F)	6.02	
Ethyl acetoacetate (71°F)	15.9	
Ethyl acetoneoxalate (66°F)	16.1	
Ethyl alcohol (Eahanol) (68°F)	25.7	
Ethyl aclohol (Ethanol) (77°F)	24.55	
Ethylamine (70°F)	6.3	
Ethylaniline (68°F)	5.9	
Ethyl amyl ether (68°F)	4.0	
Ethyl benzene (68°F)	2.5	
Ethylbenzene (76°F)	3.0	
Ethyl benzoate (68°F)	6.0	
Ethyl benzoylacetate (68°F)	12.8	
Ethyl benzoylacetoacetate (70°F)	8.6	
Ethyl benzyl ether (68°F)	3.8	
Ethyl bromide (64°F)	4.9	
Ethyl Bromoisobutyrate (68°F)	7.9	
Ethyl Bromopropionate (68°F)	9.4	
Ethyl butyrate (66°F)	5.1	
Ethyl carbonate (68°F)	3.1	
Ethyl carbonate (121°F)	14.2	
Ethyl cellulose	2.8–3.9	
Ethyl chloroacetate (68°F)	11.6	
Ethyl chloroformate (68°F)	11.3	
Ethyl chloropropionate (68°F)	10.1	
Ethyl cinnamate (66°F)	5.3	
Ethyl cyanoacetate (68°F)	27.0	
Ethyl cyclobutane (68°F)	2.0	
Ethyl dodecanoate (68°F)	3.4	
Ethylene chloride (68°F)	10.5	
Ethylene chlorohydrin (77°F)	26.0	
Ethylene cyanide (136°F)	58.3	
Ethylene diamine (64°F)	16.0	
Ethylene diamine (18°F)	16.0	
Ethylene dichloride (68°F)	10.36	
Ethylene glycol (68°F)	37.0	
Ethylene glycol, dimethyl ether (77°F)	7.20	
Ethylene iodide	3.4	

DIELECTRIC CONSTANT—TABLE 3. Dielectric Constants of Organic Compounds (*Continued*)

Compound	Dielectric constant	Loss tangent $\tan \Delta \times 10^4$
Ethylene oxide (−1°F)	13.5	
Ethylene oxide (77°F)	14.0	
Ethylene tetrafluoride	1.9	
Ethylenechlorohydrin (75°F)	25.0	
Ethylenediamine (64°F)	16.0	
Ethyl ether (−148°F)	8.1	
Ethyl ether (−40°F)	5.7	
Ethyl ether (68°F)	4.34	
Ethyl ethoxybenzoate (70°F)	7.1	
Ethyl formate (77°F)	7.1	
Ethyl fumarate (73°F)	6.5	
Ethyl iodide (68°F)	7.4	
Ethyl isothiocyanate (68°F)	19.7	
Ethyl maleate (73°F)	8.5	
Ethyl mercaptan (68°F)	8.0	
Ethyl nitrate (68°F)	19.7	
Ethyl oleate (80°F)	3.2	
Ethyl palmitate (68°F)	3.2	
Ethyl phenylacetate (70°F)	5.4	
Ethyl propionate (68°F)	5.7	
Ethyl salicylate (70°F)	8.6	
Ethyl silicate (68°F)	4.1	
Ethyl stearate (104°F)	3.0	
Ethyl thiocyanate (68°F)	29.6	
Ethyl trichloroacetate (68°F)	7.8	
Ethyl undecanoate (68°F)	3.6	
Ethyl valerate (68°F)	4.7	
Ethyl 1-bromobutyrate (68°F)	8.0	
Ethyl 2-iodopropionate (68°F)	8.8	
Ethylpentane (68°F)	1.9	
Ethyltoluene (76°F)	2.2	
Fluorotoluene (86°F)	4.2	
Formalin	23.0	
Formamide (68°F)	84.0	
Formic acid (60°F)	58.0	
Furan (77°F)	3.0	
Furfural (68°F)	42.0	
Furfuraldehyde (68°F)	41.9	
Glycerin (68°F)	43.0	
Glycerol (77°F)	42.5	
Glycerol (32°F)	47.2	
Glycerol phthalate	3.7–4.0	
Glyceryl triacetate (70°F)	6.0	
Glycol (77°F)	37.0	
Glycol (122°F)	35.6	
Glycolic nitrile (68°F)	27.0	

DIELECTRIC CONSTANT—TABLE 3.   Dielectric Constants of Organic
Compounds (*Continued*)

Compound	Dielectric constant	Loss tangent $\tan \Delta \times 10^4$
Heptadecanone (140°F)	5.3	
n-Heptane (32°F)	1.96	
n-Heptane (68°F)	1.93	
n-Heptane (140°F)	1.87	
Heptanoic acid	2.5	
Heptanone (68°F)	11.9	
Heptanoic acid (71°F)	2.59	
Heptanoic acid (160°F)	2.6	
1-Heptene (68°F)	2.1	
Heptyl alcohol (70°F)	6.7	
n-Hexane (32°F)	1.92	
n-Hexane (68°F)	1.89	
n-Hexane (77°F)	1.89	
n-Hexane (140°F)	1.82	
Hexanol (77°F)	13.3	
Hexanone (59°F)	14.6	
cis-3-Hexene (76°F)	2.1	
trans-3-Hexene (76°F)	2.0	
Hexyl iodide (68°F)	6.6	
Hexylene (62°F)	2.0	
Hexyl iodide (68°F)	6.6	
Hydroxy-4-methy-2-pentanone (76°F)	18.2	
Hydroxymethylene camphor (86°F)	5.2	
Iodobenzene (68°F)	4.6	
Iodoheptane (22°F)	4.92	
Iodoheptane (71°F)	4.9	
Iodohexane (20°F)	5.37	
Iodohexane (68°F)	5.4	
iso-Iodohexadecane	3.5	
Iodomethane (20°F)	7.0	
Iodomethane (68°F)	7.0	
Iodooctane	4.6	
Iodooctane (24°F)	4.62	
Iodotoluene (20°F)	6.1	
Iodotoluene (68°F)	6.1	
Isoprene (77°F)	2.1	
Isoquinoline (76°F)	10.7	
Lactic acid (61°F)	22.0	
Lactronitrile (68°F)	38.4	
Limonene (68°F)	2.3	
Linoleic acid (32°F)	2.6–2.9	
Linseed oil	3.2–3.5	
Maleic anhydride (140°F)	51.0	
Malonic anhydride	51.0	
Malonic nitrile (97°F)	47.0	
Mannitol (71°F)	3.0	

DIELECTRIC CONSTANT—TABLE 3. Dielectric Constants of Organic Compounds (*Continued*)

Compound	Dielectric constant	Loss tangent $\tan \Delta \times 10^4$
Menthol (42°F)	3.95	
Menthol (107°F)	4.0	
Mesityl oxide (68°F)	15.4	
Mesitylene (68°F)	2.4	
Mesitylene	3.4	
Methane (−280°F)	1.7	
Methanol (77°F)	32.70	
Methylene iodide	5.1	
Methoxy-4-methylphenol (60°F)	11.0	
Methoxybenzene (76°F)	4.3	
2-Methoxyethanol (76°F)	16.93	
Methoxyethyl stearate (140°F)	3.4	
Methoxyphenol (82°F)	11.0	
Methoxytoluene (68°F)	3.5	
Methyl acetate (77°F)	6.7	
Methylal (68°F)	2.7	
Methyl alcohol (methanol) (−112°F)	56.6	
Methyl alcohol (methanol) (32°F)	37.5	
Methyl alcohol (methanol) (68°F)	33.1	
Methyl alcohol (methanol) (77°F)	32.6	4.5
n-Methylaniline (68°F)	6.0	
Methylaniline (68°F)	6.0	
Methyl benzoate (68°F)	6.6	
Methylbenzylamine (65°F)	4.4	
Methyl butane (68°F)	1.8	
Methyl iso-butyl ketone (62°F)	12.4	11.0
Methyl iso-butyl ketone (77°F)	13.1	
Methyl butyrate (68°F)	5.6	
Methyl chloride (77°F)	12.9	
Methyl chloroacetate (68°F)	12.9	
Methylcyclohexanol (68°F)	13.0	
Methylcyclohexanone (192°F)	18.0	
Methylcylopentane (68°F)	2.0	
Methylene iodide (70°F)	5.1	
Methyleneaceloacetate (70°F)	7.8	
Methylenemalonate (72°F)	6.6	
Methylenephenylacetate (68°F)	5.0	
Methylether, Liquid	5.0	
Methyl ether (78°F)	5.0	
Methyl ethyl ketone (72°F)	18.4	5.7
Methyl ethyl ketone (77°F)	18.51	
Methyl ethyl ketoxime (68°F)	3.4	
Methyl formate (68°F)	8.5	
Methyl heptanol (68°F)	5.3	
Methyl iodide (68°F)	7.1	
Methylhexane (68°F)	1.9	
Methylisocyanate (69°F)	29.4	
Methyl o-methoxybenzoate (70°F)	7.8	

DIELECTRIC CONSTANT—TABLE 3.   Dielectric Constants of Organic
Compounds (*Continued*)

Compound	Dielectric constant	Loss tangent $\tan \Delta \times 10^4$
Methyl p-toluate (91°F)	4.3	
2-Methyl-1-propanol (77°F)	17.7	
Methyl propionate (66°F)	5.4	
Methyl n-propyl ketone (58°F)	16.8	
Methyl n-propyl ketone (68°F)	15.45	
N-Methylpyrrolidone (77°F)	32.2	
Methyl salicylate (68°F)	9.0	
Methyl thiocyanate (68°F)	35.9	
Methyl valerate (66°F)	4.3	
Methyl 5 ketocyclohexylene (68°F)	24.0	
Methyl-1-cyclopentanol (35°F)	6.9	
Methyl-2,4-pentandeiol (86°F)	24.4	
Methyl-2-Pentanone (68°F)	13.1	
Methyloctane (69°F)	30.0	
Methylpyridine (2) (68°F)	9.8	
Methoxy-4-methyl phenol	11.0	
Mineral oil (80°F)	2.1	
Monomyristin (158°F)	6.1	
Monopalmitin (152°F)	5.3	
Monostearin (170°F)	4.9	
Morpholine (77°F)	7.3	
n-Naphthyl ethyl ether (67°F)	3.2	
Naphthalene (185°F)	2.3	
Naphthalene (68°F)	2.5	
Naphthyl ethyl ether (67°F)	3.2	
Neoprene	6.0–9.0	
o-Nitroaniline (194°F)	34.5	
p-Nitroaniline (320°F)	56.3	
Nitroanisole (68°F)	24.0	
Nitrobenzene (68°F)	35.72	
Nitrobenzene (77°F)	34.8	
Nitrobenzene (176°F)	26.3	
Nitrobenzyl alcohol (68°F)	22.0	
Nitrocellulose	6.2–7.5	
Nitroethane (68°F)	19.7	
Nitroglycerin (68°F)	19.0	
Nitromethane	22.7–39.4	
Nitromethane (68°F)	39.4	
Nitrosodimethylamine (68°F)	54.0	
o-Nitrotoluene (68°F)	27.4	
m-Nitrotoluene (68°F)	23.8	
p-Nitrotoluene (137°F)	22.2	
Nonane (68°F)	2.0	
Octadecanol (136°F)	3.4	
Octane (24°F)	1.06	
Octane (68°F)	2.0	

DIELECTRIC CONSTANT—TABLE 3. Dielectric Constants of Organic
Compounds (*Continued*)

Compound	Dielectric constant	Loss tangent $\tan \Delta \times 10^4$
iso-Octane (68°F)	1.94	
1-Octanol (68°F)	10.3	
Octanone (68°F)	10.3	
Octene (76°F)	2.1	
Octyl alcohol (64°F)	3.4	
Octyl iodide (68°F)	4.9	
Octylene (65°F)	4.1	
Oleic acid (68°F)	2.5	
Palmitic acid (160°F)	2.3	
Paraldehyde (68°F)	14.5	
Paraldehyde (77°F)	13.9	
Pentachloroethane (60°F)	3.7	
Pentadiene 1,3 (77°F)	2.3	
n-Pentane (68°F)	1.83	
Pentanol (77°F)	13.9	
Pentanone (2) (68°F)	15.4	
Pentene (1) (68°F)	2.1	
Phenanthrene (68°F)	2.8	
Phenanthrene (110°F)	2.72	
Phenanthrene (230°F)	2.7	
Phenetole (70°F)	4.5	
Phenol (118°F)	9.9	
Phenol (104°F)	15.0	
Phenol (50°F)	4.3	
Phenol ether (85°F)	9.8	
Phenoxyacetylene (76°F)	4.8	
Phenetidine (70°F)	7.3	
Phenyl acetate (68°F)	6.9	
Phenyl ether (86°F)	3.7	
Phenyl isocyanate (68°F)	8.9	
Phenyl iso-thiocyanate (68°F)	10.7	
Phenyl-1-propane (68°F)	1.7	
Phenylacetaldehyde (68°F)	4.8	
Phenylacetic (68°F)	3.0	
Phenylacetonitrile (80°F)	18.0	
Phenylethanol (68°F)	13.0	
Phenylethyl acetate (58°F)	4.5	
Phenylethylene (77°F)	2.4	
Phenylhydrazine (72°F)	7.2	
Phenylsalicylate (122°F)	6.3	
p-Phthalic acid	5.1–6.3	
iso-Phthalic acid	1.4	
Piperidine (68°F)	5.9	
Propane (liquid, 32°F)	1.6	
Propanediol (68°F)	32.0	
Propanol (177°F)	20.1	
Propene (68°F)	1.9	

DIELECTRIC CONSTANT—TABLE 3.   Dielectric Constants of Organic
Compounds (*Continued*)

Compound	Dielectric constant	Loss tangent $\tan \Delta \times 10^4$
Propionaldehyde (62°F)	18.9	
Propionic acid (58°F)	3.1	
Propionic anhydride (60°F)	18.0	
Propionitrile (68°F)	27.7	
Propyl butyrate (68°F)	4.3	
Propyl acetate (68°F)	6.3	
n-Propyl alcohol (68°F)	21.8	0.28
iso-Propyl alcohol (68°F)	18.3	0.23
iso-Propyl alcohol (77°F)	19.92	
iso-Propylamine (68°F)	5.5	
Propyl benzene (68°F)	2.4	
iso-Propyl benzene (68°F)	2.4	
Propyl bromide (68°F)	7.2	
Propyl butyrate (68°F)	4.3	
Propyl chloroformate (68°F)	11.2	
Propylene carbonate (68°F)	64.9	
Propyl ether (78°F)	3.4	
iso-Propylether (77°F)	3.9	
Propyl formate (66°F)	7.9	
Propyl nitrate (64°F)	14.2	
iso-Propyl nitrate (66°F)	11.5	
Propyl propionate (68°F)	4.7	
Propyl valerate (65°F)	4.0	
Propylene (liquid)	11.9	
Pyridine (68°F)	12.5	
Pyrrole (63°F)	7.5	
Quinoline (77°F)	9.0	
Quinoline (−292°F)	2.6	
Resorcinol	3.2	
Salicylaldehyde (68°F)	13.9	
Sorbitol (176°F)	33.5	
Stearic acid (160°F)	2.3	
Styrene (77°F)	2.4	14
Succinamide (72°F)	2.9	
Succinic acid (78°F)	2.4	
Sucrose	3.3	
Tartaric acid (68°F)	6.0	
Tartaric acid (14°F)	35.9	
Tetrabromoethane (72°F)	7.0	
Tetrachloroethylene (70°F)	2.5	
Tetrafluoroethylene	2.0	
Tetrahydrofuran (68°F)	7.58	
Tetrahydro-2-naphthol (68°F)	11.0	
Tetranitromethane (68°F)	2.2	

DIELECTRIC CONSTANT—TABLE 3. Dielectric Constants of Organic Compounds (*Continued*)

Compound	Dielectric constant	Loss tangent $\tan \Delta \times 10^4$
Tetratriacontadiene (76°F)	2.8	
Thioacetic acid (68°F)	13.0	
Thiophene (60°F)	2.8	
Thrichloroethylene (61°F)	3.4	
Tobacco	1.6–1.7	
Toluene (68°F)	2.39	0.45
o-Toluidine (64°F)	6.3	
m-Toluidine (64°F)	6.0	
p-Toluidine (130°F)	5.0	
Tolyl methyl ether (68°F)	3.5	
Tribromopropane (68°F)	6.4	
Tributylphosphate (86°F)	8.0	
1,2,4-Trichlorobenzene (77°F)	2.24	
Trichlorethylene	3.4	
Trichloroacetic acid (140°F)	4.6	
Trichloroethane	7.5	
Trichloroethylene (61°F)	3.4	
Trichlorololuene (70°F)	6.9	
Trichloropropane (76°F)	2.4	
Trichlorotoluene (69°F)	6.9	
1,1,2-Trichlorotrifluoroethane (77°F)	2.41	
Triethylamine (21°F)	3.2	
Triethylamine (68°F)	2.42	
Triethylamine (77°F)	2.45	
Trifluoroacetic acid (68°F)	8.55	
Trifluorotoluene (86°F)	9.2	
Trimethyl borate (68°F)	8.2	
Trimethyl-3-heptene (68°F)	2.2	
Trimethylamine (77°F)	2.5	
Trimethylbenzene (68°F)	2.3	
Trimethylbutane (68°F)	1.9	
Trimethylpentane	1.9	
Trimethylpentane (68°F)	2.9	
Trimethylsulfanilic acid (64°F)	89.0	
Trinitrobenzene (68°F)	2.2	
Trinitrotoluene (69°F)	22.0	
Triphenylmethane (212°F)	2.3	
Tripalmitin (140°F)	2.9	
Tristearin (158°F)	2.8	
Undecane (68°F)	2.0	
Undecanone (58°F)	8.4	
Urea (71°F)	3.5	
Valeraldehyde (58°F)	11.8	
Valeric acid (68°F)	2.6	
iso-Valeric acid (68°F)	2.6	
Valeronitrile (70°F)	17.7	

DIELECTRIC CONSTANT—TABLE 3. Dielectric Constants of Organic
Compounds (*Continued*)

Compound	Dielectric constant	Loss tangent $\tan \Delta \times 10^4$
Veratrol (73°F)	4.5	
Vinyl ether (68°F)	3.9	
Vinyl formal	3.0	
Vinylidene chloride	3.0–4.0	
o-Xylene (68°F)	2.57	
m-Xylene (68°F)	2.37	
p-Xylene (68°F)	2.3	
Xylenol	17	
Xylenol (62°F)	3.9	
Xylidine (68°F)	5.0	

DIELECTRIC CONSTANT—TABLE 4. Dielectric Constants of Petroleum,
Coal, and Their Products

Compound	Dielectric constant	Loss tangent $\tan \Delta \times 10^4$
Asphalt (75°F)	2.6	
Bunker C oil	2.6	
Coal tar	2.0–3.0	
Coal, powder, fine	2.0–4.0	
Coke	1.1–2.2	
Gasoline (70°F)	2.0	
Heavy oil	3.0	
Kerosene (70°F)	1.8	
Liquefied petroleum gas (LPG)	1.6–1.9	
Lubricating oil (68°F)	2.1–2.6	
Paraffin oil	2.19	1.0
Paraffin wax	2.1–2.5	2
Petroleum (68°F)	2.1	
Vaseline	2.08	5

DIELECTRIC CONSTANT—TABLE 5. Dielectric Constants of Polymers

Compound	Dielectric constant	Loss tangent $\tan \Delta \times 10^4$
Phenol formaldehyde resin	4.5–5.0	
Phenyl urethane	2.7	
Polyamide	2.5–2.6	
Polybutylene	2.2–2.3	
Polycaprolactam	2.0–2.5	
Polycarbonate resin	2.9–3.0	
Polyester resin	2.8–5.2	
Polyether chloride	2.9	
Polyether resin	2.8–8.1	
Polyether resin, unsaturated	2.8–5.2	
Polyethylene	2.2–2.4	2–3
Polypropylene	1.5–1.8	
Polystyrene	2.4–2.6	2–4
Polytetrafluoroethylene	2.0	2
Polyvinyl alcohol	1.9–2.0	
Polyvinyl chloride	3.4	
Polyvinyl chloride resin	5.8–6.8	
Rubber	2.8–4.6	20–280
Starch	1.7–5.0	
Urea formaldehyde	6.4–6.9	
Urea resin	6.2–9.5	
Urethane (121°F)	14.2	
Urethane (74°F)	3.2	
Urethane resin	6.5–7.1	
Vinyl alcohol resin	2.6–3.5	
Vinyl chloride resin	2.8–6.4	600

DIPOLE MOMENTS

The dipole moment is the mathematical product of the distance between the centers of charge of two atoms multiplied by the magnitude of that charge. Thus, the dipole moment (μ) of a compound or molecule is:

$$\mu = Q \times r \tag{13}$$

where Q is the electrical charge(s) that is separated by the distance r; the unit of measurement is the Debye (D).

DIPOLE MOMENTS—TABLE 1.   Dipole Moments of Inorganic
Compounds

	Dipole moment (D)
Aluminum fluoride	1.53
Ammonia	1.48
Arsenic trichloride	1.59
Arsenic trifluoride	2.59
Arsine	0.20
Barium oxide	7.95
Boron hydride	1.73
Boron trichloride	0.00
Boron trifluoride	0.00
Bromine pentafluoride	1.51
Carbon dioxide	0.00
Carbon monoxide	0.10
Cesium chloride	10.42
Cesium fluoride	10.88
Hydrogen bromide	0.82
Hydrogen chloride	1.03
Hydrogen cyanide (hydrocyanic acid) (30°C, 86°F)	2.10
Hydrogen fluoride	1.82
Hydrogen iodide	1.44
Hydrogen peroxide	2.13
Hydrogen sulfide	0.93
Lithium bromide	7.27
Lithium chloride	7.13
Lithium fluoride	6.33
Lithium hydride	5.88
Lithium iodide	7.43
Mercuric chloride	0.00
Nitric acid	2.17
Nitrogen dioxide	0.40
Nitrogen trifluoride	0.24
Nitrosyl chloride	1.90
Nitrosyl fluoride	1.81
Nitrous oxide	0.17
Ozone	0.53
Phosphorus trichloride	0.90
Phosphorus trifluoride	1.03
Potassium bromide	10.41
Potassium chloride	10.27
Potassium fluoride	8.60
Rubidium fluoride	8.55
Sodium chloride	9.00
Sulfur dioxide	1.61
Sulfuryl fluoride	1.12
Thalium chloride	4.44
Thalium fluoride	4.23
Thionyl chloride	1.45
Thionyl fluoride	1.63
Water (20°C, 68°F)	1.87

DIPOLE MOMENTS—TABLE 2. Dipole Moments of Organic Compounds

	Dipole moment (D)
Acetaldehyde	2.72
Acetic acid (25°C, 77°F)	1.74
Acetic anhydride	2.80
Acetone (20°C, 68°F)	2.69
Acetonitrile (20°C, 68°F)	3.44
Acetophenone	2.90
Acetyl chloride (47°C, 117°F)	2.72
Acetyl fluoride	2.96
Acetylene	0.00
Aniline (20°C, 68°F)	1.50
Benzaldehyde	2.76
Benzene	0.00
Benzene sulfonic acid	3.80
Benzonitrile (25°C, 77°F)	3.90
Benzyl alcohol	1.67
Bromobenzene	1.64
Bromoethylene	1.42
Bromomethane	1.81
Butene-1	0.38
Benzamide	3.90
Benzonitrile	4.42
Bromobenzene	1.70
1-Bromobutane	2.08
2-Bromobutane	2.23
1-Bromopropane	2.18
2-Bromopropane	2.21
1,2-Butadiene	0.40
1,3-Butadiene	0.00
1-Butene	0.34
2-Butene (cis)	0.30
2-Butene (trans)	0.00
n-Butyl acetate (22°C, 72°F)	1.84
n-Butyl alcohol (25°C, 77°F)	1.75
iso-Butyl alcohol (25°C, 77°F)	1.79
n-Butyl chloride (25°C, 77°F)	1.90
Carbon tetrachloride	0.00
Chloroacetylene	0.44
Chlorobenzene (25°C, 77°F)	1.54
1-Chlorobutane	2.05
2-Chlorobutane	2.04
Chloroethane	2.06
Chloroethylene	1.45
Chloroform (25°C, 77°F)	1.15
Chloromethane	1.87
Chloromethyl-benzene	1.85
Chloropentafluoroethane	0.52
1-Chloropropane	2.05
2-Chloropropane	2.17

DIPOLE MOMENTS—TABLE 2. Dipole Moments of Organic Compounds (*Continued*)

	Dipole moment (D)
1-Chloroprop-2-ene	1.90
Chlorotrifluoroethylene	1.40
Cyclohexane (20°C, 68°F)	0.00
Cyclohexene	0.55
Cyclopentane (25°C, 77°F)	0.00
Cyclopentene	0.20
Cyclopropane	0.00
o-Diaminobenzene	1.53
m-Diaminobenzene	1.81
p-Diaminobenzene	1.53
Dibutyl ether	1.17
o-Dichlorobenzene (20°C, 68°F)	2.14
o-Dichlorobenzene (25°C, 77°F)	2.25
m-Dichlorobenzene (25°C, 77°F)	1.48
p-Dichlorobenzene (25°C, 77°F)	0.00
1,1-Dichloroethane	2.06
1,2-Dichloroethane	1.19
1,1-Dichloroethylene	1.34
cis-1,2-Dicholorethylene	1.90
trans-1,2-Dichloroethylene	0.00
Dichloromethane (25°C, 77°F)	1.14
1,2-Dichlorotetrafluoroethane	0.50
Diethylamine	0.92
Diethyl ether (ethyl ether) (20°C, 68°F)	1.15
Diethyl ether (ethyl ether) (25°C, 77°F)	1.00
1,1-Difluoroethane	2.27
Dimethyl acetamide (25°C, 77°F)	3.72
Dimethylamine	1.03
Dimethyl ether	1.30
N,N-Dimethylformamide (25°C, 77°F)	3.86
Dimethyl sulfide	1.50
Dimethyl sulfoxide (25°C, 77°F)	4.10
o-Dinitrobenzene	6.00
m-Dintrobenzene	3.89
p-Dintrobenzene	0.00
1,4-Dioxane (25°C, 77°F)	0.45
Diphenyl ether	1.23
Di-n-propyl ether	1.21
Di-iso-propyl ether	1.13
Ethoxybenzene	1.45
Ethyl acetate (25°C, 77°F)	1.88
Ethyl acetate (30°C, 86°F)	1.76
Ethyl alcohol (Ethanol) (20°C, 68°F)	1.66
Ethyl alcohol (Ethanol) (25°C, 77°F)	1.70
Ethylamine	1.22
Ethylbenzene	0.59
Ethyl benzoate	2.00
Ethyl bromide	2.02

DIPOLE MOMENTS—TABLE 2. Dipole Moments of Organic
Compounds (*Continued*)

	Dipole moment (D)
Ethyl chloride (20°C, 68°F)	2.10
Ethylene dichloride (25°C, 77°F)	1.83
Ethylene glycol	2.28
Ethylene glycol dimethyl ether (glyme) (25°C, 77°F)	1.71
Ethyl ether (diethyl ether) (20°C, 68°F)	1.15
Ethyl ether (diethyl ether) (25°C, 77°F)	1.10
Ethyl fluoride	1.94
Ethyl formate	1.93
Ethyl iodide	1.91
Ethyl nitrite	2.38
Fluorobenzene	1.60
Formaldehyde	2.27
Formamide	3.73
Formic acid (22°C, 72°F)	1.52
Furan	0.66
Heptane (25°C, 77°F)	0.00
Hexafluoroethane	0.00
Hexane (25°C, 77°F)	0.08
Iodobenzene	1.42
Iodoethane	1.90
Iodomethane	1.62
Isoquinoline	2.73
Methoxybenzene	1.38
2-Methoxyethanol (25°C, 77°F)	2.04
Methyl acetate	1.72
Methyl alcohol (methanol) (20°C, 68°F)	2.87
Methylamine	1.26
Methyl bromide	1.81
Methyl t-butyl ether (25°C, 77°F)	1.32
Methyl chloride (20°C, 68°F)	1.90
Methyl ether (methoxymethane)	1.30
Methyl ethyl ketone (25°C, 77°F)	2.76
Methyl fluoride	1.85
Methyl iodide	1.62
Methyl nitrate	3.12
N-Methylphenylamine	1.68
Methyl n-propyl ketone (20°C, 68°F)	2.70
N-Methylpyrrolidone (30°C, 86°F)	4.09
Methyl salicylate (20°C, 68°F)	2.40
Nitrobenzene (25°C, 77°F)	3.90
Nitroethane	3.54
Nitromethane	3.44
o-Nitrophenylamine	4.24
m-Nitrophenylamine	4.94
p-Nitrophenylamine	6.20
iso-Octane (20°C, 68°F)	0.00
Pentachloroethane	0.92
Pentane (25°C, 77°F)	0.00

DIPOLE MOMENTS—TABLE 2.   Dipole Moments of Organic Compounds (*Continued*)

	Dipole moment (D)
iso-Pentane (2-Methylbutane)	0.13
Phenol (20°C, 68°F)	1.70
Phenoxybenzene	1.23
Phenylamine	1.53
Propane	0.08
Propanoic acid	1.75
n-Propyl alcohol (20°C, 69°F)	3.09
iso-Propyl alcohol (30°C, 86°F)	1.66
Propylamine	1.17
Propylene (propene)	0.35
Propylene carbonate (20°C, 68°F)	4.94
Propyne	0.75
Pyridine (15°C, 59°F)	2.20
Pyridine (25°C, 77°F)	2.37
Pyrrole	1.84
Quinoline (25°C, 77°F)	2.20
1,1,2,2-Tetrachloroethane	1.32
Tetrachloromethane	0.00
Tetrahydrofuran (25°C, 77°F)	1.75
Thiophene	0.55
Toluene (20°C, 68°F)	0.31
Toluene (85°C, 185°F)	0.40
Trichloromethane	1.02
1,1,1-Trifluoroethane	2.32
1,1,2-Trifluoroethane	1.58
3,3,3-Trifluorpropene	2.45
Trimethylamine	0.67
Urea	4.56
o-Xylene (25°C, 77°F)	0.45
p-Xylene	0.00

ENGINEERING FORMULAS

Air/Gas Flow, Reynold's Number

$$N_R = 2.07VDw/u \qquad (14)$$

where
D = inside diameter, in
N_R = Reynold's number
V = velocity, ft/min
w = specific weight of air, lb/ft^3
u = absolute viscosity, cP

Air/Gas Flow, Reynold's Number

$$N_R = 129VD/v \tag{15}$$

where
D = inside diameter, in
N_R = Reynold's number
V = velocity, ft/min
v = kinematic viscosity, cSt

Air/Gas Flow, Standard Cubic Feet Per Minute Flow Rate, Converted From Actual Flow Rate

$$Qs = 36Q_A \ P/T \tag{16}$$

where
P = absolute pressure, psia
Q_A = actual flow rate, acfm
Qs = standard flow rate, scfm
T = absolute temperature, °R

Air/Gas Flow Volume Change Due to Temperature Change

$$v_2 = v_1\left(T_2/T_1\right) \tag{17}$$

where
T_1 = initial absolute temperature, °R
T_2 = final absolute temperature, °R
v_1 = initial volume, ft^3
v_2 = final volume, ft^3

Antoine Constants

The Antoine constants, A, B, and C, are the constants for the Antoine equation

$$\log_{10}P = A - \frac{B}{C+T} \tag{18}$$

or

$$T = \frac{B}{A - \log_{10}P} - C \tag{19}$$

where P is the saturated vapor pressure in kilopascals (kPa) and T is the temperature in degrees Kelvin. In most cases the Antoine constants are for the liquid (liq), but in some cases the constants are for the solid (sol). The listed deviation (dev) gives the estimated average difference in degrees Kelvin between the experimental values of temperature and values calculated using the given values of A, B, and C. The range is the temperature range in degrees Kelvin over which the given values of A, B, and C are valid.

API Gravity

$$\text{Degrees API} = (141.5/\text{specific gavity}) - 131.5 \tag{20}$$

Area

1 square foot = 144 square inches
1 acre = 43,560 square feet = 4840 square yards = 0.4047 hectares
1 square mile = 640 acres
1 square yard = 9 square feet = 1296 square inches
1 hectare = 2.417 acres
1 square meter = 1550 square inches = 0.0929 square feet = 11,968 square yards

Baumé Gravity

$$\text{Degrees Baumé} = (140/\text{specific gavity}) - 130 \tag{21}$$

Capacity, Cylindrical Tanks

$$C = 0.0034 D^2 L \tag{22}$$

Density, Water

62.43 lb per cubic foot = 8.33 lb per gallon = 0.1337 cubic feet per gallon
1 cubic foot = 7.48052 (U.S.) gallons = 62.43 lb water

Density, Air

13.329 cubic feet per pound = 0.075 lb per cubic foot
1 lb per cubic foot = 177.72 cubic feet per pound
1 cubic foot per pound = 0.00563 pounds per cubic foot
1 kilogram per cubic meter = 16.017 lbs per cubic foot
1 cubic meter per kilogram = 0.0624 cubic feet per pound

Density (d), Function of Pressure and Temperature

$$d = 4.476 P/RT \tag{23}$$

where
d = density, slug/ft^3
P = absolute pressure, psia
R = gas constant, ft-lb/lb/°R
T = absolute temperature, °R

Energy

1 hp = 0.746 kilowatt = 746 watts = 2545 Btu per hour = 1.0 kVA
1 kilowatt = 3413 Btu per hour = 1.341 Hp
1 watt = 3.413 Btu per hour

Flow

1 million (U.S.) gallons per day = 694.4 gallons per minute = 1.547 cubic feet per second
1 cubic foot per minute = 62.43 lbs water per minute = 448.8 gallons per hour

Fluid Flow Pipe Diameter

$$D = 0.5(Q^2 fL/h_F)^{1/5} \tag{24}$$

where
D = pipe inside diameter, in
h_F = head loss, friction, ft (fluid)
f = friction factor
L = pipe length, ft
Q = fluid flow, gal/min

Fluid Flow Pipe Wall Thickness Required

$$T = 0.5 p D/s \tag{25}$$

where
D = inside diameter of pipe, in
p = pressure, psig
s = stress (fiber), psig
T = pipe thickness, in

Fluid Flow Velocity

$$V = 1497D^2(p_1 - p_2)/\mu L \tag{26}$$

where
D = pipe inside diameter, in
L = pipe length, ft
p = pressure, psig
V = fluid velocity, ft/sec
μ = fluid absolute viscosity, cP

Fluid Flow Velocity

$$V = 0.3208 Q/A \tag{27}$$

where
A = area of pipe, in^2
Q = fluid flow, gal/min
V = fluid velocity, ft/sec

Fluid Flow Velocity

$$V = 0.408 Q/D^2 \qquad (28)$$

where
D = pipe inside diameter, in
Q = fluid flow, gal/min
V = fluid velocity, ft/sec

Fluid Flow Velocity

$$V_1 = 8 h_s^{1/2} \qquad (29)$$

$$V_2 = 12.2 \, p^{1/2} \qquad (30)$$

where
h_S = head, static pressure, ft (fluid)
p = pressure, psig

Fluid Flow Weight

$$Q_{W1} = 25 \, AVw \qquad (31)$$

$$Q_{W2} = 19.63 \, D^2 Vw \qquad (32)$$

$$Q_{W3} = 8.02 \, wQ \qquad (33)$$

where
A = area of pipe, in^2
D = pipe inside diameter, in
Q = fluid flow, gal/min
V = fluid velocity, ft/sec
w = specific weight of fluid, lb/ft^3

Gas Density

Gas density is estimated by the following modified ideal gas equation:

$$\rho = \frac{PMw}{ZTR} \qquad (34)$$

where
  P = absolute pressure
Mw = gas molecular weight
  T = absolute temperature
  Z = compressibility factor
  R = universal gas constant

Heat Transfer by Conduction

$$H = kA(t_H - t_L)/d \qquad (35)$$

where
  A = area of contact surface, ft^2
  d = depth (thickness), in
  H = heat flow, Btu/hr
  k = conduction coeff, Btu-in/hr-ft^2, °F
('L – 'H) = temperature diff., °F

Heat Transfer by Convection

$$H = hA(t_H - t_L) \qquad (36)$$

where
  A = area of contact surface, ft^2
  H = heat flow, Btu/hr
  h = convection coeff, Btu/hr-ft^2, °F
('L – 'H) = temperature diff., °F

Heat Transfer, Linear Thermal Expansion

$$L_2 - L_1 = aL_1(t_2 - t_1) \qquad (37)$$

where
  a = coeff of linear thermal expansion, length change/unit length/°F
  L_1 = initial length, in
($L_2 - L_1$) = length change, in
  $(t_2 - t_1)$ = temperature diff., °F

Heat Transfer (or Loss) by Radiation Emission

$$H = 0.174\text{E-}08 \, e \, A T^4 \qquad (38)$$

where
A = area of contact surface, ft^2
H = heat flow, Btu/hr
T = absolute temperature, °R
e = radiation factor

Length

1 mile = 1760 yards = 5280 feet = 63,360 inches = 1.609 kilometers
1 foot = 0.3048 meters = 304.8 millimeters
1 inch = 2.54 centimeters = 25.4 millimeters
1 centimeter = 0.3937 inch
1 meter = 39.37 inches = 3.2808 feet = 1.094 yards
1 kilometer = 3281 feet = 0.6214 mile = 1094 yards
1 fathom = 6 feet = 1.828804 meters
1 furlong = 660 feet

Pressure

14.7 psi = 33.95 feet of water = 29.92 inches of mercury
1 psi = 2.307 feet of water = 2.036 inches of mercury
1 foot of water = 0.4355 psi = 62.43 lb per square foot

Pressure, Absolute (psia)

$psia_1$ = gauge pressure + 14.696 psi
$psia_2$ = 0.433 [ft (water) gauge + 33.898]
$psia_3$ = 0.491 [in (mercury) gauge + 29.92]
$psia_4$ = 0.036 [in (water) gauge + 406.77]

Reactors

Homogeneous Gas-Phase, Pressure Drop. Pressure drop as a function of temperature and concentration:

$$P_2 = P_1 \frac{C_2}{C_1} \frac{T_2}{T_1} \tag{39}$$

where
C = concentration of all species, C = moles/volume
T = absolute temperature
P = absolute pressure

Subscripts 1 and 2 refer to the inlet and outlet of a reactor length increment. T_2 is found with a heat balance for the current reactor increment, and C_2 is found from the design model of the plug flow reactor.

Fluidized-Bed Reactor, Fluidizing Velocity

$$U_{mf} = \frac{\mu}{d_P \rho} \left[33.7^2 + 0.0408 \frac{d_P^3 \rho (\rho_P - \rho) g}{\mu^2} \right] \tag{40}$$

where
U_{mf} = minimum fluidization velocity
ρ = fluid density

ρ_P = particle density
μ = fluid viscosity

Packed-Bed Reactor, Pressure Drop across the Bed

$$\frac{dP}{dW} = -\frac{G}{\rho g_c d_p}\left(\frac{1}{\varepsilon_B^3}\right)\left[\frac{150(1-\varepsilon_B)\mu}{d_p}+1.75G\right]\frac{1}{A_c\rho_c} \tag{41}$$

where
G = superficial mass velocity
ε_B = bed porosity, reactor void fraction
ρ_c = density of catalyst
d_p = catalyst-particle diameter
A_c = reactor cross-sectional area

Trickle-Bed Reactor, Catalyst Wetting Factor

$$f_e = 0.0381G_L^{0.222}G_G^{-0.083}d_p^{-0.373} \tag{42}$$

where
f_e = catalyst wetting factor
G_L = mass velocity of liquid-phase, kg/m^2-s
G_G = mass velocity of gas-phase, kg/m^2-s

Trickle-Bed Reactor, Pressure Drop across the Catalyst Bed

$$\log\left(\frac{\Delta P_{GL}}{\Delta P_L + \Delta P_G}\right) = \frac{0.416}{0.666+\log(X)^2} \qquad 0.05 < X < 30 \tag{43}$$

where
ΔP_L = liquid-phase frictional pressure drop
ΔP_G = gas-phase frictional pressure drop
ΔP_{GL} = gas-liquid-phase frictional pressure drop

X is defined as:

$$X = \left(\frac{\Delta P_L}{\Delta P_G}\right)^{0.5} \tag{44}$$

The single-phase pressure drop (ΔP_L or ΔP_G) as:

$$\Delta P = \frac{150(1-\varepsilon_B)^2}{\varepsilon_B^3}\left(\frac{u\mu}{d_p^2}\right)+\frac{1.75(1-\varepsilon_B)}{\varepsilon_B^3}\left(\frac{u^2\rho}{d_p}\right) \tag{45}$$

where
μ = fluid superficial velocity
μ = fluid viscosity
ρ = fluid density
ε_B = bed porosity (reactor-bed void fraction)

Trickle-Bed Reactor, Frictional Pressure Drop

$$\Delta P_{GL} = 200(X_G\xi)^{-1.2} + 85(X_G\xi)^{-0.5} \tag{46}$$

where

$$X_G = \frac{1}{X_L}$$

$$X_L = \frac{G_{Ls}}{G_{Gs}}\left(\frac{\rho_G}{\rho_L}\right)^{0.55}\left(\frac{\mu_G}{\mu_L}\right)^{0.111} \approx \frac{G_{Ls}}{G_{Gs}}\left(\frac{\rho_G}{\rho_L}\right)^{0.50} \approx \left(\frac{We_L}{We_G}\right)^{0.5}$$

$$\xi = \frac{Re_L^2}{(0.001 + Re_L^{15})}$$

G_{Ls} = liquid-phase mass flux, g/m²-s
G_{Gs} = gas-phase mass flux, g/m²-s
We_L = liquid-phase Weber number
We_G = gas-phase Weber number

The Weber number is defined as:

$$We_L = \frac{G_{Ls}d_p}{\rho_L\sigma} \tag{47}$$

and

$$We_G = \frac{G_{Gs}d_p}{\rho_G\sigma} \tag{48}$$

where
σ = liquid-phase surface tension

Trickle-Bed Reactor, Liquid Holdup

$$\varepsilon_L = \varepsilon_B\left(0.185a_t^{0.333}X^{0.22}\right) \tag{49}$$

where
X = liquid/gas pressure drop ratio
ε_L = liquid holdup, volume of liquid per unit volume of reactor
a_t = particle surface area per unit volume of reactor, defined as:

$$a_t = \frac{6(1-\varepsilon_B)}{d_{p*}}$$

d_{p*} = particle diameter modified to account for wall effects:

$$d_{p*} = \frac{d_p}{1+(4d_p/6d_T(1-\varepsilon_B))}$$

d_T = reactor diameter
An alternate relationship is:

$$\varepsilon_L = \varepsilon_B 10^k \tag{50}$$

k is:

$$k = 0.001 - \frac{0.42}{\xi^{0.48}}$$

$$\xi = X^{0.5}\text{Re}_L^{-0.3}\left(\frac{a_c d_h}{1-\varepsilon_B}\right)^4$$

d_h = hydraulic diameter:

$$d_h = \left(\frac{16\varepsilon_B^3}{9\pi(1-\varepsilon_B)^2}\right)^{0.33} d_p$$

Slurry Reactor, Amount of Solids That Can Be Maintained in Suspension under Specified Conditions

$$\frac{W_{max}}{\rho_L} = 6.8 \times 10^{-4} \frac{C_\mu d_T u_G \rho_G}{\mu_G}\left(\frac{\sigma_L \varepsilon_G}{u_G \mu_L}\right)^{-23}\left(\frac{\varepsilon_G u_{tp}}{u_G}\right)^{-.18}\gamma^{-3} \qquad (51)$$

where
d_T = reactor diameter
W_{max} = catalysts loading, g/cm^3
C_μ = viscosity correction factor, defined as:

$$C_\mu = 0.232 - 0.1788\log(\mu_L) + 0.1026(\log(\mu_L))^2$$

σ_L = surface tension, dyne/cm
u_{tp} = terminal settling velocity of the particles and is defined as:

1. Stoke's regime, ($\text{Re}_P < 0.4$)

$$u_{tp} = \frac{gd_p^2(\rho_p - \rho_L)}{18\mu_L}$$

2. Intermediate regime, ($0.4 < \text{Re}_P < 500$)

$$u_{tp} = \left[\frac{3.1g(\rho_p - \rho_L)}{\rho_L}\right]^{1/2}$$

3. Newton's regime, ($500 < \text{Re}_P < 200,000$)

$$u_{tp} = \left[\frac{3.1g(\rho_p - \rho_L)}{\rho_L}\right]^{1/2}$$

where Re_P is the Reynolds number based on the particle settling velocity,

$$\text{Re}_P = \frac{u_{tp}\rho_L d_p}{\mu_L} \qquad (52)$$

where
ρ_p = catalyst-particle density
The wetting factor, γ, can be assumed equal to 1.0 for most catalysts

Reynold's Number, Air/Gas Flow

$$N_R = 2.07VDw/u \qquad (53)$$

where
 D = inside diameter, in
N_R = Reynold's number
 V = velocity, ft/min
 w = specific weight of air, lb/ft^3
 u = absolute viscosity, cP

Reynold's Number, Air/Gas Flow

$$N_R = 129\,VD/v \qquad (54)$$

where
 D = inside diameter, in
N_R = Reynold's number
 V = velocity, ft/min
 v = kinematic viscosity, cSt

Speed

See **Velocity**.

Tanks, Cylindrical

Capacity

$$C = 0.0034\,D^2 L \qquad (55)$$

Volume

$$V = 0.7854\,D^2 L \qquad (56)$$

where
 C = capacity, gal
 D = inside diameter, in
 L = inside length, in
 V = volume, in^3

Temperature, Absolute °C (°K)

$$°K = °C + 273.16 \qquad (57)$$

Temperature, Absolute °F (°R)

$$°R = °F + 459.69 \qquad (58)$$

Temperature Conversion

°C, Convert from °F

$$°C = 0.5556(°F - 32) \tag{59}$$

°F, Convert from °C

$$°F = (1.8°C) + 32 \tag{60}$$

°K, Convert from °F

$$°K = 0.5556(°F + 459.6) \tag{61}$$

°R

$$deg R_1 = °F + 459.69 \tag{62}$$

$$deg R_2 = 1.8°K \tag{63}$$

Thermodynamics, Reversible Adiabatic Process, Enthalpy Change

$$h_2 - h_1 = \frac{kRT_1\left[(P_2/P_1)^{(k-1)/k} - 1\right]}{778(k-1)} \tag{64}$$

where
h = enthalpy, Btu/lb
k = adiabatic exponent for gas
P = absolute pressure, psia
R = gas constant, ft-lb/lb/°R
T = absolute temperature, °R

Thermodynamics, Reversible Adiabatic Process, Enthalpy Change

$$h_2 - h_1 = C_P(t_2 - t_1) \tag{65}$$

where
C_P = specific heat, constant pressure, Btu/lb/°F
h = enthalpy, Btu/lb
t = temperature, °F

Thermodynamics, Enthalpy

$$h = U + 0.185 Pv \tag{66}$$

where
h = enthalpy, Btu/lb
P = absolute pressure, psia
U = internal energy, Btu/lb
v = specific volume, ft^3/lb

Thermodynamics, Isobaric Process, Enthalpy Change

$$h_2 - h_1 = 0.185k\,P(v_2 - v_1)/(k-1) \tag{67}$$

where
h = enthalpy, Btu/lb
k = adiabatic exponent for gas
P = absolute pressure, psia
v = specific volume, ft³/lb

Thermodynamics, Isobaric Process, Enthalpy Change

$$h_2 - h_1 = C_P(t_2 - t_1) \tag{68}$$

where
C_P = specific heat, constant pressure, Btu/lb/°F
h = enthalpy, Btu/lb
t = temperature, °F

Thermodynamics, Isobaric Process, Entropy Change

$$s_2 - s_1 = C_P \ln(v_2/v_1) \tag{69}$$

where
C_P = specific heat, constant pressure, Btu/lb/°F
v = specific volume, ft³/lb
s = entropy, Btu/lb/°R

Thermodynamics, Isobaric Process, Entropy Change

$$s_2 - s_1 = C_P \ln(T_2/T_1) \tag{70}$$

where
C_P = specific heat, constant pressure, Btu/lb/°F
T = absolute temperature, °R
s = entropy, Btu/lb/°R

Thermodynamics, Isobaric Process, Heat Change

$$Q_2 - Q_1 = 0.185k\,P(v_2 - v_1)/(k-1) \tag{71}$$

where
k = adiabatic exponent for gas
P = absolute pressure, psia
Q = heat energy, Btu/lb
v = specific volume, ft³/lb

Thermodynamics, Isobaric Process, Heat Change

$$Q_2 - Q_1 = R(k/k-1)(T_2 - T_1)/778 \tag{72}$$

where
k = adiabatic exponent for gas
Q = heat energy, Btu/lb
T = absolute temperature, °R
R = gas constant, ft-lb/lb/°R

Thermodynamics, Isobaric Process, Internal Energy Change

$$U_2 - U_1 = 0.185\,P(v_2 - v_1)/(k-1) \tag{73}$$

where
k = adiabatic exponent for gas
P = absolute pressure, psia
U = internal energy, Btu/lb
v = specific volume, ft³/lb

Thermodynamics, Isobaric Process, Temperature T₂

$$T_2 = T_1(v_2/v_1) \tag{74}$$

where
T = absolute temperature, °R
v = specific volume, ft³/lb

Thermodynamics, Isobaric Process, Work Energy Output

$$E_W = 144\,P(v_2 - v_1) \tag{75}$$

where
E_W = work energy output, ft-lb/lb
P = absolute pressure, psia
v = specific volume, ft³/lb

Thermodynamics, Isothermal (Isodynamic) Process, Entropy Change

$$s_2 - s_1 = R\ln(P_1/P_2)/778 \tag{76}$$

where
P = absolute pressure, psia
s = entropy, Btu/lb/°R
R = gas constant, ft-lb/lb/°R

Thermodynamics, Isothermal (Isodynamic) Process, Entropy Change

$$s_2 - s_1 = R \ln(v_2/v_1)/778 \tag{77}$$

where
v = specific volume, ft^3/lb
s = entropy, Btu/lb/°R
R = gas constant, ft-lb/lb/°R

Thermodynamics, Isothermal (Isodynamic) Process, Heat Change

$$Q_2 - Q_1 = RT \ln(P_1/P_2)/778 \tag{78}$$

where
P = absolute pressure, psia
Q = heat energy, Btu/lb
R = gas constant, ft-lb/lb/°R
T = absolute temperature, °R

Thermodynamics, Isothermal (Isodynamic) Process, Heat Change

$$Q_2 - Q_1 = 0.185 P_1 v_1 \ln(v_2/v_1) \tag{79}$$

where
P = absolute pressure, psia
Q = heat energy, Btu/lb
v = specific volume, ft^3/lb

Thermodynamics, Isothermal (Isodynamic) Process, Heat Change

$$Q_2 - Q_1 = RT \ln(v_2/v_1)/778 \tag{80}$$

where
Q = heat energy, Btu/lb
R = gas constant, ft-lb/lb/°R
T = absolute temperature, °R
v = specific volume, ft^3/lb

Thermodynamics, Isovolume (Isochoric) Process, Enthalpy Change

$$h_2 - h_1 = 0.185 k v(P_2 - P_1)/(k-1) \tag{81}$$

where
h = enthalpy, Btu/lb
k = adiabatic exponent for gas
P = absolute pressure, psia
v = specific volume, ft^3/lb

Thermodynamics, Isovolume (Isochoric) Process, Entropy Change

$$s_2 - s_1 = C_V \ln(P_2/P_1) \tag{82}$$

where
C_V = specific heat at constant volume, Btu/lb/°F
P = absolute pressure, psia
s = entropy, Btu/lb/°R

Thermodynamics, Isovolume (Isochoric) Process, Entropy Change

$$s_2 - s_1 = C_V \ln(T_2/T_1) \tag{83}$$

where
C_V = specific heat at constant volume, Btu/lb/°F
s = entropy, Btu/lb/°R
T = absolute temperature, °R

Thermodynamics, Polytropic Process, Enthalpy Change

$$h_2 - h_1 = \frac{k\,R\,T_1\left[(P_2/P_1)^{(n-1)/n} - 1\right]}{778(k-1)} \tag{84}$$

where
h = enthalpy, Btu/lb
k = adiabatic exponent for gas
n = polytropic constant for gas
P = absolute pressure, psia
R = gas constant, ft-lb/lb/°R
T = absolute temperature, °R

Thermodynamics, Polytropic Process, Entropy Change

$$s_2 - s_1 = C_V[(n-k)/n]\ln(P_2/P_1) \tag{85}$$

where
C_V = specific heat at constant volume, Btu/lb/°F
k = adiabatic exponent for gas
n = polytropic constant for gas
P = absolute pressure, psia
s = entropy, Btu/lb/°R

Thermodynamics, Polytropic Process, Entropy Change

$$s_2 - s_1 = C_V[(n-k)/(n-1)]\ln(T_2/T_1) \tag{86}$$

where
C_V = specific heat at constant volume, Btu/lb/°F
k = adiabatic exponent for gas

n = polytropic constant for gas
s = entropy, Btu/lb/°R
T = absolute temperature, °R

Thermodynamics, Polytropic Process, Heat Change

$$Q_2 - Q_1 = \frac{R(n-k)T_1\left[(P_2/P_1)^{(n-1)/n} - 1\right]}{778(k-1)(n-1)} \tag{87}$$

where
k = adiabatic exponent for gas
n = polytropic constant for gas
P = absolute pressure, psia
Q = heat energy, Btu/lb
R = gas constant, ft-lb/lb/°R
T = absolute temperature, °R

Thermodynamics, Polytropic Process, Heat Change

$$Q_2 - Q_1 = C_V(t_2 - t_1)(n-k)/(n-1) \tag{88}$$

where
C_V = specific heat at constant volume, Btu/lb/°F
k = adiabatic exponent for gas
n = polytropic constant for gas
Q = heat energy, Btu/lb
t = temperature, °F

Thermodynamics, Polytropic Process, Internal Energy Change

$$U_2 - U_1 = \frac{RT_1\left[(P_2/P_1)^{(n-1)/n} - 1\right]}{778(k-1)} \tag{89}$$

where
k = adiabatic exponent for gas
n = polytropic constant for gas
P = absolute pressure, psia
R = gas constant, ft-lb/lb/°R
T = absolute temperature, °R
U = internal energy, Btu/lb

Velocity

1 mile per hour = 5280 feet per hour = 88 feet per minute = 1.467 feet per second
1 mile per hour = 0.8684 knot
1 knot = 1.1515 miles per hour = 1.8532 kilometers per hour = 1.0 nautical mile per hour
1 league = 3 miles

Velocity of Sound in Air

1128.5 feet per second = 769.4 miles per hour

Volume

1 cubic yard = 27 cubic feet = 46,656 cubic inches = 1616 pints = 764.6 liters
1 cubic foot = 1728 cubic inches
1 liter = 0.2642 (US) gallons = 2.113 pints
1 gallon (US) = 8 pints = 3.785 liters
1 cubic meter = 61,023 cubic inches = 0.02832 cubic feet = 1.3093 cubic yards
1 barrel (crude oil/petroleum) = 42 (US) gallons
1 hogshead = 63 gallons = 8.42184 cubic feet

Volume, Cylindrical Tanks

$$V = 0.7854 D^2 L$$
$$C = 0.0034 D^2 L \tag{90}$$

where
C = capacity, gal
D = inside diameter, in
L = inside length, in
V = volume, in^3

Weight

1 (U.S.) gallon of water = 8.33 lbs of water
1 lb = 16 ounces = 7000 grains = 0.4536 kilogram
1 ton = 2000 lb = 907 kilograms
1 kilogram = 2.205 lb

FLAMMABILITY PROPERTIES

The *flash point* of a substance is the lowest temperature at which a flammable liquid gives off sufficient vapor to form an ignitable mixture with air near its surface or within a vessel. The *fire point* is the temperature at which the flame becomes self-sustained and the burning continues. At the flash point, the flame does not need to be sustained. The fire point is usually a few degrees above the flash point. American Society for Testing and Materials (ASTM) test methods include procedures using a closed cup (ASTM D-56, ASTM D-93, and ASTM D-3828), which is preferred, and an open cup (ASTM D-92, ASTM D-I310). When several values are available, the lowest temperature is usually taken to ensure safe operation of the process.

The *ignition temperature* (or *ignition point*) is the minimum temperature required to initiate self-sustained combustion of a substance (solid, liquid, or gaseous) independent of external ignition sources or heat.

Flash points, lower and upper flammability limits, and autoignition temperatures are the three properties that are used to indicate safe operating limits of temperature when processing organic materials. Prediction methods are somewhat erratic, but, together with comparisons with reliable experimental values for families or similar compounds, they are valuable in setting a conservative value for each of the properties.

The most preferred flash point prediction method uses the formula of the compound, the system pressure, and vapor pressure data:

$$P^{sat} = \frac{P}{1 + 4.76(2\beta - 1)} = 0$$

$$\beta = N_C + N_S + \frac{(N_H - N_X)}{4} - \frac{N_O}{2} \tag{91}$$

Ns are the numbers of atoms of carbon (C), sulfur (S), hydrogen (H), halogens (X), and oxygen (O) in the molecule. P is the total system pressure, and P^{sat} is the vapor pressure of the compound at the flash point temperature.

The upper and lower flammability limits are the boundary-line mixtures of vapor or gas with air, which, if ignited, will propagate flame and are given in terms of percent by volume of gas or vapor in the air. Each of these limits also has a temperature at which the flammability limits are reached. The temperature corresponding to the lower-limit partial vapor pressure should equal the flash point. The temperature corresponding to the upper-limit partial vapor pressure is somewhat above the lower limit and is usually considerably below the autoignition temperature. Flammability limits are calculated at one atmosphere total pressure and are normally considered synonymous with explosive limits. Limits in oxygen, rather than air, are sometimes measured and available. Limits are generally reported at 298°K and 1 atmosphere. If the temperature or the pressure is increased, the lower limit will decrease and the upper limit will increase, giving a wider range of compositions over which flame will propagate.

The most generally applicable method for prediction of the property depends only on the molecular structure of the molecule and utilizes second-order groups to construct the molecule:

$$z_n \quad \text{or} \quad z_l = \frac{\sum (n_i f_i)}{\sum \left(\dfrac{n_i f_i}{g_i} \right)} \tag{92}$$

Two sets of f_i and g_i are available for each second-order group to cover both upper (u) and lower (l) limits (z) in volume percent units. Absolute errors of 0.15% and 2.3% for the lower an upper limits, respectively, are noted.

The autoignition temperature is the minimum temperature for a substance to initiate self-combustion in air in the absence of a spark or flame. The temperature is no lower than, and is generally considerably higher than, the temperature corresponding to the upper flammability limit. Large differences can occur in reported values determined by different procedures. The lowest reasonable value should be accepted to ensure safety. Values are also sometimes given in oxygen rather than in air.

One simple method of estimating autoignition temperatures is to compare values for a compound with other members of its homologous series on a plot versus carbon number as the temperature decreases and carbon number increases.

FLAMMABILITY PROPERTIES—TABLE 1.   Boiling Points, Flash Points, and Ignition Temperatures of Organic Compounds

Compound	Boiling point °F (°C)	Flash point, °F (°C)	Ignition point, °F (°C)
Acetal $CH_3CH(OC_2H_5)_2$ (Acetaldehydediethylacetal)	215 (102)	−5 (−21)	446 (230)
Acetaldehyde CH_3CHO (Acetic aldehyde) (Ethanal)	70 (21)	−38 (−39)	347 (175)
Acetaldehydediethylacetal		See Acetal.	
Acetaldel		See Aldol.	
Acetanilide $CH_3CONHC_6H_5$	582 (306)	337 (169) (oc)	985 ± 10 (530)
Acetic Acid, Glacial CH_3COOH	245 (118)	103 (39)	867 (463)
Acetic Acid, Isopropyl Ester		See Isopropyl Acetate.	
Acetic Acid, Methyl Ester		See Methyl Acetate.	
Acetic Acid, n-Propyl Ester		See Propyl Acetate.	
Acetic Aldehyde		See Acetaldehyde	
Acetic Anhydride $(CH_3CO)_2O$ (Ethanoic anhydride)	284 (140)	120 (49)	600 (316)
Acetic Ester		See Ethyl Acetate.	
Acetic Ether		See Ethyl Acetate.	
Acetoacetanilide $CH_3COCH_2CONHC_6H_5$		365 (185)	
o-Acetoacet Anisidide $CH_3COCH_2CONHC_6$-H_4OCH_3		325 (168)	
Acetoacetic Acid, Ethyl Ester		See Ethyl acetoacetate.	
Acetoethylamide		See N-Ethylacetamide.	
Acetone CH_3COCH_3 (Dimethyl Ketone) (2-Propanone)	133 (56)	−4 (−20)	869 (465)
Acetone Cyanohydrin $(CH_3)_2C(OH)CN$ (2-Hydroxy-2-Methyl Propionitrile)	248 (120) Decomposes	165 (74)	1270 (688)
Acetonitrile CH_3CN (Methyl Cyanide)	179 (82)	42 (6)	975 (524)

FLAMMABILITY PROPERTIES—TABLE 1. Boiling Points, Flash Points, and Ignition Temperatures of Organic Compounds (*Continued*)

Compound	Boiling point °F (°C)	Flash point, °F (°C)	Ignition point, °F (°C)
Acetonyl Acetone $(CH_2COCH_3)_2$ (2,5-Hexanedione)	378 (192)	174 (79)	920 (499)
Acetophenone $C_6H_5COCH_3$ (Phenyl Methyl Ketone)	396 (202)	170 (77)	1058 (570)
p-Acetotoluidide $CH_3CONHC_6H_4CH_3$	583 (306)	334 (168)	
Acetyl Acetone		See 2,4-Pentanedione.	
Acetyl Chloride CH_3COCl (Ethanoyl Chloride)	124 (51)	40 (4)	734 (390)
Acetylene CH : CH (Ethine) (Ethyne)	−118 (−83)	Gas	581 (305)
N-Acetyl Ethanolamine $CH_3C:ONHCH_2CH_2OH$ (N-(2-Hydroxyethyl) acetamide)	304–308 (151–153) @ 10 mm Decomposes	355 (179) (oc)	860 (460)
N-Acetyl Morpholine $CH_3CONCH_2CH_2OCH_2CH:$	Decomposes	235 (113)	
Acetyl Oxide		See Acetic Anhydride.	
Acetylphenol		See Phenyl Acetate.	
Acrolein $CH_2:CHCHO$ (Acrylic Aldehyde)	125 (52)	−15 (−26)	428 (220) Unstable
Acrylic Acid (Glacial) $CH_2CHCOOH$	287 (142)	122 (50)	820 (438)
Acrylic Aldehyde		See Acrolein.	
Acrylonitrile $CH_2:CHCN$ (Vinyl Cyanide) (Propenenitrile)	171 (77)	32 (0)	898 (481)
Adipic Acid $HOOC(CH_2)_4COOH$	509 (265) @ 100 mm	385 (196)	788 (420)
Adipic Ketone		See Cyclopentanone.	
Adiponitrile $NC(CH_2)_4CN$	563 (295)	200 (93)	

FLAMMABILITY PROPERTIES—TABLE 1. Boiling Points, Flash Points, and Ignition Temperatures of Organic Compounds (*Continued*)

Compound	Boiling point °F (°C)	Flash point, °F (°C)	Ignition point, °F (°C)
Alcohol		See Ethyl Alcohol, Methyl Alcohol.	
Aldol CH$_3$CH(OH)CH$_2$CHO (3-Hydroxybutanal) (β-Hydroxybuteraldehyde)	174–176 (79–80) @ 12 mm Decomposes @ 176 (80)	150 (66)	482 (250)
Allyl Acetate CH$_3$COCH$_2$CH:CH$_2$	219 (104)	72 (22)	705 (374)
Allyl Alcohol CH$_2$:CHCH$_2$OH	206 (97)	70 (21)	713 (378)
Allylamine CH$_2$:CHCH$_2$NH$_2$ (2-Propenylamine)	128 (53)	−20 (−29)	705 (374)
Allyl Bromide CH$_2$:CHCH$_2$Br (3-Bromopropene)	160 (71)	30 (−1)	563 (295)
Allyl Caproate CH$_3$(CH$_2$)$_4$COOCH$_2$CH:Cl (Allyl Hexanoate) (2-Propenyl Hexanoate)	367–370 (186–188)	150 (66)	
Allyl Chloride CH$_2$:CHCH$_2$Cl (3-Chloropropene)	113 (45)	−25 (−32)	737 (485)
Allyl Chlorocarbonate		See Allyl Chloroformate.	
Allyl Chloroformate CH$_2$:CHCH$_2$OCOCl (Allyl Chlorocarbonate)	223–237 (106–114)	88 (31)	
Allylene		See Propyne.	
Allyl Ether (CH$_2$:CHCH$_2$)$_2$O (Diallyl Ether)	203 (95)	20 (−7)	
Allylidene Diacetate CH$_2$:CHCH(OCOCH$_3$)$_2$	225 (107) @ 50 mm	180 (82)	
Allyl Isothiocyanate		See Mustard Oil.	
Allylpropenyl		See 1,4-Hexadiene.	
Allyl Trichloride		See 1,2,3-Trichloropropane.	

FLAMMABILITY PROPERTIES—TABLE 1. Boiling Points, Flash Points, and Ignition
Temperatures of Organic Compounds (*Continued*)

Compound	Boiling point °F (°C)	Flash point, °F (°C)	Ignition point, °F (°C)
Allyl Vinyl Ether		See Vinyl Allyl Ether.	
Alpha Methyl Pyridine		2-Picoline.	
Aminobenzene		See Aniline.	
2-Aminobiphenyl		See 2-Biphenylamine.	
1-Aminobutane		See Butylamine.	
2-Amino-1-Butanol	352	165	
$CH_3CH_2CHNH_2CH_2OH$	(178)	(74)	
1-Amino-4-Ethoxybenzene		See p-Phenetidine.	
β-Aminoethyl Alcohol		See Ethanolamine.	
Amyl Acetate	300	60	680
$CH_3COOC_5H_{11}$	(149)	(16)	(360)
(1-Pentanol Acetate)		70	
Comm.		(21)	
sec-Amyl Acetate	249	89	
$CH_3COOCH(CH_3)-$	(121)	(32)	
$(CH_2)_2CH_3$			
(2-Pentanol Acetate)			
Amyl Alcohol	280	91	572
$CH_3(CH_2)_3CH_2OH$	(138)	(33)	(300)
(1-Pentanol)			
sec-Amyl Alcohol	245	94	650
$CH_3CH_2CH_2CH(OH)CH_3$	(118)	(34)	(343)
(Diethyl Carbinol)			
Amylamine	210	30	2.2 22
$C_5H_{11}NH_2$	(99)	(−1)	
(Pentylamine)			
sec-Amylamine	198	20	
$CH_3(CH_2)_2CH(CH_3)NH_2$	(92)	(−7)	
(2-Aminopentane)			
(Methylpropylcarbinylamine)			
p-tert-Amylaniline	498–504	215	
$(C_2H_5)(CH_2)_2CC_6H_4NH_2$	(259–262)	(102)	
Amylbenzene	365	150	
$C_6H_5C_5H_{11}$	(185)	(66)	
(Phenylpentane)		(oc)	
Amyl Bromide	128–9	90	
$CH_3CH_2CH_2CH_2CH_2Br$	(53–54)	(32)	
(1-Bromopentane)	@ 746 mm		
Amyl Butyrate	365	135	
$C_5H_{11}OOCC_3H_7$	(185)	(57)	

FLAMMABILITY PROPERTIES—TABLE 1. Boiling Points, Flash Points, and Ignition
Temperatures of Organic Compounds (*Continued*)

Compound	Boiling point °F (°C)	Flash point, °F (°C)	Ignition point, °F (°C)
Amyl Carbinol		See Hexyl Alcohol.	
Amyl Chloride CH$_3$(CH$_2$)$_3$CH$_2$Cl (1-Chloropentane)	223 (106)	55 (13)	500 (260)
tert-Amyl Chloride CH$_3$CH$_2$CCl(CH$_3$)CH$_3$	187 (86)		653 (345)
Amyl Chlorides (Mixed) C$_5$H$_{11}$Cl	185–228 (85–109)	38 (3)	
Amylcyclohexane C$_5$H$_{11}$C$_6$H$_{11}$	395 (202)		462 (239)
Amylene		See 1-Pentene.	
β-Amylene-cis C$_2$H$_5$CH:CHCH$_3$ (2-Pentene-cis)	99 (37)	<–4 (<–20)	
β-Amylene-trans C$_2$H$_5$CH:CHCH$_3$ (2-Pentene-trans)	97 (36)	<–4 (<–20)	
Amylene Chloride		See 1,5-Dichloropentane.	
Amyl Ether C$_5$H$_{11}$OC$_5$H$_{11}$ (Diamyl Ether) (Pentyloxypentane)	374 (190)	135 (57)	338 (170)
Amyl Formate HCOCC$_5$H$_{11}$	267 (131)	79 (26)	
Amyl Lactate C$_2$H$_5$OCOOCH$_2$- CH(CH$_3$)C$_2$H$_5$	237–239 (114–115) @ 36 mm	175 (79)	
Amyl Laurate C$_{11}$H$_{23}$COOC$_5$H$_{11}$	554–626 (290–330)	300 (149)	
Amyl Maleate (CHCOOC$_5$H$_{11}$)$_2$	518–599 (270–315)	270 (132)	
Amyl Mercaptan C$_5$H$_{11}$SH (1-Pentanethiol)	260 (127)	65 (18)	
Amyl Mercaptans (Mixed) CH$_3$(CH$_2$)$_4$SH	176–257 (80–125)	65 (18)	
Amyl Naphthalene C$_{10}$H$_7$C$_5$H$_{11}$	550 (288)	255 (124)	
Amyl Nitrate CH$_3$(CH$_2$)$_4$NO$_3$	306–315 (153–157)	118 (48)	
Amyl Nitrite CH$_3$(CH$_2$)$_4$NO$_2$	220 (104)	410 (210)	

FLAMMABILITY PROPERTIES—TABLE 1. Boiling Points, Flash Points, and Ignition Temperatures of Organic Compounds (*Continued*)

Compound	Boiling point °F (°C)	Flash point, °F (°C)	Ignition point, °F (°C)
Amyl Oleate $C_{17}H_{33}COOC_5H_{11}$	392–464 (200–240) @ 20 mm	366 (186)	
Amyl Oxalate $(COOC_5H_{11})_2$ (Diamyl Oxalate)	464–523 (240–273)	245 (118)	
o-Amyl Phenol $C_5H_{11}C_6H_4OH$	455–482 (235–250)	219 (104)	
p-tert-Amyl Phenol		See Pentaphen.	
p-sec-Amylphenol $C_5H_{11}C_6H_4OH$	482–516 (250–269)	270 (132)	
2-(p-tert-Amylphenoxy) Ethanol $C_5H_{11}C_6H_4OCH_2CH_2OH$	567–590 (297–310)	280 (138)	
2-(p-tert-Amylphenoxy) Ethyl Laurate $C_{11}H_{23}COO(CH_2)_2O-$ $C_6H_4C_5H_{11}$	464–500 (240–260) @ 6 mm	410 (210)	
p-tert-Amylphenyl Acetate $CH_3COOC_6H_4C_5H_{11}$	507–511 (264–266)	240 (116)	
p-tert-Amylphenyl Butyl Ether $C_5H_{11}C_6H_4OC_4H_9$	540–550 (282–288)	275 (135)	
Amyl Phenyl Ether $CH_3(CH_2)_4OC_6H_5$ (Amoxybenzene)	421–444 (216–229)	185 (85)	
p-tert-Amylphenyl Methyl Ether $C_5H_{11}C_6H_4OCH_3$	462–469 (239–243)	210 (99)	
Amyl Phthalate		See Diamyl Phthalate.	
Amyl Propionate $C_2H_5COO(CH_2)_4CH_3$ (Pentyl Propionate)	275–347 (135–175)	106 (41)	712 (378)
Amyl Salicylate $HOC_6H_4COOC_5H_{11}$	512 (267)	270 (132)	
Amyl Stearate $CH_3(CH_2)_{16}COOC_5H_{11}$	680 (360)	365 (185)	
Amyl Sulfides, Mixed $C_5H_{11}S$	338–356 (170–180)	185 (85)	
Amyl Toluene $C_5H_{11}C_6H_4CH_3$	400–415 (204–213)	180 (82)	

FLAMMABILITY PROPERTIES—TABLE 1. Boiling Points, Flash Points, and Ignition Temperatures of Organic Compounds (*Continued*)

Compound	Boiling point °F (°C)	Flash point, °F (°C)	Ignition point, °F (°C)
Amyl Xylyl Ether $C_5H_{11}OC_6H_3(CH_3)_2$	480–500 (249–260)	205 (96)	
Aniline $C_6H_5NH_2$ (Aminobenzene) (Phenylamine)	364 (184)	158 (70)	1139 (615)
Aniline Hydrochloride $C_6H_5NH_2HCl$	473 (245)	380 (193)	
2-Anilinoethanol $C_6H_5NHCH_2CH_2OH$ (β-Anilinoethanol Ethoxyaniline) (β-Hydroxyethylaniline)	547 (286)	305 (152)	
β-Anilinoethanol		See 2-Anilinoethanol.	
Ethoxyaniline			
o-Anisaldehyde		See o-Methoxy Benzaldehyde.	
o-Anisidine $H_2NC_6H_4OCH_3$ (2-Methoxyaniline)	435 (224)	244 (118)	
Anisole $C_6H_5OCH_3$ (Methoxybenzene) (Methyl Phenyl Ether)	309 (154)	125 (52)	887 (475)
Anol		See Cyclohexanol.	
Anthracene $(C_6H_4CH)_2$	644 (340)	250 (121)	1004 (540)
Anthraquinone $C_6H_4(CO)_2C_6H_4$	716 (380)	365 (185)	
Asphalt (Petroleum Pitch)	>700 (>371)	400+ 204+)	905 (485)
Aziridine		See Ethyleneimine.	
Azobisisobutyronitrile $N:CC(CH_3)_2N:NC(CH_3)_2C:N$	Decomposes	147 (64)	
Benzaldehyde C_6H_5CHO (Benzenecarbonal)	355 (179)	145 (63)	377 (192)
Benzedrine $C_6H_5CH_2CH(CH_3)NH_2$ (1-Phenyl Isopropyl Amine)	392 (200)	<212 <100)	

FLAMMABILITY PROPERTIES—TABLE 1. Boiling Points, Flash Points, and Ignition Temperatures of Organic Compounds (*Continued*)

Compound	Boiling point °F (°C)	Flash point, °F (°C)	Ignition point, °F (°C)
Benzene	176	12	928
C_6H_6	(80)	(−11)	(498)
(Benzol)			
Benzine		See Petroleum Ether.	
Benzocyclobutene	306	95	477
	(152)	(35)	(247)
Benzoic Acid	482	250	1058
C_6H_5COOH	(250)	(121)	(570)
Benzol		See Benzene.	
p-Benzoquinone	Sublimes	100–200	1040
$C_6H_4O_2$		(38–93)	(560)
(Quinone)			
Benzotrichloride	429	260	412
$C_6H_5CCl_3$	(221)	(127)	(211)
(Toluene, α, α, α-Trichloro)			
(Phenyl Chloroform)			
Benzotrifluoride	216	54	
$C_6H_5CF_3$	(102)	(12)	
Benzoyl Chloride	387	162	
C_6H_5COCl	(197)	(72)	
(Benzene Carbonyl Chloride)			
Benzyl Acetate	417	195	860
$CH_3COOCH_2C_6H_5$	(214)	(90)	(460)
Benzyl Alcohol	403	200	817
$C_6H_5CH_2OH$	(206)	(93)	(436)
(Phenyl Carbinol)			
Benzyl Benzoate	614	298	896
$C_6H_5COOCH_2C_6H_5$	(323)	(148)	(480)
Benzyl Butyl Phthalate	698	390	
$C_4H_9COOC_6H_4COOCH_2-$	(370)	(199)	
C_6H_5			
(Butyl Benzyl Phthalate)			
Benzyl Carbinol		See Phenethyl Alcohol.	
Benzyl Chloride	354	153	1085
$C_6H_5CH_2Cl$	(179)	(67)	(585)
(α-Chlorotoluene)			
Benzyl Cyanide	452	235	
$C_6H_5CH_2CN$	(233.5)	(113)	
(Phenyl Acetonitrile)			
(α-Tolunitrile)			
N-Benzyldiethylamine	405–420	170	
$C_6H_5CH_2N(C_2H_5)_2$	(207–216)	(77)	

FLAMMABILITY PROPERTIES—TABLE 1. Boiling Points, Flash Points, and Ignition Temperatures of Organic Compounds (*Continued*)

Compound	Boiling point °F (°C)	Flash point, °F (°C)	Ignition point, °F (°C)
Benzyl Ether		See Dibenzyl Ether.	
Benzyl Mercaptan $C_6H_5CH_2SH$ (α-Toluenethiol)	383 (195)	158 (70)	
Benzyl Salicilate $OHC_6H_4COOCH_2C_6H_5$ (Salycilic Acid Benzyl Ester)	406 (208)	>212 (>100)	
Bicyclohexyl $[CH_2(CH_2)_4CH]_2$ (Dicyclohexyl)	462 (239)	165 (74)	473 (245)
Biphenyl $C_6H_5C_6H_5$ (Diphenyl) (Phenylbenzene)	489 (254)	235 (113)	1004 (540)
2-Biphenylamine $NH_2C_6H_4C_6H_5$ (2-Aminobiphenyl)	570 (299)	842 (450)	
Bromobenzene C_6H_5Br (Phenyl Bromide)	313 (156)	124 (51)	1049 (565)
1-Bromo Butane		See Butyl Bromide.	
4-Bromodiphenyl $C_6H_5C_6H_4Br$	592 (311)	291 (144)	
Bromoethane		See Ethyl Bromide.	
Bromomethane		See Methyl Bromide.	
1-Bromopentane		See Amyl Bromide.	
3-Bromopropene		See Allyl Bromide.	
o-Bromotoluene $BrC_6H_4CH_3$	359 (182)	174 (79)	
p-Bromotoluene $BrC_6H_4CH_3$	363 (184)	185 (85)	
1,3-Butadiene $CH_2:CHCH:CH_2$	24 (−4)	Gas	788 (420)
Butadiene Monoxide $CH_2:CHCHOCH_2$ (Vinylethylene Oxide)	151 (66)	<−58 (<−50)	
Butanal		See Butyraldehyde.	
Butanal Oxime		See Butyraldoxime.	
Butane $CH_3CH_2CH_2CH_3$	31 (−1)	−76 (−60)	550 (287)

FLAMMABILITY PROPERTIES—TABLE 1. Boiling Points, Flash Points, and Ignition Temperatures of Organic Compounds (*Continued*)

Compound	Boiling point °F (°C)	Flash point, °F (°C)	Ignition point, °F (°C)
1,3-Butanediamine NH$_2$CH$_2$CH$_2$CHNH$_2$CH$_3$	289–302 (143–150)	125 (52)	
1,2-Butanediol CH$_3$CH$_2$CHOHCH$_2$OH (1,2-Dihydroxybutane) (Ethylethylene Glycol)	381 (194)	104 (40)	
1,3-Butanediol		See β-Butylene Glycol.	
1,4-Butanediol HOCH$_2$CH$_2$CH$_2$CH$_2$OH	442 (228)	250 (121)	
2,3-Butanediol CH$_3$CHOHCHOHCH$_3$	363 (184)	756 (402)	
2,3-Butanedione CH$_3$COCOCH$_3$ (Diocetyl)	190 (88)	80 (27)	
1-Butanethiol CH$_3$CH$_2$CH$_2$CH$_2$SH (Butyl Mercaptan)	208 (98)	35 (2)	
2-Butanethiol C$_4$H$_9$SH (sec-Butyl Mercaptan)	185 (85)	−10 (−23)	
1-Butanol		See Butyl Alcohol.	
2-Butanol		See sec-Butyl Alcohol.	
2-Butanone		See Methyl Ethyl Ketone.	
2-Butenal		See Crotonaldehyde.	
1-Butene CH$_3$CH$_2$CH:CH$_2$ (α-Butylene)	21 (−6)		725 (385)
2-Butene-cis CH$_3$CH:CHCH$_3$	38.7 (4)		617 (325)
2-Butene-trans CH$_3$CH:CHCH$_3$ (β-Butylene)	−34 (1)		615 (324)
Butenediol HOCH$_2$CH:CHCH$_2$OH (2-Butene-1,4-Diol)	286–300 (141–149) @ 20 mm	263 (128)	
2-Butene-1,4-Diol		See Butenediol.	
2-Butene Nitrile		See Crotononitrile.	
Butoxybenzene		See Butyl Phenyl Ether.	
1-Butoxybutane		See Dibutyl Ether.	
2,β-Butoxyethoxyethyl Chloride C$_4$H$_9$CH$_2$CH$_2$OCH$_2$CH$_2$Cl	392–437 (200–225)	190 (88)	

FLAMMABILITY PROPERTIES—TABLE 1. Boiling Points, Flash Points, and Ignition Temperatures of Organic Compounds (*Continued*)

Compound	Boiling point °F (°C)	Flash point, °F (°C)	Ignition point, °F (°C)
1-(Butoxyethoxy)-2- **Propanol** $CH_3CH(OH)CH_2OC_2H_4$- $OC_2H_4C_2H_5$	445 (229)	250 (121)	509 (265)
β-Butoxyethyl Salicylate $OCH_6H_4COOCH_2CH_2OC_4$	367–378 (186–192)	315 (157)	
N-Butyl Acetamide $CH_3CONHC_4H_9$	455–464 (235–240)	240 (116)	
N-Butylacetanilide $CH_3(CH_2)_3N(C_6H_5)COCH_3$	531–538 (277–281)	286 (141)	
Butyl Acetate $CH_3COOC_4H_9$ (Butylethanoate)	260 (127)	72 (22)	797 (425)
sec-Butyl Acetate $CH_3COOCH(CH_3)C_2H_5$	234 (112)	88 (31)	
Butyl Acetoacetate $CH_3COCH_2COO(CH_2)_3CH_3$	417 (214)	185 (85)	
Butyl Acetyl Ricinoleate $C_{17}H_{32}(OCOCH_3)$- $(COOC_4H_9)$	428 (220)	230 (110)	725 (385)
Butyl Acrylate $CH_2:CHCOOC_4H_9$	260 (127) Polymerizes	84 (29)	559 (292)
Butyl Alcohol $CH_3(CH_2)_2CH_2OH$ (1-Butanol) (Propylcarbinol) (Propyl Methanol)	243 (117)	98 (37)	650 (343)
sec-Butyl Alcohol $CH_3CH_2CHOHCH_3$ (2-Butanol) (Methyl Ethyl Carbinol)	201 (94)	75 (24)	761 (405)
tert-Butyl Alcohol $(CH_3)_2COHCH_3$ (2-Methyl-2-Propanol) (Trimethyl Carbinol)	181 (83)	52 (11)	892 (478)
Butylamine $C_4H_9NH_2$ (1-Amino Butane)	172 (78)	10 (−12)	594 (312)

FLAMMABILITY PROPERTIES—TABLE 1.　Boiling Points, Flash Points, and Ignition
Temperatures of Organic Compounds (*Continued*)

Compound	Boiling point °F (°C)	Flash point, °F (°C)	Ignition point, °F (°C)
sec-Butylamine	145	16	
$CH_3CH_2CH(NH_2)CH_3$	(63)	(−9)	
tert-Butylamine	113		716
$(CH_3)_3C:NH_2$	(45)		(380)
Butylamine Oleate		150	
$C_{17}H_{33}COONH_3C_4H_9$		(66)	
tert-Butylaminoethyl	200–221	205	
Methacrylate	(93–105)	(96)	
$(CH_3)_3CNHC_2H_4$-			
$OOCC(CH_3):CH_2$			
N-Butylaniline	465	225	
$C_6H_5NHC_4H_9$	(241)	(107)	
Butylbenzene	356	160	770
$C_6H_5C_4H_9$	(180)	(71)	(410)
sec-Butylbenzene	344	126	784
$C_6H_5CH(CH_3)C_2H_5$	(173)	(52)	(418)
tert-Butylbenzene	336	140	842
$C_6H_5C(CH_3)_3$	(169)	(60)	(450)
Butyl Benzoate	482	225	
$C_6H_5COOC_4H_9$	(250)	(107)	
2-Butylbiphenyl	−554	>212	806
$C_6H_5C_6H_4C_4H_9$	(−290)	(>100)	(430)
Butyl Bromide	215	65	509
$CH_3(CH_2)_2CH_2Br$	(102)	(18)	(265)
(1-Bromo Butane)			
Butyl Butyrate	305	128	
$CH_3(CH_2)_2COOC_4H_9$	(152)	(53)	
Butylcarbamic Acid, Ethyl Ester		See N-Butylurethane.	
tert-Butyl Carbinol	237	98	
$(CH_3)_3CCH_2OH$	(114)	(37)	
(2,2-Dimethyl-1-Propanol)			
Butyl Carbitol		See Diethylene Glycol Monobutyl Ether.	
4-tert-Butyl Catechol	545	266	
$(OH)_2C_6H_3C(CH_3)_3$	(285)	(130)	
Butyl Chloride	170	15	464
C_4H_9Cl	(77)	(−9)	(240)
(1-Chlorobutane)			
sec-Butyl Chloride	155	<32	
$CH_3CHClC_2H_5$	(68)	(<0)	
(2-Chlorobutane)			

FLAMMABILITY PROPERTIES—TABLE 1. Boiling Points, Flash Points, and Ignition
Temperatures of Organic Compounds (*Continued*)

Compound	Boiling point °F (°C)	Flash point, °F (°C)	Ignition point, °F (°C)
tert-Butyl Chloride $(CH_3)_3CCl$ (2-Chloro-2-Methyl-Propane)	124 (51)	<32 (<0)	
4-tert-Butyl-2- Chlorophenol $ClC_6H_3(OH)C(CH_3)_3$	453–484 (234–251)	225 (107)	
tert-Butyl-m-Cresol $C_6H_3(C_4H_9)(CH_3)OH$	451–469 (233–243)	116 (47)	
p-tert-Butyl-o-Cresol $(OH)C_6H_3CH_3C(CH_3)_3$	278–280 (137–138)	244 (118)	
Butylcyclohexane $C_4H_9C_6H_{11}$ (1-Cyclohexylbutane)	352–356 (178–180)		475 (246)
sec-Butylcyclohexane $CH_3CH_2CH(CH_3)C_6H_{11}$ (2-Cyclohexylbutane)	351 (177)		531 (277)
tert-Butylcyclohexane $(CH_3)_3CC_6H_{11}$	333–336 (167–169)		648 (342)
N-Butylcyclohexylamine $C_6H_{11}NH(C_4H_9)$	409 (209)	200 (93)	
Butylcyclopentane $C_4H_9C_5H_9$	314 (157)	480 (250)	
Butyl Ether	See Dibutyl Ether.		
Butylethylacetaldehyde	See 2-Ethylhexanal.		
Butyl Ethylene	See 1-Hexene.		
Butyl Ethyl Ether	See Ethyl Butyl Ether.		
Butyl Formate $HCOOC_4H_9$ (Butyl Methanoate) (Formic Acid, Butyl Ester)	225 (107)	64 (18)	612 (322)
Butyl Glycolate $CH_2OHCOOC_4H_9$	~356 (~180)	142 (61)	
tert-Butyl Hydroperoxide $(CH_3)_3COOH$		<80 (<27)	
n-Butyl Isocyanate $CH_3(CH_2)_3NCO$ (Butyl Isocyanate)	235 (113)	66 (19)	
Butyl Isovalerate $C_4H_9OOCCH_2CH(CH_3)_2$	302 (150)	127 (53)	
Butyl Lactate $CH_3CH(OH)COOC_4H_9$	320 (160)	160 (71)	720 (382)
Butyl Mercaptan	See 1-Butanethiol.		

FLAMMABILITY PROPERTIES—TABLE 1. Boiling Points, Flash Points, and Ignition
Temperatures of Organic Compounds (*Continued*)

Compound	Boiling point °F (°C)	Flash point, °F (°C)	Ignition point, °F (°C)
tert-Butyl Mercaptan		See 2-Methyl-2-Propanethiol.	
Butyl Methacrylate CH$_2$:C(CH$_3$)- COO(CH$_2$)$_3$CH$_3$	325 (163)	126 (52)	
Butyl Methanoate		See Butyl Formate.	
N-Butyl Monoethanolamine C$_4$H$_9$NHC$_2$H$_4$OH	378 (192)	170 (77)	
Butyl Naphthalene C$_4$H$_9$C$_{10}$H$_7$		680 (360)	
Butyl Nitrate CH$_3$(CH$_2$)$_3$ONO$_2$	277 (136)	97 (36)	
2-Butyloctanol C$_6$H$_{13}$CH(C$_4$H$_9$)CH$_2$OH	486 (252)	230 (110)	
Butyl Oleate C$_{17}$H$_{33}$COOC$_4$H$_9$	440.6– 442.4 (227–228) @ 15 mm	356 (180)	
Butyl Oxalate (COOC$_4$H$_9$)$_2$ (Butyl Ethanedioate)	472 (244)	265 (129) (oc)	
tert-Butyl Peracetate diluted with 25% of benzene CH$_3$CO(O$_2$)C(CH$_3$)$_3$	Explodes on heating.	<80 (<27)	
tert-Butyl Perbenzoate C$_6$H$_5$COOOC(CH$_3$)$_3$	Explodes on heating.	>190 (>88)	
tert-Butyl Peroxypivalate diluted with 25% of mineral spirits (CH$_3$)$_3$COOCOC(CH$_3$)$_3$	Explodes on heating.	>155 (>68)	
β-(p-tert-Butyl Phenoxy) Ethanol (CH$_3$)$_3$CC$_6$H$_4$OCH$_2$CH$_2$OH	293–313 (145–156)	248 (120)	
β-(p-tert-Butylphenoxy) Ethyl Acetate (CH$_3$)$_3$CC$_6$H$_6$OCH$_2$- CH$_2$OCOCH$_3$	579–585 (304–307)	324 (162)	
Butyl Phenyl Ether CH$_3$(CH$_2$)$_3$OC$_6$H$_5$ (Butoxybenzene)	410 (210)	180 (82)	
4-tert-Butyl-2-Phenylphenol C$_6$H$_5$C$_6$H$_3$OHC(CH$_3$)$_3$	385–388 (196–198)	320 (160)	

FLAMMABILITY PROPERTIES—TABLE 1. Boiling Points, Flash Points, and Ignition Temperatures of Organic Compounds (*Continued*)

Compound	Boiling point °F (°C)	Flash point, °F (°C)	Ignition point, °F (°C)
Butyl Propionate $C_2H_5COOC_4H_9$	295 (146)	90 (32)	799 (426)
Butyl Ricinoleate $C_{18}H_{33}O_3C_4H_9$	790 (421)	230 (110)	
Butyl Sebacate $[(CH_2)_4COOC_4H_9]_2$	653 (345)	353 (178)	
Butyl Stearate $C_{17}H_{35}COOC_4H_9$	650 (343)	320 (160)	671 (355)
tert-Butylstyrene	426 (219)	177 (81)	
tert-Butyl Tetralin $C_4H_9C_{10}H_{11}$		680 (360)	
Butyl Trichlorosilane $CH_3(CH_2)_3SiCl_3$	300 (149)	130 (54)	
N-Butylurethane $CH_3(CH_2)_3NHCOOC_2H_5$ (Butylcarbamic Acid, Ethyl Ester) (Ethyl Butylcarbamate)	396–397 (202–203)	197 (92)	
Butyl Vinyl Ether		See Vinyl Butyl Ether.	
2-Butyne $CH_3C \vdots CCH_3$ (Crotonylene)	81 (27)	−4 (<−20)	
Butyraldehyde $CH_3(CH_2)_2CHO$ (Butanal) (Butyric Aldehyde)	169 (76)	−8 (−22)	425 (218)
Butyraldol $C_8H_{16}O_2$	280 (138) @ 50 mm	165 (74)	
Butyraldoxime C_4H_8NOH (Butanal Oxime)	306 (152)	136 (58)	
Butyric Acid $CH_3(CH_2)_2COOH$	327 (164)	161 (72)	830 (443)
Butyric Acid, Ethyl Ester		See Ethyl Butyrate.	
Butyric Aldehyde		See Butyraldehyde.	
Butyric Anhydride $[CH_3(CH_2)_2CO]_2O$	388 (196)	180 (54)	535 (279)
Butyric Ester		See Ethyl Butyrate.	

FLAMMABILITY PROPERTIES—TABLE 1. Boiling Points, Flash Points, and Ignition Temperatures of Organic Compounds (*Continued*)

Compound	Boiling point °F (°C)	Flash point, °F (°C)	Ignition point, °F (°C)
Butyrolactone CH₂CH₂CH₂COO	399 (204)	209 (98)	
Butyrone		See 4-Heptanone.	
Butyronitrile CH₃CH₂CH₂CN	243 (117)	76 (24)	935 (501)
Caproic Acid (CH₃)(CH₂)₄COOH (Hexanoic Acid)	400 (204)	215 (102)	716 (380)
Carbolic Acid		See Phenol.	
Carbon Bisulfide		See Carbon Disulfide.	
Carbon Disulfide CS₂ (Carbon Bisulfide)	115 (46)	−22 (−30)	194 (90)
Cetane		See Hexadecane.	
Chloroacetic Acid CH₂ClCOOH	372 (189)	259 (126)	>932 (>500)
Chloroacetophenone C₆H₅COCH₂Cl (Phenacyl Chloride)	477 (247)	244 (118)	
2-Chloro-4,6-di-tert-Amylphenol (C₅H₁₁)₂C₆H₂ClOH	320–354 (160–179) @ 22 mm	250 (121)	
Chloro-4-tert-Amylphenol C₅H₁₁C₆H₃ClOH	487–509 (253–265)	225 (107)	
2-Chloro-4-tert-Amyl-Phenyl Methyl Ether C₅H₁₁C₆H₃ClOCH₃	518–529 (270–276)	230 (110)	
p-Chlorobenzaldehyde ClC₆H₄CHO	417 (214)	190 (88)	
Chlorobenzene C₆H₅Cl (Chlorobenzol) (Monochlorobenzene) (Phenyl Chloride)	270 (132)	82 (28)	1099 (593)
Chlorobenzol		See Chlorobenzene.	
o-Chlorobenzotrifluoride ClC₆H₄CF₃ (o-Chloro-α,α,α-trifluorotoluene)	306 (152)	138 (59)	

FLAMMABILITY PROPERTIES—TABLE 1. Boiling Points, Flash Points, and Ignition Temperatures of Organic Compounds (*Continued*)

Compound	Boiling point °F (°C)	Flash point, °F (°C)	Ignition point, °F (°C)
Chlorobutadiene		See 2-Chloro-1,3-Butadiene.	
2-Chloro-1,3-Butadiene $CH_2:CCl:CH:CH_2$ (Chlorobutadiene) (Chloroprene)	138 (59)	−4 (−20)	
1-Chlorobutane		See Butyl Chloride.	
2-Chlorobutene-2 $CH_3CCl:CHCH_3$	143–159 (62–71)	−3 (−19)	
Chlorodinitrobenzene		See Dinitrochlorobenzene.	
Chloroethane		See Ethyl Chloride.	
2-Chloroethanol CH_2ClCH_2OH (2-Chloroethyl Alcohol) (Ethylene Chlorohydrin)	264–266 (129–130)	140 (60)	797 (425)
2-Chloroethyl Acetate $CH_3COOCH_2CH_2Cl$	291 (144)	151 (66)	
2-Chloroethyl Alcohol		See 2-Chloroethanol.	
Chloro-4-Ethylbenzene $C_2H_5C_6H_4Cl$	364 (184)	147 (64)	
Chloroethylene		See Vinyl Chloride.	
2-Chloroethyl Vinyl Ether		See Vinyl 2-Chloroethyl Ether.	
2-Chloroethyl-2-Xenyl Ether $C_6H_5C_6H_4OCH_2CH_2Cl$	613 (323)	320 (160)	
1-Chlorohexane $CH_3(CH_2)_4CH_2Cl$ (Hexyl Chloride)	270 (132)	95 (35)	
Chloroisopropyl Alcohol		See 1-Chloro-2-Propanol.	
Chloromethane		See Methyl Chloride.	
1-Chloro-2-Methyl Propane		See Isobutyl Chloride.	
1-Chloronaphthalene $C_{10}H_7Cl$	505 (263)	250 (121)	>1036 (>558)
2-Chloro-5-Nitrobenzotrifluoride $C_6H_3CF_3(2\text{-}Cl, 5\text{-}NO_2)$ (2-Chloro-α,α,α-Trifluoro-5-Nitrotoluene)	446 (230)	275 (135)	
1-Chloro-1-Nitroethane $C_2H_4NO_2Cl$	344 (173)	133 (56)	

FLAMMABILITY PROPERTIES—TABLE 1. Boiling Points, Flash Points, and Ignition Temperatures of Organic Compounds (*Continued*)

Compound	Boiling point °F (°C)	Flash point, °F (°C)	Ignition point, °F (°C)
1-Chloro-1-Nitropropane $CHNO_2ClC_2H_5$	285 (141)	144 (62)	
2-Chloro-2-Nitropropane $CH_3CNO_2ClCH_3$	273 (134)	135 (57)	
1-Chloropentane		See Amyl Chloride.	
β-Chlorophenetole $C_6H_5OCH_2CH_2Cl$ (β-Phenoxyethyl Chloride)	306–311 (152–155)	225 (107)	
o-Chlorophenol ClC_6H_4OH	347 (175)	147 (64)	
p-Chlorophenol C_6H_4OHCl	428 (220)	250 (121)	
2-Chloro-4-Phenylphenol $C_6H_5C_6H_3ClOH$	613 (323)	345 (174)	
Chloroprene		See 2-Chloro-1,3-Butadiene.	
1-Chloropropane		See Propyl Chloride.	
2-Chloropropane		See Isopropyl Chloride.	
2-Chloro-1-Propanol $CH_3CHClCH_2OH$ (β-Chloropropyl Alcohol) (Propylene Chlorohydrin)	271–273 (133–134)	125 (52)	
1-Chloro-2-Propanol $CH_2ClCHOHCH_3$ (Chloroisopropyl Alcohol) (sec-Propylene Chlorohydrin)	261 (127)	125 (52)	
1-Chloro-1-Propene		See 1-Chloropropylene.	
3-Chloropropene		See Allyl Chloride.	
α-Chloropropionic Acid $CH_3CHClCOOH$	352–374 (178–190)	225 (107)	932 (500)
3-Chloropropionitrile $ClCH_2CH_2CN$	348.8 (176) Decomposes	168 (76)	
2-Chloropropionyl Chloride	230 (110)	88 (31)	
β-Chloropropyl Alcohol		See 2-Chloro-1-Propanol.	
1-Chloropropylene $CH_3CH:CHCl$ (1-Chloro-1-Propene)	95–97 (35–36)	<21 (<–6)	

FLAMMABILITY PROPERTIES—TABLE 1. Boiling Points, Flash Points, and Ignition Temperatures of Organic Compounds (*Continued*)

Compound	Boiling point °F (°C)	Flash point, °F (°C)	Ignition point, °F (°C)
2-Chloropropylene $CH_3CCl:CH_2$ (β-Chloropropylene) (2-Chloropropene)	73 (23)	<−4 (<−20)	
2-Chloropropylene Oxide		See Epichlorohydrin.	
γ-Chloropropylene Oxide		See Epichlorohydrin.	
Chlorotoluene $C_6H_4ClCH_3$ (Tolyl Chloride)	320 (160)	126 (52)	
α-Chlorotoluene		See Benzyl Chloride.	
Chlorotrifluoroethylene		See Trifluorochloroethylene.	
2-Chloro-α,α,α-Trifluoro-5-Nitrotoluene		See 2-Chloro-5-Nitrobenzotrifluoride.	
o-Chloro-α,α,α-Trifluorotoluene		See o-Chlorobenzotrifluoride.	
Coal Oil		See Fuel Oil No. 1.	
Coal Tar Light Oil		<80 (<27)	
Coal Tar Pitch		405 (207)	
Creosote Oil	382–752 (194–400)	165 (74)	637 (336)
o-Cresol $CH_3C_6H_4OH$ (Cresylic Acid) (o-Hydroxytoluene) (o-Methyl Phenol)	376 (191)	178 (81)	1110 (599)
p-Cresyl Acetate $CH_3C_6H_4OCOCH_3$ (p-Tolyl Acetate)		195 (91)	
Cresyl Diphenyl Phosphate $(C_6H_5O)_2[(CH_3)_2C_6H_4O]\text{-}PO_4$	734 (390)	450 (232)	
Cresylic Acid		See o-Cresol.	
Crotonaldehyde $CH_3CH:CHCHO$ (2-Butenal) (Crotonic Aldehyde) (Propylene Aldehyde)	216 (102)	55 (13)	450 (232)
Crotonic Acid $CH_3CH:CHCOOH$		372 (189)	190 (88)	745 (396)

FLAMMABILITY PROPERTIES—TABLE 1.   Boiling Points, Flash Points, and Ignition
Temperatures of Organic Compounds (*Continued*)

Compound	Boiling point °F (°C)	Flash point, °F (°C)	Ignition point, °F (°C)
Crotononitrile CH$_3$CH:CHCN (2-Butenenitrile)	230– 240.8 (110–116)	<212 (<100)	
Crotonyl Alcohol CH$_3$CH:CHCH$_2$OH (2-Buten-1-ol) (Crotyl Alcohol)	250 (121)	81 (27)	660 (349)
1-Crotyl Bromide CH$_3$CH:CHCH$_2$Br (1-Bromo-2-Butene)			
1-Crotyl Chloride CH$_3$CH:CHCH$_2$Cl (1-Chloro-2-Butene)			
Cumene C$_6$H$_5$CH(CH$_3$)$_2$ (Cumol) (2-Phenyl Propane) (Isopropyl Benzene)	306 (152)	96 (36)	795 (424)
Cumene Hydroperoxide C$_6$H$_5$C(CH$_3$)$_2$OOH	Explodes on heating.	175 (79)	
Cyanamide NH$_2$CN	500 (260) Decomposes	286 (141)	
2-Cyanoethyl Acrylate CH$_2$CHCOOCH$_2$CH$_2$CN	Polymerizes	255 (124)	
N-(2-Cyanoethyl) Cyclohexylamine C$_6$H$_{11}$NHC$_2$H$_4$CN		255 (124)	
Cyclamen Aldehyde (CH$_3$)$_2$CHC$_6$H$_4$CH(CH$_3$)CH$_2$- CHO (Methyl Para-Isopropyl Phenyl Propyl Aldehyde)		190 (88)	
Cyclobutane C$_4$H$_8$ (Tetramethylene)	55 (13)		
1,5,9-Cyclododecatriene C$_{12}$H$_{18}$	448 (231)	160 (71)	
Cycloheptane CH$_2$(CH$_2$)$_5$CH$_2$	246 (119)	<70 (<21)	

FLAMMABILITY PROPERTIES—TABLE 1. Boiling Points, Flash Points, and Ignition
Temperatures of Organic Compounds (*Continued*)

Compound	Boiling point °F (°C)	Flash point, °F (°C)	Ignition point, °F (°C)
Cyclohexane C_6H_{12} (Hexahydrobenzene) (Hexamethylene)	179 (82)	−4 (−20)	473 (245)
1,4-Cyclohexane Dimethanol $C_8H_{16}O_2$	525 (274)	332 (167)	600 (316)
Cyclohexanethiol $C_6H_{11}SH$ (Cyclohexylmercaptan)	315–319 (157–159)	110 (43)	
Cyclohexanol $C_6H_{11}OH$ (Anol) (Hexolin) (Hydralin)	322 (161)	154 (68)	572 (300)
Cyclohexanone $C_6H_{10}O$ (Pimelic Ketone)	313 (156)	111 (44)	788 (420)
Cyclohexene $CH_2CH_2CH_2CH_2CH:CH$	181 (83)	<20 (<−7)	471 (244)
3-Cyclohexene-1- Carboxaldehyde		See 1,2,3,6- Tetrahydrobenzaldehyde.	
Cyclohexenone C_6H_8O	313 (156)	93 (34)	
Cyclohexyl Acetate $CH_3CO_2C_6H_{11}$ (Hexolin Acetate)	350 (177)	136 (58)	635 (335)
Cyclohexylamine $C_6H_{11}NH_2$ (Aminocyclohexane) (Hexahydroaniline)	274 (134)	88 (31)	560 (293)
Cyclohexylbenzene $C_6H_5C_6H_{11}$ (Phenylcyclohexone)	459 (237)	210 (99)	
Cyclohexyl Chloride $CH_2(CH_2)_4CHCl$ (Chlorocyclohexane)	288 (142)	90 (32)	
Cyclohexylcyclohexanol $C_6H_{11}C_6H_{10}OH$	304–313 (151–156)	270 (132)	
Cyclohexyl Formate $CH_2(CH_2)_4HCOOCH$	324 (162)	124 (51)	

FLAMMABILITY PROPERTIES—TABLE 1. Boiling Points, Flash Points, and Ignition
Temperatures of Organic Compounds (*Continued*)

Compound	Boiling point °F (°C)	Flash point, °F (°C)	Ignition point, °F (°C)
Cyclohexylmethane		See Methylcyclohexane.	
o-Cyclohexylphenol	298	273	
$C_6H_{11}C_6H_4OH$	(148)	(134)	
	@ 10 mm		
Cyclohexyltrichlorosilane	406	196	
$C_6H_{11}SiCl_3$	(208)	(91)	
1,5-Cyclooctadiene	304	95	
C_8H_{10}	(151)	(35)	
Cyclopentane	121	<20	682
C_5H_{10}	(49)	(<−7)	(361)
Cyclopentene	111	−20	743
$CH:CHCH_2CH_2CH_2$	(44)	(−29)	(395)
Cyclopentanol	286	124	
$CH_2(CH_2)_3CHOH$	(141)	(51)	
Cyclopentanone	267	79	
$OCCH_2CH_2CH_2CH_2$	(131)	(26)	
(Adipic Ketone)			
Cyclopropane	−29		928
$(CH_2)_3$	(−34)		(498)
(Trimethylene)			
p-Cymene	349	117	817
$CH_3C_6H_4CH(CH_3)_2$ Tech.	(176)	(47)	(436)
(4-Isopropyl-1-Methyl		127	833
Benzene)		(53)	(445)
Decahydronaphthalene	382	136	482
$C_{10}H_{18}$	(194)	(58)	(250)
(Decalin)			
Decahydronaphthalene-trans	369	129	491
$C_{10}H_{18}$	(187)	(54)	(255)
Decalin		See Decahydronaphthalene.	
Decane	345	115	410
$CH_3(CH_2)_8CH_3$	(174)	(46)	(210)
Decanol	444.2	180	550
$CH_3(CH_2)_8CH_2OH$	(229)	(82)	(288)
(Decyl Alcohol)			
1-Decene	342	<131	455
$CH_3(CH_2)_7CH:CH_2$	(172)	(<55)	(235)
Decyl Acrylate	316	441	
$CH_3(CN_2)_9OCOCH:CH_2$	(158)	(227)	
	@ 50 mm		

FLAMMABILITY PROPERTIES—TABLE 1. Boiling Points, Flash Points, and Ignition Temperatures of Organic Compounds (*Continued*)

Compound	Boiling point °F (°C)	Flash point, °F (°C)	Ignition point, °F (°C)
Decyl Alcohol		See Decanol.	
Decylamine	429	210	
$CH_3(CH_2)_9NH_2$	(221)	(99)	
(1-Aminodecane)			
Decylbenzene	491–536	225	
$C_{10}H_{21}C_6H_5$	(255–280)	(107)	
tert-Decylmercaptan	410–424	190	
$C_{10}H_{21}SH$	(210–218)	(88)	
Decylnaphthalene	635–680	350	
$C_{10}H_{21}C_{10}H_7$	(335–360)	(177)	
Decyl Nitrate	261	235	
$CH_3(CH_2)_9ONO_2$	(127)	(113)	
	@ 11 mm		
Diacetone Alcohol	328	148	1118
$CH_3COCH_2C(CH_3)_2OH$	(164)		
Diacetyl		See 2,3-Butanedione.	
Diallyl Ether		See Allyl Ether.	
Diallyl Phthalate	554	330	
$C_6H_4(CO_2C_3H_5)_2$	(290)	(166)	
1,3-Diaminobutane		See 1,3-Butanediamine.	
1,3-Diamino-2-Propanol	266	270	
$NH_2CH_2CHOHCH_2NH_2$	(130)	(132)	
1,3-Diaminopropane		See 1,3-Propanediamine.	
Diamylamine	356	124	
$(C_5H_{11})_2NH$	(180)	(51)	
Diamylbenzene	491–536	225	
$(C_5H_{11})_2C_6H_4$	(255–280)	(107)	
Diamylbiphenyl	687–759	340	
$C_5H_{11}(C_6H_4)_2C_5H_{11}$	(364–404)	(171)	
(Diaminodiphenyl)			
Di-tert-Amylcyclohexanol	554–572	270	
$(C_5H_{11})_2C_6H_9OH$	(290–300)	(132)	
Diamyidlphenyl		See Diamylbiphenyl.	
Diamylene	302	118	
$C_{10}H_{20}$	(150)	(48)	
Diamyl Ether		See Amyl Ether.	
Diamyl Maleate	505–572	270	
$(CHCOOC_5H_{11})_2$	(263–300)	(132)	
Diamyl Naphthalene	624	315	
$C_{10}H_6(C_5H_{11})_2$	(329)	(159)	
2,4-Diamylphenol	527	260	
$(C_5H_{11})_2C_6H_3OH$	(275)	(127)	

FLAMMABILITY PROPERTIES—TABLE 1.   Boiling Points, Flash Points, and Ignition
Temperatures of Organic Compounds (*Continued*)

Compound	Boiling point °F (°C)	Flash point, °F (°C)	Ignition point, °F (°C)
Di-tert-Amylphenoxy Ethanol $C_6H_3(C_5H_{11})_2OC_2H_4OH$	615 (324)	300 (149)	
Diamyl Phthalate $C_6H_4(COOC_5H_{11})_2$ (Amyl Phthalate)	475–490 (246–254) @ 50 mm	245 (118)	
Diamyl Sulfide $(C_5H_{11})_2S$	338–356 (170–180)	185 (85)	
o-Dianisldine $[NH_2(OCH_3)C_6H_3]_2$ (o-Dimethoxybenzidine)		403 (206)	
Dibenzyl Ether $(C_6H_5CH_2)_2O$ (Benzyl Ether)	568 (298)	275 (135)	
Dibutoxy Ethyl Phthalate $C_6H_4(COOC_2H_4OC_4H_9)_2$	437 (225)	407 (208) (oc)	
Dibutoxymethane $CH_2(OC_4H_9)_2$	330–370 (166–188)	140 (60)	
Dibutoxy Tetraglycol $(C_4H_9OC_2H_4OC_2H_4)_2O$ (Tetraethylene Glycol Dibutyl Ether)	635 (335)	305 (152)	
N,N-Dibutylacetamide $CH_3CON(C_4H_9)_2$	469–482 (243–250)	225 (107)	
Dibutylamine $(C_4H_9)_2NH$	322 (161)	117 (47)	
Di-sec-Butylamine $[C_2H_5(CH_3)CH]_2NH$	270–275 (132–135)	75 (24)	
Dibutylaminoethanol $(C_4H_9)_2NC_2H_4OH$	432 (222)	200 (93)	
1-Dibutylamino-2-Propanol		See Dibutylisopropanolamine.	
N,N-Dibutylanlline $C_6H_5N(CH_2CH_2CH_2CH_3)_2$	505–527 (263–275)	230 (110)	
Di-tert-Butyl-p-Cresol $C_6H_2(C_4H_9)_2(CH_3)OH$	495–511 (257–266)	261 (127)	
Dibutyl Ether $(C_4H_9)_2O$ (1-Butoxybutane) (Butyl Ether)	286 (141)	77 (25)	382 (194)
2,5-Di-tert-Butylhydroquinone $[C(CH_3)_3]_2C_6H_2(OH)_2$ (DTBHQ)		420 (216)	790 (421)

FLAMMABILITY PROPERTIES—TABLE 1. Boiling Points, Flash Points, and Ignition Temperatures of Organic Compounds (*Continued*)

Compound	Boiling point °F (°C)	Flash point, °F (°C)	Ignition point, °F (°C)
Dibutyl Isophthalate $C_6H_4(CO_2C_4H_9)_2$		322 (161)	
N,N¹-Di-sec-Butyl-p-Phenylenediamine $C_6H_4[-NHCH(CH_3)-CH_2CH_3]_2$		270 (132)	625 (329)
Dibutylisopropanolamine $CH_3CHOHCH_2N(C_4H_9)_2$	444 (229)	205 (96)	
Dibutyl Maleate $(-CHCO_2C_4H_9)_2$	Decomposes	285 (141)	
Dibutyl Oxalate $C_4H_9OOCCOOC_4H_9$	472 (244)	220 (104)	
Di-tert-Butyl Peroxide $(CH_3)_3COOC(CH_3)_3$	231 (111)	65 (18)	
Dibutyl Phthalate $C_6H_4(CO_2C_4H_9)_2$ (Dibutyl-o-Phthatate)	644 (340)	315 (157)	757 (402)
n-Dibutyl Tartrate $(COOC_4H_9)_2(CHOH)_2$ (Dibutyl-d-2,3-Dihydroxybutanedioate)	650 (343)	195 (91)	544 (284)
N,N-Dibutyltoluene-sulfonamide $CH_3C_6H_4SO_3N(C_4H_9)_2$	392 (200) @ 10 mm	330 (166)	
Dicaproate	See Triethylene Glycol.		
Dicapryl Phthalate $C_6H_4[COOCH(CH_3)C_6H_{13}]_2$	441–453 (227–234) @ 4.5 mm	395 (202)	
Dichloroacetyl Chloride $CHCl_2COCl$ (Dichloroethanoyl Chloride)	225–226 (107–108)	151 (66)	
3,4-Dichloroaniline $NH_2C_6H_3Cl_2$	522 (272)	331 (166)	
o-Dichlorobenzene $C_6H_4Cl_2$ (o-Dichlorobenzol)	356 (180)	151 (66)	1198 (648)
p-Dichlorobenzene $C_6H_4Cl_2$	345 (174)	150 (66)	

FLAMMABILITY PROPERTIES—TABLE 1.　Boiling Points, Flash Points, and Ignition
Temperatures of Organic Compounds (*Continued*)

Compound	Boiling point °F (°C)	Flash point, °F (°C)	Ignition point, °F (°C)
o-Dichlorobenzol		See o-Dichlorobenzene.	
2,3-Dichlorobutadiene-1,3 $CH_2:C(Cl)C(Cl):CH_2$	212 (100)	50 (10)	694 (368)
1,2-Dichlorobutane $CH_3CH_2CHClCH_2Cl$		527 (275)	
1,4-Dichlorobutane $CH_2ClCH_2CH_2CH_2Cl$	311 (155)	126 (52)	
2,3-Dichlorobutane $CH_3CHClCHClCH_3$	241–253 (116–123)	194 (90)	
1,3-Dichloro-2-Butene $CH_2ClCH:CClCH_3$	262 (128)	80 (27)	
3,4-Dichlorobutene-1 $CH_2ClCHClCHCH_2$	316 (158)	113 (45)	
1,3-Dichlorobutene-2 $CH_2ClCH:CClCH_3$	258 (126)	80 (27)	
Dichlorodimethylsilane		See Dimethyldichlorosilane.	
1,1-Dichloroethane		See Ethylidene Dichloride.	
1,2-Dichloroethane		See Ethylene Dichloride.	
Dichloroethanoyl Chloride		See Dichloroacetyl Chloride.	
1,1-Dichloroethylene		See Vinylidene Chloride.	
Dichloroisopropyl Ether $ClCH_2CH(CH_3)OCH$- $(CH_3)CH_2Cl$ [Bis (β-Chloroisopropyl) Ether]	369 (187)	185 (85)	
2,2-Dichloro Isopropyl Ether $[ClCH_2CH(CH_3)]_2O$ [Bis (2-Chloro-1-Mothylethyl) Ether]	369 (187)	185 (85)	
Dichloromethane		See Methylene Chloride.	
1,1-Dichloro-1-Nitro Ethane $CH_3CCl_2NO_2$	255 (124)	168 (76)	
1,1-Dichloro-1-Nitro Propane $C_2H_5CCl_2NO_2$	289 (143)	151 (66)	
1,5-Dichloropentane $CH_2Cl(CH_2)_3CH_2Cl$ (Amylene Chloride) (Pentamethylene Dichloride)	352–358 (178–181)	>80 (>27)	

FLAMMABILITY PROPERTIES—TABLE 1. Boiling Points, Flash Points, and Ignition Temperatures of Organic Compounds (*Continued*)

Compound	Boiling point °F (°C)	Flash point, °F (°C)	Ignition point, °F (°C)
2,4-Dichlorophenol $Cl_2C_6H_3OH$	410 (210)	237 (114)	
1,2-Dichloropropane		See Propylene Dichloride.	
1,3-Dichloro-2-Propanol $CH_2ClCHOHCH_2Cl$	346 (174)	165 (74)	
1,3-Dichloropropene $CHCl:CHCH_2Cl$	219 (104)	95 (35)	
2,3-Dichloropropene CH_2CClCH_2Cl	201 (94)	59 (15)	
α,β-Dichlorostyrene $C_6H_5CCl:CHCl$		225 (107)	
Dicyclohexyl		See Bicyclohexyl.	
Dicyclohexylamine $(C_6H_{11})_2NH$	496 (258)	>210 (>99)	
Dicyclopentadiene $C_{10}H_{12}$	342 (172)	90 (32)	937 (503)
Didecyl Ether $(C_{10}H_{21})_2O$ (Decyl Ether)		419 (215)	
Diesel Fuel Oil No. 1-D		100 Min. (38)	
Diesel Fuel Oil No. 2-D		125 Min. (52)	
Diesel Fuel Oil No. 4-D		130 Min. (54)	
Diethanolomine $(HOCH_2CH_2)_2NH$	514 (268)	342 (172)	1224 (662)
1,2-Diethoxyethane		See Diethyl Glycol.	
Diethylacetaldehyde		See 2-Ethylbutyraldehyde.	
Diethylacetic Acid		See 2-Ethylbutyric Acid.	
N,N-Diethyl-acetoacetamide $CH_3COCH_2CON(C_2H_5)_2$	Decomposes	250 (121)	
Diethyl Acetoacetate $CH_3COC(C_2H_5)_2COOC_2H_5$	412–424 (211–218) Decomposes	170 (77)	
Diethylamine $(C_2H_5)_2NH$	134 (57)	−9 (−23)	594 (312)

FLAMMABILITY PROPERTIES—TABLE 1. Boiling Points, Flash Points, and Ignition Temperatures of Organic Compounds (*Continued*)

Compound	Boiling point °F (°C)	Flash point, °F (°C)	Ignition point, °F (°C)
2-Diethyl (Amino) Ethanol		See N,N-Diethylethanolamine.	
2-(Diethylamino) Ethyl Acrylate CH_2:$CHCOOCH_2CH_2$-$HN(CH_3CH_2)_2$	Decomposes	195 (91)	
3-(Diethylamino)-Propylamine $(C_2H_5)_2NCH_2CH_2CH_2NH_2$ (N,N-Diethyl-1,3-Propanediamine)	337 (169)	138 (59)	
N,N-Diethylaniline $C_6H_5N(C_2H_5)_2$ (Phenyldiethylamine)	421 (216)	185 (85)	1166 (630)
o-Diethyl Benzene $C_6H_4(C_2H_5)_2$	362 (183)	135 (57)	743 (395)
m-Diethyl Benzene $C_6H_4(C_2H_5)_2$	358 (181)	133 (56)	842 (450)
p-Diethyl Benzene $C_6H_4(C_2H_5)_2$	358 (181)	132 (55)	806 (430)
N,N-Diethyl-1,3-Butanediamine $C_2H_5NHCH_2CH_2CH$-$N(C_2H_5)CH_3$ [1,3-Bis(ethylamino) Buiane]	354–365 (179–185)	115 (46)	
D1-2-Ethylbutyl Phthalate $C_6H_4[COOCH_2CH(C_2H_5)_2]_2$	662 350	381 (194)	
Diethyl Carbamyl Chloride $(C_2H_5)_2NCOCl$	369–374 (187–190)	325–342 (163–172)	
Diethyl Carbinol		See sec-Amyl Alcohol.	
Diethyl Carbonate $(C_2H_5)_2CO_3$ (Ethyl Carbonate)	259 (126)	77 (25)	
Diethylcyclohexane $C_{10}H_{20}$	344 (173)	120 (49)	464 (240)
1,3-Diethyl-1,3-Diphenyl Urea $[(C_2H_5)(C_6H_5)N]_2CO$	620 (327)	302 (150)	
Diethylene Diamine	299 (150)	144 (62)	
Diethylene Dioxide		See p-Dioxane.	
Diethylene Glycol $O(CH_2CH_2OH)_2$ (2,2-Dihydroxyethyl Ether)	472 (244)	255 (124)	435 (224)
Diethylene Glycol Methyl Ether $CH_3OC_2H_4OC_2H_4OH$ (2-(2-Methoxyethoxy) Ethanol)	379 (193)	205 (96)	465 (240)

FLAMMABILITY PROPERTIES—TABLE 1. Boiling Points, Flash Points, and Ignition Temperatures of Organic Compounds (*Continued*)

Compound	Boiling point °F (°C)	Flash point, °F (°C)	Ignition point, °F (°C)
Diethylene Glycol Methyl Ether Acetate $CH_3COOC_2H_4OC_2H_4OCH_3$	410 (210)	180 (82)	
Diethylene Glycol Monobutyl Ether $C_4H_9OCH_2CH_2OCH_2CH_2OH$	448 (231)	172 (78)	400 (204)
Diethylene Glycol Monoethyl Ether Acetate $C_4H_9O(CH_2)_2O(CH_2)_2-OOCCH_3$	476 (247)	240 (116)	570 (298.9)
Diethylene Glycol Monoethyl Ether $CH_2OHCH_2OCH_2-CH_2OC_2H_5$	396 (202)	201 (94)	400 (204)
Diethylene Glycol Monoethyl Ether Acetate $C_2H_5O(CH_2)_2O(CH_2)_2-OOCCH_3$	424 (218)	225 (107)	680 (360)
Diethylene Glycol Monoisobutyl Ether $(CH_3)_2CHCH_2O(CH_2)_2-O(CH_2)_2OH$	422–437 (217–225)	222 (106)	452–485 (233–252)
Diethylene Glycol Monomethyl Ether $CH_3O(CH_2)O(CH_2)_2OH$	381 (194)	205 (96)	
Diethylene Glycol Mono-Methyl Ether Formal $CH_2(CH_3OCH_2CH_2OCH_2-CH_2O)_2$	581 (305)	310 (154)	
Diethylene Glycol Phthalate $C_6H_4[COO(CH_2)_2OC_2H_5]_2$		343 (173)	
Diethylene Oxide		See Tetrahydrofuran.	
Diethylene Triamine $NH_2CH_2CH_2NHCH_2CH_2NH_2$	404 (207)	208 (98)	676 (358)
N,N-Diethylethanolamine $(C_2H_5)_2NC_2H_4OH$ (2-(Diethylamino) Ethanol)	324 (162)	140 (60)	608 (320)
Diethyl Ether		See Ethyl Ether.	
N,N-Diethylethylene-diamine $(C_2H_5)_2NC_2H_4NH_2$	293 (145)	115 (46)	
Diethyl Fumarate $C_2H_5OCOCH:CHCOOC_2H_5$	442 (217)	220 (104)	
Diethyl Glycol $(C_2H_5OCH_2)_2$ (1,2-Diethoxyethane)	252 (122)	95 (35)	401 (205)

FLAMMABILITY PROPERTIES—TABLE 1. Boiling Points, Flash Points, and Ignition
Temperatures of Organic Compounds (*Continued*)

Compound	Boiling point °F (°C)	Flash point, °F (°C)	Ignition point, °F (°C)
Diethyl Ketone $C_2H_5COC_2H_5$ (3-Pentanone)	217 (103)	55 (13)	842 (450)
N,N-Diethyllauramide $C_{11}H_{23}CON(C_2H_5)_2$	331–351 (166–177) @ 2 mm	>150 (>66)	
Diethyl Maleate $(-CHCO_2C_2H_3)_2$	438 (226)	250 (121)	662 (350)
Diethyl Malonate $CH_2(COOC_2H_3)_2$ (Ethyl Malonate)	390 (199)	200 (93)	
Diethyl Oxide		See Ethyl Ether.	
3,3-Diethylpentane $CH_3CH_2C(C_2H_5)_2CH_2CH_3$	295 (146)	554 (290)	
Diethyl Phthalate $C_6H_4(COOC_2H_5)_2$	565 (296)	322 (161)	855 (457)
p-Diethyl Phthalate		See Diethyl Terephthalate.	
N,N-Diethylstearamide $C_{17}H_{35}CON(C_2H_5)_2$	246–401 (119–205) @ 1 mm	375 (191)	
Diethyl Succinate $(CH_2COOCH_2CH_3)_2$	421 (216)	195 (90)	
Diethyl Sulfate $(C_2H_5)_2SO_4$ (Ethyl Sulfate)	Decomposes, giving Ethyl Ether	220 (104)	817 (436)
Diethyl Tartrate $CHOHCOO(C_2H_5)_2$	536 (280)	200 (93)	
Diethyl Terephthalate $C_6H_4(COOC_2H_5)_2$ (p-Diethyl Phthalate)	576 (302)	243 (117)	
3,9-Diethyl-6-tridecanol		See Heptadecanol.	
Diglycol Chlortormate $O:(CH_2CH_2OCOCl)_2$	256–261 (124–127) @ 5 mm	295 (146)	
Diglycol Chlorohydrin $HOCH_2CH_2OCH_2CH_2Cl$	387 (197)	225 (107)	
Diglycol Diacetate $(CH_3COOCH_2CH_2)_2O$	482 (250)	255 (124)	

FLAMMABILITY PROPERTIES—TABLE 1. Boiling Points, Flash Points, and Ignition Temperatures of Organic Compounds (*Continued*)

Compound	Boiling point °F (°C)	Flash point, °F (°C)	Ignition point, °F (°C)
Diglycol Dilevulleate $(CH_2CH_2OOC-$ $(CH_2)_2COCH_3)_2:O$		340 (171)	
Diglycol Laurate $C_{16}H_{32}O_4$	559–617 (293–325)	290 (143)	
Dihexyl		See Dodecane.	
Dihexylamine $[CH_3(CH_2)_5]_2NH$	451–469 (233–243)	220 (104)	
Dihexyl Ether		See Hexyl Ether.	
Dihydropyran $CH_2CH_2CH_2:CHCHO$	186 (86)	0 (−18)	
o-Dihydroxybenione $C_6H_4(OH)_2$ (Pyrocalechol)	473 (245)	260 (127)	
p-Dihydroxybenione $C_6H_4(OH)_2$ (Hydroquinone)	547 (286)	329 (165)	959 (515)
1,2-Dihydroxybenione		See 1,2-Butanediol.	
2,2-Dihydroxyethyl Ether		See Diethylene Glycol.	
2,5-Dihydroxyhexane		See 2,5-Hexanediol.	
Diisobutylamine $[(CH_3)_2CHCH_2]_2NH$ [Bis(β-Methylpropyl) Amine]	273–286 (134–141)	85 (29)	
Diisobutyl Carbinol $[(CH_3)_2CHCH_2]_2CHOH$ (Nonyl Alcohol)	353 (178)	165 (74)	
Diisobutylene		See 2,4,4-Trimethyl-1-Pentene.	
Diisobutylene $(CH_3)_3CCH_2C(CH_3):CH_2$ (2,4,4-Trimethy-H$_2$-Pentane)	214 (101)	23 (−5)	736 (391)
Diisobutyl Ketone $[(CH_3)_2CHCH_2]_2CO$ (2,6-Dimethyl-4 Heptanone) (Isovalerone)	335 (168)	120 (49)	745 (396)
Diisobutyl Phthalate $C_6H_4[COOCH_2OH(CH_3)_2]_2$	321 (327)	365 (185)	810 (432)
Diisodecyl Adipoia $C_{10}H_{21}O_2C(CH_2)_2CO_2-C_{10}H_{21}$	660 (349)	225 (107)	
Diisodecyl Phthalate $C_6H_4(COOC_{10}H_{21})_2$	182 (250)	450 (232)	755 (402)
Diisooctyl Phthalate $(C_8H_{17}COO)_2C_2H_4$	398 (370)	450 (232)	

FLAMMABILITY PROPERTIES—TABLE 1. Boiling Points, Flash Points, and Ignition Temperatures of Organic Compounds (*Continued*)

Compound	Boiling point °F (°C)	Flash point, °F (°C)	Ignition point, °F (°C)
Diisopropanolamine [CH$_3$CH(OH)-CH$_2$]$_2$NH	480 (249)	260 (127)	705 (374)
Diisopropyl		See 2,3-Dimethylbutane.	
Diisopropylamine [(CH$_3$)$_2$CH]$_2$NH	183 (84)	30 (−1)	600 (316)
Diisopropyl Benzene [(CH$_3$)$_2$CH]$_2$C$_6$H$_4$	401 (205)	170 (77)	840 (449)
N,N-Diisopropyl-ethanolamine [(CH$_3$)$_2$CH]$_2$NC$_2$H$_4$OH	376 (191)	175 (79)	
Diisopropyl Ether		See Isopropyl Ether.	
Diisopropyl Maleate (CH$_3$)$_2$CHOCOCH: CHCOOCH(CH$_3$)$_2$	444 (229)	220 (104)	
Diisopropylmethanol		See 2,4-Dimethyl-3-Pentanol.	
Diisopropyl Peroxydicarbonate (CH$_3$)$_2$CHOCOOCO- OCH(CH$_3$)$_2$	Explodes on heating.		
Diketene CH$_2$:CCH$_2$C(O)O ⎿_____⏌ (Vinylaceto-β-Lactone)	261 (127)	93 (34)	
2,5-Dimethoxyaniline NH$_2$C$_6$H$_3$(OCH$_3$)$_2$	518 (270)	302 (150)	735 (391)
2,5-Dimethoxy chlorobenzene C$_8$H$_9$ClO$_2$	460–467 (238–242)	243 (117)	
1,2-Dimethoxyethane		See Ethylene Glycol Dimethyl Ether.	
Dimethoxyethyl Phthalate C$_6$H$_4$(COOCH$_2$CH$_2$OCH$_3$)$_2$ (Bis(2-methoxyethyl) Phthalate)	644 (340)	410 (210)	750 (399)
Dimethoxymethane		See Methylal.	
Dimethoxy Tetraglycol CH$_3$OCH$_2$(CH$_2$- OCH$_2$)$_3$CH$_2$OCH$_3$ (Tetraethylene Glycol Dimethyl Ether)	528 (276)	285 (141)	
Dimethylacetamide (CH$_3$)$_2$NC:OCH$_3$ (DMAC)	330 (165)	158 (70)	914 (490)
Dimethylamine (CH$_3$)$_2$NH	45 (7)	Gos	752 (400)

FLAMMABILITY PROPERTIES—TABLE 1. Boiling Points, Flash Points, and Ignition
Temperatures of Organic Compounds (*Continued*)

Compound	Boiling point °F (°C)	Flash point, °F (°C)	Ignition point, °F (°C)
1,2-Dimethylbenzene		See o-Xylene.	
1,3-Dimethylbenzene		See m-Xylene.	
1,4-Dimethylbenzene		See p-Xylene.	
Dimethylbenzylcarbinyl Acetate $C_6H_5CH_2C(CH_3)_2OOCCH_3$ (alpha, alpha-Dimethyl-phenethyl Acelate)		205 (96)	
2,2-Dimethylbutane $(CH_3)_3CCH_2CH_3$ (Neohexane)	122 (50)	−54 (−48)	761 (405)
2,3-Dimethylbutane $(CH_3)_2CHCH(CH_3)_2$ (Diisopropyl)	136 (58)	−20 (−29)	761 (405)
1,3-Dimethylbutanol		See Methyl Isobutyl Carbinol.	
2,3-Dimethyl-1-Butene $CH_3CH(CH_3)C(CH_3):CH_2$	133 (56)	<−4 (<−20)	680 (360)
2,3-Dimethyl-2-Butene $CH_3C(CH_3):C(CH_3)_2$	163 (73)	<−4 (<−20)	753 (401)
1,3-Dimethylbutyl Acetate $CH_3COOCH(CH_3)CH_2$- $CH(CH_3)_2$	284–297 (140–147)	113 (45)	
1,3-Dimethylbutylamine $CH_3CHNH_2(CH_2)CH(CH_3)_2$ (2-Amino-4-Methylpeniane)	223–228 (106–109)	55 (13)	
Dimethyl Carbinol		See Isopropyl Alcohol.	
Dimethyl Carbonate		See Methyl Carbonate.	
Dimethyl Chloracetal $ClCH_2CH(OCH_3)_2$	259–270 (126–132)	111 (44)	450 (232)
Dimethylcyanamide $(CH_3)_2NCN$	320 (160)	160 (71)	
1,2-Dimethylcyclohexane $(CH_3)_2C_6H_{10}$	260 (127)		579 (304)
1,3-Dimethylcyclohexane $(CH_3)_2C_6H_{10}$ (Hexahydroxylene)	~256 (124)	~50 (10)	583 (306)
1,4-Dimethylcyclohexane $(CH_3)_2C_6H_{10}$ (Hexahydroxylol)	248 (120)	52 (11)	579 (304)
1,4-Dimethylcyclohexane-cis $C_6H_{10}(CH_3)_2$	255 (124)	61 (16)	
1,4-Dimethylcyclohexane-trans $C_6H_{10}(CH_3)_2$	246 (119)	51 (11)	

FLAMMABILITY PROPERTIES—TABLE 1. Boiling Points, Flash Points, and Ignition Temperatures of Organic Compounds (*Continued*)

Compound	Boiling point °F (°C)	Flash point, °F (°C)	Ignition point, °F (°C)
Dimethyl Decalin $C_{10}H_{16}(CH_2)_2$	455 (235)	184 (84)	455 (235)
Dimethyldichlorosilane $(CH_3)_2SiCl_2$ (Dichlorodimethylsilane)	158 (70)	<70 (<21)	
Dimethyldioxane $CH_3CHCH_2OCH_2(CH_3)CHO$	243 (117)	75 (24)	
1,3-Dimethyl-1,3-Diphenylcyclobutane $(C_6H_5CCH_3)_2(CH_2)_2$	585–588 (307–309)	289 (143)	
Dimethylene Oxide		See Ethylene Oxide.	
Dimethyl Ether		See Methyl Ether.	
Dimethyl Ethyl Carbinol		See 2-Methyl-2-Butanol.	
2,4-Dimethyl-3-Ethylpentane $CH_3CH(CH_3)CH(CH_2H_5)$ $CH(CH_3)_2$ (3-Ethyl-2,4-Dimethylpentane)	279 (137)	734 (390)	
N,N-Dimethylformamide $HCON(CH_3)_2$	307 (153)	136 (58)	833 (445)
2,5-Dimethylfuran $OC(CH_3):CHCH:C(CH_3)$	200 (93)	45 (7)	
Dimethyl Glycol Phthalate $C_6H_4[COO(CH_2)_2OCH_3]_2$	446 (230)	369 (187)	
3,3-Dimethylheptane $CH_3(CH_2)_3C(CH_3)_2CH_2CH_3$	279 (137)	617 (325)	
2,6-Dimethyl-4-Heptanone		See Diisobutyl Ketone.	
2,3-Dimethylhexane $CH_3CH(CH_3)CH(CH_3)$- $C_2H_5CH_3$	237 (114)	45 (7)	820 (438)
2,4-Dimethylhexane $CH_3CH(CH_3)CH(CH_3)$- $C_2H_5CH_3$	229 (109)	50 (10)	
Dimethyl Hexynol $C_4H_9CCH_3(OH)C:CH$ (3,5-Dimethyl-1-Hexyn-3-ol)	302 (150)	135 (57)	

FLAMMABILITY PROPERTIES—TABLE 1. Boiling Points, Flash Points, and Ignition Temperatures of Organic Compounds (*Continued*)

Compound	Boiling point °F (°C)	Flash point, °F (°C)	Ignition point, °F (°C)
1,1-Dimethylhydrazine $(CH_3)_2NNH_2$ (Dimethylhydrazine, Unsymmetrical)	145 (63)	5 (−15)	480 (249)
Dimethylisophthalate $CH_3OOCC_6H_4COOCH_3$		280 (138)	
N,N-Dimethyliso-propanolamine $(CH_3)_2NCH_2CH(OH)CH_3$	257 (125)	95 (35)	
Dimethyl Ketone		See Acetone.	
Dimethyl Maleate $(-CHCOOCH_3)_2$	393 (201)	235 (113)	
2,6-Dimethylmorpholine $CH(CH_3)CH_2OCH_2CH(CH_3)NH$	296 (147)	112 (44)	
2,3-Dimethyloctane $CH_3(CH_2)_4CH(CH_3)-CH(CH_3)CH_3$	327 (164)	<131 (<55)	437 (225)
3,4-Dimethyloctane $C_3H_7CH(CH_3)CH(CH_3)C_3H_7$	324 (162)	<131 (<55)	
2,3-Dimethylpentaldehyde $CH_3CH_2CH(CH_3)CH-(CH_3)CHO$	293 (145)	94 (34)	
2,3-Dimethylpentane $CH_3CH(CH_3)CH-(CH_3)CH_2CH_3$	194 (90)	<20 (<−7)	635 (335)
2,4-Dimethylpentane $(CH_3)_2CHCH_2CH(CH_3)_2$	177 (81)	10 (−12)	
2,4-Dimethyl-3-Pentanol $(CH_3)_2CHCHOHCH(CH_3)_2$ (Diisopropylmethanol)	284 (140)	120 (49)	
Dimethyl Phthalate $C_6H_4(COOCH_3)_2$	540 (282)	295 (146)	915 (490)
Dimethylpiperazine-cis $C_6H_{14}N_2$	329 (165)	155 (68)	
2,2-Dimethylpropane $(CH_3)_4C$ (Neopentane)	49 (9)		842 (450)
2,2-Dimethy-1-Propanol		See tert-Butyl Carbinol.	
2,5-Dimethylpyrazine $CH_3C:CHN:C(CH_3)CH:N$	311 (155)	147 (64)	

FLAMMABILITY PROPERTIES—TABLE 1. Boiling Points, Flash Points, and Ignition Temperatures of Organic Compounds (*Continued*)

Compound	Boiling point °F (°C)	Flash point, °F (°C)	Ignition point, °F (°C)
Dimethyl Sebacate $[-(CH_2)_4COOCH_3]_2$ (Methyl Sebacate)	565 (296)	293 (145)	
Dimethyl Sulfate $(CH_3)_2SO_4$ (Methyl Sulfate)	370 (188)	182 (83)	370 (188)
Dimethyl Sulfide $(CH_3)_2S$	99 (37)	<0 (<−18)	403 (206)
Dimethyl Sulfoxide $(CH_3)_2SO$	372 (189)	203 (95) (oc)	419 (215)
Dimethyl Terephthalate $C_6H_4(COOCH_3)_2$ (Dimethyl-1,4-Benzene- Dicarboxylate) (DMT)	543 (284)	308 (153)	965 (518)
2,4-Dinitroaniline $(NO_2)_2C_6H_3NH_2$		435 (224)	
1,2-Dinitro Benzol $C_6H_4(NO_2)_2$ (o-Dinitrobenzene)	604 (318)	302 (150)	
Dinitrochlorobenzene $C_6H_3Cl(NO_2)_2$ (Chlorodinitrobenzene)	599 (315)	382 (194)	
2,4-Dinitrotoluene $(NO_2)_2C_6H_3CH_3$	572 (300)	404 (207)	
Dioctyl Adipate $[-(CH_2)_2COOCH_2- CH(C_2H_5)C_4-H_9]_2$ [Bis(2-Ethylhexyl) Adipate] [Di(2-Ethylhexyl) Adipate]	680 (360)	402 (206)	710 (377)
Dioctyl Azelate $(CH_2)_7[COOCH_2CH(C_2H_5)- C_4H_9]_2$ (Bis(2-Ethylhexyl) Azelate) (Di(2-Ethylhexyl) Azelate)	709 (376)	440 (227)	705 (374)
Dioctyl Ether $(C_8H_{17})_2O$ (Octyl Ether)	558 (292)	>212 (>100)	401 (205)
Dioctyl Phthalate $C_6H_4[CO_2CH_2- CH(C_2H_5)C_4H_9]_2$ [Di(2-Ethylhexyl) Phthalate] [Bis(2-Ethylhexyl) Phthalate]		420 (215)	735 (390)

FLAMMABILITY PROPERTIES—TABLE 1. Boiling Points, Flash Points, and Ignition Temperatures of Organic Compounds (*Continued*)

Compound	Boiling point °F (°C)	Flash point, °F (°C)	Ignition point, °F (°C)
p-Dioxane OCH$_2$CH$_2$OCH$_2$CH$_2$	214 (101)	54 (12)	356 (180)
(Diethylene Dioxide)			
Dioxolane OCH$_2$CH$_2$OCH$_2$	165 (74)	35 (2)	
Dipe ntene C$_{10}$H$_{16}$ (Cinene) (Limonene)	339 (170)	113 (45)	458 (237)
Diphenyl		See Biphenyl.	
Diphenylamine (C$_6$H$_5$)$_2$NH (Phenylaniline)	575 (302)	307 (153)	1173 (634)
1,1-Diphenylbutane (C$_6$H$_5$)$_2$CHC$_3$H$_7$	561 (294)	>212 (>100)	851 (455)
1,3-Diphenyl-2-buten-1-one		See Dypnone.	
Diphenyldichlorosllane (C$_6$H$_5$)$_2$SiCl$_2$	581 (305)	288 (142)	
Diphenyldodecyl Phosphite (C$_6$H$_5$O)$_2$POC$_{10}$H$_{21}$		425 (218)	
1,1-Diphenylethane (uns) (C$_6$H$_5$)$_2$CHCH$_3$	546 (286)	>212 (>100)	824 (440)
1,2-Diphenylethane (sym) C$_6$H$_5$CH$_2$CH$_2$C$_6$H$_5$	544 (284)	264 (129)	896 (480)
Diphenyl Ether		See Diphenyl Oxide.	
Diphenylmethane (C$_6$H$_5$)$_2$CH$_2$ (Ditane)	508 (264)	266 (130)	905 (485)
Diphenyl Oxide (C$_6$H$_5$)$_2$O (Diphenyl Ether)	496 (258)	239 (115)	1144 (618)
1,1-Diphenylpentane (C$_6$H$_5$)$_2$CHC$_4$H$_9$	586 (308)	>212 (>100)	824 (440)
1,1-Diphenylpropane CH$_3$CH$_2$CH(C$_6$H$_5$)$_2$	541 (283)	>212 (>100)	860 (460)
Diphenyl Phthalate C$_6$H$_4$(COOC$_6$H$_5$)$_2$	761 (405)	435 (224)	

FLAMMABILITY PROPERTIES—TABLE 1. Boiling Points, Flash Points, and Ignition Temperatures of Organic Compounds (*Continued*)

Compound	Boiling point °F (°C)	Flash point, °F (°C)	Ignition point, °F (°C)
Dipropylamine $(C_3H_7)_2NH$	229 (109)	63 (17)	570 (299)
Dipropylene Glycol $(CH_3CHOHCH_2)_2O$	449 (232)	250 (121)	
Dipropylene Glycol Methyl Ether $CH_3OC_3H_6OC_3H_6OH$	408 (209)	186 (86)	
Dipropyl Ether		See n-Propyl Ether.	
Dipropyl Ketone		See 4-Heptanone.	
Ditane		See Diphenylmethane.	
Ditridecyl Phthalate $C_6H_4(COOC_{13}H_{27})_2$	547 (286) @ 5 mm	470 (243)	
Divinyl Acetylene $(\vdots CCH:CH_2)_2$ (1,5-Hexadien-3-yne)	183 (84)	<−4 (<−20)	
Divinylbenzene $C_6H_4(CH:CH_2)_2$	392 (200)	169 (76)	
Divinyl Ether $(CH_2:CH)_2O$ (Ethenylaxyethene) (Vinyl Ether)	83 (28)	<−22 (<−30)	680 (360)
Dodecane $CH_3(CH_2)_{10}CH_3$ (Dihexyl)	421 (216)	165 (74)	397 (203)
1-Dodecanethiol $CH_3(CH_2)_{11}SH$ (Dodecyl Mercaptan) (Lauryl Mercaptan)	289 (143) @ 15 mm	262 (128)	
1-Dodecanol $CH_3(CH_2)_{11}OH$ (Louryl Alcohol)	491 (255)	260 (127)	527 (275)
Dodecyl Bromide		See Lauryl Bromide.	
Dodecylene (α) $C_{16}H_{21}CH:CH_2$ (1-Dodecane)	406 (208)	<212 (<100)	491 (255)

FLAMMABILITY PROPERTIES—TABLE 1. Boiling Points, Flash Points, and Ignition Temperatures of Organic Compounds (*Continued*)

Compound	Boiling point °F (°C)	Flash point, °F (°C)	Ignition point, °F (°C)
Dodecyl Mercaptan		See 1-Dodecanethiol.	
tert-Dodecyl Mercaptan $C_{12}H_{25}SH$	428–451 (220–233)	205 (96)	
4-Dodecyloxy-2-Hydroxy- Benzophenone $C_{25}H_{34}O_3$		498 (254)	715 (379)
Dodecyl Phenol $C_{12}H_{25}C_6H_4OH$	597–633 (314–334)	325 (163) (oc)	
Dypnone $C_6H_5COCH:C(CH_3)C_6H_5$ (1,3-Diphenyl-2- Buten-1-one)	475 (246) @ 50 mm	350 (177)	
Eicosane $C_{20}H_{42}$	651 (344)	>212 (>100)	450 (232)
Epichlorohydrin CH_2CHOCH_2Cl ⌞――――⌟ (2-Chloropropylene Oxide) (γ-Chloropropylene Oxide)	239 (115)	88 (31)	772 (411)
1,2-Epoxyethane		See Ethylene Oxide.	
Erythrene		See 1,3-Butadiene.	
Ethanal		See Acetaldehyde.	
Ethane CH_3CH_3	−128 (−89)		882 (472)
1,2-Ethanediol		See Ethylene Glycol.	
1,2-Ethanediol Diformate $HCOOCH_2CH_2OOCH$ (Ethylene Formate) (Ethylene Glycol Diformate) (Glycol Diformate)	345 (174)	200 (93)	
Ethanethiol		See Ethyl Mercaptan.	
Ethanoic Acid		See Acetic Acid.	
Ethanoic Anhydride		See Acetic Anhydride.	
Ethanol		See Ethyl Alcohol.	
Ethanolamine $NH_2CH_2CH_2OH$ (2-Amino Ethanol) (β-Aminoethyl Alcohol)	342 (172)	186 (86)	770 (410)
Ethanoyl Chloride		See Acetyl Chloride.	
Ethene		See Ethylene.	
Ethenyl Ethanoate		See Vinyl Acetate.	

FLAMMABILITY PROPERTIES—TABLE 1. Boiling Points, Flash Points, and Ignition
Temperatures of Organic Compounds (*Continued*)

Compound	Boiling point °F (°C)	Flash point, °F (°C)	Ignition point, °F (°C)
Ethenyloxyethene		See Divinyl Ether.	
Ether		See Ethyl Ether.	
Ethine		See Acetylene.	
Ethoxyacetylene $C_2H_5OC:CH$	124 (51)	<20 (<–7)	
Ethoxybenzene $C_6H_5OC_2H_5$ (Ethyl Phenyl Ether) (Phenetole)	342 (172)	145 (63)	
2-Ethoxy-3,4-Dihydro-2-Pyran $C_7H_{12}O_2$	289 (143)	111 (44)	
2-Ethoxy Ethanol		See Ethylene Glycol Monoethyl Ether.	
2-Ethoxyethyl Acetate $CH_3COOCH_2CH_2OC_2H_5$ (Ethyl Glycol Acetate)	313 (156)	117 (47)	716 (380)
3-Ethoxypropanal $C_2H_5OC_2H_4CHO$ (3-Ethoxypropionaldehyde)	275 (135)	100 (38)	
1-Ethoxypropane		See Ethyl Propyl Ether.	
3-Ethoxypropionaldehyde $C_2H_5OCH_2CH_2CHO$	275 (135)	100 (38)	
3-Ethoxypropionic Acid $C_2H_5OCH_2CH_2COOH$	426 (219)	225 (107)	
Ethoxytriglycol $C_2H_5O(C_2H_4O)_3H$ (Triethylene Glycol, Ethyl Ether)	492 (256)	275 (135)	
Ethyl Abietale $C_{19}H_{29}COOC_2H_5$	662 (350)	352 (178)	
N-Ethylacetamide $CH_3CONHC_2H_5$ (Acetoethylamide)	401 (205)	230 (110)	
N-Ethyl Acetanilide $CH_3CON(C_2H_5)(C_6H_5)$	400 (204)	126 (52)	
Ethyl Acetate $CH_3COOC_2H_5$ (Acetic Ester) (Acetic Ether) (Ethyl Ethanoate)	171 (77)	24 (–4)	800 (426)
Ethyl Acetoacetate $C_2H_5CO_2CH_2COCH_3$ (Acetoacetic Acid, Ethyl Ester) (Ethyl 3-Oxobutanoate)	356 (180)	135 (57)	563 (295)

FLAMMABILITY PROPERTIES—TABLE 1. Boiling Points, Flash Points, and Ignition Temperatures of Organic Compounds (*Continued*)

Compound	Boiling point °F (°C)	Flash point, °F (°C)	Ignition point, °F (°C)
Ethyl Acetyl Glycolate $CH_3COOCH_2COOC_2H_5$ (Ethyl Glycolate Acetate)	−365 (−185)	180 (82)	
Ethyl Acrylate $CH_2{:}CHCOOC_2H_5$	211 (99)	50 (10)	702 (372)
Ethyl Alcohol C_2H_5OH (Grain Alcohol, Ethanol)	173 (78)	55 (13)	685 (363)
Ethylamine $C_2H_5NH_2$ 70% aqueous solution (Aminoethane)	62 (17)	<0 (<−18)	725 (385)
Ethyl Amino Ethanol $C_2H_5NHC_2H_4OH$ [2-(Ethylamino)ethanol]	322 (161)	160 (71)	
Ethylaniline $C_2H_5NH(C_6H_5)$	401 (205)	185 (85)	
Ethylbenzene $C_2H_5C_6H_5$ (Ethylbenzol) (Phenylethane)	277 (136)	70 (21)	810 (432)
Ethyl Benzoate $C_6H_5COOC_2H_5$	414 (212)	190 (88)	914 (490)
Ethylbenzol	See Ethylbenzene.		
Ethyl Bromide C_2H_5Br (Bromoethane)	100 (38)	None	952 (511)
Ethyl Bromoacetate $BrCH_2COOC_2H_5$	318 (159)	118 (48)	
2-Ethylbutanol	See 2-Ethylbutyraldehyde.		
Ethyl Butanoate	See Ethyl Butyrate.		
2-Ethyl-1-Butanol	See 2-Ethylbutyl Alcohol.		
2-Ethyl-1-Butene $(C_2H_5)_2C{:}CH_2$	144 (62)	<−4 (<−20)	599 (315)
3-(2-Ethylbutoxy) Propionic Acid $CH_3CH_2CH(C_2H_5)CH_2—OCH_2CH_2COOH$	392 (200) @ 100 mm	280 (138)	
2-Ethylbutyl Acetate $CH_3COOCH_2CH(C_2H_5)_2$	324 (162)	130 (54)	

FLAMMABILITY PROPERTIES—TABLE 1. Boiling Points, Flash Points, and Ignition Temperatures of Organic Compounds (*Continued*)

Compound	Boiling point °F (°C)	Flash point, °F (°C)	Ignition point, °F (°C)
2-Ethylbutyl Acrylate CH$_2$:CHCOOCH$_2$CH— (C$_2$H$_5$)C$_2$H$_5$	180 (82) @ 10 mm	125 (52)	
2-Ethylbutyl Alcohol (C$_2$H$_5$)$_2$CHCH$_2$OH (2-Ethyl-1-Butanol)	301 (149)	135 (57) (oc)	
Ethylbutylamine CH$_3$CH$_2$CH$_2$CH$_2$— NHCH$_3$CH$_2$	232 (111)	64 (18)	
Ethyl Butylcarbamate		See N-Butylurethane.	
Ethyl Butyl Carbonate (C$_2$H$_5$)(C$_4$H$_9$)CO$_3$	275 (135)	122 (50)	
Ethyl Butyl Ether C$_2$H$_5$OC$_4$H$_9$ (Butyl Ethyl Ether)	198 (92)	40 (4)	
2-Ethyl Butyl Glycol (C$_2$H$_5$)$_2$CHCH$_2$OC$_2$H$_4$OH [2-(2-Ethylbutoxy)ethanol]	386 (197)	180 (82)	
Ethyl Butyl Ketone C$_2$H$_5$CO(CH$_2$)$_3$CH$_3$ (3-Heptanone)	299 (148)	115 (46)	
2-Ethyl-2-Butyl-1,3-Propanediol HOCH$_2$C(C$_2$H$_5$)(C$_4$H$_9$)— CH$_2$OH	352 (178) @ 50 mm	280 (138)	
2-Ethylbutyraldehyde (C$_2$H$_5$)$_2$CHCHO (Diethyl Acetaldehyde) (2-Ethylbutanal)	242 (117)	70 (21)	
Ethyl Butyrate CH$_3$CH$_2$CH$_2$COOC$_2$H$_5$ (Butyric Acid, Ethyl Ester) (Butyric Ester) (Ethyl Butanoate)	248 (120)	75 (24)	865 (463)
2-Ethylbutyric Acid (C$_2$H$_5$)$_2$CHCOOH (Diethyl Acetic Acid)	380 (193)	210 (99)	752 (400)
2-Ethylcaproaldehyde		See 2-Ethylhexanal.	
Ethyl Caproate C$_5$H$_{11}$COOC$_2$H$_5$ (Ethyl Hexoate) (Ethyl Hexanoate)	333 (167)	120 (49)	
Ethyl Caprylate CH$_3$(CH$_2$)$_6$COOC$_2$H$_5$ (Ethyl Octoate) (Ethyl Octanoate)	405–408 (207–209)	175 (79)	

FLAMMABILITY PROPERTIES—TABLE 1. Boiling Points, Flash Points, and Ignition
Temperatures of Organic Compounds (*Continued*)

Compound	Boiling point °F (°C)	Flash point, °F (°C)	Ignition point, °F (°C)
Ethyl Carbonate		See Diethyl Carbonate.	
Ethyl Chloride C_2H_5Cl (Chloroethane) (Hydrochloric Ether) (Muriatic Ether)	54 (12)	−58 (−50)	966 (519)
Ethyl Chloroacetate $ClCH_2COOC_2H_5$	295 (146)	147 (64)	
Ethyl Chlorocarbonate		See Ethyl Chloroformate.	
Ethyl Chloroformate $ClCOOC_2H_5$ (Ethyl Chlorocarbonate) (Ethyl Chloromethanoate)	201 (94)	61 (16)	932 (500)
Ethyl Chloromethanoate		See Ethyl Chloroformate.	
Ethyl Crotonate $CH_3CH{:}CHCOOC_2H_5$	282 (139)	36 (2)	
Ethyl Cyanoacetate $CH_2CNCOOC_2H_5$	401–408 (205–209)	230 (110)	
Ethylcyclobutane $C_2H_5C_4H_7$	160 (71)	<4 (<−16)	410 (210)
Ethylcyclohexane $C_2H_5C_6H_{11}$	269 (132)	95 (35)	460 (238)
N-Ethylcyclohexylamine $C_6H_{11}NHC_2H_5$		86 (30)	
Ethylcyclopentane $C_2H_5C_5H_9$	218 (103)	<70 (<21)	500 (260)
Ethyl Decanoate $C_9H_{19}COOC_2H_5$ (Ethyl Caprate)	469 (243)	>212 (>100)	
N-Ethyldiethanolamine $C_2H_5N(C_2H_4OH)_2$	487 (253)	280 (138)	
Ethyl Dimethyl Methane		See Isopentane.	
Ethylene $H_2C{:}CH_2$ (Ethene)	−155 (−104)		842 (450)
Ethylene Acetate		See Glycol Diacetate.	
Ethylene Carbonate OCH_2CH_2OCO	351 (177) @ 100 mm	290 (143)	

FLAMMABILITY PROPERTIES—TABLE 1. Boiling Points, Flash Points, and Ignition Temperatures of Organic Compounds (*Continued*)

Compound	Boiling point °F (°C)	Flash point, °F (°C)	Ignition point, °F (°C)
Ethylene Chlorohydrin		See 2-Chloroethanol.	
Ethylene Cyanohydrin $CH_2(OH)CH_2CN$ (Hydracrylonitrile)	445 (229) Decomposes	265 (129)	
Ethylenediamine $H_2NCH_2CH_2NH_2$ Anydrous 76%	241 (116) 239–252 (115–122)	104 (40) 150	725 (385) (66)
Ethylene Dichloride CH_2ClCH_2Cl (1,2-Dichloroethone)	183 (84)	56 (13)	775 (413)
2,2-Ethylenedioxydiethanol		See Triethylene Glycol.	
Ethylene Formate		See 1,2-Ethanediol Diformate.	
Ethylene Glycol HOC_2H_4OH (1,2-Ethanediol) (Glycol)	387 (197)	232 (111)	748 (398)
Ethylene Glycol n-Butyl Ether $HOCH_2CH_2OC_4H_9$	340 (171)	150 (66)	
Ethylene Glycol Diacetate		See Glycol Diacetate.	
Ethylene Glycol Dibutyl Ether $C_4H_9OC_2H_4OC_4H_9$	399 (204)	185 (85)	
Ethylene Glycol Diethyl Ether $C_2H_5OCH_2CH_2OC_2H_5$	251 (122)	95 (35)	406
Ethylene Glycol Diformate		See 1,2-Ethanediol Diformate.	
Ethylene Glycol Dimethyl Ether $CH_3O(CH_2)_2OCH_3$ (1,2-Dimethoxyethane)	174 (79) @ 630 mm	29 (–2)	395 (202)
Ethylene Glycol Ethylbutyl Ether $(C_2H_5)_2CHCH_2OCH_2CH_2OH$	386 (197)	180 (85)	
Ethylene Glycol Ethylhexyl Ether $C_4H_9CH(C_2H_5)CH_2OCH_2—CH_2OH$	442 (228)	230 (110)	
Ethylene Glycol Isopropyl Ether $(CH_3)_2CHOCH_2CH_2OH$	289 (143)	92 (33)	
Ethylene Glycol Monoacetate $CH_2OHCH_2OOCCH_3$ (Glycol Monoacetate)	357 (181)	215 (102)	

FLAMMABILITY PROPERTIES—TABLE 1. Boiling Points, Flash Points, and Ignition Temperatures of Organic Compounds (*Continued*)

Compound	Boiling point °F (°C)	Flash point, °F (°C)	Ignition point, °F (°C)
Ethylene Glycol Monoacrylate $CH_2:CHCOOC_2H_4CH$ (2-Hydroxyethylacrylate)	410 (210)	220 (104) (oc)	
Ethylene Glycol Monobenzyl Ether $C_6H_5CH_2OCH_2CH_2OH$	493 (256)	265 (129)	665 (352)
Ethylene Glycol Monobutyl Ether $C_4H_9O(CH_2)_2OH$ (2-Butoxyethanol)	340 (171)	143 (62)	460 (238)
Ethylene Glycol Monobutyl Ether Acetate $C_4H_9O(CH_2)_2OOCCH_3$	377 (192)	160 (71)	645 (340)
Ethylene Glycol Monoethyl Ether $HOCH_2CH_2OC_2H_5$ (2-Ethoxyethanol)	275 (135)	110 (43)	455 (235)
Ethylene Glycol Monoethyl Ether Acetate $CH_3COOCH_2CH_2OC_2H_5$ (Cellosolve Acetate)	313 (156)	124 (52)	715 (379)
Ethylene Glycol Monoisobutyl Ether $(CH_3)_2CHCH_2OCH_2CH_2OH$	316–323 (158–162)	136 (58)	540 (282)
Ethylene Glycol Monomethyl Ether $CH_3OCH_2CH_2OH$ (2-Methoxyethanol)	255 (124)	102 (39)	545 (285)
Ethylene Glycol Monomethyl Ether Acetal $CH_3CH(OCH_2CH_2OCH_3)_2$	405 (207)	200 (93)	
Ethylene Glycol Monomethyl Ether Acetate $CH_3O(CH_2)_2OOCCH_3$	293 (145)	120 (49)	740 (392)
Ethylene Glycol Monomethyl Ether Formal $CH_2(OCH_2CH_2OCH_3)_2$	394 (201)	155 (68)	
Ethylene Glycol Phenyl Ether $C_6H_5OC_2H_4OH$ (2-Phenoxyethanol)	473 (245)	260 (127)	

FLAMMABILITY PROPERTIES—TABLE 1. Boiling Points, Flash Points, and Ignition Temperatures of Organic Compounds (*Continued*)

Compound	Boiling point °F (°C)	Flash point, °F (°C)	Ignition point, °F (°C)
Ethylene Oxide CH_2OCH_2 ⌞____⌟ (Dimethylene Oxide) (1,2-Epoxyethane) (Oxirane)	51 (11)	−20	1058 with No Air
Ethylenimine $NHCH_2CH_2$ ⌞____⌟ (Aziridine)	132 (56)	12 (−11)	608 (320)
Ethyl Ethanoate		See Ethyl Acetate.	
N-Ethylethanolomine $C_2H_5NHC_2H_4OH$	322 (161)	160 (71)	
Ethyl Ether $C_2H_5OC_2H_5$ (Diethyl Ether) (Diethyl Oxide) (Ether) (Ethyl Oxide)	95 (35)	−49 (−45)	356 (180)
Ethylethylene Glycol		See 1,2-Butanediol.	
Ethyl Fluoride C_2H_5F (1-Fluoroethane)	−36 (−38)		
Ethyl Formate $HCO_2C_2H_5$ (Ethyl Methanoate) (Formic Acid, Ethyl Ester)	130 (54)	−4 (−20)	851 (455)
Ethyl Formate (ortho) $(C_2H_5O)_3CH$ (Triethyl Orthoformate)	291 (144)	86 (30)	
Ethyl Glycol Acetate		See 2-Ethoxyethyl Acetate.	
2-Ethylhexaldehyde		See 2-Ethylhexanal.	
2-Ethylhexanal $C_4H_9CH(C_2H_5)CHO$ (Butylethylacelaldehyde) (2-Ethylcaproaldehyde) (2-Ethylhexaldehyde)	325 (163)	112 (44)	375 (190)
2-Ethyl-1,3-Hexanediol $C_3H_7CH(OH)CH-$ $(C_2H_5)CH_2OH$	472 (244)	260 (127)	680 (360)
2-Ethylhexanoic Acid $C_4H_9CH(C_2H_5)COOH$ (2-Ethyl Hexoic Acid)	440 (227)	245 (118)	700 (371)

FLAMMABILITY PROPERTIES—TABLE 1. Boiling Points, Flash Points, and Ignition
Temperatures of Organic Compounds (*Continued*)

Compound	Boiling point °F (°C)	Flash point, °F (°C)	Ignition point, °F (°C)
2-Ethylhexanol $C_4H_9CH(C_2H_5)CH_2OH$ (2-Ethlhexyl Alcohol) (Octyl Alcohol)	359 (182)	164 (73)	448 (231)
2-Ethylhexenyl		See 2-Ethyl-3-Propylacrolein.	
2-Ethylhexoic Acid		See 2-Ethylhexanoic Acid.	
2-Ethylhexyl Acetate $CH_3COOCH_2CH(C_2H_5)C_4H_9$ (Octyl Acetate)	390 (199)	160 (71)	515 (268)
2-Ethylhexyl Acrylate $CH:CHCOOCH_2CH—$ $(C_2H_5)C_4H_9$	266 (130) @ 50 mm	180 (82)	485 (252)
2-Ethylhexylamine $C_4H_9CH(C_2H_5)CH_2NH_2$	337 (169)	140 (60)	
N-2-(Ethylhexyl) Anlline $C_6H_5NHCH_2CH(C_2H_5)C_4H_9$	379 (193) @ 50 mm	325 (163)	
2-Ethylhexyl Chloride $C_4H_9CH(C_2H_5)CH_2Cl$	343 (173)	140 (60)	
N-(2-Ethylhexyl)- cyclohexylamine $C_6H_{11}NH[CH_2CH—$ $(C_2H_5)C_4H_9]$	342 (172) @ 50 mm	265 (129)	
2-Ethylhexyl Ether $[C_4H_9CH(C_2H_5)CH_2]_2O$	517 (269)	235 (113)	
1,1-Ethylidene Dichloride CH_3CHCl_2 (1,1-Dichloroethane)	135–138 (57–59)	2 (−17)	
1,2-Ethylidene Dichloride $ClCH_2CH_2Cl$	183 (84)	55 (13)	824 (440)
Ethyl Isobutyrate $(CH_3)_2CHCOOC_2H_5$	230 (110)	<70 (<21)	
2-Ethylisohexanol $(CH_3)_2CHCH_2CH(C_2H_5)—$ CH_2OH (2-Ethyl Isohexyl Alcohol) (2-Ethyl-4-Methyl Pentanol)	343–358 (173–181)	158 (70)	600 (316)
Ethyl Lactate $CH_3CHOHCOOC_2H_5$ Tech.	309 (154)	115 (46) 131 (55)	752 (400)

FLAMMABILITY PROPERTIES—TABLE 1. Boiling Points, Flash Points, and Ignition Temperatures of Organic Compounds (*Continued*)

Compound	Boiling point °F (°C)	Flash point, °F (°C)	Ignition point, °F (°C)
Ethyl Malonate		See Diethyl Malonate.	
Ethyl Mercaptan C_2H_5SH (Ethanethiol) (Ethyl Sulfhydrate)	95 (35)	<0 (<−18)	572 (300)
Ethyl Methacrylate $CH_2{:}C(CH_3)COOC_2H_5$ (Ethyl Methyl Acrylate)	239–248 (115–120)	68 (20)	
Ethyl Methanoate		See Ethyl Formate.	
Ethyl Methyl Acrylate		See Ethyl Methacrylate.	
Ethyl Methyl Ether		See Methyl Ethyl Ether.	
7-Ethyl-2-Methyl-4-Hendecanol $C_4H_9CH(C_2H_5)C_2H_4$—$CHOHCH_2CH(CH_3)_2$	507 (264)	285 (141)	
Ethyl Methyl Ketone		See Methyl Ethyl Ketone	
4-Ethylmorpholine $CH_2CH_2OC_2H_4NCH_2CH_3$	280 (138)	90 (32)	
1-Ethylnaphthalene $C_{10}H_7C_2H_5$	496 (258)		896 (480)
Ethyl Nitrate $CH_3CH_2ONO_2$ (Nitric Ether)	190 (88)	50 (10)	
Ethyl Nitrite C_2H_5ONO (Nitrous Ether)	63 (17)	−31 (−35)	194 (90)
3-Ethyloctane $C_5H_{11}CH(C_2H_5)C_2H_5$	333 (167)		446 (230)
4-Ethyloctane $C_4H_9CH(C_2H_5)C_3H_7$	328 (164)		445 (229)
Ethyl Oxalate $(COOC_2H_5)_2$ (Oxalic Ether) (Diethyl Oxalate)	367 (186)	168 (76)	
Ethyl Oxide		See Ethyl Ether.	
p-Ethylphenol $HOC_6H_4C_2H_5$	426 (219)	219 (104)	
Ethyl Phenylacetate $C_6H_5CH_2COOC_2H_5$	529 (276)	210 (99)	

FLAMMABILITY PROPERTIES—TABLE 1. Boiling Points, Flash Points, and Ignition
Temperatures of Organic Compounds (*Continued*)

Compound	Boiling point °F (°C)	Flash point, °F (°C)	Ignition point, °F (°C)
Ethyl Phenyl Ether		See Ethoxybenzene.	
Ethyl Phenyl Ketone $C_2H_5COC_6H_5$ (Propiophenone)	425 (218)	210 (99)	
Ethyl Phthalyl Ethyl Glycolate $C_2H_5OCOC_6H_4OCO—$ $CH_2OCOC_2H_5$	608 (320)	365 (185)	
Ethyl Propenyl Ether $CH_3CH:CHOCH_2CH_3$	158 (70)	>19 (>−7)	
Ethyl Proplonate $C_2H_5COOC_2H_5$	210 (99)	54 (12)	824 (440)
2-Ethyl-3-Propylacrolein $C_3H_7CH:C(C_2H_5)CHO$ (2-Ethylhexenal)	347 (175)	155 (68)	
2-Ethyl-3-Propylacrylic Acid $C_3H_7CH:C(C_2H_5)COOH$	450 (232)	330 (166)	
Ethyl Propyl Ether $C_2H_5OC_3H_7$ (1-Ethoxypropane)	147 (64)	<−4 (<−20)	
m-Ethyltoluene $CH_3C_6H_4C_2H_5$ (1-Methyl-3-Ethylbenzene)	322 (161)		896 (480)
o-Ethyltoluene $CH_3C_6H_4C_2H_5$ (1-Methyl-2-Ethylbenzene)	329 (165)		824 (440)
p-Ethyltoluene $CH_3C_6H_4C_2H_5$ (1-Methyl-4-Ethylbenzene)	324 (162)		887 (475)
Ethyl p-Toluene Sulfonamide $C_7H_7SO_2NHC_2H_5$	208 (98) @ 745 mm	260 (127)	
Ethyl p-Toluene Sulfonate $C_7H_7SO_3C_2H_5$	345 (174)	316 (158)	

FLAMMABILITY PROPERTIES—TABLE 1. Boiling Points, Flash Points, and Ignition Temperatures of Organic Compounds (*Continued*)

Compound	Boiling point °F (°C)	Flash point, °F (°C)	Ignition point, °F (°C)
Ethyl Vinyl Ether		See Vinyl Ethyl Ether.	
Ethyne		See Acetylene.	
Fluorobenzene	185	5	
C_6H_5F	(85)	(−15)	
Formal		See Methylal.	
Formalin		See Formaldehyde.	
Formaldehyde	−3	Gas	795
HCHO	(−19)	185	(424)
37% Methanol-free	214	(85)	
	(101)		
37%, 15% Methanol		122	
(Formalin)		(50)	
(Methylene Oxide)			
Formamide	410	310	
$HCONH_2$	(210)	(154)	
	Decomposes		
Formic Acid	213	156	1004
HCOOH	(101)	(69)	(539)
90% Solution		122	813
		(50)	(434)
Formic Acid, Butyl Ester		See Butyl Formate.	
Formic Acid, Ethyl Ester		See Ethyl Formate.	
Formic Acid, Methyl Ester		See Methyl Formate.	
Fuel Oil No. 1	304–574	100–162	410
(Kerosene)	(151–301)	(38–72)	(210)
(Range Oil)			
Fuel Oil No. 2		126–204	494
		(52–96)	(257)
Fuel Oil No. 4		142–240	505
		(61–116)	(263)
Fuel Oil No. 5			
Light		156–336	
Heovy		(69–169)	
		160–250	
		(71–121)	
Fuel Oil No. 6		150–270	765
		(66–132)	(407)

FLAMMABILITY PROPERTIES—TABLE 1. Boiling Points, Flash Points, and Ignition Temperatures of Organic Compounds (*Continued*)

Compound	Boiling point °F (°C)	Flash point, °F (°C)	Ignition point, °F (°C)
2-Furaldehyde		See Furfural.	
Furan	88	<32	
CH:CHCH:CHO	(31)	(<0)	
(Furfuran)			
Furfural	322	140	600
OCH:CHCH:CHCHO	(161)	(60)	(316)
(2-Furaldehyde)			
(Furfuraldehyde)			
(Furol)			
Furfuraldehyde		See Furfural.	
Furfuran		See Furan.	
Furfuryl Acetate	356–367	185	
OCH:CHCH:CCH$_2$OOCCH$_3$	(180–186)	(85)	
Furfuryl Alcohol	340	167	915
OCH:CHCH:CCH$_2$OH	(171)	(75)	(491)
		(oc)	
Furfurylamine	295	99	
C$_4$H$_3$OCH$_2$NH$_2$	(146)	(37)	
Furol		See Furfural.	
Fusel Oil		See Isoamyl Alcohol.	
Gas Oil	500–700	150+	640
	(260–371)	(66+)	(338)
Gasoline	100–400	−45	
C$_5$H$_{12}$ to C$_9$H$_{20}$	(38–204)	(−43)	
56–60 Octane		−45	536
73 Octane		(−43)	(280)
92 Octane		−36	853
100 Octane		(−38)	(456)
Gasoline		−50	824
100–130 (Aviation Grade)		(−46)	(440)
Gasoline		−50	880
115–145 (Aviation Grade)		(−46)	(471)
Gasoline (Casinghead)		0	
		(−18)	
Glycerine	340	390	698
HOCH$_2$CHOHCH$_2$OH	(171)	(199)	(370)
(Glycerol)			

FLAMMABILITY PROPERTIES—TABLE 1. Boiling Points, Flash Points, and Ignition Temperatures of Organic Compounds (*Continued*)

Compound	Boiling point °F (°C)	Flash point, °F (°C)	Ignition point, °F (°C)
α,β-Glycerine Dichlorohydrin $CH_2ClCHClCH_2OH$	360 (182)	200 (93)	
Glycerol	See Glycerine.		
Glyceryl Triacetate $(C_3H_5)(OOCCH_3)_3$ (Triacelin)	496 (258)	280 (138)	812 (433)
Glyceryl Tributyrate $C_3H_5(OOCC_3H_7)_3$ (Tributyrin) (Butyrin) (Glycerol Tributyrate)	597 (314)	356 (180)	765 (407)
Glyceryl Trinitrate	See Nitroglycerine.		
Glyceryl Tripropionate $(C_2H_5COO)_3C_3H_5$ (Tripropionin)	540 (282)	332 (167)	790 (421)
Glycidyl Acrylate $CH_2{:}CHCOOCH_2CHCH_2O$	135 (57) @ 2 mm	141 (61)	779 (415)
Glycol	See Ethylene Glycol.		
Glycol Diacetate $(CH_2OOCCH_3)_2$ (Ethylene Acetate) (Ethylene Glycol Diaceate)	375 (191)	191 (88)	900 (482)
Glycol Dichloride	See Ethylene Dichloride.		
Glycol Diformate	See 1,2-Ethanediol Diformate.		
Glycol Dimercaptoacetate $(HSCH_2C{:}OOCH_2{-})_2$ (GDMA)	280 (138) 1.2 mm	396 (202)	
Glycol Monoacetate	See Ethylene Glycol Monoacetate.		
Grain Alcohol	See Ethyl Alcohol.		
Hendecane $CH_3(CH_2)_9CH_3$ (Undecane)	384 (196)	149 (65)	
Heptadecanol $C_4H_9CH(C_2H_5)C_2H_4{-}$ $CH(OH)C_2H_4CH(C_2H_5)_2$ (3,9-Diethyl-6-Tridecanol)	588 (309)	310 (154)	
Heptane $CH_3(CH_2)_5CH_3$	209 (98)	25 (−4)	399 (204)

FLAMMABILITY PROPERTIES—TABLE 1. Boiling Points, Flash Points, and Ignition Temperatures of Organic Compounds (*Continued*)

Compound	Boiling point °F (°C)	Flash point, °F (°C)	Ignition point, °F (°C)
2-Heptanol $CH_3(CH_2)_4CH(OH)CH_3$	320 (160)	160 (71)	
3-Heptanol $CH_3CH_2CH(OH)C_4H_9$	313 (156)	140 (60)	
3-Heptanone		See Ethyl Butyl Ketone.	
4-Heptanone $(C_3H_7)_2CO$ (Butyrone) (Dipropyl Ketone)	290 (143)	120 (49)	
1-Heptene		See Heptylene.	
3-Heptene (mixed cis and trans) $C_3H_7CH:CHC_2C_5$ (3-Heptylene)	203 (95)	21 (−6)	
Heptylamine $CH_3(CH_2)_6NH_2$ (1-Aminoheptane)	311 (155)	130 (54)	
Heptylene $C_5H_{11}CH:CH_2$ (1-Heptene)	201 (94)	<32 (<0)	500 (260)
Heptylene-2-trans $C_4H_9CH:CHCH_3$ (2-Heptene-trans)	208 (98)	<32 (<0)	
Hexachlorobutadiene $CCl_2:CClCCl:CCl_2$			1130 (610)
Hexachloro Diphenyl Oxide $(C_6H_2Cl_3)_2O$ [Bis(Trichlorophenyl) Ether]			1148 (620)
Hexadecane $CH_3(CH_2)_{14}CH_3$ (Cetane)	549 (287)	>212 (>100)	396 (202)
tert-Hexadecanethiol $C_{16}H_{33}SH$ (Hexadecyl-tert-Mercaptan)	298–307 (148–153) @ 11 mm	265 (129)	
Hexadecylene-1 $CH_3(CH_2)_{13}CH:CH_2$ (1-Hexadecene)	525 (274)	>212 (>100)	464 (240)
Hexadecyltrichiorosilane $C_{16}H_{33}SiCl_3$	516 (269)	295 (146)	
2,4-Hexadienal $CH_3CH:CHCH:CHC(O)H$	339 (171)	154 (68)	

FLAMMABILITY PROPERTIES—TABLE 1. Boiling Points, Flash Points, and Ignition
Temperatures of Organic Compounds (*Continued*)

Compound	Boiling point °F (°C)	Flash point, °F (°C)	Ignition point, °F (°C)
1,4-Hexadiene CH$_3$CH:CHCH$_2$CH:CH$_2$ (Allylpropenyl)	151 (66)	−6 (−21)	
Hexanal CH$_3$(CH$_2$)$_4$CHO (Caproaldehyde) (Hexaldehyde)	268 (131)	90 (32)	
Hexane CH$_3$(CH$_2$)$_4$CH$_3$ (Hexyl Hydride)	156 (69)	−7 (−22)	437 (225)
1,2-Hexanediol		See Hexylene Glycol.	
2,5-Hexanediol CH$_3$CH(OH)CH$_2$— CH$_2$CH(OH)CH$_3$ (2,5-Dihydroxyhexane)	429 (221)	230 (110)	
2,5-Hexanedione		See Acetonyl Acetone.	
1,2,6-Hexanetriol HOCH$_2$CH(OH)— (CH$_2$)$_3$CH$_2$OH	352 (178) @ 5 mm	375 (191)	
Hexanoic Acid		See Caproic Acid.	
1-Hexanol		See Hexyl Alcohol.	
2-Hexanone		See Methyl Butyl Ketone.	
3-Hexanone C$_2$H$_5$COC$_3$H$_7$ (Ethyl n-Propyl Ketone)	253 (123)	95 (35)	
1-Hexene CH$_2$:CH(CH$_2$)$_3$CH$_3$ (Butyl Ethylene)	146 (63)	<20 (<−7)	487 (253)
2-Hexene-cis C$_3$H$_7$CH:CHCH$_3$	156 (69)	<−4 (<−20)	
3-Hexenol-cis CH$_3$CH$_2$CH:CHCH$_2$CH$_2$OH (3-Hexen-l-ol) (Leaf Alcohol)	313 (156)	130 (54)	
Hexyl Acetate (CH$_3$)$_2$CH(CH$_2$)$_3$OOCCH$_3$ (Methylamyl Acetate)	285 (141)	113 (45)	

FLAMMABILITY PROPERTIES—TABLE 1. Boiling Points, Flash Points, and Ignition
Temperatures of Organic Compounds (*Continued*)

Compound	Boiling point °F (°C)	Flash point, °F (°C)	Ignition point, °F (°C)
Hexyl Alcohol $CH_3(CH_2)_4CH_2OH$ (Amyl Carbinol) (1-Hexanol)	311 (155)	145 (63)	
sec-Hexyl Alcohol $C_4H_9CH(OH)CH_3$ (2-Hexanol)	284 (140)	136 (58)	
Hexylamine $CH_3(CH_2)_5NH_2$	269 (132)	85 (29)	
Hexyl Chloride		See 1-Chlorohexane.	
Hexyl Cinnamic Aldehyde $C_6H_{13}C(CHO):CHC_6H_5$ (Hexyl Cinnamaldehyde)	486 (252)	>212 (>100)	
Hexylene Glycol $CH_2OHCHOH(CH_2)_3CH_3$ (1,2-Hexanediol)	385 (196)	215 (102)	
Hexyl Ether $C_6H_{13}OC_6H_{13}$ (Dihexyl Ether)	440 (227)	170 (77)	365 (185)
Hexyl Methacrylate $C_6H_{13}OOCC(CH_3):CH_2$	388–464 (198–240)	180 (82)	
Hydracrylonitrile		See Ethylene Cyanohydrin.	
Hydralin		See Cyclohexanol.	
Hydroquinone $C_6H_4(OH)_2$ (Quinol) (Hydroquinol)	547 (286)	329 (165)	960 (516)
Hydroquinone Di- **(β-Hydroxyethyl) Ether** $C_6H_4(—OCH_2CH_2OH)_2$	365–392 @ 0.3 mm (185–200)	435 (224)	875 (468)
Hydroquinone Monomethyl **Ether** $CH_3OC_6H_4OH$ (4-Methoxy Phenol) (Para-Hydroxyanisole)	475 (246)	270 (132)	790 (421)

FLAMMABILITY PROPERTIES—TABLE 1. Boiling Points, Flash Points, and Ignition
Temperatures of Organic Compounds (*Continued*)

Compound	Boiling point °F (°C)	Flash point, °F (°C)	Ignition point, °F (°C)
o-Hydroxybenzaldehyde		See Salicylaldehyde.	
3-Hydroxybutanal		See Aldol.	
β-Hydroxybutyraldehyde		See Aldol.	
Hydroxycitronellal (CH₃)₂C(OH)(CH₂)₃— CH(CH₃)CH₂CHO (Citronellal Hydrate) (3,7-Dimethyl-7-Hydroxyoctanal)	201–205 (94–96) @ 1 mm	>212 (>100)	
N-(2-Hydroxyethyl)-acetamide		See N-Acetyl Ethanolamine.	
2-Hydroxyethyl Acrylate (HEA)	410 (210)	214 (101)	1.8 @ 100°C
β-Hydroxyethylaniline		See 2-Anilinoethanol.	
N-(2-Hydroxyethyl) Cyclohexylamine C₆H₁₁NHC₂H₄OH		249 (121)	
(2-Hydraxyethyl)-Ethylenediamine CH₂OHCH₂NHCH₂CH₂NH₂	460–464 (238–240)	275 (135)	
4-(2-Hydroxyethyl) Morpholine C₂H₄OC₂H₄NC₂H₄OH	437 (225)	210 (99)	
1-(2-Hydroxyethyl) Piperazine HOCH₂CH₂— NCH₂CH₂NHCH₂CH₂	475 (246)	255 (124)	
n-(2-Hydroxyethyl) Propylenediamine CH₃CH(NHC₂H₄OH)CH₂NH₂	465 (241)	260 (127)	
4-Hydroxy-4-Methyl-2-Pentanone		See Diacetone Alcohol.	
2-Hydroxy-2-methyl-propionitrile		See Acetone Cyanohydrin.	
Hydroxypropyl Acrylate		See Propylene Glycol Monoacrylate.	
o-Hydroxytoluene		See o-Cresol.	
Ionone Alpha (α-Ionone) C(CH₃)₂CH₂CH₂CH:C(CH₃)— CHCH:CHC(CH₃):O (α-Cyclocitrylideneacetone) [4-(2,6,6-Trimethyl-2-Cyclohexen-1-yl)-3-Buten-2-one]	259–262 (126–128) @ 12 mm	>212 (>100)	

FLAMMABILITY PROPERTIES—TABLE 1. Boiling Points, Flash Points, and Ignition
Temperatures of Organic Compounds (*Continued*)

Compound	Boiling point °F (°C)	Flash point, °F (°C)	Ignition point, °F (°C)
Ionone Beta (β-Ionone)	284	>212	
C(CH₃)₂CH₂CH₂CH₂—	(140)	(>100)	
C(CH₃):CCHCHC(CH₃):O	@ 18 mm		
(β-Cyclocitrylidene-acetone)			
[4-(2,6,6-Trimethyl-1-			
Cyclohexen-1-yl)-3-			
Buten-2-one]			
Isoamyl Acetate	290	77	680
CH₃COOCH₂CH₂CH(CH₃)₂	(143)	(25)	(360)
(Banana Oil)			
(3-Methyl-1-Butanol Acetate)			
(2-Methyl Butyl Ethanoate)			
Isoamyl Alcohol	270	109	662
(CH₃)₂CHCH₂CH₂OH	(132)	(43)	(350)
(Isobutyl Carbinol)			
(Fusel Oil)			
(3-Methyl-1-Butanol)			
tert-Isoamyl Alcohol		See 2-Methyl-2-Butanol.	
Isoamyl Butyrate	352	138	
C₃H₇CO₂(CH₂)₂CH(CH₃)₂	(178)	(59)	
(Isopentyl Butyrate)			
Isoamyl Chloride	212	<70	
(CH₃)₂CHCH₂CH₂Cl	(100)	(<21)	
(1-Chloro-3-Methylbutane)			
Isobornyl Acetate	428–435	190	
C₁₀H₁₇OOCCH₃	(220–224)	(88)	
Isobutane	11		860
(CH₃)₃CH	(−12)		(460)
(2-Methylpropane)			
Isobutyl Acetate	244	64	790
CH₃COOCH₂CH(CH₃)₂	(118)	(18)	(421)
(β-Methyl Propyl Ethanoate)			
Isobutyl Acrylate	142–145	86	800
(CH₃)₂CHCH₂OOCCH:CH₂	(61–63)	(30)	(427)
	@ 15 mm		
Isobutyl Alcohol	225	82	780
(CH₃)₂CHCH₂OH	(107)	(28)	(415)
(Isopropyl Carbinol)			
(2-Methyl-1-Propanol)			
Isobutylamine	150	15	712
(CH₃)₂CHCH₂NH₂	(66)	(−9)	(378)
Isobutylbenzene	343	131	802
(CH₃)₂CHCH₂C₆H₅	(173)	(55)	(427)
Isobutyl Butyrate	315	122	
C₃H₇CO₂CH₂CH(CH₃)₂	(157)	(50)	

FLAMMABILITY PROPERTIES—TABLE 1.   Boiling Points, Flash Points, and Ignition
Temperatures of Organic Compounds (*Continued*)

Compound	Boiling point °F (°C)	Flash point, °F (°C)	Ignition point, °F (°C)
Isobutyl Carbinol		See Isoamyl Alcohol.	
Isobutyl Chloride $(CH_3)_2CHCH_2Cl$ (1-Chloro-3-Methyl-propane)	156 (69)	<70 (<21)	
Isobutylcyclohexane $(CH_3)_2CHCH_2C_6H_{11}$	336 (169)		525 (274)
Isobutylene		See 2-Methylpropene.	
Isobutyl Formate $HCOOCH_2CH(CH_3)_2$	208 (98)	<70 (<21)	608 (320)
Isobutyl Heptyl Ketone $(CH_3)_2CHCH_2COCH_2—$ $CH(CH_3)CH_2CH(CH_3)_2$ (2,6,8-Trimethyl-4-Non-anone)	412–426 (211–219)	195 (91)	770 (410)
Isobutyl Isobutyrate $(CH_3)_2CHCOOCH_2—$ $CH(CH_3)_2$	291–304 (144–151)	101 (38)	810 (432)
Isobutyl Phenylacetate $(CH_3)_2CHCH_2OOCCH_2C_6H_5$	477 (247)	>212 (>100)	
Isobutyl Phosphate $PO_4(CH_2CH(CH_3)_2)_3$ (Triisobutyl Phosphate)	302 (150) @ 20 mm	275 (135)	
Isobutyl Vinyl Ether		See Vinyl Isobutyl Ether.	
Isobutyraldehyde $(CH_3)_2CHCHO$ (2-Methylpropanal)	142 (61)	−1 (−18)	385 (196)
Isobutyric Acid $(CH_3)_2CHCOOH$	306 (152)	132 (56)	900 (481)
Isobutyric Anhydride $[(CH_3)_2CHCO]_2O$	360 (182)	139 (59)	625 (329)
Isobutyronitrile $(CH_3)_2CHCN$ (2-Methylpropanenitrile) (Isopropylcyanide)	214–216 (101–102)	47 (8)	900 (482)
Isodecaldehyde $C_9H_{19}CO$	387 (197)	185 (85)	
Isodecane $C_7H_{15}CH(CH_3)_2$ (2-Methylnonane)	333 (167)		410 (210)
Isodecanoic Acid $C_9H_{19}COOH$	489 (254)	300 (149)	

FLAMMABILITY PROPERTIES—TABLE 1. Boiling Points, Flash Points, and Ignition Temperatures of Organic Compounds (*Continued*)

Compound	Boiling point °F (°C)	Flash point, °F (°C)	Ignition point, °F (°C)
Isoevgenol $(CH_3CHCH)C_6H_3OHOCH_3$ (1-Hydroxy-2 Methoxy- 4-Propenylbanzene)	514 (268)	>212 (>100)	
Isoheptane $(CH_3)_2CHC_4H_9$ (2-Methylhexane) (Ethyl- isobutylmelhane)	194 (90)	<0 (−18)	
tert-Isohexyl Alcohol $C_2H_5(CH_3)C(OH)C_2H_5$ (3-Methyl-3-Pentanol)	252 (122)	115 (46)	
Isooctane $(CH_3)_2CHCH_4C(CH_3)_3$ (2,2,4-Trimethylpentane)	210 (99)	40 (4.5)	784 (418)
Isooctyl Alcohol $C_7H_{15}CH_2OH$ (Isooctanol)	83–91 (182–195)	180 (82)	
Isooctyl Nitrate $C_8H_{17}NO_3$	106–109 (41–43) @ 1 mm	205 (96)	
Isooctyl Vinyl Ether		See Vinyl Isooctyl Ether.	
Isopentaldehyde $(CH_3)_2CHCH_2CHO$	250 (121)	48 (9)	
Isopentane $(CH_3)_2CHCH_2CH_3$ (2-Methylbutane) (Ethyl Dimethyl Methane)	82 (28)	<−60 (<−51)	788 (420)
Isopentanoic Acid $(CH_3)_2CHCH_2COOH$ (Isovaleric Acid)	361 (183)		781 (416)
Isophorone $COCHC(CH_3)CH_2C(CH_3)_2CH_2$	419 (215)	184 (84)	860 (460)
Isophthaloyl Chloride $C_6H_4(COCl)_2$ (m-Phthalyl Dichloride)	529 (276)	356 (180)	

FLAMMABILITY PROPERTIES—TABLE 1. Boiling Points, Flash Points, and Ignition
Temperatures of Organic Compounds (*Continued*)

Compound	Boiling point °F (°C)	Flash point, °F (°C)	Ignition point, °F (°C)
Isoprene $CH_2:C(CH_3)CH:CH_2$ (2-Methyl-1,3-Butadiene)	93 (34)	−65 (−54)	743 (395)
Isopropanol		See Isopropyl Alcohol.	
Isopropenyl Acetate $CH_3COOC(CH_3):CH_2$ (1-Methylvinyl Acetate)	207 (97)	60 (16)	808 (431)
Isopropenyl Acetylene $CH_2:C(CH_3)C:CH$	92 (33)	<19 (<−7)	
2-Isopropoxypropane		See Isopropyl Ether.	
3-Isopropoxyproplonitrile $(CH_3)_2CHOCH_2CH_2CN$	149 (65) @ 10 mm	155 (68)	
Isopropyl Acetate $(CH_3)_2CHOOCCH_3$	194 (90)	35 (2)	860 (460)
Isopropyl Alcohol $(CH_3)_2CHOH$ (Isopropanol) (Dimethyl Carbinol) (2-Propanol) 87.9% iso	181 (83)	53 (12) 57 (14)	750 (399)
Isopropylamine $(CH_3)_2CHNH_2$	89 (32)	−35 (−37)	756 (402)
Isopropylbenzene		See Cumene.	
Isopropyl Benzoate $C_6H_5COOCH(CH_3)_2$	426 (219)	210 (99)	
Isopropyl Bicyclohexyl $C_{15}H_{28}$	530–541 (277–283)	255 (124)	446 (230)
2-Isopropylbiphenyl $C_{15}H_{16}$	518 (270)	285 (141)	815 (435)
Isopropyl Carbinol		See Isobutyl Alcohol.	
Isopropyl Chloride $(CH_3)_2CHCl$ (2-Chloropropane)	95 (35)	−26 (−32)	1100 (593)
Isopropylcyclohexane $(CH_3)_2CHC_6H_{11}$ (Hexahydrocumene) (Normanthane)	310 (154.5)		541 (283)

FLAMMABILITY PROPERTIES—TABLE 1. Boiling Points, Flash Points, and Ignition Temperatures of Organic Compounds (*Continued*)

Compound	Boiling point °F (°C)	Flash point, °F (°C)	Ignition point, °F (°C)
Isopropylcyclohexylamine $C_6H_{11}NHCHC_2H_6$		93 (34)	
Isopropyl Ether $(CH_3)_2CHOCH(CH_3)_2$ (2-Isopropoxypropane) (Diisopropyl Ether)	156 (69)	−18 (−28)	830 (443)
Isopropylethylene		See 3-Methyl-1-Butene.	
Isopropyl Formate $HCOOCH(CH_3)_2$ (Isopropyl Methanoate)	153 (67)	22 (−6)	905 (485)
4-Isopropylheptane $C_3H_7CH(C_3H_7)C_3H_7$ (m-Dihydroxybenzene)	155 (68)		491 (255)
Isopropyl-2-Hydroxypropanoate		See Isopropyl Lactate.	
Isopropyl Lactate $CH_3CHOHCCOCH(CH_3)_2$ (Isopropyl-2-Hydroxypropionate)	331–334 (166–168)	130 (54)	
Isopropyl Methanoate		See Isopropyl Formate.	
4-Isopropyl-1-Methyl Benzene		See p-Cymene.	
Isopropyl Vinyl Ether		See Vinyl Isopropyl Ether.	
Isovalerone		See Diisobutyl Ketone.	
Jet Fuel Jet A and Jet A-1	400–550 (204–288)	110–150 (43–66)	
Jet Fuel Jet B		−10 to +30 (−23 to −1)	
Jet Fuel JP-4		−10 to +30 (−23 to −1)	464 (240)
Jet Fuel JP-5		95–145 (35–63)	475 (246)
Jet Fuel JP-6	250 (121)	100 (38)	446 (230)
Kerosene		See Fuel Oil No. 1.	
Lactonitrile $CH_3CH(OH)CN$	361 (183)	171 (77)	
Lanolin (Wool Grease)		460 (238)	833 (445)

FLAMMABILITY PROPERTIES—TABLE 1.   Boiling Points, Flash Points, and Ignition
Temperatures of Organic Compounds (*Continued*)

Compound	Boiling point °F (°C)	Flash point, °F (°C)	Ignition point, °F (°C)
Lard Oil (Commercial or Animal)		395 (202)	833 (445)
No. 1		440 (227)	
Lard Oil (Pure)		500 (260)	
No. 2		419 (215)	
Mineral		404 (207)	
Lauryl Alcohol	See 1-Dodecanol.		
Lauryl Bromide CH₃(CH₂)₁₀CH₂Br (Dodecyl Bromide)	356 (180) @ 45 mm	291 (144)	
Lauryl Mercaptan	See 1-Dodecanethiol.		
Linalool (CH₃)₂C:CHCH₂CH₂C(CH₃)— OHCA:CH₂ (3,7-Dimethyl-1,6- Octadiene-3-01)	383–390 (195–199)	160 (71)	
Linseed Oil	600+ (316+)	432 (222)	650 (343)
Lubricating Oil (Paraffin Oil, includes Motor Oil)	680 (360)	300–450 (149–232)	500–700 (260– 371)
Lubricating Oil, Spindle (Spindle Oil)		169 (76)	478 (248)
Lubricating Oil, Turbine (Turbine Oil)		400 (204)	700 (371)
Lynalyl Acetate (CH₃)₂C:CHCH₂CH₂— C(—OOCCH₃)CH:CH₂ (Bergamol)	226–230 (108–110)	185 (85)	
Maleic Anhydride (COCH)₂O	396 (202)	215 (102)	890 (477)
Marsh Gas	See Methane.		

FLAMMABILITY PROPERTIES—TABLE 1. Boiling Points, Flash Points, and Ignition
Temperatures of Organic Compounds (*Continued*)

Compound	Boiling point °F (°C)	Flash point, °F (°C)	Ignition point, °F (°C)
2-Mercaptoethanol HSCH$_2$CH$_2$OH	315 (157)	165 (74)	
Mesitylene		See 1,3,5-Trimethylbenzene.	
Mesityl Oxide (CH$_3$)$_2$CCHCOCH$_3$	266 (130)	87 (31)	652 (344)
Metaldehyde (C$_2$H$_4$O)$_4$	subl. 233–240 (112–116)	97 (36)	
α-Methacrolein		See 2-Methylpropenal.	
Methacrylic Acid CH$_2$:C(CH$_3$)COOH	316 (158)	171 (77)	154 (68)
Methacrylonitrile C$_4$H$_5$N	194 (90)	34 (1.1)	
Methallyl Alcohol CH$_2$C(CH$_3$)CH$_2$OH	237 (114)	92 (33)	
Methallyl Chloride CH$_2$C(CH$_3$)CH$_2$Cl	162 (72)	11 (−12)	
Methane CH$_4$ (Marsh Gas)	−259 (−162)		999 (537)
Methanol		See Methyl Alcohol.	
Methanethiol		See Methyl Mercaptan.	
o-Methoxybenzaldehyde CH$_3$OC$_6$H$_4$CHO (o-Anisaldehyde)	275 (135)	104 (40)	
Methoxybenzene		See Anisole.	
3-Methoxybutanol CH$_3$CH(OCH$_3$)CH$_2$CH$_2$OH	322 (161)	165 (74)	
3-Methoxybutyl Acetate CH$_3$OCH(CH$_3$)CH$_2$CH$_2$- OOCCH$_3$ (Butoxyl)	275–343 (135–173)	170 (77)	
3-Methoxybutyraldehyde CH$_3$CH(OCH$_3$)CH$_2$CHO (Aldol Ether)	262 (128)	140 (60)	
2-Methoxyethanol		See Ethylene Glycol Monomethyl Ether.	
2-Methoxyethyl Acrylate C$_2$H$_3$COOC$_2$H$_4$OCH$_3$	142 (61) @ 17mm	180 (82)	

FLAMMABILITY PROPERTIES—TABLE 1.   Boiling Points, Flash Points, and Ignition
Temperatures of Organic Compounds (*Continued*)

Compound	Boiling point °F (°C)	Flash point, °F (°C)	Ignition point, °F (°C)
Methoxy Ethyl Phthalate (Methox)	376–412 (191–211)	275 (135)	
3-Methoxypropionitrile $CH_3OC_2H_4CN$	320 (160)	149 (65)	
3-Methoxypropylamine $CH_3OC_3H_6NH_2$	241 (116)	90 (32)	
Methoxy Triglycol $CH_3O(C_2H_4O)_3H$ (Triethylene Glycol, Methyl Ether)	480 (249)	245 (118)	
Methoxytriglycol Acetate $CH_3COO(C_2H_4O)_3CH_3$	266 (130)	260 (127)	
Methyl Abietate $C_{19}H_{29}COOCH_3$ (Abalyn)	680–689 (360–365) Decomposes	356 (180)	
Methyl Acetate CH_3COOCH_3 (Acetic Acid Methyl Ester) (Methyl Acetic Ester)	140 (60)	14 (−10) (454)	850 3.1 16
Methyl Acetic Ester		See Methyl Acetate.	
Methyl Acetoacetate $CH_3CO_2CH_2COCH_3$	338 (170)	170 (77)	536 (280)
p-Methyl Acetophenone $CH_3C_6H_4COCH_3$ (Methyl-p-Tolyl Ketone) (p-Acetotoluene)	439 (226)	205 (96)	
Methylacetylene		See Propyne.	
Methyl Acrylate $CH_2{:}CHCOOCH_3$	176 (80)	27 (−3)	875 (468)
Methylal $CH_3OCH_2OCH_3$ (Dimethoxymethane) (Formal)	111 (44)	−26 (−32)	459 (237)
Methyl Alcohol CH_3OH (Methanol) (Wood Alcohol)	147 (64)	52 (11)	867 (464)

FLAMMABILITY PROPERTIES—TABLE 1.   Boiling Points, Flash Points, and Ignition
Temperatures of Organic Compounds (*Continued*)

Compound	Boiling point °F (°C)	Flash point, °F (°C)	Ignition point, °F (°C)
Methylamine CH$_3$NH$_2$	21 (−6)		806 4 (430)
2-(Methylamino) Ethanol		See N-Methylethanolamine.	
Methylamyl Acetate		See Hexyl Acetate.	
Methylamyl Alcohol		See Methyl Isobutyl Carbinol.	
Methyl Amyl Ketone CH$_3$CO(CH$_2$)$_4$CH$_3$ 2-Heptanone	302 (150)	102 (39)	740 (393)
2-Methylaniline		See o-Toluidine.	
4-Methylaniline		See p-Toluidine.	
Methyl Anthranilate H$_2$NC$_6$H$_4$CO$_2$CH$_3$ (Methyl-ortho-Amino Benzoate) (Nevoli Oil, Artificial)	275 @ 15 mm (135)	>212 (>100)	
Methylbenzene		See Toluene	
Methyl Benzoate C$_6$H$_5$COOCH$_3$ (Niobe Oil)	302 (150)	181 (83)	
α-Methylbenzyl Alcohol		See Phenyl Methyl Carbinol.	
α-Methylbenzylamine C$_6$H$_5$CH(CH$_3$)NH$_2$	371 (188)	175 (79)	
α-Methylbenzyl Dimethyl Amine C$_6$H$_5$CH(CH$_3$)N(CH$_3$)$_2$	384 (196)	175 (79)	
α-Methylbenzyl Ether C$_6$H$_5$CH(CH$_3$)OCH- (CH$_3$)C$_6$H$_5$	548 (287)	275 (135)	
2-Methylbiphenyl C$_6$H$_5$C$_6$H$_4$CH$_3$	492 (255)	280 (137)	936 (502)
Methyl Borate B(OCH$_3$)$_3$ (Trimethyl Borate)	156 (69)	<80 (<27)	
Methyl Bromide CH$_3$Br (Bromomethane)	38.4 (4)	999 (537)	

FLAMMABILITY PROPERTIES—TABLE 1. Boiling Points, Flash Points, and Ignition Temperatures of Organic Compounds (*Continued*)

Compound	Boiling point °F (°C)	Flash point, °F (°C)	Ignition point, °F (°C)
2-Methyl-1,3-Butadiene		See Isoprene.	
2-Methylbutane		See Isopentane.	
3-Methyl-2-Butanethiol $C_5H_{11}SH$ (sec-Isoamyl Mercaptan)	230 (110)	37 (3)	
2-Methyl-1-Butanol $CH_3CH_2CH(CH_3)CH_2OH$	262 (128)	122 (50)	725 (385)
2-Methyl-2-Butanol $CH_3CH_2(CH_3)_2COH$ (tert-Isoamyl Alcohol) (Dimethyl Ethyl Carbinol)	215 (102)	67 (19)	819 (437)
3-Methyl-1-Butanol		See Isoamyl Alcohol.	
3-Methyl-1-Butanol Acetate		See Isoamyl Acetate.	
2-Methyl-1-Butene $CH_2:C(CH_3)CH_2CH_3$	88 (31)	<20 (<−7)	
2-Methyl-2-Butene $(CH_3)_2C:CCHCH_3$ (Trimethylethylene)	101 (38)	<20 (<−7)	
3-Methyl-1-Butene $(CH_3)_2CHCH:CH_2$ (Isopropylethylene)	68 (20)	<20 (<−7)	689 (365)
N-Methylbutylamine $CH_3CH_2CH_2CH_2NHCH_3$	196 (91)	55 (13)	
2-Methyl Butyl Ethanoate		See Isoamyl Acetate.	
Methyl Butyl Ketone $CH_3CO(CH_2)_3CH_3$ (2-Hexanone)	262 (128)	77 (25)	795 (423)
3-Methyl Butynol $(CH_3)_2C(OH)C:CH$	218 (103)	77 (25)	
2-Methylbutyraldehyde $CH_3CH_2CH(CH_3)CHO$	198–199 (92–93)	49 (9)	
Methyl Butyrate $CH_3OOCCH_2CH_2CH_3$	215 (102)	57 (14)	
Methyl Carbonate $CO(OCH_3)_2$ (Dimethyl Carbonate)	192 (89)	66 (19) (oc)	
Methyl Cellosolve Acetate $CH_3COOC_2H_4OCH_3$ (2-Methoxyethyl Acetate)	292 (144)	~111 (~44)	

FLAMMABILITY PROPERTIES—TABLE 1. Boiling Points, Flash Points, and Ignition
Temperatures of Organic Compounds (*Continued*)

Compound	Boiling point °F (°C)	Flash point, °F (°C)	Ignition point, °F (°C)
Methyl Chloride CH_3Cl (Chloromethane)	−11 (−24)	−50	1170 (632)
Methyl Chloroacetate $CH_2ClCOOCH_3$ (Methyl Chloroethanoate)	266 (130)	135 (57)	
Methyl Chloroethanoate		See Methyl Chloroacetate.	
Methyl-p-Cresol $CH_3C_6H_4OCH_3$ (p-Methylanisole)		140 (60)	
Methyl Cyanide		See Acetonitrile.	
Methylcyclohexane $CH_2(CH_2)_4CHCH_3$ (Cyclohexylmethane) (Hexahydrotoluene)	214 (101)	25 (−4)	482 (250)
2-Methylcyclohexanol $C_7H_{13}OH$	329 (165)	149 (65)	565 (296)
3-Methylcyclohexonol $CH_3C_6H_{10}OH$		158 (70)	563 (295)
4-Methylcyclohexanol $C_7H_{13}OH$	343 (173)	158 (70)	563 (295)
Methylcyclohexanone $C_7H_{12}O$	325 (163)	118 (48)	
4-Methylcyclohexene $CH:CHCH_2CH(CH_3)CH_2CH_2$	217 (103)	30 (−1)	
Methylcyclohexyl Acetate $C_9H_{16}O_2$	351–381 (177–194)	147 (64)	
Methyl Cyclopentadiene C_6H_8	163 (73)	120 (49)	833 (445)
Methylcyclopentane C_6H_{12}	161 (72)	<20 (<−7)	496 (258)
2-Methyldecane $CH_3(CH_2)_7CH(CH_3)_2$	374 (190)		437 (225)
Methyldichlorosilane CH_3HSiCl_2	106 (41)	15 (−9)	>600 (316)
N-Methyldiethanolamine $CH_3N(C_2H_4OH)_2$	464 (240)	260 (127)	
1-Methyl-3,5-Diethyl-benzene $(CH_3)C_6H_3(C_2H_5)_2$ (3,5-Diethyltoluene)	394 (201)		851 (455)

FLAMMABILITY PROPERTIES—TABLE 1.   Boiling Points, Flash Points, and Ignition
Temperatures of Organic Compounds (*Continued*)

Compound	Boiling point °F (°C)	Flash point, °F (°C)	Ignition point, °F (°C)
Methyl Dihydroabietate $C_{19}H_{31}COOCH_3$	689–698 (365–370)	361 (183)	
Methylene Chloride CH_2Cl_2 (Dichloromethane)	104 (40)	None	1033 (556)
Methylenedianiline $H_2NC_6H_4CH_2C_6H_4NH_2$ (MDA) (p,p'-Diaminodi- Phenylmethane)	748–750 (398–399) @ 78 mm	428 (220)	
Methylene Diisocyanate $CH_2(NCO)_2$		185 (85)	
Methylene Oxide		See Formaldehyde.	
N-Methylethanolamine $CH_3NHCH_2CH_2CH$ (2-(Methylamino) Ethanol)	319 (159)	165 (74)	
Methyl Ether $(CH_3)_2O$ (Dimethyl Ether) (Methyl Oxide)	−11 (−24)	Gas	662 (350)
Methyl Ethyl Carbinol		See sec-Butyl Alcohol.	
2-Methyl-2-Ethyl- **1,3-Dioxolane** $(CH_3)(C_2H_5)COCH_2CH_2O$	244 (118)	74 (23)	
Methyl Ethylene Glycol		See Propylene Glycol.	
Methyl Ethyl Ether $CH_3OC_2H_5$ (Ethyl Methyl Ether)	51 (11)	−35 (−37)	374 (190)
2-Methyl-4-Ethylhexane $(CH_3)_2CHCH_2CH(C_2H_5)_2$ (4-Ethyl-2-Methylhexane)	273 (134)	<70 (<21)	536 (280)
3-Methyl-4-Ethylhexane $C_2H_5CH(CH_3)CH(C_2H_5)_2$ (3-Ethyl-4-Methylhexane)	284 (140)	75 (24)	
Methyl Ethyl Ketone $C_2H_5COCH_3$ (2-Butanone) (Ethyl Methyl Ketone)	176 (80)	16 (−9)	759 (404)
Methyl Ethyl Ketoxime $CH_3C(C_2H_5):HOH$	306–307 (152–153)	156–170 (69–77)	
2-Methyl-3-Ethylpentane $(CH_3)_2CHCH(C_2H_5)_2$ (3-Ethyl-2-Methylpentane)	241 (116)	<70 (<21)	860 (460)

FLAMMABILITY PROPERTIES—TABLE 1. Boiling Points, Flash Points, and Ignition
Temperatures of Organic Compounds (*Continued*)

Compound	Boiling point °F (°C)	Flash point, °F (°C)	Ignition point, °F (°C)
2-Methyl-5-Ethyl-piperidine NHCH(CH$_3$)CH$_2$CH$_2$CH- ⎿———————— (C$_2$H$_5$)CH$_2$ ⎿——⌐	326 (163)	126 (52)	
2-Methyl-5-Ethylpyridine N:C(CH$_3$)CH:CHC(C$_2$H$_5$):CH ⎿————————————⌐	353 (178)	155 (68)	
Methyl Formate CH$_3$OOCH (Formic Acid, Methyl Ether)	90 (32)	−2 (−19)	840 (449)
2-Methylfuran C$_4$H$_3$OCH$_3$ (Sylvan)	144–147 (62–64)	−22 (−30)	
Methyl Glycol Acetate CH$_2$OHCHOHCH$_2$CO$_1$CH$_3$ (Propylene Glycol Acetate)		111 (44)	
Methyl Heptolocyl Ketone C$_{17}$H$_{35}$COCH$_3$	329 (165) @ 3 mm	255 (124)	
Methylheptenone (CH$_3$)$_2$C:CH(CH$_2$)$_2$COCH$_3$ (6-Methyl-5-Hepten-2-one)	343–345 (173–174)	135 (57)	
Methyl Heptine Carbonate CH$_3$(CH$_2$)$_4$C:CCOOCH$_3$ (Methyl 2-Octynoate)		190 (88)	
Methyl Heptyl Ketone C$_7$H$_{15}$COCH$_4$ (5-Methyl-2-Octanone)	361–383 (183–195)	140 (60)	680 (360)
2-Methylhexane (CH$_3$)$_2$CH(CH$_2$)$_3$CH$_3$	194 (90)	<0 (<−18)	536 (280)
3-Methylhexane CH$_3$CH$_2$CH(CH$_3$)CH$_2$CH$_2$CH$_3$	198 (92)	25 (−4)	536 (280)
Methyl Hexyl Ketone CH$_3$COC$_6$H$_{13}$ (2-Octanone) (Octanone)	344 (173.5)	125 (52)	
Methyl-3-Hydroxybutyrate CH$_3$CHOHCH$_2$COOCH$_3$	347 (175)	180 (82)	
Methyl Ionone C$_{14}$H$_{22}$O (Irone)	291 (144) @ 16 mm	>212 (>100)	

FLAMMABILITY PROPERTIES—TABLE 1. Boiling Points, Flash Points, and Ignition Temperatures of Organic Compounds (*Continued*)

Compound	Boiling point °F (°C)	Flash point, °F (°C)	Ignition point, °F (°C)	
Methyl Isoamyl Ketone $CH_3COCH_2CH_2CH(CH_3)_2$	294 (146)	96 (36)	375 (191)	
Methyl Isobutyl Carbinol $CH_3CHOHCH_2CHCH_3CH_3$ (1,3-Dimethylbutanol) (4-Methyl-2-Pentanol) (Methylamyl Alcohol)	266–271 (130–133)	106 (41)		
Methylisobutylcarbinol Acetate		See 4-Methyl-2-Pentanol Acetate.		
Methyl Isobutyl Ketone $CH_3COCH_2CH(CH_3)_2$ (Hexone) (4-Methyl-2-Pentanone)	244 (118)	64 (18)	840 (448)	
Methyl Isopropenyl Ketone $CH_2COC:CH_2(CH_3)$	208 (98)			
Methyl Isocyanate CH_3NCO (Methyl Carbonimide)	102 (39)	19 (−7)	994 (534)	
Methyl Iso Eugenol $CH_3CH:CHC_6H_3(OCH_3)_2$ (Propenyl Guaiacol)	504–507 (262–264)	>212 (>100)		
Methyl Lactate $CH_3CHOHCOOCH_3$	293 (145)	121 (49)	725 (385) 212 (100)	2.2 @
Methyl Mercaptan CH_3SH (Methanethiol)	42.4 (6)			
β-Methyl Mercapto-propionaldehyde $CH_3SC_2H_4CHO$ (3-(Methylthio) Propionalde-hyde)	~329 (~165)	142 (61)	491 (255)	
Methyl Methacrylate $CH_2:C(CH_3)COOCH_3$	212 (100)	50 (10)		
Methyl Methanoate		See Methyl Formate.		
4-Methylmorpholine $C_2H_4OC_2H_4NCH_3$	239 (115)	75 (24)		
1-Methylnaphthalene $C_{10}H_7CH_3$	472 (244)		984 (529)	
Methyl Nonyl Ketone $C_9H_{19}COCH_3$	433 (223)	192 (89)		

FLAMMABILITY PROPERTIES—TABLE 1. Boiling Points, Flash Points, and Ignition Temperatures of Organic Compounds (*Continued*)

Compound	Boiling point °F (°C)	Flash point, °F (°C)	Ignition point, °F (°C)
Methyl Oxide		See Methyl Ether.	
Methyl Pentadecyl Ketone $C_{15}H_{31}COCH_3$	313 (156) @ 3 mm	248 (120)	
2-Methyl-1,3-Pentadiene $CH_2{:}C(CH_3)CH{:}CHCH_3$	169 (76)	<−4 (<−20)	
4-Methyl-1,3-Pentadiene $CH_2{:}CHCH_2{:}C(CH_3)_2$	168 (76)	−30 (−34)	
Methylpentaldehyde $CH_3CH_2CH_2C(CH_3)HCHO$ |⎣_____⎦ (Methyl Pentanal)	243 (117)	68 (20)	
Methyl Pentanal		See Methylpentaldehyde.	
2-Methylpentane $(CH_3)_2CH(CH_2)_2CH_3$ (Isohexane)	140 (60)	<20 (<−7)	583 (306)
3-Methylpentane $CH_3CH_2CH(CH_3)CH_2CH_3$	146 (63)	<20 (<−7)	532 (278)
2-Methyl-1,3-Pentanediol $CH_3CH_2CH(OH)$- $CH(CH_3)CH_2OH$	419 (215)	230 (110)	
2-Methyl-2,4-Pentanediol $(CH_3)_2C(OH)CH_2CH$- $(OH)CH_3$	385 (196)	205 (96)	
2-Methylpentanoic Acid $C_3H_7CH(CH_3)COOH$	381 (194)	225 (107)	712 (378)
2-Methyl-1-Pentanol $CH_3(CH_2)_2CH(CH_3)CH_2OH$	298 (148)	129 (54)	590 (310)
4-Methyl-2-Pentanol		See Methyl Isobutyl Carbinol.	
4-Methyl-2-Pentanol Acetate $CH_3COOCH(CH_3)CH_2$- $CH(CH_3)_2$ (Methylisobutylcarbinol Acetate)	295 (146)	110 (43)	660 (349)
4-Methyl-2-Pentanone		See Methyl Isobutyl Ketone.	
2-Methyl-1-Pentene $CH_2{:}C(CH_3)CH_2CH_2CH_3$	143 (62)	<20 (<−7)	572 (300)
4-Methyl-1-Pentene $CH_2{:}CHCH_2CH(CH_3)_2$	129 (54)	<20 (<−7)	572 (300)
2-Methyl-2-Pentene $(CH_3)_2C{:}CHCH_2CH_3$	153 (67)	<20 (<−7)	
4-Methyl-2-Pentene $CH_3CH{:}CHCH(CH_3)_2$	133–137 (56–58)	<20 (<−7)	

FLAMMABILITY PROPERTIES—TABLE 1.   Boiling Points, Flash Points, and Ignition
Temperatures of Organic Compounds (*Continued*)

Compound	Boiling point °F (°C)	Flash point, °F (°C)	Ignition point, °F (°C)
3-Methyl-1-Pentynol $(C_2H_5)(CH_3)C(OH)C:CH$	250 (121)	101 (38)	
o-Methyl Phenol		See o-Cresol.	
Methyl Phenylacetate $C_6H_5CH_2COOCH_3$	424 (218)	195 (91)	
Methylphenyl carbinol $C_6H_5CH(CH_3)OH$ (α-Methylbenzyl Alcohol) (Styralyl Alcohol) (sec-Phenethyl Alcohol)	399 (204)	200 (93)	
Methyl Phenyl Carbinyl Acetate $C_6H_5CH(CH_3)OOCH_3$ (α-Methyl-Benzyl Acetate) (Styrolyl Acetate) (sec-Phenylethyl Acetate) (Phenyl Methylcarbinyl Acetate)		195 (91)	
Methyl Phenyl Ether		See Anisole.	
Methyl Phthalyl Ethyl Glycolate $CH_3COOC_6H_4COO-CH_2COOC_2H_5$	590 (310)	380 (193)	
1-Methyl Piperazine $CH_3NCH_2CH_2NHCH_2CH_2$	280 (138)	108 (42)	
2-Methylpropanal		See Isobutyraldehyde.	
2-Methylpropane		See Isobutane.	
2-Methyl-2-Propanethiol $(CH_3)_3CSH$ (tert-Butyl Mercaptan)	149–153 (65–67)	<−20 (<−29)	
2-Methyl Propanol-1		See Isobutyl Alcohol.	
2-Methyl-2-Propanol		See tert-Butyl Alcohol.	
2-Methylpropenal $CH_2:C(CH_3)CHO$ (Methacrolein) (α-Methyl Acrolein)	154 (68)	35 (2)	
2-Methylpropene $CH_2:C(CH_3)CH_3$ (γ-Butylene) (Isobutylene)	20 (−7)		869 (465)
Methyl Propionate $CH_3COOCH_2CH_3$	176 (80)	28 (−2)	876 (469)

FLAMMABILITY PROPERTIES—TABLE 1.   Boiling Points, Flash Points, and Ignition
Temperatures of Organic Compounds (*Continued*)

Compound	Boiling point °F (°C)	Flash point, °F (°C)	Ignition point, °F (°C)
Methyl Propyl Acetylene $CH_3C_2H_4ClCCH_3$ (2-Hexyne)	185 (85)	<14 (<–10)	
Methyl Propyl Carbinol $CH_3CHOHC_3H_7$ (2-Pentanol)	247 (119)	105 (41)	
Methylpropylcarbinylumine		See sec-Amylamine.	
Methyl n-Propyl Ether $CH_3OC_3H_7$	102 (39)	<–4 (<–20)	
Methyl Propyl Ketone $CH_3COC_3H_7$ (2-Pentanone)	216 (102)	45 (7)	846 (452)
2-Methylpyrazine $N{:}C(CH_3)CH{:}NCH{:}CH$		122 (50)	
2-Methyl Pyridine		See 2-Picoline.	
Methylpyrrole $N(CH_3)CH{:}CHCH{:}CH$	234 (112)	61 (16)	
Methylpyrrolidine $CH_3NC_4H_5$	180 (82)	7 (–14)	
1-Methyl-2-Pyrrolidone $CH_3NCOCH_2CH_2CH_2$ (N-Methyl-2-Pyrrolidone)	396 (202)	204 (96)	655 (346)
Methyl Salicylate $HOC_6H_4COOCH_3$ (Oil of Wintergreen) (Gaultheria Oil) (Betula Oil) (Sweet-Birch Oil)	432 (222)	205 (96)	850 (454)
Methyl Stearate $C_{17}H_{35}COOCH_3$	421 (216)	307 (153)	
α-Methylstyrene **1-Methylethenyl Benzene** **1-Methyl-1-phenylethene**	329–331 (165–166)	129 (54)	1066 (574)

FLAMMABILITY PROPERTIES—TABLE 1. Boiling Points, Flash Points, and Ignition Temperatures of Organic Compounds (*Continued*)

Compound	Boiling point °F (°C)	Flash point, °F (°C)	Ignition point, °F (°C)
Methyl Sulfate		See Dimethyl Sulfate.	
2-Methyltetrahydrofuran $C_4H_7OCH_3$	176 (80)	12 (−11)	
Methyl Toluene Sulfonate $CH_3C_6H_4SO_3CH_3$	315 (157) @ 8 mm	306 (152)	
Methyltrichlorosilane CH_3SiCl_3 (Methyl Silico Chloroform) (Trichloromethylsilane)	151 (66)	15 (−9)	>760 (>404)
Methyl Undecyl Ketone $C_{11}H_{23}COCH_3$ (2-Tridecanone)	248 (120)	225 (107)	
1-Methylvinyl Acetate		See Isopropenyl Acetate.	
Methyl Vinyl Ether		See Vinyl Methyl Ether.	
Methyl Vinyl Ketone $CH_3COCH:CH_2$	177 (81)	20 (−7)	915 (491)
Mineral Wax		See Wax, Ozocerite.	
Morpholine $OC_2H_4NHCH_2CH_2$	262 (128)	98 (37)	555 (290)
Mustard Oil $C_3H_5N:C:S$ (Allyl Isothiocyanate)	304 (151)	115 (46)	
Naphtha, Coal		107 (42)	531 (277)
Naphtha, Petroleum		See Petroleum Ether.	
Naphtha V.M. & P., 50° Flash (10)	240–290 (116–143)	50 (10)	450 (232)
Naphtha V.M. & P., High Flash	280–350 (138–177)	85 (29)	450 (232)
Naphtha V.M. & P., Regular	212–320 (100–160)	28 (−2)	450 (232)

FLAMMABILITY PROPERTIES—TABLE 1. Boiling Points, Flash Points, and Ignition Temperatures of Organic Compounds (*Continued*)

Compound	Boiling point °F (°C)	Flash point, °F (°C)	Ignition point, °F (°C)
Naphthalene $C_{10}H_8$	424 (218)	174 (79)	979 (526)
β-Naphthol $C_{10}H_7OH$ (β-Hydroxy Naphthalene) (2-Naphthol)	545 (285)	307 (153)	
1-Naphthylamine $C_{10}H_7NH_2$	572 (300)	315 (157)	
Nechexane		See 2,2-Dimethylbutane.	
Neopentone		See 2,2-Dimethylpropane.	
Neopentyl Glycol $HOCH_2C(CH_3)_2CH_2OH$ (2,2-Dimethyl 1,3 Propanediol)	410 (210)	265 (129)	750 (399)
Nicoline $C_{10}H_{14}N_2$	475 (246)		471 (244)
Niobe Oil		See Methyl Benzoate.	
Nitric Ether		See Ethyl Nitrate.	
p-Nitroaniline $NO_2C_6H_4NH_2$	637 (336)	390 (199)	
Nitrobenzene $C_6H_5NO_2$ (Nitrobenzol) (Oil of Mirbane)	412 (211)	190 (88)	900 (482)
1,3-Nitrobenzotrifluoride $C_6H_4NO_2CF_3$ (α,μ,α-Trifluoronitrotoluene	397 (203)	217 (103)	
Nitrobenzol		See Nitrobenzene.	
Nitrobiphenyl $C_6H_5C_6H_4NO_2$	626 (330)	290 (143)	
p-Nitrochlorobenzene $C_6H_4ClNO_2$ (1-Chloro-4-Nitrobenzene)	468 (242)	261 (127)	

FLAMMABILITY PROPERTIES—TABLE 1. Boiling Points, Flash Points, and Ignition Temperatures of Organic Compounds (*Continued*)

Compound	Boiling point °F (°C)	Flash point, °F (°C)	Ignition point, °F (°C)
Nitrocyclohexane $CH_2(CH_2)_4CHNO_2$	403 (206) Decomposes	190 (88)	
Nitroethane $C_2H_5NO_2$	237 (114)	82 (28)	778 (414)
Nitroglycerine $C_3H_5(NO_3)_3$ (Glyceryl Trinitrate)	502 (261) Explodes	Explodes	518 (270)
Nitromethane CH_3NO_2	214 (101)	95 (35)	785 (418)
1-Nitronaphthalene $C_{10}H_7NO_2$	579 (304)	327 (164)	
1-Nitropropane $CH_3CH_2CH_2NO_2$	268 (131)	96 (36)	789 (421)
2-Nitropropane $CH_3CH(NO_2)CH_3$ (sec-Nitropropane)	248 (120)	75 (24)	802 (428)
sec-Nitropropane		See 2-Nitropropane.	
m-Nitrotoluene $C_6H_4CH_3NO_2$	450 (232)	223 (106)	
o-Nitrotoluene $C_6H_4CH_3NO_2$	432 (222)	223 (106)	
p-Nitrotoluene $HO_2C_6H_4CH_3$	461 (238)	223 (106)	
2-Nitro-p-toludine $CH_3C_6H_3(NH_2)NO_2$		315 (157)	
Nitrous Ether		See Ethyl Nitrite.	
Nonadecane $CH_3(CH_2)_{17}CH_3$	628 (331)	>212 (>100)	446 (230)
Nonane C_9H_{20}	303 (151)	88 (31)	401 (205)
Nonane (iso) $C_6H_{13}CH(CH_3)_2$ (2-Methyloctane)	290 (143)		428 (220)
Nonane $C_5H_{11}CH(CH_3)C_2H_5$ (3-Methyloctane)	291 (144)		428 (220)
Nonane $C_4H_9CH(CH_3)C_3H_7$ (4-Methyloctane)	288 (142)		437 (225)

FLAMMABILITY PROPERTIES—TABLE 1.   Boiling Points, Flash Points, and Ignition Temperatures of Organic Compounds (*Continued*)

Compound	Boiling point °F (°C)	Flash point, °F (°C)	Ignition point, °F (°C)
Nonene C_9H_{18} (Nonylene)	270–290 (132–143)	78 (26)	
Nonyl Acetate $CH_2COOC_9H_{19}$	378 (192)	155 (68)	
Nonyl Alcohol		See Diisobutyl Carbinol.	
Nonylbenzene $C_9H_{19}C_6H_5$	468–486 (242–252)	210 (99)	
tert-Nonyl Mercaptan $C_9H_{19}SH$	370–385 (188–196)	154 (68)	
Nonylnaphthalene $C_9H_{19}C_{10}H_7$	626–653 (330–345)	<200 (<93)	
Nonylphenol $C_6H_4(C_9H_{19})OH$	559–567 (293–297)	285 (141)	
2,5-Norbornadiene C_7H_8 (NBD)	193 (89)	–6 (–21)	
Octadecane $C_{18}H_{38}$	603 (317)	>212 (>100)	441 (227)
Octadecylene α $CH_3(CH_2)_{15}CH:CH_2$ (1-Octadecene)	599 (315)	>212 (>100)	482 (250)
Octadecyltrichlorosilane $C_{18}H_{37}SiCl_3$ (Trichlorooctadecylsilane)	716 (380)	193 (89)	
Octadecyl Vinyl Ether		See Vinyl Octodecyl Ether.	
Octane $CH_3(CH_2)_6CH_3$	258 (126)	56 (13)	403 (206)
1-Octanethiol $C_8H_{17}SH$ (n-Octyl Mercapian)	390 (199)	156 (69)	
1-Octanol		See Octyl Alcohol.	
2-Octanol $CH_3CHOH(CH_2)_5CH_3$	363 (184)	190 (88)	
1-Octene $CH_2:C_7H_{14}$	250 (121)	70 (21)	446 (230)

FLAMMABILITY PROPERTIES—TABLE 1. Boiling Points, Flash Points, and Ignition
Temperatures of Organic Compounds (*Continued*)

Compound	Boiling point °F (°C)	Flash point, °F (°C)	Ignition point, °F (°C)
Octyl Acetate		See 2-Ethylhexyl Acetate.	
Octyl Alcohol $CH_3(CH_2)_6CH_2OH$ (1-Octanol)	381 (194)	178 (81)	
Octylamine $CH_3(CH_2)_6CH_2NH_2$ (1-Aminooctane)	338 (170)	140 (60)	
tert-Octylamine $(CH_3)_3CCH_2C(CH_3)_2NH_2$ (1,1,3,3-Tetramethyl- butylamine)	284 (140)	91 (33)	
Octyl Chloride $CH_3(CH_2)_7Cl$	359 (182)	158 (70)	
Octylene Glycol $(CH_3(CH_2)_2CHOH)_2$	475 (246)	230 (110)	635 (335)
tert-Octyl Mercaptan $C_8H_{17}SH$	318–329 (159–165)	115 (46) (oc)	
p-Octylphenyl Salicylate $C_{21}H_{26}O_3$		420 (216)	780 (416)
Oil of Mirbane		See Nitrobenzene.	
Oil of Wintergreen		See Methyl Salicylate.	
Oleic Acid $C_8H_{17}CH:CH(CH_2)_7COOH$ (Red Oil) Distilled	547 (286)	372 (189) 364 (184)	685 (363)
Oxalic Ether		See Ethyl Oxalate.	
Oxirane		See Ethylene Oxide.	
Paraffin Oil (See also Lubricating Oil)		444 (229)	
Paraformaldehyde $HO(CH_2O)_nH$		158 (70)	572 (300)
Paraldehyde $(CH_3CHO)_3$	255 (124)	96 (36)	460 (238)

1FLAMMABILITY PROPERTIES—TABLE 1. Boiling Points, Flash Points, and Ignition Temperatures of Organic Compounds (*Continued*)

Compound	Boiling point °F (°C)	Flash point, °F (°C)	Ignition point, °F (°C)
1,2,3,4,5-Pentamethyl Benzene $C_6H(CH_3)_5$ (Pentamethylbenzene)	449 (232)	200 (93)	800 est (427)
Pentamethylene Dichloride		See 1,5-Dichloropentane.	
Pentamethylene Glycol		See 1,5-Pentanediol.	
Pentamethylene Oxide $O(CH_2)_4CH_2$ ⌞_____⌟ (Tetrahydropyran)	178 (81)	−4 (−20)	
Pentanal		See Valeraldehyde.	
Pentane $CH_3(CH_2)_3CH_3$	97 (36)	<−40 (<−40)	500 (260)
1,5-Pentanediol $HO(CH_2)_5OH$ (Pentamethylene Glycol)	468 (242)	265 (129)	635 (335)
2,4-Pentanedione $CH_3COCH_2COCH_3$ (Acetyl Acetone)	284 (140)	93 (34)	644 (340)
Pentanoic Acid C_4H_9COOH (Valeric Acid)	366 (186)	205 (96)	752 (400)
1-Pentanol		See Amyl Alcohol.	
2-Pentanol		See Methyl Propyl Carbinol.	
3-Pentanol $CH_3CH_2CH(OH)CH_2CH_3$ (tert-n-Amyl Alcohol)	241 (116)	105 (41)	815 (435)
1-Pentanol Acetate		See Amyl Acetate.	
2-Pentanol Acetate		See sec-Amyl Acetate.	
2-Pentanone		See Methyl Propyl Ketone.	
3-Pentanone		See Diethyl Ketone.	
Pentaphen $C_5H_{11}C_6H_4OH$ (p-tert-Amyl Phenol)	482 (250)	232 (111)	
1-Pentene $CH_3(CH_2)_2CH:CH_2$ (Amylene)	86 (30)	0 (−18)	527 (275)
1-Pentene-cis		See β-Amylene-cis.	
2-Pentene-trans		See β-Amylene-trans.	
Pentylamine		See Amylamine.	

FLAMMABILITY PROPERTIES—TABLE 1. Boiling Points, Flash Points, and Ignition
Temperatures of Organic Compounds (*Continued*)

Compound	Boiling point °F (°C)	Flash point, °F (°C)	Ignition point, °F (°C)
Pentyloxypentane		See Amyl Ether.	
Pentyl Propionate		See Amyl Propionate.	
1-Pentyne $HC_1CC_3H_7$ (n-Propyl Acetylene)	104 (40)	<−4 (<−20)	
Perchloroethylene $Cl_2C{=}CCl_2$ (Tetrachloroethylene)	250 (121)	None	None
Perhydrophenanthrene $C_{14}H_{24}$ (Tetradecahydro Phenanthrene)	187–192 (86–89)		475 (246)
Petroleum, Crude Oil		20–90 (−7 to 32)	
Petroleum Ether (Benzine) (Petroleum Naphtha)	95–140 (35–60)	<0 (<−18)	550 (288)
Petroleum Pitch		See Asphalt.	
β-Pheliandrene $CH_2{:}CCH{:}CHCH[CH(CH_3)_2]—$ $\underline{\quad\quad\quad\quad\quad\quad}$ CH_2CH_2 $\underline{\quad}$ (p-Mentha-1(7), 2-Diene)	340 (171)	120 (49)	
Phenanthrene $(C_6H_4CH)_2$ (Phenanthrin)	644 (340)	340 (171)	
Phenethyl Alcohol $C_6H_5CH_2CH_2OH$ (Benzyl Carbinol) (Phenylethyl Alcohol)	430 (221)	205 (96)	
o-Phenetidine $H_2NC_6H_4OC_2H_5$ (2-Ethoxyaniline) (o-Amino-Phenetole)	442–446 (228–230)	239 (115)	

FLAMMABILITY PROPERTIES—TABLE 1. Boiling Points, Flash Points, and Ignition Temperatures of Organic Compounds (*Continued*)

Compound	Boiling point °F (°C)	Flash point, °F (°C)	Ignition point, °F (°C)
p-Phenetidine $C_2H_5OC_0H_4NH_2$ (1-Amino-4-Ethoxy-benzene) (p-Aminophenetole)	378–484 (192–251)	241 (116)	
Phenetole		See Ethoxybenzene.	
Phenol C_6H_5OH (Carbolic Acid)	358 (181)	175 (79)	1319 (715)
2-Phenoxyethanol		See Ethylene Glycol, Phenyl Ether.	
Phenoxy Ethyl Alcohol $C_6H_5O(CH_2)_2OH$ (2-Phenoxyethanol) (Phenyl Cellosolve)	468 (242)	250 (121)	
N-(2-Phenoxyethyl) Anlline $C_6H_5O(CH_2)_3NHC_6H_5$	396 (202)	338 (170)	
β-Phenoxyethyl Chloride		See β-Chlorophenetole.	
Phenylacetaldehyde $C_6H_5CH_2CHO$ (α-Toluic Aldehyde)	383 (195)	160 (71)	
Phenyl Acetate $CH_3COOC_6H_5$ (Acetylphenol)	384 (196)	176 (80)	
Phenylocetic Acid $C_6H_5CH_2COOH$ (α-Toluic Acid)	504 (262)	>212 (>100)	
Phenylamine		See Aniline.	
N-Phenylaniline		See Diphenylamine.	
Phenylbenzene		See Biphenyl.	
Phenyl Bromide		See Bromobenzene.	
Phenyl Carbinol		See Benzyl Alcohol.	
Phenyl Chloride		See Chlorobenzene.	
Phenyicyclohexane		See Cyclohexylbenzene.	
Phenyl Didecyl Phosphite $(C_6H_5O)P(OC_{10}H_{21})_2$		425 (218)	
N-Phenyldiethanolamine $C_6H_5N(C_2H_4OH)_2$	376 (191)	385 (196)	730 (387)

FLAMMABILITY PROPERTIES—TABLE 1. Boiling Points, Flash Points, and Ignition Temperatures of Organic Compounds (*Continued*)

Compound	Boiling point °F (°C)	Flash point, °F (°C)	Ignition point, °F (°C)
Phenyidiethylamine		See N,N-Diethylaniline.	
o-Phenylenediamine $NH_2C_6H_4NH_2$ (1,2-Diaminobenzene)	513 (267)	313 (156)	
Phenylethane		See Ethylbenzene.	
N-Phenylethanolamine $C_6H_5NHC_2H_4OH$	545 (285)	305 (152)	
Phenylethyl Acetate (β) $C_6H_5CH_2CH_2OOCCH_3$	435 (224)	230 (110)	
Phenylethyl Alcohol		See Phenethyl Alcohol.	
Phenylethylene		See Styrene.	
N-Phenyl-N-Ethyl-ethanolamine $C_6H_5N(C_2H_5)C_2H_4OH$	514 (268) @ 740 mm	270 (132) (oc)	685 (362)
Phenylhydrazine $C_6H_5NHNH_2$	Decomposes	190 (88)	
Phenylmethane		See Toluene.	
Phenylmethyl Ethanol Amine $C_6H_5N(CH_3)C_2H_4OH$ (2-(N-Methylaniline)-Ethanol)	378 (192) @ 100 mm	280 (138)	
Phenyl Methyl Ketone		See Acetophenone.	
4-Phenylmorpheline $C_6H_5NC_2H_4OCH_2CH_2$	518 (270)	220 (104) (oc)	
Phenylpentane		See Amylbenzene.	
o-Phenylphenol $C_6H_5C_6H_4OH$	547 (286)	255 (124)	986 (530)
Phenylpropane		See Propylbenzene.	
2-Phenylpropane		See Cumene.	
Phenylpropyl Alcohol $C_6H_5(CH_2)_3OH$ (Hydrocinnamic Alcohol) (3-Phenyl-l-propanol) (Phenylethyl Carbinol)	426 (219)	212 (100)	

FLAMMABILITY PROPERTIES—TABLE 1. Boiling Points, Flash Points, and Ignition Temperatures of Organic Compounds (*Continued*)

Compound	Boiling point °F (°C)	Flash point, °F (°C)	Ignition point, °F (°C)
Phenyl Propyl Aldehyde $C_6H_5CH_2CH_2CHO$ (3-Phenylpropionaldehyde) (Hydrocinnamic Aldehyde)		205 (96)	
Phenyl Toluene o $C_6H_5C_6H_4CH_3$ (2-Methylbiphenyl)	500 (260)	>212 (>100)	923 (495)
Phorone $(CH_3)_2CCHCOCHC(CH_3)_2$	388 (198)	185 (85)	
Phosphine PH_3	−126 (−88)		212 (100)
Phthalic Acid $C_6H_4(COOH)_2$	552 (289)	334 (168)	
Phthalic Anhydride $C_6H_4(CO)_2O$	543 (284)	305 (152)	1058 (570)
m-Phthalyl Dichloride		See Isophthaloyl Chloride.	
2-Picoline $CH_3C_5H_4N$ (2-Methylpyridine)	262 (128)	102 (39) (oc)	1000 (538)
4-Picoline $CH_3C_5H_4N$ (4-Methylpyridine)	292 (144)	134 (57)	
Pinane $C_{10}H_{18}$	336 (151)		523 (273)
α-Pinene $C_{10}H_{16}$	312 (156)	91 (33)	491 (255)
Pine Oil Steam Distilled	367–439 (186–226)	172 (78) 138 (59)	
Pine Pitch	490 (254)	285 (141)	
Pine Tar	208 (98)	130 (54)	671 (355)
Pine Tar Oil (Wood Tar Oil)		144 (62)	
Piperazine HNCH₂CH₂NHCH₂CH₂	294 (146)	178 (81) (oc)	
Piperidine $(CH_2)_5NH$ (Hexahydropyridine)	223 (106)	61 (16)	

FLAMMABILITY PROPERTIES—TABLE 1.   Boiling Points, Flash Points, and Ignition
Temperatures of Organic Compounds (*Continued*)

Compound	Boiling point °F (°C)	Flash point, °F (°C)	Ignition point, °F (°C)
Polyamyl Naphthalene Mixture of Polymers	667–747 (353–397)	360 (182)	
Polyethylene Glycols $OH(C_2H_5O)_nC_2H_4OH$		360–550 (182–287)	
Polyoxyethylene Lauryl Ether $C_{12}H_{25}O(OCH_2CH_2)_nOH$		>200 (>93)	
Polypropylene Glycols $OH(C_3H_6O)_nC_3H_4OH$	Decomposes	365 (185)	
Polyvinyl Alcohol Mixture of Polymers		175 (79)	
Potasslum Xanthate $KS_2C\text{-}OC_2H_5$	392 (200) Decomposes	205 (96)	
Propanal CH_3CH_2CHO (Propionaldehyde)	120 (49)	−22 (−30)	405 (207)
Propane $CH_3CH_2CH_3$	−44 (−42)		842 (450)
1,3-Propanediamine $NH_2CH_2CH_2CH_2NH_2$ (1,3-Diaminopropane) (Trimethylenediamine)	276 (136)	75 (24)	
1,2-Propanediol		See Propylene Glycol.	
1,3-Propanediol		See Trimethylene Glycol.	
1-Propanol		See Propyl Alcohol.	
2-Propanol		See Isopropyl Alcohol.	
2-Propanone		See Acetone.	
Propanoyl Chloride		See Propionyl Chloride.	
Propargyl Alcohol HC_1CCH_2OH (2-Propyn-1-ol)	239 (115)	97 (36)	
Propargyl Bromide HC_1CCH_2Br (3-Bromopropyne)	192 (89)	50 (10)	615 (324)
Propene		See Propylene.	
2-Propenylamine		See Allylamine.	
Propenyl Ethyl Ether $CH_3CH:CHOCH_2CH_3$	158 (70)	<20 (<−7)	

FLAMMABILITY PROPERTIES—TABLE 1. Boiling Points, Flash Points, and Ignition Temperatures of Organic Compounds (*Continued*)

Compound	Boiling point °F (°C)	Flash point, °F (°C)	Ignition point, °F (°C)
β-Propiolactone $C_3H_4O_2$	311 (155)	165 (74)	
Propionaldehyde		See Propanal.	
Propionic Acid CH_3CH_2COOH	297 (147)	126 (52)	870 (465)
Propionic Anhydride $(CH_3CH_2CO)_2O$	336 (169)	145 (63)	545 (285)
Propionic Nitrile CH_3CH_2CN (Propionitrile)	207 (97)	36 (2)	
Propionic Chloride CH_3CH_2COCl (Propanoyl Chloride)	176 (80)	54 (12)	
Propyl Acetate $C_3H_7OOCCH_3$ (Acetic Acid, n-Propyl Ester)	215 (102)	55 (13)	842 (450)
Propyl Alcohol $CH_3CH_2CH_2OH$ (1-Propanol)	207 (97)	74 (23)	775 (412)
Propylamine $CH_3(CH_2)_2NH_2$	120 (49)	−35 (−37)	604 (318)
Propylbenzene $C_3H_7C_6H_5$ (Phenylpropane)	319 (159)	86 (30)	842 (450)
2-Propylbiphenyl $C_6H_5C_6H_4C_3H_7$	~536 (~280)	>212 (>100)	833 (445)
n-Propyl Bromide C_3H_7Br (1-Bromopropane)	160 (71)		914 (490)
n-Propyl Butyrate $C_3H_7COOC_3H_7$	290 (143)	99 (37)	
Propyl Carbinol		See Butyl Alcohol.	
Propyl Chloride C_3H_7Cl	115 (46)	<0 (<−18)	968 (520)
Propyl Chlorothiolformate C_3H_7SCOCl	311 (155)	145 (63)	
Propylcyclohexane $H_7C_3C_6H_{11}$	313–315 (156–157)		478 (248)
Propylcyclopentane $C_3H_7C_5H_9$ (1-Cyclopentylpropane)	269 (131)		516 (269)

FLAMMABILITY PROPERTIES—TABLE 1.   Boiling Points, Flash Points, and Ignition
Temperatures of Organic Compounds (*Continued*)

Compound	Boiling point °F (°C)	Flash point, °F (°C)	Ignition point, °F (°C)
Propylene $CH_2:CHCH_3$ (Propene)	−53 (−47)	Gas	851 (455)
Propylene Aldehyde		See Crotonaldehyde.	
Propylene Carbonate $OCH_2CH_2CH_2OCO$ ⌞_____⌟	468 (242)	275 (135)	
Propylene Chlorohydrin		See 2-Chloro-1-Propanol.	
sec-Propylene Chlorohydrin		See 1-Chloro-2-Propanol.	
Propylenedlamine $CH_3CH(NH_2)CH_2NH_2$	246 (119)	92 (33) (oc)	780 (416)
Propylene Dichloride $CH_3CHClCH_2Cl$ (1,2-Dichloropropane)	205 (96)	60 (16)	1035 (557)
Propylene Glycol $CH_3CHOHCH_2OH$ (Methyl Ethylene Glycol) (1,2-Propanediol)	370 (188)	210 (99)	700 (371)
Propylene Glycol Acetate		See Methyl Glycol Acetate.	
Propylene Glycol Isopropyl Ether	283 (140)	110 (43)	
Propylene Glycol Methyl Ether $CH_3OCH_2CHOHCH_3$ (1-Methoxy-2-propanol)	248 (120)	90 (32)	
Propylene Glycol Methyl Ether Acetate (99% Pure)	295 (146)	108 (42)	
Propylene Glycol Monoacylate $CH_2:CHCOO(C_3H_6)OH$ (Hydroxypropyl Acrylate)	410 (210)	207 (97)	
Propylene Oxide OCH_2CHCH_3 ⌞____⌟	94 (35)	−35 (−37)	840 (449)
n-Propyl Ether $(C_3H_7)_2O$ (Dipropyl Ether)	194 (90)	70 (21)	370 (188)
Propyl Formate $HCOOC_3H_7$	178 (81)	27 (−3)	851 (455)
Propyl Methanol		See Butyl Alcohol.	
Propyl Nitrate $CH_3CH_2CH_2NO_3$	231 (111)	68 (20)	347 (175)

FLAMMABILITY PROPERTIES—TABLE 1. Boiling Points, Flash Points, and Ignition Temperatures of Organic Compounds (*Continued*)

Compound	Boiling point °F (°C)	Flash point, °F (°C)	Ignition point, °F (°C)
Propyl Proplonate $CH_3CH_2COOCH_2CH_2CH_3$	245 (118)	175 (79)	
Propyltrichlorosilane $(C_3H_7)SiCl_3$	254 (123.5)	98 (37)	
Propyne CH_3C_1CH (Allylene) (Methylacetylene)	−10 (−23)		
Pseudocumene		See 1,2,4-Trimethylbenzene.	
Pyridine $CH < (CHCH)_2 > N$	239 (115)	68 (20)	900 (482)
Pyrrole $(CHCH)_2NH$ (Azole)	268 (131)	102 (39)	
Pyrrolidine $NHCH_2CH_2CH_2CH_2$ (Tetrahydropyrrole)	186–189 (86–87)	37 (3)	
2-Pyrrolidene $NHCOCH_2CH_2CH_2$	473 (245)	265 (129)	
Quinoline $C_6H_4N:CHCH:CH$	460 (238)		896 (480)
Range Oil	See Fuel Oil No. 1.		
Rape Seed Oil (Colza Oil)		325 (163)	836 (447)
Resorcinol $C_6H_4(OH)_2$ (Dihydroxybenzol)	531 (277)	261 (127)	1126 (608)
Rhodinol $CH_2:C(CH_3)(CH_2)_3CH—$ $(CH_3)(CH_2)_2OH$	237–239 (114–115) @ 12 mm	>212 (>100)	
Rosin Oil	>680 (>360)	266 (130)	648 (342)

FLAMMABILITY PROPERTIES—TABLE 1. Boiling Points, Flash Points, and Ignition Temperatures of Organic Compounds (*Continued*)

Compound	Boiling point °F (°C)	Flash point, °F (°C)	Ignition point, °F (°C)
Salicylaldehyde HOC_6H_4CHO (o-Hydroxybenzaldehyde)	384 (196)	172 (78)	
Salicylic Acid HOC_6H_4COOH	Sublimes @ 169 (76)	315 (157)	1004 (540)
Safrole $C_3H_5C_6H_3O_2CH_2$ (4-allyl-1,2-Mathylene-dioxy-benzene)	451 (233)	212 (100)	
Santatol $C_{15}H_{24}O$ (Arheol)	~572 (~300)	>212 (>100)	
Sesame Oil		491 (255)	
Soy Bean Oil		540 (282)	833 (445)
Sperm Oil No. 1 **No.2**		428 (220) 460 (238)	586 (308)
Stearic Acid $CH_3(CH_2)_{16}COOH$	726 (386)	385 (196)	743 (395)
Stearyl Alcohol $CH_3(CH_2)_{17}OH$ (1-Ocladecanol)	410 (210) @ 15 mm		842 (450)
Styrene $C_6H_5CH:CH_2$ (Cinnamene) (Phenylethylene) (Vinyl Benzene)	295 (146)	88 (31)	914 (490)
Styrene Oxide $C_6H_5CHOCH_2$		165 (74)	929 (498)
Succinonitrile $NCCH_2CH_2CN$ (Ethylene Dicyanide)	509–513 (265–267)	270 (132)	

FLAMMABILITY PROPERTIES—TABLE 1. Boiling Points, Flash Points, and Ignition Temperatures of Organic Compounds (*Continued*)

Compound	Boiling point °F (°C)	Flash point, °F (°C)	Ignition point, °F (°C)
Sulfolane $CH_2(CH_2)_3SO_2$ (Tetrahydrothiophene-1,1-Dioxide) (Tetramethylune Sulfone)	545 (285)	350 (177)	
Tartaric Acid (d, l) $(CHOHCO_2H)_2$		410 (210) (oc)	797 (425)
Terephthalle Acid $C_6H_4(COOH)_2$ (para-Phthalic Acid) (Benzene-para-Dicarboxylic Acid)	Sublimes above 572 (300)	500 (260)	925 (496)
Terephthaloyl Chloride $C_6H_4(COCl)_2$ (Terephthalyl Dichloride) (p-Phthalyl Dichloride) (1,4-Benzenedicarbonyl Chloride)	498 (259)	356 (180)	
o-Terphenyl $(C_6H_5)_2C_6H_4$	630 (332)	325 (163)	
m-Terphenyl $(C_6H_5)_2C_6H_4$	685 (363)	375 (191)	
Terpineol $C_{10}H_{17}OH$ (Terpilenol)	417–435 (214–224)	195 (91)	
Terpinyl Acetate $C_{10}H_{17}OOCCH_3$	428 (220)	200 (93)	
Tetraamylbenzene $(C_5H_{11})_4C_6H_2$	608–662 (320–350)	295 (146)	
1,1,2,2-Tetrabromoethane $CHBr_2CHBr_2$ (Acetylene Tetrabromide)	275 (135)		635 (335)
1,2,4,5-Tetrachlorobenzene $C_6H_{12}Cl_4$	472 (245)	311 (155)	

FLAMMABILITY PROPERTIES—TABLE 1.   Boiling Points, Flash Points, and Ignition
Temperatures of Organic Compounds (*Continued*)

Compound	Boiling point °F (°C)	Flash point, °F (°C)	Ignition point, °F (°C)
Tetradecane $CH_3(CH_2)_{12}CH_3$	487 (253)	212 (100)	392 (200)
Tetradecanol $C_{14}H_{29}OH$	507 (264)	285 (141) (oc)	
1-Tetradecene $CH_2{:}CH(CH_2)_{11}CH_3$	493 (256)	230 (110)	455 (235)
tert-Tetradecyl Mercaptan $C_{14}H_{29}SH$	496–532 (258–278)	250 (121)	
Tetraethoxypropane $(C_2H_5O)_4C_3H_4$	621 (327)	190 (88)	
Tetra (2-Ethylbutyl) Silicate $[C_2H_5CH(C_2H_5)CH_2O]_4Si$	460 (238) @ 50 mm	335 (168)	
Tetraethylene Glycol $HOCH_2(CH_2OCH_2)_3CH_2OH$	Decomposes	360 (182)	
Tetraethylene Glycol, Dimethyl Ether		See Dimethoxy Tetraglycol.	
Tetraethylene Pentamine $H_2N(C_2H_4NH)_3C_2H_4NH_2$	631 (333)	325 (163)	610 (321)
Tetra (2-Ethylhexyl) Silicate $[C_4H_9CH(C_2H_5)CH_2O]_4Si$		390 (199)	
Tetrafluoroethylene $F_2C{:}CF_2$ (TFE) (Perfluoroethylene)	−105 (−76)		392 (200)
1,2,3,6-Tetrahydro- benzaldehyde $CH_2CH{:}CHCH_2CH_2CHCHO$	328 (164)	135 (57)	
(3-Cyclohexene-1-Car- boxaldehyde)			

FLAMMABILITY PROPERTIES—TABLE 1. Boiling Points, Flash Points, and Ignition
Temperatures of Organic Compounds (*Continued*)

Compound	Boiling point °F (°C)	Flash point, °F (°C)	Ignition point, °F (°C)
endo-Tetrahydrodicyclo-pentadiene $C_{10}H_{16}$ (Tricyclodecane)	379 (193)		523 (273)
Tetrahydrofuran $OCH_2CH_2CH_2CH_2$ (Diethylene Oxide) (Tetramethylene Oxide)	151 (66)	6 (−14)	610 (321)
Tetrahydrofurfuryl Alcohol $C_4H_7OCH_2OH$	352 (178) @ 743 mm	167 (75)	540 (282)
Tetrahydrofurfuryl Oleale $C_4H_7OCH_2OOCC_{17}H_{33}$	392–545 (200–285) @ 16 mm	390 (199)	
Tetrahydronaphthalene $C_6H_2(CH_3)_2C_2H_4$ (Tetralin)	405 (207)	160 (71)	725 (385)
Tetrahydropyran		See Pentamethylene Oxide.	
Tetraphydropyran-2-Methanol $OCH_2CH_2CH_2CH_2CHCH_2OH$	368 (187)	200 (93)	
Tetrahydropyrrole		See Pyrrolidine.	
Tetralin		See Tetrahydronophthalene.	
1,1,3,3-Tetramethoxy-propane $[(CH_3O)_2CH]_2CH_2$	361 (183)	170 (77)	
1,2,3,4-Tetramethyl-benzene 95% $C_6H_2(CH_3)_4$ (Prohnitene)	399–401 (204–205)	166 (74)	800 est. (427)
1,2,3,5-Tetramethyl-benzene 85.5% $C_6H_2(CH_3)_4$ (Isodurene)	387–389 (197–198)	160 (71)	800 est. (427)
1,2,4,5-Tetramethyl-benzene 95% $C_6H_2(CH_3)_4$ (Durene)	385 (196)	130 (54)	
Tetramethylene		See Cyclobutane	
Tetramethyleneglycol $CH_2OH(CH_2)_2CH_2OH$	230 (110)	734 (390)	

FLAMMABILITY PROPERTIES—TABLE 1. Boiling Points, Flash Points, and Ignition Temperatures of Organic Compounds (*Continued*)

Compound	Boiling point °F (°C)	Flash point, °F (°C)	Ignition point, °F (°C)
Tetramethylene Oxide		See Tetrahydrofuran.	
Tetramethyl Lead, Compounds $Pb(CH_3)_4$		100 (38)	
2,2,3,3-Tetramethyl Pentane $(CH_3)_3CC(CH_3)_2CH_2CH_3$	273 (134)	<70 (<21)	806 (430)
2,2,3,4-Tetramethyl-pentane $(CH_3)_3CCH(CH_3)CH(CH_3)_2$	270 (132) 172	<70 (<21) <70	
Thialdine $SCH(CH_3)SCH(CH_3)NHCHCH_3$	Decomposes	200 (93)	
2,2-Thiodiethanol $(HOCH_2CH_2)_2S$ (Thiodiethylene Glycol)	540 (282)	320 (160)	
Thiodiethylene Glycol		See 2,2-Thiodiethanol.	
Thiodiglycol $(CH_2CH_2OH)_2S$ (Thiodiethylene Glycol) (Beta-bis-Hydroxyethyl Sulfide) (Dihydroxyethyl Sulfide)	541 (283)	320 (160)	568 (298)
Thiophene $SCH:CHCH:CH$	184 (84)	30 (−1)	
1,4-Thioxane $O(CH_2CH_2)_2S$ (1,4-Oxathiane)	300 (149)	108 (42)	
Toluene $C_6H_5CH_3$ (Methylbenzene) (Phenylmethane) (Toluol)	231 (111)	40 (4)	896 (480)
Toluene-2,4-Diisocyanate $CH_3C_6H_3(NCO)_2$	484 (251)	260 (127)	
p-Toluenesulfonic Acid $C_6H_4(SO_3H)(CH_3)$	295 (140) @ 20 mm	363 (184)	

FLAMMABILITY PROPERTIES—TABLE 1. Boiling Points, Flash Points, and Ignition
Temperatures of Organic Compounds (*Continued*)

Compound	Boiling point °F (°C)	Flash point, °F (°C)	Ignition point, °F (°C)
Toluhydroquinone $C_6H_3(OH)_2CH_3$ (Methylhydroquinone)	545 (285)	342 (172)	875 (468)
o-Toluidine $CH_3C_6H_4NH_2$ (2-Methylaniline)	392 (200)	185 (85)	900 (482)
p-Toluidine $CH_3C_6H_4NH_2$ (4-Mothylaniline)	392 (200)	188 (87)	900 (482)
Toluol		See Toluene.	
m-Tolydiethanolamine $(HOC_2H_4)_2NC_6H_4CH_3$ (MTDEA)	400 (204)	740 (393)	0.6
2,4-Tolylene Diisocyanate		See Toluene-2,4-Diisocyanate.	
o-Tolyl Phosphate		See Tri-o-Cresyl Phosphate.	
o-Tolyl p-Toluene Sulfonate $C_{14}H_{14}O_3S$		363 (184)	
Transformer Oil (Tronsil Oil)		295 (146)	
Triacetin		See Glyceryl Triacetate.	
Triamylamine $(C_5H_{11})_3N$	453 (234)	215 (102)	
Triamylbenzene $(C_5H_{11})_3C_6H_3$	575 (302)	270 (132)	
Tributylamine $(C_4H_9)_3N$	417 (214)	187 (86)	
Tri-n-Butyl Borate $B(OC_4H_9)_3$	446 (230)	200 (93)	
Tributyl Citrate $C_3H_4(OH)(COOC_4H_9)_3$	450 (232)	315 (157)	695 (368)
Tributyl Phosphate $(C_4H_9)_3PO_4$	560 (293)	295 (146)	
Tributylphosphine $(C_4H_9)_3P$	473 (245)		392 (200)
Tributyl Phosphite $(C_4H_9)_3PO_3$	244–250 (118–121) @ 7 mm	248 (120)	
1,2,4-Trichlorobenzene $C_6H_3Cl_3$	415 (213)	222 (105)	1060 (571)

FLAMMABILITY PROPERTIES—TABLE 1. Boiling Points, Flash Points, and Ignition Temperatures of Organic Compounds (*Continued*)

Compound	Boiling point °F (°C)	Flash point, °F (°C)	Ignition point, °F (°C)
1,1,1-Trichloroethane CH_3CCl_3 (Methyl Chloroform)	165 (74)		
Trichloroethylene $ClHC:CCl_2$	188 (87)		788 (420)
1,2,3-Trichloropropane $CH_2ClCHClCH_2Cl$ (Allyl Trichloride) (Glyceryl Trichlorohydrin)	313 (156)	160 (71)	
Trichlorosilane $HSiCl_3$	89 (32)	7 (−14)	
Tri-o-Cresyl Phosphate $(CH_3C_6H_4)_3PO_4$ (o-Tolyl Phosphate)	770 (410) Decomposes	437 (225)	725 (385)
Tridecanol $CH_3(CH_2)_{12}OH$	525 (274)	250 (121)	
2-Tridecanone		See Methyl Undecyl Ketone.	
Tridecyl Acrylate $CH_2:CHCOOC_{13}H_{27}$	302 (150) @ 10 mm	270 (132)	
Tridecyl Alcohol $C_{12}H_{25}CH_2OH$ (Tridecanol)	485–503 (252–262)	180 (82)	
Tridecyl Phosphite $(C_{10}H_{21}O)_3P$	356 (180) @ 0.1 mm	455 (235)	
Triethanolamine $(CH_2OHCH_2)_3N$ (2,2′,2″-Nitrilotriethanol)	650 (343)	354 (179)	
1,1,3-Triethoxyhexane $CH(OC_2H_5)_2CH_2CH$- $(OC_2H_5)C_8H_7$	271 (133) @ 50 mm Decomposes @ 760 mm	210 (99)	
Triethylamine $(C_2H_5)_3N$	193 (89)	16 (−7)	480 (249)
1,2,4-Triethylbenzene $(C_2H_5)_3C_6H_3$	423 (217)	181 (83)	

FLAMMABILITY PROPERTIES—TABLE 1. Boiling Points, Flash Points, and Ignition Temperatures of Organic Compounds (*Continued*)

Compound	Boiling point °F (°C)	Flash point, °F (°C)	Ignition point, °F (°C)
Triethyl Cltrate $HOC(CH_2CO_2C_2H_5)$- $CO_2H_2H_5$	561 (294)	303 (151)	
Triethylene Glycol $HOCH_2(CH_2OCH_2)_2CH_2OH$ (Dicaproate) (2,2-Ethylenedioxy- diethanol)	546 (286)	350 (177)	700 (371)
Triethylene Glycol Diacetate $CH_3COO(CH_2CH_2O)_3$- $COCH_3$ (TDAC)	572 (300)	345 (174)	
Triethylene Glycol, Dimethyl Ether $CH_3(OCH_2)_3OCH_3$	421 (216)	232 (111)	
Triethylene Glycol, Ethyl Ether		See Ethoxytriglycol.	
Triethylene Glycol, Methyl Ether		See Methoxy Triglycol.	
Triethyleneglycol Monobutyl Ether $C_4H_9O(C_2H_4O)_3H$	270 (132)	290 (143)	
Triethylenetetramine $N_2NCH_2(CH_2NHCH_2)_2$- CH_2NH_2	532 (278)	275 (135)	640 (338)
Triethyl Phosphate $(C_2H_5)_3PO_4$ (Ethyl Phosphate)	408–424 (209–218)	240 (115)	850 (454)
Trifluorochloroethylene $CF_2:CFCl$ (R-1113) (Chlorotrifluoroethylene)	−18 (−28)		
Triglycol Dichloride $ClCH_2(CH_3OCH_2)_2CH_2Cl$	466 (241)	250 (121)	
Trihexyl Phosphite $(C_6H_{13})_3PO_3$	275–286 (135–141) @ 2 mm	320 (160)	
Triisopropanolamine $[(CH_3)_2COH]_3N$ (1,1′,1″-Nitrolotri-2-propanol)	584 (307)	320 (160)	608 (320)

FLAMMABILITY PROPERTIES—TABLE 1. Boiling Points, Flash Points, and Ignition Temperatures of Organic Compounds (*Continued*)

Compound	Boiling point °F (°C)	Flash point, °F (°C)	Ignition point, °F (°C)
Triisopropylbenzene $C_6H_3(CH_3CHCH_3)_3$	495 (237)	207 (97)	
Triisopropyl Borate $(C_3H_7O)_3B$	288 (142)	82 (28)	
Triiauryl Trithiophosphite $[CH_3(CH_2)_{11}S]_3P$		398 (203)	
Trimethylamine $(CH_3)_3N$	38 (3)		374 (190)
1,2,3-Trimethylbenzene $C_6H_3(CH_3)_3$ (Hemellitol)	349 (176)	111 (44)	878 (470)
1,2,4-Trimethylbenzene $C_6H_3(CH_3)_3$ (Pseudocumene)	329 (165)	112 (44)	932 (500)
1,3,5-Trimethylbenzene $C_6H_3(CH_3)_3$ (Mesitylene)	328 (164)	122 (50)	1039 (559)
Trimethyl Borate		See Methyl Borate.	
2,2,3-Trimethylbutane $(CH_3)_3C(CH_3)CHCH_3$ (Triptane—an isomer of Heptane)	178 (81)	<32 (<0)	774 (412)
2,3,3-Trimethyl-1-Butene $(CH_3)_3CC(CH_3):CH_2$ (Heplylene)	172 (78)	<32 (<0)	707 (375)
Trimethyl Carbinol		See tert-Butyl Alcohol.	
Trimethylchlorosiiane $(CH_3)_3SiCl$	135 (57)	−18 (−28)	
1,3,5-Trimethylcyclohexane $(CH_3)_3C_6H_9$ (Hexahydromesitylene)	283 (139)		597 (314)
Trimethylcyclohexanol $CH(OH)CH_2C(CH_3)_2$- $CH_2CH(CH_3)CH_2$	388 (198)	165 (74)	

FLAMMABILITY PROPERTIES—TABLE 1. Boiling Points, Flash Points, and Ignition Temperatures of Organic Compounds (*Continued*)

Compound	Boiling point °F (°C)	Flash point, °F (°C)	Ignition point, °F (°C)
3,3,5-Trimethyl-1-Cyclohexanol $CH_2CH(CH_3)CH_2C(CH_3)_2$- CH_2CHOH	388 (198)	190 (88)	
Trimethylene		See Cyclopropane.	
Trimethylenediamine		See 1,3-Propanediamine.	
Trimethylene Glycol $HO(CH_2)_3OH$ (1,3-Propanediol)	417 (214)		752 (400)
Trimethylethylene		See 2-Methyl-2-Butene.	
2,5,5-Trimethylheptane $C_2H_5C(CH_3)_2(CH_2)_2$- $CH(CH_3)_2$	304 (151)	<131 (<55)	527 (275)
2,2,5-Trimethylhexane $(CH_3)_3C(CH_2)_2CH(CH_3)_2$	255 (124)	55 (13) (oc)	
3,5,5-Trimethylhexanol $CH_3C(CH_3)_2CH_2CH$- $(CH_3)CH_2CH_2OH$	381 (194)	200 (93)	
2,4,8-Trimethyl-6-Nonanol $C_4H_9CH(OH)C_7H_{15}$ (2,6,8-Trimethyl-4-nonanol)	491 (255)	199 (93)	
2,6,8-Trimethyl-4-Nonanol $(CH_3)_2CHCH_2CH(OH)CH_2$- $CH(CH_3)CH_2CH(CH_3)_2$	438 (226)	200 (93)	
2,6,8-Trimethyl-4-Nonanone $(CH_3)_2CHCH_2CH(CH_3)CH_2$- $COCH_2CH(CH_3)_2$	425 (218)	195 (91)	
2,2,4-Trimethylpentane $(CH_3)_3CCH_2CH(CH_3)_2$	211 (99)	10 (−12)	779 (415)
2,3,3-Trimethylpentane $CH_3CH_2C(CH_3)_2CH(CH_3)_2$	239 (115)	<70 (<21)	797 (425)
2,2,4-Trimethyl-1,3-Pentanediol $(CH_3)_2CHCH(OH)C(CH_3)_2$- CH_2OH	419–455 (215–235)	235 (113)	655 (346)
2,2,4-Trimethyl pentanediol Diisobutyrate $C_{16}H_{30}O_4$	536 (280)	250 (121)	795 (424)

FLAMMABILITY PROPERTIES—TABLE 1. Boiling Points, Flash Points, and Ignition Temperatures of Organic Compounds (*Continued*)

Compound	Boiling point °F (°C)	Flash point, °F (°C)	Ignition point, °F (°C)
2,2,4-Trimethyl-1,3-Pentanediol Isobutyrate $(CH_3)_2CHCH(OH)C(CH_3)_2$- $CH_2OOCCH(CH_3)_2$	356–360 125 mm (180–182)	248 (120)	740 (393)
2,2,4-Trimethylpentanediol Isobutyrate Benzoate $C_{19}H_{28}O_4$	167 (75) @ 10 mm	325 (163)	
2,3,4-Trimethyl-1-Pentene $H_2C{:}C(CH_3)CH(CH_3)$- $CH(CH_3)_2$	214 (101)	<70 (<21)	495 (257)
2,4,4-Trimethyl-1-Pentene $CH_2{:}C(CH_3)CH_2C(CH_3)_3$ (Diisobutylene)	214 (101)	23 (–5)	736 (391)
2,4,4-Trimethyl-2-Pentene $CH_3CH{:}C(CH_3)C(CH_3)_3$	221 (105)	35 (2) (oc)	581 (305)
3,4,4-Trimethyl-2-Pentene $(CH_3)_3CC(CH_3){:}CHCH_3$	234 (112)	<70 (<21)	617 (325)
Trimethyl Phosphite $(CH_3O)_3P$	232–234 (111–112)	130 (54)	
Trioctyl Phosphite $(C_8H_{17}O)_3P$ [Tris (2-Ethylhexyl) Phosphite]	212 (100) @ 0.01 mm	340 (171)	
Trioxane $OCH_2OCH_2OCH_2$ (ring)	239 (115) Sublimes	113 (45)	777 (414)
Triphenylmethane $(C_6H_5)_3CH$	678 (359)	>212 (>100)	
Triphenyl Phosphate $(C_6H_5)_3PO_4$	750 (399)	428 (220)	
Triphenylphosphine		See Triphenylphosphorus.	
Triphenyl Phosphite $(C_6H_5O)_3PO_3$	311–320 (155–160) @ 0.1 mm	425 (218)	
Triphenylphosphorus $(C_6H_5)_3P$ (Triphenylphosphine)	711 (377)	356 (180)	—
Tripropylamine $(CH_3CH_2CH_2)_3N$	313 (156)	105 (41)	

FLAMMABILITY PROPERTIES—TABLE 1. Boiling Points, Flash Points, and Ignition Temperatures of Organic Compounds (*Continued*)

Compound	Boiling point °F (°C)	Flash point, °F (°C)	Ignition point, °F (°C)
Tripropylene C_9H_{18} (Propylene Trimer)	271–288 (133–142)	75 (24)	
Tripropylene Glycol $H(OC_3H_6)_3OH$	514 (268)	285 (141)	
Tripropylene Glycol Methyl Ether $HO(C_3H_6O)_2C_3H_6OCH_3$	470 (243)	250 (121)	
Tris (2-Ethylhexyl) Phosphite		See Trioctyl Phosphite.	
Tung Oil (China Wood Oil)		552 (289)	855 (457)
Turkey Red Oil		476 (247)	833 (445)
Turpentine	300 (149)	95 (35)	488 (253)
Undecane		See Hendecane.	
2-Undecanol $C_4H_9CH(C_2H_5)C_2H_4$- $CH(OH)CH_3$	437 (225)	235 (113)	
Valeraldehyde $CH_3(CH_2)_3CHO$ (Pentanal)	217 (103)	54 (12)	432 (222)
Valeric Acid		See Pentanoic Acid.	
Vinyl Acetate $CH_2:CHOOCCH_3$ (Ethenyl Ethanoate)	161 (72)	18 (−8)	756 (402)
Vinylaceto-β-Lactone		See Diketene.	
Vinyl Acetylene $CH_2:CHC:CH$ (1-Buten-3-yne)	41 (5)		
Vinyl Allyl Ether $CH_2:CHOCH_2CH_2O$- $(CH_2)_3CH_3$ (Allyl Vinyl Ether)	153 (67)	<68 (<20)	
Vinylbenzene		See Styrene.	
Vinylbenzylchloride $ClCH_2H_6H_4CH:CH_2$	444 (229)	220 (104)	

FLAMMABILITY PROPERTIES—TABLE 1. Boiling Points, Flash Points, and Ignition Temperatures of Organic Compounds (*Continued*)

Compound	Boiling point °F (°C)	Flash point, °F (°C)	Ignition point, °F (°C)
Vinyl Bromide	60 (15.8)	None	986 (530)
Vinyl Butyl Ether $CH_2:CHOCH_4H_9$ (Butyl Vinyl Ether)	202 (94)	15 (−9)	437 (255)
Vinyl Butyrate $CH_2:CHOCOC_3H_7$	242 (117)	68 (20)	
Vinyl 2-Chloroethyl Ether $CH_2:CHOCH_2CH_2Cl$ (2-Chloroethyl Vinyl Ether)	228 (109)	80 (27)	
Vinyl Chloride CH_2CHCl (Chloroethylene)	7 (−14)	−108.4 (−78)	882 (472)
Vinyl Crotonate $CH_2:CHOCOCH:CHCH_3$	273 (134)	78 (26)	
Vinyl Cyanide		See Acrylonitrile.	
4-Vinyl Cyclohexene C_8H_{12}	266 (130)	61 (16)	517 (269)
Vinyl Ether		See Divinyl Ether.	
Vinyl Ethyl Alcohol $CH_2:CH(CH_2)_2OH$ (3-Buten-1-ol)	233 (112)	100 (38)	
Vinyl Ethyl Ether $CH_2:CHOC_2H_5$ (Ethyl Vinyl Ether)	96 (36)	<−50 (<−46)	395 (202)
Vinyl 2-Ethylhexoate $CH_2:CHOCOCH(C_2H_5)C_4H_9$	365 (185)	165 (74)	
Vinyl 2-Ethylhexyl Ether $C_{10}H_{20}O$ (2-Ethylhexyl Vinyl Ether)	352 (178)	135 (57)	395 (202)
2-Vinyl-5-Ethylpyridine $N:C(CH:CH_2)CH:CH-$ $\overline{}$ $C(C_2H_5):CH$ $\underline{}$	248 (120) @50mm	200 (93)	
Vinyl Fluoride $CH_2:CHF$	−97.5 (−72)		
Vinylidene Chloride $CH_2:CCl_2$ (1,1-Dichloroethylene)	89 (32)	−19 (−28)	1058 (570)

FLAMMABILITY PROPERTIES—TABLE 1.   Boiling Points, Flash Points, and Ignition
Temperatures of Organic Compounds (*Continued*)

Compound	Boiling point °F (°C)	Flash point, °F (°C)	Ignition point, °F (°C)
Vinylidene Fluoride $CH_2:CF_2$	−122.3 (−86)		
Vinyl Isobutyl Ether $CH_2:CHOCH_2CH(CH_3)CH_3$ (Isobutyl Vinyl Ether)	182 (83)	15 (−9)	
Vinyl Isooctyl Ether $CH_2:CHO(CH_2)_5CH(CH_3)_2$ (Isooctyl Vinyl Ether)	347 (175)	140 (60)	
Vinyl Isopropyl Ether $CH_2:CHOCH(CH_3)_2$ (Isopropyl Vinyl Ether)	133 (56)	−26 (−32)	522 (272)
Vinyl 2-Methoxyethyl Ether $CH_2:CHOC_2H_4OCH_3$ (1-Methoxy-2-Vinyloxyethane)	228 (109)	64 (18)	
Vinyl Methyl Ether $CH_2:CHOCH_3$ (Methyl Vinyl Ether)	43 (6)		549 (287)
Vinyl Octadecyl Ether $CH_2:CHO(CH_2)_{17}CH_3$ (Octadecyl Vinyl Ether)	297–369 (147–187) @ 5 mm	350 (177)	
Vinyl Propionate $CH_2:CHOCOC_2H_5$	203 (95)	34 (1)	
1-Vinylpyrrolidone $CH_2:CHNCOCH_2CH_2CH_2$ $\llcorner_____\lrcorner$ (Vinyl-2-Pyrrolidone)	205 (96) @ 14 mm	209 (98)	
Vinyl-2-Pyrrolidone		See 1-Vinylpyrrolidone.	
Vinyl Trichlorosilane $CH_2:CHSiCl_3$	195 (91)	70 (21)	
Wax, Microcrystalline		>400 (>204)	
Wax, Ozocerite (Mineral Wax)		236 (113)	
Wax, Paraffin	>700 (>371)	390 (199)	473 (245)

FLAMMABILITY PROPERTIES—TABLE 1. Boiling Points, Flash Points, and Ignition Temperatures of Organic Compounds (*Continued*)

Compound	Boiling point °F (°C)	Flash point, °F (°C)	Ignition point, °F (°C)
White Tar		See Naphthalene.	
Wood Alcohol		See Methyl Alcohol.	
Wood Tar Oil		See Pine Tar Oil.	
Wool Grease		See Lanolin.	
m-Xylene $C_6H_4(CH_3)_2$ (1,3-Dimethylbenzene)	282 (139)	81 (27)	982 (527)
o-Xylene $C_6H_4(CH_3)_2$ (1,2-Dimethylbenzene) (o-Xylol)	292 (144)	90 (32)	867 (463)
p-Xylene $C_6H_4(CH_3)_2$ (1,4-Dimethylbenzene)	281 (138)	81 (27)	984 (528)
o-Xylidine $C_6H_3(CH_3)_2NH_2$ (o-Dimethylaniline)	435 (224)	206 (97)	
o-Xylol		See o-Xylene.	

FLAMMABILITY PROPERTIES—TABLE 2.
Flammability Limits of Inorganic Compounds in Air

	Limits of Flammability	
Compound	Lower volume %	Upper volume %
Ammonia	15.50	27.00
Carbon monoxide	12.50	74.20
Carbonyl sulfide	11.90	28.50
Cyanogen	6.60	42.60
Hydrocyanic acid	5.60	40.00
Hydrogen	4.00	74.20
Hydrogen sulfide	4.30	45.50

FLAMMABILITY PROPERTIES—TABLE 3.
Flammability Limits of Organic Compounds in Air

| | Limits of Flammability | |
| | Lower, volume % | Upper, volume % |
Compound		
Acetaldehyde	3.97	57.00
Acetic acid	5.40	20.00
Acetone	2.55	12.80
Acetylene	2.50	80.00
Allyl alcohol	2.50	18.00
Allyl bromide	4.36	7.25
Allyl chloride	3.28	11.15
n-Amyl acetate	1.10	7.50
n-Amyl alcohol	1.19	10.00
iso-Amyl alcohol	1.20	9.00
n-Amyl chloride	1.60	8.63
n-Amylene	1.42	8.70
Benzene	1.40	7.10
n-Butane	1.86	8.41
iso-Butane	1.80	8.44
Butene-1	1.65	9.95
Butene-2	1.75	9.70
n-Butyl acetate	1.39	7.55
n-Butyl alcohol	1.45	11.25
iso-Butyl alcohol	1.68	9.80
n-Butyl chloride	1.85	10.10
iso-Butyl chloride	2.05	8.75
Carbon disulfide	1.25	50.00
Crotonic aldehyde	2.12	15.50
Cyclohexane	1.26	7.75
Cyclopropane	2.40	10.40
n-Decane	0.77	5.35
Diethylamine	1.77	10.10
Diethyl ether	1.85	36.50
Diethyl peroxide	2.34	
Dimethylamine	2.80	14.40
2,3-Dimethylpentane	1.12	6.75
2,2-Dimethylpropane	1.38	7.50
1,4-Dioxane	1.97	22.25
Divinyl ether	1.70	27.00
Ethane	3.00	12.50
Ethyl acetate	2.18	11.40
Ethyl alcohol	3.28	18.95
Ethylamine	3.55	13.95
Ethyl bromide	6.75	11.25
Ethyl chloride	4.00	14.80
Ethylene	2.75	28.60
Ethylene dichloride	6.20	15.90
Ethylene oxide	3.00	80.00
Ethyl formate	2.75	16.40

FLAMMABILITY PROPERTIES—TABLE 3.

Flammability Limits of Organic Compounds in Air
(*Continued*)

	Limits of Flammability	
Compound	Lower, volume %	Upper, volum %
Ethyl nitrite	3.01	50.00
Ethyl nitrate	3.80	
Ethyl nitrite	3.01	50.00
Furfural	2.10	19.30
n-Heptane	1.10	6.70
n-Hexane	1.18	7.40
Methane	5.00	15.00
Methyl acetate	3.15	15.60
Methyl alcohol	6.72	36.50
Methylamine	4.95	20.75
Methyl bromide	13.50	14.50
Methyl iso-butyl ketone	1.35	7.60
Methyl chloride	8.25	18.70
Methylcyclohexane	1.15	6.70
Methyl ethyl ether	2.00	10.00
Methyl ethyl ketone	1.81	9.50
Methyl formate	5.05	22.70
Methyl iso-propyl ketone	1.55	8.15
n-Nonane	0.83	2.90
n-Octane	0.95	6.50
Paraldehyde	1.30	
n-Pentane	1.40	7.80
iso-Pentane (2-Methylbutane)	1.32	7.60
Propane	2.12	9.35
n-Propyl acetate	1.77	8.00
iso-Propyl acetate	1.78	7.80
n-Propyl alcohol	2.15	13.50
iso-Propyl alcohol	2.02	11.80
Propylamine	2.01	10.35
n-Propyl chloride	2.60	11.10
Propylene	2.00	11.20
Propylene dichloride	3.40	14.50
Propylene oxide	2.00	22.00
Pyridine	1.81	12.40
Toluene	1.27	6.75
Triethylamine	1.25	7.90
Trimethylamine	2.00	11.60
Turpentine	0.80	
Vinyl chloride	4.00	21.70
o-Xylene	1.00	6.00
m-Xylene	1.10	7.00
p-Xylene	1.10	7.00

GLASS AND SILICA

GLASS AND SILICA—TABLE 1. Properties of Glass and Silica

	Pyroceram	96% silica	Borosilicate	Glass lining
Specific gravity, 77°F	2.60	2.18	2.23	2.56
Water absorption, %	0.00	0.00	0.00	
Gas permeability	Gastight	Gastight	Gastight	
Softening temperature, °F (°C)	2282 (1250)	2732 (1500)	1508 (1820)	
Specific heat, 77°F Btu/(lb·°F)[J/(kg·K)]	0.185 (775)	0.178 (746)	0.186 (779)	
Mean specific heat (77–752°F)	0.230	0.224	0.233	
Thermal conductivity, mean temperature, 77°F, Btu/(ft²·h·°F)/in [W/(m·K)]	25.2 (3.6)		7.5 (1.1)	
Linear thermal expansion, per °F (77–572°F); (per °C), ×10⁻⁶	3.2 (5.8)	0.44 (0.79)	1.8 (3.2)	
Modulus of elasticity, kip/in² (MPa) × 10³	17.3 (119)	9.6 (66)	9.5 (66)	6–9 (40–60)
Poisson's ratio	0.245	0.17	0.20	
Modulus of rupture, kip/in²	20 (140)	5–9 (35–63)	6–10 (42–70)	
Knoop hardness, 100 g	698	532	481	480
Knoop hardness, 500 g	619	477	442	
Adhesion strength kip/in² (MPa)				5–10 (35–70)
Maximum operating temperature, °F (°C)				500 (260)
Thermal shock resistance, temperature difference, °F (°C)				305 (152)

GRAPHITE AND SILICON CARBIDE

GRAPHITE AND SILICON CARBIDE—TABLE 1. Properties of Graphite and Silicon Carbide

	Graphite	Impervious graphite	Impervious silicon carbide
Specific gravity	1.4–1.8	1.75	3.10
Tensile strength, lbf/in² (MPa)	400–1,400 (3–10)	2,600 (18)	20,650 (143)
Compressive strength, lbf/in² (MPa)	2,000–6,000 (14–42)	10,500 (72)	150,000 (1,000)
Flexural strength, lbf/in² (MPa)	750–3,000 (5–21)	4,700 (32)	
Modulus of elasticity (×10⁵), lbf/in² (MPa)	0.5–1.8 (0.3–12 × 10⁴)	2.3 (1.6 × 10⁴)	56 (39 × 10⁴)
Thermal expansion, in/(in·°F × 10⁻⁶) [mm/(mm·°C)]	0.7–2.1 (1.3–3.8)	2.5 (4.5)	1.80 (3.4)
Thermal conductivity, Btu/[h·ft²)(°F/ft)][W/(m·K)]	15–97 (85–350)	85 (480)	60 (340)
Maximum working temperature (inert atmosphere), °F(°C)	5,000 (2,800)	350 (180)	4,200 (2,300)
Maximum working temperature (oxidizing atmosphere), °F(°C)	660 (350)	350 (180)	3,000 (1,650)

Source: Carborundum Co. Courtesy of National Association of Corrosion Engineers.

See also **MASS TRANSFER**.

HEAT TRANSFER—TABLE 1. Normal Total Emissivity of Various Surfaces

A. Metals and Their Oxides

Surface	t, °F*	Emissivity*
Aluminum		
Highly polished plate, 98.3% pure	440–1070	0.039–0.057
Polished plate	73	0.040
Rough plate	78	0.055
Oxidized at 1110°F	390–1110	0.11–0.19
Aluminum-surfaced roofing	100	0.216
Calorized surfaces, heated at 1110°F.		
Copper	390–1110	0.18–0.19
Steel	390–1110	0.52–0.57
Brass		
Highly polished:		
73.2% Cu, 26.7% Zn	476–674	0.028–0.031
62.4% Cu, 36.8% Zn, 0.4% Pb,	494–710	0.033–0.037
0.3% Al		
82.9% Cu, 17.0% Zn	530	0.030
Hard rolled, polished:		
But direction of polishing visible	70	0.038
But somewhat attacked	73	0.043
But traces of stearin from polish left on	75	0.053

Surface	t, °F*	Emissivity*
Sheet steel, strong rough oxide layer	75	0.80
Dense shiny oxide layer	75	0.82
Cast plate		
Smooth	73	0.80
Rough	73	0.82
Cast iron, rough, strongly oxidized	100–480	0.95
Wrought iron, dull oxidized	70–680	0.94
Steel plate, rough	100–700	0.94–0.97
High temperature alloy steels (see Nickel Alloys)		
Molten metal		
Cast iron	2370–2550	0.29
Mild steel	2910–3270	0.28
Lead		
Pure (99.96%), unoxidized	260–440	0.057–0.075
Gray oxidized	75	0.281
Oxidized at 390°F	390	0.63
Mercury	32–212	0.09–0.12

Material	Temperature, °F	Emissivity
Polished	100–600	0.096
Rolled plate, natural surface	72	0.06
Rubbed with coarse emery	72	0.20
Dull plate	120–660	0.22
Oxidized by heating at 1110°F	390–1110	0.61–0.59
Chromium; see Nickel Alloys for Ni—Cr steels	100–1000	0.08–0.26
Copper		
Carefully polished electrolytic copper	176	0.018
Commercial, emeried, polished, but pits remaininig	66	0.030
Commercial, scraped shiny but not mirror-like	72	0.072
Polished	242	0.023
Plate, heated long time, covered with thick oxide layer	77	0.78
Plate heated at 1110°F	390–1110	0.57
Cuprous oxide	1470–2010	0.66–0.54
Molten copper	1970–2330	0.16–0.13
Gold		
Pure, highly polished	440–1160	0.018–0.035
Iron and steel		
Metallic surfaces (or very thin oxide layer):		
Electrolytic iron, highly polished	350–440	0.052–0.064
Polished iron	800–1880	0.144–0.377
Iron freshly emeried	68	0.242
Cast iron, polished	392	0.21
Wrought iron, highly polished	100–480	0.28
Cast iron, newly turned	72	0.435
Polished steel casting	1420–1900	0.52–0.56
Molybdenum filament	1340–4700	0.096–0.292
Monel metal, oxidized at 1110°F	390–1110	0.41–0.46
Nickel		
Electroplated on polished iron, then polished	74	0.045
Technically pure (98.9% Ni, + Mn), polished	440–710	0.07–0.087
Electroplated on pickled iron, not polished	68	0.11
Wire	368–1844	0.096–0.186
Plate, oxidized by heating at 1110°F	390–1110	0.37–0.48
Nickel oxide	1200–2290	0.59–0.86
Nickel alloys		
Chromnickel	125–1894	0.64–0.76
Nickelin (18-32 Ni; 55–68 Cu; 20 Zn), gray oxidized	70	0.262
KA-2S alloy steel (8% Ni; 18% Cr), light silvery, rough, brown, after heating	420–914	0.44–0.36
After 42hr, heating at 980°F:	420–980	0.62–0.73
NCT-3 alloy (20% Ni; 25% Cr.), brown, splotched, oxidized from service	420–980	0.90–0.97
NCT-6 alloy (60% Ni; 12% Cr), smooth, black, firm adhesive oxide coat from service	520–1045	0.89–0.82
Platinum		
Pure, polished plate	440–1160	0.054–0.104
Strip	1700–2960	0.12–0.17
Filament	80–2240	0.036–0.192
Wire	440–2510	0.073–0.182
Silver		

HEAT TRANSFER—TABLE 1. Normal Total Emissivity of Various Surfaces (*Continued*)

A. Metals and Their Oxides (*Continued*)

Surface	t, °F*	Emissivity*	Surface	t, °F*	Emissivity*
Ground sheet steel	1720–2010	0.55–0.61	Polished, pure	440–1160	0.0198–0.0324
Smooth sheet iron	1650–1900	0.55–0.60	Polished	100–700	0.0221–0.0312
Cast iron, turned on lathe	1620–1810	0.60–0.70	Steel, see Iron.		
Oxidized surfaces:			Tantalum filament	2420–5430	0.194–0.31
Iron plate, pickled, then rusted red	68	0.612	Tin—bright tinned iron sheet	76	0.043 and 0.064
Completely rusted	67	0.685	Tungsten		
Rolled sheet steel	70	0.657	Filament, aged	80–6000	0.032–0.35
Oxidized iron	212	0.736	Filament	6000	0.39
Cast iron, oxidized at 1100°F	390–1110	0.64–0.78	Zinc		
Steel, oxidized at 1100°F	390–1110	0.79	Commercial, 99.1% pure, polished	440–620	0.045–0.053
Smooth oxidized electrolytic iron	260–980	0.78–0.82	Oxidized by heating at 750°F.	750	0.11
Iron oxide	930–2190	0.85–0.89	Galvanized sheet iron, fairly bright	82	0.228
Rough ingot iron	1700–2040	0.87–0.95	Galvanized sheet iron, gray oxidized	75	0.276

B. Refractories, Building Materials, Paints, and Miscellaneous

Surface	t, °F*	Emissivity*	Surface	t, °F*	Emissivity*
Asbestos			Carbon		
Board	74	0.96	T-carbon (Gebr. Siemens) 0.9% ash	260–1160	0.81–0.79
Paper	100–700	0.93–0.945	(this started with emissivity at 260°F.		
Brick			of 0.72, but on heating changed to		
Red, rough, but no gross irregularities	70	0.93	values given)		
Silica, unglazed, rough	1832	0.80	Carbon filament	1900–2560	0.526
Silica, glazed, rough	2012	0.85	Candle soot	206–520	0.952
Grog brick, glazed	2012	0.75	Lampblack-waterglass coating	209–362	0.959–0.947
See Refractory Materials below					

Surface	°F	Emissivity
Same		
Thin layer on iron plate	260–440	0.957–0.952
Thick coat	69	0.927
Lampblack, 0.003 in. or thicker	68	0.967
Enamel, white fused, on iron	100–700	0.945
Glass, smooth	66	0.897
Gypsum, 0.02 in. thick on smooth or blackened plate	72	0.937
	70	0.903
Marble, light gray, polished	72	0.931
Oak, planed	70	0.895
Oil layers on polished nickel (lube oil)	68	
Polished surface, alone		0.045
+0.001-in. oil		0.27
+0.002-in. oil		0.46
+0.005-in. oil		0.72
Infinitely thick oil layer		0.82
Oil layers on aluminum foil (linseed oil)		
Al foil	212	0.087
+1 coat oil	212	0.561
+2 coats oil	212	0.574
Paints, lacquers, varnishes		
Snowhite enamel varnish or rough iron plate	73	0.906
Black shiny lacquer, sprayed on iron	76	0.875
Black shiny shellac on tinned iron sheet	70	0.821
Black matte shellac	170–295	0.91
Black lacquer	100–200	0.80–0.95
Flat black lacquer	100–200	0.96–0.98
White lacquer	100–200	0.80–0.95
Oil paints, sixteen different, all colors	212	0.92–0.96
Aluminum paints and lacquers		
10% Al, 22% lacquer body, on rough or smooth surface	212	0.52
26% Al, 27% lacquer body, on rough or smooth surface	212	0.3
Other Al paints, varying age and Al content	212	0.27–0.67
Al lacquer, varnish binder, on rough plate	70	0.39
Al paint, after heating to 620°F.	300–600	0.35
Paper, thin		
Pasted on tinned iron plate	66	0.924
On rough iron plate	66	0.929
On black lacquered plate	66	0.944
Plaster, rough lime	50–190	0.91
Porcelain, glazed	72	0.924
Quartz, rough, fused	70	0.932
Refractory materials, 40 different	1110–1830	poor radiators 0.65]–0.75 / 0.70 ; good radiators 0.80]–0.85 / 0.85]–0.90
Roofing paper	69	0.91
Rubber		
Hard, glossy plate	74	0.945
Soft, gray, rough (reclaimed)	76	0.859
Serpentine, polished	74	0.900
Water	32–212	0.95–0.963

* When two temperatures and two emissivities are given, they correspond, first to first and second to second, and linear interpolation is permissible. °C = (°F − 32)/1.8.

HEAT TRANSFER—TABLE 2. Emissivity ε_G of $H_2O:CO_2$ Mixtures

Limited range for furnaces, valid over 25-fold range of $p_{w+c}L$, 0.046–1.15 m atm (0.15–3.75 ft. atm)

p_w/p_c	0		½		1		2		3		∞	
$\dfrac{p_w}{p_w+p_c}$	0		⅓(0.3–0.42)		½(0.42–0.5)		⅔(0.6–0.7)		¾(0.7–0.8)		1	
	CO₂ only		corresponding to (CH)ₓ, covering coal, heavy oils, pitch		corresponding to (CH₂)ₓ, covering distillate oils, paraffins, olefines		corresponding to CH₄, covering natural gas and refinery gas		corresponding to (CH₆)ₓ, covering future high H₂ fuels		H₂O only	

Constants b and n of Eq., $\varepsilon_G T = b(pL - 0.015)^n$, pL = m atm, $T = K$

$T,\ K$	b	n	b	n	b	n	b	n	b	n	b	n
1000	188	0.209	384	0.33	416	0.34	444	0.34	455	0.35	416	0.400
1500	252	0.256	448	0.38	495	0.40	540	0.42	548	0.42	548	0.523
2000	267	0.316	451	0.45	509	0.48	572	0.51	594	0.52	632	0.640

Constants b and n of Eq., $\varepsilon_G T = b(pL - 0.05)^n$, pL = ft. atm, $T = °R$

$T,\ °R$	b	n	b	n	b	n	b	n	b	n	b	n
1800	264	0.209	467	0.33	501	0.34	534	0.34	541	0.35	466	0.400
2700	335	0.256	514	0.38	555	0.40	591	0.42	600	0.42	530	0.523
3600	330	0.316	476	0.45	519	0.48	563	0.51	577	0.52	532	0.640

Full range, valid over 2000-fold range of $p_{w+c}L$, 0.005–10.0 m atm (0.016–32.0 ft. atm)

Constants of Eq., $\log_{10}\varepsilon_G T_G = a_0 + a_1\log pL + a_2\log^2 pL + a_3\log^3 pL$

$\dfrac{p_w}{p_c}$	$\dfrac{p_w}{p_w+p_c}$	pL = m. atm, $T = K$					pL = ft., atm, $T = °R$				
		T, K	a_0	a_1	a_2	a_3	$T, °R$	a_0	a_1	a_2	a_3
0	0	1000	2.2661	0.1742	-0.0390	0.0040	1800	2.4206	0.2176	-0.0452	0.0040
		1500	2.3954	0.2203	-0.0433	0.00562	2700	2.5248	0.2695	-0.0521	0.00562
		2000	2.4104	0.2602	-0.0651	-0.00155	3600	2.5143	0.3621	-0.0627	-0.00155
½	⅓	1000	2.5754	0.2792	-0.0648	0.0017	1800	2.6691	0.3474	-0.0674	0.0017
		1500	2.6451	0.3418	-0.0685	-0.0043	2700	2.7074	0.4091	-0.0618	-0.0043
		2000	2.6504	0.4279	-0.0674	-0.0120	3600	2.6686	0.4879	-0.0489	-0.0120
1	½	1000	2.6090	0.2799	-0.0745	-0.0006	1800	2.7001	0.3563	-0.0736	-0.0006
		1500	6.6862	0.3450	-0.0816	-0.0039	2700	2.7423	0.4561	-0.0756	-0.0039
		2000	2.7029	0.4440	-0.0859	-0.0135	3600	2.7081	0.5210	-0.0650	-0.0135
2	⅔	1000	2.6367	0.2723	-0.0804	0.0030	1800	2.7296	0.3577	-0.0850	0.0030
		1500	2.7178	0.3386	-0.0990	-0.0030	2700	2.7724	0.4384	-0.0944	-0.0030
		2000	2.7482	0.4464	-0.1086	-0.0139	3600	2.7461	0.5474	-0.0871	-0.0139
3	¾	1000	2.6432	0.2715	-0.0816	-0.0052	1800	2.7359	0.3599	-0.0896	0.0052
		1500	2.7257	0.3355	-0.0981	0.0045	2700	2.7811	0.4403	-0.1051	0.0045
		2000	2.7592	0.4372	-0.1122	-0.0065	3600	2.7599	0.5478	-0.1021	-0.0065
∞	1	1000	2.5995	0.3015	-0.0961	0.0119	1800	2.6720	0.4102	-0.1145	0.0119
		1500	2.7083	0.3969	-0.1309	0.00123	2700	2.7238	0.5330	-0.1328	0.00123
		2000	2.7709	0.5099	-0.1646	-0.0165	3600	2.7215	0.6666	-0.1391	-0.0165

Note: $p_w/(p_w + p_c)$ of ⅓, ½, ⅔, and ¾ may be used to cover the ranges 0.2–0.4, 0.4–0.6, 0.6–0.7, and 0.7–0.8, respectively, with a maximum error in ε_G of 5 percent at pL = 6.5 m atm, less at lower pLs. Linear interpolation reduces the error generally to less than 1 percent. Linear interpolation or extrapolation on T introduces an error generally below 2 percent, less than the accuracy of the original data.

HEAT TRANSFER—TABLE 3. Total Emissivities of Some Gases

Temperature	1000°R			1600°R			2200°R			2800°R		
P_xL, (atm)(ft)	0.01	0.1	1.0	0.01	0.1	1.0	0.01	0.1	1.0	0.01	0.1	1.0
NH_3	0.047	0.20	0.61	0.020	0.120	0.44	0.0057	0.051	0.25	(0.001)	(0.015)	(0.14)
SO_2	0.020	0.13	0.28	0.013	0.090	0.32	0.0085	0.051	0.27	0.0058	0.043	0.20
CH_4	0.020	0.060	0.15	0.023	0.072	0.194	0.022	0.070	0.185	0.019	0.059	0.17
CO	0.011	0.031	0.061	0.022	0.057	0.10	0.022	0.050	0.080	(0.012)	(0.035)	(0.050)
NO	0.0046	0.018	0.060	0.0046	0.021	0.070	0.0019	0.010	0.040	0.00078	0.004	0.025
HCl	0.00022	0.00079	0.0020	0.00036	0.0013	0.0033	0.00037	0.0014	0.0036	0.00029	0.0010	0.0027

HEAT TRANSFER—TABLE 4. Rules for Diffusivities

	D_i magnitude		D_i range		Comments
Continuous phase	m²/s	cm²/s	m²/s	cm²/s	
Gas at atmospheric pressure	10^{-5}	0.1	10^{-4}–10^{-6}	1–10^{-2}	Accurate theories exist, generally within ±10%; $D_iP \cong$ constant; $D_i \propto T^{1.66 \text{ to } 2.0}$
Liquid	10^{-9}	10^{-5}	10^{-8}–10^{-10}	10^{-4}–10^{-6}	Approximate correlations exist, generally within ±25%
Liquid occluded in solid matrix	10^{-10}	10^{-6}	10^{-8}–10^{-12}	10^{-4}–10^{-8}	Hard cell walls: $D_{eff}/D_i = 0.1$ to 0.2. Soft cell walls: $D_{eff}/D_i = 0.3$ to 0.9
Polymers and glasses	10^{-12}	10^{-8}	10^{-10}–10^{-14}	10^{-6}–10^{-10}	Approximate theories exist for dilute and concentrated limits; strong composition dependence
Solid	10^{-14}	10^{-10}	10^{-10}–10^{-34}	10^{-6}–10^{-30}	Approximate theories exist; strong temperature dependence

HEAT TRANSFER—TABLE 5. Correlations of Diffusivities for Gases

Equation	Error
1. Binary Mixtures—Low Pressure—Nonpolar	
$D_{AB} = \dfrac{0.001858 T^{3/2} M_{AB}^{1/2}}{P \sigma_{AB}^2 \Omega_D}$	7.3%
$D_{AB} = \dfrac{\left(0.0027 - 0.0005 M_{AB}^{1/2}\right) T^{3/2} M_{AB}^{1/2}}{P \sigma_{AB}^2 \Omega_D}$	7.0%
$D_{AB} = \dfrac{0.001 T^{1.75} M_{AB}^{1/2}}{P\left[(\Sigma v)_A^{1/3} + (\Sigma v)_B^{1/3}\right]^2}$	5.4%
2. Binary Mixtures—Low Pressure—Polar	
$D_{AB} = \dfrac{0.001858 T^{3/2} M_{AB}^{1/2}}{P \sigma_{AB}^2 \Omega_D}$	9.0%
3. Self-Diffusivity—High Pressure	
$D_{AA} = \dfrac{10.7 \times 10^{-5} T_r}{\beta p_r}\ (p_r \leq 1.5)$	5%
$D_{AA} = \dfrac{0.77 \times 10^{-5} T_r}{p_r \delta}\ (p_r \leq 1)$	0.5%
$D_{AA} = \dfrac{(0.007094 G + 0.001916)^{2.5} T_r}{\delta},\ [p_r > 1, G < 1]$	17%
4. Supercritical Mixtures	
$D_{AB} = \dfrac{1.23 \times 10^{-10} T}{\mu^{0.799} V_{C_A}^{0.49}}$	5%
$D_{AB} = 5.152 D_0 T_r \dfrac{\left(p^{-0.667} - 0.4510\right)\left(1 + M_A/M_B\right) R}{\left(1 + (V_{cB}/V_{cA})^{0.333}\right)^2}$	10%

HUMIDITY

HUMIDITY—TABLE 1. Maintenance of Constant Humidity

Solid phase	Max. temp., °C.	% Humidity
$H_3PO_4 \cdot \frac{1}{2} H_2O$	24.5	9
$ZnCl_2 \cdot \frac{1}{2} H_2O$	20	10
$KC_2H_3O_2$	168	13
$LiCl \cdot H_2O$	20	15
$KC_2H_3O_2$	20	20

HUMIDITY—TABLE 1 Maintenance of Constant Humidity (*Continued*)

Solid phase	Max. temp., °C.	% Humidity
KF	100	22.9
NaBr	100	22.9
$CaCl_2 \cdot 6H_2O$	24.5	31
$CaCl_2 \cdot 6H_2O$	20	32.3
$CaCl_2 \cdot 6H_2O$	18.5	35
CrO_3	20	35
$CaCl_2 \cdot 6H_2O$	10	38
$CaCl_2 \cdot 6H_2O$	5	39.8
$K_2CO_3 \cdot 2H_2O$	24.5	43
$K_2CO_3 \cdot 2H_2O$	18.5	44
$Ca(NO_3)_2 \cdot 4H_2O$	24.5	51
$NaHSO_4 \cdot H_2O$	20	52
$Mg(NO_3)_2 \cdot 6H_2O$	24.5	52
$NaClO_3$	100	54
$Ca(NO_3)_2 \cdot 4H_2O$	18.5	56
$Mg(NO_3)_2 \cdot 6H_2O$	18.5	56
$NaBr \cdot 2H_2O$	20	58
$Mg(C_2H_3O_2)_2 \cdot 4H_2O$	20	65
$NaNO_2$	20	66
$(NH_4)_2SO_4$	108.2	75
$(NH_4)_2SO_4$	20	81
$NaC_2H_3O_2 \cdot 3H_2O$	20	76
$Na_2S_2O_3 \cdot 5H_2O$	20	78
NH_4Cl	20	79.5
NH_4Cl	25	79.3
NH_4Cl	30	77.5
KBr	20	84
Tl_2SO_4	104.7	84.8
$KHSO_4$	20	86
$Na_2CO_3 \cdot 10H_2O$	24.5	87
K_2CrO_4	20	88
$NaBrO_3$	20	92
$Na_2CO_3 \cdot 10H_2O$	18.5	92
$Na_2SO_4 \cdot 10H_2O$	20	93
$Na_2HPO_4 \cdot 12H_2O$	20	95
NaF	100	96.6
$Pb(NO_3)_2$	20	98
$TlNO_3$	100.3	98.7
TlCl	100.1	99.7

HYDROGEN BONDING

The *hydrogen bond* is formed by the electrostatic interaction between the hydrogen atom and an electronegative atom, such as oxygen, nitrogen, or fluorine.

HYDROGEN BONDING—TABLE 1. Hydrogen Bonding Classification of Chemical Families

Class	Chemical family			
H-Bonding, Strongly Associative (HBSA)	Water Primary amides Secondary amides	Polyacids Dicarboxylic acids Monohydroxy acids	Polyphenols Oximes Hydroxylamines	Amino alcohols Polyols
H-Bond Acceptor-Donor (HBAD)	Phenols Aromatic acids Aromatic amines Alpha H nitriles	Imines Monocarboxylic acids Other monoacids Peracids	Alpha H nitros Azines Primary amines Secondary amines	n-Alcohols Other alcohols Ether alcohols
H-Bond Acceptor (HBA)	Acyl chlorides Acyl fluorides Hetero nitrogen aromatics Hetero oxygen aromatics	Tertiary amides Tertiary amines Other nitriles Other nitros Isocyanates Peroxides	Aldehydes Anhydrides Cyclo ketones Aliphatic ketones Esters Ethers	Aromatic esters Aromatic nitriles Aromatic ethers Sulfones Sulfolanes
π-Bonding Acceptor (π-HBA)	Alkynes Alkenes	Aromatics Unsaturated esters		
H-Bond Donor (HBD)	Inorganic acids Active H chlorides	Active H fluorides Active H iodides	Active H bromides	
Non-Bonding (NB)	Paraffins Nonactive H chlorides	Nonactive H fluorides Sulfides	Nonactive H iodides Disulfides	Nonactive H bromides Thiols

HYDROGEN BONDING—TABLE 1. Hydrogen Bonding classification of Chemical Families (*Continued*)

H-Bonding classes	Deviations from Raoult's Law	
	Type of deviations	Comments
HBSA + NB	Alway positive dev., HBSA + NB often limited miscibility	H-bonds broken by interactions
HBAD + NB		
HBA + HBD	Always negative dev.	H-bonds formed by interactions
HBSA + HBD	Always positive deviations, HBSA + HBD often limited miscibility	H-bonds broken and formed: dissociation of
HBAD + HBD		HBSA or HBAD liquid most important effect
HBSA + HBSA	Usually positive deviations: some give maximum-boiling azeotropes	H-bonds broken and formed
HBSA + HBAD		
HBSA + HBA		
HBAD + HBAD		
HBAD + HBA		
HBA + HBA	Ideal, quasi-ideal systems; always positive or no deviations; azeotropes,	No H-bonding involved
HBA + NB	if any, minimum-boiling	
HBD + HBD		
HBD + NB		
NB + NB		

Note: π-HBA is *enhanced* version of HBA.

226

INFRARED ABSORPTION

Infrared (IR) absorption spectroscopy is the measurement of the wavelength and intensity of the absorption of mid-infrared light by a sample. Mid-infrared light (2.5–$50\,\mu$m, 4000–$200\,\text{cm}^{-1}$) is energetic enough to excite molecular vibrations to higher energy levels. The wavelengths of many infrared absorption bands are characteristic of specific types of chemical bonds, and infrared spectroscopy finds its greatest utility for qualitative analysis of organic and organometallic molecules. Infrared spectroscopy is used to confirm the identity of a particular compound and as a tool to help determine the structure of a molecule.

Significant for the identification of the source of an absorption band are *intensity* (weak, medium, or strong), *shape* (broad or sharp), *position* (cm^{-1}) in the spectrum. Characteristic examples are provided in Table 1 to assist the user in becoming familiar with the intensity and shape absorption bands for representative absorptions.

INFRARED ABSORPTION—TABLE 1. Infrared Absorption Frequencies

Bond	Compound type	Frequency range, cm^{-1} (wave number)
O—H	Monomeric alcohols and phenols	3640–3160 (s, br) stretch
O—H	Hydrogen-bonded alcohols and phenols	3600–3200 (b) stretch
N—H	Amines, amides	3560-3320 (m) stretch; non-hydrogen bonded
N—H	Amines	3400–3100 (m) stretch; hydrogen bonded
C—H	Alkynes	3333–3267 (s) stretch
C—H	Aromatic rings	3100–3000 (m) stretch
C—H	Alkanes	3080–3020 (m) stretch
O—H	Carboxylic acids	3000–2500 (b) stretch
C—H	Alkanes	2960–2850 (s) stretch
C≡N	Nitriles	2260–2220 (v) stretch
C≡C	Alkynes	2260–2100 (w, sh) stretch
C—H	Phenyl ring substitution	2000–1600 (w)
C=O	Aldehydes, ketones, carboxylic acids, esters	1760–1670 (s) stretch
C=C	Alkenes	1680–1640 (m, w) stretch
NO$_2$	Nitro-compounds	1660–1500 (s) asymmetrical stretch
N—H	Amines	1650–1580 (m) bend
C=C	Aromatic rings	1600, 1500 (w) stretch
C—H	Alkanes	1470–1350 (v) scissoring and bending
NO$_2$	Nitro-compounds	1390–1260 (s) symmetrical stretch
C—H	CH$_3$ Deformation	1380 (m or w, can be a doublet)—isopropyl, t-butyl
C—N	Amines	1340–1020 (m) stretch
C—O	Alcohols, ethers, carboxylic acids, esters	1260–1000 (s) stretch
C—H	Alkenes	1000–675 (s) bend
C—H	Phenyl ring substitution	870–675 (s) bend
C—H	Alkynes	700–610 (b) bend
C—Cl	Chloro-compounds	600–800 (m, w)
C—Br	Bromo-compounds	500–600 (m, w)
C—I	Iodo-compounds	500 (n w)

v = variable, m = medium, s = strong, br = broad, w = weak; wave number = 1/wavelength.

LIQUID–LIQUID EXTRACTION

LIQUID–LIQUID EXTRACTION—TABLE 1.　Selected List of Ternary Systems
Component A = feed solvent, component B = solute, and component S = extraction solvent. K_1 is the partition ratio in weight-fraction solute y/x for the tie line of lowest solute concentration reported. Ordinarily, K will approach unity as the solute concentration is increased.

Component B	Component S	Temp., °C	K_1
A = cetane			
Benzene	Aniline	25	1.290
n-Heptane	Aniline	25	0.0784
A = cottonseed oil			
Oleic acid	Propane	85	0.150
		98.5	0.1272
A = cyclohexene			
Benzene	Furfural	25	0.680
Benzene	Nitromethane	25	0.397
A = docosane			
1,6-Diphenylhexane	Furfural	45	0.980
		80	1.100
		115	1.062
A = dodecane			
Methylnaphthalene	β,β'-Iminodipropionitrile	ca. 25	0.625
Methylnaphthalene	β,β'-Oxydipropionitrile	ca. 25	0.377
A = ethylbenzene			
Styrene	Ethylene glycol	25	0.190
A = ethylene glycol			
Acetone	Amyl acetate	31	1.838
Acetone	n-Butyl acetate	31	1.940
Acetone	Cyclohexane	27	0.508
Acetone	Ethyl acetate	31	1.850
Acetone	Ethyl butyrate	31	1.903
Acetone	Ethyl propionate	31	2.32
A = furfural			
Trilinolein	n-Heptane	30	47.5
		50	21.4
		70	19.5
Triolein	n-Heptane	30	95
		50	108
		70	41.5
A = glycerol			
Ethanol	Benzene	25	0.159
Ethanol	Carbon tetrachloride	25	0.0667
A = n-heptane			
Benzene	Ethylene glycol	25	0.300
		125	0.316
Benzene	β,β'-thiodipropionitrile	25	0.350
Benzene	Triethylene glycol	25	0.351
Cyclohexene	Aniline	25	0.0815

LIQUID–LIQUID EXTRACTION — TABLE 1. Selected List of Ternary Systems (*Continued*)

Component *B*	Component *S*	Temp., °C	K_1
Cyclohexene	Benzyl alcohol	0	0.107
		15	0.267
Cyclohexene	Dimethylformamide	20	0.1320
Cyclohexene	Furfural	30	0.0635
Ethylbenzene	Dipropylene glycol	25	0.329
Ethylbenzene	β,β'-Oxydipropionitrile	25	0.180
Ethylbenzene	β,β'-Thiodipropionitrile	25	0.100
Ethylbenzene	Triethylene glycol	25	0.140
Methylcyclohexene	Aniline	25	0.087
Toluene	Aniline	0	0.577
		13	0.477
		20	0.457
		40	0.425
Toluene	Benzyl alcohol	0	0.694
Toluene	Dimethylformamide	0	0.667
		20	0.514
Toluene	Dipropylene glycol	25	0.331
Toluene	Ethylene glycol	25	0.150
Toluene	Propylene carbonate	20	0.732
Toluene	β,β'-Thiodipropionitrile	25	0.150
Toluene	Triethylene glycol	25	0.289
m-Xylene	β,β'-Thiodipropionitrile	25	0.050
o-Xylene	β,β'-Thiodipropionitrile	25	0.150
p-Xylene	β,β'-Thiodipropionitrile	25	0.080
A = *n*-hexane			
Benzene	Ethylenediamine	20	4.14
A = *neo*-hexane		15	0.1259
Cyclopentane	Aniline	25	0.311
A = methyleyclohexane			
Toluene	Methylperfluorooctanoate	10	0.1297
		25	0.200
A = *iso*-octane			
Benzene	Furfural	25	0.833
Cyclohexane	Furfural	25	0.1076
n-Hexane	Furfural	30	0.083
A = perfluoroheptane			
Perfluoroclic oxide	Carbon tetrachloride	30	0.1370
Perfluoroclic oxide	*n*-Heptane	30	0.329
A = perfluoro-*n*-hexane			
n-Hexane	Benzene	30	6.22
n-Hexane	Carbon disulfide	25	6.50
A = perfluorotri-*n*-butylamine			
Iso-octane	Nitroethane	25	3.59
		31.5	2.36
		33.7	4.56
A = toluene			
Acetone	Ethylene glycol	0	0.286
		24	0.326

LIQUID–LIQUID EXTRACTION—TABLE 1. Selected List of Ternary Systems (*Continued*)

Component *B*	Component *S*	Temp., °C	K_1
A = triethylene glycol			
α-Picoline	Methylcyclohexane	20	3.87
α-Picoline	Diisobutylene	20	0.445
α-Picoline	Mixed heptanes	20	0.317
A = triolein			
Oleic acid	Propane	85	0.138
A = water			
Acetaldehyde	*n*-Amyl alcohol	18	1.43
Acetaldehyde	Benzene	18	1.119
Acetaldehyde	Furfural	16	0.967
Acetaldehyde	Toluene	17	0.478
Acetaldehyde	Vinyl acetate	20	0.560
Acetic acid	Benzene	25	0.0328
		30	0.0984
		40	0.1022
		50	0.0558
		60	0.0637
Acetic acid	I-Butanol	26.7	1.613
Acetic acid	Butyl acetate	30	0.705
			0.391
Acetic acid	Caproic acid	25	0.349
Acetic acid	Carbon tetrachloride	27	0.1920
		27.5	0.0549
Acetic acid	Chloroform	*ca.* 25	0.178
		25	0.0865
		56.8	0.1573
Acetic acid	Creosote oil	34	0.706
Acetic acid	Cyciohexanol	26.7	1.325
Acetic acid	Diisobutyl ketone	25–26	0.284
Acetic acid	Di-*n*-butyl ketone	25–26	0.379
Acetic acid	Diisopropyl carbinol	25–26	0.800
Acetic acid	Ethyl acetate	30	0.907
Acetic acid	2-Ethylbutyric acid	25	0.323
Acetic acid	2-Ethylhexoic acid	25	0.286
Acetic acid	Ethylidene diacetate	25	0.85
Acetic acid	Ethyl propionate	28	0.510
Acetic acid	Fenchone	25–26	0.310
Acetic acid	Furfural	26.7	0.787
Acetic acid	Heptadecanol	25	0.312
		50	0.1623
Acetic acid	3-Heptanol	25	0.828
Acetic acid	Hexalin acetate	25–26	0.520
Acetic acid	Hexane	31	0.0167
Acetic acid	Isoamyl acetate	25–26	0.343
Acetic acid	Isophorone	25–26	0.858
Acetic acid	Isopropyl ether	20	0.248
		25–26	0.429
Acetic acid	Methyl acetate		1.273
Acetic acid	Methyl butyrate	30	0.690

LIQUID–LIQUID EXTRACTION—TABLE 1.   Selected List of Ternary Systems (*Continued*)

Component *B*	Component *S*	Temp., °C	K_1
Acetic acid	Methyl cyclohexanone	25–26	0.930
Acetic acid	Methylisobutyl carbinol	30	1.058
Acetic acid	Methylisobutyl ketone	25	0.657
		25–26	0.755
Acetic acid	Monochlorobenzene	25	0.0435
Acetic acid	Octyl acetate	25–26	0.1805
Acetic acid	*n*-Propyl acetate		0.638
Acetic acid	Toluene	25	0.0644
Acetic acid	Trichloroethylene	27	0.140
		30	0.0549
Acetic acid	Vinyl acetate	28	0.294
Acetone	Amyl acetate	30	1.228
Acetone	Benzene	15	0.940
		30	0.862
		45	0.725
Acetone	*n*-Butyl acetate		1.127
Acetone	Carbon tetrachlonde	30	0.238
Acetone	Chloroform	25	1.830
		25	1.720
Acetone	Dibutyl ether	25–26	1.941
Acetone	Diethyl ether	30	1.00
Acetone	Ethyl acetate	30	1.500
Acetone	Ethyl butyrate	30	1.278
Acetone	Ethyl propionate	30	1.385
Acetone	*n*-Heptane	25	0.274
Acetone	*n*-Hexane	25	0.343
Acetone	Methyl acetate	30	1.153
Acetone	Methylisobutyl ketone	25–26	1.910
Acetone	Monochlorobenzene	25–26	1.000
Acetone	Propyl acetate	30	0.243
Acetone	Tetrachloroethane	25–26	2.37
Acetone	Tetrachloroethylene	30	0.237
Acetone	1.1,2-Trichloroethune	25	1.467
Acetone	Toluene	25–26	0.835
Acetone	Vinyl acetate	20	1.237
		25	3.63
Acetone	Xylene	25–26	0.659
Allyl alcohol	Diallyl ether	22	0.572
Aniline	Benzene	25	14.40
		50	15.50
Aniline	*n*-Heptane	25	1.425
		50	2.20
Aniline	Methylcyclohexene	25	2.05
		50	3.41
Aniline	Nitrobenzene	25	18.89
Aniline	Toluene	25	12.91
Aniline hydrochloride	Aniline	25	0.0540
Benzoic acid	Methylisobutyl ketone	26.7	76.9*

LIQUID–LIQUID EXTRACTION—TABLE 1. Selected List of Ternary Systems (*Continued*)

Component *B*	Component *S*	Temp., °C	K_1
iso-Butanol	Benzene	25	0.989
iso-Butanol	1,1,2,2-Tetrachloroethane	25	1.80
iso-Butanol	Tetrachloroethylene	25	0.0460
n-Butanol	Benzene	25	1.263
		35	2.12
n-Butanol	Toluene	30	1.176
tert-Butanol	Benzene	25	0.401
tert-Butanol	*tert*-Butyl hypochlorite	0	0.1393
		20	0.1487
		40	0.200
		60	0.539
tert-Butanol	Ethyl acetate	20	1.74
2-Butoxyethanol	Methylethyl ketone	25	3.05
2,3-Butylene glycol	*n*-Butanol	26	0.597
		50	0.893
2,3-Butylene glycol	Butyl acetate	26	0.0222
		50	0.0326
2,3-Butylene glycol	Butylene glycol diacetate	26	0.1328
		75	0.565
2,3-Butylene glycol	Methylvinyl carbinol acetate	26	0.237
		50	0.351
		75	0.247
n-Butylamine	Monochlorobenzene	25	1.391
l-Butyraidehyde	Ethyl acetate	37.8	41.3
Butyric acid	Methyl butyrate	30	6.75
Butyric acid	Methylisobutyl carbinol	30	12.12
Cobaltous chloride	Dioxane	25	0.0052
Cupric sulfate	*n*-Butanol	30	0.000501
Cupric sulfate	*sec*-Butanol	30	0.00702
Cupric sulfate	Mixed peutanols	30	0.000225
p-Cresol	Methylnaphthalene	35	9.89
Diacetone alcohol	Ethylbenzene	25	0.335
Diacetone alcohol	Styrene	25	0.445
Dichloroacetic acid	Monochlorobenzene	25	0.0690
1,4-Dioxane	Benzene	25	1.020
Ethanol	*n*-Amyl alcohol	25–26	0.598
Ethanol	Benzene	25	0.1191
		25	0.0536
Ethanol	*n*-Butanol	20	3.00
Ethanol	Cyclohexane	25	0.0157
Ethanol	Cyclohexene	25	0.0244
Ethanol	Dibutyl ether	25–26	0.1458
Ethanol	Di-*n*-propyl ketone	25–26	0.592
Ethanol	Ethyl acetate	0	0.0263
		20	0.500
		70	0.455
Ethanol	Ethyl isovalerate	25	0.392
Ethanol	Heptadecanol	25	0.270
Ethanol	*n*-Heptane	30	0.274
Ethanol	3-Heptanol	25	0.783
Ethanol	*n*-Hexane	25	0.00212

LIQUID–LIQUID EXTRACTION—TABLE 1.   Selected List of Ternary Systems (*Continued*)

Component *B*	Component *S*	Temp., °C	K_1
Ethanol	*n*-Hexanol	28	1.00
Ethanol	*sec*-Octanol	28	0.825
Ethanol	Toluene	25	0.01816
Ethylene glycol	Trichloroethylene	25	0.0682
Ethylene glycol	*n*-Amyl alcohol	20	0.1159
Ethylene glycol	*n*-Butanol	27	0.412
Ethylene glycol	Furfural	25	0.315
Ethylene glycol	*n*-Hexanol	20	0.275
Ethylene glycol	Methylethyl ketone	30	0.0527
Formic acid	Chloroform	25	0.00445
		56.9	0.0192
Formic acid	Methylisobutyl carbinol	30	1.218
Furfural	*n*-Butane	51.5	0.712
		79.5	0.930
Furfural	Methylisobutyl ketone	25	7.10
Furfural	Toluene	25	5.64
Hydrogen chloride	*iso*-Amyl alcohol	25	0.170
Hydrogen chloride	2,6-Dimethyl-4-heptanol	25	0.266
Hydrogen chloride	2-Ethyl-1-butanol	25	0.534
Hydrogen chloride	Ethylbutyl ketone	25	0.01515
Hydrogen chloride	3-Heptanol	25	0.0250
Hydrogen chloride	1-Hexanol	25	0.345
Hydrogen chloride	2-Methyl-1-butanol	25	0.470
Hydrogen chloride	Methylisobutyl ketone	25	0.0273
Hydrogen chloride	2-Methyl-1-pentanol	25	0.502
Hydrogen chloride	2-Methyl-2-pentanol	25	0.411
Hydrogen chloride	Methylisopropyl ketone	25	0.0814
Hydrogen chloride	1-Octanol	25	0.424
Hydrogen chloride	2-Octanol	25	0.380
Hydrogen chloride	1-Pentanol	25	0.257
Hydrogen chloride	Pentanols (mixed)	25	0.271
Hydrogen chloride	Methylisobutyl ketone	25	0.370
Lactic acid	*iso*-Amyl alcohol	25	0.352
Methanol	Benzene	25	0.01022
Methanol	*n*-Butanol	0	0.600
		15	0.479
		30	0.510
		45	1.260
		60	0.682
Methanol	*p*-Cresol	35	0.313
Methanol	Cyclohexane	25	0.0156
Methanol	Cyclohexane	25	0.01043
Methanol	Ethyl acetate	0	0.0589
		20	0.238
Methanol	*n*-Hexanol	28	0.565
Methanol	Methylnaphthalene	25	0.025
		35	0.0223
Methanol	*sec*-Octanol	28	0.584
Methanol	Phenol	25	1.333
Methanol	Toluene	25	0.0099

LIQUID–LIQUID EXTRACTION—TABLE 1. Selected List of Ternary Systems (*Continued*)

Component B	Component S	Temp., °C	K_1
Methanol	Trichloroethylene	27.5	0.0167
Methyl-*n*-butyl ketone	*n*-Butanol	37.8	53.4
Methylethyl ketone	Cyclohexane	25	1.775
		30	3.60
Methylethyl ketone	Gasoline	25	1.686
Methylethyl ketone	*n*-Heptane	25	1.548
Methylethyl ketone	*n*-Hexane	25	1.775
		37.8	2.22
Methylethyl ketone	2-Methyl furan	25	84.0
Methylethyl ketone	Monochlorobenzene	25	2.36
Methylethyl ketone	Naphtha	26.7	0.885[†]
Methylethyl ketone	1,1,2-Trichloroethane	25	3.44
Methylethyl ketone	Trichloroethylene	25	3.27
Methylethyl ketone	2,2,4-Trimethylpentane	25	1.572
Nickelous chloride	Dioxane	25	0.0017
Nicotine	Carbon tetrachloride	25	9.50
Phenol	Methylnaphthalene	25	7.06
α-Picoline	Benzene	20	8.75
α-Picoline	Diisobutylene	20	1.360
α-Picoline	Heptanes (mixed)	20	1.378
α-Picoline	Methylcyclohexane	20	1.00
iso-Propanol	Benzene	25	0.276
iso-Propanol	Carbon tetrachloride	20	1.405
iso-Propanol	Cyclohexane	25	0.0282
iso-Propanol	Cyclohexene	15	0.0583
		25	0.0682
		35	0.1875
iso-Propanol	Diisopropyl ether	25	0.406
iso-Propanol	Ethyl acetate	0	0.200
		20	1.205
iso-Propanol	Tetrachloroethylene	25	0.388
iso-Propanol	Toluene	25	0.1296
n-Propanol	*iso*-Amyl alcohol	25	3.34
n-Propanol	Benzene	37.8	0.650
n-Propanol	*n*-Butanol	37.8	3.61
n-Propanol	Cyclohexane	25	0.1553
		35	0.1775
n-Propanol	Ethyl acetate	0	1.419
		20	1.542
n-Propanol	*n*-Heptane	37.8	0.540
n-Propanol	*n*-Hexane	37.8	0.326
n-Propanol	*n*-Propyl acetate	20	1.55
		35	2.14
n-Propanol	Toluene	25	0.299
Propionic acid	Benzene	30	0.598
Propionic acid	Cyclohexane	31	0.1955
Propionic acid	Cyclohexene	31	0.303
Propionic acid	Ethyl acetate	30	2.77
Propionic acid	Ethyl butyrate	26	1.470
Propionic acid	Ethyl propionate	28	0.510

LIQUID–LIQUID EXTRACTION—TABLE 1.   Selected List of Ternary Systems (*Continued*)

Component *B*	Component *S*	Temp., °C	K_1
Propionic acid	Hexanes (mixed)	31	0.186
Propionic acid	Methyl butyrate	30	2.15
Propionic acid	Methylisobutyl carbinol	30	3.52
Propionic acid	Methylisobutyl ketone	26.7	1.949
Propionic acid	Monochlorobenzene	30	0.513
Propionic acid	Tetrachloroethylene	31	0.167
Propionic acid	Toluene	31	0.515
Propionic acid	Trichloroethylene	30	0.496
Pyridine	Benzene	15	2.19
		25	3.00
		25	2.73
		45	2.49
		60	2.10
Pyridine	Monochlorobenzene	25	2.10
Pyridine	Toluene	25	1.900
Pyridine	Xylene	25	1.260
Sodium chloride	*iso*-Butanol	25	0.0182
Sodium chloride	*n*-Ethyl-*sec*-butyl amine	32	0.0563
Sodium chloride	*n*-Ethyl-*tert*-butyl amine	40	0.1792
Sodium chloride	2-Ethylhexyl amine	30	0.187
Sodium chloride	1-Methyldiethyl amine	39.1	0.0597
Sodium chloride	1-Methyldodecyl amine	30	0.693
Sodium chloride	*n*-Methyl-1,3-dimethylbutyl amine	30	0.0537
Sodium chloride	1-Methyloctyl amine	30	0.589
Sodium chloride	*tert*-Nonyl amine	30	0.0318
Sodium chloride	1,1,3,3-Tetramethyl butyl amine	30	0.072
Sodium hydroxide	*iso*-Butanol	25	0.00857
Sodium nitrate	Dioxane	25	0.0246
Succinic acid	Ethyl ether	15	0.220
		30	0.198
		25	0.1805
Trimethyl amine	Benzene	25	0.857
		70	2.36

* Concentrations in lb.-moles/cu. ft.
† Concentrations in volume fraction.

MASS TRANSFER

See also **HEAT TRANSFER**.

MASS TRANSFER—TABLE 1. Mass Transfer Correlations for a Single Flat Plate or Disk—Transfer to or from Plate to Fluid

Situation	Correlation	Comments E = Empirical, S = Semiempirical, T = Theoretical
A. Laminar, local, flat plate, forced flow	$N_{Sh,x} = \dfrac{k'x}{D} = 0.323(N_{Re,x})^{1/2}(N_{Sc})^{1/3}$ Coefficient 0.332 is a better fit.	[T] Low M.T. rates. Low mass-flux, constant property systems. $N_{Sh,s}$ is local k. use with arithmetic difference in concentration. Coefficient 0.323 is Blasius' approximate solution. $N_{Re,x} = \dfrac{xu_\infty\rho}{\mu}$, $x =$ length along plate $N_{Re,L} = \dfrac{Lu_\infty\rho}{\mu}$, 0.664 (Polhausen) is a better fit for $N_{Sc}^x > 0.6$, $N_{Re,x} < 3 \times 10^5$.
Laminar, average, flat plate, forced flow	$N_{Sh,\text{avg}} = \dfrac{k_m'L}{D} = 0.646(N_{Re,L})^{1/2}(N_{Sc})^{1/3}$ k_m' is mean mass-transfer coefficient for dilute systems.	
j-factors	$j_D = j_H = \dfrac{f}{2} = 0.664(N_{Re,L})^{-1/2}$	[S] Analogy. $N_{sc} = 1.0$, $f =$ drag coefficient. j_D is defined in terms of k_m'.
B. Laminar, local, flat plate, blowing or suction and forced flow	$N_{Sh,x} = \dfrac{k'x}{D} = (\text{Slope})_{y=0}(N_{Re,x})^{1/2}(N_{Sc})^{1/3}$	[T] Blowing is positive. Other conditions as above. $\dfrac{u_o}{u_\infty}\sqrt{N_{Re,x}}$ $\quad$ 0.6 $\quad$ 0.5 $\quad$ 0.25 $\quad$ 0.0 $\quad$ −2.5 $(\text{Slope})_{y=0}$ $\quad$ 0.01 $\quad$ 0.06 $\quad$ 0.17 $\quad$ 0.332 $\quad$ 1.64
C. Laminar, local, flat plate, natural convection vertical plate	$N_{Sh,x} = \dfrac{k'x}{D} = 0.508 N_{Sc}^{1/2}(0.952 + N_{Sc})^{-1/4} N_{Gr}^{1/4}$	[T] Low MT rates. Dilute systems, $\Delta\rho/\rho \ll 1$. $N_{Gr}N_{Sc} < 10^8$. Use with arithmetic concentration difference. $x =$ length from plate bottom. $N_{Gr} = \dfrac{gx^3}{(\mu/\rho)^2}\left(\dfrac{\rho_\infty}{\rho_0} - 1\right)$

D. Laminar, stationary disk	$N_{Sh} = \dfrac{k'd_{disk}}{D} = \dfrac{8}{\pi}$	[T] Stagnant fluid. Use arithmetic concentration difference.
Laminar, spinning disk	$N_{Sh} = \dfrac{k'd_{disk}}{D} = 0.879 N_{Re}^{1/2} N_{Sc}^{1/3}$	[T] Asymptotic solution for large N_{Sc}. $N_{Re} < \sim 10^4$. $\mu = \omega d_{disk/2}$, ω = rotational speed, rad/s. Rotating disks are often used in electrochemical research.
E. Laminar, inclined, plate	$N_{Sh,avg} = 0.783 N_{Re,film}^{1/9} N_{Sc}^{1/3} \left(\dfrac{x^3 \rho^2 g \sin\alpha}{\mu^2} \right)^{2/9}$ $N_{Sh,avg} = \dfrac{k'_m x}{D}$ $\delta_{film} = \left(\dfrac{3\mu Q}{w\rho g \sin\alpha} \right)^{1/3}$ = film thickness	[T] Constant-property liquid film with low mass-transfer rates. Use arithmetic concentration difference. $N_{Re,film} = \dfrac{4Q\rho}{\mu^2} < 2000$ ω = width of plate, δ_f = film thickness, α = angle of inclination, x = distance from start soluble surface. Newtonian fluid. Solute does not penetrate past region of linear velocity profile. Differences between theory and experiment.
F. Turbulent, local flat plate, forced flow	$N_{Sh,x} = \dfrac{k'x}{D} = 0.0292 N_{Re,x}^{0.8}$	[S] Low mass-flux with constant property system. Use with arithmetic concentration difference. $N_{Sc} = 1.0$, $N_{Re,x} > 10^5$ Based on Prandtl's 1/7-power velocity law, $\dfrac{u}{u_\infty} = \left(\dfrac{y}{\delta} \right)^{1/7}$
Turbulent, average, flat plate, forced flow	$N_{Sh,avg} = \dfrac{k'L}{D} = 0.0365 N_{Re,L}^{0.8}$, average coefficient	

MASS TRANSFER—TABLE 1. Mass Transfer Correlations for a Single Flat Plate or Disk—Transfer to or from Plate to Fluid (*Continued*)

Situation	Correlation	Comments E = Empirical, S = Semiempirical, T = Theoretical
G. Laminar and turbulent, flat plate, forced flow	$j_D = j_H = \dfrac{f}{2} = 0.037 N_{Re,L}^{-0.2}$	Chilton-Colburn analogies, $N_{Sc} = 1.0$, (gases), $f =$ drag coefficient. Corresponds to item 5-21-F and refers to same conditions. $8000 < N_{Re} < 300{,}000$. Can apply analogy, $j_D = f/2$, to entire plate (including laminar portion) if average values are used.
H. Laminar and turbulent, flat plate, forced flow	$N_{Sh,avg} = 0.037 N_{Sc}^{1/3} (N_{Re,L}^{0.8} - 15{,}500)$ to $N_{Re,L} = 320{,}000$ $N_{Sh,avg} = 0.037 N_{Sc}^{1/3}$ $\times \left(N_{Re,L}^{0.8} - N_{Re,Cr}^{0.8} + \dfrac{0.664}{0.037} N_{Re,Cr}^{1/2} \right)$ in range 3×10^5 to 3×10^6.	[E] Use arithmetic concentration difference. $N_{Sh,avg} = \dfrac{k_m' L}{D}$, $N_{Sc} > 0.5$ Entrance effects are ignored. $N_{Re,Cr}$ is transition laminar to turbulent.
I. Turbulent, local flat plate, natural convection, vertical plate	$N_{Sh,x} = \dfrac{k_x' x}{D} = 0.0299 N_{Cr}^{2/5} N_{Sc}^{7/15}$ $\times (1 + 0.494 N_{Sc}^{2/3})^{-2/5}$	[S] Low solute concentration and low transfer rates. Use arithmetic concentration difference. $N_{Cr} = \dfrac{g x^3}{(\mu/\rho)^2} \left(\dfrac{\rho_\infty}{\rho_0} - 1 \right)$ $N_{Cr} > 10^{10}$
Turbulent, average, flat plate, natural convection, vertical plate	$N_{Sh,avg} = 0.0249 N_{Cr}^{2/5} N_{Sc}^{7/15} \times (1 + 0.494 N_{Sc}^{2/3})^{-2/5}$	

J. Turbulent, vertical plate

$$N_{Sh,avg} = \frac{k'_m x}{D} = 0.327 N_{Re,film}^{2/9} N_{Sc}^{1/3} \left(\frac{x^3 \rho^2 g}{\mu^2}\right)^{2/9}$$

$$\delta_{film} = 0.172 \left(\frac{Q^2}{w^2 g}\right)^{1/3}$$

Assumes laminar boundary layer is small fraction of total.

$$N_{Sh,avg} = \frac{k'_m L}{D}$$

$$N_{Re,film} = \frac{4Q\rho}{w\mu^2} > 2360$$

Solute remains laminar sublayer.

K. Turbulent, spinning disk

$$N_{Sh} = \frac{k' d_{disk}}{D} = 5.6 N_{Re}^{1.1} N_{Sc}^{1/3}$$

[E] Use arithmetic concentration difference.
$6 \times 10^5 < N_{Re} < 2 \times 10^6$
$120 < N_{Sc} < 1200$
$u = \omega d_{disk}/2$ where ω = rotational speed, radians/s.
$N_{Re} = \rho \omega d^2/2\mu$.

L. Mass transfer to a flat plate membrane in a stirred vessel

$$N_{Sh} = \frac{k' d_{tank}}{D} = a N_{Re}^b N_{Sc}^c$$

a depends on system. $a = 0.0443$ [73, 165]; b is often 0.65–0.70 [110]. If

$$N_{Re} = \frac{\omega d_{tank}^2 \rho}{\mu}$$

$b = 0.785$ [73]. c is often 0.33 but other values have been reported [110].

[E] Use arithmetic concentration difference. ω = stirrer speed, radians/s. Useful for laboratory dialysis, R.O., U.F., and microfiltration systems.

MASS TRANSFER—TABLE 2. Mass Transfer Correlations for Falling Films with a Free Surface in Wetted Wall Columns—Transfer between Gas and Liquid

Situation	Correlation	Comments E = Empirical, S = Semiempirical, T = Theoretical
A. Laminar, vertical wetted wall column	$N_{Sh,avg} = \dfrac{k'_m x}{D} \approx 3.41 \dfrac{x}{\delta_{film}}$ (first term of infinite series) $\delta_{film} = \left(\dfrac{3\mu Q}{w\rho g}\right)^{1/3}$ = film thickness ω = film width (circumference in column)	[T] Low rates M.T. Use with log mean concentration difference. Parabolic velocity distribution in films. $N_{Re,film} = \dfrac{4Q\rho}{w\mu} < 20$ Derived for flat plates, used for tubes if $r_{tube}\left(\dfrac{\rho g}{2\sigma}\right)^{1/2} > 3.0.$ σ = surface tension If $N_{Re,film} > 20$, surface waves and rates increase. An approximate solution $D_{apparent}$ can be used. Ripples are suppressed with a wetting agent good to $N_{Re} = 1200$.
B. Turbulent, vertical wetted wall column	$N_{Sh,avg} = \dfrac{k'_m d_t}{D} = 0.023 N_{Re}^{0.83} N_{Sc}^{0.44}$ A coefficient 0.0163 has also been reported using $N_{Re'}$, where $v = v$ of gas relative to liquid film.	[E] Use with log mean concentration difference for correlations in B and C. N_{Re} is for gas. N_{Sc} for vapor in gas. $2000 < N_{Re} \leq 35{,}000$, $0.6 \leq N_{Sc} \leq 2.5$. Use for gases, d_t = tube diameter.
C. Turbulent, vertical wetted wall column with ripples	$N_{Sh,avg} = \dfrac{k'_m d_t}{D} = 0.00814 N_{Re}^{0.83} N_{Sc}^{0.44} \left(\dfrac{4Q\rho}{w\mu}\right)^{0.15}$ $N_{Sh,avg} = \dfrac{k'_m d_t}{D} = 0.023 N_{Re}^{0.8} N_{Sc}^{1/3}$	[E] For gas systems with rippling. Fits B for $\left(\dfrac{4Q\rho}{w\mu}\right) = 1000$ $30 \leq \left(\dfrac{4Q\rho}{w\mu}\right) < 1200$ [E] "Rounded" approximation to include ripples. Includes solid-liquid mass-transfer data to find $1/3$ coefficient on N_{Sc}. May use $N_{Re}^{0.83}$. Use for liquids.
D. Rectification in vertical wetted wall column with turbulent vapor flow, Johnstone and Pigford correlation	$N_{Sh,avg} = \dfrac{k'_G d_{col} p_{BM}}{D_v p} = 0.0328(N'_{Re})^{0.77} N_{Sc}^{0.33}$ $3000 < N'_{Re} < 40{,}000$, $0.5 < N_{Sc} < 3$ $N'_{Re} = \dfrac{d_{col} v_{rel} \rho_v}{\mu_v}$, v_{rel} = gas velocity relative to liquid film = $\dfrac{3}{2} u_{avg}$ in film	[E] Use logarithmic mean driving force at two ends of column. Based on four systems with gas-side resistance only. p_{BM} = logarithmic mean partial pressure of nondiffusing species B in binary mixture. p = total pressure Modified form is used for structured packings.

MASS TRANSFER—TABLE 3. Mass Transfer Correlations for Flow in Pipes and Ducts—Transfer Is from Wall to Fluid

Situation	Correlation	Comments E = Empirical, S = Semiempirical, T = Theoretical
A. Tubes, laminar, fully developed parabolic velocity profile, developing concentration profile, constant wall concentration	$N_{Sh} = \dfrac{k'd_t}{D} = 3.66 + \dfrac{0.0668(d_t/x)N_{Re}N_{Sc}}{1+0.04[(d_t/x)N_{Re}N_{Sc}]^{2/3}}$	[T] Use log mean concentration difference. For $$\frac{x/d_t}{N_{Re}N_{Sc}} < 0.10 \cdot N_{Re} < 2100.$$ x = distance from tube entrance. Good agreement with experiment at values $$10^4 > \frac{\pi}{4}\frac{d_t}{x}N_{Re}N_{Sc} > 10$$
B. Tubes, fully developed concentration profile	$N_{Sh} = \dfrac{k'd_t}{D} = 3.66$	$\dfrac{x/d_t}{N_{Re}N_{Sc}} > 0.1$
C. Tubes, approximate solution	$N_{Sh,x} = \dfrac{k'd_t}{D} = 1.077\left(\dfrac{d_t}{x}\right)^{1/3}(N_{Re}N_{Sc})^{1/3}$ $N_{Sh,avg} = \dfrac{k'd_t}{D} = 1.615\left(\dfrac{d_t}{L}\right)^{1/3}(N_{Re}N_{Sc})^{1/3}$	[T] For arithmetic concentration difference. $$\frac{W}{\rho Dx} > 400$$ Leveque's approximation: Concentration BL is thin. Assume velocity profile is linear. High mass velocity. Fits liquid data well.

Situation	Correlation	Comments E = Empirical, S = Semiempirical, T = Theoretical
D. Tubes, laminar, uniform plug velocity, developing concentration profile, constant wall concentration	$$N_{Sh,avg} = \frac{1}{2}\frac{d_t}{L}N_{Re}N_{Sc}\left[\frac{1-4\sum\limits_{j=1}^{\infty}a_j^{-2}\exp\left(\dfrac{-2a_j^{-2}(x/r_t)}{N_{Re}N_{Sc}}\right)}{1+4\sum\limits_{j=1}^{\infty}a_j^{-2}\exp\left(\dfrac{-2a_j^{2}(x/r_t)}{N_{Re}N_{Sc}}\right)}\right]$$ Graetz solution for heat transfer written for M.T.	[T] Use arithmetic concentration difference. Fits gas data well, for $\dfrac{W}{Dpx} < 50$ (fit is fortuitous). $N_{Sh,avg} = (k'_m d_t)/D$. $a_1 = 2.405$, $a_2 = 5.520$, $a_3 = 8.654$, $a_4 = 11.792$, $a_5 = 14.931$.
E. Laminar, fully developed parabolic velocity profile, constant mass flux at wall	$$N_{Sh,x} = \left[\frac{11}{48} - \frac{1}{2}\sum_{j=1}^{\infty}\frac{\exp[-\lambda_j^2(x/r_t)/(N_{Re}N_{Sc})]}{C_j\lambda_j^4}\right]^{-1}$$ j λ_j^2 c_j 1 25.68 7.630×10^{-3} 2 83.86 2.058×10^{-3} 3 174.2 0.901×10^{-3} 4 296.5 0.487×10^{-3} 5 450.9 0.297×10^{-3}	[T] Use log mean concentration difference. $N_{Re} < 2100$ $N_{Sh,x} = \dfrac{k'd_t}{D}$ $N_{Re} = \dfrac{vd_t\rho}{\mu}$
F. Laminar, alternate	$$N_{Sh} = 4.36 + \frac{0.023(d_t/L)N_{Re}N_{Sc}}{1+0.0012(d_t/L)N_{Re}N_{Sc}}$$	[T] $N_{Sh} = \dfrac{k'd_t}{D}$ Use log mean concentration difference. $N_{Re} < 2100$
G. Laminar, fully developed concentration and velocity profile	$$N_{Sh} = \frac{k'd_t}{D} = \frac{48}{11} = 4.3636$$	[T] Use log mean concentration difference. $N_{Re} < 2100$

H. Vertical tubes, laminar flow, forced and natural convection	$N_{Sh,avg} = 1.62 N_{Cz}^{1/3} \left[1 \pm 0.0742 \dfrac{\left(N_{Cr} N_{Sc} d/L\right)^{3/4}}{N_{Cz}} \right]^{-1/3}$	[T] Approximate solution. Use minus sign if forced and natural convection oppose each other. $N_{Cr} = \dfrac{N_{Re} N_{Sc} d}{L}$ $N_{Gr} = \dfrac{g\Delta\rho d^3}{\rho v^2}$ Good agreement with experiment.
I. Tubes, laminar, RO systems	$N_{Sh,avg} = \dfrac{k'_m d_1}{D} = 1.632 \left(\dfrac{u d_1^2}{DL}\right)^{1/3}$	Use arithmetic concentration difference. Thin concentration polarization layer, not fully developed. $N_{Re} < 2000$, L = length tube.
J. Tubes and parallel plates, laminar RO	Graphical solutions for concentration polarization. Uniform velocity through walls.	[T]
K. Parallel plates, laminar, parabolic velocity, developing concentration profile, constant wall concentration	Graphical solution	[T] Low transfer rates.
L. 5-23-K, fully developed	$N_{Sh} = \dfrac{k'(2h)}{D} = 7.6$	[T] h = distance between plates. Use log mean concentration difference. $\dfrac{N_{Re} N_{Sc}}{x/(2h)} < 20$
M. Parallel plates, laminar, parabolic velocity, developing concentration profile, constant mass flux at wall	Graphical solution	[T] Low transfer rates.
N. 5-23-M, fully developed	$N_{Sh} = \dfrac{k'(2h)}{D} = 8.23$	[T] Use log mean concentration difference. $\dfrac{N_{Re} N_{Sc}}{x/(2h)} < 20$

MASS TRANSFER—TABLE 3. Mass Transfer Correlations for Flow in Pipes and Ducts—Transfer Is from Wall to Fluid (*Continued*)

Situation	Correlation	Comments E = Empirical, S = Semiempirical, T = Theoretical
O. Laminar flow, vertical parallel plates, forced and natural convection	$N_{Sh,avg} = 1.47 N_{Cz}^{1/3} \left[1 \pm 0.0989 \dfrac{(N_{Cr} N_{Sc} h/L)^{3/4}}{N_{Cz}} \right]^{-1/3}$	[T] Approximate solution. Use minus sign if forced and natural convection oppose each other. $$N_{Cz} = \frac{N_{Re} N_{Sc} h}{L}$$ $$N_{Cr} = \frac{g\Delta\rho h^3}{\rho v^2}$$ Good agreement with experiment.
P. Parallel plates, laminar, RO systems	$N_{Sh,avg} = \dfrac{k'(2H_p)}{D} = 2.354 \left(\dfrac{uH_p^2}{DL} \right)^{1/3}$	Thin concentration polarization layer. Short tubes, concentration profile not fully developed. Use arithmetic concentration difference.
Q. Tubes, turbulent	$N_{Sh,avg} = \dfrac{k'_m d_t}{D} = 0.023 N_{Re}^{0.83} N_{Sc}^{1/3}$	[E] Use with log mean concentration difference at two ends of tube. $2100 < N_{Re} < 35{,}000$ $0.6 < N_{Sc} < 3000$
R. Tubes, turbulent	$N_{Sh,avg} = \dfrac{k'_m d_t}{D} = 0.023 N_{Re}^{0.83} N_{Sc}^{0.44}$	[E] Evaporation of liquids. Use with log mean concentration difference. See item above. Better fit for gases. $2000 < N_{Re} < 35{,}000$ $0.6 < N_{Sc} < 2.5$.
S. Tubes, turbulent	$N_{Sh} = \dfrac{k' d_t}{D} = 0.0096 N_{Re}^{0.913} N_{Sc}^{0.346}$	[E] $430 < N_{Se} < 100{,}000$. Dissolution data. Use for high N_{Sc}.

T. Tubes, turbulent, smooth tubes, Reynolds analogy	$N_{Sh} = \dfrac{k'd_t}{D} = \left(\dfrac{f}{2}\right)N_{Re}N_{Sc}$ f = Fanning friction faction	[T] Use arithmetic concentration difference. N_{Sc} near 1.0 Turbulent core extends to wall. Of limited utility.
U. Tubes, turbulent, smooth tubes, Chilton-Colburn analogy	$j_D = j_H = \dfrac{f}{2}$ If $\dfrac{f}{2} = 0.023 N_{Re}^{-0.2}$, $j_D = \dfrac{N_{Sh}}{N_{Re}N_{Sc}^{1/3}} = 0.023 N_{Re}^{-0.2}$ $N_{Sh} = \dfrac{k'd_t}{D}$ $j_D = j_H = f(N_{Re}, \text{geometry and B.C.})$	[T] Use log-mean concentration difference. Relating j_D to $f/2$ approximate. N_{Pr} and N_{Sc} near 1.0. Low concentration. Results about 20% lower than experiment. $3 \times 10^4 < N_{Re} < 10^6$ [E] Good over wide ranges.
V. Tubes, turbulent, smooth tubes, constant surface concentration, Prandtl analogy	$N_{Sh} = \dfrac{k'd_t}{D} = \dfrac{(f/2)N_{Re}N_{Sc}}{1+5\sqrt{f/2}(N_{Sc}-1)}$ $\dfrac{f}{2} = 0.04 N_{Re}^{-0.25}$	[T] Use arithmetic concentration difference. Improvement over Reynolds analogy. Best for N_{Sc} near 1.0.
W. Tubes, turbulent, smooth tubes, constant surface concentration, Von Karman analogy	$N_{Sh} = \dfrac{(f/2)N_{Re}N_{Sc}}{1+5\sqrt{f/2}\left\{(N_{Sc}-1)+\ln\left[1+\dfrac{5}{6}(N_{Sc}-1)\right]\right\}}$ $\dfrac{f}{2} = 0.04 N_{Re}^{-0.25}$	[T] Use arithmetic concentration difference. $N_{Sh} = k'd_t/D$. Improvement over Prandtl, $N_{Sc} < 25$.

MASS TRANSFER—TABLE 3. Mass Transfer Correlations for Flow in Pipes and Ducts—Transfer Is from Wall to Fluid (*Continued*)

Situation	Correlation	Comments E = Empirical, S = Semiempirical, T = Theoretical
X. Tubes, turbulent, smooth tubes, constant surface concentration	For $0.5 < N_{Sc} < 10$: $$N_{Sh,avg} = 0.0097\, N_{Re}^{9/10} N_{Sc}^{1/2} \times (1.10 + 0.44 N_{Sc}^{-1/3} - 0.70 N_{Sc}^{1/6})$$ For $10 < N_{Sc} < 1000$: $$N_{Sh,avg} = \frac{0.0097 N_{Re}^{9/10} N_{Sc}^{1/2}(1.10 + 0.44 N_{Sc}^{-1/3} - 0.70 N_{Sc}^{-1/6})}{1 + 0.064 N_{Sc}^{1/2}(1.10 + 0.44 N_{Sc}^{-1/3} - 0.70 N_{Sc}^{-1/6})}$$ For $N_{Sc} > 1000$: $$N_{Sh,avg} = 0.0102 N_{Re}^{9/10} N_{Sc}^{1/3}$$	[S] Use arithmetic concentration difference. Based on partial fluid renewal and an infrequently replenished thin fluid layer for high N_{Sc}. Good fit to available data. $$N_{Re} = \frac{u_{bulk} d_t}{\nu}$$ $$N_{Sh,avg} = \frac{k'_{avg} d_t}{D}$$
Y. Turbulent flow, tubes	$$N_{St} = \frac{N_{Sh}}{N_{Pe}} = \frac{N_{Sh}}{N_{Re} N_{Sc}} = 0.0149 N_{Re}^{-0.12} N_{Sc}^{-2/3}$$	[E] Smooth pipe data. Data fits within 4% except at $N_{Sc} > 20{,}000$, where experimental data is underpredicted. $N_{Sc} > 100, 10^5 > N_{Re} > 2100$
Z. Turbulent flow, noncircular ducts	Use correlations with $$d_{eq} = \frac{4 \text{ cross-sectional area}}{\text{wetted perimeter}}$$ Parallel plates: $$d_{eq} = 4 \frac{2hw}{2w + 2h}$$	Can be suspect for systems with sharp corners.

MASS TRANSFER—TABLE 4. Mass Transfer Correlations for Flow Past Submerged Objects

Situation	Correlations	Comments E = Empirical, S = Semiempirical, T = Theoretical
A. Single sphere	$$N_{Sh} = \frac{k'_G \, p_{BLM} RT d_s}{PD} = \frac{2r}{r - r_s}$$ r/r_s: 2, 5, 10, 50 (asymptotic limit), ∞ N_{Sh}: 4.0, 2.5, 2.22, 2.04, 2.0	[T] Use with log mean concentration difference. r = distance from sphere, r_s, d_s = radius and diameter of sphere No convection.
B. Single sphere, creeping flow with forced convention	$$N_{Sh} = \frac{k'd}{D} = \left[4.0 + 1.21(N_{Re}N_{Sc})^{2/3}\right]^{1/2}$$ $$N_{Sh} = \frac{k'd}{D} = a(N_{Re} - N_{Sc})^{1/3}$$ $a = 1.01, 1.0, \text{ or } 0.991$	[T] Use with log mean concentration difference. Average over sphere. Numerical calculations. $(N_{Re}N_{Sc}) < 10,000 \; N_{Re} < 1.0$. Constant sphere diameter. Low mass-transfer rates. [T] Fit to above ignoring molecular diffusion. $1000 < (N_{Re}N_{Sc}) < 10,000.$
C. Single spheres, molecular diffusion, and forced convection, low flow rates	$$N_{sh} = 2.0 + A N_{Re}^{1/2} N_{Sc}^{1/3}$$ $A = 0.5 \text{ to } 0.62$ $A = 0.60.$ $A = 0.95.$ $A = 0.95.$ $A = 0.544.$	[E] Use with log mean concentration difference. Average over sphere. Frössling Eq. ($A = 0.552$), $2 \le N_{Re} \le 800$, $0.6 \le N_{sc} \le 2.7$. N_{Sh} lower than experimental at high N_{Re}. [E] $2 \le N_{Re} \le 200, 0.6 \le N_{sc} \le 2.5$. [E] Liquids $\le N_{Re} \le 2000$. Graph in Ref. 146, p. 217–218. [E] $100 \le N_{Re} \le 700; 1200 \le N_{Sc} \le 1525$. [E] Use with arithmetic concentration difference. $N_{Sc} = 1; 50 \le N_{Re} \le 350$.

MASS TRANSFER – TABLE 4. Mass Transfer Correlations for Flow Past Submerged Objects (*Continued*)

Situation	Correlations	Comments E = Empirical, S = Semiempirical, T = Theoretical
D.	$$N_{Sh} = \frac{k'd_s}{D} = 2.0 + 0.575 N_{Re}^{1/2} N_{Sc}^{0.35}$$	[E] Use with log mean concentration difference. $N_{Sc} \leq 1$, $N_{Re} < 1$.
E.	$$N_{Sh} = \frac{k'd_s}{D} = 2.0 + 0.552 N_{Re}^{0.53} N_{Sc}^{1/3}$$	[E] Use with log mean concentration difference. $1.0 < N_{Re} \leq 48{,}000$ Cases; $0.6 \leq N_{Sc} \leq 2.7$.
F. Single spheres, forced concentration, any flow rate	$$N_{Sh} = \frac{k'_L d_s}{D} = 2.0 + 0.59 \left[\frac{E^{1/3} d_p^{4/3} \rho}{\mu} \right]^{0.57} N_{Sc}^{1/3}$$ Energy dissipation rate per unit mass of fluid (ranges $570 < N_{Sc} < 1420$); $$E = \left(\frac{C_{Dr}}{2} \right) \left(\frac{v_r^3}{d_p} \right) \frac{\text{m}^2}{\text{s}^3}$$ $$2 < \left(\frac{E^{1/3} d_p^{4/3} \rho}{\mu} \right) < 63{,}000$$	[S] Correlated large amount of data and compares to published data. v_r = relative velocity between fluid and sphere, m/s, C_{Dr} = drag coefficient for single particle fixed in fluid at velocity v_r.
G. Single spheres, forced convection, high flow rates, ignoring molecular diffusion	$$N_{Sh} = \frac{k'd_s}{D} = 0.347 N_{Re}^{0.62} N_{Sc}^{1/3}$$	[E] Use with arithmetic concentration difference. Liquids, $2000 < N_{Re} < 17{,}000$.
	$$N_{Sh} = \frac{k'd_s}{D} = 0.33 N_{Re}^{0.6} N_{Sc}^{1/3}$$	[E] $1500 \leq N_{Re} \leq 12{,}000$.
	$$N_{Sh} = \frac{k'd_s}{D} = 0.43 N_{Re}^{0.56} N_{Sc}^{1/3}$$	[E] $200 \leq N_{Re} \leq 4 \times 10^4$ "air" $\leq N_{Sc} \leq$ "water."
	$$N_{Sh} = \frac{k'd_s}{D} = 0.692 N_{Re}^{0.514} N_{Sc}^{1/3}$$	[E] $500 \leq N_{Re} \leq 5000$.
H. Single cylinders, perpendicular flow	$$N_{Sh} = \frac{k'd_s}{D} = A N_{Re}^{1/2} N_{Sc}^{1/3}, \; A = 0.82$$	[E] $100 < N_{Re} \leq 3500$, $N_{Sc} = 1560$.
	$A = 0.74$.	[E] $120 < N_{Re} \leq 6000$, $N_{Sc} = 2.44$.

	$A = 0.582$ $j_D = 0.600(N_{Re})^{-0.487}$ $N_{Sh} = \dfrac{k' d_{cyl}}{D}$	[E] $300 \le N_{Re} \le 7600$, $N_{Sc} = 1200$. [E] Use with arithmetic concentration difference. $50 \le N_{Re} \le 50{,}000$; gases, $0.6 \le N_{Sc} \le 2.6$; liquids; $1000 \le N_{Sc} \le 3000$. Data scatter ±30%.
I.	Can use $j_D = j_H$. Graphical correlation.	[E] Used with linear concentration difference.
J. Rotating cylinder in an infinite liquid, no forced flow	$j'_D = \dfrac{k'}{v} N_{Sc}^{0.644} = 0.0791 N_{Re}^{-0.30}$ Results presented graphically to $N_{Re} = 241{,}000$. $N_{Re} = \dfrac{v d_{cyl}\mu}{\rho}$ where $v = \dfrac{\omega d_{cyl}}{2}$ = peripheral velocity	$N_{Sh} = \dfrac{k' d_s}{D}$ [E] Used with arithmetic concentration difference. $112 < N_{Re} \le 100{,}000$. $835 < N_{Sc} < 11{,}490$ k' = mass-transfer coefficient, cm/s; ω = rotational speed, radian/s. Useful geometry in electrochemical studies.
K. Oblate spheroid, forced convection	$j_D = \dfrac{N_{Sh}}{N_{Re} N_{Sc}^{1/3}} = 0.74 N_{Re}^{-0.5}$ $N_{Re} = \dfrac{d_{ch} v \rho}{\mu}, d_{ch} = \dfrac{\text{total surface area}}{\text{perimeter normal to flow}}$ e.g., for cube with side length a, $d_{ch} = 1.27a$. $N_{Sh} = \dfrac{k' d_{ch}}{D}$	[E] Used with arithmetic concentration difference. $120 \le N_{Re} \le 6000$; standard deviation 2.1%. Eccentricities between 1:1 (spheres) and 3:1. Shape is often approximated by drops.
L. Other objects, including prisms, cubes, hemispheres, spheres, and cylinders; forced convection	$j_D = 0.692 N_{Re,p}^{-0.486}, N_{Re,p} = \dfrac{v d_{ch}\rho}{\mu}$	[E] Used with arithmetic concentration difference. $500 \le N_{Re,ji} \le 5000$. Turbulent. Agrees with cylinder and oblate spheroid results. ±15%. Assumes molecular diffusion and natural convection are negligible.

MASS TRANSFER—TABLE 4. Mass Transfer Correlations for Flow Past Submerged Objects (*Continued*)

Situation	Correlations	Comments E = Empirical, S = Semiempirical, T = Theoretical
M. Other objects, molecular diffusion limits	$N_{Sh} = \dfrac{k' d_{ch}}{D} = A$ Spheres and cubes $A = 2$, tetrahedrons $A = 2\sqrt{6}$ octahedrons $2\sqrt{2}$.	[T] Use with arithmetic concentration difference. Hard to reach limits in experiments.
N. Shell side of microporous hollow fiber module for solvent extraction	$N_{Sh} = \beta[d_h(1 - \varphi)/L]N_{Re}^{0.6}N_{Sc}^{0.33}$ $N_{Sh} = \dfrac{\overline{K} d_h}{D}$ $\beta = 5.8$ for hydrophobic membrane. $\beta = 6.1$ for hydrophilic membrane.	[E] Used with logarithmic mean concentration difference. $N_{Re} = \dfrac{d_h v \rho}{\mu}$, $\overline{K}$ = overall mass-transfer coefficient d_h = hydraulic diameter $= \dfrac{4 \times \text{cross-sectional area of flow}}{\text{wetted perimeter}}$ φ = pickling fraction of shell side. L = module length. Based on area of contact according to inside or outside diameter of tubes depending on location of interface between aqueous and organic phases. Can also be applied to gas–liquid systems with liquid on shell side.

MASS TRANSFER—TABLE 5. Mass Transfer Correlations for Drops and Bubbles

Conditions	Correlations	Comments E = Empirical, S = Semiempirical, T = Theoretical
A. Single liquid drop in immiscible liquid, drop formation, discontinuous (drop) phase coefficient	$$\hat{k}_{df} = A\left(\frac{\rho_d}{M_d}\right)_{av}\left(\frac{D_d}{\pi t_f}\right)^{1/2}$$ $$A = \frac{24}{7} \text{ (penetration theory)}$$ $$A = 1.31 \text{ (semiempirical value)}$$ $$A = \left[\frac{24}{7}(0.8624)\right] \text{ (extension by fresh surface elements)}$$	[T,S] Use arithmetic mole fraction difference. Fits some, but not all, data. Low mass transfer rate. M_d = mean molecular weight of dispersed phase; t_f = formation time of drop. k_{Ld} = mean dispersed liquid phase M.T. coefficient kmole/[s·m² (mole fraction)].
B.	$$\hat{k}_{df} = 0.0432 \times \frac{d_p}{t_f}\left(\frac{\rho_d}{M_d}\right)_{av}\left(\frac{u_o}{d_p g}\right)^{0.089}\left(\frac{d_p^2}{t_f D_d}\right)^{-0.334}\left(\frac{\mu_d}{\sqrt{\rho_d d_p \sigma g_c}}\right)^{-0.601}$$	[E] Use arithmetic mole fraction difference. Based on 23 data points for 3 systems. Average absolute deviation 26%. Use with surface area of drop after detachment occurs. u_o = velocity through nozzle; σ = interfacial tension.
C. Single liquid drop in immiscible liquid, drop formation, continuous phase coefficient	$$\hat{k}_{df} = 4.6\left(\frac{\rho_c}{M_c}\right)_{av}\sqrt{\frac{D_c}{\pi t_f}}$$	[T] Use arithmetic mole fraction difference. Based on rate of bubble growth away from fixed orifice. Approximately three times too high compared to experiments.
D.	$$k_{L,c} = 0.386 \times \left(\frac{\rho_c}{M_c}\right)_{av}\left(\frac{D_c}{t_f}\right)^{0.5}\left(\frac{\rho_c \sigma g_c}{\Delta\rho g t_f \mu_c}\right)^{0.407}\left(\frac{g t_f^2}{d_p}\right)^{0.148}$$	[E] Average absolute deviation 11% for 20 data points for 3 systems.
E. Single liquid drop in immiscible liquid, free rise or fall, discontinuous phase coefficient, stagnant drops	$$k_{L,d,m} = \frac{-d_p}{6t} \times \left(\frac{\rho_d}{M_d}\right)_{av}\ln\left\{\frac{6}{\pi^2}\sum_{j=1}^{\infty}\frac{1}{j^2}\exp\left[\left(\frac{-D_d j^2 \pi^2 t}{(d_p/2)^2}\right)\right]\right\}$$	[T] Use with log mean mole fraction differences based on ends of column t = rise time. No continuous phase resistance. Stagnant drops are likely if drop is very viscous, quite small, or is coated with surface active agent. $k_{L,d,m}$ = mean dispersed liquid M.T. coefficient.

MASS TRANSFER—TABLE 5. Mass Transfer Correlations for Drops and Bubbles (Continued)

Conditions	Correlations	Comments E = Empirical, S = Semiempirical, T = Theoretical
F.	$\hat{k}_{L,d,m} = \dfrac{-d_p}{6t} \times \left(\dfrac{\rho_d}{M_d}\right)_{av} \ln\left[1 - \dfrac{\pi D_d^{1/2} t^{1/2}}{d_p/2}\right]$	[S] Approximation for fractional extractions less than 50%.
G. Continuous phase coefficient, stagnant drops spherical	$N_{Sh} = \dfrac{k_{l,c,m}d_c}{D_c} = 0.74\left(\dfrac{\rho_c}{M_c}\right)_{av} N_{Re}^{1/2}(N_{Sc})^{1/3}$	[E] $N_{Re} = \dfrac{v_s d_p \rho_c}{\mu_c}$. v_s = slip velocity between drop and continuous phase.
H. Oblate spheroid	$N_{Sh} = \dfrac{k_{l,c,m}d_3}{D_c} = 0.74\left(\dfrac{\rho_c}{M_c}\right)_{av}(N_{Re3})^{1/2}(N_{Sc,c})^{1/3}$	[E] Used with log mean mole fraction. Differences based on ends of extraction column; 100 measured values ±2% deviation. Based on area oblate spheroid. $N_{Re3} = \dfrac{v_s d_3 \rho_c}{\mu_c}$ v_s = slip velocity, $d_3 = \dfrac{\text{total drop surface area}}{\text{perimeter normal to flow}}$
I. Single liquid drop in immiscible liquid, Free rise or fall, discontinuous phase coefficient, circulating drops	$k_{dr,circ} = -\dfrac{d_p}{6\theta}\ln\left[\dfrac{3}{8}\sum_{j=1}^{\infty}B_j^2\exp\left(-\dfrac{\lambda_j 64 D_d\theta}{d_p^2}\right)\right]$	[T] Use with arithmetic concentration difference. θ = drop residence time $k'_{L,d,circ}$ is m/s.

Eigenvalues for Circulating Drop

$k_d d_p/D_d$	λ_1	λ_2	λ_3	B_1	B_2	B_3
3.20	0.262	0.424		1.49	0.107	
10.7	0.680	4.92		1.49	0.300	
26.7	1.082	5.90	15.7	1.49	0.495	0.205
107	1.484	7.88	19.5	1.39	0.603	0.384
320	1.60	8.62	21.3	1.31	0.583	0.391
∞	1.656	9.08	22.2	1.29	0.596	0.386

J.

$$\hat{k}_{L,d,circ} = -\frac{d_p}{6\theta}\left(\frac{\rho_d}{M_d}\right)_{av}\ln\left[1-\left(\frac{R^{1/2}\pi D_d^{1/2}\theta^{1/2}}{d_p/2}\right)\right]$$

[E] Used with mole fractions for extraction less than 50%, $R \approx 2.25$.

K.

$$N_{Sh} = \frac{\hat{k}_{L,d,circ}d_p}{D_d} = 31.4\left(\frac{\rho_d}{M_f}\right)_{av}\left(\frac{4D_d t}{d_p^2}\right)^{-0.34}N_{Sc,d}^{-0.125}\left(\frac{d_p v_s^2 \rho_c}{\sigma g_c}\right)^{-0.37}$$

[E] Used with log mean mole fraction difference. d_v = diameter of sphere with same volume as drop. $856 \leq N_{Sc} \leq 79{,}800$, $2.34 \leq \sigma \leq 4.8$ dynes/cm.

L. Liquid drop in immiscible liquid, free rise or fall, continuous phase coefficient, circulating single drops

$$N_{Sh,c} = \frac{k'_{L,c}d_p}{D_d} = \left[2 + 0.463N_{Re,drop}^{0.484}N_{Sc,c}^{0.339}\left(\frac{d_p g^{1/3}}{D_c^{2/3}}\right)^{0.072}\right]F$$

$$F = 0.281 + 1.615K + 3.73K^2 - 1.874K$$

$$K = N_{Re,drop}^{1/8}\left(\frac{\mu_c}{\mu_d}\right)^{1/4}\left(\frac{\mu_c v_s}{\sigma g_c}\right)^{1/6}$$

[E] Used as an arithmetic concentration difference.

$$N_{Re,drop} = \frac{d_p v_s \rho_c}{\mu_c}$$

Solid sphere form with correction factor F.

M. Circulating, single drop

$$N_{Sh} = \frac{k_{L,c}d_p}{D_c} = 0.6\left(\frac{\rho_c}{M_c}\right)_{av}N_{Re,drop}^{1/2}N_{Sc,c}^{1/2}$$

[E] Used as an arithmetic concentration difference. Low σ.

N. Circulating swarm of drops

$$k_{L,c} = 0.725\left(\frac{\rho_c}{M_c}\right)_{av}N_{Re,drop}^{-0.43}N_{Sc,c}^{-0.58}v_2(1-\phi_d)$$

[E] Used as an arithmetic concentration difference. Low σ, disperse-phase holdup of drop swarm. ϕ_d = volume fraction dispersed phase.

O. Liquid drops in immiscible liquid, free rise or fall, discontinuous phase coefficient, oscillating drops

$$N_{Sh} = \frac{k_{L,d,osc}d_p}{D_d} = 0.32\left(\frac{\rho_d}{M_d}\right)_{av}\left(\frac{4D_d t}{d_p^2}\right)^{-0.14}N_{Re,drop}^{0.68}N_{Sc,c}^{0.10}\left(\frac{\sigma^3 g_c^3 \rho_c^2}{g\mu_c^4 \Delta\rho}\right)^{0.10}$$

[E] Used with a log mean mole fraction difference. Based on ends of extraction column.

$$N_{Re,drop} = \frac{d_p v_s \rho_c}{\mu_c}$$

d_p = diameter of sphere with volume of drop. Average absolute deviation from data, 10.5%. $411 \leq N_{Re} \leq 3114$. Low interfacial tension (3.5–5.8 dynes), $\mu_c < 1.35$ centipoises.

MASS TRANSFER—TABLE 5. Mass Transfer Correlations for Drops and Bubbles (*Continued*)

Conditions	Correlations	Comments E = Empirical, S = Semiempirical, T = Theoretical
P.	$$k_{L,d,osc} = \frac{0.00375 v_s}{1 + \mu_d / \mu_c}$$	[T] Use with log mean concentration difference. Based on end of extraction column. No continuous phase resistance. $k_{L,d,osc}$ in cm/s, v_s = drop velocity relative to continuous phase.
Q. Single liquid drop in immiscible liquid, range rigid to fully circulating	Rigid drops: $10^4 < N_{Pe,c} < 10^6$ $$N_{Sh,c,rigid} = \frac{k_c d_p}{D_c} = 2.43 + 0.774 N_{Re}^{0.5} N_{Sc}^{0.33} + 0.0103 N_{Re} N_{Sc}^{0.33}$$ Circulating drops: $10 < N_{Re} < 1200, 190 < N_{Sc} < 241,000,$ $10^3 < N_{Pe,s} < 10^6$ $$N_{Sh,c,fully,\,circular} = \left[\frac{2}{\pi^{0.5}}\right] N_{Pe,c}^{0.33}$$ Drops in intermediate range: $$\frac{N_{Sh,c} - N_{Sh,c,rigid}}{N_{Sh,c,fully\,circular} - N_{Sh,c,rigid}} = 1 - \exp\left[-(4.18 \times 10^{-3}) N_{Pe,c}^{0.42}\right]$$	[E] Allows for slight effect of wake.
R. Coalescing drops in immiscible liquid, discontinuous phase coefficient	$$\hat{k}_{d,coal} = 0.173 \frac{d_p}{t_f} \left(\frac{\rho_d}{M_d}\right)_{av} \left(\frac{\mu_d}{\rho_d D_d}\right)^{-1.115} \times \left(\frac{\Delta\rho g d_p^2}{\sigma g_c}\right)^{1.302} \left(\frac{v_s^2 t_f}{D_d}\right)^{0.146}$$	[E] Used with log mean mole fraction difference. 23 data points. Average absolute deviation 25%. t_f = formation time.

254

Situation	Equation	Comments
S. Continuous phase coefficient	$$\hat{k}_{c,coal} = 5.959 \times 10^{-4}\left(\frac{\rho}{M}\right)_{av} \times \left(\frac{D_c}{t_f}\right)^{0.5}\left(\frac{\rho_d \mu_s^3}{g\mu_c}\right)^{0.332}\left(\frac{d_p^2 \rho_c \rho_d v_s^3}{\mu_d \sigma g_c}\right)^{0.525}$$	[E] Used with log mean mole fraction difference. 20 data points. Average absolute deviation 22%.
T. Single liquid drops in gas, gas side coefficient	$$\frac{\hat{k}_g M_g d_p P}{D_{gas}\rho_g} = 2 + A N_{Re,g}^{1/2} N_{Sc\text{-}g}^{1/3}$$ $A = 0.552$ or 0.60.	[E] Used for spray drying (arithmetic partial pressure difference). $$N_{Re,g} = \frac{d_p \rho_g v_s}{\mu_g},\; v_s = \text{slip velocity between drop}$$ and gas stream. Sometimes written with: $$\frac{M_g P}{\rho_g} = RT$$
U. Single water drop in air, liquid side coefficient	$$k_L = 2\left(\frac{D_L}{\pi t}\right)^{1/2}, \text{short contact times}$$ $$k_L = 10\frac{D_L}{d_p}, \text{long contact times}$$	[T] Use arithmetic concentration difference. Penetration theory. t = contact time of drop. Gives plot for $k_G a$ also. Air-water system.
V. Single bubbles of gas in liquid, continuous phase coefficient, very small bubbles	$$N_{Sh} = \frac{k_c' d_b}{D_c} = 1.0\left(N_{Re}N_{Sc}\right)^{1/3}$$	[T] Solid-sphere. $d_b < 0.1$ cm, k_c' is average over entire surface of bubble.
W. Medium to large bubbles	$$N_{sh} = \frac{k_c' d_b}{D_c} = 1.13\left(N_{Re}N_{Sc}\right)^{1/2}$$	[T] Use arithmetic concentration difference. Droplet equation: $d_b > 0.5$ cm.

MASS TRANSFER—TABLE 5. Mass Transfer Correlations for Drops and Bubbles (*Continued*)

Conditions	Correlations	Comments E = Empirical, S = Semiempirical, T = Theoretical
X.	$$N_{Sh} = \dfrac{k'_c d_b}{D_c} = 1.13(N_{Re}N_{Sc})^{1/2}\left[\dfrac{d_b}{0.45+0.2d_b}\right]$$	[S] Use arithmetic concentration difference. Modification of above (W), $d_b > 0.5$ cm. $500 \le N_{Re} \le 8000$. No effect SAA for $d_p > 0.6$ cm.
Y. Rising small bubbles of gas in liquid, continuous phase	$$N_{Sh} = \dfrac{k'_c d_b}{D_c} = 2 + 0.31(N_{Gr})^{1/3}\,N_{Sc}^{1/3},\ d_b < 0.25\ \text{cm}$$	[E] Use with arithmetic concentration difference. $$N_{Ra} = \dfrac{d_b^3|\rho_G - \rho_L|g}{\mu_L D_L} = \text{Raleigh number}$$ Note that $N_{Re} = N_{Gr}N_{Sc}$. Valid for single bubbles or swarms. Independent of agitation as long as bubble size is constant.
Z. Large bubbles	$$N_{Sh} = \dfrac{k'_c d_b}{D_c} = 0.42(N_{Gr})^{1/3}\,N_{Sc}^{1/2},\ d_b < 0.25\ \text{cm}$$ $$\dfrac{\text{Interfacial area}}{\text{volume}} = a = \dfrac{6H_g}{d_b}$$	[E] Use with arithmetic concentration difference. H_g = fractional gas holdup, volume gas/total volume. For large bubbles, k'_c is independent of bubble size and independent of agitation or liquid velocity. Resistance is entirely in liquid phase for most gas–liquid mass transfer.

MASS TRANSFER—TABLE 6. Mass Transfer Correlations for Particles, Drops, and Bubbles in Agitated Systems

Situation	Correlation	Comments E = Empirical, S = Semiempirical, T = Theoretical
A. Solid particles suspended in agitated vessel containing vertical baffles, continuous phase coefficient	$$\frac{K'_{LT}d_p}{D} = 2 + 0.6 N_{Re,T}^{1/2} N_{Sc}^{1/3}$$ Replace v_{slip} with v_T = terminal velocity. Calculate Stokes' law terminal velocity $$v_{Ts} = \frac{d_p^2 |\rho_p - \rho_c| g}{18\mu_c}$$ and correct: | $N_{Re,Ts}$	1	10	100	1,000	10,000	100,000
---	---	---	---	---	---	---
v_T/v_n	0.9	0.65	0.37	0.17	0.07	0.023 | Approximate: $K_L = 2k_{LT}$	[S] Use log mean concentration difference. Modified Frossling equation: $$N_{Re,Ts} = \frac{v_{Ts}d_p\rho_v}{\mu_c}$$ (Reynolds number based on Stokes' law.) $$N_{Re,T} = \frac{v_T d_p \rho_c}{\mu_c}$$ (terminal velocity Reynolds number) k'_L almost independent of d_p. Harriott suggests different correction procedures. Range k_L/k'_{LT} is 1.5 to 8.0.
B.	Graphical comparisons experiments and correlations.	[E,S] For spheres. Includes transpiration effects and changing diameters.
C. Solid, neutrally buoyant particles, continuous phase coefficient	$$N_{Sh} = \frac{k'_L d_p}{D} = 2 + 0.47 N_{Re,p}^{0.62} N_{Sc}^{0.36}\left(\frac{d_{imp}}{d_{tank}}\right)^{0.17}$$ Graphical comparisons are in Ref. 109, p. 116	[E] Use log mean concentration difference. Density unimportant if particles are close to neutrally buoyant. [E] E = energy dissipation rate per unit mass fluid $$= \frac{pg_c}{V_{tank}\rho_c}, \; p = \text{power}$$ $$N_{Re,p} = \frac{E^{1/3}d_p^{4/3}}{\nu}$$ Also used for drops. Geometric effect (d_{imp}/d_{tank}) is usually unimportant.

MASS TRANSFER—TABLE 6. Mass Transfer Correlations for Particles, Drops, and Bubbles in Agitated Systems (Continued)

Situation	Correlation	Comments E = Empirical, S = Semiempirical, T = Theoretical
D. Small particles	$N_{Sh} = 2 + 0.52 N_{Re,p}^{0.52} N_{Sc}^{1/3}, N_{Re,p} < 1.0$	[E] Terms same as above.
E. Solid particles with significant density difference	$N_{Sh} = \dfrac{k'_L d_p}{D} = 2 + 0.44\left(\dfrac{d_p v_{slip}}{\nu}\right)^{1/2} N_{Sc}^{0.38}$	[E] Use log mean concentration difference. N_{Sh} standard deviation 11.1% v_{slip} calculated by methods given in reference.
F. Small solid particles, gas bubbles or liquid drops, $d_p < 2.5$ mm	$N_{Sh} = \dfrac{k'_L d_p}{D} = 2 + 0.31\left[\dfrac{d_p^3 |\rho_p - \rho_\ell|}{\mu_c D}\right]^{1/3}$	[E] Use log mean concentration difference. $g = 9.80665$ m/s². Second term RHS is free-fall or rise term.
G. Highly agitated systems; solid particles, drops, and bubbles; continuous phase coefficient	$k'_L N_{Sc}^{2/3} = 0.13\left[\dfrac{(P/V_{tank})\mu_c g_c}{\rho_c^2}\right]^{1/4}$	[E] Use arithmetic concentration difference. Use when gravitational forces overcome by agitation. Up to 60% deviation. Correlation prediction is low (P/V_{tank}) = power dissipated by agitator per unit volume liquid.
H. Liquid drops in baffled tank with flat six-blade turbine	$k'_c a = 2.621 \times 10^{-3} \dfrac{(ND)^{1/2}}{d_{imp}}$ $\times \phi^{0.304}\left(\dfrac{d_{imp}}{d_{tank}}\right)^{1.552} N_{Re}^{1.929} N_{Oh}^{1.025}$	[E] Use arithmetic concentration difference. Studied for five systems. $N_{Re} = d_{imp}^2 N\rho_\ell/\mu_c, N_{Oh} = \mu_c/(\rho_c d_{imp}\sigma)^{1/2}$ ϕ = volume fraction dispersed phase. N = impeller speed (revolutions/time). For $d_{tank} = h_{tank}$, average absolute deviation 23.8%.
I. Liquid drops in baffled tank, low volume fraction dispersed phase	$N_{Sh} = \dfrac{k'_c d_p}{D} = 1.237 \times 10^{-3} N_{Sc}^{1/3} N_{Fr}^{2/3}$ $\times N_{Fr}^{5/12}\left(\dfrac{d_{imp}}{d_p}\right)\left(\dfrac{d_p}{D_{tank}}\right)^{1/2}\left(\dfrac{\rho_d d_p^2}{\sigma}\right)^{3/4} \phi^{-1/2}$	[E] 180 runs, 9 systems, $\phi = 0.01$. k_c is time-averaged. Use arithmetic concentration difference. $N_{Re} = \left(\dfrac{d_{imp}^2 N_{Se}}{\mu_c}\right), N_{Fr} = \left(\dfrac{d_{imp}N^2}{g}\right)$

Stainless steel flat six-blade turbine. Thank had four baffles.
Correlation recommended for $\phi \le 0.06$ [Ref. 156] $a = 6\phi/d_{32}$, where d_{32} is Sauter mean diameter when 33% mass transfer has occurred.

d_p = particle or drop diameter; σ = interfacial tension.
N/m; ϕ = volume fraction dispersed phase; a = interfacial volume, l/m; and $k_c \alpha D_c^{2/3}$ implies rigid drops.
Negligible drop coalescence.
Average absolute deviation—19.71%.

J. Gas bubble swarms in sparged tank reactors

$$k'_L a \left(\frac{v}{g^2}\right)^{1/3} = C\left[\frac{P/V_L}{\rho(vg^4)^{1/3}}\right]^{7a}\left[\frac{qc}{V_L}\left(\frac{v}{g^2}\right)^{1/3}\right]^{b}$$

Rushton turbines; $C = 7.94 \times 10^{-4}$, $a = 0.62$, $b = 0.23$.
Intermig impellers: $C = 5.89 \times 10^{-4}$, $a = 0.62$, $b = 0.19$.

[E] Use arithmetic concentration difference. Done for biological system, O_2 transfer. $h_{tank}/D_{tank} = 2.1$; P = power, kW. V_L = liquid volume, m^3, q_G = gassing rate, m^3/s, $k'_L a$ = s^{-1}. Since a = m^2/m^3, v = kinematic viscosity, m^2/s. Low viscosity system. Better fit claimed with q_G/V_L than with u_G.

K.

$$k'_L a = 2.6 \times 10^{-2}\left(\frac{P}{V_L}\right)^{0.4} u_G^{0.5}$$

[E] Use arithmetic concentration difference. Ion free water $V_L < 2.6$, u_G = superficial gas velocity in m/s. $500 < P/V_L < 10,000$. P/V_L = watts/m^3, V_L = liquid volume, m^3.

L.

$$k'_L a = 2.0 \times 10^{-3}\left(\frac{P}{V_L}\right)^{0.7} u_G^{0.2}$$

[E] Use arithmetic concentration difference. Water with ions. $0.002 < V_L < 4.4$, $500 < P/V_L < 10,000$.

M. Baffled tank with standard blade Rushton impeller

$$k'_L a = 93.37\left(\frac{P}{V_L}\right)^{0.76} u_G^{0.45}$$

[E] Air-water.
$0.005 < u_G < 0.025$, $3.83 < N < 8.33$, $400 < P/V_L < 7000$
$h = D_{tank} = 0.305$ or 0.610m. V_G = gas volume, m^3, N = stirrer speed, rpm. Method assumes perfect liquid mixing.

MASS TRANSFER—TABLE 6. Mass Transfer Correlations for Particles, Drops, and Bubbles in Agitated Systems (*Continued*)

Situation	Correlation	Comments E = Empirical, S = Semiempirical, T = Theoretical
N.	$k'_L a \dfrac{d_{imp}^2}{D} = 7.57 \left[\dfrac{\mu_{eff}}{\rho D}\right]^{0.5} \left[\dfrac{\mu_G}{\mu_{eff}}\right]^{-0.694}$ $\times \left[\dfrac{d_{imp}^2 N \rho_L}{\mu_{eff}}\right]^{1.11} \left(\dfrac{u_G d}{\sigma}\right)^{0.447}$ d_{imp} = impeller diameter, m; D = diffusivity, m²/s	[E] Use arithmetic concentration difference. CO_2 into aqueous carboxyl polyethylene. μ_{eff} = effective viscosity from power law model, Pa·s, σ = surface tension liquid, N/m.
O. Bubbles	$\dfrac{k'_L a d_{imp}^2}{D} = 0.060 \left(\dfrac{d_{imp}^2 N_P}{\mu_{eff}}\right) \left(\dfrac{d_{imp}^2 N^2}{g}\right)^{0.19} \left(\dfrac{\mu_{eff} u_G}{\sigma}\right)^{0.6}$	[E] Use arithmetic concentration difference. O_2 into aqueous glycerol solutions. O_2 into aqueous millet jelly solutions.
P. Gas bubble swarm in sparged stirred tank reactor with solids present	$\dfrac{k'_L a}{(k'_L a)_0} = 1 - 3.54(\varepsilon_s - 0.03)$ $300 \le P/V_{rs} < 10{,}000\,\text{W/m}^2, \; 0.03 \le \varepsilon_s \le 0.12$ $0.34 \le u_C \le 4.2\,\text{cm/s}, \; 5 < \mu_L < 75\,\text{Pa·s}$	[E] Use arithmetic concentration difference. Solids are glass beads, $d_v = 320\,\mu m$. ε_s = solids holdup m³/m³ liquid, $(k_L a)_0$ = mass transfer in absence of solids. Ionic salt solution—noncoalescing.
Q.	$\dfrac{k'_L a}{(k'_L a)_0} = 1 - \varepsilon_s$	[E] Use arithmetic concentration difference. Variety of solids, $d_p > 150\,\mu m$ (glass, amberlite, polypropylene). Tap water. Slope very different than item *P*. Coalescence may have occurred.

MASS TRANSFER—TABLE 7. Mass Transfer Correlations for Fixed and Fluidized Beds

Transfer is to or from particles

Situation	Correlation	Comments E = Empirical, S, = Semiempirical, T = Theoretical
A. Heat or mass transfer in packed bed for gases and liquids	$j_D = j_H = 0.91 \Psi N_{Re}^{-0.51}$, $0.01 < N_{Re} < 50$ Equivalent $N_{Sh} = 0.91 \Psi N_{Re}^{0.49} N_{Sc}^{1/3}$ $j_D = j_H = 0.61 \Psi N_{Re}^{-0.41}$, $50 < N_{Re} < 1000$ Equivalent $N_{Sh} = 0.61 \Psi N_{Re}^{0.59} N_{Sc}^{1/3}$	[E] Different constants and shape factors reported in other references. Evaluate terms at film temperature or composition. $N_{Sh} = \dfrac{k'd_s}{D}$, $j_D = \dfrac{N_{Sh}}{N_{Re} N_{Sc}^{1/3}}$ $N_{Re} = \dfrac{v_{super}\rho}{\mu \Psi a}$, $v_{super} = $ superficial velocity $a = \dfrac{\text{surface area}}{\text{volume}} = 6(1-\varepsilon)/d_p$
(shape factor, Ψ)	<table><tr><td>particle</td><td>sphere</td><td>cylinder</td></tr><tr><td>1.00</td><td>0.91</td><td>0.81</td></tr></table>	For spheres, $d_p = $ diameter. For nonspherical: $d_p = 0.567 \sqrt{\text{Part. Surf. Area}}$ Results are from too-short beds—use with caution.
B. For gases, fixed and fluidized beds, Cupta and Thodos correlation	$j_H = j_D = \dfrac{2.06}{\varepsilon N_{Re}^{0.573}}$, $90 \le N_{Re} \le A$ Equivalent: $N_{Sh} = j_D = \dfrac{2.06}{\varepsilon} N_{Re}^{0.425} N_{Sc}^{1/3}$ For other shapes: $\dfrac{\varepsilon j_D}{(\varepsilon j_D)_{sphere}} = 0.79 \text{(cylinder) or } 0.71 \text{(cude)}$ Graphical results are available for N_{Re} from 1900 to 10,300.	[E] For spheres. $N_{Re} = \dfrac{v_{super} d_p \rho}{\mu}$ $A = 2453$ [Ref. 151], $A = 4000$ [Ref. 100]. For $N_{Re} > 1900$, $j_H = 1.05 j_D$. Heat transfer result is in absence of radiation. $N_{Sh} = \dfrac{k'd_s}{D}$

261

MASS TRANSFER—TABLE 7. Mass Transfer Correlations for Fixed and Fluidized Beds (*Continued*)
Transfer is to or from particles

Situation	Correlation	Comments E = Empirical, S, = Semiempirical, T = Theoretical
C. For gases, for fixed beds. Petrovic and Thodos correlation	$N_{Sh} = \dfrac{0.357}{\varepsilon} N_{Re}^{0.641} N_{Sc}^{1/3}$	[E] Packed spheres, deep beds, $3 < N_{Re} < 900$ can be extrapolated to $N_{Re} < 2000$. Corrected for axial dispersion with axial Peclet number = 2.0. Prediction is low at low N_{Re}, N_{Re}.
D. For gases and liquids, fixed and fluidized beds	$j_D = \dfrac{0.4548}{\varepsilon N_{Re}^{-0.4069}},\ 10 \le N_{Re} \le 2000$ $j_D = \dfrac{N_{Sh}}{N_{Re} N_{Sc}^{1/3}},\ N_{Sh} = \dfrac{k' d_s}{D}$	[E] Packed spheres, deep bed. Average deviation ± 20%, $N_{Re} = d_p v_{super}\rho/\mu$. Can use for fluidized beds. $10 \le N_{Re} \le 4000$.
E. For gases, fixed beds	$j_D = \dfrac{0.499}{\varepsilon N_{Re}^{0.382}}$	[E] Data on sublimination of naphthalene spheres dispersed in inert beads. $0.1 < N_{Re} < 100$, $N_{Sc} = 2.57$. Correlation coefficient = 0.978.
F. For liquids, fixed bed, Wilson and Geankoplis correlation	$j_D = \dfrac{1.09}{\varepsilon N_{Re}^{2/3}},\ 0.0016 < N_{Re} < 55$ $165 \le N_{Sc} \le 70,6000,\ 0.35 < \varepsilon < 0.75$ Equivalent: $N_{Sh} = \dfrac{1.09}{\varepsilon} N_{Re}^{1/3} N_{Sc}^{1/3}$ $j_D = \dfrac{0.25}{\varepsilon N_{Re}^{0.31}},\ 55 < N_{Re} < 1500,\ 165 \le N_{Sc} \le 10,690$ Equivalent: $N_{Sh} = \dfrac{0.25}{\varepsilon} N_{Re}^{0.89} N_{Sc}^{1/3}$	[E] Beds of spheres, $N_{Re} = \dfrac{d_p V_{super}\rho}{\mu}$ Deep beds. $N_{Sh} = \dfrac{k' d_s}{D}$

G. For liquids, fixed beds, Ohashi et al. correlation

$$N_{Sh} = \frac{k'd}{D} = 2 + 0.51\left(\frac{E^{1/3}d_p^{4/3}\rho}{\mu}\right)^{0.60} N_{Sc}^{1/3}$$

E = Energy dissipation rate per unit mass of fluid

$$= 50(1-\varepsilon)\varepsilon^2 C_{Do}\left(\frac{v_r^3}{d_p}\right), \text{m}^2/\text{s}^3$$

$$= \left[\frac{50(1-\varepsilon)C_D}{\varepsilon}\right]\left(\frac{v_{super}^3}{d_p}\right)$$

General form:

$$N_{Sh} = 2 + K\left(\frac{E^{1/3}D_p^{4/3}\rho}{\mu}\right)^a N_{Sc}^\beta$$

applies to single particles, packed beds, two-phase tube flow, suspended bubble columns, and stirred tanks with different definitions of E.

[S] Correlates large amount of published data. Compares numbers of correlations, v_r = relative velocity, m/s. In packed bed, $v_r = v_{super}/\varepsilon$.

C_{Dc} = single particle drag coefficient at v_{super} calculated from $C_{Dc} = AN_{Re_f}^{-m}$

N_{Re}	A	m
0 to 5.8	24	1.0
5.8 to 500	10	0.5
>500	0.44	0

Ranges for packed bed:
$0.001 < N_{Re} < 1000$
$505\ N_{Sc} < 70,600$

$$0.2 < \frac{E^{1/3}d_p^{4/3}\rho}{\mu} < 4600$$

Compares different situations versus general correlation.

H. For liquids, fixed and fluidized beds

$$\varepsilon j_D = \frac{1.1068}{N_{Re}^{0.78}}, 1.0 < N_{Re} \leq 10$$

$$\varepsilon j_D = \frac{N_{Sh}}{N_{Re}N_{Sc}^{1/3}}, N_{Sh} = \frac{k'd_s}{D}$$

[E] Spheres:

$$N_{Re} = \frac{d_p v_{super}\rho}{\mu}$$

MASS TRANSFER—TABLE 7. Mass Transfer Correlations for Fixed and Fluidized Beds (*Continued*)

Transfer is to or from particles

Situation	Correlation	Comments E = Empirical, S. = Semiempirical, T = Theoretical
I. For gases and liquids, fixed and fluidized beds, Dwivedi and Upadhyay correlation	$\varepsilon j_D = \dfrac{0.765}{N_{Re}^{0.82}} + \dfrac{0.365}{N_{Re}^{0.386}}$ Gases: $10 \leq N_{Re} \leq 15{,}000$. Liquids: $0.01 \leq N_{Re} \leq 15{,}000$.	[E] Deep beds of spheres, $j_D = \dfrac{N_{Sh}}{N_{Re} N_{Sc}^{1/3}}$ $N_{Re} = \dfrac{d_p \upsilon_{super} \rho}{\mu}, \; N_{Sh} = \dfrac{k'ds}{D}$ Based on 20 gas studies and 17 liquid studies.
J. For gases and liquids, fixed bed	$j_D = 1.17 N_{Re}^{-0.415}, \; 10 \leq N_{Re} \leq 2500$ $j_D = \dfrac{k'}{\upsilon_{av}} \dfrac{p_{BM}}{P} N_{Sc}^{2/3}$ Comparison with other results is shown.	[E] Spheres: $N_{Re} = \dfrac{d_p \upsilon_{super} \rho}{\mu}$ Variation in packing that changes ε not allowed for. Extensive data referenced. $0.5 < N_{Sc} < 15{,}000$.
K. For liquids, fixed and fluidized beds, Rahman and Streat correlation	$N_{Sh} = \dfrac{0.86}{\varepsilon} N_{Re} N_{Sc}^{1/3}, \; 2 \leq N_{Re} \leq 25$	[E] Can be extrapolated to $N_{Re} = 2000$. $N_{Re} = d_p \upsilon_{super} \rho/\mu$. Done for neutralization of ion exchange resin.
L. For liquids and gases, Ranz and Marshall correlation	$N_{Sh} = \dfrac{k'd}{D} = 2.0 + 0.6 N_{Sc}^{1/3} N_{Re}^{1/2}$ $N_{Re} = \dfrac{d_p \upsilon_{super} \rho}{\mu}$	[E] Based on freely falling, evaporating spheres. Has been applied to packed beds. Prediction is low compared to experimental data for packed beds. Limit of 2.0 at low N_{Re} is too high. Not corrected for axial dispersion.

M. For liquids and gases. Wakao and Funazkri correlation

$$N_{Sh} = 2.0 + 1.1 N_{Sc}^{1/3} N_{Re}^{0.6}, \quad 3 < N_{Re} < 10{,}000$$

$$N_{Sh} = \frac{k'_{film} d_p}{D} t$$

[E] $N_{Re} = \dfrac{\rho_l v_{super}\rho}{\mu}$

Correlate 20 gas studies and 16 liquid studies.
Correlated for axial dispersion with:

$$\frac{e D_{axial}}{D} = 10 + 0.5 N_{Sc} N_{Re}$$

D_{axial} is axial dispersion coefficient.

N. Liquid fluidized beds

$$N_{Sh} = \frac{2\xi/\varepsilon^m + \left[\dfrac{(2\xi/\varepsilon^m)(1-\varepsilon)^{1/2}}{[1-(1-\varepsilon)^{1/3}]^2} - 2\right]\tan h(\xi/\varepsilon^m)}{\dfrac{\xi/\varepsilon^m}{1-(1-\varepsilon)^{1/2}} - \tan h(\xi/\varepsilon^m)}$$

where

$$\xi = \left[\frac{1}{(1-\varepsilon)^{1/3}} - 1\right]\frac{\alpha}{2} N_{Sc}^{1/3} N_{Re}^{1/2}$$

This simplifies to:

$$N_{Sh} = \frac{\varepsilon^{1-2m}}{(1-\varepsilon)^{1/3}}\left[\frac{1}{(1-\varepsilon)^{1/3}}-1\right]\frac{\alpha^2}{2} N_{Re} N_{Sc}^{2/3}$$

$$(N_{Re} < 0.1)$$

[S] Modification of theory to fit experimental data. For spheres, $m = 1$, $N_{Re} > 2$.

$$N_{Sh} = \frac{k'_L d_p}{D} \qquad N_{Re} = \frac{v_{super} d_p \rho}{\mu}$$

$m = 1$ for $N_{Re} > 2$; $m = 0.5$ for $N_{Re} < 1.0$; $\varepsilon = $ voidage; $\alpha = $ const.
Best fit data is $\alpha = 0.7$.

O. Liquid fluidized beds

$$N_{Sh} = 0.250 N_{Re}^{0.023} N_{Ga}^{0.306}\left(\frac{\rho_s - \rho}{\rho}\right)^{0.282} N_{Sc}^{0.410}$$

$$(\varepsilon < 0.85)$$

$$N_{Sh} = 0.304 N_{Re}^{-0.067} N_{Ga}^{0.332}\left(\frac{\rho_s - \rho}{\rho}\right)^{0.297} N_{Se}^{0.404}$$

$$(\varepsilon > 0.85)$$

This can be simplified (with slight loss in accuracy at high ε) to

$$N_{Sh} = 0.245 N_{Ga}^{0.323}\left(\frac{\rho_s - \rho}{\rho}\right)^{0.300} N_{Sc}^{0.400}$$

[E] Correlate amount of data from literature. Compare large number of published correlations.

$$N_{Sh} = \frac{k'_L d_p}{D}, \quad N_{Re} = \frac{d_p \rho v_{super}}{\mu}$$

$$N_{Ga} = \frac{d_p^3 \rho^2 g}{\mu^2}, \quad N_{Sc} = \frac{\mu}{\rho D}$$

$1.6 < N_{Re} < 1320, \; 2470 < N_{Ga} < 4.42 \times 10^n$

$0.27 < \dfrac{\rho_s - \rho}{\rho} < 1.114, \; 305 < N_{Sc} < 1595$

Predicts very little dependence of N_{Sh} on velocity.

MASS TRANSFER — TABLE 7. Mass Transfer Correlations for Fixed and Fluidized Beds (*Continued*)
Transfer is to or from particles

Situation	Correlation	Comments E = Empirical, S. = Semiempirical, T = Theoretical
P. Liquid film flowing over solid particles with air present, trickle bed reactors, fixed bed	$N_{Sh} = \dfrac{k_1}{aD} = 1.8 N_{Re}^{1/2} N_{Sc}^{1/3}$, $0.013 < N_{Re} < 12.6$ two-phases, liquid trickle, no forced flow of gas. $N_{Sh} = 0.8 N_{Re}^{1/2} N_{Sc}^{1/3}$, one-phase, liquid only.	$[E] N_{Re} = \dfrac{L}{a\mu}$ L = superficial liquid flow rate, kg/m²s. a = surface area/col, volume, m²/m³. Irregular granules of benzoic acid, $0.29 \le d_p \le$ 1.45 cm.
Q. Supercritical fluids in packed bed	$\dfrac{N_{Sh}}{(N_{Sc}N_{Gr})^{1/4}} = 0.1813 \left(\dfrac{N_{Re}^2 N_{Sc}^{1/3}}{N_{Gr}} \right)^{1/4} \left(N_{Re}^{1/2} N_{Sc}^{1/3} \right)^{3/4}$ $+ 1.2149 \left[\left(\dfrac{N_{Re}^2 N_{Sc}^{1/3}}{N_{Gr}} \right)^{3/4} - 0.01649 \right]^{1/3}$	[E] Natural and forced convection, $4 < N_{Re} < 135$.
R. Supercritical fluids in packed bed	$\dfrac{N_{Sh}}{(N_{Sc}N_{Gr})^{1/4}} = 0.5265 \left(\dfrac{N_{Re}^{1/2} N_{Sc}^{1/3}}{(N_{Sc}N_{Gr})^{1/4}} \right)^{1.6808}$ $+ 2.48 \left[\left(\dfrac{N_{Re}^2 N_{Sc}^{1/3}}{N_{Gr}} \right)^{0.6439} - 0.8768 \right]^{1.553}$	[E] Natural and forced convection. $0.3 < N_{Re} < 135$. Improvement of correlation in Q.

Note: For $N_{Re} < 3$ convective contributions which are not included may become important. Use with logarithmic concentration difference (integrated form) or with arithmetic concentration difference (differential form).

Situation	Correlation	Comments E = Empirical, S = Semiempirical, T = Theoretical
A. Absorption, counter-current, liquid-phase coefficient H_L, Sherwood and Holloway correlation for random packings	$H_L = a_L\left(\dfrac{L}{\mu_L}\right)^n N_{Sc,L}^{0.5}, \quad L = \text{lb/hr ft}^2$ $H_L = \dfrac{L_M}{\hat{k}_L a}$ L_M = lbmoles/hr ft², $\hat{k}_L$ = lbmoles/hr ft², a = ft²/ft³, μ_L in lb/(hr ft).	[E] From experiments on desorption of sparingly soluble gases from water. Equation is dimensional. A typical value of n is 0.3 p. 633 has constants in kg, m², and s units for use in 5-28-A and B with k_G in kgmole/s m² and $\hat{k}_L$ in kgmole/s m² (kgmole/m³).

Ranges for 5-28-B (G and L).

Packing	a_G	b	c	G	L	a_L	n
Raschig rings							
3/8 inch	2.32	0.45	0.47	200–500	500–1500	0.00182	0.46
1	7.00	0.39	0.58	200–800	400–500	0.010	0.22
1	6.41	0.32	0.51	200–600	500–4500	—	—
2	3.82	0.41	0.45	200–800	500–4500	0.0125	0.22
Berl saddles							
1/2 inch	32.4	0.30	0.74	200–700	500–1500	0.0067	0.28
1/2	0.811	0.30	0.24	200–800	400–4500	—	—
1	1.97	0.36	0.40	200–800	400–4500	0.0059	0.28
1.5	5.05	0.32	0.45	200–1000	400–4500	0.0062	0.28

Range for 5-28-A is $400 < L < 150{,}000\,\text{lb/hr ft}^2$

Situation	Correlation	Comments
B. Absorption counter-current, gas-phase coefficient H_G, for random packing	$H_G = \dfrac{a_G(G)^b N_{Sc,v}^{0.5}}{(L)^c}, \quad G = \text{lb/hr ft}^2$ $H_G = \dfrac{G_M}{\hat{k}_G a}$ G_M = lbmoles/hr ft², $\hat{k}_G$ = lbmoles/hr ft².	[E] Based on ammonia-water-air data in Fellinger's 1941 MIT thesis.

MASS TRANSFER—TABLE 8. Mass Transfer Correlations for Packed Two-Phase Contactors—Absorption, Distillation, Cooling Towers and Extractors (Packing Is Inert) (*Continued*)

Situation	Correlation	Comments E = Empirical, S = Semiempirical, T = Theoretical
C. Absorption, counter-current, gas-liquid individual coefficients and interfacial area, Shulman data for random packings	$$\frac{k_G N_{Sc,v}^{2/3}}{G_M} = 1.195 \left[\frac{d_p G}{\mu_G(1-\varepsilon_{Lo})} \right]^{-0.36}$$ $$\frac{\hat{k}_L d_p}{D_L} = 25.1 \left(\frac{d_p L}{\mu_L} \right)^{0.45} N_{Sc,L}^{0.5}$$	[E] Compared napthalene sublimination to aqueous absorption to obtain $\hat{k}_G$, a, and $\hat{k}_L$ separately. Raschig rings and Berl saddles. d_p = diameter of sphere with same surface area as packing piece. ε_{Lv} = operating void space = $\varepsilon - \phi_{Lt}$, where ε = void fraction w/o liquid, and ϕ_{Lt} = liquid holdup. Same definition as 5-28-A and B. Onda et al. correlation is preferred. G = $\rho_G v_{super,gas}$
D. Absorption and distillation, counter-current, gas and liquid individual coefficients and wetted surface area, Onda et al. correlation for random packings	$$\frac{k_G' RT}{a_p D_C} = A \left(\frac{G}{a_p \mu_C} \right)^{0.1} N_{Sc,G}^{1/3} (a_p d_p')^{-2.0}$$ $A = 5.23$ for packing $\geq 1/2$ inch $(0.012\,\text{m})$ $A = 2.0$ for packing $< 1/2$ inch $(0.012\,\text{m})$ $k_G' = \text{lbmoles/hr·ft}^2\text{·atm [kgmol/s·m}^2 \text{ (N/m}^2)]$ $$k_L' \left(\frac{\rho_L}{\mu_L g} \right)^{1/3} = 0.0051 \left(\frac{L}{a_w \mu_L} \right)^{2/3} N_{Sc,L}^{-1/2} (a_p d_p')^{0.4}$$ $k_L' = \text{lbmoles/hr·ft}^2 \text{ (lbmoles/ft}^3) \text{ [kgmoles/s·m}^2 \text{ (kgmoles/m}^3)]$ $$\frac{a_w}{a_p} = 1 - \exp \left\{ \begin{array}{l} -1.45 \left(\frac{\sigma_c}{\sigma} \right)^{0.75} \left(\frac{L}{a_p \mu_L} \right)^{0.1} \\ \times \left(\frac{L^2 a_p}{\rho_L^2 g} \right)^{-0.05} \left(\frac{L}{\rho_L \sigma a_p} \right)^{0.2} \end{array} \right\}$$	[E] Gas absorption and desorption from water and organics plus vaporization of pure liquids for Raschig rings, saddles, spheres, and rods. d_p' = nominal packing size, a_p = dry packing surface area/volume, a_w = wetted packing surface area/volume. Equations are dimensionally consistent, so any set of consistent units can be used. σ = surface tension, dynes/cm.

268

E. Distillation and absorption, counter-current, random packings, modification of Onda correlation, Bravo and Fair correlation

Critical surface tensions, $\sigma_c = 61$ (ceramic), 75 (steel), 33 (polyethylene), 40 (PVC), 56 (carbon) dynes/cm.

$$4 < \frac{L}{a_w \mu_L} < 400$$

$$5 < \frac{G}{a_p \mu_c} < 1000$$

Most data ± 20% of correlation, some ± 50%.

Use Onda's correlations (5-28-D) for k'_G and k'_L. Calculate:

$$H_G = \frac{G}{k'_G a_e P M_G}, \quad H_L = \frac{L}{k'_L a_e \rho_L}$$

$$H_{OG} = H_G + \lambda H_L$$

where

$$\lambda = \frac{m}{L_L / G_M}$$

Using

$$a_e = 0.498 a_p \left(\frac{\sigma^{0.3}}{Z^{0.4}} \right) \left(N_{Ca, L} N_{ReG} \right)^{0.302}$$

where

$$N_{ReG} = \frac{6G}{a_p \mu_c}$$

$$N_{Ca,L} = \frac{L \mu_L}{\rho_L \sigma g_c} \quad \text{(dimensionless)}$$

[E] Uses Bolles & Fair data base to determine new effective area a_c to use with Onda et al. correlation. P = Total pressure, atm; M_G = gas, molecular weight; m = local slope of equilibrium curve; L_{Mf}/G_M = slope operating line; Z = height of packing in feet.

Equation for a_c is dimensional. Fit to data for effective area quite good for distillation. Good for absorption at low values of $(N_{Ca,L} \times N_{ReG})$, but correlation is too high at higher values of $(N_{Ca,L} \times N_{ReG})$.

F. Absorption, co-current downward flow, random packings

Air-oxygen-water results correlated by $k'_L a = 0.12 E_L^{0.5}$. Extended to other systems.

$$k'_L a = 0.12 E_L^{0.5} \left(\frac{D_L}{2.4 \times 10^5} \right)^{0.3}$$

$$E_L = \left(\frac{\Delta p}{\Delta L} \right)_{2\text{-phase}} v_L$$

[E] Based on oxygen transfer from water to air 77°F. Liquid film resistance controls. (D_{water} @ 77°F = 2.4×10^{-5}). Equation is dimensional. Data was for thin-walled polyethylene Raschig rings. Correlation also fit data for spheres. Fit ±25%.

269

MASS TRANSFER—TABLE 8. Mass Transfer Correlations for Packed Two-Phase Contactors—Absorption, Distillation, Cooling Towers and Extractors (Packing Is Inert) *(Continued)*

Situation	Correlation	Comments E = Empirical, S = Semiempirical, T = Theoretical
	$k_L a = s^{-1}$ $D_L = cm/s$ $E_L = ft, lbf/s \cdot ft^3$ $v_L = $ superficial liquid velocity, ft/s $\dfrac{\Delta p}{\Delta L} = $ pressure loss in two-phase flow $= lbf/ft^2 \, ft$ $k_G' a = 2.0 + 0.91 E_G^{2/3}$ for NH_3 $E_g = \left(\dfrac{\Delta p}{\Delta L}\right)_{2\text{-phase}} v_g$ $v_g = $ superficial gas velocity, ft/s.	[E] Ammonia absorption into water from air at 70°F. Gas-film resistance controls. Thin-walled polyethylene Raschig rings and 1-inch Intalox saddles. Fit ±25%. Terms defined as above.
G. Absorption, striping, distillation, counter-current, H_L and H_G, random packings. Cornell et al. correlation, and Bolles and Fair correlation	For Raschig rings, Berl saddles, and spiral tile: $H_L = \dfrac{\phi G_{\text{flood}} N_{Sc,L}^{0.3}}{3.28} \left(\dfrac{Z}{3.05}\right)^{0.15}$ $H_G = \dfrac{A \Psi (d'_{\text{col}})^m \, Z^{0.33} N_{Sc,C}^{0.5}}{\left[L \left(\dfrac{\mu_L}{\mu_{\text{water}}}\right)^{0.15} \left(\dfrac{\rho_{\text{water}}}{\rho_L}\right)^{1.25} \left(\dfrac{\sigma_{\text{water}}}{\sigma_L}\right)^{0.8} \right]^n}$ $A = 0.017$ (rings) or 0.029 (saddles) $d'_{\text{col}} = $ column diameter in m (if diameter > 0.6 m, used $d'_{\text{col}} = 0.6$) $m = 1.24$ (rings) or 1.11 (saddles) $n = 0.6$ (rings) or 0.5 (saddles) Ψ is given in Range: $25 < \Psi < 190$ m.	[E] $Z = $ packed height, m of each section with its own liquid distribution. $L = $ liquid rate, kg/(sm^2), $\mu_{\text{water}} = 1.0$ Pa·s, $\rho_{\text{water}} = 1000$ kg/m^3, $\sigma_{\text{water}} = 72.8$ mN/m (72.8 dynes/cm). H_G and H_L will vary from location to location. Design each section of packing separately.

H. Distillation and absorption. Counter-current flow. Structured packings. Gauze-type with triangular flow channels, Bravo, Rocha, and Fair correlation

Equivalent channel:

$$d_{eq} = Bh\left[\frac{1}{B+2S} + \frac{1}{2S}\right]$$

Use modified correlation for wetted wall column:

$$N_{Sh,v} = \frac{k'_v d_{eq}}{D_v} = 0.0338 N_{Re,v}^{0.8} N_{Sc,v}^{0.333}$$

$$N_{Re,v} = \frac{d_{eq}\rho_v(U_{v,\text{eff}}+U_{L,\text{eff}})}{\mu_v}$$

where effective velocities

$$U_{v,\text{eff}} = \frac{U_{v,\text{super}}}{\epsilon \sin\theta}$$

$$U_{L,\text{eff}} = \frac{3\Gamma}{2\rho_L}\left(\frac{\rho_L^2 g}{3\mu_L\Gamma}\right)^{0.333}, \quad \Gamma = \frac{L}{Per}$$

$$Per = \frac{\text{Perimeter}}{\text{Area}} = \frac{4s+2B}{Bh}$$

Calculate k'_L from penetration model (use time for liquid to flow distance s).

$$k'_L = 2(D_L U_{L,\text{eff}}/\pi S)^{1/2}.$$

[T] Check of 132 data points showed average deviation 14.6% from theory. Johnstone and Pigford correlation has exponent on N_{Re} rounded to 0.8. Assume gauze packing is completely wet. Thus, $a_{\text{eff}} = a_p$ to calculate H_G and H_L. Same approach may be used generally applicable to sheet-metal packings, but they will not be completely wet and need to estimate transfer area.

L = liquid flux, kg/sm².
G = vapor flux, kg/sm².

$$H_G = \frac{G}{k_v a_p \rho_v}, \quad H_L = \frac{L}{k'_L a_p \rho_L}$$

MASS TRANSFER—TABLE 8. Mass Transfer Correlations for Packed Two-Phase Contactors—Absorption, Distillation, Cooling Towers and Extractors (Packing Is Inert) (*Continued*)

Situation	Correlation	Comments E = Empirical, S = Semiempirical, T = Theoretical
I. High-voidage packings, cooling towers, splash-grid packings	$$\frac{(Ka)_H V_{tower}}{L} = 0.07 + A'N'\left(\frac{L}{G_a}\right)^{-n'}$$ A' and n' depend on deck type (Ref. 107), $0.060 \leq A' \leq 0.135$, $0.46 \leq n' \leq 0.62$.	[E] General form. G_a = lb dry air/hr·ft^2. L = lb/h·ft^2, N' = number of deck levels. $(Ka)_H$ = overall enthalpy transfer coefficient $$= \text{lb/(h)(ft}^3)\left(\frac{\text{lb water}}{\text{lb dry air}}\right)$$ V_{tower} = tower volume, ft^3/ft^2. If normal packings are used, use absorption mass-transfer correlations.
J. Liquid-liquid extraction, packed towers	Use k values for drops. Enhancement due to packing is at most 20%. Packing decreases drop size and increases interfacial area.	[E]
K. Liquid-liquid extraction in Rotating-disc contactor (RDC)	$$\frac{k_{c,RDC}}{k_c} = 1.0 + 2.44\left(\frac{N}{N_{Cr}}\right)^{2.5}$$ $$N_{Cr} = 7.6 \times 10^{-4}\left(\frac{\sigma}{d_{drop}\mu_c}\right)\left(\frac{H}{D_{tank}}\right)$$ $$\frac{k_{d,RDC}}{k_d} = 1.0 + 1.825\left(\frac{N}{N_{Cr}}\right)\frac{H}{D_{tank}}$$	k_c, k_d are for drops. N = impeller speed Breakage occurs when $N > N_{Cr}$. Maximum enhancement before breakage was factor of 2.0. H = compartment height, D_{tank} = tank diameter, σ = interfacial tension, N/m. Done in 0.152 and 0.600 m RDC.
L. Liquid-liquid extraction, stirred tanks		[E]

MELTING POINTS

MELTING POINTS—TABLE 1.   Melting Points of Organic Compounds

(a) Derivatives of Alcohols

	3,5-Dinitro-benzoate $\theta_{C,m}/°C$		3,5-Dinitro-benzoate $\theta_{C,m}/°C$
Methanol	109	2-Methylpropan-2-ol	142
Ethanol	94	Pentan-1-ol	46
Propan-1-ol	75	Hexan-1-ol	61
Propan-2-ol	122	Phenylmethanol	113
Butan-1-ol	64	Cyclohexanol	113
2-Methylpropan-1-ol	88	Ethane-1,2-diol (glycol)	169*
Butan-2-ol	76		

(b) Derivatives of Phenols

	3,5-Dinitro benzoate $\theta_{C,m}/°C$	4-Methyl-benzene-sulphonate $\theta_{C,m}/°C$		3,5-Dinitro-benzoate $\theta_{C,m}/°C$	4-Methyl-benzene sulphonate $\theta_{C,m}/°C$
Phenol	146	96	Benzene-1,2-diol	152*	—
2-Methylphenol	138	55	Benzene-1,3-diol	201*	81*
3-Methylphenol	165	51	Benzene-1,4-diol	317*	159*
4-Methylphenol	189	70	2-Nitrophenol	155	83
Naphthalen-1-ol	217	88	3-Nitrophenol	159	113
Naphthalen-2-ol	210	125	4-Nitrophenol	188	97

(c) Derivatives of Aldehydes and Ketones

	2,4-Dinitro-phenyl-hydrazone $\theta_{C,m}/°C$		2,4-Dinitro-phenyl-hydrazone $\theta_{C,m}/°C$
Methanal	166	Propanone	126
Ethanal	168	Butanone	116
Propanal	155	Pentan-3-one	156
Butanal	126	Pentan-2-one	144
Benzaldehyde	237	Heptan-4-one	75
2-Hydroxybenzaldehyde	252 dec.	Phenylethanone	250
Ethanedial	327	Diphenylmethanone	239
Trichloroethanal	131	Cyclohexanone	162

MELTING POINTS—TABLE 1. Melting Points of Organic Compounds (*Continued*)

(d) Derivatives of Amines

	Ethanoyl derivative $\theta_{C,m}/°C$	Benzoyl derivative $\theta_{C,m}/°C$	4-Methyl-benzene-sulphonyl derivative $\theta_{C,m}/°C$
Methylamine	28	80	75
Ethylamine	205*	69	62
Propylamine	47	85	52
Butylamine	229‡	70	65
(Phenylmethyl)amine	60	105	116
Phenylamine	114	163	103
Cyclohexylamine	104	147	87
2-Methylphenylamine	112	143	110
3-Methylphenylamine	66	125	114
4-Methylphenylamine	152	158	118
Dimethylamine	116‡	42	87
Diethylamine	186‡	42	60
Diphenylamine	103	180	142

* Disubstituted derivative.
† Boiling temperature.

MELTING POINTS—TABLE 2. Melting Points of the *n*-Paraffins

	Melting point	
Number of carbon atoms	°C	°F
1	−182	−296
2	−183	−297
3	−188	−306
4	−138	−216
5	−130	−202
6	−95	−139
7	−91	−132
8	−57	−71
9	−54	−65
10	−30	−22
11	−26	−15
12	−10	14
13	−5	23
14	6	43
15	10	50
16	18	64
17	22	72
18	28	82
19	32	90
20	36	97
30	66	151
40	82	180

MELTING POINTS—TABLE 2. Melting Points of the *n*-Paraffins (*Continued*)

Number of carbon atoms	Melting point	
	°C	°F
50	92	198
60	99	210

MELTING POINTS—TABLE 3. Melting Temperatures of Common Alloys

	UNS	Melting range	
		°F	°C
Aluminum alloy AA11000	A91100	1190–1215	640–660
Aluminum alloy AA5052	A95052	1125–1200	610–650
Aluminum cast alloy 43	A24430	1065–1170	570–630
Copper	C11000	1980	1083
Red brass	C23000	1810–1880	990–1025
Admiralty brass	C44300	1650–1720	900–935
Muntz Metal	C28000	1650–1660	900–905
Aluminum bronze D	C61400	1910–1940	1045–1060
Ounce metal	C83600	1510–1840	854–1010
Manganese bronze	C86500	1583–1616	862–880
90-10 copper nickel	C70600	2010–2100	1100–1150
70-30 copper nickel	C71500	2140–2260	1170–1240
Carbon steel, AISI 1020	G10200	2760	1520
Gray cast iron	F10006	2100–2200	1150–1200
4-6 Cr, $\frac{1}{2}$ Mo Street	S50100	2700–2800	1480–1540
Stainless steel, AISI 410	S41000	2700–2790	1480–1530
Stainless steel, AISI 446	S44600	2600–2750	1430–1510
Stainless steel, AISI 304	S30400	2550–2650	1400–1450
Stainless steel, AISI 310	S31000	2500–2650	1400–1450
Stainless steel, ACI HK	J94224	2550	1400
Nickel alloy 200	N02200	2615–2635	1440–1450
Nickel alloy 400	N04400	2370–2460	1300–1350
Nickel alloy 600	N06600	2470–2575	1350–1410
Nickel-molybdenum alloy B-2	N10665	2375–2495	1300–1370
Nickel-molybdenum alloy C-276	N10276	2420–2500	1320–1370
Titanium, commercially pure	R50250	3100	1705
Titanium alloy T1-6A1-4V	R56400	2920–3020	1600–1660
Magnesium alloy AZ 31B	M11311	1120–1170	605–632
Magnesium alloy HK 31A	M13310	1092–1204	589–651
Chemical lead		618	326
50-50 solder	L05500	361–421	183–216
Zinc	Z13001	787	420
Tin	Z13002	450	232
Zirconium	R60702	3380	1860
Molybdenum	R03600	4730	2610
Tantalum	R05200	5425	2996

NATURAL GAS

NATURAL GAS—TABLE 1. Composition of Natural Gas from a Petroleum Well

Category	Component	Amount, %
Paraffinic	Methane (CH_4)	70–98
	Ethane (C_2H_6)	1–10
	Propane (C_3H_8)	Trace–5
	Butane (C_4H_{10})	Trace–2
	Pentane (C_5H_{12})	Trace–1
	Hexane (C_6H_{14})	Trace–0.5
	Heptane and higher (C_7+)	None–trace
Cyclic	Cyclopropane (C_3H_6)	Traces
	Cyclohexane (C_6H_{12})	Traces
Aromatic	Benzene (C_6H_6), others	Traces
Nonhydrocarbon	Nitrogen (N_2)	Trace–15
	Carbon dioxide (CO_2)	Trace–1
	Hydrogen sulfide (H_2S)	Trace occasionally
	Helium (He)	Trace–5
	Other sulfur and nitrogen compounds	Trace occasionally
	Water (H_2O)	Trace–5

NATURAL GAS—TABLE 2. Variations of Natural Gas Composition with Source

	Type of gas field			Natural gas separated from crude oil		
				Ventura*		
Component	Dry gas, Los Medanos,* mole %	Sour gas, Jumping Pound,† mole %	Gas condensate, Paloma,* mole %	400 lb, mole %	50 lb, mole %	Vapor, mole %
Hydrogen sulfide	0	3.3	0	0	0	0
Carbon dioxide	0	6.7	0.68	0.30	0.68	0.81
Nitrogen and air	0.8	0	0	0	—	2.16
Methane	95.8	84.0	74.55	89.57	81.81	69.08
Ethane	2.9	3.6	8.28	4.65	5.84	5.07
Propane	0.4	1.0	4.74	3.60	6.46	8.76
Isobutane	0.1	0.3	0.89	0.52	0.92	2.14
n-Butane	Trace	0.4	1.93	0.90	2.26	5.20
Isopentane	0		0.75	0.19	0.50	1.42
n-Pentane	0		0.63	0.12	0.48	1.41
Hexane	0	0.7	1.25			
Heptane	0			0.15	1.05	4.13
Octane	0		6.30			
Nonane	0					
	100.0	100.0	100.00	100.00	100.00	100.00

* California.
† Canada.

NATURAL GAS—TABLE 3. Molecular Weights, Boiling Points, and Densities of Hydrocarbon Gases That Occur in Natural Gas

Gas	Molecular weight	Boiling point 1 atm, °C (°F)	Density at 60°F (15.6°C), 1 atm Real gas	
			g/litre	Relative to air = 1
Methane	16.043	−161.5 (−258.7)	0.6786	0.5547
Ethylene	28.054	−103.7 (−154.7)	1.1949	0.9768
Ethane	30.068	−88.6 (−127.5)	1.2795	1.0460
Propylene	42.081	−47.7 (−53.9)	1.8052	1.4757
Propane	44.097	−42.1 (−43.8)	1.8917	1.5464
1,2-Butadiene	54.088	10.9 (51.6)	2.3451	1.9172
1,3-Butadiene	54.088	−4.4 (24.1)	2.3491	1.9203
1-Butene	56.108	−6.3 (20.7)	2.4442	1.9981
cis-2-Butene	56.108	3.7 (38.7)	2.4543	2.0063
trans-2-Butene	56.108	0.9 (33.6)	2.4543	2.0063
iso-Butene	56.104	−6.9 (19.6)	2.4442	1.9981
n-Butane	58.124	−5.4 (31.1)	2.5320	2.0698
iso-Butane	58.124	−11.7 (10.9)	2.5268	2.0656

NATURAL GASOLINE

NATURAL GASOLINE—TABLE 1 Composition of Natural Gasoline from Natural Gas

Reid vapor pressure	Ventura gasoline plant			Ten-section gasoline plant, 22 psia
	38 psia	60 psia	100 psia	
Ethane	Trace	0.5	0.7	0
Propane	1.1	16.0	43.8	0
Isobutane	19.0	16.0	10.7	0.2
n-Butane	41.0	34.7	23.0	22.7
Isopentane	13.2	11.2	7.4	24.1
n-Pentane	11.3	9.5	6.3	21.0
Hexane	6.8	5.7	3.8	12.6
Heptane	5.3	4.4	2.9	13.7
Octane	1.2	1.0	0.7	4.1
Nonane	1.1	1.0	0.7	1.2
Decane	Trace	Trace	Trace	0.4
	100.0	100.0	100.0	100.0

NUCLEAR MAGNETIC RESONANCE

Nuclear magnetic resonance (NMR) is a characterization technique in which a sample is immersed in a magnetic field and subjected to radio waves. These radio waves encourage the nuclei of the molecule to oscillate, and these molecular movements are detected on a receiver. Different nuclei oscillate at different frequencies that translate into chemical shifts that are dependent on the location of the nucleus under study.

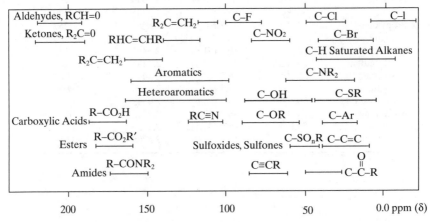

NUCLEAR MAGNETIC RESONANCE—FIGURE 1.   Carbon Chemical Shift Ranges*
* For samples in $CDCl_3$ solution. The δ scale is relative to TMS as $\delta = 0$.

NUCLEAR MAGNETIC RESONANCE—TABLE 1.   Carbon Chemical Shifts, ppm

Compound	X = F	X = Cl	X = Br	X = I	X = OMe
CH_3X	75	25	10	−21	60
CH_2X_2	110	54	22	−54	103
CHX_3	119	77.5	12	−140	115
CX_4	94	97	−29	−293	121

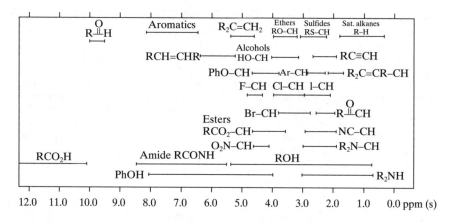

NUCLEAR MAGNETIC RESONANCE—FIGURE 2. Proton Chemical Shift Ranges*
* For samples in CDCl₃ solution. The δ scale is relative to TMS as $\delta = 0$.

NUCLEAR MAGNETIC RESONANCE—TABLE 2. Proton Chemical Shifts, ppm

Cpd./Sub	X = Cl	X = Br	X = I	X = OR	X = SR
CH_3X	3.0	2.7	2.1	3.1	2.1
CH_2X_2	5.3	5.0	3.9	4.8	3.8
CHX_3	7.3	7.0	5.7	5.2	

PETROLEUM

Petroleum (*crude oil*) is the term used to describe myriad hydrocarbon-rich fluids that have accumulated in subterranean reservoirs. Petroleum varies dramatically in color, odor, and flow properties that reflect the diversity of its origin.

Petroleum may be called *light* or *heavy* in reference to the amount of its low-boiling constituents and relative density (specific gravity). Likewise, odor is used to distinguish between *sweet* (low-sulfur) and *sour* (high-sulfur) petroleum. Viscosity indicates the ease of (or, more correctly, the resistance to) flow.

Although not directly derived from composition, the terms *light* or *heavy* or *sweet* and *sour* provide convenient terms for use in descriptions. For example, *light petroleum* (often referred to as *conventional petroleum*) is usually rich in low-boiling constituents and waxy molecules, whereas *heavy petroleum* contains greater proportions of higher-boiling, more aromatic, and heteroatom-containing (N-, O-, S-, and metal-containing) constituents. *Heavy petroleum* is more viscous than conventional petroleum and requires enhanced methods for recovery.

Bitumen is *near solid* or *solid* and cannot be recovered by enhanced oil recovery methods.

Conventional (light) petroleum is composed of hydrocarbons and smaller amounts of organic compounds of nitrogen, oxygen, and sulfur as well as still smaller amounts of compounds containing metallic constituents, particularly vanadium, nickel, iron, and copper. The processes by which petroleum was formed dictate that petroleum composition varies and is *site specific*, thus leading to a wide variety of compositional differences. The term *site specific* is intended to convey that petroleum composition is dependent on regional and local variations of the proportion of the various precursors that went into the formation of the *protopetroleum* as well as variations in temperature and pressure to which the precursors were subjected.

PETROLEUM—TABLE 1. Variation in Composition and Properties of Petroleum

Crude oil	Specific gravity	API gravity	Residuum >1000°F
U.S. Domestic			
California	0.858	33.4	23.0
Oklahoma	0.816	41.9	20.0
Pennsylvania	0.800	45.4	2.0
Texas	0.827	39.6	15.0
Texas	0.864	32.3	27.9
Foreign			
Bahrain	0.861	32.8	26.4
Iran	0.836	37.8	20.8
Iraq	0.844	36.2	23.8
Kuwait	0.860	33.0	31.9
Saudi Arabia	0.840	37.0	27.5
Venezuela	0.950	17.4	33.6

PETROLEUM—TABLE 2.   Analysis of Petroleum and Petroleum Products

Product	Gravity, °API	Specific gravity at 60°F	Wt lb/gal	High-heat value, Btu/lb*	Ultimate analysis, %				
					C	H	S	N	O
California crude	22.8	0.917	7.636	18,910	84.00	12.70	0.75	1.70	1.20
Kansas crude	22.1	0.921	7.670	19,130	84.15	13.00	1.90	0.45	
Oklahoma crude (east)	31.3	0.869	7.236	19,502	85.70	13.11	0.40	0.30	
Oklahoma crude (west)	31.0	0.871	7.253	19,486	85.00	12.90	0.76		
Pennsylvania crude	42.6	0.813	6.769	19,505	86.06	13.88	0.06	0.00	0.00
Texas crude	30.2	0.875	7.286	19,460	85.05	12.30	1.75	0.70	0.00
Wyoming crude	31.5	0.868	7.228	19,510					
Mexican crude	13.6	0.975	8.120	18,755	83.70	10.20	4.15		
Gasoline	67.0	0.713	5.935	—	84.3	15.7			
Gasoline	60.0	0.739	6.152	20,750	84.90	14.76	0.08		
Gasoline-benzene blend	46.3	0.796	6.627	—	88.3	11.7			
Kerosene	41.3	0.819	6.819	19,810					
Gas oil	32.5	0.863	7.186	19,200					
Fuel oil (Mex.)	11.9	0.987	8.220	18,510	84.02	10.06	4.93		
Fuel oil (midcontinent)	27.1	0.892	7.428	19,376	85.62	11.98	0.35	0.50	0.60
Fuel oil (Calif.)	16.7	0.9554	7.956	18,835	84.67	12.36	1.16		

* Btu/lb × 2.328 = kJ/kg.

PETROLEUM—TABLE 3.   Properties of Petroleum Products

	Molecular weight	Specific gravity	Boiling point, °F	Ignition temperature, °F	Flash point, °F	Flammability limits in air, % v/v
Benzene	78.1	0.879	176.2	1040	12	1.35–6.65
n-Butane	58.1	0.601	31.1	761	−76	1.86–8.41
iso-Butane	58.1		10.9	864	−117	1.80–8.44
n-Butene	56.1	0.595	21.2	829	Gas	1.98–9.65
iso-Butene	56.1		19.6	869	Gas	1.8–9.0
Diesel fuel	170–198	0.875			100–130	
Ethane	30.1	0.572	−127.5	959	Gas	3.0–12.5
Ethylene	28.0		−154.7	914	Gas	2.8–28.6
Fuel oil No. 1		0.875	304–574	410	100–162	0.7–5.0
Fuel oil No. 2		0.920		494	126–204	
Fuel oil No. 4	198.0	0.959		505	142–240	
Fuel oil No. 5		0.960			156–336	
Fuel oil No. 6		0.960			150	
Gasoline	113.0	0.720	100–400	536	−45	1.4–7.6
n-Hexane	86.2	0.659	155.7	437	−7	1.25–7.0
n-Heptane	100.2	0.668	419.0	419	25	1.00–6.00
Kerosene	154.0	0.800	304–574	410	100–162	0.7–5.0
Methane	16.0	0.553	−258.7	900–1170	Gas	5.0–15.0
Naphthalene	128.2		424.4	959	174	0.90–5.90
Neohexane	86.2	0.649	121.5	797	−54	1.19–7.58
Neopentane	72.1		49.1	841	Gas	1.38–7.11
n-Octane	114.2	0.707	258.3	428	56	0.95–3.2
iso-Octane	114.2	0.702	243.9	837	10	0.79–5.94
n-Pentane	72.1	0.626	97.0	500	−40	1.40–7.80
iso-Pentane	72.1	0.621	82.2	788	−60	1.31–9.16
n-Pentene	70.1	0.641	86.0	569	—	1.65–7.70
Propane	44.1		−43.8	842	Gas	2.1–10.1
Propylene	42.1		−53.9	856	Gas	2.00–11.1
Toluene	92.1	0.867	321.1	992	40	1.27–6.75
Xylene	106.2	0.861	281.1	867	63	1.00–6.00

PETROLEUM—TABLE 4. Dielectric Constants of
Hydrocarbons and Petroleum Products

Material	Temperature		Dielectric constant
	°C	°F	
n-Hexane	0	32	1.918
	20	68	1.890
	60	140	1.817
n-Heptane	0	32	1.958
	20	68	1.930
	60	140	1.873
Benzene	10	50	2.296
	20	68	2.283
	60	140	2.204
Cyclohexane	20	68	2.055
Petroleum products			
Gasoline	20	68	1.8–2.0
Kerosene	20	68	2.0–2.2
Lubricating oil	20	68	2.1–2.6

PETROLEUM—TABLE 5. Dielectric Constants of Petroleum,
Coal, and Their Products

Compound	Dielectric constant	Loss tangent $\tan \Delta \times 10^4$
Asphalt (75°F)	2.6	
Bunker C oil	2.6	
Coal tar	2.0–3.0	
Coal, powder, fine	2.0–4.0	
Coke	1.1–2.2	
Gasoline (70°F)	2.0	
Heavy oil	3.0	
Heavy oil, bunker C	2.6	
Kerosene (70°F)	1.8	
Liquefied petroleum gas (LPG)	1.6–1.9	
Lubricating oil (68°F)	2.1–2.6	
Paraffin oil	2.19	1.0
Paraffin wax	2.1–2.5	2
Petroleum (68°F)	2.1	
Vaseline	2.08	5
Wax, paraffin	2.1–2.5	2

PETROLEUM—TABLE 6. Fuel Oil Analysis

Composition, %	No. 1 fuel oil (41.5° API)	No. 2 fuel oil (33° API)	No. 4 fuel oil (23.2° API)	Low sulfur, No. 6 F.O. (12.6° API)	High sulfur, No. 6 (15.5° API)
Carbon	86.4	87.3	86.47	87.26	84.67
Hydrogen	13.6	12.6	11.65	10.49	11.02
Oxygen	0.01	0.04	0.27	0.64	0.38
Nitrogen	0.003	0.006	0.24	0.28	0.18
Sulfur	0.09	0.22	1.35	0.84	3.97
Ash	<0.01	<0.01	0.02	0.04	0.02
C/H Ratio	6.35	6.93	7.42	8.31	7.62

Note: The C/H ratio is a weight ratio.

Heat of Combustion of Petroleum Fuels

The *heat value* (hhv) of petroleum products is determined by combustion in a bomb with oxygen under pressure (ASTM D240). It may also be calculated, in products free from impurities, by the formula

$$Q_v = 22,320 - 3,780d^2$$

in which Q_v is the hhv at constant volume in Btu/lb and d is the specific gravity at 60/60°F.

The low heat value at constant pressure Q_p may be calculated by the relation

$$Q_p = Q_v - 90.8H$$

where H is the weight percentage of hydrogen and can be obtained from the relation

$$H = 26 - 15d$$

Typical heats of combustion of petroleum oils free from water, ash, and sulfur vary (within an estimated accuracy of 1%) with the API gravity (i.e., with the "heaviness" or "lightness" of the material).

The heat value should be corrected when the oil contains sulfur by using an hhv of 4050 Btu/lb for sulfur (ASTM D1405).

The *specific heat c* of petroleum products of specific gravity d and at temperature t (°F) is given by the equation

$$c = (0.388 + 0.00045t)/\sqrt{d}$$

The *heat of vaporization L* (Btu/lb) may be calculated from the equation

$$L = (110.9 - 0.09t)/d$$

The heat of vaporization per gallon (measured at 60°F) is

$$8.34Ld = 925 - 0.75t$$

indicating that the heat of vaporization per gallon depends only on the temperature of vaporization t and varies over the range of 450 for the heavier products to 715

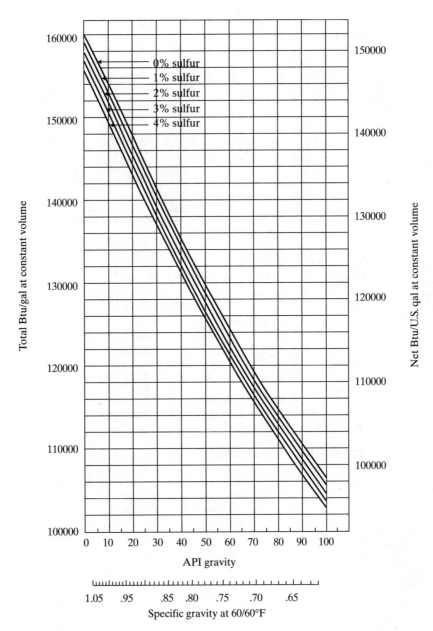

PETROLEUM—FIGURE 1. Heat of combustion of petroleum fuels. To convert Btu/U.S. gal to kJ/m³, multiply by 278.7.

for gasoline. These data have an estimated accuracy within 10% when the vaporization is at constant temperature and at pressures below 50 lb/in² without chemical change.

PETROLEUM—TABLE 7. Relationship of Heat Content to API Gravity

Deg API at 60°F	Density, lb/gal*	High heat value at constant volume Q_v, Btu		Low heat value at constant pressure Q_p, Btu	
		Per lb	Per gal	Per lb	Per gal
10	8.337	18,540	154,600	17,540	146,200
20	7.787	19,020	148,100	17,930	139,600
30	7.305	19,420	141,800	18,250	133,300
40	6.879	19,750	135,800	18,510	127,300
50	6.500	20,020	130,100	18,720	121,700
60	6.160	20,260	124,800	18,900	116,400
70	5.855	20,460	119,800	19,020	112,500
80	5.578	20,630	115,100	19,180	107,000

* Btu/lb $\times$ 2.328 = kJ/kg; Btu/gal $\times$ 279 = kJ/m^3.

PETROLEUM—TABLE 8. Latent Heat of Vaporization of Petroleum Products

Product	Gravity, °API	Average boiling temp, °F	Heat of vaporization	
			Btu/lb	Btu/gal
Gasoline	60	280	116	715
Naphtha	50	340	103	670
Kerosine	40	440	86	595
Fuel oil	30	580	67	490

PETROLEUM—TABLE 9. Properties of Conventional Crude Oil, Synthetic Crude Oil, and Bitumen

Property	Athabasca bitumen	Synthetic crude oil	Conventional crude oil
Specific gravity	1.03	0.85	0.85–0.90
Viscosity, cP			
38C/100°F	750,000	210	<210
100C/212°F	11,300		
Pour point, °F	>50	−35	ca. −20
Elemental analysis (wt. %):			
Carbon	83.0	86.3	86.0
Hydrogen	10.6	13.4	13.5
Nitrogen	0.5	0.02	0.2
Oxygen	0.9	0.00	<0.5
Sulfur	4.9	0.03	<2.0
Ash	0.8	0.0	0.0
Nickel (ppm)	250	0.01	<10.0
Vanadium (ppm)	100	0.01	<10.0
Fractional composition (wt. %)			
Asphaltenes (pentane)	17.0	0.0	<10.0

PETROLEUM—TABLE 9. Properties of Conventional Crude Oil, Synthetic Crude Oil, and Bitumen (*Continued*)

Property	Athabasca bitumen	Synthetic crude oil	Conventional crude oil
Resins	34.0	0.0	<20.0
Aromatics	34.0	40.0	>30.0
Saturates	15.0	60.0	>30.0
Carbon residue (wt. %)			
Conradson	14.0	<0.5	<10.0

PETROLEUM—TABLE 10. Relationship of the Refractive Index to the Dielectric Constant of Hydrocarbons and Petroleum Products

Material	Refractive Index n	n^2	Dielectric Constant
Benzene	1.501	2.253	2.283
Cyclohexane	1.427	2.036	2.055
n-Hexane	1.375	1.890	1.890
n-Heptane	1.388	1.926	1.933
Kerosene	1.449	2.100	2.135
Paraffin oil	1.481	2.193	2.195
Vaseline	1.480	2.190	2.078

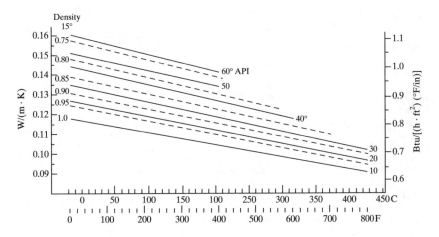

PETROLEUM—FIGURE 2. Thermal conductivity of petroleum liquids. The solid lines refer to density expressed as degrees API; the broken lines refer to relative density at 288 K (15°C). (K = [°F + 459.7]/1.8)

PHYSICAL CONSTANTS

PHYSICAL CONSTANTS—TABLE 1.   Table of Physical Constants

Quantity	Symbol	Value
Permeability of vacuum	μ_0	$4\pi \times 10^{-7}\,\mathrm{H\,m^{-1}}$
		$= 12.5663706144 \times 10^{-7}\,\mathrm{H\,m^{-1}}$
Speed of light in vacuum	c	$2.99792458(1.2)\,\mathrm{m\,s^{-1}}$
Permittivity of vacuum	$\varepsilon_0 = (\mu_0 c^2)^{-1}$	$8.85418782(7) \times 10^{-12}\,\mathrm{F\,m^{-1}}$
Fine structure constant, $\mu_0 c e^2 / 2h$	α	$0.0072973506(60)$
	α^{-1}	$137.03604(11)$
Elementary charge	e	$1.6021892(46) \times 10^{-19}\,\mathrm{C}$
Planck constant	h	$6.626176(36) \times 10^{-34}\,\mathrm{J\,Hz^{-1}}$
	$\hbar = h/2\pi$	$1.0545887(57) \times 10^{-34}\,\mathrm{J\,s}$
Avogadro constant	N_A	$6.022045(31) \times 10^{23}\,\mathrm{mol^{-1}}$
Atomic mass unit	$1\,\mathrm{u} = (10^{-3}\,\mathrm{kg\,mol^{-1}})/N_A$	$1.6605655(86) \times 10^{-27}\,\mathrm{kg}$
Electron rest mass	m_e	$0.9109534(47) \times 10^{-30}\,\mathrm{kg}$
		$5.4858026(21) \times 10^{-4}\,\mathrm{u}$
Muon rest mass	m_μ	$1.883566(11) \times 10^{-28}\,\mathrm{kg}$
		$0.11342920(26)\,\mathrm{u}$
Proton rest mass	m_p	$1.6726485(86) \times 10^{-27}\,\mathrm{kg}$
		$1.007276470(11)\,\mathrm{u}$
Neutron rest mass	m_n	$1.6749543(86) \times 10^{-27}\,\mathrm{kg}$
		$1.008665012(37)\,\mathrm{u}$
Ratio, proton mass to electron mass	m_p/m_e	$1836.15152(70)$
Ratio, muon mass to electron mass	m_μ/m_e	$206.76865(47)$
Specific electron charge	e/m_e	$1.7588047(49) \times 10^{11}\,\mathrm{C\,kg^{-1}}$
Faraday constant	$F = N_A e$	$9.648456(27) \times 10^4\,\mathrm{C\,mol^{-1}}$
Magnetic flux quantum	$\Phi_0 = h/2e$	$2.0678506(54) \times 10^{-15}\,\mathrm{Wb}$
	h/e	$4.135701(11) \times 10^{-15}\,\mathrm{J\,Hz^{-1}\,C^{-1}}$
Josephson frequency-voltage ratio	$2e/h$	$483.5939(13)\,\mathrm{THz\,V^{-1}}$
Quantum of circulation	$h/2m_e$	$3.6369455(60) \times 10^{-4}\,\mathrm{J\,Hz^{-1}\,kg^{-1}}$
	h/m_e	$7.273891(12) \times 10^{-4}\,\mathrm{J\,Hz^{-1}\,kg^{-1}}$
Rydberg constant	R_∞	$1.097373177(83) \times 10^7\,\mathrm{m^{-1}}$
Bohr radius	$a_0 = \alpha/4\pi R_\infty$	$0.52917706(44) \times 10^{-10}\,\mathrm{m}$
Electron Compton wavelength	$\lambda_C = \alpha^2/2R_\infty$	$2.4263089(40) \times 10^{-12}\,\mathrm{m}$
	$\lambdabar_C = \lambda_C/2\pi = \alpha a_0$	$3.8615905(64) \times 10^{-13}\,\mathrm{m}$
Classical electron radius	$r_e = \mu_0 e^2/4\pi m_e = \alpha \lambdabar_C$	$2.8179380(70) \times 10^{-15}\,\mathrm{m}$
Electron g-factor	$\frac{1}{2}g_e = \mu_e/\mu_B$	$1.0011596567(35)$
Muon g-factor	$\frac{1}{2}g_\mu$	$1.00116616(31)$

PHYSICAL CONSTANTS—TABLE 1. Table of Physical Constants (*Continued*)

Quantity	Symbol	Value
Proton moment in nuclear magnetons	μ_p/μ_N	2.7928456(11)
Bohr magneton	$\mu_B = e\hbar/2m_e$	$9.274078(36) \times 10^{-24}\,\mathrm{J\,T^{-1}}$
Nuclear magneton	$\mu_N = e\hbar/2m_p$	$5.050824(20) \times 10^{-27}\,\mathrm{J\,T^{-1}}$
Electron magnetic moment	μ_e	$9.284832(36) \times 10^{-24}\,\mathrm{J\,T^{-1}}$
Proton magnetic moment	μ_p	$1.4106171(55) \times 10^{-26}\,\mathrm{J\,T^{-1}}$
Proton magnetic moment in Bohr magnetons	μ_p/μ_B	$1.521032209(16) \times 10^{-3}$
Ratio, electron to proton magnetic moment	μ_e/μ_p	658.2106880(66)
Ratio, Muon moment to proton moment	μ_μ/μ_p	3.1833402(72)
Muon magnetic moment	μ_μ	$4.490474(18) \times 10^{-26}\,\mathrm{J\,T^{-1}}$
Proton gyromagnetic ratio	γ_p	$2.6751987(75) \times 10^{8}\,\mathrm{s^{-1}\,T^{-1}}$
Diamagnetic shielding factor, spherical H_2O sample	$1 + \sigma(H_2O)$	1.000025637(67)
Proton gyromagnetic ratio (uncorrected)	γ_p'	$2.6751301(75) \times 10^{8}\,\mathrm{s^{-1}\,T^{-1}}$
	$\gamma_p'/2\pi$	$42.57602(12)\,\mathrm{MHz\,T^{-1}}$
Proton moment in nuclear magnetons (uncorrected)	μ_p'/μ_N	2.7927740(11)
Proton Compton wavelength	$\lambda_{C,p} = h/m_p c$	$1.3214099(22) \times 10^{-15}\,\mathrm{m}$
	$\lambda_{C,p} = \lambda_{C,p}/2\pi$	$2.1030892(36) \times 10^{-16}\,\mathrm{m}$
Neutron Compton wavelength	$\lambda_{C,n} = h/m_n c$	$1.3195909(22) \times 10^{-15}\,\mathrm{m}$
	$\lambda_{C,n} = \lambda_{C,n}/2\pi$	$2.1001941(35) \times 10^{-16}\,\mathrm{m}$
Molar gas constant	R	$8.31441(26)\,\mathrm{J\,mol^{-1}\,K^{-1}}$
Molar volume, ideal gas ($T_0 = 273.15\,\mathrm{K}, p_0 = 1\,\mathrm{atm}$)	$V_m = RT_0/p_0$	$0.02241383(70)\,\mathrm{m^3\,mol^{-1}}$
Boltzmann constant	$k = R/N_A$	$1.380662(44) \times 10^{-23}\,\mathrm{J\,K^{-1}}$
Stefan-Boltzmann constant	$\sigma = (\pi^2/60)k^4/h^3 c^2$	$5.67032(71) \times 10^{-8}\,\mathrm{W\,m^{-2}\,K^{-4}}$
First radiation constant	$c_1 = 2\pi h c^2$	$3.741832(20) \times 10^{-16}\,\mathrm{W\,m^2}$
Second radiation constant	$c_2 = hc/k$	$0.01438786(45)\,\mathrm{m\,K}$
Gravitational constant	G	$6.6720(41) \times 10^{-11}\,\mathrm{N\,m^2\,kg^{-2}}$

REFRACTIVE INDEX

The refractive index (n) of a substance is the ratio of the velocity of light in a vacuum to the velocity of the light in the substance:

$$n = V_v/V \tag{93}$$

where V_v is the velocity of light in a vacuum, and V is the velocity of light in the substance.

REFRACTIVE INDEX — TABLE 1.
Refractive Indices of Hydrocarbons

Hydrocarbon	n_D^{20}
Pentane	1.3579
Hexane	1.3749
Heptane	1.3876
Octane	1.3975
Nonane	1.4054
Decane	1.4119
Nonadecane	1.4409
Eicosane	1.4425
Cyclopentane	1.4064
Cyclohexane	1.4266
Cycloheptane	1.4449
Benzene	1.5011
Toluene	1.4961
Ethylbenzene	1.4959
Propylbenzene	1.4920
Tetralin	1.5461
Decalin	1.4811

* n_D^{20}: Refractive index at 20°C (68°F) at the wavelength of the sodium D line.

REFRACTIVE INDEX — TABLE 2. Relationship of the Refractive Index to the Dielectric Constant of Hydrocarbons and Petroleum Products

Material	Refractive index n	n^2	Dielectric constant
Benzene	1.501	2.253	2.283
Cyclohexane	1.427	2.036	2.055
n-Hexane	1.375	1.890	1.890
n-Heptane	1.388	1.926	1.933
Kerosene	1.449	2.100	2.135
Paraffin oil	1.481	2.193	2.195
Vaseline	1.480	2.190	2.078

REFRACTIVE INDEX—TABLE 3.
Refractive Indices of Minerals

Mineral name	Refractive index
Actinolite	1.618–1.641
Adularia moonstone	1.525
Adventurine feldspar	1.532–1.542
Adventurine quartz	1.544–1.533
Agalmatoite	1.55
Agate	1.544–1.553
Albite feldspar	1.525–1.536
Albite moonstone	1.535
Alexandrite	1.745–1.759
Almandine garnet	1.76–1.83
Almandite garnet	1.79
Amazonite feldspar	1.525
Amber	1.540
Amblygonite	1.611–1.637
Amethyst	1.544–1.553
Anatase	2.49–2.55
Andalusite	1.634–1.643
Andradite garnet	1.82–1.89
Anhydrite	1.571–1.614
Apatite	1.632–1.648
Apophyllite	1.536
Aquamarine	1.577–1.583
Aragonite	1.530–1.685
Augelite	1.574–1.588
Axinite	1.675–1.685
Azurite	1.73–1.838
Barite	1.636–1.648
Barytocalcite	1.684
Benitoite	1.757–1.8
Beryl	1.577–1.60
Beryllonite	1.553–1.562
Brazilianite	1.603–1.623
Brownite	1.567–1.576
Calcite	1.486–1.658
Cancrinite	1.491–1.524
Cassiterite	1.997–2.093
Celestite	1.622–1.631
Cerussite	1.804–2.078
Ceylanite	1.77–1.80
Chalcedony	1.53–1.539
Chalybite	1.63–1.87
Chromite	2.1
Chrysoberyl	1.745
Chrysocolla	1.50
Chrysoprase	1.534
Citrine	1.55
Clinozoisite	1.724–1.734

REFRACTIVE INDEX—TABLE 3.
Refractive Indices of Minerals (*Continued*)

Mineral name	Refractive index
Colemanite	1.586–1.614
Coral	1.486–1.658
Cordierite	1.541
Corundum	1.766–1.774
Crocoite	2.31–2.66
Cuprite	2.85
Danburite	1.633
Demantoid garnet	1.88
Diamond	2.417–2.419
Diopsite	1.68–1.71
Dolomite	1.503–1.682
Dumortierite	1.686–1.723
Ekanite	1.60
Elaeolite	1.532–1.549
Emerald	1.576–1.582
Enstatite	1.663–1.673
Epidote	1.733–1.768
Euclase	1.652–1.672
Fibrolite	1.659–1.680
Fluorite	1.434
Gaylussite	1.517
Glass	1.44–1.90
Grossular garnet	1.738–1.745
Hambergite	1.559–1.631
Hauynite	1.502
Hematite	2.94–3.22
Hemimorphite	1.614–1.636
Hessonite garnet	1.745
Hiddenite	1.655–1.68
Howlite	1.586–1.609
Hypersthene	1.67–1.73
Idocrase	1.713–1.72
Iolite	1.548
Ivory	1.54
Jadeite	1.66–1.68
Jasper	1.54
Jet	1.66
Kornerupine	1.665–1.682
Kunzite	1.655–1.68
Kyanite	1.715–1.732
Labradorite feldspar	1.565
Lapis gem	1.50

REFRACTIVE INDEX—TABLE 3.
Refractive Indices of Minerals (*Continued*)

Mineral name	Refractive index
Lazulite	1.615–1.645
Leucite	1.5085
Magnesite	1.515–1.717
Malachite	1.655–1.909
Meerschaum	1.53. . . . none
Microcline feldspar	1.525
Moldavite	1.50
Moss agate	1.54–1.55
Natrolite	1.48–1.493
Nephrite	1.60–1.63
Nephrite jade	1.600–1.627
Obsidian	1.48–1.51
Oligoclase feldspar	1.539–1.547
Olivine	1.672
Onyx	1.486–1.658
Opal	1.45
Orthoclase feldspar	1.525
Painite	1.787–1.816
Pearl	1.52–1.69
Periclase	1.74
Peridot	1.654–1.69
Peristerite	1.525–1.536
Petalite	1.502–1.52
Phenakite	1.65–1.67
Phosgenite	2.117–2.145
Prase	1.54–1.533
Prasiolite	1.54–1.553
Prehnite	1.61–1.64
Proustite	2.79–3.088
Purpurite	1.84–1.92
Pyrite	1.81
Pyrope	1.74
Quartz	1.55
Rhodizite	1.69
Rhodochrisite	1.60–1.82
Rhodolite garnet	1.76
Rhodonite	1.73–1.74
Rock crystal	1.544–1.553
Ruby	1.76–1.77
Rutile	2.61–2.90
Sanidine	1.522
Sapphire	1.76–1.77
Scapolite	1.54–1.56

REFRACTIVE INDEX—TABLE 3.

Refractive Indices of Minerals (*Continued*)

Mineral name	Refractive index
Scapolite (yellow)	1.555
Scheelite	1.92–1.934
Serpentine	1.555
Shell	1.53–1.686
Sillimanite	1.658–1.678
Sinhalite	1.699–1.707
Smaragdite	1.608–1.63
Smithsonite	1.621–1.849
Sodalite	1.483
Spessartite garnet	1.81
Spinel	1.712–1.736
Sphalerite	2.368–2.371
Sphene	1.885–2.05
Spodumene	1.65–1.68
Staurolite	1.739–1.762
Steatite	1.539–1.589
Stichtite	1.52–1.55
Sulfur	1.96–2.248
Taaffeite	1.72
Tantalite	2.24–2.41
Tanzanite	1.691–1.70
Thomsonite	1.531
Tiger eye	1.544–1.553
Topaz (white)	1.638
Topaz (blue)	1.611
Topaz (pink, yellow)	1.621
Tourmaline	1.616–1.652
Tremolite	1.60–1.62
Tugtupite	1.496–1.50
Turquoise	1.61–1.65
Turquoise gem	1.61
Ulexite	1.49–1.52
Uvarovite	1.87
Variscite	1.55–1.59
Vivianite	1.580–1.627
Wardite	1.59–1.599
Willemite	1.69–1.72
Witherite	1.532–1.68
Wulfenite	2.300–2.40
Zincite	2.01–1.03
Zircon	1.801–2.01
Zirconia (cubic)	2.17
Zoisite	1.695

REFRACTIVE INDEX — TABLE 4. Refractive Indices of Organic Compounds (at 20°C under otherwise stated)

Substance	Formula	Density, g/ml	Refractive Index
Acenaphthene	$C_{12}H_{10}$	1.220	1.6048/98.8°
Acetaldehyde	C_2H_4O	0.788/16°	1.3316
Acetamide	C_2H_5ON	1.159	1.4274/78°
Acetanilide	C_8H_9ON	1.21/4°	
Acetic acid	$C_2H_4O_2$	1.0492	1.3718
Acetic anhydride . . .	$C_4H_6O_3$	1.0850/15°	1.3904
Acetone	C_3H_6O	0.787/25°	1.3620/15°
Acetonitrile	C_2H_3N	0.7828	1.3460
Acetophenone	C_8H_8O	1.0329/15°	1.5342/19°
Acetyl chloride	C_2H_3OCl	1.1051	1.3898
Acetylene	C_2H_2	0.61/–80°	
Adipic acid	$C_6H_{10}O_4$	1.366	
Alloxan + $_4H_2O$	$C_4H_{10}O_8N_2$		
Allyl alcohol	C_3H_6O	0.8573/15°	1.4135
p-Aminobenzoic acid	$C_7H_7O_2N$		
2-Aminopyridine . . .	$C_5H_6N_2$		
n-Amyl alcohol	$C_5H_{12}O$	0.8154	1.414/13°
act-Amyl alcohol . . .	$C_5H_{12}O$	0.816	
sec-Amyl alcohol . . .	$C_5H_{12}O$	0.8103	1.4053
tert-Amyl alcohol . . .	$C_5H_{12}O$	0.809	1.4045
Aniline	C_6H_7N	1.026/15°	1.5863
Aniline hydrochloride	C_6H_8NCl	1.222/4°	
Anisole	C_7H_8O	0.9925/25°	1.5150/22°
Anthracene	$C_{14}H_{10}$	1.243	
Anthraquinone	$C_{14}H_8O$	1.419/4°	
Azobenzene	$C_{12}H_{10}N_2$		
Benzaldehyde	C_7H_6O	1.0504/15°	1.5463/17.6°
Benzene	C_6H_6	0.8790	1.5011
Benzoic acid	$C_7H_6O_2$	1.2659/15°	1.5397/15°
Benzoic anhydride . .	$C_{14}H_{10}O_3$	1.1989/15°	1.5767/15°
Benzoin	$C_{14}H_{12}O_2$		
Benzonitrile	C_7H_5N	1.0093/15°	1.5289
Benzophenone (a) . .	$C_{13}H_{10}O$	1.085/50°	
Benzoquinone	$C_6H_4O_2$		
Benzoyl chloride	C_7H_5OCl	1.212	1.5537
Benzoyl peroxide . . .	$C_{14}H_{10}O_4$		
Benzyl alcohol	C_7H_8O	1.049/15°	1.5396
Benzyl benzoate	$C_{14}H_{12}O_2$	1.114/18°	1.5681/21°
Benzyl chloride	C_7H_7Cl	1.0983	1.5415/15°
Benzyl cinnamate . . .	$C_{16}H_{14}O_2$		
Borneol (DL)	$C_{10}H_{18}O$	1.01	
a-Bromonaphthalene	$C_{10}H_7Br$	1.4888/16.5°	1.6601/16.5°
Bromobenzene	C_6H_5Br	1.4978/15°	1.5625/15°
Bromoform	$CHBr_3$	2.900/15°	1.6005/15°
n-Butane	C_4H_{10}	0.5788 (at sat. pressure)	
n-Butyl alcohol	$C_4H_{10}O$	0.8098	1.3993
iso-Butyl alcohol . . .	$C_4H_{10}O$	0.8169	1.3968/17.5°
sec-Butyl alcohol . . .	$C_4H_{10}O$	0.808	1.3949/25°
tert-Butyl alcohol . . .	$C_4H_{10}O$	0.7887	1.3878
n-Butyl chloride	C_4H_9Cl	0.9074/0	1.4015

REFRACTIVE INDEX—TABLE 4.   Refractive Indices of Organic Compounds (*Continued*)

Substance	Formula	Density, g/ml	Refractive Index
n-Butyric acid	$C_4H_8O_2$	0.9587	1.3991
iso-Butyric acid	$C_4H_8O_2$	0.950	
Camphene (DL)	$C_{10}H_{16}$	0.879	1.4402/80°
Camphor (D)	$C_{10}H_{16}O$	0.992/10°	
Carbitol (Diethyleneglycol-monomethylether)	$C_6H_{14}O_3$	0.9902	
Carbon disulphide	CS_2	1.2927/0°	1.6276
Carbon tetrabromide	CBr_4	2.9109/99.5°	
Carbon tetrachloride	CCl_4	1.6320/0°	1.4607
Cellosolve (Glycolmonoethylether)	$C_4H_{10}O_2$	0.9311	
Chloral hydrate	$C_2H_3O_2Cl_3$	1.9081	
Chloroacetic acid	$C_2H_3O_2Cl$	1.39/75°	1.4297/65°
Chlorobenzene	C_6H_5Cl	1.066	1.5248
Chloroform	$CHCl_3$	1.4985/15°	1.4467
Cholesterol	$C_{27}H_{46}O$	1.067	
Cineol (Eucalyptol)	$C_{10}H_{18}O$	0.9267	1.4584/18°
Cinnamic acid (trans)	$C_9H_8O_2$	1.247	
Cinnamyl alcohol	$C_9H_{10}O$	1.0440	1.5819
Citric acid	$C_6H_8O_7$	1.542/18°	
o-Cresol	C_7H_8O	1.051	1.5372/40°
m-Cresol	C_7H_8O	1.035	1.5406
p-Cresol	C_7H_8O	1.035	1.5316
Cumene	C_9H_{12}	0.8615	1.4909
Cyclohexane	C_6H_{12}	0.7786	1.4262
Cyclohexanol	$C_6H_{12}O$	0.9624	1.4656/22°
Cyclohexanone	$C_6H_{10}O$	0.9478	1.4507
Cyclohexene	C_6H_{10}	0.8108	1.4467
p-Cymene	$C_{10}H_{14}$	0.8766	1.5006
cis-Decalin	$C_{10}H_{18}$	0.8963	1.4811
trans-Decalin	$C_{10}H_{18}$	0.8703/18°	1.4697/18°
Dibenzyl	$C_{14}H_{14}$	0.995	
n-Dibutyl phthalate	$C_{16}H_{22}O_4$	1.0465	
Diethylamine	$C_4H_{11}N$	0.7108/18°	1.3873/18°
Difluorodichloro-methane (Freon 12)	CCl_2F_2		
Difluoromonochloro-methane (Freon 22)	$CHClF_2$		
Dimethylamine	C_2H_7N	0.6804/0°	1.350/17°
Dimethylaniline	$C_8H_{11}N$	0.9557	1.5582
Dioxane	$C_4H_8O_2$	1.0338	1.4224
Diphenyl	$C_{12}H_{10}$	1.180/0°	1.5852/79°
Diphenylamine	$C_{12}H_{11}N$	1.159	
Epichlorhydrin	C_3H_5OCl	1.180	1.4420/11.6°
Ethane	C_2H_6		
Ethanolamine	C_2H_7ON	1.022	1.4539
di-Ethanolamine	$C_4H_{11}O_2N$	1.0966	1.4776
tri-Ethanolamine	$C_6H_{15}O_3N$	1.1242	1.4852

REFRACTIVE INDEX—TABLE 4. Refractive Indices of Organic Compounds (*Continued*)

Substance	Formula	Density, g/ml	Refractive Index
Ether (diethyl)	$C_4H_{10}O$	0.714/20°	1.3538
Ethyl acetate	$C_4H_8O_2$	0.9245	1.3701/25°
Ethyl acetoacetate	$C_6H_{10}O_3$	1.0282	1.4209/16°
Ethyl alcohol	C_2H_6O	0.7893	1.3610/20.5°
Ethylamine	C_2H_7N	0.7057/0°	
Ethylbenzene	C_8H_{10}	0.8669	1.4959
Ethyl benzoate	$C_9H_{10}O_2$	1.0509/15°	1.5068/17.3°
Ethyl bromide	C_2H_5Br	1.4555	1.4239
Ethyl chloride	C_2H_5Cl	0.9214/0°	
Ethylene	C_2H_4		
Ethylenediamine	$C_2H_8N_2$	0.902/15°	1.4540/26.1°
Ethylene dibromide	$C_2H_4Br_2$	2.1785	1.5379
Ethylene dichloride	$C_2H_4Cl_2$	1.2521	1.4443
Ethylene glycol	$C_2H_6O_2$	1.1155	1.4274
Ethylene oxide	C_2H_4O	0.877/7°	1.3597/7°
Ethyl formate	$C_3H_6O_2$	0.9168	1.3598
Ethyl iodide	C_2H_5I	1.9133/30°	1.5168/15°
Ethyl mercaptan	C_2H_6S	0.8315/25°	1.4351
Ethyl nitrate	$C_2H_5O_3N$	1.109	1.3853
Ethyl nitrite	$C_2H_5O_2N$	0.900/15°	
Ethyl oxalate	$C_6H_{10}O_4$	1.0785	1.4101
Ethyl salicylate	$C_9H_{10}O_3$	1.131	1.5226
Ethyl sulphate	$C_4H_{10}O_4S$	1.180/18°	1.4010/18°
Eugenol	$C_{10}H_{12}O_2$	1.0620/25°	1.5439/19°
Fluorescein	$C_{20}H_{12}O_5$		
Fluorobenzene	C_6H_5F	1.0236	1.4677
Formaldehyde	CH_2O	0.815/−20°	
Formamide	CH_3ON	1.1334	1.4472
Formic acid	CH_2O_2	1.220	1.3714
Fructose	$C_6H_{12}O_6$	1.598	
Fumaric acid	$C_4H_4O_4$	1.635	
Furfural	$C_5H_4O_2$	1.1594	1.5261
Furfuryl alcohol	$C_5H_6O_2$	1.1282/23°	1.4852
Furan	C_4H_4O	0.9644/0°	1.4216
Glucose	$C_6H_{12}O_6$	1.544/25°	
Glycerol	$C_3H_8O_3$	1.2604/17.5°	1.4730
Glyceryl trioleate	$C_{57}H_{104}O_6$	0.8992/50°	1.4561/60°
Glyceryl tripalmitate	$C_{51}H_{98}O_6$	0.8752/70°	1.4381/80°
Glyceryl tristearate	$C_{57}H_{110}O_6$	0.8559/90°	1.4385/80°
Glycine	$C_2H_5O_2N$		
Guaiacol	$C_7H_8O_2$	1.1287/21.4°	
n-Heptane	C_7H_{16}	0.6838	1.3877
Hexachlorotethane	C_2Cl_6	2.091	
Hexamine	$C_6H_{12}N_4$		
n-Hexane	C_6H_{14}	0.6594	1.3749
Hippuric acid	$C_9H_9O_3N$	1.371	
Hydroquinone	$C_6H_6O_2$	1.358	
Indene	C_9H_8	0.996	1.5766
Iodoform	CHI_3	4.008	

REFRACTIVE INDEX—TABLE 4.   Refractive Indices of Organic Compounds (*Continued*)

Substance	Formula	Density, g/ml	Refractive Index
Isobutane	C_4H_{10}	0.5572 (at sat. press.)	
Isopentane	C_5H_{12}	0.6192	1.3538
isoprene	C_5H_8	0.6806	1.4194
Isooctane	C_8H_{18}	0.6919	1.3915
Isoquinoline	C_9H_7N	1.099	1.6223/25°
Lactic acid	$C_3H_6O_3$	1.2485	1.4414
Lactose + H_2O	$C_{12}H_{24}O_1$	1.525	
Maleic acid	$C_4H_4O_4$	1.5920	
Maleic anhydride	$C_4H_2O_3$	0.934	
Malonic acid	$C_3H_4O_4$	1.631/15°	
Maltose + H_2O	$C_{12}H_{24}O_1$	1.540	
Menthol (L)	$C_{10}H_{20}O$	0.903/15°	
Mesitylene	C_9H_{12}	0.8652	1.4994
Metaldehyde	$(C_2H_4O)_n$		
Methane	CH_4		
Methyl acetate	$C_3H_6O_2$	0.9280	1.3593/20°
Methyl alcohol	CH_4O	0.7910	1.3276/25°
Methylamine	CH_5N	0.699/−10.8°	
Methylaniline	C_7H_9N	0.9891	1.5702/21.2°
Methyl anthranilate	$C_8H_9O_2N$	1.1682/18.6°	
Methyl benzoate	$C_8H_8O_2$	1.0937/15°	1.5205/15°
Methyl bromide	CH_3Br	1.732/0°	
Methyl carbonate	$C_3H_6O_3$	1.0694	1.3687
Methyl chloride	CH_3Cl	0.991/−25°	
Methylene bromide	CH_2Br_2	2.8098/15°	
Methylene chloride	CH_2Cl_2	1.3348/15°	1.4237
Methyl ethyl ketone	C_4H_8O	0.8054	1.3814/15°
Methyl formate	$C_2H_4O_2$	0.9867/15°	1.344
Methyl iodide	CH_3I	2.251/30°	1.5293/21°
Methyl methacrylate	$C_5H_8O_2$	0.936	1.413
Methyl sulphate	$C_2H_6O_4S$	1.3348/15°	1.3874
Methyl salicylate	$C_8H_8O_3$	1.1787/25°	1.538/18.1°
Monofluorotrichloromethane (Freon 11)	CCl_3F	1.494/17°	
Morpholine	C_4H_9ON	0.9994	1.4545
Naphthalene	$C_{10}H_8$	1.14	1.5822/100°
α-Naphthol	$C_{10}H_8O$	1.099/99°	1.6206/98.7°
β-Naphthol	$C_{10}H_8O$	1.272	
α-Naphthylamine	$C_{10}H_9N$	1.1196/25°	1.6703/51°
β-Naphthylamine	$C_{10}H_9N$	1.0614/98°	1.6493/98°
Nicotine (L)	$C_{10}H_{14}N_2$	1.0097	1.5280
Nitrobenzene	$C_6H_5O_2N$	1.1732/25°	1.5530
Nitroethane	$C_2H_5O_2N$	1.050	1.3916
Nitromethane	CH_3O_2N	1.137	1.3818
1-Nitropropane	$C_3H_7O_2N$	1.001	1.4015
2-Nitropropane	$C_3H_7O_2N$	0.990	1.3941
n-Octane	C_8H_{18}	0.7025	1.3974
n-Octyl alcohol	$C_8H_{18}O$	0.8270	1.4292
Oleic acid	$C_{18}H_{34}O_2$	0.898	1.4582
Oxalic acid	$C_2H_2O_4$		

REFRACTIVE INDEX—TABLE 4.   Refractive Indices of Organic Compounds (*Continued*)

Substance	Formula	Density, g/ml	Refractive Index
Palmitic acid	$C_{16}H_{32}O_2$	0.8527/62°	1.4339/60°
Paraformaldehyde	$(CH_2O)_n$		
Paraldehyde	$C_6H_{12}O_3$	0.9943	1.4049
n-Pentane	C_5H_{12}	0.6262	1.3575
Phosgene	$COCl_2$		
Phenanthrene	$C_{14}H_{10}$	1.17	1.6567/129°
Phenol	C_6H_6O	1.073	1.5425/40.6°
Phthalic acid	$C_8H_6O_4$	1.593	
Phthalic anhydride	$C_8H_4O_3$	1.527/4°	
Phthalimide	$C_8H_5O_2N$		
α-Picoline	C_6H_7N	0.9443	1.5010
β-Picoline	C_6H_7N	0.9566	1.5068
γ-Picoline	C_6H_7N	0.9548	1.5058
Picric acid	$C_6H_3O_7N_3$	1.763	
Picryl chloride	$C_6H_2O_6N_3Cl$	1.797	
Pinene (Turpentine)	$C_{10}H_{16}$	0.861	1.4685/15°
Piperidine	$C_5H_{11}N$	0.8606	1.4530
Propane	C_3H_8		
n-Propyl acetate	$C_5H_{10}O_2$	0.887	1.3844
n-Propyl alcohol	C_3H_8O	0.8035	1.3850
iso-Propyl alcohol	C_3H_8O	0.7855	1.3776
Propylene	C_3H_6	0.5139 (at sat. press.)	
Pyridine	C_5H_5N	0.9831	1.5102
Pyrocatechol	$C_6H_6O_2$	1.344	
Pyrogallol	$C_6H_6O_3$		
Quinhydrone	$C_{12}H_{10}O_4$	1.401	
Quinoline	C_9H_7N	1.095	1.6269
Resorcinol	$C_6H_6O_2$	1.285/15°	
Salicylic acid	$C_7H_6O_3$	1.443	
Stearic acid	$C_{18}H_{36}O_2$	0.9408	1.4335/70°
Styrene	C_8H_8	0.9060	1.5469
Succinic acid	$C_4H_6O_4$	1.564/15°	
Succinic anhydride	$C_4H_4O_3$	1.234	
Sucrose	$C_{12}H_{22}O_{11}$	1.588/15°	
Sylvan (2-Methylfuran)	C_5H_6O	0.916	
Tartaric acid (*meso-*)	$C_4H_6O_6$	1.666	
Tartaric acid (racemic) + H_2O	$C_4H_8O_7$	1.697	
Tartaric acid (D)	$C_4H_6O_6$	1.7598	
Tartaric acid (L)	$C_4H_6O_6$	1.7598	
Tetralin	$C_{10}H_{12}$		1.5453/17°
Thiophen	C_4H_4S	1.0644	1.5287
Thiourea	CH_4N_2S	1.405	
Thymol	$C_{10}H_{14}O$	0.969	
Toluene	C_7H_8	0.8670	1.4969
o-Toluidine	C_7H_9N	1.0035	1.5688
m-Toluidine	C_7H_9N	0.987/25°	1.5686
p-Toluidine	C_7H_9N	0.961/50°	1.5532/59.1°
Trichloroethylene	C_2HCl_3	1.4597/15°	1.4782
Tri-*o*-cresyl phosphate	$C_{21}H_{21}O_4P$		

Substance	Formula	Density, g/ml	Refractive Index
Tri-*p*-cresyl phosphate	$C_{21}H_{21}O_4P$		
Triethylamine	$C_6H_{15}N$	0.7495/0°	1.4003
Trimethylamine	C_3H_9N	0.6709/0°	
Trinitrotoluene	$C_7H_5O_6N_3$	1.654	
Triphenylmethane	$C_{19}H_{16}$		
Urea	CH_4ON_2	1.335	
Uric acid	$C_5H_4O_3N_4$	1.893	
n-Valeric acid	$C_5H_{10}O_2$	0.942	1.4086
iso-Valeric acid	$C_5H_{10}O_2$	0.937/15°	1.4018/22.4°
Vanillin	$C_8H_8O_3$		
o-Xylene	C_8H_{10}	0.8802	1.5054
m-Xylene	C_8H_{10}	0.8642	1.4972
p-Xylene	C_8H_{10}	0.8611	1.4958

REFRACTIVE INDEX—TABLE 5. Refractive Indices of Polymers

Polymer Name	Refractive Index (20°C, 68°F)
Acetal homopolymer	1.48
Acrylics	1.49–1.52
Allyl diglycol carbonate	1.50
Cellulose acetate	1.46–1.50
Cellulose acetate butyrate	1.46–1.49
Cellulose ester	1.47–1.50
Cellulose nitrate	1.49–1.51
Cellulose propionate	1.46–1.49
Chlorotrifluoroethylene (CTFE)	1.42
Diallyl isophthalate	1.57
Epoxies	1.55–1.65
Ethyl cellulose	1.47
Fluorinated ethylene-propylene	1.34
Methylpentene polymer	1.485
Nylon	1.52–1.53
Phenol formaldehyde	1.50–1.70
Phenoxy polymer	1.60
Polyacetal	1.48
Polyallomer	1.49
Polyallyl methacrylate	1.52
Polyamide nylon 6/6	1.53
Polyamide nylon 11	1.52
Polybutylene	1.50
Polycarbornate	1.57–1.59
Poly(cyclohexyl methacrylate)	1.51
Poly(diallyl phthalate)	1.57
Polyester	1.53–1.58
Poly(ester-styrene)	1.54–1.57

Polymer Name	Refractive Index (20°C, 68°F)
Polyethylene (low density)	1.51
Polyethylene (medium density)	1.52
Polyethylene (high density)	1.54
Polyethylene dimethacrylate	1.51
Poly(ethylene terephthalate)	1.57–1.58
Poly(methyl-α-chloroacrylate)	1.52
Poly(methyl methacrylate)	1.49
Polypropylene	1.49
Poly(propyl methacrylate)	1.48
Polystyrene	1.57–1.60
Polysulfone	1.63
Poly(tetrafluoroethylene) (PTFE)	1.35
Poly(trifluorochloroethylene)	1.43
Poly(trifluoroethylene)	1.35–1.37
Poly(vinyl alcohol)	1.49–1.53
Poly(vinyl acetal)	1.48
Poly(vinyl acetate)	1.46–1.47
Poly(vinyl butyral)	1.49
Poly(vinyl chloride)	1.52–1.55
Poly(vinyl cyclohexene dioxide)	1.53
Poly(vinyl formal)	1.60
Poly(vinyl naphthalene)	1.68
Poly(vinylidene chloride)	1.60–1.63
Poly(vinylidene fluoride)	1.42
Silicone polymer	1.43
Styrene acrylonitrile copolymer	1.56–1.57
Styrene butadiene thermoplastic	1.52–1.55
Styrene methacrylate copolymer	1.53
Urea formaldehyde	1.54–1.58
Urethane	1.50–1.60

TABLE 6. Refractivity Intercept for Hydrocarbon Types

Material	Refractivity intercept
Paraffins	1.0461
Saturated monocyclics	1.0400
Saturated polycyclics	1.0285
Olefins	1.0521
Diolefins	1.0592
Conjugated diolefins	1.0877
Cycloolefins	1.0461
Conjugated cyclodiolefins	1.0643
Aromatics	1.0627

SI UNITS AND CONVERSION FACTORS

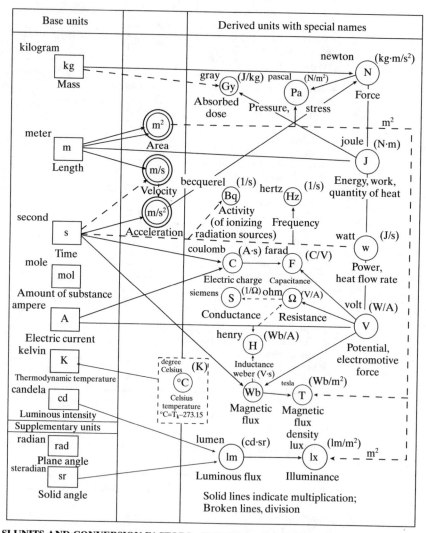

SI UNITS AND CONVERSION FACTORS—FIGURE 1. Relationship of SI Units

SI UNITS AND CONVERSION FACTORS—TABLE 1. SI Base and
Supplementary Quantities and Units

Quantity or "dimension"	SI unit	SI unit symbol ("abbreviation") Use roman (upright) type
Base quantity or "dimension"		
length	meter	m
mass	kilogram	kg
time	second	s
electric current	ampere	A
thermodynamic temperature	kelvin	K
amount of substance	mole*	mol
luminous intensity	candela	cd
Supplementary quantity or "dimension"		
plane angle	radian	rad
solid angle	steradian	sr

   * When the mole is used, the elementary entities must be specified; they may be atoms, molecules, ions, electrons, other particles, or specified groups of such particles.

SI UNITS AND CONVERSION FACTORS—TABLE 2. Derived Units of SI That
Have Special Names

Quantity	Unit	Symbol	Formula
Frequency (of a periodic phenomenon)	hertz	Hz	l/s
Force	newton	N	$(kg \cdot m)/s^2$
Pressure, stress	pascal	Pa	N/m^2
Energy, work, quantity of heat	joule	J	$N \cdot m$
Power, radiant flux	watt	W	J/s
Quantity of electricity, electric charge	coulomb	C	$A \cdot s$
Electric potential, potential difference, electromotive force	volt	V	W/A
Capacitance	farad	F	C/V
Electric resistance	ohm	Ω	V/A
Conductance	siemens	S	A/V
Magnetic flux	weber	Wb	$V \cdot s$
Magnetic-flux density	tesla	T	Wb/m^2
Inductance	henry	H	Wb/A
Luminous flux	lumen	lm	$cd \cdot sr$
Illuminance	lux	lx	lm/m^2
Activity (of radionuclides)	becquerel	Bq	l/s
Absorbed dose	gray	Gy	J/kg

SI UNITS AND CONVERSION FACTORS—TABLE 3. Additional Common Derived Units of SI

Quantity	Unit	Symbol
Acceleration	meter per second squared	m/s^2
Angular acceleration	radian per second squared	rad/s^2
Angular velocity	radian per second	rad/s
Area	square meter	m^2
Concentration (of amount of substance)	mole per cubic meter	mol/m^3
Current density	ampere per square meter	A/m^2
Density, mass	kilogram per cubic meter	kg/m^3
Electric-charge density	coulomb per cubic meter	C/m^3
Electric-field strength	volt per meter	V/m
Electric-flux density	coulomb per square meter	C/m^2
Energy density	joule per cubic meter	J/m^3
Entropy	joule per kelvin	J/K
Heat capacity	joule per kelvin	J/K
Heat-flux density, irradiance	watt per square meter	W/m^2
Luminance	candela per square meter	cd/m^2
Magnetic-field strength	ampere per meter	A/m
Molar energy	joule per mole	J/mol
Molar entropy	joule per mole-kelvin	$J/(mol \cdot K)$
Molar-heat capacity	joule per mole-kelvin	$J/(mol \cdot K)$
Moment of force	newton-meter	$N \cdot m$
Permeability	henry per meter	H/m
Permittivity	farad per meter	F/m
Radiance	watt per square-meter-steradian	$W/(m^2 \cdot sr)$
Radiant intensity	watt per steradian	W/sr
Specific-heat capacity	joule per kilogram-kelvin	$J/(kg \cdot K)$
Specific energy	joule per kilogram	J/kg
Specific entropy	joule per kilogram-kelvin	$J/(kg \cdot K)$
Specific volume	cubic meter per kilogram	m^3/kg
Surface tension	newton per meter	N/m
Thermal conductivity	watt per meter-kelvin	$W/(m \cdot K)$
Velocity	meter per second	m/s
Viscosity, dynamic	pascal-second	$Pa \cdot s$
Viscosity, kinematic	square meter per second	m^2/s
Volume	cubic meter	m^3
Wave number	1 per meter	$1/m$

SI UNITS AND CONVERSION FACTORS—TABLE 4. SI Prefixes

Multiplication factor	Prefix	Symbol
1 000 000 000 000 000 000 = 10^{18}	exa	E
1 000 000 000 000 000 = 10^{15}	peta	P
1 000 000 000 000 = 10^{12}	tera	T
1 000 000 000 = 10^{9}	giga	G
1 000 000 = 10^{6}	mega	M
1 000 = 10^{3}	kilo	k
100 = 10^{2}	hecto*	h
10 = 10^{1}	deka*	da
0.1 = 10^{-1}	deci*	d
0.01 = 10^{-2}	centi	c
0.001 = 10^{-3}	milli	m
0.000 001 = 10^{-6}	micro	μ
0.000 000 001 = 10^{-9}	nano	n
0.000 000 000 001 = 10^{-12}	pico	p
0.000 000 000 000 001 = 10^{-15}	femto	f
0.000 000 000 000 000 001 = 10^{-18}	atto	a

* Generally to be avoided.

SI UNITS AND CONVERSION FACTORS—TABLE 5. Conversion Factors: U.S. Customary and Commonly Used Units to SI Units

Quantity	Customary or commonly used unit	SI unit	Alternate SI unit	Conversion factor; multiply customary unit by factor to obtain SI unit	
		Space,† time			
Length	naut mi	km		1.852*	E + 00
	mi	km		1.609 344*	E + 00
	chain	m		2.011 68*	E + 01
	link	m		2.011 68*	E – 01
	fathom	m		1.828 8*	E + 00
	yd	m		9.144*	E – 01
	ft	m		3.048*	E – 01
		cm		3.048*	E + 01
	in	mm		2.54*	E + 01
	in	cm		2.54	E + 00
	mil	μm		2.54*	E + 01
Length/length	ft/mi	m/km		1.893 939	E – 01
Length/volume	ft/U.S. gal	m/m^3		8.051 964	E + 01
	ft/ft^3	m/m^3		1.076 391	E + 01
	ft/bbl	m/m^3		1.917 134	E + 00
Area	mi^2	km^2		2.589 988	E + 00
	section	ha		2.589 988	E + 02
	acre	ha		4.046 856	E – 01
	ha	m^2		1.000 000*	E + 04
	yd^2	m^2		8.361 274	E – 01

SI UNITS AND CONVERSION FACTORS—TABLE 5.   Conversion Factors: U.S.
Customary and Commonly Used Units to SI Units (*Continued*)

Quantity	Customary or commonly used unit	SI unit	Alternate SI unit	Conversion factor; multiply customary unit by factor to obtain SI unit	
		Space,† time			
	ft^2	m^2		9.290304*	E − 02
	in^2	mm^2		6.4516*	E + 02
		cm^2		6.4516*	E + 00
Area/volume	ft^2/in^3	m^2/cm^3		5.699291	E − 03
	ft^2/ft^3	m^2/m^3		3.280840	E + 00
Volume	cubem	km^3		4.168182	E + 00
	acre·ft	m^3		1.233482	E + 03
		ha·m		1.233482	E − 01
	yd^3	m^3		7.645549	E − 01
	bbl	m^3		1.589873	E − 01
	(42 U.S. gal)	m^3		2.831685	E − 02
	ft^3	dm^3	L	2.831685	E + 01
	U.K. gal	m^3		4.546092	E − 03
		dm^3	L	4.546092	E + 00
	U.S. gal	m^3		3.785412	E − 03
		dm^3	L	3.785412	E + 00
	U.K. qt	dm^3	L	1.136523	E + 00
	U.S. qt	dm^3	L	9.463529	E − 01
	U.S. pt	dm^3	L	4.731765	E − 01
	U.K. fl oz	cm^3		2.841307	E + 01
	U.S. fl oz	cm^3		2.957353	E + 01
	in^3	cm^3		1.638706	E + 01
Volume/length (linear displacement)	bbl/in	m^3m		6.259342	E + 00
	bbl/ft	m^3/m		5.216119	E − 01
	ft^3/ft	m^3/m		9.290304*	E − 02
	U.S. gal/ft	m^3/m		1.241933	E − 02
		L/m		1.241933	E + 01
Plane angle	rad	rad		1	
	deg (°)	rad		1.745329	E − 02
	min (′)	rad		2.908882	E − 04
	sec (″)	rad		4.848137	E − 06
Solid angle	sr	sr		1	
Time	year	a		1	
	week	d		7.0*	E + 00
	h	s		3.6*	E + 03
		min		6.0*	E + 01
	min	s		6.0*	E + 01
		h		1.666667	E − 02
	mµs	ns		1	
		Mass, amount of substance			
Mass	U.K. ton	Mg	t	1.016047	E + 00
	U.S. ton	Mg	t	9.071847	E − 01

SI UNITS AND CONVERSION FACTORS—TABLE 5.　Conversion Factors: U.S. Customary and Commonly Used Units to SI Units (*Continued*)

Quantity	Customary or commonly used unit	SI unit	Alternate SI unit	Conversion factor; multiply customary unit by factor to obtain SI unit	
		Space,† time			
	U.K. cwt	kg		5.080 234	E + 01
	U.S. cwt	kg		4.535 924	E + 01
	lbm	kg		4.535 924	E − 01
	oz (troy)	g		3.110 348	E + 01
	oz (av)	g		2.834 952	E + 01
	gr	mg		6.479 891	E + 01
Amount of	lbm·mol	kmol		4.535 924	E − 01
substance	std m³ (0°C, 1 atm)	kmol		4.461 58	E − 02
	std ft³ (60°F, 1 atm)	kmol		1.195 30	E − 03
		Enthalpy, calorific value, heat, entropy, heat capacity			
Calorific value,	Btu/lbm	MJ/kg		2.326 000	E − 03
enthalpy		kJ/kg	J/g	2.326 000	E + 00
(mass basis)		kWh/kg		6.461 112	E − 04
	cal/g	kJ/kg	J/g	4.184*	E + 00
	cal/lbm	J/kg		9.224 141	E + 00
Caloric value,	kcal/(g·mol)	kJ/kmol		4.184*	E + 03
enthalpy	Btu/(lb·mol)	kJ/kmol		2.326 000	E + 00
(mole basis)					
Calorific value	Btu/U.S. gal	MJ/m³	kJ/dm³	2.787 163	E − 01
(volume basis		kJ/m³		2.787 163	E + 02
—solids and		kWh/m³		7.742 119	E − 02
liquids)	Btu/U.K. gal	MJ/m³	kJ/dm³	2.320 800	E − 01
		kJ/m³		2.320 800	E + 02
	Btu/ft³	kWh/m³		6.446 667	E − 02
		MJ/m³	kJ/dm³	3.725 895	E − 02
		kJ/m³		3.725 895	E + 01
		kWh/m³		1.034 971	E − 02
	cal/mL	MJ/m³		4.184*	E + 00
	(ft·lbf)/U.S. gal	kJ/m³		3.581 692	E − 01
Calorific value	cal/mL	kJ/m³	J/dm³	4.184*	E + 03
(volume	kcal/m³	kJ/m³	J/dm³	4.184*	E + 00
basis—gases)	Btu/ft³	kJ/m³	J/dm³	3.725 895	E + 01
		kWh/m³		1.034 971	E − 02
Specific	Btu/(lbm·°R)	kJ/(kg·K)	J/(g·K)	4.186 8*	E + 00
entropy	cal/(g·K)	kJ/(kg·K)	J/(g·K)	4.184*	E + 00
	kcal/(kg·°C)	kJ/(kg·K)	J/(g·K)	4.184*	E + 00
Specific-heat	kWh/(kg·°C)	kJ/(kg·K)	J/(g·K)	3.6*	E + 03
capacity	Btu/(lbm·°F)	kJ/(kg·K)	J/(g·K)	4.186 8*	E + 00
(mass basis)	kcal/(kg·°C)	kJ/(kg·K)	J(g·K)	4.184*	E + 00
Specific-heat	Btu/(lb·mol·°F)	kJ/(kmol·K)		4.186 8*	E + 00
capacity	cal/(g·mol·°C)	kJ/(kmol·K)		4.184*	E + 00
(mole basis)					

SI UNITS AND CONVERSION FACTORS—TABLE 5.   Conversion Factors: U.S.
Customary and Commonly Used Units to SI Units (*Continued*)

Quantity	Customary or commonly used unit	SI unit Space,† time	Alternate SI unit	Conversion factor; multiply customary unit by factor to obtain SI unit
		Temperature, pressure, vacuum		
Temperature (absolute)	°R	K		5/9
	K	K		1
Temperature (traditional)	°F	°C		5/9(°F − 32)
Temperature (difference)	°F	K, °C		5/9
Pressure	atm (760 mmHg at 0°C or 14,696 psi)	MPa		1.013 250* E − 01
		kPa		1.013 250* E + 02
		bar		1.013 250* E + 00
	bar	MPa		1.0* E − 01
		kPa		1.0* E + 02
	mmHg (0°C) = torr	MPa		6.894 757 E − 03
		kPa		6.894 757 E + 00
		bar		6.894 757 E − 02
	μmHg (0°C)	kPa		3.376 85 E + 00
	μ bar	kPa		2.488 4 E − 01
	mmHg = torr (0°C)	kPa		1.333 224 E − 01
	cmH_2O (4°C)	kPa		9.806 38 E − 02
	lbf/ft² (psf)	kPa		4.788 026 E − 02
	mHg (0°C)	Pa		1.333 224 E − 01
	bar	Pa		1.0* E + 05
	dyn/cm²	Pa		1.0* E − 01
Vacuum, draft	inHg (60°F)	kPa		3.376 85 E + 00
	inH_2O (39.2°F)	kPa		2.490 82 E − 01
	inH_2O (60°F)	kPa		2.488 4 E − 01
	mmHg (0°C) = torr	kPa		1.333 224 E − 01
	cmH_2O (4°C)	kPa		9.806 38 E − 02
Liquid head	ft	m		3.048* E − 01
	in	mm		2.54* E + 01
		cm		2.54* E + 00
Pressure drop/length	psi/ft	kPa/m		2.262 059 E + 01
		Density, specific volume, concentration, dosage		
Density	lbm/ft³	kg/m³		1.601 846 E + 01
		g/m³		1.601 846 E + 04
	lbm/U.S. gal	kg/m³		1.198 264 E + 02
		g/cm³		1.198 264 E − 01
	lbm/U.K. gal	kg/m³		9.977 633 E + 01
	lbm/ft³	kg/m³		1.601 846 E + 01
		g/cm³		1.601 846 E − 02

SI UNITS AND CONVERSION FACTORS—TABLE 5. Conversion Factors: U.S. Customary and Commonly Used Units to SI Units (*Continued*)

Quantity	Customary or commonly used unit	SI unit Space,† time	Alternate SI unit	Conversion factor; multiply customary unit by factor to obtain SI unit	
	g/cm^3	kg/m^3		1.0*	E + 03
	lbm/ft^3	kg/m^3		1.601 846	E + 01
Specific volume	ft^3/lbm	m^3/kg		6.242 796	E − 02
		m^3/g		6.242 796	E − 05
	ft^3/lbm	dm^3/kg		6.242 796	E + 01
	U.K. gal/lbm	dm^3/kg	cm^3/g	1.002 242	E + 01
	U.S. gal/lbm	dm^3/kg	cm^3/g	8.345 404	E + 00
Specific volume (mole basis)	$L/(g{\cdot}mol)$	$m^3/kmol$		1	
	$ft^3/(lb{\cdot}mol)$	$m^3/kmol$		6.242 796	E − 02
Specific volume	bbl/U.S. ton	m^3/t		1.752 535	E − 01
	bbl/U.K. ton	m^3/t		1.564 763	E − 01
Yield	bbl/U.S. ton	dm^3/t	L/t	1.752 535	E + 02
	bbl/U.K. ton	dm^3/t	L/t	1.564 763	E + 02
	U.S. gal/U.S. ton	dm^3/t	L/t	4.172 702	E + 00
	U.S. gal/U.K. ton	dm^3/t	L/t	3.725 627	E + 00
Concentration (mass/mass)	wt %	kg/kg		1.0*	E − 02
		g/kg		1.0*	E + 01
	wt ppm	mg/kg		1	
Concentration (mass/ volume)	lbm/bbl	kg/m^3	g/dm^3	2.853 010	E + 00
	g/U.S. gal	kg/m^3		2.641 720	E − 01
	g/U.K. gal	kg/m^3	g/L	2.199 692	E − 01
	lbm/1000 U.S. gal	g/m^3	mg/dm^3	1.198 264	E + 02
	lbm/1000 U.K. gal	g/m^3	mg/dm^3	9.977 633	E + 01
	gr/U.S. gal	g/m^3	mg/dm^3	1.711 806	E + 01
	gr/ft^3	mg/m^3		2.288 351	E + 03
	lbm/1000 bbl	g/m^3	mg/dm^3	2.853 010	E + 00
	mg/U.S. gal	g/m^3	mg/dm^3	2.641 720	E − 01
	$gr/100 ft^3$	mg/m^3		2.288 351	E + 01
Concentration (volume/ volume)	ft^3/ft^3	m^3/m^3		1	
	bbl/(acre·ft)	m^3/m^3		1.288 931	E − 04
	vol%	m^3/m^3		1.0*	E − 02
	U.K. gal/ft^3	dm^3/m^3	L/m^3	1.605 437	E + 02
	U.S. gal/ft^3	dm^3/m^3	L/m^3	1.336 806	E + 02
	mL/U.S. gal	dm^3/m^3	L/m^3	2.641 720	E − 01
	mL/U.K. gal	dm^3/m^3	L/m^3	2.199 692	E − 01
	vol ppm	cm^3/m^3		1	
		dm^3/m^3	L/m^3	1.0*	E − 03
	U.K. gal/1000 bbl	cm^3/m^3		2.859 403	E + 01
	U.S. gal/1000 bbl	cm^3/m^3		2.380 952	E + 01
	U.K. pt/1000 bbl	cm^3/m^3		3.574 253	E + 00
Concentration (mole/ volume)	(lb·mol)/U.S. gal	$kmol/m^3$		1.198 264	E + 02
	(lb·mol)/U.K. gal	$kmol/m^3$		9.977 644	E + 01
	$(lb{\cdot}mol)/ft^3$ std ft^3	$kmol/m^3$		1.601 846	E + 01
	(60°F, 1 atm)/bbl	$kmol/m^3$		7.518 21	E − 03

SI UNITS AND CONVERSION FACTORS—TABLE 5. Conversion Factors: U.S. Customary and Commonly Used Units to SI Units (*Continued*)

Quantity	Customary or commonly used unit	SI unit Space,† time	Alternate SI unit	Conversion factor; multiply customary unit by factor to obtain SI unit	
Concentration (volume/ mole)	U.S. gal/1000 std ft³ (60°F/60°F) bbl/million std ft³ (60°F/60°F)	dm³/kmol dm³/kmol	L/kmol L/kmol	3.166 91 1.330 10	E + 00 E − 01

	Facility throughput, capacity				
Throughput (mass basis)	U.K. ton/year	t/a		1.016 047	E + 00
	U.S. ton/year	t/a		9.071 847	E − 01
	U.K. ton/day	t/d		1.016 047	E + 00
		t/h		4.233 529	E − 02
	U.S. ton/day	t/d		9.071 847	E − 01
		t/h		3.779 936	E − 02
	U.K. ton/h	t/h		1.016 047	E + 00
	U.S. ton/h	t/h		9.071 847	E − 01
	lbm/h	kg/h		4.535 924	E − 01
Throughput (volume basis)	bbl/day	t/a		5.803 036	E + 01
		m³/d		1.589 873	E − 01
	ft³/day	m³/h		1.179 869	E − 03
	bbl/h	m³/h		1.589 873	E − 01
	ft³/h	m³/h		2.831 685	E − 02
	U.K. gal/h	m³/h		4.546 092	E − 03
		L/s		1.262 803	E − 03
	U.S. gal/h	m³/h		3.785 412	E − 03
		L/s		1.051 503	E − 03
	U.K. gal/min	m³/h		2.727 655	E − 01
		L/s		7.576 819	E − 02
	U.S. gal/min	m³/h		2.271 247	E − 01
		L/s		6.309 020	E − 02
Throughput (mole basis)	(lbm·mol)/h	kmol/h		4.535 924	E − 01
		kmol/s		1.259 979	E − 04

	Flow rate				
Flow rate (mass basis)	U.K. ton/min	kg/s		1.693 412	E + 01
	U.S. ton/min	kg/s		1.511 974	E + 01
	U.K. ton/h	kg/s		2.822 353	E − 01
	U.S. ton/h	kg/s		2.519 958	E − 01
	U.K. ton/day	kg/s		1.175 980	E − 02
	U.S. ton/day	kg/s		1.049 982	E − 02
	million lbm/year	kg/s		5.249 912	E + 00
	U.K. ton/year	kg/s		3.221 864	E − 05
	U.S. ton/year	kg/s		2.876 664	E − 05
	lbm/s	kg/s		4.535 924	E − 01
	lbm/min	kg/s		7.559 873	E − 03
	lbm/h	kg/s		1.259 979	E − 04

SI UNITS AND CONVERSION FACTORS—TABLE 5.   Conversion Factors: U.S. Customary and Commonly Used Units to SI Units (*Continued*)

Quantity	Customary or commonly used unit	SI unit Space,† time	Alternate SI unit	Conversion factor; multiply customary unit by factor to obtain SI unit	
Flow rate	bbl/day	m³/d		1.589 873	E − 01
(volume		L/s		1.840 131	E − 03
basis)	ft³/day	m³/d		2.831 685	E − 02
		L/s		3.277 413	E − 04
	bbl/h	m³/s		4.416 314	E − 05
		L/s		4.416 314	E − 02
	ft³/h	m³/s		7.865 791	E − 06
		L/s		7.865 791	E − 03
	U.K. gal/h	dm³/s	L/s	1.262 803	E − 03
	U.S. gal/h	dm³/s	L/s	1.051 503	E − 03
	U.K. gal/min	dm³/s	L/s	7.576 820	E − 02
	U.S. gal/min	dm³/s	L/s	6.309 020	E − 02
	ft³/min	dm³/s	L/s	4.719 474	E − 01
	ft³/s	dm³/s	L/s	2.831 685	E + 01
Flow rate	(lb·mol)/s	kmol/s		4.535 924	E − 01
(mole basis)	(lb·mol)/h	kmol/s		1.259 979	E − 04
	million scf/D	kmol/s		1.383 45	E − 02
Flow rate/length	lbm/(s·ft)	kg/(s·m)		1.488 164	E + 00
(mass basis)	lbm/(h·ft)	kg/(s·m)		4.133 789	E − 04
Flow rate/length	U.K. gal/(min·ft)	m²/s	m³/(s·m)	2.485 833	E − 04
(volume	U.S. gal/(min·ft)	m²/s	m³/(s·m)	2.069 888	E − 04
basis)	U.K. gal/(h·in)	m²/s	m³/(s·m)	4.971 667	E − 05
	U.S. gal/(h·in)	m²/s	m³/(s·m)	4.139 776	E − 05
	U.K. gal/(h·ft)	m²/s	m³/(s·m)	4.143 055	E − 06
	U.S. gal/(h·ft)	m²/s	m³/(s·m)	3.449 814	E − 06
Flow rate/area	lbm/(s·ft²)	kg/(s·m²)		4.882 428	E + 00
(mass basis)	lbm/(h·ft²)	kg/(s·m²)		1.356 230	E − 03
Flow rate/area	ft³/(s·ft²)	m/s	m³/(s·m²)	3.048*	E − 01
(volume	ft³/(min·ft²)	m/s	m³/(s·m²)	5.08*	E − 03
basis)	U.K. gal/(h·in²)	m/s	m³/(s·m²)	1.957 349	E − 03
	U.S. gal/(h·in²)	m/s	m³/(s·m²)	1.629 833	E − 03
	U.K. gal/(min·ft²)	m/s	m³/(s·m²)	8.155 621	E − 04
	U.S. gal/(min·ft²)	m/s	m³/(s·m²)	6.790 972	E − 04
	U.K. gal/(h·ft²)	m/s	m³/(s·m²)	1.359 270	E − 05
	U.S. gal/(h·ft²)	m/s	m³/(s·m²)	1.131 829	E − 05

	Energy, work, power				
Energy work	therm	MJ		1.055 056	E + 02
		kJ		1.055 056	E + 05
		kWh		2.930 711	E + 01
	U.S. tonf·mi	MJ		1.431 744	E + 01
	hp·h	MJ		2.684 520	E + 00
		kJ		2.684 520	E + 03
		kWh		7.456 999	E − 01

SI UNITS AND CONVERSION FACTORS—TABLE 5. Conversion Factors: U.S.
Customary and Commonly Used Units to SI Units (*Continued*)

Quantity	Customary or commonly used unit	SI unit Space,† time	Alternate SI unit	Conversion factor; multiply customary unit by factor to obtain SI unit	
	ch·h or CV·h	MJ		2.647 780	E + 00
		kJ		2.647 780	E + 03
		kWh		7.354 999	E − 01
	kWh	MJ		3.6*	E + 00
		kJ		3.6*	E + 03
	Chu	kJ		1.899 101	E + 00
		kWh		5.275 280	E − 04
	Btu	kJ		1.055 056	E + 00
		kWh		2.930 711	E − 04
	kcal	kJ		4.184*	E + 00
	cal	kJ		4.184*	E − 03
	ft·lbf	kJ		1.355 818	E − 03
	lbf·ft	kJ		1.355 818	E − 03
	J	kJ		1.0*	E − 03
	$(lbf·ft^2)/s^2$	kJ		4.214 011	E − 05
	erg	J		1.0*	E − 07
Impact energy	kgf·m	J		9.806 650*	E + 00
	lbf·ft	J		1.355 818	E + 00
Surface energy	erg/cm^2	mJ/m^2		1.0*	E + 00
Specific-impact energy	$(kgf·m)/cm^2$	J/cm^2		9.806 650*	E − 02
	$(lbf·ft)/in^2$	J/cm^2		2.101 522	E − 03
Power	million Btu/h	MW		2.930 711	E − 01
	ton of refrigeration	kW		3.516 853	E + 00
	Btu/s	kW		1.055 056	E + 00
	kW	kW		1	
	hydraulic horsepower—hhp	kW		7.460 43	E − 01
	hp (electric)	kW		7.46*	E − 01
	hp [(550 ft·lbf)/s]	kW		7.456 999	E − 01
	ch or CV	kW		7.354 999	E − 01
	Btu/min	kW		1.758 427	E − 02
	(ft·lbf)/s	kW		1.355 818	E − 03
	kcal/h	W		1.162 222	E + 00
	Btu/h	W		2.930 711	E − 01
	(ft·lbf)/min	W		2.259 697	E − 02
Power/area	$Btu/(s·ft^2)$	kW/m^2		1.135 653	E + 01
	$cal/(h·cm^2)$	kW/m^2		1.162 222	E − 02
	$Btu/(h·ft^2)$	kW/m^2		3.154 591	E − 03
Heat-release rate, mixing power	hp/ft^3	kW/m^3		2.633 414	E + 01
	$cal/(h·cm^3)$	kW/m^3		1.162 222	E + 00
	$Btu/(s·ft^3)$	kW/m^3		3.725 895	E + 01
	$Btu/(h·ft^3)$	kW/m^3		1.034 971	E − 02
Cooling duty (machinery)	Btu/(bhp·h)	W/kW		3.930 148	E − 01

SI UNITS AND CONVERSION FACTORS—TABLE 5. Conversion Factors: U.S. Customary and Commonly Used Units to SI Units (*Continued*)

Quantity	Customary or commonly used unit	SI unit Space,† time	Alternate SI unit	Conversion factor; multiply customary unit by factor to obtain SI unit	
Specific fuel consumption (mass basis)	lbm/(hp·h)	mg/J kg/kWh	kg/MJ	1.689 659 6.082 774	E – 01 E – 01
Specific fuel consumption (volume basis)	m³/kWh U.S. gal/(hp·h) U.K. pt/(hp·h)	dm³/MJ dm³/MJ dm³/MJ	mm³/J mm³/J mm³/J	2.777 778 1.410 089 2.116 806	E + 02 E + 00 E – 01
Fuel consumption	U.K. gal/mi U.S. gal/mi mi/U.S. gal mi/U.K. gal	dm³/100 km dm³/100 km km/dm³ km/dm³	L/100 km L/100 km km/L km/L	2.824 807 2.352 146 4.251 437 3.540 064	E + 02 E + 02 E – 01 E – 01
Velocity (linear), speed	knot mi/h ft/s ft/min ft/h ft/day in/s in/min	km/h km/h m/s cm/s m/s mm/s mm/s m/d mm/s mm/s		1.852* 1.609 344* 3.048* 3.048* 5.08* 8.466 667 3.527 778 3.048* 2.54* 4.233 333	E + 00 E + 00 E – 01 E + 01 E – 03 E – 02 E – 03 E – 01 E + 01 E – 01
Corrosion rate	in/year (ipy) mil/year	mm/a mm/a		2.54* 2.54*	E + 01 E – 02
Rotational frequency	r/min	r/s rad/s		1.666 667 1.047 198	E – 02 E – 01
Acceleration (linear)	ft/s²	m/s² cm/s²		3.048* 3.048*	E – 01 E + 01
Acceleration (rotational)	rpm/s	rad/s²		1.047 198	E – 01
Momentum	(lbm·ft)/s	(kg·m)/s		1.382 550	E – 01
Force	U.K. tonf U.S. tonf kgf (kp) lbf dyn	kN kN N N mN		9.964 016 8.896 443 9.806 650* 4.448 222 1.0	E + 00 E + 00 E + 00 E + 00 E – 02
Bending moment, torque	U.S. tonf·ft kgf·m lbf·ft lbf·in	kN·m N·m N·m N·m		2.711 636 9.806 650* 1.355 818 1.129 848	E + 00 E + 00 E + 00 E – 01
Bending moment/ length	(lbf·ft)/in (lbf·in)/in	(N·m)/m (N·m)/m		5.337 866 4.448 222	E + 01 E + 00

SI UNITS AND CONVERSION FACTORS—TABLE 5. Conversion Factors: U.S. Customary and Commonly Used Units to SI Units (*Continued*)

Quantity	Customary or commonly used unit	SI unit Space,† time	Alternate SI unit	Conversion factor; multiply customary unit by factor to obtain SI unit	
Moment of inertia	lbm·ft^2	kg·m^2		4.214011	E − 02
Stress	U.S. tonf/in^2	MPa	N/mm^2	1.378951	E + 01
	kgf/mm^2	MPa	N/mm^2	9.806650*	E + 00
	U.S. tonf/ft^2	MPa	N/mm^2	9.576052	E − 02
	lbf/in^2 (psi)	MPa	N/mm^2	6.894757	E − 03
	lbf/ft^2 (psf)	kPa		4.788026	E − 02
	dyn/cm^2	Pa		1.0*	E − 01
Mass/length	lbm/ft	kg/m		1.488164	E + 00
Mass/area structural loading, bearing capacity (mass basis)	U.S. ton/ft^2	Mg/m^2		9.764855	E + 00
	lbm/ft^2	kg/m^2		4.882428	E + 00
Miscellaneous transport properties					
Diffusivity	ft^2/s	m^2/s		9.290304*	E − 02
	m^2/s	mm^2/s		1.0*	E + 06
	ft^2/h	m^2/s		2.58064*	E − 05
Thermal resistance	(°C·m^2·h)/kcal	(K·m^2)/kW		8.604208	E + 02
	(°F·ft^2·h)/Btu	(K·m^2)/kW		1.761102	E + 02
Heat flux	Btu/(h·ft^2)	kW/m^2		3.154591	E − 03
Thermal conductivity	(cal·cm)/(s·cm^2·°C)	W/(m·K)		4.184*	E + 02
	(Btu·ft)/(h·ft^2·°F)	W/(m·K)		1.730735	E + 00
		(kJ·m)/(h·m^2·K)		6.230646	E + 00
	(kcal·m)/(h·m^2·°C)	W/(m·K)		1.162222	E + 00
	(Btu·in)/(h·ft^2·°F)	W/(m·K)		1.442279	E − 01
	(cal·cm)/(h·cm^2·°C)	W/(m·K)		1.162222	E − 01
Heat-transfer coefficient	cal/(s·cm^2·°C)	kW/(m^2·K)		4.184*	E + 01
	Btu/(s·ft^2·°F)	kW/(m^2·K)		2.044175	E + 01
	cal/(h·cm^2·°C)	kW/(m^2·K)		1.162222	E − 02
	Btu/(h·ft^2·°F)	kW/(m^2·K)		5.678263	E − 03
		kJ/(h·m^2·K)		2.044175	E + 01
	Btu/(h·ft^2·°R)	kW/(m^2·K)		5.678263	E − 03
	kcal/(h·m^2·°C)	kW/(m^2·K)		1.162222	E − 03
Volumetric heat-transfer coefficient	Btu/(s·ft^3·°F)	kW/(m^3·K)		6.706611	E + 01
	Btu/(h·ft^3·°F)	kW/(m^3·K)		1.862947	E − 02

SI UNITS AND CONVERSION FACTORS—TABLE 5.   Conversion Factors: U.S. Customary and Commonly Used Units to SI Units (*Continued*)

Quantity	Customary or commonly used unit	SI unit Space,† time	Alternate SI unit	Conversion factor; multiply customary unit by factor to obtain SI unit	
Surface tension	dyn/cm	mN/m		1	
Viscosity (dynamic)	(lbf·s)/in²	Pa·s	(N·s)/m²	6.894757	E + 03
	(lbf·s)/ft²	Pa·s	(N·s)/m²	4.788026	E + 01
	(kgf·s)/m²	Pa·s	(N·s)/m²	9.806650*	E + 00
	lbm/(ft·s)	Pa·s	(N·s)/m²	1.488164	E + 00
	(dyn·s)/cm²	Pa·s	(N·s)/m²	1.0*	E − 01
	cP	Pa·s	(N·s)/m²	1.0*	E − 03
	lbm/(ft·h)	Pa·s	(N·s)/m²	4.133789	E − 04
Viscosity (kinematic)	ft²/s	m²/s		9.290304*	E − 02
	in²/s	mm²/s		6.4516*	E − 02
	m²/h	mm²/s		2.777778	E + 02
	ft²/h	m²/s		2.58064*	E − 05
	cSt	mm²/s		1	
Permeability	darcy	μm²		9.869233	E − 01
	millidarcy	μm²		9.869233	E − 04
Thermal flux	Btu/(h·ft²)	W/m²		3.152	E + 00
	Btu/(s·ft²)	W/m²		1.135	E + 04
	cal/(s·cm²)	W/m²		4.184	E + 04
Mass-transfer coefficient	(lb·mol)/[h·ft² (lb·mol/ft³)] (g·mol)/[s·m² (g·mol/L)]	m/s m/s		8.467 1.0	E − 05 E + 01

Electricity, magnetism					
Admittance	S	S		1	
Capacitance	μF	μF		1	
Charge density	C/mm³	C/mm³		1	
Conductance	S	S		1	
	℧ (mho)	S		1	
Conductivity	S/m	S/m		1	
	℧/m	S/m		1	
	m℧/m	mS/m		1	
Current density	A/mm²	A/mm²		1	
Displacement	C/cm²	C/cm²		1	
Electric charge	C	C		1	
Electric current	A	A		1	
Electric-dipole moment	C·m	C·m		1	

SI UNITS AND CONVERSION FACTORS—TABLE 5.   Conversion Factors: U.S. Customary and Commonly Used Units to SI Units (*Continued*)

Quantity	Customary or commonly used unit	SI unit Space,† time	Alternate SI unit	Conversion factor; multiply customary unit by factor to obtain SI unit
Electric-field strength	V/m	V/m		1
Electric flux	C	C		1
Electric polarization	C/cm^2	C/cm^2		1
Electric potential	V mV	V mV		1 1
Electromagnetic moment	A·m^2	A·m^2		1
Electromotive force	V	V		1
Flux of displacement	C	C		1
Frequency	cycles/s	Hz		1
Impedance	Ω	Ω		1
Linear-current density	A/mm	A/mm		1
Magnetic-dipole moment	Wb·m	Wb·m		1
Magnetic-field strength	A/mm Oe gamma	A/mm A/m A/m		1 7.957 747 E + 01 7.957 747 E − 04
Magnetic flux	mWb	mWb		1
Magnetic-flux density	mT G gamma	mT T nT		1 1.0* E − 04 1
Magnetic induction	mT	mT		1
Magnetic moment	A·m^2	A·m^2		1
Magnetic polarization	mT	mT		1
Magnetic potential difference	A	A		1
Magnetic-vector potential	Wb/mm	Wb/mm		1
Magnetization	A/mm	A/mm		1
Modulus of admittance	S	S		1
Modulus of impedance	Ω	Ω		1

SI UNITS AND CONVERSION FACTORS—TABLE 5.   Conversion Factors: U.S. Customary and Commonly Used Units to SI Units (*Continued*)

Quantity	Customary or commonly used unit	SI unit Space,† time	Alternate SI unit	Conversion factor; multiply customary unit by factor to obtain SI unit	
Mutual inductance	H	H		1	
Permeability	μH/m	μH/m		1	
Permeance	H	H		1	
Permitivity	μF/m	μF/m		1	
Potential difference	V	V		1	
Quantity of electricity	C	C		1	
Reactance	Ω	Ω		1	
Reluctance	H^{-1}	H^{-1}		1	
Resistance	Ω	Ω		1	
Resistivity	Ω·cm	Ω·cm		1	
	Ω·m	Ω·m		1	
Self-inductance	mH	mH		1	
Surface density of change	mC/m^2	mC/m^2		1	
Susceptance	S	S		1	
Volume density of charge	C/mm^3	C/mm^3		1	

Acoustics, light, radiation					
Absorbed dose	rad	Gy		1.0*	E − 02
Acoustical energy	J	J		1	
Accoutical intensity	W/cm^2	W/m^2		1.0*	E + 04
Acoustical power	W	W		1	
Sound pressure	N/m^2	N/m^2		1.0*	
Illuminance	fc	lx		1.076 391	E + 01
Illumination	fc	lx		1.076 391	E + 01
Irradiance	W/m^2	W/m^2		1	
Light exposure	fc·s	lx·s		1.076 391	E + 01
Luminance	cd/m^2	cd/m^2		1	
Luminous efficacy	lm/W	lm/W		1	

Quantity	Customary or commonly used unit	SI unit Space,† time	Alternate SI unit	Conversion factor; multiply customary unit by factor to obtain SI unit	
Luminous exitance	lm/m^2	lm/m^2		1	
Luminous flux	lm	lm		1	
Luminous intensity	cd	cd		1	
Radiance	W/m^2·sr	W/m^2·sr		1	
Radiant energy	J	J		1	
Radiant flux	W	W		1	
Radiant intensity	W/sr	W/sr		1	
Radiant power	W	W		1	
Wavelength	Å	nm		1.0*	E − 01
Capture unit	10^{-3} cm^{-1}	m^{-1}		1.0*	E + 01
			10^{-3} cm^{-1}	1	
	m^{-1}	m^{-1}		1	
Radioactivity	Ci	Bq		3.7*	E + 10

* An asterisk indicates that the conversion factor is exact.

† Conversion factors for length, area, and volume are based on the international foot. The international foot is longer by 2 parts in 1 million than the U.S. Survey foot (land-measurement use).

Note: The following unit symbols are used in the table:

Unit symbol	Name	Unit symbol	Name
A	ampere	lm	lumen
a	annum (year)	lx	lux
Bq	becquerel	m	meter
C	coulomb	min	minute
cd	candela	′	minute
Ci	curie	N	newton
d	day	naut mi	U.S. nautical mile
°C	degree Celsius	Oe	oersted
°	degree	Ω	ohm
dyn	dyne	Pa	pascal
F	farad	rad	radian
fc	footcandle	r	revolution
G	gauss	S	siemens
g	gram	s	second
gr	grain	″	second
Gy	gray	sr	steradian
H	henry	St	stokes
h	hour	T	tesla
ha	hectare	t	tonne
Hz	hertz	V	volt
J	joule	W	watt
K	kelvin	Wb	weber
L, ℓ, l	liter		

SI UNITS AND CONVERSION FACTORS—TABLE 6. Metric Conversion Factors as Exact Numerical Multiples of SI Units
The first two digits of each numerical entry represent a power of 10. For example, the entry "−02 2.54" expresses the fact that 1 in = 2.54 × 10^{-2} m

To convert from	To	Multiply by
abampere	ampere	+01 1.00
abcoulomb	coulomb	+01 1.00
abfarad	farad	+09 1.00
abhenry	henry	−09 1.00
abmho	mho	+09 1.00
abohm	ohm	−09 1.00
abvolt	volt	−08 1.00
acre	$meter^2$	+03 4.046 856
ampere (international of 1948)	ampere	−01 9.998 35
angstrom	meter	−10 1.00
are	$meter^2$	+02 1.00
astronomical unit	meter	+11 1.495 978
atmosphere	$newton/meter^2$	+05 1.013 25
bar	$newton/meter^2$	+05 1.00
barn	$meter^2$	−28 1.00
barrel (petroleum 42 gal)	$meter^3$	−01 1.589 873
barye	$newton/meter^2$	−01 1.00
British thermal unit (ISO/TC 12)	joule	+03 1.055 06
British thermal unit (International Steam Table)	joule	+03 1.055 04
British thermal unit (mean)	joule	+03 1.055 87
British thermal unit (thermochemical)	joule	+03 1.054 350
British thermal unit (39°F)	joule	+03 1.059 67
British thermal unit (60°F)	joule	+03 1.054 68
bushel (U.S.)	$meter^3$	−02 3.523 907
cable	meter	+02 2.194 56
caliber	meter	−04 2.54
calorie (International Steam Table)	joule	+00 4.186 8
calorie (mean)	joule	+00 4.190 02
calorie (thermochemical)	joule	+00 4.184
calorie (15°C)	joule	+00 4.185 80
calorie (20°C)	joule	+00 4.181 90
calorie (kilogram, International Steam Table)	joule	+03 4.186 8
calorie (kilogram, mean)	joule	+03 4.190 02
calorie (kilogram, thermochemical)	joule	+03 4.184
carat (metric)	kilogram	−04 2.00
Celsius (temperature)	kelvin	$t_k = t_c$ $+ 273.15$
centimeter of mercury (0°C)	$newton/meter^2$	+03 1.333 22
centimeter of water (4°C)	$newton/meter^2$	+01 9.806 38
chain (engineer's)	meter	+01 3.048
chain (surveyor's or Gunter's)	meter	+01 2.011 68
circular mil	$meter^2$	−10 5.067 074
cord	$meter^3$	+00 3.624 556
coulomb (international of 1948)	coulomb	−01 9.998 35
cubit	meter	−01 4.572
cup	$meter^3$	−04 2.365 882
curie	disintegration/second	+10 3.70

SI UNITS AND CONVERSION FACTORS—TABLE 6. Metric Conversion Factors as Exact Numerical Multiples of SI Units (*Continued*)

The first two digits of each numerical entry represent a power of 10. For example, the entry "−02 2.54" expresses the fact that 1 in = 2.54 × 10⁻⁴ m

To convert from	To	Multiply by
day (mean solar)	second (mean solar)	+04 8.64
day (sidereal)	second (mean solar)	+04 8.616 409
degree (angle)	radian	−02 1.745 329
denier (international)	kilogram/meter	−07 1.111 111
dram (avoirdupois)	kilogram	−03 1.771 845
dram (troy or apothecary)	kilogram	−03 3.887 934
dram (U.S. fluid)	meter³	−06 3.696 691
dyne	newton	−05 1.00
electron volt	joule	−19 1.602 10
erg	joule	−07 1.00
Fahrenheit (temperature)	kelvin	$t_K = (5/9)(t_F + 459.67)$
Fahrenheit (temperature)	Celsius	$t_c = (5/9)(t_F - 32)$
farad (international of 1948)	farad	−01 9.995 05
faraday (based on carbon 12)	coulomb	+04 9.648 70
faraday (chemical)	coulomb	+04 9.649 57
faraday (physical)	coulomb	+04 9.652 19
fathom	meter	+00 1.828 8
fermi (femtometer)	meter	−15 1.00
fluid ounce (U.S.)	meter³	−05 2.957 352
foot	meter	−01 3.048
foot (U.S. survey)	meter	−01 3.048 006
foot of water (39.2°F)	newton/meter²	+03 2.988 98
footcandle	lumen/meter²	+01 1.076 391
footlambert	candela/meter²	+00 3.426 259
furlong	meter	+02 2.011 68
gal (galileo)	meter/second²	−02 1.00
gallon (U.K. liquid)	meter³	−03 4.546 087
gallon (U.S. dry)	meter³	−03 4.404 883
gallon (U.S. liquid)	meter³	−03 3.785 411
gamma	tesla	−09 1.00
gauss	tesla	−04 1.00
gilbert	ampere turn	−01 7.957 747
gill (U.K.)	meter³	−04 1.420 652
gill (U.S.)	meter³	−04 1.182 941
grad	degree (angular)	−01 9.00
grad	radian	−02 1.570 796
grain	kilogram	−05 6.479 891
gram	kilogram	−03 1.00
hand	meter	−01 1.016
hectare	meter²	+04 1.00
henry (international of 1948)	henry	+00 1.000 495
hogshead (U.S.)	meter³	−01 2.384 809
horsepower (550 ft lbf/s)	watt	+02 7.456 998
horsepower (boiler)	watt	+03 9.809 50
horsepower (electric)	watt	+02 7.46

SI UNITS AND CONVERSION FACTORS—TABLE 6. Metric Conversion Factors as Exact Numerical Multiples of SI Units (*Continued*)
The first two digits of each numerical entry represent a power of 10. For example, the entry "–02 2.54" expresses the fact that 1 in = 2.54×10^{-4} m

To convert from	To	Multiply by
horsepower (metric)	watt	+02 7.354 99
horsepower (U.K.)	watt	+02 7.457
horsepower (water)	watt	+02 7.460 43
hour (mean solar)	second (mean solar)	+03 3.60
hour (sidereal)	second (mean solar)	+03 3.590 170
hundredweight (long)	kilogram	+01 5.080 234
hundredweight (short)	kilogram	+01 4.535 923
inch	meter	–02 2.54
inch of mercury (32°F)	newton/meter^2	+03 3.386 389
inch of mercury (60°F)	newton/meter^2	+03 3.376 85
inch of water (39.2°F)	newton/meter^2	+02 2.490 82
inch of water (60°F)	newton/meter^2	+02 2.4884
joule (international of 1948)	joule	+00 1.000 165
kayser	l/meter	+02 1.00
kilocalorie (International Steam Table)	joule	+03 4.186 74
kilocalorie (mean)	joule	+03 4.190 02
kilocalorie (thermochemical)	joule	+03 4.184
kilogram mass	kilogram	+00 1.00
kilogram-force (kgf)	newton	+00 9.806 65
kilopond-force	newton	+00 9.806 65
kip	newton	+03 4.448 221
knot (international)	meter/second	–01 5.144 444
lambert	candela/meter^2	+04 1/π
lambert	candela/meter^2	+03 3.183 098
langley	joule/meter^2	+04 4.184
lbf (pound-force, avoirdupois)	newton	+00 4.448 221
lbm (pound-mass, avoirdupois)	kilogram	–01 4.535 923
league (British nautical)	meter	+03 5.559 552
league (international nautical)	meter	+03 5.556
league (statute)	meter	+03 4.828 032
light-year	meter	+15 9.460 55
link (engineer's)	meter	–01 3.048
link (surveyor's or Gunter's)	meter	–01 2.011 68
liter	meter^3	–03 1.00
lux	lumen/meter^2	+00 1.00
maxwell	weber	–08 1.00
meter	wavelengths Kr 86	+06 1.650 763
micrometer	meter	–06 1.00
mil	meter	–05 2.54
mile (U.S. statute)	meter	+03 1.609 344
mile (U.K. nautical)	meter	+03 1.853 184
mile (international nautical)	meter	+03 1.852
mile (U.S. nautical)	meter	+03 1.852
millibar	newton/meter^2	+02 1.00
millimeter of mercury (0°C)	newton/meter^2	+02 1.333 224
minute (angle)	radian	–04 2.908 882
minute (mean solar)	second (mean solar)	+01 6.00

SI UNITS AND CONVERSION FACTORS—TABLE 6.   Metric Conversion Factors as Exact Numerical Multiples of SI Units (*Continued*)
The first two digits of each numerical entry represent a power of 10. For example, the entry "–02 2.54" expresses the fact that 1 in = 2.54 × 10^{-4} m

To convert from	To	Multiply by
minute (sidereal)	second (mean solar)	+01 5.983 617
month (mean calendar)	second (mean solar)	+06 2.628
nautical mile (international)	meter	+03 1.852
nautical mile (U.S.)	meter	+03 1.852
nautical mile (U.K.)	meter	+03 1.853 184
oersted	ampere/meter	+01 7.957 747
ohm (international of 1948)	ohm	+00 1.000 495
ounce-force (avoirdupois)	newton	–01 2.780 138
ounce-mass (avoirdupois)	kilogram	–02 2.834 952
ounce-mass (troy or apothecary)	kilogram	–02 3.110 347
ounce (U.S. fluid)	meter^3	–05 2.957 352
pace	meter	–01 7.62
parsec	meter	+16 3.083 74
pascal	newton/meter^2	+00 1.00
peck (U.S.)	meter^3	–03 8.809 767
pennyweight	kilogram	–03 1.555 173
perch	meter	+00 5.0292
phot	lumen/meter^2	+04 1.00
pica (printer's)	meter	–03 4.217 517
pint (U.S. dry)	meter^3	–04 5.506 104
pint (U.S. liquid)	meter^3	–04 4.731 764
point (printer's)	meter	–04 3.514 598
poise	(newton-second)/meter^2	–01 1.00
pole	meter	+00 5.0292
pound-force (lbf avoirdupois)	newton	+00 4.448 221
pound-mass (lbm avoirdupois)	kilogram	–01 4.535 923
pound-mass (troy or apothecary)	kilogram	–01 3.732 417
poundal	newton	–01 1.382 549
quart (U.S. dry)	meter^3	–03 1.101 220
quart (U.S. liquid)	meter^3	–04 9.463 529
rad (radiation dose absorbed)	joule/kilogram	–02 1.00
Rankine (temperature)	kelvin	$t_k = (5/9)t_R$
rayleigh (rate of photon emission)	l/second-meter^2	+10 1.00
rhe	meter^2/(newton-second)	+01 1.00
rod	meter	+00 5.0292
roentgen	coulomb/kilogram	–04 2.579 76
rutherford	disintegration/second	+06 1.00
second (angle)	radian	–06 4.848 136
second (ephemeris)	second	+00 1.000 000
second (mean solar)	second (ephemeris)	Consult American Ephemeris and Nautical Almanac
second (sidereal)	second (mean solar)	–01 9.972 695
section	meter^2	+06 2.589 988
scruple (apothecary)	kilogram	–03 1.295 978

SI UNITS AND CONVERSION FACTORS—TABLE 6. Metric Conversion Factors as Exact Numerical Multiples of SI Units (*Continued*)
The first two digits of each numerical entry represent a power of 10. For example, the entry "−02 2.54" expresses the fact that 1 in = 2.54 × 10⁻⁴ m

To convert from	To	Multiply by
shake	second	−08 1.00
skein	meter	+02 1.097 28
slug	kilogram	+01 1.459 390
span	meter	−01 2.286
statampere	ampere	−10 3.335 640
statcoulomb	coulomb	−10 3.335 640
statfarad	farad	−12 1.112 650
stathenry	henry	+11 8.987 554
statmho	mho	−12 1.112 650
statohm	ohm	+11 8.987 554
statute mile (U.S.)	meter	+03 1.609 344
statvolt	volt	+02 2.997 925
stere	meter³	+00 1.00
stilb	candela/meter²	+04 1.00
stoke	meter²/second	−04 1.00
tablespoon	meter³	−05 1.478 676
teaspoon	meter³	−06 4.928 921
ton (assay)	kilogram	−02 2.916 666
ton (long)	kilogram	+03 1.016 046
ton (metric)	kilogram	+03 1.00
ton (nuclear equivalent of TNT)	joule	+09 4.20
ton (register)	meter³	+00 2.831 684
ton (short, 2000 lb)	kilogram	+02 9.071 847
tonne	kilogram	+03 1.00
torr (0°C)	newton/meter²	+02 1.333 22
township	meter²	+07 9.323 957
unit pole	weber	−07 1.256 637
volt (international of 1948)	volt	+00 1.000 330
watt (international of 1948)	watt	+00 1.000 165
yard	meter	−01 9.144
year (calendar)	second (mean solar)	+07 3.1536
year (sidereal)	second (mean solar)	+07 3.155 815
year (tropical)	second (mean solar)	+07 3.155 692
year 1900, tropical, Jan., day 0, hour 12	second (ephemeris)	+07 3.155 692
year 1900, tropical, Jan., day 0, hour 12	second	+07 3.155 692

SOLUBILITY PARAMETERS

SOLUBILITY PARAMETERS—TABLE 1. Hansen Solubility Parameters of Polymers*

Polymer	$\delta/MPa^{1/2}$			
	δ_d	δ_p	δ_h	R
Cellulose acetate	18.6	12.7	11.0	7.6
Chlorinated polypropylene	20.3	6.3	5.4	10.6
Epoxy	20.4	12.0	11.5	12.7
Isoprene elastomer	16.6	1.4	−0.8	9.6
Cellulose nitrate	15.4	14.7	8.8	11.5
Polyamide, thermoplastic	17.4	−1.9	14.9	9.6
Poly(isobutylene)	14.5	2.5	4.7	12.7
Poly(ethylmethacrylate)	17.6	9.7	4.0	10.6
Poly(methyl methacrylate)	18.6	10.5	7.5	8.6
Polystyrene	21.3	5.8	4.3	12.7
Poly(vinyl acetate)	20.9	11.3	9.6	13.7
Poly(vinyl butyral)	18.6	4.4	13.0	10.6
Poly(vinyl chloride)	18.2	7.5	8.3	3.5
Saturated polyester	21.5	14.9	12.3	16.8

* $\delta_t^2 = \delta_d^2 + \delta_p^2 + \delta_h^2$
where δ_t^2 = total Hildebrand parameter, δ_d^2 = dispersion component, δ_p^2 = polar component, and δ_h^2 = hydrogen bonding component.

SOLUBILITY PARAMETERS—TABLE 2.
Hildebrand Solubility Parameters of
Polymers

Compound	Solubility parameter $\delta/MPa^{1/2}$
Poly(ethylene)	7.9
Poly(ethylene terephthalate)	10.7
Poly(hexamethylene adipamide)	13.6
Poly(isobutylene)	8.1
Poly(methylmethacrylate)	9.3
Poly(styrene)	9.1
Poly(tetrafluoroethylene)	6.2
Poly(vinyl acetate)	9.6
Poly(vinyl chloride)	9.7
Poly(vinylidene chloride)	12.2

SOLUBILITY PARAMETERS—TABLE 3. Hildebrand
Solubility Parameters of Common Organic Solvents

Compound	Solubility Parameter δ/Mpa$^{1/2}$	Solubility parameter SI
Acetone	9.8	19.7
Acetonitrile	11.9	24.2
Amyl acetate	8.5	17.1
Benzene	9.2	18.7
n-Butyl alcohol	11.3	28.7
Butyl cellosolve	11.9	21.9
Carbon disulfide	10.2	20.5
Carbon tetrachloride	8.7	18.0
Cellosolve acetate	9.6	19.1
Chloroform	9.4	18.7
m-Cresol	13.0	27.2
Cyclohexane	8.2	16.8
Decafluorobutane	5.3	10.6
Diacetone alcohol	10.2	20.0
Diethyl ether	7.6	15.4
Di-isobutyl ketone	7.8	15.8
Dimethylformamide	12.1	24.8
Dimethyl sulfoxide	12.9	26.4
1,4-Dioxane	10.2	20.5
n-Dodecane	8.0	16.0
Ethyl acetate	9.1	18.2
Ethyl alcohol (ethanol)	12.9	26.2
2-Ethyl butanol	10.5	21.0
Ethylene carbonate	14.7	29.8
Ethylene dichloride	9.8	20.2
Ethylene glycol	16.31	34.9
2-Ethyl hexanol	9.5	19.2
Formic acid	13.6	27.6
Glycerol	21.1	36.2
n-Heptane	7.4	15.3
4-Heptanone	8.9	18.2
n-Hexane	7.2	14.9
Methyl alcohol (methanol)	14.3	29.7
Methyl cellosolve	10.8	22.2
Methylene chloride	9.9	20.2
Methyl ethyl ketone	9.3	19.3
Methyl hexyl ketone	8.9	18.2
Morpholine	10.5	22.1
Neopentane	6.4	12.9
Nitroethane	11.1	22.4
Nitropropane	10.4	21.0
n-Pentane	7.0	14.4
n-Propyl alcohol	11.8	24.9
iso-Propyl alcohol	11.8	24.9
Propylene carbonate	13.3	27.0
Propylene glycol	14.8	30.7
Tetrahydrofuran	9.5	18.5
Toluene	8.9	18.3
1,1,1-Trichloroethane	8.6	15.8
Trichloroethylene	9.3	18.7
Turpentine	6.4	16.6
Xylene	8.9	18.2

SOLUBILITY PRODUCT CONSTANTS

SOLUBILITY PRODUCT CONSTANTS—TABLE 1. Solubility Product Constants (K_{sp}) at 25°C

Compound	Formula	K_{sp}
Aluminum hydroxide	$Al(OH)_3$	4.6×10^{-33}
Aluminum phosphate	$AlPO_4$	6.3×10^{-19}
Barium carbonate	$BaCO_3$	5.1×10^{-9}
Barium chromate	$BaCrO_4$	2.2×10^{-10}
Barium fluoride	BaF_2	1.0×10^{-6}
Barium hydroxide	$Ba(OH)_2$	5.0×10^{-3}
Barium iodate	$Ba(IO_3)_2$	1.5×10^{-9}
Barium oxalate	BaC_2O_4	2.3×10^{-8}
Barium sulfate	$BaSO_4$	1.1×10^{-10}
Barium sulfite	$BaSO_3$	8.0×10^{-7}
Barium thiosulfate	BaS_2O_3	1.6×10^{-5}
Bismuth hydroperoxide	$BiOOH$	4×10^{-10}
Bismuth oxychloride	$BiOCl$	1.8×10^{-31}
Bismuth sulfide	Bi_2S_3	1.0×10^{-97}
Cadmium carbonate	$CdCO_3$	5.2×10^{-12}
Cadmium hydroxide	$Cd(OH)_2$	2.5×10^{-14}
Cadmium iodate	$Cd(IO_3)_2$	2.3×10^{-8}
Cadmium sulfide	CdS	8.0×10^{-27}
Calcium chromate	$CaCrO_4$	7.1×10^{-4}
Calcium dichromate	CaC_2O_4	4.0×10^{-9}
Calcium fluoride	CaF_2	5.3×10^{-9}
Calcium fluorapatite	$Ca_5(PO_4)_3F$	1.0×10^{-60}
Calcium hydrogen phosphate	$CaHPO_4$	1.0×10^{-7}
Calcium hydroxyapatite	$Ca_5(PO_4)_3OH$	1.0×10^{-36}
Calcium hydroxide	$Ca(OH)_2$	5.5×10^{-6}
Calcium iodate	$Ca(IO_3)_2$	7.1×10^{-7}
Calcium oxalate hydrate	$CaC_2O_4 . H_2O$	1.96×10^{-8}
Calcium phosphate	$Ca_3(PO_4)_2$	1.0×10^{-26}
Calcium sulfate	$CaSO_4$	9.1×10^{-6}
Calcium sulfite	$CaSO_3$	6.8×10^{-8}
Calcium sulfate	$CaSO_4$	1.2×10^{-6}
Chromium (II) hydroxide	$Cr(OH)_2$	2×10^{-16}
Chromium (III) hydroxide	$Cr(OH)_3$	6.3×10^{-31}
Cobalt (II) carbonate	$CoCO_3$	1.4×10^{-13}
Cobalt (III) hydroxide	$Co(OH)_3$	1.6×10^{-44}
Cobalt sulfide	CoS	2.0×10^{-25}
Copper (I) chloride	$CuCl$	1.2×10^{-6}
Copper (I) cyanide	$CuCN$	3.2×10^{-20}
Copper (I) iodide	CuI	1.1×10^{-12}
Copper (I) sulfide	Cu_2S	2.5×10^{-48}
Copper (II) arsenate	$Cu_3(AsO_4)_2$	7.6×10^{-36}
Copper (II) carbonate	$CuCO_3$	1.4×10^{-10}
Copper (II) chromate	$CuCrO_4$	3.6×10^{-6}
Copper (II) ferrocyanide	$Cu_2[Fe(CN)_6]$	1.3×10^{-16}

SOLUBILITY PRODUCT CONSTANTS—TABLE 1. Solubility Product Constants (K_{sp}) at 25°C (*Continued*)

Compound	Formula	K_{sp}
Copper (II) hydroxide	$Cu(OH)_2$	2.2×10^{-20}
Copper (II) sulfide	CuS	6.0×10^{-37}
Copper (II) thiocyanate	$Cu(SCN)_2$	4.0×10^{-14}
Iron (II) carbonate	$FeCO_3$	3.2×10^{-11}
Iron (II) hydroxide	$Fe(OH)_2$	8.0×10^{-16}
Iron (III) hydroxide	$Fe(OH)_3$	4.0×10^{-38}
Iron (II) sulfide	FeS	6×10^{-19}
Iron (III) arsenate	$FeAsO_4$	5.7×10^{-21}
Iron (III) ferrocyanide	$Fe_4[Fe(CN)_6]_3$	3.3×10^{-41}
Iron (III) hydroxide	$Fe(OH)_3$	4×10^{-38}
Iron (III) phosphate	$FePO_4$	1.3×10^{-22}
Lead (II) arsenate	$Pb_3(AsO_4)_2$	4.0×10^{-36}
Lead (II) azide	$Pb(N_3)_2$	2.5×10^{-9}
Lead (II) bromate	$Pb(BrO_3)_2$	7.9×10^{-6}
Lead (II) bromide	$PbBr_2$	4.0×10^{-5}
Lead (II) carbonate	$PbCO_3$	7.4×10^{-14}
Lead (II) chloride	$PbCl_2$	1.6×10^{-5}
Lead (II) chromate	$PbCrO_4$	2.8×10^{-13}
Lead (II) fluoride	PbF_2	2.7×10^{-8}
Lead (II) hydroxide	$Pb(OH)_2$	1.2×10^{-5}
Lead (II) iodate	$Pb(IO_3)_2$	2.6×10^{-13}
Lead (II) iodide	PbI_2	7.1×10^{-9}
Lead sulfate	$PbSO_4$	1.6×10^{-8}
Lead sulfide	PbS	8.0×10^{-28}
Lithium carbonate	Li_2CO_3	2.5×10^{-2}
Lithium fluoride	LiF	3.8×10^{-3}
Lithium phosphate	Li_3PO_4	3.2×10^{-9}
Magnesium ammonium phosphate	$MgNH_4PO_4$	2.5×10^{-13}
Magnesium arsenate	$Mg_3(AsO_4)_2$	2.1×10^{-20}
Magnesium carbonate	$MgCO_3$	3.5×10^{-8}
Magnesium dichromate	MgC_2O_4	8.6×10^{-5}
Magnesium fluoride	MgF_2	3.7×10^{-8}
Magnesium hydroxide	$Mg(OH)_2$	1.8×10^{-11}
Magnesium oxalate	MgC_2O_4	7.0×10^{-7}
Magnesium phosphate	$Mg_3(PO_4)_2$	1.0×10^{-25}
Manganese (II) carbonate	$MnCO_3$	1.8×10^{-11}
Manganese hydroxide	$Mn(OH)_2$	1.9×10^{-13}
Manganese sulfide	MnS	2.5×10^{-13}
Mercury (I) bromide	$HgBr$	5.6×10^{-23}
Mercury (I) chloride	$HgCl$	5.0×10^{-13}
Mercury (I) chromate	Hg_2CrO_4	2.0×10^{-9}
Mercury (I) cyanide	$Hg(CN)$	5.0×10^{-40}
Mercury (I) iodide	HgI	4.5×10^{-29}
Mercury (I) sulfate	Hg_2SO_4	7.4×10^{-7}
Mercury (I) sulfide	Hg_2S	1.0×10^{-47}
Mercury (I) thiocyanate	$Hg(SCN)$	3.0×10^{-20}
Mercury (II) thiocyanate	$Hg(SCN)_2$	2.8×10^{-20}

Compound	Formula	K_{sp}
Nickel (II) carbonate	$NiCO_3$	6.6×10^{-9}
Nickel (II) hydroxide	$Ni(OH)_2$	2.0×10^{-15}
Nickel (II) sulfide	NiS	1.0×10^{-24}
Scandium fluoride	ScF_3	4.2×10^{-18}
Scandium hydroxide	$Sc(OH)_3$	4.2×10^{-18}
Silver arsenate	Ag_3AsO_4	1.0×10^{-22}
Silver acetate	$AgC_2H_3O_2$	2.0×10^{-3}
Silver azide	AgN_3	2.0×10^{-8}
Silver benzoate	$AgC_7H_5O_2$	2.5×10^{-5}
Silver bromate	$AgBrO_3$	5.5×10^{-5}
Silver bromide	$AgBr$	5.0×10^{-13}
Silver carbonate	Ag_2CO_3	8.1×10^{-12}
Silver chloride	$AgCl$	1.8×10^{-10}
Silver chromate	Ag_2CrO_4	1.1×10^{-12}
Silver cyanide	$AgCN$	1.2×10^{-16}
Silver iodate	$AgIO_3$	3.1×10^{-8}
Silver iodide	AgI	8.3×10^{-17}
Silver nitrite	$AgNO_2$	6.0×10^{-4}
Silver oxalate	$Ag_2C_2O_4$	3.6×10^{-11}
Silver phosphate	Ag_3PO_4	1.3×10^{-20}
Silver sulfate	Ag_2SO_4	1.4×10^{-5}
Silver sulfide	Ag_2S	6.0×10^{-51}
Silver sulfite	$AgSO_3$	1.5×10^{-14}
Silver thiocyanate	$AgSCN$	1.0×10^{-12}
Strontium carbonate	$SrCO_3$	1.1×10^{-10}
Strontium chromate	$SrCrO_4$	2.2×10^{-5}
Strontium fluoride	SrF_2	2.5×10^{-9}
Strontium oxalate	SrC_2O_4	4.0×10^{-7}
Strontium sulfate	$SrSO_4$	3.2×10^{-7}
Strontium sulfite	$SrSO_3$	4.0×10^{-8}
Thallium (I) bromate	$TlBrO_3$	1.7×10^{-4}
Thallium (I) bromide	$TlBr$	3.4×10^{-6}
Thallium (I) chloride	$TlCl$	1.7×10^{-4}
Thallium (I) chromate	Tl_2CrO_4	9.8×10^{-15}
Thallium (I) iodate	$TlIO_3$	3.1×10^{-6}
Thallium (I) iodide	TlI	6.5×10^{-8}
Thallium (I) sulfide	Tl_2S	6.0×10^{-22}
Thallium (I) Thiocyanate	$TlSCN$	1.6×10^{-4}
Thallium (III) hydroxide	$Tl(OH)_3$	6.3×10^{-46}
Tin (II) hydroxide	$Sn(OH)_2$	1.4×10^{-28}
Tin (II) sulfide	SnS	1.0×10^{-26}
Zinc carbonate	$ZnCO_3$	1.4×10^{-11}
Zinc cyanide	$Zn(CN)_2$	3.0×10^{-16}
Zinc hydroxide	$Zn(OH)_2$	3.3×10^{-17}
Zinc iodate	$Zn(IO_3)_2$	3.9×10^{-6}
Zinc oxalate	ZnC_2O_4	2.7×10^{-8}
Zinc phosphate	$Zn_3(PO_4)_2$	9.0×10^{-33}
Zinc sulfide	ZnS	1.6×10^{-23}

SPECIFIC GRAVITY

Density is defined as the mass of a unit volume of material at a specified temperature and has the dimensions of grams per cubic centimeter (a close approximation to grams per milliliter). Density is measured at a standard temperature, mostly 15.6 or 20°C (60 or 68°F) and is often written in the form of d_4^{20} to indicate that it was measured at 20°C and in pycnometer calibrated with water at 4°C (39°F), that is, with a medium of density equal to 1.0000.

Specific gravity is the ratio of the mass of a volume of the substance to the mass of the same volume of water and is dependent on two temperatures, those at which the masses of the sample and the water are measured. When the water temperature is 4°C (39°F), the specific gravity is equal to the density in the centimeter-gram-second (cgs) system, because the volume of 1 g of water at that temperature is, by definition, 1 ml. Thus the density of water, for example, varies with temperature, and its specific gravity at equal temperatures is always unity. The standard temperatures for a specific gravity in the petroleum industry in North America are 60/60°F (15.6/15.6°C).

In the petroleum industry, the use of density or specific gravity has largely been replaced by the American Petroleum Institute (API) gravity is the preferred property. It is one criterion that is used in setting prices for petroleum. In the United States, the API gravity has been used for the classification of a reservoir and the accompanying tax and royalty consequences. This property was derived from the Baumé scale:

$$\text{degrees Baumé} = 140/\text{sp gr @ } 60/60°F - 130 \tag{94}$$

However, a considerable number of hydrometers calibrated according to the Baumé scale were soon found to be in error by a consistent amount, and this led to the adoption of the equation:

$$\text{degrees API} = 141.5/\text{sp gr @ } 60/60°F - 131.5 \tag{95}$$

SPECIFIC GRAVITY—TABLE 1. Specific Gravity of Inorganic Compounds

Compound	Formula	Molecular weight	Specific gravity
Actinium			
Bromide	$AcBr_3$	466.7	5.85
Chloride	$AcCl_3$	333.4	4.81
Fluoride	AcF_3	284.0	7.88
Oxide	Ac_2O_3	502.0	9.19
Aluminum			
Bromide	$AlBr_3$	266.7	2.64[10]
Carbide	Al_4C_3	143.9	2.36
Chloride	$AlCl_3$	133.3	2.44
Fluoride	AlF_3	84.0	2.88
Hydroxide	$Al(OH)_3$	78.0	2.42
Iodide	AlI_3	407.7	3.95
Nitrate	$Al(NO_3)_3 \cdot 9H_2O$	375.1	1.72
Nitride	AlN	41.0	3.13

SPECIFIC GRAVITY — TABLE 1. Specific Gravity of Inorganic Compounds (*Continued*)

Compound	Formula	Molecular weight	Specific gravity
Oxide	Al_2O_3	102.0	3.97
Phosphate	$AlPO_4$	122.0	2.57
Silicate	Al_2SiO_5	162.0	3.21
Sulfate	$Al_2(SO_4)_3$	342.2	2.71
Sulfide	Al_2S_3	150.2	2.02
Americium			
Oxide IV	AmO_2	275.1	11.7
Ammonium			
Bromide	NH_4Br	98.0	2.43
Carbonate	$(NH_4)_2CO_3 \cdot H_2O$	114.1	
Chlorate	NH_4ClO_3	101.5	1.80
Chloride	NH_4Cl	53.5	1.53
Chromate	$(NH_4)_2CrO_4$	152.1	1.89
Fluoride	NH_4F	37.0	1.015
Iodate	NH_4IO_3	192.9	3.31
Iodide	NH_4I	144.9	2.86
Nitrate	NH_4NO_3	80.0	1.725
Nitrite	NH_4NO_2	64.0	1.69
Oxalate	$(NH_4)_2C_2O_4 \cdot H_2O$	142.1	1.50
Perchlorate	NH_4ClO_4	117.5	1.95
Hydrogen Phosphate	$(NH_4)_2HPO_4$	132.1	1.62
Dihydrogen Phosphate	$NH_4H_2PO_4$	115.0	1.80
Sulfate	$(NH_4)_2SO_4$	132.1	1.77
Hydrogen sulfide	NH_4HS	51.1	1.17
Thiocyanate	HN_4SCN	76.1	1.30
Antimony			
Bromide III	$SbBr_3$	361.5	4.15
Chloride III	$SbCl_3$	228.1	3.14
Chloride V	$SbCl_5$	299.0	2.34
Fluoride III	SbF_3	178.8	4.38
Fluoride V	SbF_5	216.7	2.99
Hydride III	SbH_3	124.8	2.26^{-25}
Iodide III	SbI_3	502.5	4.85
Iodide V	SbI_5	756.3	
Oxide III	Sb_2O_3	291.5	5.67
Oxide V	Sb_2O_5	323.5	3.78
Oxychloride III	$SbOCl$	173.2	
Sulfate III	$Sb_2(SO_4)_3$	531.7	3.62
Sulfide III	Sb_2S_3	339.7	4.64
Sulfide V	Sb_2S_5	403.8	4.12
Arsenic			
Acid, ortho	$H_3AsO_4 \cdot {}^1\!/_2H_2O$	151.0	2.0–2.5
Bromide III	$AsBr_3$	314.7	3.54
Chloride III	$AsCl_3$	181.3	2.16
Chloride V	$AsCl_5$	252.2	
Fluoride III	AsF_3	131.9	2.70
Fluoride V	AsF_5	169.9	7.71 g/l
Hydride III	AsH_3	77.9	2.70
Iodide III	AsI_3	455.6	4.39

SPECIFIC GRAVITY—TABLE 1. Specific Gravity of Inorganic Compounds (*Continued*)

Compound	Formula	Molecular weight	Specific gravity
Iodide V	AsI_5	709.5	3.93
Oxide III	As_2O_3	197.2	3.86
Oxide V	As_2O_5	229.9	4.09
Sulfide II	As_2S_2	214.0	3.20
Sulfide III	As_2S_3	246.0	3.43
Sulfide V	As_2S_5	310.2	
Barium			
Bromate	$Ba(BrO_3)_2 \cdot H_2O$	411.2	3.99
Bromide	$BaBr_2$	297.2	4.78
Carbide	BaC_2	161.4	3.75
Carbonate	$BaCO_3$	197.4	4.43
Chlorate	$Ba(ClO_3)_2 \cdot H_2O$	322.3	3.18
Chloride	$BaCl_2$	208.3	3.86^{24}
Chromate	$BaCrO_4$	253.3	4.50^{15}
Fluoride	BaF_2	175.3	4.89
Hydride	BaH_2	139.4	4.21^0
Hydroxide	$Ba(OH)_2 \cdot 8H_2O$	315.5	2.18^{16}
Iodide	BaI_2	391.2	5.15
Nitrate	$Ba(NO_3)_2$	261.4	3.24
Oxalate	BaC_2O_4	225.4	2.66
Oxide	BaO	153.3	5.72
Perchlorate	$Ba(ClO_4)_2$	336.2	3.2
Sulfate	$BaSO_4$	233.4	4.50^{15}
Sulfide	BaS	169.4	4.25^{15}
Titanate	$BaTiO_3$	233.3	6.0/5.8
Beryllium			
Bromide	$BeBr_2$	168.8	3.47
Carbide	Be_2C	30.0	1.90^{15}
Chloride	$BeCl_2$	79.9	1.90
Fluoride	BeF_2	47.0	1.99
Hydroxide	$Be(OH)_2$	43.0	1.91
Iodide	BeI_2	262.8	4.33
Nitrate	$Be(NO_3)_2 \cdot 3H_2O$	187.1	1.56
Nitride	Be_3N_2	55.1	2.71
Oxide	BeO	25.0	3.03
Sulfate	$BeSO_4$	105.1	2.43
Sulfate	$BeSO_4 \cdot 4H_2O$	177.1	1.71^{10}
Bismuth			
Bromide III	$BiBr_3$	448.7	5.60
Chloride III	$BiCl_3$	315.4	4.75
Fluoride III	BiF_3	266.0	8.75
Hydroxide III	$Bi(OH)_3$	260.0	4.36
Iodide III	BiI_3	589.7	5.64
Nitrate III	$Bi(NO_3)_3 \cdot 5H_2O$	485.1	2.83
Nitrate, Basic III	$BiO(NO_3) \cdot H_2O$	305.0	4.93
Oxide III	Bi_2O_3	466.0	8.9
Oxide IV	$Bi_2O_4 \cdot 2H_2O$	518.0	5.6
Oxide V	Bi_2O_5	498.0	5.10
Oxychloride III	$BiOCl$	260.5	7.72

SPECIFIC GRAVITY—TABLE 1. Specific Gravity of Inorganic Compounds (*Continued*)

Compound	Formula	Molecular weight	Specific gravity
Phosphate III	$BiPO_4$	304.0	6.32
Sulfate III	$Bi_2(SO_4)_3$	706.1	5.08
Sulfide III	Bi_2S_3	514.2	6.82
Boron			
Arsenate	$BAsO_4$	149.7	3.64
Boric Acid	H_3BO_3	61.8	1.44
Bromide	BBr_3	250.5	2.65
Carbide	B_4C	55.3	2.52
Chloride	BCl_3	117.2	1.43
Diborane	B_2H_6	27.7	0.447^{-112}
Fluoride	BF_3	67.8	1.58
Iodide	BI_3	391.6	3.35
Nitride	BN	24.8	2.30
Oxide	B_2O_3	69.6	1.84
Sulfide	B_2S_3	117.8	1.55
Bromine			
Chloride I	$BrCl$	115.4	
Fluoride I	BrF	98.9	
Fluoride III	BrF_3	136.9	2.49
Fluoride V	BrF_5	174.9	2.47
Hydride I	HBr	80.9	2.16^{-67}
Cadmium			
Bromide	$CdBr_2$	272.2	5.19
Carbonate	$CdCO_3$	172.4	4.26
Chloride	$CdCl_2$	228.4	4.05
Fluoride	CdF_2	150.4	6.64
Hydroxide	$Cd(OH)_2$	146.4	4.79
Iodide	CdI_2	366.2	5.67
Nitrate	$Cd(NO_3)_2 \cdot 4H_2O$	308.5	2.46
Oxide	CdO	128.4	8.15
Sulfate	$CdSO_4$	208.5	4.69
Sulfate	$3CdSO_4 \cdot 8H_2O$	769.6	3.09
Sulfide	CdS	144.5	4.82
Calcium			
Bromate	$CaBrO_3 \cdot H_2O$	313.9	3.33
Bromide	$CaBr_2 \cdot 6H_2O$	308.0	2.30
Carbide	CaC_2	64.1	2.22^{18}
Carbonate	$CaCO_3$	100.1	2.93
Chloride	$CaCl_2$	111.0	2.15
Chloride	$CaCl_2 \cdot 6H_2O$	219.1	1.71^{25}
Chromate	$CaCrO_4 \cdot 2H_2O$	192.1	
Fluoride	CaF_2	78.1	3.18
Hydride	CaH_2	42.1	1.9
Hydroxide	$Ca(OH)_2$	74.1	2.24
Iodide	CaI_2	293.9	3.96
Nitrate	$Ca(NO_3)_2$	164.1	2.50
Nitride	Ca_3N_2	148.3	2.63^{17}
Oxalate	CaC_2O_4	128.1	2.2
Oxide	CaO	56.1	3.3

SPECIFIC GRAVITY—TABLE 1. Specific Gravity of Inorganic Compounds (*Continued*)

Compound	Formula	Molecular weight	Specific gravity
Perchlorate	$Ca(ClO_4)_2$	239.0	2.65
Peroxide	CaO_2	72.1	2.92
Sulfate	$CaSO_4$	136.1	2.96
Sulfate	$CaSO_4 \cdot 2H_2O$	172.2	2.32
Sulfide	CaS	72.1	2.59
Carbon			
Dioxide	CO_2	44.0	1.107^{-37}
Disulfide	CS_2	76.1	1.261
Monoxide	CO	28.0	0.793
Oxybromide	$COBr_2$	187.8	2.44
Oxychloride	$COCl_2$ (Phosgene)	98.9	1.38
Oxysulfide	COS	60.1	1.24^{-87}
Cerium			
Bromide III	$CeBr_3$	380.0	5.18
Chloride III	$CeCl_3$	246.5	3.92^0
Fluoride III	CeF_3	197.1	6.16
Iodate IV	$Ce(IO_3)_4$	839.7	
Iodide III	CeI_3	520.8	2.27
Molybdate III	$Ce_2(MoO_4)_3$	760.0	4.83
Nitrate III	$Ce(NO_3)_3 \cdot 6H_2O$	434.2	
Oxide III	Ce_2O_3	328.2	6.86
Oxide IV	CeO_2	172.1	7.13
Sulfate III	$Ce_2(SO_4)_3$	568.4	3.91
Sulfide	Ce_2S_3	376.4	5.19
Cesium			
Bromide	$CsBr$	212.8	4.44
Carbonate	Cs_2CO_3	325.8	
Chloride	$CsCl$	168.4	3.99
Fluoride	CsF	151.9	4.12
Hydroxide	$CsOH$	149.9	3.68
Iodide	CsI	259.8	4.51
Iodide III	CsI_3	513.7	4.47
Nitrate	$CsNO_3$	194.9	3.69
Oxide	Cs_2O	281.8	4.25
Perchlorate	$CsClO_4$	232.4	3.33^4
Periodate	$CsIO_4$	323.8	4.26
Peroxide	Cs_2O_2	297.8	4.47
Sulfate	Cs_2SO_4	361.9	4.23
Superoxide	CsO_2	164.9	3.77
Trioxide	Cs_2O_3	313.8	4.25
Chlorine			
Dioxide	ClO_2	67.5	
Fluoride	ClF	54.5	1.67^{-108}
Trifluoride	ClF_3	92.5	1.77^{13}
Monoxide	Cl_2O	86.9	3.01
Hydrochloric Acid	HCl	36.5	1.19^{-85}
Perchloric Acid	$HClO_4$	100.5	1.77
Chromium			
Bromide II	$CrBr_2$	211.8	4.36

SPECIFIC GRAVITY—TABLE 1. Specific Gravity of Inorganic Compounds (*Continued*)

Compound	Formula	Molecular weight	Specific gravity
Carbide III	Cr_3C_2	180.0	6.68
Chloride II	$CrCl_2$	122.9	2.88
Chloride III	$CrCl_3$	158.4	2.76
Fluoride II	CrF_2	90.0	4.11
Fluoride III	CrF_3	109.0	3.78
Iodide II	CrI_2	305.8	5.20
Nitrate III	$Cr(NO_3)_3$	238.0	
Nitride III	CrN	66.0	6.14
Oxide II	CrO	68.0	
Oxide III	Cr_2O_3	152.0	5.21
Oxide IV	CrO_2	84.0	
Oxide VI	CrO_3	100.0	2.70
Phosphate III	$CrPO_4 \cdot 6H_2O$	255.1	2.12
Sulfate III	$Cr_2(SO_4)_3 \cdot 18H_2O$	716.5	1.86
Sulfide II	CrS	84.1	4.09
Sulfide III	Cr_2S_3	200.2	3.97
Cobalt			
Bromide II	$CoBr_2$	218.8	4.91
Chlorate II	$Co(ClO_3)_2 \cdot 6H_2O$	333.9	1.92
Chloride II	$CoCl_2$	129.8	3.36
Fluoride II	CoF_2	96.9	4.46
Fluoride III	CoF_3	115.9	3.88
Hydroxide II	$Co(OH)_2$	92.9	3.60
Iodate II	$Co(IO_3)_2$	408.7	5.00
Iodide II	CoI_2	312.7	5.68
Nitrate II	$Co(NO_3)_2 \cdot 6H_2O$	291.0	1.87
Oxide II	CoO	74.9	6.43
Oxide III	Co_2O_3	165.9	5.18
Oxide II–III	Co_3O_4	240.8	6.07
Perchlorate II	$Co(ClO_4)_2$	257.8	3.33
Sulfate II	$CoSO_4$	155.0	3.71
Sulfate II	$CoSO_4 \cdot 7H_2O$	281.1	1.95
Sulfide II	CoS	91.0	5.95
Sulfide III	Co_2S_3	214.1	4.8
Copper			
Bromide I	$CuBr$	143.5	5.05
Bromide II	$CuBr_2$	223.4	4.72
Carbonate, Basic II	$2CuCO_3 \cdot Cu(OH)_2$	344.7	3.88
Chloride I	$CuCl$	99.0	4.14
Chloride II	$CuCl_2$	134.5	3.39
Chloride II	$CuCl_2 \cdot 2H_2O$	170.5	2.54
Fluoride II	$CuF_2 \cdot 2H_2O$	137.6	4.23
Hydroxide I	$CuOH$	80.6	
Hydroxide II	$Cu(OH)_2$	97.6	3.37
Iodide I	CuI	190.5	5.62
Nitrate II	$Cu(NO_3)_2 \cdot 3H_2O$	241.6	2.32
Oxide I	Cu_2O	143.1	6.0
Oxide II	CuO	79.5	6.32
Sulfate II	$CuSO_4$	159.6	3.60

SPECIFIC GRAVITY—TABLE 1.   Specific Gravity of Inorganic Compounds (*Continued*)

Compound	Formula	Molecular weight	Specific gravity
Sulfate I	$CuSO_4 \cdot 5H_2O$	249.7	2.28
Sulfide I	Cu_2S	159.1	5.6
Sulfide II	CuS	95.6	4.64
Thiocyanate I	$CuSCN$	121.6	2.85
Curium			
Bromide III	$CmBr_3$	488	6.87
Chloride III	$CmCl_3$	353	5.81
Fluoride III	CmF_3	304	9.70
Fluoride IV	CmF_4	323	7.49
Iodide III	CmI_3	628	6.37
Dysprosium			
Bromide	$DyBr_3$	402.3	4.78
Chloride	$DyCl_3$	268.9	3.67
Fluoride	DyF_3	219.5	7.46
Iodide	DyI_3	543.2	3.21
Nitrate	$Dy(NO_3)_3 \cdot 5H_2O$	438.6	
Oxide	Dy_2O_3	373.0	8.15
Sulfate	$Dy_2(SO_4)_3 \cdot 8H_2O$	757.3	
Erbium			
Bromide	$ErBr_3$	407.1	4.93
Chloride	$ErCl_3$	273.6	
Fluoride	ErF_3	224.3	7.81
Iodide	ErI_3	548.0	3.28
Oxide	Er_2O_3	382.6	8.64
Sulfate	$Er_2(SO_4)_3$	622.7	3.68
Sulfide	Er_2S_3	263.5	6.21
Europium			
Bromide II	$EuBr_2$	311.8	
Bromide III	$EuBr_3$	391.7	5.40
Chloride II	$EuCl_2$	222.9	
Chloride III	$EuCl_3$	258.3	4.89
Fluoride II	EuF_2	190.0	6.50
Fluoride III	EuF_3	209.0	6.79
Iodide II	EuI_2	405.8	5.5
Iodide III	EuI_3	532.7	
Oxide III	Eu_2O_3	351.9	7.42
Sulfate III	$Eu_2(SO_4)_3 \cdot 8H_2O$	736.2	375 ($-8H_2O$)
Fluorine			
Dioxide	F_2O_2	70.0	1.45^{-57}
Hydride	HF	20.0	$0.991^{-19.9}$
Oxide	F_2O	54.0	1.90^{-224}
Gadolinium			
Bromide	$GdBr_3$	397.0	4.57
Chloride	$GdCl_3$	263.6	4.52^0
Fluoride	GdF_3	214.3	7.05
Iodide	GdI_3	538.0	3.14
Nitrate	$Gd(NO_3)_3 \cdot 6H_2O$	451.4	2.33
Oxide	Gd_2O_3	362.5	7.41

SPECIFIC GRAVITY—TABLE 1. Specific Gravity of Inorganic Compounds (*Continued*)

Compound	Formula	Molecular weight	Specific gravity
Sulfate	$Gd_2(SO_4)_3$	602.7	4.14
Sulfide	Gd_2S_3	410.7	6.15
Gallium			
Arsenide III	GaAs	144.6	5.35
Bromide III	$GaBr_3$	309.5	3.69
Chloride II	Ga_2Cl_4	281.3	
Chloride III	$GaCl_3$	176.0	2.47
Fluoride III	GaF_3	126.7	4.47
Iodide III	GaI_3	450.4	4.15
Oxide I	Ga_2O	155.4	4.77
Oxide III	Ga_2O_3	187.4	5.88
Sulfide I	Ga_2S	171.5	4.2
Sulfide III	Ga_2S_3	235.6	3.7
Germanium			
Bromide IV	$GeBr_4$	392.2	3.13
Chloride IV	$GeCl_4$	214.4	1.87
Fluoride IV	GeF_4	148.6	2.46^{-37}
Hydride IV	GeH_4 (Germane)	76.6	1.52^{-142}
Iodide IV	GeI_4	580.2	4.32
Oxide II	GeO	88.6	1.83
Oxide IV	GeO_2	104.6	4.70
Sulfide II	GeS	104.7	4.01
Sulfide IV	GeS_2	136.7	2.94^{14}
Gold			
Bromide I	AuBr	276.9	7.90
Bromide III	$AuBr_3$	436.7	
Chloride I	AuCl	232.4	7.4
Chloride III	$AuCl_3$	303.3	3.9
Hydroxide III	$Au(OH)_3$	248.0	
Iodide I	AuI	323.9	8.25
Iodide III	AuI_3	577.7	
Sulfate III	$Au_2(SO_4)_3 \cdot H_2O$	490.5	
Sulfide I	Au_2S	426.0	
Sulfide III	Au_2S_3	490.1	8.75
Hafnium			
Bromide	$HfBr_4$	498.1	
Carbide	HfC	190.5	12.7
Chloride	$HfCl_4$	320.3	
Fluoride	HfF_4	254.5	7.13
Iodide	HfI_4	686.1	
Nitride	HfN	192.5	13.9
Oxide	HfO_2	210.5	10.0
Sulfide	HfS_2	242.6	6.0
Holmium			
Bromide	$HoBr_3$	404.7	4.86
Chloride	$HoCl_3$	271.3	
Fluoride	HoF_3	221.9	7.83
Iodide	HoI_3	545.6	3.24
Oxide	Ho_2O_3	377.9	8.35

SPECIFIC GRAVITY—TABLE 1. Specific Gravity of Inorganic Compounds (*Continued*)

Compound	Formula	Molecular weight	Specific gravity
Hydrogen			
Bromide	HBr	80.9	2.16^{-67}
Chloride	HCl	36.5	1.19^{-85}
Fluoride	HF	20.0	$0.991^{19.9}$
Iodide	HI	127.9	2.80^{-35}
Oxide	H_2O	18.0	1.00
Oxide-Deutero	$2H_2O$	20.0	1.104
Peroxide	H_2O_2	34.0	1.442
Selenide	H_2Se	31.0	2.12^{-42}
Sulfide	H_2S	34.1	0.96^{-60}
Telluride	H_2Te	129.9	2.57^{-20}
Indium			
Bromide I	InBr	194.7	4.98
Bromide III	$InBr_3$	354.5	4.75
Chloride I	InCl	150.3	4.2
Chloride III	$InCl_3$	221.2	3.46
Fluoride III	InF_3	171.8	4.39
Iodide I	InI	241.7	5.31
Iodide III	InI_3	495.5	4.69
Oxide III	In_2O_3	277.6	7.18
Sulfate III	$In_2(SO_4)_3$	517.8	3.44
Sulfide III	In_2S_3	325.8	4.90
Iodine			
Bromide I	IBr	206.8	4.42
Chloride I, α	ICl	162.4	3.18
Chloride I, β	ICl	162.4	3.24
Chloride III	ICl_3	233.3	3.19
Fluoride V	IF_5	221.9	3.5
Fluoride VII	IF_7	259.9	2.8^6
Oxide IV	I_2O_4	317.8	4.2
Oxide V	I_2O_5	333.8	4.80
Iodic Acid	HIO_3	175.9	4.63
Hydrogen Iodide	HI	127.9	2.80^{-35}
Iridium			
Bromide III	$IrBr_3 \cdot 4H_2O$	504.0	
Bromide IV	$IrBr_4$	511.8	
Chloride III	$IrCl_3$	298.6	5.30
Chloride IV	$IrCl_4$	334.0	
Fluoride VI	IrF_6	306.2	6.0
Iodide III	IrI_3	572.9	
Iodide IV	IrI_4	699.8	
Oxide IV	IrO_2	224.2	11.7
Sulfide IV	IrS_4	256.3	8.43
Iron			
Arsenide	FeAs	130.8	7.83
Arsenide, di-	$FeAs_2$	205.7	7.38
Bromide II	$FeBr_2$	215.7	4.64
Bromide III	$FeBr_3 \cdot 6H_2O$	403.7	
Carbide	Fe_3C	179.6	7.4

SPECIFIC GRAVITY—TABLE 1.   Specific Gravity of Inorganic Compounds (*Continued*)

Compound	Formula	Molecular weight	Specific gravity
Carbonate II	$FeCO_3$	115.9	3.84
Chloride II	$FeCl_2$	126.8	2.98
Chloride III	$FeCl_3$	162.2	2.90
Fluoride III	FeF_3	112.9	3.18
Hydroxide II	$Fe(OH)_2$	89.9	3.4
Hydroxide III	$Fe(OH)_3$	106.9	3.9
Iodide II	FeI_2	309.7	5.31
Nitrate II	$Fe(NO_3)_3 \cdot 6H_2O$	288.0	
Nitrate III	$Fe(NO_3)_3 \cdot 9H_2O$	404.0	1.68
Nitride	Fe_2N	125.7	6.35
Oxide II	FeO	71.9	6.04
Oxide III	Fe_2O_3	159.7	5.25
Oxide II–III	Fe_3O_4	231.6	5.21
Phosphate III	$FePO_4 \cdot 2H_2O$	186.9	2.87
Phosphide	Fe_2P	142.7	6.56
Sulfate II	$FeSO_4 \cdot 7H_2O$	278.0	1.90
Sulfate III	$Fe_2(SO_4)_3$	399.9	3.10
Sulfate II, Ammonium	$(NH_4)_2Fe(SO_4) \cdot 6H_2O$	392.2	1.86
Sulfide II	FeS	87.9	4.76
Sulfide III	Fe_2S_3	207.9	4.3
Sulfide, di	FeS_2	120.0	5.00
Lanthanum			
Bromate	$La(BrO_3)_3 \cdot 9H_2O$	684.8	
Bromide	$LaBr_3$	378.6	5.07
Chloride	$LaCl_3$	245.3	3.84
Fluoride	LaF_3	195.9	5.94
Iodide	LaI_3	519.6	2.25
Molybdate	$La_2(MoO_4)_3$	757.6	4.77
Oxide	La_2O_3	325.8	6.51
Sulfate	$La_2(SO_4)_3$	566.0	3.60
Sulfide	La_2S_3	374.0	4.91
Lead			
Acetate II	$Pb(C_2H_3O_2)_2$	325.3	3.25
Acetate IV	$Pb(C_2H_3O_2)_4$	443.4	2.23
Arsenate II	$Pb_3(AsO_4)_2$	899.4	7.80
Bromide II	$PbBr_2$	367.0	6.67
Carbonate II	$PbCO_3$	267.2	6.60
Chloride II	$PbCl_2$	278.1	5.85
Chloride IV	$PbCl_4$	349.0	3.18
Chromate II	$PbCrO_4$	323.2	6.12
Fluoride II	PbF_2	245.2	8.37
Hydroxide II	$Pb(OH)_2$	241.2	
Iodate II	$Pb(IO_3)_2$	557.0	6.16
Iodide II	PbI_2	461.0	6.16
Molybdate II	$PbMoO_4$	367.2	6.92
Nitrate II	$Pb(NO_3)_2$	331.2	4.53
Oxide II	PbO	223.2	9.36
Oxide IV	PbO_2	239.2	9.38
Oxide II–IV	Pb_3O_4	685.6	9.1

SPECIFIC GRAVITY—TABLE 1. Specific Gravity of Inorganic Compounds (*Continued*)

Compound	Formula	Molecular weight	Specific gravity
Phosphate, III	$Pb_3(PO_4)_2$	811.6	6.99
Sulfate II	$PbSO_4$	303.3	6.34
Sulfide II	PbS	239.3	7.58
Tungstate II	$PbWO_4$	455.1	8.46
Lithium			
Aluminum Hydride	$LiAlH_4$	37.9	0.917
Bromide	$LiBr$	86.9	3.47
Carbonate	Li_2CO_3	73.9	2.11
Chloride	$LiCl$	42.4	2.07
Fluoride	LiF	25.9	2.63
Hydride	LiH	8.0	0.82
Hydroxide	$LiOH$	24.0	2.54
Iodide	LiI	133.9	4.06
Nitrate	$LiNO_3$	68.9	2.38
Oxide	Li_2O	29.9	2.10
Peroxide	Li_2O_2	45.9	2.36
Perchlorate	$LiClO_4$	160.4	2.43
Phosphate	Li_3PO_4	115.8	2.54
Sulfate, α	Li_2SO_4	109.9	2.21
Sulfide	Li_2S	45.9	1.66
Lutetium			
Bromide	$LuBr_3$	414.7	5.17
Chloride	$LuCl_3$	281.3	3.98
Fluoride	LuF_3	232.0	8.33
Iodide	LuI_3	555.7	3.39
Oxide	Lu_2O_3	397.9	9.42
Magnesium			
Aluminate	$MgO \cdot Al_2O_3$	142.3	3.6
Bromide	$MgBr_2$	184.1	3.72
Carbonate	$MgCO_3$	84.3	3.04
Chloride	$MgCl_2$	95.2	2.33
Fluoride	MgF_2	62.3	3.13
Hydroxide	$Mg(OH)_2$	58.3	2.36
Iodide	MgI_2	278.2	4.2
Nitrate	$Mg(NO_3)_2 \cdot 6H_2O$	256.4	1.46
Oxide	MgO	40.3	3.65
Silicide	Mg_2Si	76.7	1.94
Silicate, m	$MgSiO_3$	100.4	3.18
Silicate, o	Mg_2SiO_4	140.7	3.21
Sulfate	$MgSO_4$	120.4	2.66
Sulfide	MgS	56.4	2.80
Manganese			
Bromide II	$MnBr_2$	214.8	4.38
Carbonate II	$MnCO_3$	114.9	3.12
Chloride II	$MnCl_2$	125.9	2.98
Fluoride II	MnF_2	92.9	3.98
Iodide II	MnI_2	308.8	5.0
Oxide II	MnO	70.9	5.44
Oxide III	Mn_2O_3	157.9	4.50

SPECIFIC GRAVITY—TABLE 1.   Specific Gravity of Inorganic Compounds (*Continued*)

Compound	Formula	Molecular weight	Specific gravity
Oxide IV	MnO_2	86.9	5.03
Oxide II–IV	Mn_3O_4	228.8	4.86
Potassium Permanganate	$KMnO_4$	158.0	2.70
Silicide	$MnSi$	83.0	5.90
Sulfate II	$MnSO_4$	151.0	3.25
Sulfide II	MnS	87.0	3.99
Mercury			
Bromide I	Hg_2Br_2	561.1	7.31
Bromide II	$HgBr_2$	360.4	5.92
Chloride I	Hg_2Cl_2	472.1	7.15
Chloride II	$HgCl_2$	271.5	5.53
Cyanide II	$Hg(CN)_2$	252.7	4.00
Fluoride I	Hg_2F_2	439.2	8.73
Fluoride II	HgF_2	238.6	8.95
Iodide I	Hg_2I_2	655.0	7.70
Iodide II	HgI_2	454.4	6.27
Nitrate I	$Hg_2(NO_3)_2 \cdot 2H_2O$	561.2	4.79
Nitrate II	$Hg(NO_3)_2 \cdot \frac{1}{2}H_2O$	333.6	4.39
Oxide I	Hg_2O	417.2	9.8
Oxide II	HgO	216.6	11.1
Sulfate I	Hg_2SO_4	497.3	7.56
Sulfate II	$HgSO_4$	296.7	6.47
Sulfide II	HgS	232.7	8.10
Molybdenum			
Carbide II	Mo_2C	203.9	8.9
Carbide IV	MoC	108.0	8.40
Chloride II	$MoCl_2$	166.9	3.71
Chloride III	$MoCl_3$	202.3	3.58
Chloride V	$MoCl_5$	273.2	2.93
Fluoride VI	MoF_6	202.9	2.55
Iodide II	MoI_2	349.8	5.28
Molybdic Acid	$H_2MoO_4 \cdot 4H_2O$	180.0	3.12
Oxide IV	MoO_2	127.9	6.47
Oxide VI	MoO_3	143.9	4.50
Silicide IV	$MoSi_2$	152.1	6.31
Sulfide IV	MoS_2	160.1	4.80
Neodymium			
Bromide	$NdBr_3$	384.0	5.35
Chloride	$NdCl_3$	250.6	4.17
Fluoride	NdF_3	201.2	
Iodide	NdI_3	524.9	2.34
Oxide	Nd_2O_3	336.5	7.24
Sulfide	Nd_2S_3	384.7	5.18
Neptunium			
Bromide II	$NpBr_3$	476.7	6.62
Chloride III	$NpCl_3$	343.4	5.58
Chloride IV	$NpCl_4$	378.8	4.92
Fluoride III	NpF_3	294.0	9.12
Fluoride VI	NpF_6	351.0	5.00

SPECIFIC GRAVITY—TABLE 1. Specific Gravity of Inorganic Compounds (*Continued*)

Compound	Formula	Molecular weight	Specific gravity
Iodide III	NpI_3	617.7	6.82
Oxide IV	NpO_2	269.0	11.1
Nickel			
Arsenide	$NiAs$	133.6	7.57
Bromide II	$NiBr_2$	218.5	4.64
Carbonyl	$Ni(CO)_4$	170.7	1.32
Chloride II	$NiCl_2$	129.6	3.55
Fluoride II	NiF_2	96.7	4.63
Hydroxide II	$Ni(OH)_2$	92.7	4.15
Iodide II	NiI_2	312.5	5.83
Nitrate II	$Ni(NO_3)_2 \cdot 6H_2O$	290.8	2.05
Oxide II	NiO	74.7	7.45
Phosphide	Ni_2P	148.4	6.31
Sulfate II	$NiSO_4$	154.8	3.68
Sulfide II	NiS	90.8	5.5
Niobium			
Bromide	$NbBr_5$	492.5	4.44
Carbide	NbC	104.9	7.82
Chloride	$NbCl_5$	270.2	2.75
Fluoride	NbF_5	187.9	3.29
Iodide	NbI_5	727.4	5.11
Oxide	Nb_2O_5	265.8	4.46
Nitrogen			
Ammonia	NH_3	17.0	0.681^{-33}
Hydrazine	N_2H_4	32.0	1.01
Hydrazoic Acid	HN_3	43.0	1.09
Hydroxylamine	NH_2OH	33.0	1.20
Nitric Acid	HNO_3	63.0	1.50
Chloride	NCl_3	120.4	1.65
Fluoride	NF_3	71.0	1.54^{-129}
Iodide	NI_3	394.7	
Oxide I (nitrous-)	N_2O	44.0	0.784
Oxide II (nitric-)	NO	30.0	1.269^{-150}
Oxide III (tri-)	N_2O_3	76.0	1.45^2
Oxide IV (per-)	NO_2	46.0	1.45
Oxide V (penta-)	N_2O_5	108.0	1.64^{18}
Sulfide II	N_4S_4	184.3	2.24^{18}
Nitrosyl Chloride	$NOCl$	65.5	1.42^{-12}
Nitrosyl Fluoride	NOF	49.0	1.80^{-72}
Nitryl Chloride	NO_2Cl	81.5	1.32^{14}
Osmium			
Chloride IV	$OsCl_4$	332.0	
Fluoride V	OsF_5	285.2	
Fluoride VI	OsF_6	304.2	
Fluoride VIII	OsF_8	342.2	3.87
Iodide IV	OsI_4	697.8	
Oxide IV	OsO_2	222.2	7.91
Oxide VIII	OsO_4	254.1	4.91
Sulfide IV	OsS_2	254.3	9.47

SPECIFIC GRAVITY—TABLE 1.   Specific Gravity of Inorganic Compounds (*Continued*)

Compound	Formula	Molecular weight	Specific gravity
Oxygen			
Fluoride	OF_2	54.0	1.90^{-224}
Ozone	O_3	48.0	3.03^{-80}
Palladium			
Bromide II	$PdBr_2$	266.6	5.17
Chloride II	$PdCl_2$	177.3	4.0^{18}
Fluoride II	PdF_2	144.4	5.80
Iodide II	PdI_2	360.2	6.00
Oxide II	PdO	122.4	8.31
Sulfide II	PdS	138.5	6.60
Phosphorus			
Hypophosphorous Acid	H_3PO_2	66.0	1.49
Phosphoric Acid	H_3PO_4	98.0	1.87
Phosphorous Acid	H_3PO_3	82.0	1.65
Bromide III	PBr_3	270.7	2.85^{15}
Bromide V	PBr_5	430.5	
Chloride III	PCl_3	137.3	1.57
Chloride V	PCl_5	208.3	1.6
Fluoride III	PF_3	88.0	
Fluoride V	PF_5	126.0	
Hydride (Phosphine)	PH_3	34.0	0.746^{-90}
Iodide III	PI_3	411.7	4.18
Oxide III	P_4O_6	219.9	2.13
Oxide IV	PO_2	63.0	2.54
Oxide V	P_2O_5	142.0	2.30
Oxybromide V	$POBr_3$	286.7	2.77^{82}
Oxychloride V	$POCl_3$	153.4	1.67
Oxyfluoride	POF_3	104.0	
Sulfide	P_4S_7	348.4	2.19
Sulfide V	P_2S_5	222.3	2.03
Thiobromide V	$PSBr_3$	302.8	2.85
Thiochloride V	$PSCl_3$	169.4	1.63
Platinum			
Bromide II	$PtBr_2$	354.9	6.65
Bromide IV	$PtBr_4$	514.8	5.69
Chloride II	$PtCl_2$	260.0	5.87
Chloride IV	$PtCl_4$	336.9	4.30
Fluoride IV	PtF_4	271.2	
Fluoride VI	PtF_6	309.1	
Hydroxide II	$Pt(OH)_2$	229.1	
Hydroxide IV	$Pt(OH)_4$	263.1	
Iodide II	PtI_2	448.9	6.40
Oxide II	PtO	211.1	14.9
Oxide IV	PtO_2	227.1	10.2
Sulfate IV	$Pt(SO_4)_2 \cdot 4H_2O$	459.4	
Sulfide II	PtS	227.2	10.1
Sulfide III	Pt_2S_3	486.6	5.52
Sulfide IV	PtS_2	259.2	7.66

SPECIFIC GRAVITY—TABLE 1.   Specific Gravity of Inorganic Compounds (*Continued*)

Compound	Formula	Molecular weight	Specific gravity
Plutonium			
Bromide III	$PuBr_3$	481.7	6.83
Carbide IV	PuC	256.0	13.5
Chloride III	$PuCl_3$	346.4	5.70
Fluoride III	PuF_3	299.0	9.32
Fluoride IV	PuF_4	318.0	7.00
Fluoride VI	PuF_6	356.0	4.86
Iodide III	PuI_3	622.7	6.92
Nitride III	PuN	256.0	14.2
Oxide IV	PuO_2	274.0	11.5
Polonium			
Bromide IV	$PoBr_4$	529.7	
Chloride II	$PoCl_2$	281.0	6.50
Chloride IV	$PoCl_4$	351.9	
Oxide IV	PoO_2	242.0	8.96
Potassium			
Bromate	$KBrO_3$	167.0	3.24
Bromide	KBr	119.0	2.76
Carbonate	K_2CO_3	138.2	2.43
Chlorate	$KClO_3$	122.6	2.32
Chloride	KCl	74.6	1.98
Cyanide	KCN	65.1	1.52
Dichromate	$K_2Cr_2O_7$	294.2	2.69
Ferrocyanide	$K_4[Fe(CN)_6] \cdot 3H_2O$	422.4	1.85
Fluoride	KF	58.1	2.48
Hydroxide	KOH	56.1	2.04
Iodate	KIO_3	214.0	3.99
Iodide	KI	166.0	3.13
Nitrate	KNO_3	101.1	2.11
Oxide	K_2O	94.2	2.32
Perchlorate	$KClO_4$	138.6	2.52
Periodate	KIO_4	230.0	3.62
Permanganate	$KMnO_4$	158.0	2.70
Peroxide	K_2O_2	110.2	2.40
Phosphate, o	K_3PO_4	212.3	2.26
Sulfate	K_2SO_4	174.3	2.66
Sulfide	K_2S	110.3	1.80
Superoxide	KO_2	71.1	2.14
Thiocyanate	$KSCN$	97.2	1.89
Praseodymium			
Bromide	$PrBr_3$	380.6	5.26
Chloride	$PrCl_3$	247.3	4.02
Fluoride	PrF_3	197.9	6.14
Iodide	PrI_3	521.6	2.31
Oxide	Pr_2O_3	329.8	7.07
Sulfate	$Pr_2(SO_4)_3 \cdot 8H_2O$	714.1	2.83
Sulfide	Pr_2S_3	378.0	5.24

SPECIFIC GRAVITY—TABLE 1. Specific Gravity of Inorganic Compounds (*Continued*)

Compound	Formula	Molecular weight	Specific gravity
Protoctinium			
Bromide IV	$PaBr_4$	470.9	
Chloride IV	$PaCl_4$	372.9	4.72
Fluoride IV	PaF_4	307.1	6.36
Iodide III	PaI_3	611.8	
Oxide IV	PaO_2	263.1	
Radium			
Bromide	$RaBr_2$	385.8	5.78
Chloride	$RaCl_2$	296.1	4.91
Sulfate	$RaSO_4$	322.1	
Rhenium			
Bromide III	$ReBr_3$	425.9	
Chloride III	$ReCl_3$	292.6	
Chloride V	$ReCl_5$	363.5	4.9
Fluoride IV	ReF_4	262.5	5.38
Fluoride VI	ReF_6	300.2	3.62^{19}
Fluoride VII	ReF_7	319.2	
Oxide IV	ReO_2	218.2	11.4
Oxide VI	ReO_3	234.2	6.9–7.4
Oxide VII	Re_2O_7	484.4	8.2
Oxybromide VII	ReO_3Br	314.1	
Oxychloride VII	ReO_3Cl	269.7	3.87
Sulfide IV	ReS_2	250.4	7.51
Sulfide VII	Re_2S_7	596.9	4.87
Rhodium			
Chloride III	$RhCl_3$	209.3	
Fluoride III	RhF_3	159.9	5.38
Hydroxide III	$Rh(OH)_3$	155.9	
Oxide III	Rh_2O_3	253.8	8.20
Oxide IV	RhO_2	134.9	
Sulfide III	Rh_2S_3	302.0	6.40
Rubidium			
Bromate	$RbBrO_3$	213.4	3.68
Bromide	$RbBr$	165.4	3.36
Carbonate	Rb_2CO_3	231.0	3.47
Chloride	$RbCl$	120.9	2.76
Fluoride	RbF	104.5	3.56
Hydroxide	$RbOH$	102.5	3.20
Iodide	RbI	212.4	3.55
Nitrate	$RbNO_3$	147.5	3.13
Oxide	Rb_2O	187.0	3.72
Perchlorate	$RbClO_4$	189.4	3.01
Peroxide	Rb_2O_2	202.9	3.65
Sulfate	Rb_2SO_4	267.0	3.61
Sulfide	Rb_2S	203.0	2.91
Superoxide	RbO_2	117.5	3.05
Ruthenium			
Chloride III	$RuCl_3$	207.4	3.11
Fluoride V	RuF_5	196.1	2.96

SPECIFIC GRAVITY—TABLE 1. Specific Gravity of Inorganic Compounds (*Continued*)

Compound	Formula	Molecular weight	Specific gravity
Oxide IV	RuO_2	133.1	6.97
Oxide VIII	RuO_4	165.1	3.29
Sulfide IV	RuS_2	165.2	6.99
Samarium			
Bromate III	$Sm(BrO_3)_3 \cdot 9H_2O$	696.2	
Bromide II	$SmBr_2$	310.2	5.1
Bromide III	$SmBr_3$	390.1	5.40
Chloride II	$SmCl_2$	221.3	3.69
Chloride III	$SmCl_3$	256.7	4.46
Fluoride II	SmF_2	188.4	
Fluoride III	SmF_3	207.4	6.64
Iodide II	SmI_2	404.2	
Iodide III	SmI_3	531.1	3.14
Nitrate III	$Sm(NO_3)_3 \cdot 6H_2O$	444.5	2.38
Oxide III	Sm_2O_3	348.7	7.43
Sulfate III	$Sm_2(SO_4)_3 \cdot 8H_2O$	733.0	2.93
Sulfide III	Sm_2S_3	396.9	5.83
Scandium			
Bromide	$ScBr_3$	284.7	3.91
Chloride	$ScCl_3$	151.3	2.4
Fluoride	ScF_3	102.0	
Iodide	ScI_3	425.7	
Nitrate	$Sc(NO_3)_3$	231.0	
Oxide	Sc_2O_3	137.9	3.86
Sulfate	$Sc_2(SO_4)_3$	378.1	2.58
Selenium			
Bromide I	Se_2Br_2	317.7	3.60^{15}
Bromide IV	$SeBr_4$	398.6	4.03^{-78}
Chloride I	Se_2Cl_2	228.8	$2.91^{17.5}$
Chloride IV	$SeCl_4$	220.8	3.80
Fluoride IV	SeF_4	154.9	2.77
Fluoride VI	SeF_6	192.9	2.26^{-35}
Hydride II	H_2Se	81.0	2.00^{-42}
Oxide IV	SeO_2	111.0	3.95
Oxide VI	SeO_3	127.0	
Oxybromide	$SeOBr_2$	254.8	3.38^{50}
Oxychloride	$SeOCl_2$	165.9	2.44^{16}
Oxyfluoride	$SeOF_2$	133.0	2.67
Selenic Acid	H_2SeO_4	145.0	3.00
Selenous Acid	H_2SeO_3	129.0	3.00
Silicon			
Bromide	$SiBr_4$	347.7	2.77
Carbide	SiC	40.1	3.17
Chloride	$SiCl_4$	169.9	1.52^0
Fluoride	SiF_4	104.1	1.59^{-78}
Hydride (silane)	SiH_4	32.1	0.68^{-185}
Hydride (disilane)	Si_2H_6	62.2	0.69^{-25}
Hydride (trisilane)	Si_3H_8	92.3	0.73^0
Iodide	SiI_4	535.7	4.2

SPECIFIC GRAVITY—TABLE 1.   Specific Gravity of Inorganic Compounds (*Continued*)

Compound	Formula	Molecular weight	Specific gravity
Nitride	Si_3N_4	140.3	3.44
Oxide II	SiO	44.1	2.18
Oxide IV (amorph)	SiO_2	60.1	2.63
Oxychloride	Si_2OCl_6	284.9	
Sulfide	SiS_2	92.2	1.875
Silver			
Bromate	$AgBrO_3$	235.8	5.21
Bromide	AgBr	187.8	6.48
Carbonate	Ag_2CO_3	257.8	6.08
Chlorate	$AgClO_3$	191.3	4.43
Chloride	AgCl	143.3	5.56
Cyanide	AgCN	133.9	3.95
Fluoride	AgF	126.9	5.85
Iodate	$AgIO_3$	282.8	5.53
Iodide	AgI	234.8	5.67
Nitrate	$AgNO_3$	169.9	4.35
Nitrite	$AgNO_2$	153.9	4.45
Oxide	Ag_2O	231.8	7.22
Perchlorate	$AgClO_4$	207.4	2.81
Phosphate, o	Ag_3PO_4	418.6	6.37
Sulfate	Ag_2SO_4	311.8	5.45
Sulfide	Ag_2S	247.8	7.32
Telluride	Ag_2Te	343.4	8.32
Thiocyanate	AgSCN	166.0	
Sodium			
Bicarbonate	$NaHCO_3$	84.0	2.16
Bromate	$NaBrO_3$	150.9	3.34
Bromide	NaBr	102.9	3.21
Carbonate	Na_2CO_3	106.0	2.53
Chlorate	$NaClO_3$	106.4	2.5
Chloride	NaCl	58.4	2.17
Cyanide	NaCN	49.0	1.86
Fluoride	NaF	42.0	2.78
Hydride	NaH	24.0	1.36
Hydroxide	NaOH	40.0	2.13
Iodate	$NaIO_3$	197.9	4.28
Iodide	NaI	149.9	3.67
Nitrate	$NaNO_3$	85.0	2.26
Nitrite	$NaNO_2$	69.0	2.17
Oxide	Na_2O	62.0	2.27
Perchlorate	$NaClO_4$	122.4	2.50
Periodate	$NaIO_4$	213.9	4.17
Peroxide	Na_2O_2	78.0	2.60
Phosphate, o	Na_3PO_4	163.9	2.54
Silicate, m	Na_2SiO_3	122.1	2.4
Sulfate	Na_2SO_4	142.1	2.68
Sulfide	Na_2S	78.1	1.86
Sulfite	Na_2SO_3	126.1	2.63
Thiosulfate	$Na_2S_2O_3$	158.1	1.67

SPECIFIC GRAVITY—TABLE 1.   Specific Gravity of Inorganic Compounds (*Continued*)

Compound	Formula	Molecular weight	Specific gravity
Strontium			
Bromide	$SrBr_2$	247.5	4.22
Carbonate	$SrCO_3$	147.6	3.74
Chloride	$SrCl_2$	158.5	3.05
Fluoride	SrF_2	125.6	4.24
Hydride	SrH_2	89.6	3.27
Hydroxide	$Sr(OH)_2$	121.7	3.63
Iodate	$Sr(IO_3)_2$	437.4	5.04
Iodide	SrI_2	341.4	4.55
Nitrate	$Sr(NO_3)_2$	211.7	2.99
Oxide	SrO	103.6	4.7
Peroxide	SrO_2	119.6	4.71
Sulfate	$SrSO_4$	183.7	3.96
Sulfide	SrS	119.7	3.70
Sulfur			
Bromide I	S_2Br_2	224.0	2.64
Chloride I	S_2Cl_2	135.0	1.68
Chloride II	SCl_2	103.0	1.62
Chloride IV	SCl_4	173.9	
Fluoride I	S_2F_2	102.1	1.5^{-100}
Fluoride VI	SF_6	146.0	1.88^{-51}
Hydride	H_2S	34.1	0.96^{-60}
Oxide IV	SO_2	64.1	1.434
Oxide VI	SO_3	80.1	1.97
Pyrosulfuric Acid	$H_2S_2O_7$	178.1	1.89
Sulfuric Acid	H_2SO_4	98.1	1.841
Sulfuryl Chloride	SO_2Cl_2	135.0	1.67
Thionyl Bromide	$SOBr_2$	207.9	2.68
Thionyl Chloride	$SOCl_2$	119.0	1.64
Tantalum			
Bromide	$TaBr_5$	580.5	4.99
Carbide	TaC	193.0	13.9
Chloride	$TaCl_5$	358.2	3.76
Fluoride	TaF_5	275.9	4.74
Iodide	TaI_5	815.4	5.80
Nitride	TaN	194.9	14.4
Oxide	Ta_2O_5	441.9	8.0
Tellurium			
Bromide II	$TeBr_2$	287.4	5.24
Bromide IV	$TeBr_4$	447.3	4.31
Chloride II	$TeCl_2$	198.5	7.05
Chloride IV	$TeCl_4$	269.4	3.26
Fluoride VI	TeF_6	241.6	4.00^{-191}
Hydride	H_2Te	129.6	2.68^{-12}
Iodide IV	TeI_4	635.2	5.05
Oxide IV	TeO_2	159.6	5.67/5.91
Oxide VI	TeO_3	175.6	5.08
Telluric Acid, o	H_2TeO_6	229.7	3.16

SPECIFIC GRAVITY—TABLE 1. Specific Gravity of Inorganic Compounds (*Continued*)

Compound	Formula	Molecular weight	Specific gravity
Terbium			
Bromide	$TbBr_3$	398.6	4.67
Chloride	$TbCl_3$	265.3	4.35
Fluoride	TbF_3	215.9	7.24
Iodide	TbI_3	539.6	3.16
Nitrate	$Tb(NO_3)_3 \cdot 6H_2O$	453.0	
Oxide	Tb_2O_3	365.8	7.81
Thallium			
Bromide I	$TlBr$	284.3	7.54
Carbonate I	Tl_2CO_3	468.8	7.11
Chloride I	$TlCl$	239.8	7.00
Chloride III	$TlCl_3$	310.8	
Fluoride	TlF	223.4	8.36
Hydroxide I	$TlOH$	221.4	
Iodide I	TlI	331.3	7.3/7.1
Nitrate I	$TlNO_3$	266.4	5.55
Oxide I	Tl_2O	424.7	10.36
Oxide III	Tl_2O_3	456.7	9.65
Sulfate I	Tl_2SO_4	504.8	6.77
Sulfide I	Tl_2S	440.8	8.46
Thorium			
Bromide	$ThBr_4$	551.7	5.67
Carbide	ThC_2	256.1	8.96
Chloride	$ThCl_4$	373.9	4.60
Fluoride	ThF_4	308.0	6.19
Iodide	ThI_4	739.7	6.00
Oxide	ThO_2	264.0	9.69
Sulfate	$Th(SO_4)_2$	424.2	4.22
Sulfide	ThS_2	296.2	7.36
Thulium			
Bromide	$TmBr_3$	408.7	5.02
Chloride	$TmCl_3$	275.2	
Fluoride	TmF_3	225.9	7.97
Iodide	TmI_3	549.6	3.32
Oxide	Tm_2O_3	385.9	8.77
Tin			
Bromide II	$SnBr_2$	278.5	5.12
Bromide IV	$SnBr_4$	438.4	3.35
Chloride II	$SnCl_2$	189.6	3.95
Chloride IV	$SnCl_4$	260.5	2.23
Fluoride II	SnF_2	156.7	
Fluoride IV	SnF_4	194.7	4.78
Hydride	SnH_4	122.7	
Iodide II	SnI_2	372.5	5.28
Iodide IV	SnI_4	626.3	4.70
Oxide II	SnO	143.7	6.45
Oxide IV	SnO_2	150.7	6.95
Sulfide II	SnS	150.8	5.08
Sulfide IV	SnS_2	182.8	4.49

SPECIFIC GRAVITY—TABLE 1. Specific Gravity of Inorganic Compounds (*Continued*)

Compound	Formula	Molecular weight	Specific gravity
Titanium			
Bromide IV	$TiBr_4$	367.6	3.42
Carbide IV	TiC	59.9	4.92
Chloride II	$TiCl_2$	118.8	3.13
Chloride III	$TiCl_3$	154.3	2.64
Chloride IV	$TiCl_4$	189.7	1.73
Fluoride IV	TiF_4	123.9	2.79
Iodide IV	TiI_4	555.5	4.40
Nitride	TiN	61.9	5.21
Oxide II	TiO	63.9	4.89
Oxide IV	TiO_2	79.9	3.84
Sulfide IV	TiS_2	112.0	3.28
Tungsten			
Bromide V	WBr_5	583.4	
Carbide II	W_2C	379.7	16.1
Carbide IV	WC	195.9	15.7
Chloride V	WCl_5	361.1	3.88
Chloride VI	WCl_6	396.6	3.52
Fluoride VI	WF_6	297.8	3.44[15]
Oxide IV	WO_2	215.9	12.1
Oxide VI	WO_3	231.9	7.16
Sulfide IV	WS_2	248.0	7.5
Tungstic Acid	H_2WO_4	250.0	5.5
Uranium			
Bromide III	UBr_3	477.8	6.53
Bromide IV	UBr_4	557.7	5.55
Carbide	UC	250.0	13.63
Carbide	UC_2	262.0	11.68
Chloride III	UCl_3	344.4	5.51
Chloride IV	UCl_4	379.9	4.87
Fluoride IV	UF_4	314.1	6.70
Fluoride VI	UF_6	352.1	5.06
Nitride	UN	252.0	14.31
Oxide IV	UO_2	270.1	10.96
Oxide VI	UO_3	286.1	8.34
Oxide IV–VI	U_3O_8	842.2	8.39
Uranyl Acetate	$UO_2(C_2H_3O_2)_2 \cdot 6H_2O$	422.1	2.89
Uranyl Nitrate	$UO_2(NO_3)_2 \cdot 6H_2O$	502.1	2.81
Vanadium			
Carbide IV	VC	62.9	5.77
Chloride IV	VCl_4	192.7	1.87
Fluoride III	VF_3	107.9	3.36
Fluoride V	VF_5	145.9	2.18
Iodide II	VI_2	304.7	5.44
Oxide III	V_2O_3	149.9	4.82
Oxide IV	VO_2	82.9	4.65
Oxide V	V_2O_5	181.9	3.36
Oxychloride V	$VOCl_3$	173.3	1.83
Sulfide II	VS	83.0	4.20

SPECIFIC GRAVITY—TABLE 1.   Specific Gravity of Inorganic Compounds (*Continued*)

Compound	Formula	Molecular weight	Specific gravity
Xenon			
Fluoride II	XeF_2	169.3	4.3
Fluoride IV	XeF_4	207.3	4.1
Fluoride VI	XeF_6	245.3	3.6
Oxide VI	XeO_3	179.3	4.6
Ytterbium			
Bromide III	$YbBr_3$	412.8	5.10
Chloride II	$YbCl_2$	244.0	5.08
Chloride III	$YbCl_3$	279.3	
Fluoride III	YbF_3	230.0	8.17
Iodide II	YbI_2	426.9	5.40
Iodide III	YbI_3	553.8	3.33
Oxide III	Yb_2O_3	394.1	9.17
Sulfate III	$Yb_2(SO_4)_3$	634.3	3.79
Yttrium			
Bromide	YBr_3	328.6	3.95
Chloride	YCl_3	195.3	2.67
Fluoride	YF_3	145.9	5.07
Iodide	YI_3	469.6	
Oxide	Y_2O_3	225.8	4.84
Sulfate	$Y_2(SO_4)_3$	466.0	2.61
Zinc			
Acetate	$Zn(C_2H_3O_2)_2$	183.5	1.84
Bromide	$ZnBr_2$	225.2	4.20
Carbonate	$ZnCO_3$	125.4	4.42
Chloride	$ZnCl_2$	136.3	2.91
Fluoride	ZnF_2	103.4	4.90
Hydroxide	$Zn(OH)_2$	99.4	3.05
Iodide	ZnI_2	319.2	4.74
Nitrate	$Zn(NO_3)_2 \cdot 6H_2O$	297.5	2.07
Oxide	ZnO	81.4	5.61
Sulfate	$ZnSO_4$	161.4	3.54
Sulfide	ZnS	97.5	4.04
Zirconium			
Bromide	$ZrBr_4$	410.9	
Carbide	ZrC	103.2	6.73
Chloride	$ZrCl_4$	233.1	2.80
Fluoride	ZrF_4	167.2	4.43
Iodide	ZrI_4	598.8	
Nitride	ZrN	105.2	7.09
Oxide	ZrO_2	123.2	5.73

SPECIFIC HEAT

The *specific heat* is the amount of heat per unit mass required to raise the temperature by 1°C. The relationship between heat and temperature change is usually expressed in the form shown below, where c is the specific heat. The relationship does not apply if a phase change is encountered, because the heat added or removed during a phase change does not change the temperature.

$$Q = cm \ \Delta T \tag{96}$$

that is, heat added is equal to the specific heat multiplied by the mass (weight) multiplied by the temperature difference ($\Delta T = t_{final} - t_{initial}$)

SPECIFIC HEAT—TABLE 1.   Specific Heats of the Elements

Atomic number	Element	Specific heat, J/g·K	Atomic number	Element	Specific heat, J/g·K
1	H	14.304	2	He	5.193
3	Li	3.582	4	Be	1.825
5	B	1.026	6	C	0.709
7	N	1.042	8	O	0.920
9	F	0.824	10	Ne	1.030
11	Na	1.230	12	Mg	1.020
13	Al	0.900	14	Si	0.7
15	P	0.769	16	S	0.710
17	Cl	0.480	18	Ar	0.520
19	K	0.757	20	Ca	0.647
21	Sc	0.568	22	Ti	0.523
23	V	0.489	24	Cr	0.449
25	Mn	0.480	26	Fe	0.449
27	Co	0.421	28	Ni	0.444
29	Cu	0.385	30	Zn	0.388
31	Ga	0.371	32	Ge	0.320
33	As	0.330	34	Se	0.320
35	Br	0.226	36	Kr	0.248
37	Rb	0.363	38	Sr	0.300
39	Y	N/A	40	Zr	0.278
41	Nb	0.250	42	Mo	0.250
43	Tc	0.240	44	Ru	0.238
45	Rh	0.242	46	Pd	0.244
47	Ag	0.232	48	Cd	0.233
49	In	0.233	50	Sn	0.228
51	Sb	0.207	52	Te	0.202
53	I	0.145	54	Xe	0.158
55	Cs	0.240	56	Ba	0.204
57	La	0.190	58	Ce	0.190
59	Pr	0.190	60	Nd	0.190
61	Pm	N/A	62	Sm	0.197
63	Eu	0.182	64	Gd	0.230
65	Tb	0.180	66	Dy	0.173
67	Ho	0.165	68	Er	0.168
69	Tm	0.160	70	Yb	0.155
71	Lu	0.150	72	Hf	0.140
73	Ta	0.140	74	W	0.130
75	Re	0.137	76	Os	0.130

SPECIFIC HEAT—TABLE 1. Specific Heats of the Elements (*Continued*)

Atomic number	Element	Specific heat, J/g·K	Atomic number	Element	Specific heat, J/g·K
77	Ir	0.130	78	Pt	0.130
79	Au	0.128	80	Hg	0.140
81	Tl	0.129	82	Pb	0.129
83	Bi	0.122	84	Po	N/A
85	At	N/A	86	Rn	0.094
87	Fr	N/A	88	Ra	0.094
89	Ac	0.120	90	Th	0.113
91	Pa	N/A	92	U	0.120
93	Np	0.120	94	Pu	0.130
95	Am	0.110	96	Cm	N/A
97	Bk	N/A	98	Cf	N/A
99	Es	N/A	100	Fm	N/A
101	Md	N/A	102	No	N/A
103	Lr	N/A	104	Rf	N/A
105	Db	N/A	106	Sg	N/A
107	Bh	N/A	108	Hs	N/A
109	Mt	N/A	110	Uun	N/A
111	Uuu	N/A	112	Uub	N/A
114	Uuq	N/A	116	Uuh	N/A
118	Uuo	N/A			

SPECIFIC HEAT—TABLE 2. Specific Heats of Various Inorganic Materials

Substance	Specific heat, J/kg·K	Specific heat, c, cal/g·K or Btu/lb·F	Molar C, J/mol K
Brass	0.380	0.092	—
Cement		0.430	
Glass	0.840	0.200	—
Granite	0.790	0.190	—
Hydroxyapatite		0.210	
Ice (−5°C)	0.210		
Ice (−10°C)	2.050	0.490	36.9
Lead	0.128	0.031	26.4
Marble		0.860	
Mercury	0.140	0.033	28.3
Silver	0233	0.056	24.9
Steel		0.450	
Water	4.186	1.000	75.2
Wood	1.700		
Zinc oxide		0.095	
Zinc phosphate		0.122	

SPECIFIC HEAT—TABLE 3. Specific Heats of Organic Compounds

Compound	Formula	Temperature, °C	Specific heat, cal/g °C
Acetic acid	$C_2H_4O_2$	−200 to +25	$0.330 + 0.00080t$
Acetone	C_3H_6O	−210 to −80	$0.540 + 0.0156t$
Aminobenzoic acid (*o*-)	$C_7H_7NO_2$	85 to mp	$0.254 + 0.00136t$
(*m*-)	$C_7H_7NO_2$	120 to mp	$0.253 + 0.00122t$
(*p*-)	$C_7H_7NO_2$	128 to mp	$0.287 + 0.00088t$
Aniline	C_6H_7N		0.741
Anthracene	$C_{14}H_{10}$	50	0.308
		100	0.350
		150	0.382
Anthraquinone	$C_{14}H_8O_2$	0 to 270	$0.258 + 0.00069t$
Apiol	$C_{12}H_{14}O_4$	10	0.299
Azobenzene	$C_{12}H_{10}N_2$	28	0.330
Benzene	C_6H_6	−250	0.0399
		−225	0.0908
		−200	0.124
		−150	0.170
		−100	0.227
		−50	0.299
		0	0.375
Benzoic acid	$C_7H_6O_2$	20 to mp	$0.287 + 0.00050t$
Benzophenone	$C_{13}H_{10}O$	−150	0.115
		−100	0.172
		−50	0.220
		0	0.275
		+20	0.303
Betol	$C_{17}H_{12}O_3$	−150	0.129
		−100	0.167
		0	0.248
		+50	0.308
Bromoiodobenzene (*o*-)	C_6H_4BrI	−50 to 0	$0.143 + 0.00025t$
(*m*-)	C_6H_4BrI	−75 to −15	0.143
(*p*-)	C_6H_4BrI	−40 to 50	$0.116 + 0.00032t$
Bromonaphthalene (*β*-)	$C_{10}H_7Br$	41	0.260
Bromophenol	C_6H_5BrO	32	0.263
Camphene	$C_{10}H_{16}$	35	0.380
Capric acid	$C_{10}H_{20}O_2$	8	0.695
Caprylic acid	$C_8H_{16}O_2$	−2	0.628
Carbon tetrachloride	CCl_4	−240	0.013
		−200	0.081
		−160	0.131
		−120	0.162
		−80	0.182
		−40	0.201
Chloral alcoholate	$C_4H_7Cl_3O_2$	78	0.509
hydrate	$C_2H_3Cl_3O_2$	32	0.213
Chloroacetic acid	$C_2H_3ClO_2$	60	0.363
Chlorobenzoic acid (*o*-)	$C_7H_5ClO_2$	80 to mp	$0.228 + 0.00084t$
(*m*-)	$C_7H_5ClO_2$	94 to mp	$0.232 + 0.00073t$
(*p*-)	$C_7H_5ClO_2$	180 to mp	$0.242 + 0.00055t$

SPECIFIC HEAT—TABLE 3. Specific Heats of Organic Compounds (*Continued*)

Compound	Formula	Temperature, °C	Specific heat, cal/g °C
Chlorobromobenzene (*o*-)	C_6H_4BrCl	−34	0.192
(*m*-)	C_6H_4BrCl	−52	0.150
(*p*-)	C_6H_4BrCl	−40	0.150
Cyanamide	CH_2N_2	20	0.547
Cyanuric acid	$C_3H_3N_3O_3$	40	0.318
Dextrin	$(C_6H_{10}O_5)x$	0 to 90	$0.291 + 0.00096t$
Dextrose	$C_6H_{12}O_6$	−250	0.016
		−200	0.077
		−100	0.160
		0	0.277
		20	0.300
Dibenzyl	$C_{14}H_{14}$	28	0.363
Dibromobenzene (*o*-)	$C_6H_4Br_2$	−36	0.248
(*m*-)	$C_6H_4Br_2$	−25	0.134
(*p*-)	$C_6H_4Br_2$	−50 to +50	$0.139 + 0.00038t$
Dichloroacetic acid	$C_2H_2Cl_2O_2$		0.406
Dichlorobenzene (*o*-)	$C_6H_4Cl_2$	−48.5	0.185
(*m*-)	$C_6H_4Cl_2$	−52	0.186
(*p*-)	$C_6H_4Cl_2$	−50 to +53	$0.219 + 0.0021t$
Dicyandiamide	$C_2H_4N_4$	0 to 204	0.456
Dihydroxybenzene (*o*-)	$C_6H_6O_2$	−163 to mp	$0.278 + 0.00098t$
(*m*-)	$C_6H_6O_2$	−160 to mp	$0.269 + 0.00118t$
(*p*-)	$C_6H_6O_2$	−250	0.025
		−240	0.038
		−220	0.061
		−200	0.081
		−150 to mp	$0.268 + 0.00093t$
Di-iodobenzene (*o*-)	$C_6H_4I_2$	−50 to +15	$0.109 + 0.00026t$
(*m*-)	$C_6H_4I_2$	−52 to −42	$0.100 + 0.00026t$
(*p*-)	$C_6H_4I_2$	−50 to +80	$0.101 + 0.00026t$
Dimethyl oxalate	$C_4H_6O_4$	10 to 50	$0.212 + 0.0044t$
Dimethylpyrene	$C_7H_8O_2$	50	0.368
Dinitrobenzene (*o*-)	$C_6H_4N_2O_4$	−160 to mp	$0.252 + 0.00083t$
(*m*-)	$C_6H_4N_2O_4$	−160 to mp	$0.248 + 0.00077t$
(*p*-)	$C_6H_4N_2O_4$	119 to mp	$0.259 + 0.00057t$
Diphenyl	$C_{12}H_{10}$	40	0.385
Diphenylamine	$C_{12}H_{11}N$	26	0.337
Dulcitol	$C_6H_{14}O_6$	20	0.282
Erythritol	$C_4H_{10}O_4$	60	0.351
Ethyl alcohol	C_2H_6O (crystalline)	−190	0.232
		−180	0.248
		−160	0.282
		−140	0.318
		−130	0.376
	(vitreous)	−190	0.260
		−180	0.296
		−175	0.380
		−170	0.399
Ethylene glycol	$C_2H_6O_2$	−190 to −40	$0.366 + 0.00110t$

SPECIFIC HEAT—TABLE 3. Specific Heats of Organic Compounds (*Continued*)

Compound	Formula	Temperature, °C	Specific heat, cal/g °C
Formic acid	CH_2O_2	−22	0.387
		0	0.430
Glutaric acid	$C_5H_8O_4$	20	0.299
Glycerol	$C_3H_8O_3$	−265	0.009
		−260	0.022
		−250	0.047
		−220	0.085
		−200	0.115
		−100	0.217
		0	0.330
Hexachloroethane	C_2Cl_6	25	0.174
Hexadecane	$C_{16}H_{34}$		0.495
Hydroxyacetanilide	$C_8H_9NO_2$	41 to mp	$0.249 + 0.00154t$
Iodobenzene	C_6H_5I	40	0.191
Isopropyl alcohol	C_3H_8O	−200 to −160	$0.051 + 0.00165t$
Lactose	$C_{12}H_{22}O_{11}$	20	0.287
	$C_{12}H_{22}O_{11} \cdot H_2O$	20	0.299
Lauric acid	$C_{12}H_{24}O_2$	−30 to +40	$0.430 + 0.000027t$
Levoglucosane	$C_6H_{10}O_5$	40	0.607
Levulose	$C_6H_{12}O_6$	20	0.275
Malonic acid	$C_3H_4O_4$	20	0.275
Maltose	$C_{12}H_{22}O_{11}$	20	0.320
Mannitol	$C_6H_{14}O_6$	0 to 100	$0.313 + 0.00025t$
Melamine	$C_3H_6N_6$	40	0.351
Myristic acid	$C_{14}H_{28}O_2$	0 to 35	$0.381 + 0.00545t$
Naphthalene	$C_{10}H_5$	−130 to mp	$0.281 + 0.00111t$
Naphthol (α-)	$C_{10}H_5O$	50 to mp	$0.240 + 0.00147t$
(β-)	$C_{10}H_8O$	61 to mp	$0.252 + 0.00128t$
Naphthylamine (α-)	$C_{10}H_9N$	0 to 50	$0.270 + 0.0031t$
Nitroaniline (o-)	$C_6H_6N_2O_2$	−160 to up	$0.269 + 0.000920t$
(m-)	$C_6H_6N_2O_2$	−160 to mp	$0.275 + 0.000946t$
(p-)	$C_6H_6N_2O_2$	−160 to mp	$0.276 + 0.001000t$
Nitrobenzoic acid (o-)	$C_7H_5NO_4$	−163 to mp	$0.256 + 0.00085t$
(m-)	$C_7H_5NO_4$	66 to mp	$0.258 + 0.00091t$
(p-)	$C_7H_5NO_4$	−160 to mp	$0.247 + 0.00077t$
Nitronaphthalene	$C_{10}H_7NO_2$	0 to 55	$0.236 + 0.00215t$
Oxalic acid	$C_2H_2O_4$	−200 to +50	$0.259 + 0.00076t$
	$C_2H_2O_4 \cdot 2H_2O$	−200	0.117
		−100	0.239
		0	0.338
		+50	0.385
		100	0.416
Palmitic acid	$C_{16}H_{32}O_2$	−180	0.167
		−140	0.208
		−100	0.251
		−50	0.306
		0	0.382
		+20	0.430

SPECIFIC HEAT—TABLE 3.　Specific Heats of Organic Compounds (*Continued*)

Compound	Formula	Temperature, °C	Specific heat, cal/g °C
Phenol	C_6H_6O	14 to 26	0.561
Phthalic acid	$C_8H_6O_4$	20	0.232
Picric acid	$C_6H_3N_3O_7$	−100	0.165
		0	0.240
		+50	0.263
		100	0.297
		120	0.332
Propionic acid	$C_3H_6O_2$	−33	0.726
Propyl alcohol (*n*-)	C_3H_8O	−200	0.170
		−175	0.363
		−150	0.471
		−130	0.497
Pyrotartaric acid	$C_6H_8O_4$	20	0.301
Quinhydrone	$C_{12}H_{10}O_4$	−250	0.017
		−225	0.061
		−200	0.098
		−100	0.191
		0	0.256
Quinone	$C_6H_4O_2$	−250	0.031
		−225	0.082
		−200	0.113
		−150 to mp	$0.282 + 0.00083t$
Stearic acid	$C_{18}H_{36}O_2$	15	0.399
Succinic acid	$C_4H_6O_4$	0 to 160	$0.248 + 0.00153t$
Sucrose	$C_{12}H_{22}O_{11}$	20	0.299
Sugar (cane)	$C_{12}H_{22}O_{11}$	22 to 51	0.301
Tartaric acid	$C_4H_6O_6$	36	0.287
Tartaric acid	$C_4H_6O_6 \cdot H_2O$	−150	0.112
		−100	0.170
		−50	0.231
		0	0.308
		+50	0.366
Tetrachloroethylene	C_2Cl_4	−40 to 0	$0.198 + 0.00018t$
Tetryl	$C_7H_5N_5O_8$	−100	0.182
		−50	0.199
		0	0.212
		+100	0.236
1 Tetryl + 1 picric acid	$C_{13}H_8N_8O_{15}$	−100 to +100	$0.253 + 0.00072t$
1 Tetryl + 2 TNT	$C_{21}H_{15}N_{11}O_{20}$	−100	0.172
		0	0.280
		+50	0.325
Thymol	$C_{10}H_{14}O$	0 to 49	$0.315 + 0.0031t$
Toluic acid (*o*-)	$C_8H_8O_2$	54 to mp	$0.277 + 0.00120t$
(*m*-)	$C_8H_8O_2$	54 to mp	$0.239 + 0.00195t$
(*p*-)	$C_8H_8O_2$	130 to mp	$0.271 + 0.00106t$
Toluidine (*p*-)	C_7H_9N	0	0.337
		20	0.387
		40	0.440

SPECIFIC HEAT—TABLE 3. Specific Heats of Organic Compounds (*Continued*)

Compound	Formula	Temperature, °C	Specific heat, cal/g °C
Trichloroacetic acid	$C_2HCl_3O_2$	solid	0.459
Trimethyl carbinol	$C_4H_{10}O$	−4	0.559
Trinitrotoluene	$C_7H_5N_3O_6$	−100	0.170
		−50	0.253
		0	0.311
		+100	0.385
Trinitroxylene	$C_8H_7N_3O_6$	−185 to +23	0.241
		20 to 50	0.423
Triphenylmethane	$C_{19}H_{16}$	0 to 91	$0.189 + 0.0027t$
Urea	CH_4N_2O	20	0.320

Recalculated from "International Critical Tables," Volume 5, pp. 101–105.

SPECIFIC HEAT—TABLE 4 Specific Heats of the Various Phases of Water

Phase	$J\,g^{-1}\,°C^{-1}$	$J\,kg^{-1}\,K^{-1}$
Gas	2.02	2.02×10^3
Liquid	4.184	4.184×10^3
Solid	2.06	2.06×10^3

THERMAL CONDUCTIVITY

The *thermal conductivity* is a measure of the effectiveness of a material as a thermal insulator. The energy transfer rate through a substance is proportional to the temperature gradient across the substance and the cross-sectional area of the body. In the limit of infinitesimal thickness and temperature difference, the fundamental law of heat conduction is:

$$Q = \lambda\,A\,DT/dx \qquad (97)$$

where Q is the heat flow, A is the cross-sectional area, dT/dx is the temperature/thickness gradient, and λ is the thermal conductivity.

A substance with a large thermal conductivity value is a good conductor of heat; one with a small thermal conductivity value is a poor heat conductor (i.e., a good insulator).

Thermal conductivity describes the ease with which conductive heat can flow through a vapor, liquid, or solid layer of a substance. It is defined as the proportionality constant in Fourier's law of heat conduction in units of energy·length/time·area·temperature (e.g., W/m K).

Gases

For *pure component, low pressure* (<350 kPa) *hydrocarbon* gases:

$$k_C = 4.45 \times 10^{-7} T_r \frac{C_p}{\lambda} \qquad (98)$$

For these hydrocarbons above reduced temperatures of 1.0 and for other hydrocarbons at any temperature:

$$k_C = 10^{-7}(14.52T_r - 5.14)^{2/3}\left(\frac{C_p}{\lambda}\right) \tag{99}$$

In these equations,

$$\lambda = T_c^{1/6}M^{1/2}\left(\frac{101.325}{P_c}\right)^{2/3} \tag{100}$$

where k_C = vapor thermal conductivity, W/m·K
T_r = reduced temperature, T/T_c
T = temperature, K
T_c = critical temperature, K
C_p = heat capacity at constant pressure, J/kmol·K
M = molecular weight
P_c = critical pressure, kPa

C_p may be assumed to be the ideal gas heat capacity, C_p^o. Average errors can be expected to be less than 5%.

For *pure nonhydrocarbon* gases at law pressures (up through 1 atm), the following equations may be used at temperature T (K):

Monatomic gases:
$$k_C = 2.5\frac{\eta_c C_v}{M} \tag{101}$$

Linear molecules:
$$k_C = \frac{\eta_c}{M}\left(1.30C_v + 14644.0 - \frac{2928.8}{T_r}\right) \tag{102}$$

Nonlinear molecules:
$$k_C = \frac{\eta_c}{M}(1.15C_v + 16903.36) \tag{103}$$

where k_c = vapor thermal conductivity, W/m·K
η_c = vapor viscosity, Pa·s
C_v = heat capacity at constant volume, J/kmol·K
M = molecular weight
T_r = reduced temperature, T/T_c
T_c = critical temperature, K

C_o may be calculated as $C_p^o - R$, where C_p^o is the ideal gas heat capacity in J/kmol K and R is the gas constant, 8314 J/kmol·K. Average errors are in the 8–10% range but may be higher for polar compounds. This method should not be used for molecules that associate (e.g., organic acids).

For pure gases *above atmospheric pressure*:

$$k_c = k_C' + \frac{A \times 10^{-4}\left(e^{Rpr} + C\right)}{\left(\frac{T_r^{1/6}M^{1/2}}{P_c^{2/3}}\right)Z_e^5} \tag{104}$$

$\rho_r < 0.5$	A = 2.702	B = 0.535	C = −1.000
$0.5 < \rho_r > 2.0$	A = 2.528	B = 0.670	C = −1.069
$2.0 < \rho_r > 2.8$	A = 0.574	B = 1.155	C = 2.016

where k_c = vapor thermal conductivity at the temperature T (K) and pressure P
 of interest, W/m·K

 k'_c = vapor thermal conductivity at T and atmospheric pressure, W/m K

 ρ_r = reduced density = V_r/V

 V_r = critical molar volume, m³/kmol

 V = molar volume at T and P, m³/kmol

 T_r = critical temperature, K

 M = molecular weight

 P_r = critical pressure, MPa

 Z_r = critical compressibility factor = $P_r V_r / R T_r$

 R = gas constant = 0.008314 MPa m³/kmol·K

Errors in the range of 5–6% are typical with this method but may be higher for branched compounds.

The thermal conductivity of *low pressure* (1 atm or less) *gas mixtures* can be determined from the relation:

$$k_m = \sum_{i=1}^{n} \frac{y_i k_i}{\sum_{j-1}^{n} y_j A_{ij}} \qquad (105)$$

where k_m = mixture thermal conductivity, W/m·K

 n = number of components

 $y_{i,j}$ = mole fraction of component i or j in the vapor mixture

 k_i = thermal conductivity of pure component i at the temperature of
 interest

The binary interaction parameter A_{ij} is obtained by:

$$A_{ij} = \frac{1}{4} \left\{ 1 + \left[\frac{\mu_i}{\mu_j} \left(\frac{M_j}{M_i} \right)^{3/4} \left(\frac{T + S_i}{T + S_j} \right) \right]^{1/2} \right\}^2 \left(\frac{T + S_{ij}}{T + S_i} \right) \qquad (106)$$

where $\mu_{i,j}$ = vapor viscosity of pure component i or j at the temperature T of
 interest and low pressure, Pa·s

 $M_{i,j}$ = molecular weight of pure component i or j

 T = temperature, K

 $S_{ij} = S_{ji}$

 C = 1.0 except when either or both components i and j are very polar;
 then $C = 0.73$

 $S_{i,j}$ = 79 K for helium, hydrogen, and neon

 $T_{bi,j}$ = normal boiling temperature of pure component i or j, K

$$S_{ji} = C \left(S_i S_j \right)^{1/2} \qquad (107)$$

$$S_{i,j} = 1.5 T_{bi,j} \qquad (108)$$

Expected errors for this method are 4–5%. At *higher pressures*, a pressure correction (Eq. 104) may be used.

Liquids

For *pure component hydrocarbon* liquids at reduced temperatures between 0.25 and 0.8 and at pressures below 3.4 MPa:

$$k_L = C \rho M^n \left[\frac{3 + 20(1 - T_r)^{2/3}}{3 + 20\left(1 - \dfrac{293.15}{T_c}\right)^{2/3}} \right] \qquad (109)$$

where k_L = liquid thermal conductivity, W/m K
    M = molecular weight
    ρ = molar density at 293.15 K, kmol/m³
    T_r = reduced temperature, T/T_c
    T = temperature, K
    T_c = critical temperature, K

For unbranched, straight chain hydrocarbons, $n = 1.001$ and $C = 1.811 \times 10^{-4}$.
  For branched and cyclic hydrocarbons, $n = 0.7717$ and $C = 4.407 \times 10^{-4}$.
  Average errors are 5% when this equation is used. For pressures greater than 3.4 MPa, the thermal conductivity (Eq. 109) may be corrected. The correction factor is the ratio of conductivity factors F/F', where F is at the desired temperature and higher pressure, and F' is at the same temperature and lower pressure (usually atmospheric). The conductivity factors are calculated from:

$$F = 17.77 + 0.065 P_r - 7.764 T_r - \frac{2.054 T_r^2}{e^{0.2 Pr}} \qquad (110)$$

where T_r = reduced temperature
    P_r = reduced pressure, P/P_c
    P = pressure, MPa
    P_c = critical pressure, MPa

The average error in the pressure correction alone is typically 3%.
  For pure component hydrocarbon liquids *above the normal boiling point and all pressures*:

$$k_L = \frac{\alpha \, e^{\beta p_r}}{\lambda} \qquad p < 10{,}000 \, \text{kPa} \qquad (111)$$

$$k_L = \frac{2.596 \times 10^{-4} \, P_r^{1.6} + \alpha \, e^{\beta p_r}}{\lambda} \qquad p > 10{,}000 \, \text{kPa} \qquad (112)$$

where k_L = liquid thermal conductivity at the temperature T (K) and pressure P (kPa) of interest, W/m K
    $\alpha = 0.0112 \, \beta^{-3.322}$
    $\beta = 0.40 + 0.986 e^{-0.64\lambda}$
    $\lambda = T_c^{1/6} M^{1/2} \left(\dfrac{101.325}{P_c} \right)^{2/3}$
    ρ_r = reduced density = V_c/V
    V_c = critical molar volume, m³/kmol
    V = molar volume at T and P, m³/kmol
    T_c = critical temperature, K
    M = molecular weight
    P_c = critical pressure, kPa
    P_r = reduced pressure, P/P_c

Average errors can be expected to be on the order of 10%.

The thermal conductivity of *pure component nonhydrocarbon* liquids may be estimated for silicon compounds, at reduced temperatures between 0.3 and 0.8 and at pressures below 3.5 MPa:

$$k_L = \left(\frac{abc}{m}\right)\left(\frac{(1-T_r)^{0.38}}{T_r^{1/6}}\right) \tag{113}$$

where k_L = liquid thermal conductivity, W/m K
$\quad T_r$ = reduced temperature, T/T_c
$\quad T$ = temperature, K
$\quad T_r$ = critical temperature, K
$\quad a$ = constant parameter
$\quad b$ = constant parameter = funtion of normal boiling temperature T_b, K
$\quad c$ = constant parameter = function of critical temperature T_c, K
$\quad m$ = constant parameter = function of molecular weight M

Values of a, b, c, and m for various compound classes are available (Table 3). Average errors are about 8%.

For pure component nonhydrocarbon liquids *for which Eq. 113 is not applicable*, the method of Missenard may be used at temperature T (K) and below pressures of 3.5 MPa:

$$k_L = k_L^r\left[\frac{3+20(1-T_r)^{2/3}}{3+20\left(1-\dfrac{273.15}{T_c}\right)^{2/3}}\right] \tag{114}$$

$$k_L^r = 2.656 \times 10^{-7}\frac{(T_b\rho^r)^{1/2}C_\mu^r}{M^{1/2}N^{1/4}} \tag{115}$$

where k_L = liquid thermal conductivity, W/m K
$\quad k_L^r$ = liquid thermal conductiviy at 273.15 K, W/m K
$\quad T_r$ = reduced temperature, T/T_c
$\quad T_c$ = critical temperature, K
$\quad T_b$ = normal boiling temperature, K
$\quad \rho^r$ = molar density at 273.15 K, kmol/m^3
$\quad C_p^r$ = molar heat capacity at 273.15 K, J/kmol K
$\quad M$ = molecular weight
$\quad N$ = number of atoms in the molecule

Errors on the order of 8% can be expected.

For *pressures greater than 3.5 MPa*, a correction factor may be used to obtain the thermal conductivity of pure component nonhydrocarbon liquids. Thus:

$$k_L = k_L'\left[0.98 + 0.0079P_rT_r^{1.4} + 0.63T_r^{1.2}\left(\frac{P_r}{30+P_r}\right)\right] \tag{116}$$

where k_L = liquid thermal conductivity at the desired temperature T (K) and pressure P (MPa), W/m K
$\quad k_L'$ = liquid thermal conductivity at T and pressure of 0.1 MPa, W/m K
$\quad P_r$ = reduced pressure, P/P_c
$\quad P_c$ = critical pressure, MPa
$\quad T_r$ = reduced temperature, T/T_c
$\quad T_c$ = critical temperature, K

Average errors are in the range of 5–20%.

THERMAL CONDUCTIVITY–TABLE 1. Thermal Conductivity: Conversion Factors

	$\dfrac{\text{Watt}}{\text{cm·K}}$	$\dfrac{\text{Watt}}{\text{m·K}}$	$\dfrac{\text{Watt·in}}{\text{in}^2\text{·R}}$	$\dfrac{\text{Cal·cm}}{\text{cm}^2\text{·sec·K}}$	$\dfrac{\text{Kcal·m}}{\text{m}^2\text{·hr·K}}$	$\dfrac{\text{Cal·in}}{\text{in}^2\text{·se·°R}}$	$\dfrac{\text{Btu·in}}{\text{in}^2\text{·sec·°R}}$	$\dfrac{\text{Btu·in}}{\text{in}^2\text{·hr·°R}}$	$\dfrac{\text{Btu·ft}}{\text{ft}^2\text{·hr·°R}}$	$\dfrac{\text{Btu·in}}{\text{ft}^2\text{·hr·°R}}$
$\dfrac{\text{Watt}}{\text{cm·K}}$	= 1.000	100.0	1.411	0.2390	86.04	0.3373	1.338×10^{-3}	4.818	57.82	693.8
$\dfrac{\text{Watt}}{\text{m·K}}$	= 1.000×10^{-2}	1.000	1.411×10^{-2}	2.390×10^{-3}	0.8604	3.373×10^{-3}	1.338×10^{-5}	4.818×10^{-2}	0.5782	6.938
$\dfrac{\text{Watt·in}}{\text{in}^2\text{·R}}$	= 0.7087	70.87	1.000	0.1694	60.97	0.2390	9.485×10^{-4}	3.414	40.97	491.7
$\dfrac{\text{Cal·cm}}{\text{cm}^2\text{·sec·K}}$	= 4.184	418.4	5.904	1.000	360.0	1.411	5.600×10^{-3}	20.16	241.9	2.903
$\dfrac{\text{Kcal·m}}{\text{m}^2\text{·hr·K}}$	= 1.162×10^{-2}	1.162	1.640×10^{-2}	2.778×10^{-3}	1.000	3.920×10^{-3}	1.555×10^{-5}	5.600×10^{-2}	0.6720	8.064
$\dfrac{\text{Cal·in}}{\text{in}^2\text{·se·°R}}$	= 2.965	296.5	4.184	0.7087	255.1	1.000	3.968×10^{-3}	14.29	171.4	2.057
$\dfrac{\text{Btu·in}}{\text{in}^2\text{·sec·°R}}$	= 747.2	7.472×10^{4}	1054	178.6	6.429×10^{4}	252.0	1.000	3600	4.320×10^{4}	5.184×10^{5}
$\dfrac{\text{Btu·in}}{\text{in}^2\text{·hr·°R}}$	= 0.2075	20.75	0.2929	4.961×10^{-2}	67.86	7.000×10^{-2}	2.778×10^{-4}	1.000	12.00	144.0
$\dfrac{\text{Btu·ft}}{\text{ft}^2\text{·hr·°R}}$	= 1.730×10^{-2}	1.730	2.441×10^{-2}	4.134×10^{-3}	0.488	5.833×10^{-3}	2.315×10^{-5}	8.333×10^{-2}	1.000	12.00
$\dfrac{\text{Btu·in}}{\text{ft}^2\text{·hr·°R}}$	= 1.441×10^{-3}	0.1441	2.034×10^{-3}	3.445×10^{-4}	0.1240	4.861×10^{-4}	1.929×10^{-6}	6.944×10^{-3}	8.333×10^{-2}	1.000

THERMAL CONDUCTIVITY—TABLE 2. Values of Constant
Parameters in Equation 113 for Various Compound Classes

Compound class	a	b	c	m
Acids*	0.00319	$T_b^{6/5}$	$T_c^{-1/6}$	$M^{1/2}$
Alcohols, phenols	0.00339	$T_b^{6/5}$	$T_c^{-1/6}$	$M^{1/2}$
Esters[†]	0.0415	$T_b^{6/5}$	$T_c^{-1/6}$	M
Ethers	0.0385	$T_b^{6/5}$	$T_c^{-1/6}$	M
Halides[‡]	0.494	1.0	$T_c^{1/6}$	$M^{1/2}$
Refrigerants R20–R23	0.562	1.0	$T_c^{1/6}$	$M^{1/2}$
Ketones	0.00383	$T_b^{6/5}$	$T_c^{-1/6}$	$M^{1/2}$
Alkoxysilanes	0.00482	$T_b^{6/5}$	$T_c^{-1/6}$	$M^{1/2}$
Alkyl-(aryl)-chlorosilanes	0.6510	1.0	$T_c^{1/6}$	$M^{1/2}$

* Do not use for formic, myristic, or oleic acids.
[†] Do not use for butyl stearate.
[‡] Do not use for refrigerants R20–R23 ($CHCl_3$, $CHFCl_2$, $CHClF_2$, or CHF_3).

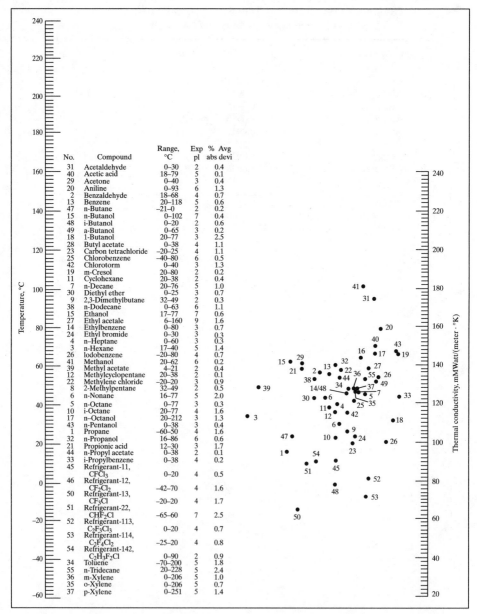

No.	Compound	Range, °C	Exp pl	% Avg abs devi
31	Acetaldehyde	0–30	2	0.4
40	Acetic acid	18–79	5	0.1
29	Acetone	0–40	3	0.4
20	Aniline	0–93	6	1.3
2	Benzaldehyde	18–68	4	0.7
13	Benzene	20–118	5	0.6
47	n-Butane	−21–0	2	0.2
15	n-Butanol	0–102	7	0.4
48	i-Butanol	0–20	2	0.6
49	a-Butanol	0–65	3	0.2
18	1-Butanol	20–77	3	2.5
28	Butyl acetate	0–38	4	1.1
23	Carbon tetrachloride	−20–25	4	1.1
25	Chlorobenzene	−40–80	6	0.5
42	Chlorotorm	0–40	3	1.3
19	m-Cresol	20–80	2	0.2
11	Cyclohexane	20–38	2	0.4
7	n-Decane	20–76	5	1.0
30	Diethyl ether	0–25	3	0.7
9	2,3-Dimethylbutane	32–49	2	0.3
38	n-Dodecane	0–63	6	1.1
15	Ethanol	17–77	7	0.6
27	Ethyl acetale	6–160	9	1.6
14	Ethylbenzene	0–80	3	0.7
24	Ethyl bromide	0–30	3	0.3
4	n–Heptane	0–60	3	0.3
3	n-Hexane	17–40	5	1.4
26	Iodobenzene	−20–80	4	0.7
41	Methanol	20–62	6	0.2
39	Methyl acetate	4–21	2	0.4
12	Methylcyclopentane	20–38	2	0.1
22	Methylene chloride	−20–20	3	0.9
8	2-Methylpentane	32–49	2	0.5
6	n-Nonane	16–77	5	2.0
5	n-Octane	0–77	3	0.3
10	i-Octane	20–77	4	1.6
17	n-Octanol	20–212	3	1.3
43	n-Pentanol	0–38	3	0.2
1	Propane	−60–50	4	1.6
32	n-Propanol	16–86	6	0.6
21	Propionic acid	12–30	3	1.7
44	n-Propyl acetate	0–38	2	0.1
33	i-Propylbenzene	0–38	4	0.2
45	Refrigerant-11, $CFCl_3$	0–20	4	0.5
46	Refrigerant-12, CF_2Cl_2	−42–70	4	1.6
50	Refrigerant-13, CF_3Cl	−20–20	4	1.7
51	Refrigerant-22, CHF_2Cl	−65–60	7	2.5
52	Refrigerant-113, $C_2F_3Cl_3$	0–20	4	0.7
53	Refrigerant-114, $C_2F_4Cl_2$	−25–20	4	0.8
54	Refrigerant-142, $C_2H_3F_2Cl$	0–90	2	0.9
34	Toluene	−70–200	5	1.8
55	n-Tridecane	20–228	5	2.4
36	m-Xylene	0–206	5	1.0
35	o-Xylene	0–206	5	0.7
37	p-Xylene	0–251	5	1.4

Temperature, °C

Thermal conductivity, mMWatt/(meter · °K)

THERMAL CONDUCTIVITY—FIGURE 1. Nomograph (right) for thermal conductivity of organie liquids

THERMAL CONDUCTIVITY—TABLE 3. Thermal Conductivity of the Elements

Element number	Element symbol	Thermal conductivity (W/m)/K 27°C, 81°F	Element number	Element symbol	Thermal conductivity (W/m)/K 27°C, 81°F
1	H	0.1815	2	He	0.152
3	Li	84.7	4	Be	200
5	B	27	6	C	155
7	N	0.02598	8	O	0.02674
9	F	0.0279	10	Ne	0.0493
11	Na	141	12	Mg	156
13	Al	237	14	Si	148
15	P	0.235	16	S	0.269
17	Cl	0.0089	18	Ar	0.0177
19	K	102.5	20	Ca	200
21	Sc	15.8	22	Ti	21.9
23	V	30.7	24	Cr	93.7
25	Mn	7.82	26	Fe	80.2
27	Co	100	28	Ni	90.7
29	Cu	401	30	Zn	116
31	Ga	40.6	32	Ge	59.9
33	As	50	34	Se	2.04
35	Br	0.122	36	Kr	0.00949
37	Rb	58.2	38	Sr	35.3
39	Y	17.2	40	Zr	22.7
41	Nb	53.7	42	Mo	138
43	Tc	50.6	44	Ru	117
45	Rh	150	46	Pd	71.8
47	Ag	429	48	Cd	96.8
49	In	81.6	50	Sn	66.6
51	Sb	24.3	52	Te	2.35
53	I	0.449	54	Xe	0.00569
55	Cs	35.9	56	Ba	18.4
57	La	13.5	58	Ce	11.4
59	Pr	12.5	60	Nd	16.5
61	Pm	17.9	62	Sm	13.3
63	Eu	13.9	64	Gd	10.6
65	Tb	11.1	66	Dy	10.7
67	Ho	16.2	68	Er	14.3

THERMODYNAMIC RELATIONSHIPS AND PROPERTIES

Examples of *thermodynamic relationships* are (1) the thermodynamic relationship between a solid and its melt when vapor pressure remains constant and (2) the thermodynamic relationship between the concentration of a solid and a solvent, other than the melt of that solid.

The thermodynamic relationships are more properly expressed by the following equations:

First law:	$dQ - dW = dU$	(117)
Enthalpy:	$H = U + pV \quad \text{or} \quad h = u + pv$	(118)
for a reversible process,	$dS = \left(\dfrac{dQ}{T}\right)_{\text{reV}} \quad \text{or} \quad dQ = TdS$	(119)
	$dW = pdV$	(120)
For a homogeneous fluid:	$Tds = du + pdv = dh - vdp$	(121)
Specific heat at constant volume:	$c_v = \left(\dfrac{du}{dT}\right)_v$	(122)
Specific heat at constant pressure:	$c_p = \left(\dfrac{dh}{dT}\right)_p$	(123)
Ratio of specific heats:	$\gamma = \dfrac{c_p}{c_v}$	(124)
Helmholtz function:	$F = U - TS \quad \text{or} \quad f = u - Ts$	(125)
Gibbs function, Gibbs free energy:	$G = H - TS \quad \text{or} \quad g = h - Ts$	(126)

Perfect gas relationships:

$pv = RT \quad \text{or} \quad p = \rho RT$	$PV = mRT$	(127)
$pv_0 = R_0 T$	$MR = R_0$	(128)
$\Delta U = mc_v (T_2 - T_1)$	$\Delta H = mc_p (T_2 - T_1)$	(129)
$\Delta S = mc_v \ln\left(\dfrac{p_2}{p_1}\right) + mc_p \ln\left(\dfrac{v_2}{v_1}\right)$		
$c_p - c_v = R$	$\dfrac{\gamma - 1}{\gamma} = \dfrac{R}{c_p}$	(130)

Continuity equation:	$\dot{m} = \rho A c$	(131)

Steady flow energy equation:

$$\frac{Q - W}{m} = h_2 - h_1 + \frac{1}{2}\left(c_2^2 - c_1^2\right) + g(z_2 - z_1) \tag{132}$$

Availability (closed system):

$$(A_1 - A_0) = (U_1 + p_0 V_1 - T_0 S_1) - (U_0 + p_0 V_0 - T_0 S_0) \tag{133}$$

Availability (flow process):

$$(B_1 - B_0) = (H_1 - T_0 S_1) - (H_0 - T_0 S_0) \tag{134}$$

Irreversibility:	$I = W_{\text{rev}} - W$	(135)

Van der Waal's equation:
$$\left(p + \frac{a}{v^2}\right)(v - b) = RT \tag{136}$$

Entropy:
$$S = k \ln p, \text{ where } k = R_0/N \tag{137}$$

Reversible (Carnot) engine efficiency:
$$1 - \frac{T_{\text{sink}}}{T_{\text{source}}} \tag{138}$$

Engine indicated power:
$$P_i = p_m v_s N_c \tag{139}$$

Maximum work of a reaction:
$$W_{\text{max}} = G_{\text{react}} - G_{\text{prod}} = R_o T \ln\left(K_p p^n\right) \tag{140}$$

THERMODYNAMIC RELATIONSHIPS AND PROPERTIES—TABLE 1.
Thermodynamic Values for the Hydrogen–Oxygen Reaction $H_2 + \frac{1}{2}O_2 \rightarrow H_2O$

Temperature, °K	Enthalpy of reaction $(\Delta H°)$, kJ/mol	Free energy of reaction $(\Delta G°)$, kJ/mol	Equilibrium cell potential $(E°)$, V
300	−241.8	−228.4	1.18
500	−243.9	−219.2	1.14
700	−245.6	−208.8	1.06
900	−247.3	−197.9	1.03
1100	−248.5	−187.0	0.97
1300	−249.4	−175.7	0.91

THERMODYNAMIC RELATIONSHIPS AND PROPERTIES—TABLE 2. Thermodynamic Data for Inorganic Compounds

Substance	State	$\Delta H_f^{\ominus}$ kJ mol^{-1}	$\Delta G_f^{\ominus}$ kJ mol^{-1}	$S^{\ominus}$ J K^{-1} mol^{-1}	$C_p^{\ominus}$ J K^{-1} mol^{-1}
Aluminium					
Al	s	0	0	28·3	24·3
Al_2O_3	s (α, corundum)	−1669	−1576	51·0	79·0
$Al(OH)_3$	am	−1273			
AlF_3	s	−1301	−1230	96	
$AlCl_3$	s	−695	−637	170	89·1
$AlBr_3$	s	−527	−505	180	102
AlI_3	s	−315	−314	200	99·2
Al_4C_3	s	−129	−121	100	
AlN	s	−241	−210	20	32
Al_2S_3	s	−509	−492	96	
$Al(NO_3)_3 \cdot 9H_2O$	s	−3754	−2930	569	
$Al_2(SO_4)_3$	s	−3435	−3092	239	259
$AlNH_4(SO_4)_2 \cdot 12H_2O$	s	−5939	−4933	697	683
$AlK(SO_4)_2 \cdot 12H_2O$	s	−6057	−5137	687	

THERMODYNAMIC RELATIONSHIPS AND PROPERTIES—TABLE 2.   Thermodynamic Data for Inorganic Compounds (*Continued*)

Substance	State	$\dfrac{\Delta H_f^{\ominus}}{\text{kJ mol}^{-1}}$	$\dfrac{\Delta G_f^{\ominus}}{\text{kJ mol}^{-1}}$	$\dfrac{S^{\ominus}}{\text{J K}^{-1}\text{mol}^{-1}}$	$\dfrac{C_p^{\ominus}}{\text{J K}^{-1}\text{mol}^{-1}}$
Antimony					
Sb	s	0	0	43·9	25·4
Sb_2O_3	s	−705	−623	123	101
Sb_2O_4	s	−895	−787	127	115
Sb_2O_5	s	−981	−839	125	118
SbF_3	s	−909			
$SbCl_3$	s	−382	−325	186	
$SbCl_5$	l	−438			
SbOCl	s	−380			
SbH_3	s	140			
Sb_2S_3	s (black)	−182	−180	127	
Sb_2S_3	am (orange)	−151			
Argon					
Ar	g	0	0	155	20·8
Arsenic					
As	s (α, grey)	0	0	35	25·0
As_4	g	149	105	290	
As_4O_6	s (octahedral)	−1314	−1152	214	191
As_2O_5	s	−915	−772	105	116
AsF_3	l	−949	−902	181	127
$AsCl_3$	l	−336	−295	234	
AsH_3	g	172			
As_2S_3	s	−150			
H_3AsO_4	s	−900			
Barium					
Ba	s	0	0	67	26·4
BaO	s	−558	−528	70·3	47·4
BaO_2	s	−630			
$Ba(OH)_2 \cdot 8H_2O$	s	−3345			
BaF_2	s	−1201	−1148	96·2	71·2
$BaCl_2$	s	−860	−811	130	75·3
$BaCl_2 \cdot 2H_2O$	s	−1461	−1296	203	155
$BaBr_2$	s	−755			
$BaBr_2 \cdot 2H_2O$	s	−1365			
BaI_2	s	−602			
$BaI_2 \cdot 2H_2O$	s	−1218			
BaH_2	s	−171			
Ba_3N_2	s	−364			
BaS	s	−444			
$BaCO_3$	s (witherite)	−1219	−1139	112	85·4
$Ba(NO_3)_2$	s	−992	−795	214	
$BaSO_4$	s	−1465	−1353	132	102
Beryllium					
Be	s	0	0	9·5	17·8
BeO	s	−611	−582	14·1	25·4
$Be(OH)_2$	s	−907			

THERMODYNAMIC RELATIONSHIPS AND PROPERTIES—TABLE 2. Thermodynamic
Data for Inorganic Compounds (*Continued*)

Substance	State	$\dfrac{\Delta H_f^{\ominus}}{\text{kJ mol}^{-1}}$	$\dfrac{\Delta G_f^{\ominus}}{\text{kJ mol}^{-1}}$	$\dfrac{S^{\ominus}}{\text{J K}^{-1}\text{mol}^{-1}}$	$\dfrac{C_p^{\ominus}}{\text{J K}^{-1}\text{mol}^{-1}}$
$BeCl_2$	s	−512			
$BeBr_2$	s	−370			
BeI_2	s	−212			
Be_3N_2	s	−568	−512	33·4	
BeS	s	−234			
$BeSO_4$	s	−1197			
Bismuth					
Bi	s	0	0	56·9	26
Bi_2O_3	s	−577	−497	152	114
$Bi(OH)_3$	s	−710			
$BiCl_3$	s	−379	−319	190	
BiOCl	s	−365	−322	86·2	
Bi_2S_3	s	−183	−165	148	
Boron					
B	s	0	0	6·5	12·0
B_2O_3	s	−1264	−1184	54·0	62·3
BF_3	g	−1111	−1093	254	50·5
BCl_3	l	−418	−379	209	
BBr_3	l	−221	−219	229	
B_2H_6	g	31	82·8	233	56·4
BN	s	−134	−114	30	25
B_2S_3	s	−238			
H_3BO_3	s	−1089	−963	89·6	82·0
Bromine					
Br_2	l	0	0	152	71·6
Br_2	g	30·7	3·1	245	36·0
BrCl	g	14·7	−0·9	240	
HBr	g	−36·2	−53·2	198	29·1
Cadmium					
Cd	s (α)	0	0	51·5	25·9
CdO	s	−255	−225	54·8	43·4
$Cd(OH)_2$	s	−558	−471	95·4	
CdF_2	s	−690	−648	110	
$CdCl_2$	s	−389	−343	118	
$CdCl_2 \cdot 2 \cdot 5H_2O$	s	−1129	−943	233	
$CdBr_2$	s	−314	−293	134	
$CdBr_2 \cdot 4H_2O$	s	−1491	−1246	312	
CdI_2	s	−201	−201	168	
Cd_3N_2	s	162			
CdS	s	−144	−141	70	
$CdCO_3$	s	−748	−670	105	
$Cd(NO_3)_2 \cdot 4H_2O$	s	−165			
$CdSO_4$	s	−926	−820	137	

THERMODYNAMIC RELATIONSHIPS AND PROPERTIES—TABLE 2. Thermodynamic Data for Inorganic Compounds (*Continued*)

Substance	State	$\Delta H_f^{\ominus}$ $\mathrm{kJ\,mol^{-1}}$	$\Delta G_f^{\ominus}$ $\mathrm{kJ\,mol^{-1}}$	$S^{\ominus}$ $\mathrm{J\,K^{-1}\,mol^{-1}}$	$C_p^{\ominus}$ $\mathrm{J\,K^{-1}\,mol^{-1}}$
Cesium					
Cs	s	0	0	82·8	31·0
Cs_2O	s	−318			
Cs_2O_2	s	−402			
CsOH	s	−407			
CsF	s	−531			
CsCl	s	−433			
CsBr	s	−394	−383	120	
CsI	s	−337	−334	130	
CsH	g	121	102	214	
Calcium					
Ca	s	0	0	41·6	26·3
CaO	s	−635	−604	40	42·8
$Ca(OH)_2$	s	−987	−897	76·2	84·5
CaF_2	s	−1214	−1162	68·9	67·0
$CaCl_2$	s	−795	−750	114	72·6
$CaCl_2 \cdot 6H_2O$	s	−2608			
$CaBr_2$	s	−675	−656	130	
CaI_2	s	−535	−530	140	
CaH_2	s	−189	−150	40	
CaC_2	s	−62·8	−67·8	70·3	62·3
Ca_3N_2	s	−432	−369	100	
CaS	s	−483	−477	56·5	
$CaCO_3$	s (calcite)	−1207	−1129	92·9	81·9
$CaCO_3$	s (aragonite)	−1207	−1128	88·7	81·2
$Ca(HCO_3)_2$	s	−1354			
$Ca(NO_3)_2$	s	−937	−742	193	
$Ca(NO_3)_2 \cdot 4H_2O$	s	−2132	−1701	340	
$Ca_3(PO_4)_2$	s (α)	−4126	−3890	241	232
$CaSO_4$	s (anhydrite)	−1433	−1320	107	99·6
$CaSO_4 \cdot 0·5H_2O$	s (α)	−1575	−1435	130	120
	s (β)	−1573	−1434	134	124
$CaSO_4 \cdot 2H_2O$	s	−2021	−1796	194	186
Carbon					
C	s (graphite)	0	0	5·7	8·6
C	s (diamond)	1·9	2·9	2·4	6·1
C	g	715	673	158	21
CO	g	−111	−137	198	29·1
CO_2	g	−394	−395	214	37·1
CF_4	g	−680	−661	262	
CCl_4	g	−107	−64·0	309	83·5
CCl_4	l	−139	−68·6	214	132
CBr_4	g	50·2	36	358	
$COCl_2$	g	−223	−210	289	60·7
CS_2	l	87·9	63·6	151	75·7
HCN	g	130	120	202	35·9
HCN	l	105	121	113	70·6
$(CN)_2$	g	308	296	242	56·9
Cerium					
Ce	s	0	0	57·7	25·9

THERMODYNAMIC RELATIONSHIPS AND PROPERTIES—TABLE 2. Thermodynamic Data for Inorganic Compounds (*Continued*)

Substance	State	$\Delta H_f^{\ominus}$ kJ mol^{-1}	$\Delta G_f^{\ominus}$ kJ mol^{-1}	$S^{\ominus}$ J K^{-1} mol^{-1}	$C_p^{\ominus}$ J K^{-1} mol^{-1}
Chlorine					
Cl_2	g	0	0	223	33·9
Cl_2O	g	76·2	93·7	266	
ClO_2	g	103	123	249	
Cl_2O_7	g	265			
ClF	g	−55·6	−56·9	218	
HCl	g	−92·3	−95·3	187	29·1
Chromium					
Cr	s	0	0	23·8	23·4
Cr_2O_3	s	−1128	−1047	81·2	119
CrO_3	s	−579			
$Cr(OH)_3$	s	−1033			
$CrCl_2$	s	−396	−356	115	70·6
$CrCl_3$	s	−563	−494	126	90·1
CrO_2Cl_2	l	−568			
Cobalt					
Co	s	0	0	28	25·6
CoO	s	−239	−213	43·9	
Co_3O_4	s	−879			
CoF_2	s	−665			
$CoCl_2$	s	−326	−282	106	78·7
$CoCl_2 \cdot 6H_2O$	s	−2130			
$CoBr_2$	s	−232			
CoI_2	s	−102			
CoS	s	−84·5	−82·8	55	
$CoCO_3$	s	−722	−650		
$Co(NO_3)_2 \cdot 6H_2O$	s	−2216			
$CoSO_4$	s	−868	−762	113	
$CoSO_4 \cdot 7H_2O$	s	−2986			
Copper					
Cu	s	0	0	33·3	24·5
Cu_2O	s	−167	−146	101	69·9
CuO	s	−155	−127	43·5	44·4
$Cu(OH)_2$	s	−448			
CuF_2	s	−531			
$CuCl$	s	−135	−119	84·5	
$CuCl_2$	s	−206			
$CuCl_2 \cdot 2H_2O$	s	−808			
$CuBr$	s	−105	−99·6	91·6	
$CuBr_2$	s	−141			
CuI	s	−67·8	−69·5	96·6	54·0
CuI_2	s	−7·1			
Cu_2S	s	−79·5	−86·2	121	76·3
CuS	s	−48·5	−49·0	66·5	47·8
$CuCO_3$	s	−595	−518	88	
$Cu(NO_3)_2$	s	−307			
$Cu(NO_3)_2 \cdot 3H_2O$	s	−1219			

THERMODYNAMIC RELATIONSHIPS AND PROPERTIES—TABLE 2. Thermodynamic Data for Inorganic Compounds (*Continued*)

Substance	State	$\dfrac{\Delta H_f^{\ominus}}{\text{kJ mol}^{-1}}$	$\dfrac{\Delta G_f^{\ominus}}{\text{kJ mol}^{-1}}$	$\dfrac{S^{\ominus}}{\text{J K}^{-1}\text{mol}^{-1}}$	$\dfrac{C_p^{\ominus}}{\text{J K}^{-1}\text{mol}^{-1}}$
Cu_2SO_4	s	−750			
$CuSO_4$	s	−770	−662	113	
$CuSO_4 \cdot H_2O$	s	−1084	−917	150	
$CuSO_4 \cdot 5H_2O$	s	−2278	−1880	305	281
Deuterium					
D_2	g	0	0	145	29·2
D_2O	g	−249	−235	198	34·3
D_2O	l	−295	−244	76·0	82·4
HDO	g	−246	−234	199	33·7
HDO	l	−290	−242	79·3	78·9
HD	g	0·2	−1·6	144	29·2
Fluorine					
F_2	g	0	0	203	31·5
F_2O	g	23	41	247	
HF	g	−269	−271	174	29·1
Gallium					
Ga	s	0	0	42·7	26·6
Ga_2O	s	−340			
Ga_2O_3	s	−1079			
$Ga(OH)_3$	s		−833		
$GaCl_3$	s	−525			
$GaBr_3$	s	−387			
GaI_3	s	−214			
GaN	s	−100			
Germanium					
Ge	s	0	0	42·4	26·1
GeO_2	am	−537			
$GeCl_4$	l	−544			
GeH_4	g			214	45·0
Gold					
Au	s	0	0	47·7	25·2
Au_2O_3	s	80·8	163	130	
AuCl	s	−35			
$AuCl_3$	s	−118			
Hafnium					
Hf	s	0	0	54·8	25·7
HfO_2	s	−1094			
Helium					
He	g	0	0	126	20·8
Hydrogen					
H_2	g	0	0	131	28·8
H	g	218	203	115	20·8
H_2O	g	−242	−229	189	33·6
H_2O	l	−286	−237	69·9	75·3

THERMODYNAMIC RELATIONSHIPS AND PROPERTIES—TABLE 2. Thermodynamic Data for Inorganic Compounds (*Continued*)

Substance	State	$\Delta H_f^{\ominus}$ kJ mol^{-1}	$\Delta G_f^{\ominus}$ kJ mol^{-1}	$S^{\ominus}$ J K^{-1} mol^{-1}	$C_p^{\ominus}$ J K^{-1} mol^{-1}
H_2O_2	g	−133			
H_2O_2	l	−188	−118	102	
Indium					
In	s	0	0	52·3	27·4
In_2O_3	s	−931			
$In(OH)_3$	s	−895	−762	100	
InCl	s	−186			
$InCl_3$	s	−537			
$InBr_3$	s	−404			
InI_3	s	−230			
InN	s	−20			
Iodine					
I_2	s	0	0	117	55·0
I_2	g	62·2	19·4	261	36·9
I_2O_5	s	−177			
ICl	g	17·6	−5·5	247	35·4
ICl_3	s	−88·3	−22·4	172	
IBr	g	40·8	3·8	259	
HI	g	25·9	1·3	206	29·2
HIO_3	s	−23·9			
Iridium					
Ir	s	0	0	36	25
IrO_2	s	−168			
$IrCl_2$	s	−179			
$IrCl_3$	s	−257			
Iron					
Fe	s	0	0	27·2	25·2
$Fe_{0·95}O$	s (wüstite)	−266	−244	54·0	
Fe_2O_3	s (haematite)	−822	−741	90·0	105
Fe_3O_4	s (magnetite)	−1117	−1015	146	
$Fe(OH)_2$	s	−568	−484	80	
$Fe(OH)_3$	s	−824			
$FeCl_2$	s	−341	−302	120	76·4
$FeCl_3$	s	−405			
$FeCl_3·6H_2O$	s	−2226			
$FeBr_2$	s	−251			
FeI_2	s	−125			
Fe_3C	s (cementite)	21	15	108	106
FeS	s (α)	−95·1	−97·6	67·4	54·8
FeS_2	s (pyrites)	−178	−167	53·1	61·9
$FeCO_3$	s (siderite)	−748	−674	92·9	82·1
$Fe(CO)_5$	l	−786			
$Fe(NO_3)_3·9H_2O$	s	−3282			
$FeSO_4$	s	−923	−815	108	
$FeSO_4·7H_2O$	s	−3007			

THERMODYNAMIC RELATIONSHIPS AND PROPERTIES—TABLE 2.   Thermodynamic
Data for Inorganic Compounds (*Continued*)

Substance	State	$\Delta H_f^{\ominus}$ kJ mol^{-1}	$\Delta G_f^{\ominus}$ kJ mol^{-1}	$S^{\ominus}$ J K^{-1} mol^{-1}	$C_p^{\ominus}$ J K^{-1} mol^{-1}
Krypton					
Kr	g	0	0	164	20·8
Lanthanum					
La	s	0	0	57·3	28
La$_2$O$_3$	s	−1916			
LaCl$_3$	s (α)	−1103			
Lead					
Pb	s	0	0	64·9	26·8
PbO	s (red)	−219	−189	67·8	
PbO	s (yellow)	−218	−188	69·4	49·0
PbO$_2$	s	−277	−219	76·6	64·4
Pb$_3$O$_4$	s	−735	−618	211	147
Pb(OH)$_2$	s	−515	−421	88	
PbF$_2$	s	−663	−620	120	
PbCl$_2$	s	−359	−314	136	77·0
PbBr$_2$	s	−277	−260	162	
PbI$_2$	s	−175	−174	177	
PbS	s	−94·3	−92·7	91·2	49·5
PbCO$_3$	s	−700	−626	131	
Pb(CH$_3$COO)$_2$·3H$_2$O	s	−1854			
Pb(C$_2$H$_5$)$_4$	l	220			
Pb(NO$_3$)$_2$	s	−449			
PbSO$_4$	s	−918	−811	147	104
Lithium					
Li	s	0	0	28·0	23·6
Li$_2$O	s	−596	−560	39·2	
Li$_2$O$_2$	s	−635			
LiOH	s	−487	−444	50	
LiF	s	−612	−584	−35·9	
LiCl	s	−409			
LiBr	s	−350			
LiI	s	−271			
LiH	g	128	105	171	
Li$_3$N	s	−198			
Li$_2$CO$_3$	s	−1216	−1133	90·4	
LiNO$_3$	s	−482			
Li$_2$SO$_4$	s	−1434			
LiAlH$_4$	s	−101			
Magnesium					
Mg	s	0	0	32·5	23·9
MgO	s	−602	−570	27	37·4
Mg(OH)$_2$	s	−925	−834	63·1	
MgF$_2$	s	−1102	−1049	57·2	61·6
MgCl$_2$	s	−642	−592	89·5	
MgCl$_2$·6H$_2$O	s	−2500	−2116	366	316
MgBr$_2$	s	−518			

THERMODYNAMIC RELATIONSHIPS AND PROPERTIES—TABLE 2. Thermodynamic Data for Inorganic Compounds (*Continued*)

Substance	State	$\dfrac{\Delta H_f^{\ominus}}{\text{kJ mol}^{-1}}$	$\dfrac{\Delta G_f^{\ominus}}{\text{kJ mol}^{-1}}$	$\dfrac{S^{\ominus}}{\text{J K}^{-1}\text{mol}^{-1}}$	$\dfrac{C_p^{\ominus}}{\text{J K}^{-1}\text{mol}^{-1}}$
MgI_2	s	−360			
Mg_3N_2	s	−461			
MgS	s	−347			
$MgCO_3$	s	−1113	−1030	65·7	75·5
$Mg(NO_3)_2$	s	−790	−588	164	
$Mg(NO_3)_2 \cdot 6H_2O$	s	−2612			
$Mg_3(PO_4)_2$	s	−4023			
$MgNH_4PO_4 \cdot 6H_2O$	s	−3686			
$MgSO_4$	s	−1278	−1174	91·6	
$MgSO_4 \cdot 7H_2O$	s	−3384			
Manganese					
Mn	s (α)	0	0	31·8	26·3
MnO	s	−385	−363	60·2	43·0
MnO_2	s	−521	−466	53·1	54·0
Mn_2O_3	s	−971			
Mn_3O_4	s	−1387	−1280	148	
$Mn(OH)_2$	am	−694	−610	88·3	
MnF_2	s	−791	−749	92·9	
$MnCl_2$	s	−482	−441	117	72·9
$MnBr_2$	s	−380			
MnI_2	s	−248			
MnS	s (green)	−204	−209	78·2	
MnS	s (red)	−199			
$MnCO_3$	s	−895	−818	85·8	
$Mn(NO_3)_2$	s	−696			
$Mn(NO_3)_2 \cdot 6H_2O$	s	−2370			
$MnSO_4$	s	−1064	−956	112	
$MnSO_4 \cdot 4H_2O$	s	−2256			
Mercury					
Hg	l	0	0	77·4	27·8
Hg	g	60·8	31·8	175	20·8
HgO	s (red)	−90·7	−58·5	72·0	45·7
HgO	s (yellow)	−90·2	−58·4	73·2	
Hg_2O	s	−91·2	−53·6		
Hg_2Cl_2	s	−265	−211	196	102
$HgCl_2$	s	−230	−177	121	76·6
Hg_2Br_2	s	−207	−179	213	
$HgBr_2$	s	−170	−162	206	
Hg_2I_2	s (yellow)	−121	−113	239	106
HgI_2	s (red)	−105			
HgI_2	s (yellow)	−103			
HgS	s (red)	−58·2	−48·8	77·8	
HgS	s (black)	−54·0	−46·2	83·3	
$Hg_2(NO_3)_2 \cdot 2H_2O$	s	−866			
$Hg(NO_3)_2 \cdot 0.5H_2O$	s	−389			
Hg_2SO_4	s	−742	−624	201	
$HgSO_4$	s	−704			

THERMODYNAMIC RELATIONSHIPS AND PROPERTIES—TABLE 2. Thermodynamic Data for Inorganic Compounds (*Continued*)

Substance	State	$\dfrac{\Delta H_f^{\ominus}}{\text{kJ mol}^{-1}}$	$\dfrac{\Delta G_f^{\ominus}}{\text{kJ mol}^{-1}}$	$\dfrac{S^{\ominus}}{\text{J K}^{-1}\text{mol}^{-1}}$	$\dfrac{C_p^{\ominus}}{\text{J K}^{-1}\text{mol}^{-1}}$
Molybdenum					
Mo	s	0	0	28·6	23·5
MoO$_3$	s	−754	−678	78·2	73·6
MoS$_2$	s	−232	−225	63·2	
Neon					
Ne	g	0	0	146	20·8
Nickel					
Ni	s	0	0	30·1	26·0
NiO	s	−244	−216	38·6	44·4
Ni(OH)$_2$	s	−538	−453	80	
NiF$_2$	s	−667			
NiCl$_2$	s	−316	−272	107	
NiCl$_2$·6H$_2$O	s	−2116	−1718	315	
NiBr$_2$	s	−227			
NiI$_2$	s	−85·8			
NiS	s	−73·2			
NiCO$_3$	s		−614		
Ni(CO)$_4$	g	−605	−587	402	
Ni(NO$_3$)$_2$	s	−428			
Ni(NO$_3$)$_2$·6H$_2$O	s	−2224			
NiSO$_4$	s	−891	−774	77·8	
NiSO$_4$·6H$_2$O	s (blue)	−2688	−2222	306	340
Niobium					
Nb	s	0	0	35	25
Nb$_2$O$_5$	s	−1938			
Nitrogen					
N$_2$	g	0	0	192	29·1
N$_2$O	g	81·6	104	220	38·7
NO	g	90·4	86·7	211	29·8
N$_2$O$_3$	g	92·9			
NO$_2$	g	33·9	51·8	240	37·9
N$_2$O$_4$	g	9·7	98·3	304	79·1
N$_2$O$_5$	g	15			
N$_2$O$_5$	s	−41·8			
NF$_3$	g	−114			
NH$_3$	g	−46·2	−16·6	193	35·7
HN$_3$	g	294	328	237	
NOCl	g	52·6	66·4	264	
NH$_4$F	s	−467			
NH$_4$Cl	s	−315	−204	94·6	84·1
NH$_4$Br	s	−270			
NH$_4$I	s	−202			
NH$_4$HS	s	−159			
NH$_4$HCO$_3$	s	−852			
NH$_4$CNO	s	−312			
NH$_4$SCN	s	−83·7			

THERMODYNAMIC RELATIONSHIPS AND PROPERTIES—TABLE 2. Thermodynamic Data for Inorganic Compounds (*Continued*)

Substance	State	$\dfrac{\Delta H_f^{\ominus}}{\text{kJ mol}^{-1}}$	$\dfrac{\Delta G_f^{\ominus}}{\text{kJ mol}^{-1}}$	$\dfrac{S^{\ominus}}{\text{J K}^{-1}\text{mol}^{-1}}$	$\dfrac{C_p^{\ominus}}{\text{J K}^{-1}\text{mol}^{-1}}$
NH_4NO_2	s	−264			
NH_4NO_3	s	−365			182
$(NH_4)_3PO_4$	s	−1681			
$(NH_4)_2HPO_4$	s	−1572			
$NH_4H_2PO_4$	s	−1451	−1215	152	
$(NH_4)_2SO_4$	s	−1179	−900	220	187
NH_4HSO_4	s	−1024			143
$(NH_4)_2S_2O_8$	s	−1659			
NH_4VO_3	s	−1051	−886	141	
N_2H_4	l	50·4			
N_2H_5Cl	s	−196			
NH_2OH	s	−107			
NH_3OHCl	s	−310			
HNO_3	l	−173	−79·9	156	110
Osmium					
Os	s	0	0	33	25
OsO_4	s (white)	−384	−295	145	
OsO_4	s (yellow)	−391	−296	124	
Oxygen					
O_2	g	0	0	205	29·4
O_3	g	142	163	238	38·2
Palladium					
Pd	s	0	0	37	26
PdO	s	−85·4			
$PdCl_2$	s	−190			
Phosphorus					
P	s (white)	0	0	44·4	23·2
P	s (red)	−18			21
P	s (black)	−43·1			
P_4	g	54·9	24·4	280	
P_4O_6	s	−1640			
P_4O_{10}	s	−3012			
PCl_3	g	−306	−286	312	
PCl_3	l	−339			
PCl_5	g	−399	−325	353	
PCl_5	s	−463			
PBr_3	g	−150	−172	348	
PBr_3	l	−199			
PBr_5	s	−280			
PI_3	s	−45·6			
$POCl_3$	g	−592	−545	325	
$POCl_3$	l	−632			
PH_3	g	9·2	18·2	210	
H_3PO_3	s	−972			
H_3PO_4	s	−1281			
HPO_3	s	−955			

THERMODYNAMIC RELATIONSHIPS AND PROPERTIES—TABLE 2. Thermodynamic Data for Inorganic Compounds (*Continued*)

Substance	State	$\dfrac{\Delta H_f^{\ominus}}{\text{kJ mol}^{-1}}$	$\dfrac{\Delta G_f^{\ominus}}{\text{kJ mol}^{-1}}$	$\dfrac{S^{\ominus}}{\text{J K}^{-1}\text{mol}^{-1}}$	$\dfrac{C_p^{\ominus}}{\text{J K}^{-1}\text{mol}^{-1}}$
Platinum					
Pt	s	0	0	41·8	26·6
$PtCl_2$	s	−148			
$PtCl_4$	s	−263			
Potassium					
K	s	0	0	63·6	29·2
K_2O	s	−362			
K_2O_2	s	−494			
K_2O_4	s	−561			
KOH	s	−426			
KF	s	−563	−533	66·6	49·1
KHF_2	s	−920	−852	104	76·9
KCl	s	−436	−408	82·7	51·5
KBr	s	−392	−379	96·4	53·6
KI	s	−328	−322	104	55·1
KH	g	126	105	198	
K_2S	s	−418			
K_2CO_3	s	−1146			
$KHCO_3$	s	−959			
KCN	s	−112			
KCNO	s	−412			
KSCN	s	−203			
KNO_2	s	−370			
KNO_3	s	−493	−393	133	96·3
KNH_2	s	−118			
KH_2PO_4	s	−1569			
K_2SO_3	s	−1117			
K_2SO_4	s	−1434	−1316	176	130
$KHSO_4$	s	−1158			
$K_2S_2O_8$	s	−1917			
$KClO_3$	s	−391	−290	143	100
$KClO_4$	s	−434	−304	151	110
$KBrO_3$	s	−332	−244	149	105
KIO_3	s	−508	−426	152	106
K_2CrO_4	s	−1383			
$K_2Cr_2O_7$	s	−2033			
$KMnO_4$	s	−813	−714	172	119
$K_3Fe(CN)_6$	s	−173			
$K_4Fe(CN)_6$	s	−523			
Radium					
Ra	s	0	0	71	28
RaO	s	−523			
Radon					
Rn	g	0	0	176	20·8
Rhenium					
Re	s	0	0	42	26
Re_2O_7	s	−1245			

THERMODYNAMIC RELATIONSHIPS AND PROPERTIES—TABLE 2. Thermodynamic Data for Inorganic Compounds (*Continued*)

Substance	State	$\dfrac{\Delta H_f^{\ominus}}{\text{kJ mol}^{-1}}$	$\dfrac{\Delta G_f^{\ominus}}{\text{kJ mol}^{-1}}$	$\dfrac{S^{\ominus}}{\text{J K}^{-1}\text{mol}^{-1}}$	$\dfrac{C_p^{\ominus}}{\text{J K}^{-1}\text{mol}^{-1}}$
Rhodium					
Rh	s	0	0	32	26
RhO	s	−90·8			
Rh_2O_3	s	−286			
$RhCl_2$	s	−150			
$RhCl_3$	s	−230			
Rubidium					
Rb	s	0	0	69·4	30·4
Rb_2O	s	−330			
Rb_2O_2	s	−426			
RbOH	s	−414			
RbF	s	−549			
RbCl	s	−431	−412	119	51·5
RbBr	s	−389	−378	108	
RbI	s	−328	−326	118	
RbH	g	140			
Ruthenium					
Ru	s	0	0	29	23
RuO_2	s	−220			
$RuCl_3$	s	−260			
Scandium					
Sc	s	0	0	38	25
$ScCl_3$	s	−924			
Selenium					
Se	s (grey)	0	0	41·8	24·9
SeO_2	s	−230			
H_2Se	g	85·8	71·1	221	
SeF_6	g	−1030			
Silicon					
Si	s	0	0	18·7	19·9
SiO_2	s (quartz)	−859	−805	41·8	44·4
SiO_2	s (crystobalite)	−858	−804	42·6	44·4
SiO_2	s (tridymite)	−857	−803	43·3	44·4
SiF_4	g	−1550	−1510	284	76·2
$SiCl_4$	g	−610	−570	331	90·8
$SiCl_4$	l	−640	−573	239	145
$SiBr_4$	l	−398			
SiI_4	s	−132			
SiH_4	g	−61·9	−39	204	42·8
SiC	s	−112	−26·1	16·5	26·6
Silver					
Ag	s	0	0	42·7	25·5
Ag_2O	s	−30·6	−10·8	122	65·6
AgF	s	−203	−185	80	
AgCl	s	−127	−110	96·1	50·8
AgBr	s	−99·5	−93·7	107	52·4

THERMODYNAMIC RELATIONSHIPS AND PROPERTIES—TABLE 2. Thermodynamic Data for Inorganic Compounds (*Continued*)

Substance	State	$\dfrac{\Delta H_f^{\ominus}}{\text{kJ mol}^{-1}}$	$\dfrac{\Delta G_f^{\ominus}}{\text{kJ mol}^{-1}}$	$\dfrac{S^{\ominus}}{\text{J K}^{-1}\,\text{mol}^{-1}}$	$\dfrac{C_p^{\ominus}}{\text{J K}^{-1}\,\text{mol}^{-1}}$
AgI	s	−62·4	−66·3	114	54·4
Ag$_2$S	s (α)	−31·8	−40·2	146	
Ag$_2$CO$_3$	s	−506	−437	167	
AgCN	s	146	164	8·37	
AgSCN	s	87·9			
AgNO$_2$	s	−44·4	19·8	128	
AgNO$_3$	s	−123	−32·2	141	93·0
Ag$_2$SO$_4$	s	−713	−616	200	131
Ag$_2$CrO$_4$	s	−712	−622	217	
Sodium					
Na	s	0	0	51·0	28·4
Na$_2$O	s	−416	−377	72·8	68·2
Na$_2$O$_2$	s	−505			
NaOH	s	−427			80·3
NaF	s	−569	−541	58·6	46·0
NaHF$_2$	s	−906			
NaCl	s	−411	−384	72·4	49·7
NaBr	s	−360			52·3
NaI	s	−288			54·4
NaH	g	125	104	188	
NaH	s	−57·3			
Na$_2$S	s	−373			
Na$_2$B$_4$O$_7$	s	−3254			
Na$_2$B$_4$O$_7$·10H$_2$O	s	−6264			
Na$_2$CO$_3$	s	−1131	−1048	136	110
Na$_2$CO$_3$·H$_2$O	s	−1430			
Na$_2$CO$_3$·10H$_2$O	s	−4082			
NaHCO$_3$	s	−948	−852	102	87·6
NaCN	s	−89·8			
NaCNO	s	−400			
NaSCN	s	−175			
Na$_2$SiO$_3$	s	−1520	−1430	114	112
NaNO$_2$	s	−359			
NaNO$_3$	s	−467	−366	116	93·0
NaNH$_2$	s	−119			
Na$_3$PO$_4$	s	−1920			
Na$_3$PO$_4$·12H$_2$O	s	−5477			
Na$_2$HPO$_4$	s	−1747			
Na$_2$HPO$_4$·12H$_2$O	s	−5299			558
NaNH$_4$HPO$_4$·4H$_2$O	s	−2856			
Na$_2$SO$_3$	s	−1090	−1002	146	
Na$_2$SO$_3$·7H$_2$O	s	−3153			
Na$_2$SO$_4$	s	−1385	−1267	150	128
Na$_2$SO$_4$·10H$_2$O	s	−4324	−3644	593	587
NaHSO$_4$	s	−1126			
Na$_2$S$_2$O$_3$	s	−1117			
Na$_2$S$_2$O$_3$·5H$_2$O	s (I)	−2602			361
NaClO$_3$	s	−359			
NaClO$_4$	s	−386			101

THERMODYNAMIC RELATIONSHIPS AND PROPERTIES—TABLE 2. Thermodynamic Data for Inorganic Compounds (*Continued*)

Substance	State	$\Delta H_f^{\ominus}$ kJ mol^{-1}	$\Delta G_f^{\ominus}$ kJ mol^{-1}	$S^{\ominus}$ J K^{-1} mol^{-1}	$C_p^{\ominus}$ J K^{-1} mol^{-1}
Na_2CrO_4	s	−1329			
$NaBH_4$	s	−183	−120	105	86·6
Strontium					
Sr	s	0	0	54·4	25
SrO	s	−590	−560	54·4	
SrO_2	s	−643			
$Sr(OH)_2{\cdot}8H_2O$	s	−3352			
SrF_2	s	−1214			
$SrCl_2$	s	−828	−781	120	
$SrCl_2{\cdot}6H_2O$	s	−2624			
$SrBr_2$	s	−716			
SrI_2	s	−567			
SrH_2	s	−177			
Sr_3N_2	s	−391			
SrS	s	−452			
$SrCO_3$	s	−1219	−1138	97·1	
$Sr(NO_3)_2{\cdot}4H_2O$	s	−2153			
$SrSO_4$	s	−1444	−1335	122	
Sulphur					
S	s (rhombic, α)	0	0	31·9	22·6
S	s (monoclinic, β)	0·3	0·1	32·6	23·6
S	g	223	182	168	23·7
S_8	g	102	49·8	430	
SO_2	g	−297	−300	248	39·8
SO_3	g	−395	−370	256	50·6
S_2Cl_2	l	−60·2			130
$SOCl_2$	l	−206			
SO_2Cl_2	l	−389			132
SF_6	g	−1100	−992	291	
H_2S	g	−20·2	−33·0	206	34·0
H_2SO_4	l	−811		138	
Tantalum					
Ta	s	0	0	41	25·3
Ta_2O_5	s	−2092	−1969	143	135
Technetium					
Tc	s	0	0	40	24
Tellurium					
Te	s	0	0	49·7	25·7
TeO_2	s	−325	−270	71·1	66·5
$TeCl_4$	s	−323			
TeF_6	g	−1320	−1220	338	
H_2Te	g	154	138	230	
Thallium					
Tl	s	0	0	64·4	26·6
Tl_2O	s	−175	−136	99·6	
TlOH	s	−238	190	72·4	
$Tl(OH)_3$	s	−513			

THERMODYNAMIC RELATIONSHIPS AND PROPERTIES—TABLE 2.   Thermodynamic
Data for Inorganic Compounds (*Continued*)

Substance	State	$\dfrac{\Delta H_f^\ominus}{\text{kJ mol}^{-1}}$	$\dfrac{\Delta G_f^\ominus}{\text{kJ mol}^{-1}}$	$\dfrac{S^\ominus}{\text{J K}^{-1}\text{mol}^{-1}}$	$\dfrac{C_p^\ominus}{\text{J K}^{-1}\text{mol}^{-1}}$
TlCl	s	−205	−185	108	
TlCl$_3$	s	−251			
TlBr	s	−172	−166	111	
TlBr$_3$	s	−247			
TlI	s (II)	−124	−124	123	
Thorium					
Th	s	0	0	56·9	32
ThO$_2$	s	−1220			85·3
Tin					
Sn	s (white)	0	0	51·5	26·4
Sn	s (grey)	2·5	4·6	44·8	25·8
SnO	s	−286	−257	56·5	44·4
SnO$_2$	s	−581	−520	52·3	52·6
Sn(OH)$_2$	s	−579	−492	96·6	
SnCl$_2$	s	−350			
SnCl$_2$·2H$_2$O	s	−945			
SnCl$_4$	l	−545	−474	259	165
SnBr$_2$	s	−266			
SnI$_2$	s	−144			
SnS	s	−77·8	−82·4	98·7	
SnS$_2$	s	−167	−159	87·4	70·3
Sn(SO$_4$)$_2$	s	−1646			
Titanium					
Ti	s	0	0	30·3	25·2
TiO$_2$	s (rutile)	−912	−853	50·2	55·1
TiCl$_4$	l	−750	−674	253	157
Tungsten					
W	s	0	0	33	25·0
WO$_3$	s (yellow)	−840	−764	83·3	81·5
Uranium					
U	s	0	0	50·3	27·5
UO$_2$	s	−1130	−1080	77·8	
UO$_3$	s	−1260	−1180	98·6	
U$_3$O$_8$	s	−3760			
UF$_4$	s	−1850	−1760	151	118
UF$_6$	g	−2110	−2030	380	
Vanadium					
V	s	0	0	29·5	24·5
V$_2$O$_3$	s	−1210	−1130	98·7	
VO$_2$	s	−720	−665	51·6	59·4
V$_2$O$_5$	s	−1560	−1440	131	130
Vcl$_2$	s	−452	−406	97·1	
Vcl$_3$	s	−573	−502	131	
Xenon					
Xe	g	0	0	170	20·8
XeO$_3$	s	400			

THERMODYNAMIC RELATIONSHIPS AND PROPERTIES—TABLE 2. Thermodynamic
Data for Inorganic Compounds (*Continued*)

Substance	State	$\Delta H_f^{\ominus}$ kJ mol^{-1}	$\Delta G_f^{\ominus}$ kJ mol^{-1}	$S^{\ominus}$ J K^{-1} mol^{-1}	$C_p^{\ominus}$ J K^{-1} mol^{-1}
XeF$_2$	g	−82·0			
XeF$_4$	s	−252	−123	145	118
XeF$_6$	g	−329			
Yttrium					
Y	s	0	0	46	26
YCl$_3$	s (γ)	−982			
Zinc					
Zn	s	0	0	41·6	25·1
ZnO	s	−348	−318	43·9	40·2
Zn(OH)$_2$	s	−642			
ZnCl$_2$	s	−416	−369	108	76·6
ZnBr$_2$	s	−327	−310	137	
ZnI$_2$	s	−209	−209	159	
ZnS	s (sphalerite)	−203	−198	57·7	45·2
Zns	s (wurtzite)	−190			
ZnCO$_3$	s	−812	−731	82·4	
Zn(NO$_3$)$_2$	s	−482			
Zn(NO$_3$)$_2$·6H$_2$O	s	−2305			
ZnSO$_4$	s	−979	−872	125	
ZnSO$_4$·7H$_2$O	s	−3075	−2560	387	392
Zirconium					
Zr	s	0	0	38·4	26
ZrO$_2$	s	−1080	−1023	50·3	
ZrCl$_4$	s	−962	−874	186	

$\Delta H_f^{\ominus}$ denotes the standard enthalpy of formation of a substance from its elements at 298°K and 101,325 N m^{-2} (1 atm) pressure.

$\Delta G_f^{\ominus}$ denotes the standard Gibbs free energy of formation of a substance from its elements at 298°K and 101,325 N m^{-2} (1 atm) pressure.

$S^{\ominus}$ denotes the standard entropy of a substance at 298°K and 101,325 N m^{-2} (1 atm) pressure.

$C_p^{\ominus}$ denotes the standard molar heat capacity of a substance at constant pressure at 298°K.

s = solid, l = liquid, g = gas, am = amorphous.

THERMODYNAMIC RELATIONSHIPS AND PROPERTIES—TABLE 3. Thermodynamic
Data for Organic Compounds

Substance	State	$\Delta H_f^{\ominus}$ kJ mol^{-1}	$\Delta G_f^{\ominus}$ kJ mol^{-1}	$S^{\ominus}$ J K^{-1} mol^{-1}	$C_p^{\ominus}$ J K^{-1} mol^{-1}
CH$_4$	g	−74·9	−50·8	186	35·3
C$_2$H$_6$	g	−84·7	−32·9	230	52·7
C$_3$H$_8$	g	−104	−23·5	270	73·6
n-C$_4$H$_{10}$	g	−125	−15·7	310	97·5
n-C$_5$H$_{12}$	g	−146	−8·2	348	120
n-C$_6$H$_{14}$	g	−167	0·2	387	143
C$_2$H$_4$	g	52·3	68·1	219	43·5

THERMODYNAMIC RELATIONSHIPS AND PROPERTIES—TABLE 3. Thermodynamic Data for Organic Compounds (*Continued*)

Substance	State	$\Delta H_f^\ominus$ kJ mol^{-1}	$\Delta G_f^\ominus$ kJ mol^{-1}	$S^\ominus$ J K^{-1} mol^{-1}	$C_p^\ominus$ J K^{-1} mol^{-1}
C$_3$H$_6$	g	20·4	62·7	267	64·0
C$_4$H$_8$ (but-1-ene)	g	1·2	72·0	307	86
C$_4$H$_8$ (cis-but-2-ene)	g	−5·7	67·1	301	79
C$_4$H$_8$ (trans-but-2-ene)	g	−10·1	64·1	296	88
C$_2$H$_2$	g	227	209	201	43·9
C$_3$H$_4$	g	185	194	248	61
C$_4$H$_6$ (buta-1,3-diene)	g	112	152	279	79·5
C$_6$H$_{12}$ (cyclohexane)	l	−156	26·8	204	154
C$_6$H$_6$	g	82·9	130	269	8·16
C$_6$H$_6$	l	49·0	125	173	136
C$_6$H$_5$CH$_3$	g	50·0	122	320	104
C$_6$H$_5$CH$_2$CH$_3$	g	29.8	131	360	128
C$_8$H$_8$ (phenylethene)	g	148	214	345	
CH$_3$Cl	g	−82·0	−58·6	234	41
CH$_2$Cl$_2$	l	−117	−63·2	179	100
CHCl$_3$	l	−132	−71·6	203	116
CH$_3$Br	g	−36	−26	246	42·7
CHBr$_3$	l	−20	3	222	
CH$_3$I	l	−8·4	20	163	
CHI$_3$	s	141			
C$_2$H$_5$Cl	g	−105	−53·1	276	63
C$_2$H$_5$Br	l	−85·4			
C$_2$H$_5$I	l	−31			
CH$_2$ = CHCl	g	31	52	264	
CH$_2$ClCH$_2$Cl	l	−166	−80·3	208	129
C$_6$H$_5$Cl	g	52·3	99	314	
(CH$_3$)$_2$O	g	−185	−114	267	66·1
CH$_3$OH	g	−201	−162	238	45·2
CH$_3$OH	l	−239	−166	127	81·6
C$_2$H$_5$OH	g	−235	−169	282	65
C$_2$H$_5$OH	l	−278	−175	161	111
C$_6$H$_5$OH	s	−163	−50·9	146	
HCHO	g	−116	−110	219	35
CH$_3$CHO	g	−166	−134	266	62·8
(CH$_3$)$_2$CO	g	−216	−152	295	74·9
HCO$_2$H	g	−363	−336	251	
HCO$_2$H	l	−409	−346	129	99·2
CH$_3$CO$_2$H	l	−487	−392	16	123
C$_6$H$_5$CO$_2$H	s	−385	−245	167	147
CH$_3$CO$_2$C$_2$H$_5$	l	−481			
CH$_3$COCl	l	−275	−208	201	
CH$_3$CONH$_2$	s	−320			
CH$_3$CN	l	53·1	100	144	
CH$_3$NH$_2$	g	−28	28	242	54.0
C$_2$H$_5$NH$_2$	g	−48·5	37	285	70
CO(NH$_2$)$_2$	s	−333	−47·1	105	93·3

$\Delta H_f^\ominus$ denotes the standard enthalpy of formation of a substance from its elements at 298°K and 101,325 N m^{-2} (1 atm) pressure.

$\Delta G_f^\ominus$ denotes the standard Gibbs free energy of formation of a substance from its elements at 298°K and 101,325 N m^{-2} (1 atm) pressure.

$S^\ominus$ denotes the standard entropy of a substance at 298°K and 101,325 N m^{-2} (1 atm) pressure.

$C_p^\ominus$ denotes the standard molar heat capacity of a substance at constant pressure at 298°K.

s = solid, l = liquid, g = gas, am=amorphous.

THERMODYNAMIC RELATIONSHIPS AND PROPERTIES—TABLE 4. Thermodynamic Properties of Moist Air (Standard Atmospheric Pressure, 29.921 in Hg)

Temp. t, °F	Saturation humidity $H_s \times 10^8$	Volume, cu ft/lb dry air			Enthalpy, Btu/lb dry air			Entropy, Btu/(°F)(lb dry air)			Condensed water			Temp. t, °F
		v_a	v_{as}	v_s	h_a	h_{as}	h_s	s_a	s_{as}	s_s	Enthalpy, h_w Btu/lb	Entropy, s_w Btu/(lb)(°F)	Vapor press., in Hg $p_s \times 10^6$	
-160	0.2120	7.520	0.000	7.520	-38.504	0.000	-38.504	-0.10300	0.00000	-0.10300	-222.00	-0.4907	0.1009	-160
-155	.3869	7.647	.000	7.647	-37.296	.000	-37.296	-0.09901	.00000	-0.09901	-220.40	-0.4853	.1842	-155
-150	.6932	7.775	.000	7.775	-36.088	.000	-36.088	-0.09508	.00000	-0.09508	-218.77	-0.4800	.3301	-150
-145	1.219	7.902	.000	7.902	-34.881	.000	-34.881	-0.09121	.00000	-0.09121	-217.12	-0.4747	.5807	-145
-140	2.109	8.029	.000	8.029	-33.674	.000	-33.674	-0.08740	.00000	-0.08740	-215.44	-0.4695	1.004	-140
-135	3.586	8.156	.000	8.156	-32.468	.000	-32.468	-0.08365	.00000	-0.08365	-213.75	-0.4642	1.707	-135
-130	6.000	8.283	.000	8.283	-31.262	.000	-31.262	-0.07997	.00000	-0.07997	-212.03	-0.4590	2.858	-130
	$H_s \times 10^7$												$p_s \times 10^5$	
-125	0.9887	8.411	.000	8.411	-30.057	.000	-30.057	-0.07634	.00000	-0.07634	-210.28	-0.4538	0.4710	-125
-120	1.606	8.537	.000	8.537	-28.852	.000	-28.852	-0.07277	.00000	-0.07277	-208.52	-0.4485	.7653	-120
-115	2.571	8.664	.000	8.664	-27.648	.000	-27.648	-0.06924	.00000	-0.06924	-206.73	-0.4433	1.226	-115
-110	4.063	8.792	.000	8.792	-26.444	.000	-26.444	-0.06577	.00000	-0.06577	-204.92	-0.4381	1.939	-110
-105	6.340	8.919	.000	8.919	-25.240	.001	-25.239	-0.06234	.00000	-0.06234	-203.09	-0.4329	3.026	-105
-100	9.772	9.046	.000	9.046	-24.037	.001	-24.036	-0.05897	.00000	-0.05897	-201.23	-0.4277	4.666	-100
	$H_s \times 10^6$												$p_s \times 10^4$	
-95	1.489	9.173	.000	9.173	-22.835	.002	-22.833	-0.05565	.00000	-0.05565	-199.35	-0.4225	0.7111	-95
-90	2.242	9.300	.000	9.300	-21.631	.002	-21.629	-0.05237	.00001	-0.05236	-197.44	-0.4173	1.071	-90
-85	3.342	9.426	.000	9.426	-20.428	.003	-20.425	-0.04913	.00001	-0.04912	-195.51	-0.4121	1.597	-85
-80	4.930	9.553	.000	9.553	-19.225	.005	-19.220	-0.04595	.00001	-0.04594	-193.55	-0.4069	2.356	-80
-75	7.196	9.680	.000	9.680	-18.022	.007	-18.015	-0.04280	.00002	-0.04278	-191.57	-0.4017	3.441	-75

THERMODYNAMIC RELATIONSHIPS AND PROPERTIES—TABLE 4. Thermodynamic Properties of Moist Air (Standard Atmospheric Pressure, 29.921 in Hg) (*Continued*)

Temp. t, °F	Saturation humidity $H_s \times 10^8$	Volume, cu ft/lb dry air			Enthalpy, Btu/lb dry air			Entropy, Btu/(°F)(lb dry air)			Condensed water			Temp. t, °F
		v_a	v_{as}	v_s	h_a	h_{as}	h_s	s_a	s_{as}	s_s	Enthalpy, Btu/lb h_w	Entropy, Btu/(lb)(°F) s_w	Vapor press., in Hg $p_s \times 10^6$	
-70	10.40	9.806	.000	9.806	-16.820	.011	-16.809	-0.03969	.00003	-0.03966	-189.56	-0.3965	4.976	-70
-65	14.91	9.932	.000	9.932	-15.617	.015	-15.602	-0.03663	.00005	-0.03658	-187.53	-0.3913	7.130	-65
	$H_s \times 10^5$												$p_s \times 10^3$	
-60	2.118	10.059	.000	10.059	-14.416	.022	-14.394	-0.03360	.00006	-0.03354	-185.47	-0.3861	1.0127	-60
-55	2.982	10.186	.000	10.186	-13.214	.031	-13.183	-0.03061	.00009	-0.03052	-183.39	-0.3810	1.4258	-55
-50	4.163	10.313	.001	10.314	-12.012	.043	-11.969	-0.02766	.00012	-0.02754	-181.29	-0.3758	1.9910	-50
-45	5.766	10.440	.001	10.441	-10.811	.060	-10.751	-0.02474	.00015	-0.02459	-179.16	-0.3707	2.7578	-45
-40	7.925	10.566	.001	10.567	-9.609	.083	-9.526	-0.02186	.00021	-0.02165	-177.01	-0.3655	3.7906	-40
-35	10.81	10.693	.002	10.695	-8.408	.113	-8.295	-0.01902	.00028	-0.01874	-174.84	-0.3604	5.1713	-35
	$H_s \times 10^4$												$p_s \times 10^2$	
-30	1.464	10.820	.002	10.822	-7.207	.154	-7.053	-0.01621	.00038	-0.01583	-172.64	-0.3552	0.70046	-30
-25	1.969	10.946	.004	10.950	-6.005	.207	-5.798	-0.01342	.00051	-0.01291	-170.42	-0.3500	.94212	-25
-20	2.630	11.073	.005	11.078	-4.804	.277	-4.527	-0.01067	.00068	-0.00999	-168.17	-0.3449	1.2587	-20
-15	3.491	11.200	.006	11.206	-3.603	.368	-3.235	-0.00796	.00089	-0.00707	-165.90	-0.3398	1.6706	-15
-10	4.606	11.326	.008	11.334	-2.402	.487	-1.915	-0.00529	.00115	-0.00414	-163.60	-0.3346	2.2035	-10
-5	6.040	11.452	.011	11.463	-1.201	.639	-0.562	-0.00263	.00149	-0.00114	-161.28	-0.3295	2.8886	-5
	$H_s \times 10^3$													
0	0.7872	11.578	.015	11.593	0.000	.835	0.835	0.00000	.00192	0.00192	-158.93	-0.3244	3.7645	0
5	1.020	11.705	.019	11.724	1.201	1.085	2.286	.00260	.00246	.00506	-156.57	-0.3193	4.8779	5

	p_s												$H_s \times 10^2$	
10	6.2858	-0.3141	-154.17	.00832	.00314	.00518	3.803	1.401	2.402	11.856	.025	11.831	1.315	10
15	8.0565	-0.3090	-151.76	.01171	.00399	.00772	5.403	1.800	3.603	11.990	.032	11.958	1.687	15
20	10.272	-0.3039	-149.31	.01527	.00504	.01023	7.106	2.302	4.804	12.126	.042	12.084	2.152	20
25	13.032	-0.2988	-146.85	.01908	.00635	.01273	8.934	2.929	6.005	12.265	.054	12.211	2.733	25
30	16.452	-0.2936	-144.36	.02315	.00796	.01519	10.915	3.709	7.206	12.406	.068	12.338	3.454	30
32	18.035	-0.2916	-143.36	.02487	.00870	.01617	11.758	4.072	7.686	12.463	.075	12.388	3.788	32
32*	18.037	0.0000	0.04	.02487	.00870	.01617	11.758	4.072	7.686	12.463	.075	12.388	3.788	32*
34	19.546	.0041	2.06	.02655	.00940	.01715	12.585	4.418	8.167	12.520	.082	12.438	4.107	34
36	0.21166	.0081	4.07	.02828	.01016	.01812	13.438	4.791	8.647	12.578	.089	12.489	4.450	36
38	.22904	.0122	6.08	.03006	.01097	.01909	14.319	5.191	9.128	12.637	.097	12.540	4.818	38
40	.24767	.0162	8.09	.03188	.01183	.02005	15.230	5.622	9.608	12.695	.105	12.590	5.213	40
42	.26763	.0202	10.09	.03376	.01275	.02101	16.172	6.084	10.088	12.755	.114	12.641	5.638	42
44	.28899	.0242	12.10	.03570	.01373	.02197	17.149	6.580	10.569	12.815	.124	12.691	6.091	44
46	.31185	.0282	14.10	.03771	.01478	.02293	18.161	7.112	11.049	12.876	.134	12.742	6.578	46
48	.33629	.0321	16.11	.03978	.01591	.02387	19.211	7.681	11.530	12.938	.146	12.792	7.100	48
50	.36240	.0361	18.11	.04192	.01711	.02481	20.301	8.291	12.010	13.001	.158	12.843	7.658	50
52	.39028	.0400	20.11	.04414	.01839	.02575	21.436	8.945	12.491	13.064	.170	12.894	8.256	52
54	.42004	.0439	22.12	.04645	.01976	.02669	22.615	9.644	12.971	13.129	.185	12.944	8.894	54
56	.45176	.0478	24.12	.04883	.02121	.02762	23.84	10.39	13.452	13.195	.200	12.995	9.575	56
58	.48558	.0517	26.12	.05131	.02276	.02855	25.12	11.19	13.932	13.261	.216	13.045	10.30	58
60	.52159	.0555	28.12	.05389	.02441	.02948	26.46	12.05	14.413	13.329	.233	13.096	11.08	60
62	.55994	.0594	30.12	.05656	.02616	.03040	27.85	12.96	14.893	13.398	.251	13.147	11.91	62
64	.60073	.0632	32.12	.05935	.02803	.03132	29.31	13.94	15.374	13.468	.271	13.197	12.80	64
66	.64411	.0670	34.11	.06225	.03002	.03223	30.83	14.98	15.855	13.539	.292	13.247	13.74	66
68	.69019	.0708	36.11	.06527	.03213	.03314	32.42	16.09	16.335	13.613	.315	13.298	14.75	68
70	.73915	.0746	38.11	.06842	.03437	.03405	34.09	17.27	16.816	13.687	.339	13.348	1.582	70
72	.79112	.0784	40.11	.07170	.03675	.03495	35.83	18.53	17.297	13.762	.364	13.398	1.697	72
74	.84624	.0821	42.10	.07513	.03928	.03585	37.66	19.88	17.778	13.841	.392	13.449	1.819	74

THERMODYNAMIC RELATIONSHIPS AND PROPERTIES—TABLE 4. Thermodynamic Properties of Moist Air (Standard Atmospheric Pressure, 29.921 inHg) (*Continued*)

Temp. t, °F	Saturation humidity $H_s \times 10^8$	Volume, cu ft/lb dry air			Enthalpy, Btu/lb dry air			Entropy, Btu/(°F)(lb dry air)			Condensed water			Temp. t, °F
		v_a	v_{as}	v_s	h_a	h_{as}	h_s	s_a	s_{as}	s_s	Enthalpy, Btu/lb h_w	Entropy, Btu/(lb)(°F) s_w	Vapor press., in Hg $p_s \times 10^6$	
76	1.948	13.499	.422	13.921	18.259	21.31	39.57	.03675	.04197	.07872	44.10	.0859	.90470	76
78	2.086	13.550	.453	14.003	18.740	22.84	41.58	.03765	.04482	.08247	46.10	.0896	.96665	78
80	2.233	13.601	.486	14.087	19.221	24.47	43.69	.03854	.04784	.08638	48.10	.0933	1.0323	80
82	2.389	13.651	.523	14.174	19.702	26.20	45.90	.03943	.05105	.09048	50.09	.0970	1.1017	82
84	2.555	13.702	.560	14.262	20.183	28.04	48.22	.04031	.05446	.09477	52.09	.1007	1.1752	84
86	2.731	13.752	.602	14.354	20.663	30.00	50.66	.04119	.05807	.09926	54.08	.1043	1.2529	86
88	2.919	13.803	.645	14.448	21.144	32.09	53.23	.04207	.06189	.10396	56.08	.1080	1.3351	88
90	3.118	13.853	.692	14.545	21.625	34.31	55.93	.04295	.06596	.10890	58.08	.1116	1.4219	90
92	3.330	13.904	.741	14.645	22.106	36.67	58.78	.04382	.07025	.11407	60.07	.1153	1.5135	92
94	3.556	13.954	.795	14.749	22.587	39.18	61.77	.04469	.07480	.11949	62.07	.1188	1.6102	94
96	3.795	14.005	.851	14.856	23.068	41.85	64.92	.04556	.07963	.12519	64.06	.1224	1.7123	96
98	4.049	14.056	.911	14.967	23.548	44.68	68.23	.04643	.08474	.13117	66.06	.1260	1.8199	98
100	4.319	14.106	.975	15.081	24.029	47.70	71.73	.04729	.09016	.13745	68.06	.1296	1.9333	100
102	4.606	14.157	1.043	15.200	24.510	50.91	75.42	.04815	.09591	.14406	70.05	.1332	2.0528	102
104	4.911	14.207	1.117	15.324	24.991	54.32	79.31	.04900	.1020	.1510	72.05	.1367	2.1786	104
	$H_s \times 10$													
106	0.5234	14.258	1.194	15.452	25.472	57.95	83.42	.04985	.1085	.1584	74.04	.1403	2.3109	106
108	.5578	14.308	1.278	15.586	25.953	61.80	87.76	.05070	.1153	.1660	76.04	.1438	2.4502	108
110	.5944	14.359	1.365	15.724	26.434	65.91	92.34	.05155	.1226	.1742	78.03	.1472	2.5966	110
112	.6333	14.409	1.460	15.869	26.915	70.27	97.18	.05239	.1302	.1826	80.03	.1508	2.7505	112
114	.6746	14.460	1.560	16.020	27.397	74.91	102.31	.05323	.1384	.1916	82.03	.1543	2.9123	114

116	.7185	14.510	1.668	16.178	27.878	79.85	107.73	.05407	.1470	.2011	84.02	.1577	3.0821	116
118	.7652	14.561	1.782	16.343	28.359	85.10	113.46	.05490	.1562	.2111	86.02	.1612	3.2603	118
120	.8149	14.611	1.905	16.516	28.841	90.70	119.54	.05573	.1659	.2216	88.01	.1646	3.3474	120
122	.8678	14.662	2.034	16.696	29.322	96.66	125.98	.05656	.1763	.2329	90.01	.1681	3.6436	122
124	.9242	14.712	2.174	16.886	29.804	103.0	132.8	.05739	.1872	.2446	92.01	.1715	3.8493	124
126	.9841	14.763	2.323	17.086	30.285	109.8	140.1	.05821	.1989	.2571	94.01	.1749	4.0649	126
128	1.048	14.813	2.482	17.295	30.766	117.0	147.8	.05903	.2113	.2703	96.00	.1783	4.2907	128
130	1.116	14.864	2.652	17.516	31.248	124.7	155.9	.05985	.2245	.2844	98.00	.1817	4.5272	130
132	1.189	14.915	2.834	17.749	31.729	133.0	164.7	.06067	.2386	.2993	100.00	.1851	4.7747	132
134	1.267	14.965	3.029	17.994	32.211	141.8	174.0	.06148	.2536	.3151	102.00	.1885	5.0337	134
136	1.350	15.016	3.237	18.253	32.692	151.2	183.9	.06229	.2695	.3318	104.00	.1918	5.3046	136
138	1.439	15.066	3.462	18.528	33.174	161.2	194.4	.06310	.2865	.3496	106.00	.1952	5.5878	138

H_s

140	0.1534	15.117	3.702	18.819	33.655	172.0	205.7	.06390	.3047	.3686	107.99	.1985	5.8838	140
142	.1636	15.167	3.961	19.128	34.136	183.6	217.7	.06470	.3241	.3888	109.99	.2018	6.1930	142
144	.1745	15.218	4.239	19.457	34.618	196.0	230.6	.06549	.3449	.4104	111.99	.2051	6.5160	144
146	.1862	15.268	4.539	19.807	35.099	209.3	244.4	.06629	.3672	.4335	113.99	.2084	6.8532	146
148	.1989	15.319	4.862	20.181	35.581	223.7	259.3	.06708	.3912	.4583	115.99	.2117	7.2051	148
150	.2125	15.369	5.211	20.580	36.063	239.2	275.3	.06787	.4169	.4848	117.99	.2150	7.5722	150
152	.2271	15.420	5.587	21.007	36.545	255.9	292.4	.06866	.4445	.5132	119.99	.2183	7.9550	152
154	.2430	15.470	5.996	21.466	37.026	273.9	310.9	.06945	.4743	.5438	121.99	.2216	8.3541	154
156	.2602	15.521	6.439	21.960	37.508	293.5	331.0	.07023	.5066	.5768	123.99	.2248	8.7701	156
158	.2788	15.571	6.922	22.493	37.990	314.7	352.7	.07101	.5415	.6125	125.99	.2281	9.2036	158
160	.2990	15.622	7.446	23.068	38.472	337.8	376.3	.07179	.5793	.6511	128.00	.2313	9.6556	160
162	.3211	15.672	8.020	23.692	38.954	363.0	402.0	.07257	.6204	.6930	130.00	.2345	10.125	162
164	.3452	15.723	8.648	24.371	39.436	390.5	429.9	.07334	.6652	.7385	132.00	.2377	10.614	164
166	.3716	15.773	9.339	25.112	39.918	420.8	460.7	.07411	.7142	.7883	134.00	.2409	11.123	166
168	.4007	15.824	10.098	25.922	40.400	454.0	494.4	.07488	.7680	.8429	136.01	.2441	11.652	168
170	.4327	15.874	10.938	26.812	40.882	490.6	531.5	.07565	.8273	.9030	138.01	.2473	12.203	170
172	.4682	15.925	11.870	27.795	41.364	531.3	572.7	.07641	.8927	.9691	140.01	.2505	12.775	172
174	.5078	15.975	12.911	28.886	41.846	576.5	618.3	.07718	.9564	1.0426	142.02	.2537	13.369	174

THERMODYNAMIC RELATIONSHIPS AND PROPERTIES — TABLE 4. Thermodynamic Properties of Moist Air (Standard Atmospheric Pressure, 29.921 in Hg) (Continued)

| Temp. t, °F | Saturation humidity $H_s \times 10^8$ | Volume, cu ft/lb dry air | | | Enthalpy, Btu/lb dry air | | | Entropy, Btu/(°F)(lb dry air) | | | Condensed water | | | Temp. t, °F |
		v_a	v_{as}	v_s	h_a	h_{as}	h_s	s_a	s_{as}	s_s	Enthalpy, Btu/lb h_w	Entropy, Btu/(lb)(°F) s_w	Vapor press., in Hg $p_s \times 10^6$	
176	.5519	16.026	14.074	30.100	42.328	627.1	669.4	.07794	1.047	1.125	144.02	.2568	13.987	176
178	.6016	16.076	15.386	31.462	42.810	684.1	726.9	.07870	1.137	1.216	146.03	.2600	14.628	178
180	.6578	16.127	16.870	32.997	43.292	748.5	791.8	.07946	1.240	1.319	148.03	.2631	15.294	180
182	.7218	16.177	18.565	34.742	43.775	821.9	865.7	.08021	1.357	1.437	150.04	.2662	15.985	182
184	.7953	16.228	20.513	36.741	44.257	906.2	950.5	.08096	1.490	1.571	152.04	.2693	16.702	184
186	.8805	16.278	22.775	39.053	44.740	1004	1049	.08171	1.645	1.727	154.05	.2724	17.446	186
188	.9802	16.329	25.427	41.756	45.222	1119	1164	.08245	1.825	1.907	156.06	.2755	18.217	188
190	1.099	16.379	28.580	44.959	45.704	1255	1301	.08320	2.039	2.122	158.07	.2786	19.017	190
192	1.241	16.430	32.375	48.805	46.187	1418	1464	.08394	2.296	2.380	160.07	.2817	19.845	192
194	1.416	16.480	37.036	53.516	46.670	1619	1666	.08468	2.609	2.694	162.08	.2848	20.704	194
196	1.635	16.531	42.885	59.416	47.153	1871	1918	.08542	3.002	3.087	164.09	.2879	21.594	196
198	1.917	16.581	50.426	67.007	47.636	2195	2243	.08616	3.507	3.593	166.10	.2910	22.514	198
200	2.295	16.632	60.510	77.142	48.119	2629	2677	.08689	4.179	4.266	168.11	.2940	23.468	200

To convert British thermal units per pound to joules per kilogram, multiply by 2326; to convert British thermal units per pound dry air-degree Fahrenheit to joules per kilogram-kelvin, multiply by 4186.8; and to convert cubic feet per pound to cubic meters per kilogram, multiply by 0.0624.

THERMODYNAMIC RELATIONSHIPS AND PROPERTIES—TABLE 5 Additive Corrections for H, h, and v When Barometric Pressure Differs from Standard Barometer

Wet-bulb temp. t_u	Sat. vapor press., in. Hg	-900 $\Delta p = +1$		900 $\Delta p = -1$		1800 $\Delta p = -2$		2700 $\Delta p = -3$		3700 $\Delta p = -4$		4800 $\Delta p = -5$		5900 $\Delta p = -6$	
		ΔH_s	Δh	ΔH_s	Δh	ΔH_s	Δh	ΔH_s	Δh	ΔH_s	Δh	ΔH_s	Δh	ΔH_s	Δh
-10	0.022	-0.10	-0.02	0.11	0.02	0.23	0.03	0.36	0.05	0.50	0.07	0.64	0.10	0.81	0.12
-8	.025	-0.12	-0.02	.12	.02	.26	.04	.40	.06	.55	.08	.72	.11	.90	.13
-6	.027	-0.13	-0.02	.14	.02	.29	.04	.44	.07	.62	.09	.80	.12	1.00	.15
-4	.030	-0.14	0.02	.15	.02	.32	.05	.50	.07	.69	.10	.89	.13	1.12	.17
-2	.034	-0.16	-0.02	.17	.02	.35	.05	.55	.08	.76	.11	.99	.15	1.24	.19
0	.038	-0.18	-0.03	.19	.03	.39	.06	.61	.09	.85	.13	1.10	.17	1.38	.21
2	.042	-0.20	-0.03	.21	.03	.44	.07	.68	.10	.94	.14	1.22	.19	1.53	.23
4	.046	-.022	-0.03	.23	.03	.48	.07	.75	.11	1.05	.16	1.36	.21	1.70	.26
6	.051	-0.24	-0.04	.26	.04	.54	.08	.83	.13	1.16	.18	1.51	.23	1.89	.29
8	.057	-0.27	-0.04	.29	.04	.59	.09	.93	.14	1.28	.19	1.67	.25	2.09	.32
10	.063	-0.30	-0.04	.32	.05	.66	.10	1.03	.16	1.42	.22	1.85	.28	2.31	.35
12	.069	-0.33	-0.05	.35	.05	.73	.11	1.13	.17	1.57	.24	2.04	.31	2.56	.39
14	.077	-0.36	-0.05	.39	.06	.81	.12	1.25	.19	1.74	.26	2.26	.34	2.82	.43
16	.085	-0.40	-0.06	.43	.06	.89	.14	1.38	.21	1.92	.29	2.49	.38	3.12	.48
18	.093	-0.44	-0.07	.47	.07	.98	.15	1.53	.23	2.12	.32	2.75	.42	3.44	.53
20	.103	-0.49	-0.08	.52	.08	1.08	.17	1.68	.26	2.33	.36	3.03	.46	3.79	.58
22	.113	-0.5	-0.08	.6	.09	1.2	.18	1.9	.29	2.6	.40	3.4	.52	4.2	.64
24	.124	-0.6	-0.09	.6	.10	1.3	.20	2.1	.32	2.8	.43	3.7	.57	4.6	.71
26	.137	-0.7	-0.10	.7	.11	1.4	.22	2.3	.35	3.1	.48	4.1	.63	5.1	.78
28	.150	-0.7	-0.11	.8	.12	1.6	.24	2.5	.38	3.4	.52	4.5	.69	5.6	.86

Approximate altitude in feet

THERMODYNAMIC RELATIONSHIPS AND PROPERTIES—TABLE 5   Additive Corrections for H, h, and v When Barometric Pressure Differs from Standard Barometer (*Continued*)

Wet-bulb temp. t_u	Sat. vapor press., in.Hg	−900 $\Delta p = +1$ ΔH_s	Δh	900 $\Delta p = -1$ ΔH_s	Δh	1800 $\Delta p = -2$ ΔH_s	Δh	2700 $\Delta p = -3$ ΔH_s	Δh	3700 $\Delta p = -4$ ΔH_s	Δh	4800 $\Delta p = -5$ ΔH_s	Δh	5900 $\Delta p = -6$ ΔH_s	Δh
30	.165	−0.8	−0.12	.8	.13	1.7	.27	2.7	.42	3.8	.58	4.9	.75	6.1	.92
32	.180	−0.9	−0.13	.9	.14	1.9	.29	3.0	.45	4.1	.63	5.3	.82	6.6	1.01
34	.197	−0.9	−0.14	1.0	.15	2.1	.32	3.2	.49	4.4	.68	5.7	.88	7.2	1.11
36	.212	−1.0	−0.15	1.1	.17	2.2	.35	3.5	.53	4.8	.74	6.2	.96	7.8	1.20
38	.229	−1.1	−0.17	1.2	.18	2.4	.37	3.8	.58	5.2	.80	6.8	1.05	8.4	1.30
40	.248	−1.2	−0.18	1.3	.20	2.6	.41	4.1	.63	5.7	.88	7.4	1.14	9.2	1.42
42	.268	−1.3	−0.20	1.4	.21	2.8	.44	4.4	.69	6.1	.94	8.0	1.23	10.0	1.54
44	.289	−1.4	−0.22	1.5	.23	3.1	.47	4.8	.74	6.7	1.04	8.7	1.34	10.8	1.67
46	.312	−1.5	−0.23	1.6	.25	3.3	.51	5.2	.80	7.2	1.11	9.4	1.45	11.7	1.81
48	.336	−1.6	−0.25	1.8	.27	3.6	.56	5.6	.87	7.8	1.21	10.2	1.58	12.6	1.95
50	.3624	−1.7	−0.27	1.9	.29	3.9	.60	6.1	.94	8.4	1.30	10.9	1.69	13.6	2.11
52	.3903	−1.9	−0.29	2.0	.32	4.2	.65	6.5	1.01	9.0	1.40	11.8	1.83	14.7	2.28
54	.4200	−2.0	−0.31	2.2	.34	4.5	.70	7.0	1.09	9.7	1.50	12.7	1.97	15.8	2.45
56	.4518	−2.2	−0.34	2.4	.37	4.9	.76	7.6	1.18	10.5	1.63	13.7	2.13	17.1	2.66
58	.4856	−2.3	−0.37	2.5	.39	5.3	.82	8.2	1.27	11.3	1.76	14.7	2.28	18.4	2.86
60	.522	−2.5	−0.40	2.7	.42	5.7	.88	8.8	1.37	12.2	1.90	15.9	2.47	19.9	3.09
62	.560	−2.7	−0.43	2.9	.46	6.1	.95	9.5	1.48	13.2	2.05	17.1	2.66	21.4	3.33
64	.601	−2.9	−0.46	3.2	.49	6.5	1.02	10.2	1.59	14.2	2.21	18.4	2.87	23.1	3.60
66	.644	−3.2	−0.50	3.4	.53	7.1	1.10	11.0	1.72	15.3	2.38	19.8	3.09	24.8	3.87
68	.690	−3.4	−0.53	3.7	.57	7.6	1.18	11.8	1.84	16.4	2.56	21.3	3.32	26.7	4.16

70	.739	-3.7	-0.57	3.9	.61	8.1	1.27	12.7	1.98	17.6	2.75	22.9	3.58	28.7	4.48
72	.791	-3.9	-0.61	4.2	.66	8.7	1.36	13.6	2.13	18.8	2.94	24.6	3.84	30.9	4.82
74	.846	-4.2	-0.66	4.6	.71	9.4	1.46	14.6	2.28	20.2	3.16	26.4	4.14	33.1	5.18
76	.905	-4.5	-0.71	4.9	.77	10.0	1.57	15.7	2.46	21.7	3.39	28.3	4.42	35.5	5.56
78	.967	-4.9	-0.76	5.2	.82	10.8	1.69	16.9	2.65	23.3	3.65	30.5	4.77	38.2	5.98
80	1.032	-5.2	-0.82	5.6	.88	11.6	1.82	18.1	2.84	25.1	3.93	32.7	5.13	41.0	6.43
82	1.102	-5.6	-0.88	6.0	.94	12.5	1.96	19.5	3.06	27.0	4.24	35.1	5.51	44.0	6.90
84	1.175	-6.0	-0.94	6.4	1.00	13.3	2.10	20.9	3.28	28.9	4.54	37.7	5.92	47.2	7.41
86	1.253	-6.4	-1.00	6.9	1.08	14.3	2.24	22.3	3.50	30.9	4.85	40.4	6.34	50.6	7.94
88	1.335	-6.9	-1.08	7.4	1.16	15.3	2.40	23.9	3.75	33.1	5.20	43.2	6.79	54.2	8.51
90	1.422	-7.4	-1.16	7.9	1.24	16.5	2.59	25.7	4.04	35.6	5.60	46.4	7.29	58.2	9.15
92	1.514	-7.9	-1.24	8.5	1.34	17.6	2.77	27.5	4.33	38.2	6.01	49.8	7.83	62.5	9.83
94	1.610	-8.5	-1.34	9.1	1.43	18.9	2.98	29.5	4.64	41.0	6.46	53.4	8.41	67.0	10.55
96	1.712	-9.1	-1.43	9.8	1.54	20.2	3.18	31.5	4.96	43.8	6.90	57.2	9.01	71.7	11.30
98	1.820	-9.7	-1.53	10.4	1.64	21.7	3.42	33.8	5.33	47.0	7.41	61.3	9.67	76.8	12.11
100	1.933	-10.4	-1.64	11.2	1.77	23.2	3.66	36.3	5.73	50.4	7.95	65.7	10.37	82.5	13.02
102	2.053	-11.1	-1.75	12.0	1.90	24.8	3.92	38.9	6.14	54.1	8.54	70.5	11.13	88.5	13.98
104	2.179	-11.9	-1.88	12.8	2.02	26.6	4.20	41.6	6.58	57.9	9.15	75.5	11.93	94.8	14.98
106	2.311	-12.8	-2.02	13.7	2.17	28.6	4.52	44.6	7.06	62.1	9.82	81.1	12.83	101.7	16.09
108	2.450	-13.7	-2.17	14.7	2.33	30.6	4.84	47.7	7.55	66.5	10.53	87.0	13.77	109.1	17.27
110	2.597	-14.7	-2.33	15.8	2.50	32.8	5.20	51.3	8.13	71.3	11.30	93.1	14.75	117.0	18.54
112	2.751	-15.7	-2.49	16.9	2.68	35.2	5.58	55.0	8.72	76.4	12.11	99.9	15.84	125.9	19.96
114	2.913	-16.9	-2.68	18.1	2.87	37.7	5.98	58.9	9.50	82.0	13.01	107.3	17.03	135.0	21.42
116	3.082	-18.0	-2.86	19.4	3.08	40.4	6.42	63.2	10.03	88.0	13.97	115.1	18.28	144.7	22.98
118	3.260	-19.3	-3.07	20.8	3.31	43.3	6.88	67.8	10.77	94.4	15.00	123.5	19.63	155.4	24.73
120	3.448	-20.7	-3.29	22.4	3.56	46.6	7.41	72.8	11.58	101.4	16.13	132.7	21.10	167.1	26.58
122	3.644	-22.2	-3.53	24.0	3.82	50.0	7.96	78.2	12.45	109.0	17.35	142.6	22.70	179.6	28.58
124	3.850	-23.8	-3.79	25.8	4.11	53.7	8.55	84.0	13.38	117.1	18.65	153.3	24.42	193.2	30.77
126	4.065	-25.6	-4.08	27.6	4.40	57.7	9.20	90.3	14.39	125.9	20.07	165.0	26.30	208.0	33.15
128	4.291	-27.5	-4.39	29.7	4.74	62.0	9.89	97.1	15.49	135.5	21.61	177.6	28.33	224.0	35.73

THERMODYNAMIC RELATIONSHIPS AND PROPERTIES—TABLE 5 Additive Corrections for H, h, and v When Barometric Pressure Differs from Standard Barometer (*Continued*)

Approximate altitude in feet

Wet-bulb temp. t_w	Sat. vapor press., in.Hg	−900 $\Delta p = +1$		900 $\Delta p = -1$		1800 $\Delta p = -2$		2700 $\Delta p = -3$		3700 $\Delta p = -4$		4800 $\Delta p = -5$		5900 $\Delta p = -6$	
		ΔH_s	Δh	ΔH_s	Δh	ΔH_s	Δh	ΔH_s	Δh	ΔH_s	Δh	ΔH_s	Δh	ΔH_s	Δh
130	4.527	−29.5	−4.71	32.0	5.11	66.7	10.64	104.5	16.68	145.9	23.29	191.4	30.55	241.5	38.55
132	4.775	−31.8	−5.08	34.4	5.50	71.8	11.47	112.6	17.99	157.2	25.11	206.3	32.96	260.6	41.63
134	5.034	−34.2	−5.47	37.1	5.93	77.4	12.37	121.4	19.41	169.6	27.12	222.7	35.60	281.4	44.99
136	5.305	−36.8	−5.89	40.0	6.40	83.4	13.34	130.9	20.94	183.1	29.30	240.5	38.48	304.2	48.67
138	5.588	−39.7	−6.36	43.2	6.92	90.0	14.41	141.4	22.64	197.8	31.67	260.1	41.65	329.3	52.73
140	5.884	−42.8	−6.86	46.5	8.45	97.3	15.59	152.8	24.48	214.0	34.29	281.6	45.12	356.8	57.17

t = dry-bulb temperature, °F.

t_w = wet-bulb temperature, °F.

p = barometric pressure, inHg.

Δp = pressure difference from standard barometer (inHg).

H = moisture content of air, gr/lb dry air.

H_s = moisture content of air saturated at wet-bulb temperature (t_w), gr/lb dry air.

ΔH = moisture-content correction of air when barometric pressure differs from standard barometer, gr/lb dry air.

ΔH_s = moisture-content correction of air saturated at wet-bulb temperature when barometric pressure differs from standard barometer, gr/lb dry air.

Note: To obtain ΔH reduce value of ΔH_s by 1 percent where $t - t_w = 24°F$ and correct proportionally when $t - t_w$ is not 24°F.

h = enthalpy of moist air, Btu/lb dry air.

Δh = enthalpy correction when barometric pressure differs from standard barometer, for saturated or unsaturated air, Btu/lb dry air.

v = volume of moist air, ft³/lb dry air.

$$= \frac{0.754(t+459.8)}{p}\left(1+\frac{H}{4360}\right).$$

THERMODYNAMIC AND THERMOPHYSICAL PROPERTIES

THERMODYNAMIC AND THERMOPHYSICAL PROPERTIES—TABLE 1. Saturated Acetone

Temperature, K	Pressure, bar	v_f, m³/kg	v_g, m³/kg	h_f, kJ/kg	h_g, kJ/kg	s_f, kJ/kg·K	s_g, kJ/kg·K	c_{pf}, kJ/kg·K	μ_f, 10^{-6} Pa·s	k_f, W/m·K	Pr
300	0.318	0.001 261	1.415	-67	466	-0.213	1.561				
310	0.482	0.001 285	0.942	-46	476	-0.144	1.540				
320	0.710	0.001 309	0.645	-22	490	-0.068	1.531				
329.3^b	1.013	0.001 333	0.456	0	506	0	1.537	2.29	232	0.141	3.77
330	1.040	0.001 335	0.448	2	506	0.003	1.521	2.29	231	0.141	3.75
340	1.52	0.001 359	0.311	25	509	0.075	1.514	2.33	212	0.137	3.61
350	2.04	0.001 383	0.237	51	529	0.150	1.516	2.38	200	0.132	3.61
360	2.74	0.001 408	0.179	78	543			2.43	187	0.128	3.55
370	3.60	0.001 435	0.138	103	554			2.48	176	0.124	3.52
380	4.52	0.001 464	0.110	127	566			2.53	165	0.119	3.51
390	5.87	0.001 495	0.0854	151	577			2.59	153	0.115	3.45
400	7.31	0.001 528	0.0684	184	588			2.65	141	0.111	3.37
410	8.94	0.001 564	0.0556	207	598			2.73	130	0.107	3.32
420	10.82	0.001 604	0.0454	231	608			2.82	119	0.103	3.26
430	13.64	0.001 647	0.0356	256	618			2.92	109	0.099	3.21
440	16.37	0.001 695	0.0292	281	625			3.03	99	0.095	3.16
450	19.42	0.001 748	0.0240	308	632			3.15	90	0.092	3.08
460	22.79	0.001 81	0.0199	337	637			3.29	80	0.088	2.99
470	27.52	0.001 88	0.0159	365	641			3.45	71	0.083	2.95
480	32.52	0.001 98	0.0130	396	638			3.76	64	0.077	3.13
490	37.73	0.002 15	0.0091								
500	43.08	0.002 46	0.0063								
508.2^c	47.61	0.003 67	0.0037								

b = normal boiling point; c = critical point.

THERMODYNAMIC AND THERMOPHYSICAL PROPERTIES — TABLE 2. Saturated Acetylene

Temperature, K	Pressure, bar	v_{cond}, m³/kg	v_g, m³/kg	h_{cond}, kJ/kg	h_g, kJ/kg	s_{cond}, kJ/(kg·K)	s_g, kJ/(kg·K)
162.0	0.101		5.081	158	983	2.967	8.062
169.3	0.203		2.644	173	994	3.039	7.889
173.9	0.304		1.805	182	999	3.095	7.797
180.0	0.507		1.116	194	1007	3.161	7.672
184.3	0.709		0.810	203	1011	3.216	7.596
189.1	1.013		0.5780	214	1015	3.272	7.511
192.4ᵗ	1.283		0.4617	221	1018	3.312	7.455
192.4ᵗ	1.283	0.00164	0.4617	378	1018	4.127	7.455
200.9	2.027	0.00165	0.3011	411	1027	4.296	7.362
209.4	3.040	0.00169	0.2074	445	1035	4.461	7.280
221.5	5.066	0.00174	0.1264	493	1046	4.684	7.180
230.4	7.093	0.00179	0.0907	528	1052	4.837	7.111
240.7	10.13	0.00186	0.0635	565	1058	4.990	7.037
253.2	15.20	0.00195	0.0420	602	1061	5.133	6.947
263.0	20.27	0.00204	0.0309	628	1061	5.231	6.878
271.6	25.33	0.00213	0.0240	654	1060	5.326	6.822
278.9	30.40	0.00223	0.0193	680	1057	5.414	6.767
284.9	35.46	0.00232	0.0159	704	1051	5.494	6.716
290.4	40.53	0.00242	0.0133	727	1041	5.576	6.658
300.0	50.66	0.00270	0.0093	778	1017	5.737	6.534
307.8	60.80	0.00335	0.0061	850	968	5.965	6.351
308.7ᶜ	62.47	0.00434	0.0043	908	908	6.158	6.158

THERMODYNAMIC AND THERMOPHYSICAL PROPERTIES—TABLE 3. Saturated Air

T, K	P_f, bar	P_g, bar	v_f, m³/kg	v_g, m³/kg	h_f, kJ/kg	h_g, kJ/kg	s_f, kJ/(kg·K)	s_g, kJ/(kg·K)	c_{pf}, kJ/(kg·K)	μ_f, 10⁻⁴ Pa·s	k_f, W/(m·K)
60			1.040.−3	5.55	−159.2	59.7	2.528	6.255		3.25	0.180
62			1.050.−3	3.73	−155.2	61.7	2.585	6.164		2.98	0.176
64	0.123	0.071	1.060.−3	2.57	−151.4	63.6	2.641	6.080		2.75	0.173
66	0.174	0.104	1.070.−3	1.82	−147.8	65.5	2.696	6.002		2.54	0.169
68	0.239	0.147	1.080.−3	1.313	−144.2	67.4	2.747	5.929		2.36	0.166
70	0.323	0.205	1.089.−3	0.968	−140.6	69.2	2.797	5.862	1.817	2.21	0.163
72	0.429	0.280	1.101.−3	0.728	−137.1	71.0	2.847	5.799	1.827	2.07	0.160
74	0.560	0.376	1.113.−3	0.556	−133.5	72.8	2.895	5.740	1.838	1.95	0.156
76	0.721	0.495	1.125.−3	0.431	−129.9	74.5	2.941	5.685	1.849	1.84	0.152
78	0.915	0.644	1.136.−3	0.339	−126.3	76.2	2.988	5.634	1.861	1.74	0.148
80	1.146	0.825	1.146.−3	0.270	−122.6	77.8	3.034	5.585	1.873	1.65	0.145
82	1.420	1.043	1.160.−3	0.217	−118.8	79.4	3.079	5.540	1.885	1.58	0.142
84	1.741	1.305	1.173.−3	0.177	−115.0	80.9	3.123	5.496	1.898	1.51	0.139
86	2.114	1.614	1.187.−3	0.145	−111.2	82.3	3.167	5.454	1.912	1.44	0.135
88	2.544	1.976	1.201.−3	0.120	−107.4	83.6	3.209	5.414	1.927	1.38	0.132
90	3.036	2.397	1.216.−3	0.1002	−103.5	84.8	3.251	5.376	1.944	1.32	0.128
92	3.596	2.884	1.231.−3	0.0843	−99.5	85.9	3.293	5.340	1.962	1.27	0.125
94	4.229	3.441	1.247.−3	0.0713	−95.5	87.0	3.335	5.304	1.982	1.23	0.121
96	4.940	4.075	1.265.−3	0.0607	−91.5	87.9	3.376	5.270	2.003	1.18	0.117
98	5.736	4.792	1.283.−3	0.0520	−87.5	88.7	3.416	5.236	2.027	1.14	0.114
100	6.621	5.599	1.302.−3	0.0447	−83.3	89.3	3.456	5.204	2.053	1.10	0.110
105	9.265	8.056	1.355.−3	0.0312	−72.8	90.2	3.553	5.124	2.137	1.02	0.102
110	12.59	11.22	1.418.−3	0.0222	−61.9	90.1	3.649	5.045	2.264	0.95	0.093
115	16.68	15.21	1.495.−3	0.0159	−50.3	88.4	3.747	4.964	2.477	0.87	0.084
120	21.61	20.14	1.596.−3	0.0115	−37.5	84.8	3.850	4.877	2.916	0.75	0.076
125	27.43	26.14	1.757.−3	0.0081	−22.0	78.2	3.969	4.776	4.585	0.42	0.067
130	34.16	33.32	2.075.−3	0.0054	0.4	66.1	4.136	4.644			
132.55^c		37.69	3.196.−3	0.0032	37.4	37.4	4.410	4.410	∞		∞

THERMODYNAMIC AND THERMOPHYSICAL PROPERTIES—TABLE 4. Compressed Air

Temperature, K

Pressure, bar	80	90	100	120	140	160	180	200	220	240	260	280	300
1 v		0.251	0.281	0.340	0.399	0.457	0.515	0.537	0.631	0.688	0.746	0.803	0.861
h		87.9	98.3	118.8	139.1	159.3	179.5	199.7	219.8	239.9	260.0	280.2	300.3
s	Mix	5.650	5.759	5.946	6.103	6.238	6.357	6.463	6.559	6.647	6.727	6.802	6.871
C_v		1.044	1.032	1.020	1.014	1.010	1.008	1.007	1.006	1.006	1.006	1.006	1.007
μ		0.064	0.071	0.085	0.097	0.109	0.121	0.133	0.144	0.154	0.165	0.175	0.185
k		0.0084	0.0093	0.0112	0.0129	0.0147	0.0164	0.0181	0.0198	0.0214	0.0231	0.0247	0.0263
5 v	0.00115	0.00122	0.0509	0.0646	0.0773	0.0895	0.102	0.114	0.125	0.137	0.149	0.160	0.172
h	−122.3	−103.3	90.6	113.6	135.3	156.4	177.1	197.7	218.1	238.5	258.8	279.1	299.4
s	3.031	3.250	5.246	5.455	5.623	5.763	5.885	5.994	6.092	6.180	6.262	6.337	6.406
C_v	1.868	1.941	1.212	1.107	1.065	1.045	1.033	1.025	1.020	1.017	1.015	1.013	1.013
μ	1.794	1.163	0.077	0.087	0.098	0.110	0.122	0.134	0.145	0.155	0.165	0.175	0.185
k	0.146	0.128	0.0103	0.0119	0.0135	0.0151	0.0168	0.0184	0.0201	0.0217	0.0234	0.0250	0.0265
10 v	0.00115	0.00121	0.00130	0.0298	0.0370	0.0436	0.0499	0.0561	0.0621	0.0681	0.0741	0.0800	0.0859
h	−122.0	−103.1	−83.2	106.2	130.2	152.5	174.1	195.2	216.1	236.7	257.3	277.8	298.3
s	3.028	3.246	3.452	5.214	5.398	5.548	5.675	5.786	5.885	5.975	6.058	6.134	6.204
C_v	1.863	1.932	2.041	1.270	1.146	1.093	1.065	1.049	1.038	1.031	1.026	1.023	1.201
μ	1.816	1.177	0.838	0.089	0.101	0.112	0.124	0.135	0.146	0.156	0.166	0.176	0.186
k	0.146	0.128	0.111	0.0126	0.0141	0.0157	0.0173	0.0189	0.0205	0.0221	0.0237	0.0253	0.0268
20 v	0.00114	0.00121	0.00129	0.0116	0.0167	0.0206	0.0241	0.0274	0.0306	0.0337	0.0368	0.0398	0.0428
h	−121.3	−102.5	−82.9	85.2	118.5	144.3	167.7	190.1	211.9	233.2	254.3	275.2	296.0
s	3.022	3.239	3.442	4.882	5.140	5.312	5.450	5.568	5.672	5.765	5.849	5.927	5.998
C_v	1.853	1.916	2.010	2.237	1.390	1.215	1.141	1.101	1.076	1.061	1.050	1.042	1.037
μ	1.859	1.205	0.857	0.098	0.106	0.116	0.127	0.137	0.148	0.158	0.168	0.178	0.187
k	0.147	0.130	0.112	0.0152	0.0157	0.0169	0.0182	0.0197	0.0212	0.0228	0.0243	0.0258	0.0273

40 v	0.00114	0.00120	0.00128	0.00153	0.0058	0.0090	0.0114	0.0131	0.0148	0.0165	0.0182	0.0198	0.0214
h	−120.0	−101.4	−82.2	−39.8	83.6	125.3	154.3	179.7	203.5	226.3	248.5	270.2	291.7
s	3.011	3.225	3.424	3.807	4.745	5.025	5.196	5.330	5.444	5.543	5.632	5.712	5.786
C_p	1.834	1.886	1.958	2.432	3.193	1.610	1.335	1.221	1.159	1.122	1.097	1.081	1.068
μ	1.943	1.261	0.896	0.516	0.132	0.129	0.135	0.144	0.154	0.163	0.172	0.182	0.191
k	0.149	0.132	0.115	0.0814	0.0460	0.0201	0.0206	0.0217	0.0229	0.0242	0.0256	0.0270	0.0284
60 v	0.00113	0.00119	0.00126	0.00147	0.00222	0.00505	0.00687	0.00833	0.00963	0.0108	0.0120	0.0131	0.0142
h	−118.6	−100.3	−81.4	−40.8	22.8	90.0	132.6	163.9	191.1	216.1	240.0	263.1	285.6
s	3.000	3.211	3.407	3.773	4.260	4.798	5.020	5.174	5.298	5.404	5.497	5.581	5.657
C_p	1.818	1.860	1.915	2.205	4.808	2.338	1.594	1.361	1.249	1.186	1.146	1.119	1.100
$*\mu$	2.028	1.318	0.936	0.559	0.277	0.153	0.149	0.154	0.161	0.169	0.178	0.186	0.195
k	0.150	0.134	0.117	0.0861	0.0480	0.0360	0.0238	0.0240	0.0248	0.0258	0.0270	0.0283	0.0296
80 v	0.00113	0.00119	0.00126	0.00145	0.00188	0.00327	0.00480	0.00601	0.00706	0.00803	0.00894	0.00981	0.0107
h	−117.2	−99.1	−80.4	−41.3	9.0	78.4	125.3	158.7	187.1	212.9	237.3	260.8	283.7
s	2.989	3.198	3.391	3.745	4.138	4.597	4.875	5.051	5.186	5.299	5.396	5.484	5.562
C_p	1.802	1.838	1.881	2.078	2.992	3.029	1.887	1.510	1.342	1.250	1.194	1.156	1.130
μ	2.12	1.38	0.977	0.597	0.356	0.194	0.167	0.166	0.170	0.177	0.184	0.191	0.200
k	0.152	0.134	0.120	0.0901	0.0599	0.0420	0.0278	0.0268	0.0269	0.0276	0.0286	0.0296	0.0308
100 v	0.00112	0.00118	0.00125	0.00142	0.00174	0.00252	0.00366	0.00467	0.00556	0.00637	0.00713	0.00785	0.00855
h	−115.8	−97.8	−79.4	−41.3	3.9	61.7	111.8	148.8	179.4	206.7	232.2	256.4	279.9
s	2.978	3.186	3.376	3.721	4.076	4.457	4.753	4.949	5.095	5.214	5.315	5.406	5.486
C_p	1.789	1.818	1.852	1.992	2.506	2.874	2.114	1.650	1.431	1.311	1.239	1.191	1.158
μ	2.21	1.44	1.02	0.631	0.405	0.249	0.193	0.181	0.181	0.185	0.191	0.198	0.205
k	0.154	0.137	0.122	0.0936	0.0669	0.0500	0.0327	0.0299	0.0293	0.0295	0.0302	0.0311	0.0320
150 v	0.00111	0.00116	0.00122	0.00137	0.00158	0.00194	0.00247	0.00309	0.00369	0.00425	0.00478	0.00529	0.00578
h	−112.2	−94.5	−76.6	−40.1	0.5	45.2	89.5	129.2	163.2	193.4	221.0	247.0	271.8
s	2.954	3.157	3.342	3.673	3.988	4.287	4.548	4.757	4.919	5.051	5.161	5.257	5.343
C_p	1.789	1.818	1.852	1.992	2.506	2.874	2.114	1.650	1.431	1.311	1.239	1.267	1.220
μ	2.44	1.60	1.13	0.709	0.490	0.349	0.266	0.229	0.215	0.211	0.212	0.215	0.220
k	0.157	1.142	0.127	0.101	0.0785	0.0588	0.0455	0.0389	0.0360	0.0348	0.0346	0.0349	0.0354

THERMODYNAMIC AND THERMOPHYSICAL PROPERTIES—TABLE 4. Compressed Air (Continued)

Temperature, K

Pressure, bar	80	90	100	120	140	160	180	200	220	240	260	280	300
200 v	0.00110	0.00115	0.00120	0.00133	0.00150	0.00174	0.00206	0.00245	0.00287	0.00328	0.00368	0.00407	0.00446
h	−108.5	−91.2	−73.6	−38.0	0.2	40.2	79.8	117.6	152.2	183.6	212.5	239.6	265.5
s	2.930	3.130	3.312	3.634	3.931	4.198	4.432	4.631	4.796	4.932	5.048	5.149	5.238
C_p	1.733	1.747	1.761	1.809	1.905	1.988	1.953	1.814	1.643	1.501	1.396	1.321	1.266
μ	2.70	1.78	1.25	0.782	0.561	0.420	0.331	0.279	0.253	0.241	0.236	0.235	0.237
k	0.161	0.146	0.132	0.107	0.0868	0.0691	0.0559	0.0476	0.0429	0.0405	0.0393	0.0389	0.0389
250 v	0.00109	0.00114	0.00119	0.00130	0.00144	0.00162	0.00186	0.00214	0.00244	0.00276	0.00307	0.00338	0.00368
h	−104.8	−87.6	−70.3	−35.4	1.3	38.9	75.8	111.7	145.6	177.1	206.6	234.3	260.8
s	2.909	3.106	3.285	3.601	3.886	4.138	4.355	4.544	4.706	4.843	4.961	5.064	5.155
C_p	1.712	1.722	1.733	1.767	1.824	1.854	1.831	1.748	1.635	1.522	1.427	1.353	1.297
μ	2.96	1.97	1.39	0.855	0.625	0.476	0.385	0.327	0.292	0.272	0.262	0.257	0.256
k	0.165	0.150	0.137	0.113	0.0935	0.0769	0.0641	0.0552	0.0495	0.0460	0.0441	0.0430	0.0426
300 v	0.00108	0.00112	0.00117	0.00127	0.00139	0.00155	0.00173	0.00195	0.00219	0.00243	0.00269	0.00294	0.00318
h	−101.0	−84.0	−67.0	−32.4	3.1	39.2	74.5	109.0	142.0	173.2	202.7	230.8	257.7
s	2.888	3.083	3.260	3.572	3.849	4.090	4.298	4.480	4.637	4.773	4.891	4.995	5.088
C_p	1.694	1.703	1.713	1.740	1.769	1.777	1.751	1.689	1.607	1.518	1.438	1.370	1.316
μ	3.24	2.18	1.53	0.932	0.687	0.529	0.433	0.370	0.329	0.303	0.288	0.280	0.276
k	0.168	0.154	0.141	0.118	0.0996	0.0836	0.0710	0.0619	0.0555	0.0514	0.0487	0.0471	0.0462
400 v		0.00110	0.00114	0.00123	0.00133	0.00145	0.00158	0.00173	0.00189	0.00206	0.00224	0.00242	0.00260
h		−76.6	−59.8	−25.9	8.3	42.4	75.8	108.5	140.1	170.5	199.7	227.8	254.8
s		3.042	3.216	3.523	3.788	4.016	4.214	4.386	4.537	4.669	4.786	4.890	4.983
C_p		1.674	1.686	1.704	1.702	1.685	1.654	1.607	1.550	1.490	1.431	1.378	1.331

	μ											
μ	2.63	1.86	1.10	0.802	0.631	0.500	0.446	0.397	0.364	0.341	0.325	0.316
k	0.161	0.149	0.127	0.110	0.0946	0.0823	0.0729	0.0660	0.0610	0.0574	0.0550	0.0533
500 v	0.00109	0.00112	0.00120	0.00128	0.00138	0.00148	0.00160	0.00173	0.00186	0.00199	0.00213	0.00227
h	−69.0	−52.3	−18.7	14.4	47.4	79.8	111.4	142.0	171.7	200.5	228.4	255.4
s	3.005	3.177	3.482	3.743	3.966	4.151	4.317	4.463	4.593	4.708	4.811	4.905
C_p	1.655	1.670	1.686	1.667	1.644	1.598	1.557	1.509	1.461	1.415	1.371	1.331
μ	3.13	2.24	1.31	0.924	0.710	0.560	0.512	0.459	0.420	0.391	0.370	0.356
k	0.167	0.156	0.135	0.119	0.104	0.0916	0.0822	0.0749	0.0694	0.0653	.0622	0.0599
600 v							0.00151	0.00161	0.00172	0.00183	0.00194	0.00205
h							116.0	146.1	175.3	203.6	231.2	258.1
s							2.263	4.406	4.533	4.646	4.749	4.842
C_p							1.525	1.480	1.438	1.398	1.361	1.327
μ								0.516	0.472	0.439	0.414	0.396
k							0.0903	0.0828	0.0769	0.0724	0.0689	0.0662
800 v								0.00147	0.00155	0.00163	0.00171	0.00179
h								157.4	185.9	213.7	240.3	267.3
s								4.318	4.442	4.553	4.653	4.745
C_p								1.445	1.406	1.372	1.342	1.314
μ										0.529	0.497	0.473
k								0.0964	0.0901	0.0850	0.0809	0.0776
1000 v										0.00151	0.00157	0.00163
h										226.4	253.2	279.5
s										4.482	4.582	4.672
C_p										1.355	1.327	1.303
μ												0.546
k										0.0961	0.0916	0.0878

THERMODYNAMIC AND THERMOPHYSICAL PROPERTIES – TABLE 4. Compressed Air (Continued)

Pressure, bar		Temperature, K												
		350	400	450	500	600	800	1000	1200	1400	1600	1800	2000	2500
1	v	1.005	1.148	1.292	1.436	1.723	2.297	2.872	3.446	4.020	4.594	5.168	5.743	7.200
	h	350.7	401.2	452.1	503.4	607.5	822.5	1046.8	1278	1515	1764	2017	2279	3011
	s	7.026	7.161	7.282	7.389	7.579	7.888	8.138	8.334	8.531	8.695	8.844	8.983	9.308
	C_v	1.009	1.014	1.021	1.030	1.051	1.099	1.141	1.175	1.207	1.248	1.286	1.337	1.665
	μ	0.208	0.230	0.251	0.270	0.306	0.370	0.424	0.473	0.527	0.584	0.637	0.689	0.818
	k	0.0301	0.0336	0.0371	0.0404	0.0466	0.0577	0.0681	0.0783	0.0927	0.106	0.120	0.137	0.222
5	v	0.201	0.230	0.259	0.288	0.345	0.460	0.575	0.690	0.805	0.920	1.034	1.149	1.438
	h	350.0	400.8	451.8	503.2	607.4	822.6	1046.9	1279	1516	1764	2017	2278	2981
	s	6.563	6.698	6.818	6.927	7.116	7.426	7.676	7.887	8.069	8.233	8.382	8.520	8.832
	C_v	1.014	1.017	1.024	1.032	1.053	1.100	1.142	1.175	1.208	1.248	1.285	1.326	1.516
	μ	0.208	0.230	0.251	0.270	0.306	0.370	0.425	0.473	0.527	0.584	0.637	0.689	0.818
	k	0.0303	0.0338	0.0372	0.0405	0.0467	0.0578	0.0681	0.0783	0.0927	0.106	0.120	0.136	0.195
10	v	0.101	0.115	0.130	0.144	0.173	0.231	0.288	0.345	0.403	0.460	0.518	0.575	0.720
	h	349.2	400.2	451.4	502.9	607.3	822.7	1047.2	1279	1516	1765	2018	2279	2974
	s	6.361	6.497	6.618	6.727	6.917	7.226	7.477	7.688	7.870	8.034	8.183	8.321	8.630
	C_v	1.019	1.021	1.027	1.034	1.055	1.100	1.142	1.175	1.208	1.248	1.284	1.324	1.481
	μ	0.209	0.231	0.252	0.271	0.306	0.370	0.425	0.473	0.527	0.584	0.637	0.689	0.817
	k	0.0305	0.0340	0.0374	0.0407	0.0469	0.0579	0.0682	0.0784	0.0927	0.106	0.120	0.135	0.187
20	v	0.0503	0.0577	0.0650	0.0723	0.0868	0.116	0.145	0.173	0.202	0.231	0.260	0.288	0.360
	h	347.7	399.1	450.7	502.4	607.2	823.0	1047.7	1280	1517	1766	2019	2279	2970
	s	6.158	6.295	6.417	6.526	6.716	7.027	7.277	7.489	7.671	7.835	7.984	8.121	8.428
	C_p	1.030	1.029	1.033	1.039	1.057	1.102	1.143	1.176	1.209	1.249	1.284	1.322	1.456
	μ	0.210	0.232	0.253	0.272	0.307	0.371	0.425	0.474	0.527	0.584	0.637	0.689	0.817
	k	0.0309	0.0344	0.0377	0.0410	0.0471	0.0581	0.0685	0.0787	0.0928	0.106	0.120	0.135	0.181

40	v	0.0252	0.0290	0.0327	0.0364	0.0438	0.0583	0.0728	0.0872	0.102	0.116	0.130	0.145	0.181
	h	344.6	397.0	449.2	501.5	606.9	823.7	1048.8	1281	1519	1768	2021	2281	2969
	s	5.950	6.090	6.212	6.323	6.515	6.826	7.077	7.289	7.473	7.636	7.785	7.922	8.229
	C_p	1.051	1.044	1.044	1.049	1.063	1.105	1.145	1.177	1.210	1.249	1.284	1.322	1.438
	μ	0.213	0.235	0.255	0.274	0.309	0.372	0.426	0.474	0.527	0.584	0.637	0.689	0.817
	k	0.0318	0.0351	0.0384	0.0416	0.0476	0.0584	0.0687	0.0789	0.0928	0.106	0.120	0.135	0.177
60	v	0.0169	0.0194	0.0220	0.0245	0.0294	0.0392	0.0489	0.0585	0.0681	0.0776	0.0872	0.0968	0.1207
	h	340.4	394.0	447.1	500.6	606.8	824.3	1050.0	1283	1521	1770	2023	2284	2969
	s	5.824	5.967	6.091	6.202	6.396	6.708	6.960	7.172	7.355	7.520	7.669	7.806	8.112
	C_p	1.072	1.059	1.055	1.057	1.069	1.108	1.147	1.178	1.210	1.249	1.286	1.322	1.430
	μ	0.217	0.237	0.257	0.275	0.310	0.373	0.427	0.475	0.527	0.584	0.637	0.689	0.817
	k	0.0328	0.0359	0.0391	0.0422	0.0481	0.0588	0.0690	0.0790	0.0929	0.106	0.120	0.134	0.176
80	v	0.0127	0.0147	0.0166	0.0185	0.0223	0.0296	0.0369	0.0442	0.0513	0.0585	0.0657	0.0729	0.0908
	h	339.0	393.1	446.5	499.8	606.7	825.1	1051.1	1284	1522	1772	2025	2285	2971
	s	5.733	5.878	6.004	6.116	6.311	6.624	6.877	7.089	7.273	7.437	7.586	7.723	8.029
	C_p	1.091	1.073	1.066	1.065	1.075	1.111	1.149	1.180	1.210	1.249	1.286	1.322	1.426
	μ	0.220	0.240	0.259	0.278	0.312	0.374	0.428	0.475	0.527	0.584	0.637	0.689	0.817
	k	0.0337	0.0368	0.0398	0.0428	0.0486	0.0592	0.0693	0.0793	0.0929	0.106	0.120	0.134	0.175
100	v	0.0102	0.0118	0.0134	0.0149	0.0180	0.0239	0.0298	0.0356	0.0413	0.0470	0.0528	0.0584	0.0729
	h	336.5	391.3	445.3	499.0	606.6	825.8	1052.4	1286	1524	1774	2027	2288	2972
	s	5.661	5.807	5.935	6.048	6.244	6.559	6.812	7.024	7.208	7.373	7.522	7.659	7.964
	C_p	1.110	1.087	1.076	1.073	1.080	1.114	1.151	1.181	1.211	1.250	1.288	1.323	1.423
	μ	0.224	0.243	0.262	0.280	0.314	0.375	0.429	0.477	0.527	0.584	0.637	0.689	0.817
	k	0.0347	0.0376	0.0405	0.0434	0.0491	0.0595	0.0696	0.0795	0.0930	0.106	0.120	0.134	0.175
150	v	0.00695	0.00806	0.00914	0.0102	0.0123	0.0163	0.0202	0.0241	0.0279	0.0317	0.0356	0.0394	0.0490
	h	330.9	387.5	442.9	497.5	606.6	827.8	1055.5	1290	1529	1779	2033	2294	2977
	s	5.525	5.677	5.807	5.922	6.121	6.439	6.693	6.906	7.092	7.256	7.405	7.543	7.848
	C_p	1.151	1.117	1.099	1.092	1.093	1.121	1.155	1.184	1.213	1.252	1.290	1.325	1.418
	μ	0.235	0.252	0.270	0.286	0.318	0.379	0.431	0.478	0.527	0.584	0.637	0.689	
	k	0.0374	0.0398	0.0424	0.0451	0.0504	0.0605	0.0703	0.0801	0.0932	0.106	0.120	0.133	

THERMODYNAMIC AND THERMOPHYSICAL PROPERTIES—TABLE 4. Compressed Air (*Continued*)

Pressure, bar		350	400	450	500	600	800	1000	1200	1400	1600	1800	2000	2500
							Temperature, K							
200	v	0.00534	0.00620	0.00702	0.00783	0.00940	0.0125	0.0154	0.0184	0.0212	0.0241	0.0269	0.0298	0.0370
	h	326.5	384.5	440.9	496.6	607.0	829.9	1058.7	1294	1533	1783	2038	2299	2982
	s	5.426	5.581	5.715	5.831	6.033	6.353	6.608	6.822	7.009	7.173	7.323	7.460	7.765
	C_p	1.184	1.141	1.119	1.108	1.104	1.128	1.160	1.187	1.214	1.254	1.292	1.326	1.415
	μ	0.248	0.262	0.278	0.293	0.324	0.382	0.434	0.481	0.528	0.585	0.638		
	k	0.0400	0.0420	0.0423	0.0467	0.0517	0.0614	0.0711	0.0808	0.0934	0.106	0.120		
250	v	0.00440	0.00509	0.00576	0.00642	0.00770	0.0102	0.0126	0.0149	0.0172	0.0195	0.0218	0.0241	0.0298
	h	323.2	382.3	439.6	496.0	607.6	832.2	1062.0	1298	1538	1789	2043	2304	2988
	s	5.348	5.506	5.641	5.760	5.963	6.286	6.542	6.757	6.944	7.108	7.258	7.396	7.701
	C_p	1.208	1.161	1.135	1.121	1.115	1.135	1.164	1.190	1.216	1.256	1.294	1.328	1.414
	μ	0.262	0.273	0.286	0.301	0.329	0.386	0.437	0.483	0.528	0.585			
	k	0.0429	0.0443	0.0462	0.0484	0.0531	0.0624	0.0718	0.0814	0.0937	0.106			
300	v	0.00379	0.00437	0.00493	0.00548	0.00656	0.00864	0.0107	0.0126	0.0145	0.0164	0.0183	0.0202	0.0250
	h	320.9	380.9	438.9	495.9	608.5	834.5	1065.3	1302	1542	1794	2049	2310	2993
	s	5.283	5.443	5.580	5.700	5.906	6.230	6.488	6.703	6.891	7.056	7.206	7.344	7.648
	C_p	1.226	1.176	1.148	1.133	1.124	1.140	1.168	1.193	1.217	1.256	1.298	1.330	1.413
	μ	0.276	0.284	0.296	0.308	0.335	0.390	0.440	0.485	0.529				
	k	0.0457	0.0466	0.0481	0.0501	0.0544	0.0634	0.0726	0.0820	0.0940				
400	v	0.00304	0.00348	0.00390	0.00432	0.00514	0.00673	0.00826	0.00977	0.0111	0.0126	0.0140	0.0155	0.0190
	h	319.1	380.0	439.0	496.8	611.0	839.4	1072.0	1310	1552	1804	2059	2321	3004
	s	5.181	5.344	5.483	5.605	5.813	6.142	6.401	6.618	6.808	6.972	7.123	7.261	7.566
	C_p	1.246	1.195	1.166	1.149	1.138	1.151	1.176	1.199	1.222	1.258	1.301	1.333	1.412
	μ	0.307	0.308	0.315	0.325	0.348	0.398	0.446	0.490					
	k	0.0513	0.0512	0.0521	0.0535	0.0571	0.0653	0.0740	0.0832					

500	v	0.00262	0.00296	0.00330	0.00364	0.00430	0.00558	0.00683	0.00804	0.00911	0.0103	0.0114	0.0126	0.0154
	h	319.9	381.3	440.8	499.1	614.3	844.6	1078.8	1318	1561	1814	2070	2332	3015
	s	5.103	5.267	5.408	5.531	5.741	6.072	6.333	6.550	6.743	6.907	7.058	7.196	7.501
	C_p	1.255	1.206	1.176	1.159	1.148	1.159	1.183	1.205	1.226	1.265	1.306	1.337	1.412
	μ	0.338	0.333	0.336	0.343	0.361	0.407	0.452	0.495					
	k	0.0568	0.0557	0.0560	0.0569	0.0598	0.0672	0.0755	0.0844					
600	v	0.00234	0.00262	0.00290	0.00318	0.00374	0.00481	0.00586	0.00689	0.00776	0.00873	0.00970	0.0107	0.0130
	h	322.6	384.2	444.0	502.6	618.5	850.1	1085.5	1326	1570	1824	2080	2343	3026
	s	5.041	5.205	5.346	5.470	5.681	6.014	6.277	6.495	6.690	6.854	7.005	7.144	7.449
	C_p	1.258	1.211	1.182	1.166	1.154	1.166	1.189	1.210	1.231	1.267	1.310	1.341	1.412
	μ	0.370	0.359	0.358	0.361	0.375	0.416	0.459	0.501					
	k	0.0620	0.0602	0.0598	0.0603	0.0625	0.0691	0.0770	0.0857					
800	v	0.00200	0.00221	0.00242	0.00263	0.00304	0.00385	0.00465	0.00544	0.00608	0.00681	0.00754	0.00826	0.0101
	h	331.6	393.8	453.4	512.3	625.8	862.0	1099.3	1341	1588	1844	2101	2365	3049
	s	4.943	5.108	5.250	5.374	5.586	5.922	6.136	6.407	6.605	6.769	6.921	7.060	7.366
	C_p	1.257	1.216	1.188	1.172	1.161	1.175	1.198	1.219	1.240	1.275	1.318	1.347	1.412
	μ	0.432	0.411	0.402	0.399	0.405	0.436	0.474	0.512					
	k	0.0718	0.0688	0.0673	0.0669	0.0679	0.0730	0.0800	0.0881					
1000	v	0.00180	0.00196	0.00213	0.00230	0.00262	0.00328	0.00392	0.00455	0.00507	0.00565	0.00624	0.00681	0.00825
	h	343.4	405.1	465.3	524.4	641.2	875.1	1113.3	1356	1606	1863	2121	2386	3071
	s	4.869	5.034	5.176	5.300	5.513	5.850	6.115	6.337	6.539	6.703	6.856	6.995	7.302
	C_p	1.254	1.217	1.192	1.175	1.164	1.179	1.204	1.225	1.248	1.283	1.325	1.354	1.413
	μ	0.494	0.463	0.446	0.438	0.435	0.456	0.489	0.524					
	k	0.0810	0.0768	0.0744	0.0733	0.0732	0.0768	0.0830	0.0906					

THERMODYNAMIC AND THERMOPHYSICAL PROPERTIES—TABLE 5. Enthalpy and Psi Functions for Ideal-Gas Air

T, K	h, kJ/kg	ψ	T, K	h, kJ/kg	ψ	T, K	h, kJ/kg	ψ
200	200.0	−0.473	650	659.8	1.339	1200	1278	2.376
210	210.0	−0.400	660	670.5	1.364	1220	1301	2.406
220	220.0	−0.329	670	681.1	1.388	1240	1325	2.435
230	230.1	−0.262	680	691.8	1.412	1260	1349	2.463
240	240.1	−0.197	690	702.5	1.436	1280	1372	2.491
250	250.1	−0.135	700	713.3	1.459	1300	1396	2.519
260	260.1	−0.076	710	724.0	1.482	1320	1420	2.547
270	270.1	−0.018	720	734.8	1.505	1340	1444	2.574
280	280.1	0.037	730	745.6	1.528	1360	1467	2.601
290	290.2	0.090	740	756.4	1.550	1380	1491	2.627
300	300.2	0.142	750	767.3	1.572	1400	1515	2.653
310	310.3	0.191	760	778.2	1.594	1420	1539	2.679
320	320.3	0.240	770	789.1	1.615	1440	1563	2.705
330	330.4	0.286	780	800.0	1.637	1460	1587	2.730
340	340.4	0.332	790	811.0	1.658	1480	1612	2.755
350	350.5	0.376	800	821.9	1.679	1500	1636	2.779
360	360.6	0.419	810	832.9	1.699	1520	1660	2.803
370	370.7	0.461	820	844.0	1.720	1540	1684	2.827
380	380.8	0.502	830	855.0	1.740	1560	1709	2.851
390	390.9	0.541	840	866.1	1.760	1580	1738	2.875
400	401.0	0.580	850	877.2	1.780	1600	1758	2.898
410	411.2	0.618	860	888.3	1.800	1620	1782	2.921
420	421.3	0.655	870	899.4	1.819	1640	1806	2.944
430	431.5	0.691	880	910.6	1.838	1660	1831	2.966
440	441.7	0.727	890	921.8	1.857	1680	1855	2.988
450	451.8	0.761	900	933.0	1.876	1700	1880	3.010
460	462.1	0.795	910	944.2	1.895	1720	1905	3.032
470	472.3	0.829	920	955.4	1.914	1740	1929	3.054
480	482.5	0.861	930	966.7	1.932	1760	1954	3.075
490	492.8	0.893	940	978.0	1.950	1780	1979	3.096
500	503.1	0.925	950	989.3	1.969	1800	2003	3.117
510	513.4	0.956	960	1000.6	1.987	1820	2028	3.138
520	523.7	0.986	970	1011.9	2.004	1840	2053	3.158
530	534.0	1.016	980	1023.3	2.022	1860	2078	3.178
540	544.4	1.045	990	1034.7	2.039	1880	2102	3.198
550	554.8	1.074	1000	1046.1	2.057	1900	2127	3.218
560	565.2	1.102	1020	1068.9	2.091	1920	2152	3.238
570	575.6	1.130	1040	1091.9	2.125	1940	2177	3.258
580	586.1	1.158	1060	1114.9	2.158	1960	2202	3.277
590	596.5	1.185	1080	1138.0	2.190	1980	2227	3.296
600	607.0	1.211	1100	1161.1	2.223	2000	2252	3.215
610	617.5	1.238	1120	1184.3	2.254	2050	2315	3.362
620	628.1	1.264	1140	1207.6	2.285	2100	2377	3.408
630	638.6	1.289	1160	1230.9	2.316	2150	2440	3.453
640	649.2	1.314	1180	1254.3	2.346	2200	2504	3.496

THERMODYNAMIC AND THERMOPHYSICAL PROPERTIES—TABLE 6. Saturated Ammonia

T, K	P, bar	v_f, m³/kg	v_g, m³/kg	h_f, kJ/kg	h_g, kJ/kg	s_f, kJ/(kg·K)	s_g, kJ/(kg·K)	c_{pf}, kJ/(kg·K)	μ_f, 10⁻⁴ Pa·s	k_f, W/(m·K)
195.5ᵗ	0.0608	1.327,–3	15.648	–1110.1	380.1	4.203	11.827	4.73	4.25	0.715
200	0.0865	1.372,–3	11.237	–1088.8	388.5	4.311	11.698	4.61	4.07	0.709
210	0.1775	1.394,–3	5.729	–1044.1	406.7	4.529	11.438	4.38	3.69	0.685
220	0.3381	1.417,–3	3.135	–1000.6	424.1	4.731	11.207	4.35	3.34	0.661
230	0.6044	1.442,–3	1.822	–957.0	440.7	4.925	11.002	4.38	3.02	0.638
240	1.0226	1.468,–3	1.115	–912.9	456.2	5.113	10.817	4.43	2.73	0.615
250	1.6496	1.495,–3	0.712	–868.2	470.6	5.294	10.650	4.48	2.45	0.592
260	2.5529	1.524,–3	0.472	–823.1	483.8	5.471	10.498	4.54	2.20	0.569
270	3.8100	1.551,–3	0.324	–777.3	495.6	5.643	10.358	4.60	1.97	0.546
280	5.5077	1.589,–3	0.228	–730.9	506.0	5.811	10.228	4.66	1.76	0.523
290	7.741	1.626,–3	0.165	–683.8	514.7	5.975	10.108	4.73	1.58	0.500
300	10.61	1.666,–3	0.121	–636.0	521.5	6.135	9.994	4.82	1.41	0.477
310	14.24	1.710,–3	0.091	–587.2	526.1	6.293	9.885	4.91	1.26	0.454
320	18.72	1.760,–3	0.069	–537.5	528.2	6.448	9.779	5.02	1.13	0.431
330	24.20	1.815,–3	0.053	–486.7	527.5	6.602	9.675	5.17	1.02	0.408
340	30.79	1.878,–3	0.0410	–434.3	523.3	6.755	9.571	5.37	0.92	0.385
350	38.64	1.952,–3	0.0319	–380.0	515.1	6.908	9.465	5.64	0.83	0.361
360	47.90	2.039,–3	0.0249	–323.2	501.8	7.063	9.354	6.04	0.75	0.337
370	58.74	2.148,–3	0.0194	–262.6	481.9	7.222	9.235	6.68	0.69	0.313
380	71.35	2.291,–3	0.0149	–196.5	452.7	7.391	9.100	7.80	0.61	0.286
390	85.98	2.499,–3	0.0113	–120.9	408.1	7.578	8.935	10.3	0.50	0.254
400	103.0	2.882,–3	0.0077	–23.5	329.0	7.813	8.694	21.	0.39	0.21
405.4ᶜ	113.0	4.255,–3	0.0043	142.7	142.7	8.216	8.216	∞	0.25	∞

THERMODYNAMIC AND THERMOPHYSICAL PROPERTIES — TABLE 7.   Saturated Argon (R740)

T, K	P, bar	v_f, m³/kg	v_g, m³/kg	h_f, kJ/kg	h_g, kJ/kg	s_f, kJ/(kg·K)	s_g, kJ/(kg·K)	c_{pf}, kJ/(kg·K)	μ_f, 10⁻⁴ Pa·s	k_f, W/(m·K)
10		5.646.−4		0.20		0.0266		0.083		
20		5.666.−4		2.20		0.1559		0.306		
30		5.707.−4		6.12		0.3129		0.466		
40		5.763.−4		11.30		0.4610		0.560		
50		5.831.−4		17.26		0.5937		0.627		
60		5.912.−4		23.85		0.7138		0.687		
70	0.082	6.008.−4	2.1800	31.08	229.08	0.8250	3.415	0.752		
80	0.406	6.125.−4	0.3918	39.07	232.88	0.9316	3.364	0.836		
83.8[t]	0.687	6.178.−4	0.2434	42.34	235.06	0.9720	3.280	0.877		
83.8[t]	0.687	7.068.−4	0.2434	71.88	235.06	1.333	3.280	1.050	2.93	
85	0.790	7.107.−4	0.2145	73.16	235.55	1.348	3.258	1.058	2.81	0.134
87.3	1.013	7.174.−4	0.1710	75.61	236.39	1.375	3.216	1.073	2.60	0.132
90	1.338	7.269.−4	0.1327	78.55	237.37	1.403	3.168	1.091	2.40	0.128
95	2.317	7.440.−4	0.0864	84.15	238.91	1.462	3.091	1.124	2.08	0.124
100	3.247	7.628.−4	0.0588	89.85	240.20	1.520	3.023	1.158	1.82	0.116
110	6.665	8.064.−4	0.0299	101.83	241.66	1.632	2.903	1.229	1.46	0.109
115	9.107	8.322.−4	0.0221	108.11	241.78	1.685	2.848	1.274	1.32	0.096
120	12.13	8.618.−4	0.0166	114.62	241.33	1.738	2.794	1.336	1.21	0.090
125	15.81	8.965.−4	0.0126	121.50	240.30	1.792	2.743	1.427	1.12	0.084
130	20.23	9.620.−4	0.0096	128.79	238.41	1.846	2.690	1.550	1.01	0.078
135	25.49	9.906.−4	0.0074	136.76	234.60	1.902	2.633	1.752	0.89	0.072
140	31.68	1.061.−3	0.0056	145.58	230.74	1.961	2.570		0.75	0.066
145	38.93	1.172.−3	0.0041	155.73	223.09	2.026	2.490		0.60	0.060
150	47.39	1.468.−3	0.0026	174.64	204.35	2.133	2.331		0.45	0.054
150.9	48.98	1.867.−3	0.0019	189.94	189.94	2.201	2.201		0.28	∞

THERMODYNAMIC AND THERMOPHYSICAL PROPERTIES—TABLE 8. Compressed Argon

						Pressure, bar					
T, K		1	100	200	300	400	500	600	700	800	900	1000
100	v	0.2035	7.420.–4	7.255.–4	7.120.–4	7.006.–4	6.907.–4	6.819.–4	6.050.–4	6.009.–4	5.976.–4	5.935.–4
	h	243.4	93.6	97.9	102.5	107.2	112.0	116.8	91.1	96.1	101.0	106.0
	s	3.299	1.494	1.464	1.438	1.414	1.393	1.372	1.037	1.026	1.016	1.007
200	v	0.4151	2.96.–3	1.430.–5	1.159.–3	1.045.–3	9.778.–4	9.312.–4	8.962.–4	8.683.–4	8.454.–4	8.260.–4
	h	296.4	250.2	217.1	209.1	207.9	209.2	211.9	215.1	218.9	223.0	227.4
	s	3.667	2.538	2.276	2.173	2.112	2.068	2.033	2.004	1.979	1.957	1.936
300	v	0.6241	5.96.–3	2.976.–3	2.071.–3	1.666.–3	1.443.–3	1.304.–3	1.207.–3	1.136.–3	1.081.–3	1.037.–3
	h	348.6	330.9	316.3	306.6	301.4	299.3	299.2	300.5	302.7	305.6	310.0
	s	3.879	2.872	2.686	2.572	2.493	2.435	2.389	2.352	2.320	2.293	2.269
400	v	0.8326	8.37.–3	4.279.–3	2.957.–3	2.322.–3	1.955.–3	1.719.–3	1.557.–3	1.435.–3	1.344.–3	1.271.–3
	h	400.7	391.3	383.6	378.4	375.2	373.8	373.8	374.8	376.6	379.2	382.0
	s	4.028	3.048	2.881	2.780	2.707	2.651	2.603	2.565	2.533	2.505	3.480
500	v	1.0409	1.062.–2	5.464.–3	3.772.–3	2.940.–3	2.448.–3	2.124.–3	1.899.–3	1.730.–3	1.607.–3	1.506.–3
	h	452.8	447.7	444.3	442.0	440.9	440.6	441.4	422.9	444.7	447.1	449.9
	s	4.145	3.174	3.018	2.924	2.854	2.801	2.755	2.718	2.685	2.658	2.633
600	v	1.2489	1.280.–2	6.589.–3	4.539.–3	3.525.–3	2.922.–3	2.522.–3	2.238.–3	2.023.–3	1.866.–3	1.736.–3
	h	504.9	502.4	501.6	501.4	501.8	503.0	504.6	506.6	508.7	511.2	513.9
	s	4.240	3.274	3.122	3.031	2.966	2.914	2.870	2.834	2.801	2.774	2.750
700	v	1.4569	1.495.–2	7.686.–3	5.281.–3	4.088.–3	3.377.–3	2.906.–3	2.570.–3	2.317.–3	2.123.–3	1.966.–3
	h	556.9	556.5	556.9	558.0	559.8	561.8	564.2	566.9	569.6	527.5	575.3
	s	4.320	3.356	3.207	3.118	3.054	3.005	2.963	2.928	2.897	2.870	2.845
800	v	1.6659	1.708.–2	8.768.–3	6.011.–3	4.640.–3	3.822.–3	3.280.–3	2.893.–3	2.603.–3	2.376.–3	2.196.–3
	h	609.9	609.8	611.0	612.9	615.2	618.1	621.2	624.5	627.8	631.3	634.8
	s	4.389	3.427	3.279	3.191	3.129	3.081	3.039	3.005	2.975	2.948	2.924
900	v	1.8739	1.920.–2	9.841.–3	6.732.–3	5.183.–3	4.259.–3	3.646.–3	3.209.–3	2.881.–3	2.626.–3	2.423.–3
	h	661.0	662.7	664.6	667.2	670.1	673.3	676.8	680.7	684.4	688.3	692.3
	s	4.451	3.490	3.342	3.255	3.193	3.145	3.105	3.071	3.042	3.016	2.992
1000	v	2.0819	2.131.–2	1.091.–2	7.448.–3	5.723.–3	4.692.–3	4.008.–3	3.520.–3	3.156.–3	2.872.–3	2.645.–3
	h	713.1	715.4	717.9	720.9	724.3	727.8	731.5	735.6	739.8	744.1	748.5
	s	4.506	3.545	3.398	3.312	3.250	3.203	3.163	3.129	3.100	3.074	3.051

THERMODYNAMIC AND THERMOPHYSICAL PROPERTICS—TABLE 9. Liquid–Vapor Equilibrium Data for the Argon-Nitrogen-Oxygen System*

Liquid mole fraction		Vapor mole fraction			Temperature, °R	Relative volatility			Pressure activity coefficient			Enthalpy, Btu/(lb·mol)		Heat capacity, Btu/(lb·mol·°R)	
$N_2/N_2 + O_2$	Ar	N_2	Ar	O_2		N_2/Ar	N_2/O_2	Ar/O_2	N_2	Ar	O_2	Liquid	Vapor	Liquid	Vapor
						Pressure, 1 atm									
0.	0.	0.	0.	1.0000	162.4	2.575	4.010	1.557	1.118	1.165	0.999	−1841.	1093.	13.2	7.406
0.	0.01	0.	0.0154	0.9845	162.3	2.581	4.007	1.553	1.117	1.161	1.000	−1844.	1087.	13.1	7.374
0.	0.02	0.	0.0306	0.9694	162.2	2.586	4.004	1.548	1.115	1.158	1.000	−1847.	1082.	13.1	7.342
0.	0.03	0.	0.0456	0.9544	162.1	2.592	4.001	1.544	1.113	1.155	1.000	−1850.	1076.	13.1	7.311
0.	0.04	0.	0.0603	0.9397	162.0	2.597	3.998	1.540	1.112	1.151	1.001	−1852.	1071.	13.0	7.281
0.	0.05	0.	0.0748	0.9253	161.9	2.602	3.995	1.535	1.110	1.148	1.001	−1855.	1066.	13.0	7.251
0.	0.07	0.	0.1031	0.8970	161.7	2.613	3.989	1.526	1.107	1.142	1.002	−1860.	1056.	12.9	7.192
0.	0.10	0.	0.1439	0.8561	161.5	2.629	3.979	1.513	1.103	1.132	1.003	−1868.	1041.	12.9	7.107
0.	0.20	0.	0.2687	0.7313	160.7	2.682	3.941	1.469	1.091	1.104	1.010	−1893.	997.	12.6	6.847
0.	0.40	0.	0.4796	0.5204	159.4	2.786	3.852	1.382	1.076	1.058	1.034	−1938.	924.	11.9	6.406
0.	0.60	0.	0.6605	0.3395	158.5	2.888	3.746	1.297	1.075	1.026	1.072	−1978.	862.	11.3	6.026
0.	0.80	0.	0.8293	0.1707	157.7	2.991	3.632	1.214	1.087	1.008	1.127	−2015.	807.	10.7	5.669
0.	0.90	0.	0.9136	0.0865	157.5	3.042	3.572	1.174	1.099	1.003	1.162	−2032.	779.	10.4	5.491
0.10	0.	0.3135	0.	0.6865	157.7	2.621	4.111	1.568	1.103	1.168	1.012	−1834.	1060.	13.2	7.410
0.10	0.01	0.3095	0.0119	0.6786	157.6	2.626	4.106	1.563	1.102	1.164	1.012	−1837.	1057.	13.1	7.386
0.10	0.02	0.3056	0.0237	0.6707	157.6	2.631	4.100	1.558	1.100	1.161	1.012	−1839.	1053.	13.1	7.361
0.10	0.03	0.3017	0.0354	0.6630	157.6	2.636	4.095	1.554	1.099	1.157	1.013	−1842.	1049.	13.1	7.337
0.10	0.04	0.2978	0.0470	0.6553	157.6	2.641	4.090	1.549	1.098	1.154	1.013	−1844.	1045.	13.1	7.313
0.10	0.05	0.2939	0.0585	0.6476	157.5	2.645	4.085	1.544	1.096	1.151	1.013	−1846.	1042.	13.0	7.289
0.10	0.07	0.2863	0.0812	0.6325	157.5	2655	4.074	1.534	1.094	1.144	1.014	−1851.	1034.	13.0	7.242
0.10	0.10	0.2752	0.1145	0.6103	157.4	2.669	4.058	1.520	1.090	1.135	1.015	−1858.	1024.	12.9	7.173
0.10	0.20	0.2399	0.2207	0.5394	157.2	2.717	4.003	1.473	1.080	1.106	1.022	−1882.	990.	12.5	6.951
0.10	0.40	0.1759	0.4170	0.4072	157.0	2.812	3.887	1.382	1.070	1.061	1.045	−1926.	928.	11.9	6.540

0.10	0.60	0.1169	0.6036	0.2795	1.029	156.9	2.906	3.766	1.296	1.072	1.082	−1969.	871.	11.3	6.147
0.10	0.80	0.0595	0.7933	0.1471	1.009	156.9	3.001	3.640	1.213	1.086	1.134	−2009.	813.	10.7	5.746
0.10	0.90	0.0303	0.8937	0.0762	1.004	157.1	3.048	3.576	1.173	1.099	1.166	−2029.	783.	10.4	5.534
0.20	0.	0.5095	0.	0.4905	1.171	154.0	2.641	4.155	1.573	1.085	1.026	−1814.	1035.	13.2	7.422
0.20	0.01	0.5042	0.0096	0.4861	1.168	154.0	2.646	4.149	1.568	1.084	1.026	−1816.	1032.	13.2	7.402
0.20	0.02	0.4990	0.0192	0.4818	1.164	154.0	2.651	4.143	1.563	1.083	1.026	−1819.	1029.	13.1	7.382
0.20	0.03	0.4938	0.0288	0.4775	1.161	154.1	2.655	4.137	1.558	1.082	1.027	−1821.	1027.	13.1	7.362
0.20	0.04	0.4886	0.0383	0.4731	1.158	154.1	2.660	4.131	1.553	1.081	1.027	−1824.	1024.	13.1	7.342
0.20	0.05	0.4834	0.0477	0.4688	1.154	154.1	2.665	4.125	1.548	1.080	1.028	−1826.	1021.	13.0	7.322
0.20	0.07	0.4732	0.0666	0.4602	1.148	154.1	2.674	4.112	1.538	1.078	1.030	−1831.	1016.	13.0	7.283
0.20	0.10	0.4580	0.0946	0.4474	1.139	154.2	2.688	4.094	1.523	1.075	1.036	−1839.	1008.	12.9	7.224
0.20	0.20	0.4083	0.1866	0.4051	1.110	154.3	2.735	4.032	1.474	1.068	1.058	−1863.	981.	12.6	7.031
0.20	0.40	0.3123	0.3680	0.3197	1.064	154.8	2.829	3.907	1.381	1.062	1.093	−1911.	930.	11.9	6.648
0.20	0.60	0.2162	0.5550	0.2288	1.032	155.4	2.921	3.779	1.294	1.068	1.140	−1958.	877.	11.3	6.252
0.20	0.80	0.1144	0.7602	0.1254	1.011	156.2	3.009	3.647	1.212	1.086	1.169	−2004.	819.	10.7	5.817
0.20	0.90	0.0593	0.8744	0.0663	1.006	156.7	3.052	3.580	1.173	1.099		−2026.	787.	10.4	5.574
0.40	0.	0.7333	0.	0.2667	1.187	148.7	2.629	4.124	1.569	1.050	1.065	−1748.	997.	13.3	7.452
0.40	0.01	0.7226	0.0070	0.2651	1.183	148.8	2.634	4.119	1.564	1.049	1.065	−1751.	996.	13.3	7.437
0.40	0.02	0.7172	0.0140	0.2635	1.179	148.8	2.640	4.114	1.558	1.049	1.065	−1754.	994.	13.2	7.422
0.40	0.03	0.7118	0.0210	0.2619	1.176	148.9	2.645	4.108	1.553	1.048	1.066	−1757.	992.	13.2	7.407
0.40	0.04	0.7064	0.0280	0.2602	1.172	148.9	2.650	4.103	1.548	1.048	1.066	−1760.	991.	13.2	7.392
0.40	0.05		0.0350	0.2586	1.169	149.0	2.656	4.098	1.543	1.047	1.066	−1763.	989.	13.1	7.377
0.40	0.07	0.6956	0.0491	0.2553	1.162	149.1	2.667	4.087	1.533	1.047	1.067	−1770.	986.	13.1	7.347
0.40	0.10	0.6794	0.0703	0.2503	1.152	149.3	2.683	4.072	1.517	1.046	1.071	−1779.	981.	13.0	7.301
0.40	0.20	0.6244	0.1426	0.2331	1.122	149.9	2.737	4.018	1.468	1.044	1.088	−1810.	964.	12.6	7.145
0.40	0.40	0.5075	0.2977	0.1948	1.074	151.2	2.841	3.907	1.375	1.048	1.116	−1871.	928.	11.9	6.811
0.40	0.60	0.3743	0.4776	0.1482	1.039	152.8	2.939	3.788	1.289	1.062	1.154	−1932.	885.	11.3	6.432
0.40	0.80	0.2121	0.7010	0.0870	1.016	154.8	3.025	3.657	1.209	1.084	1.177	−1991.	829.	10.7	5.944
0.40	0.90	0.1141	0.8382	0.0477	1.008	155.9	3.063	3.586	1.171	1.099		−2020.	794.	10.4	5.651
0.60	0.	0.8569	0.	0.1431	1.218	144.9	2.575	3.993	1.551	1.024	1.126	−1663.	970.	13.6	7.483
0.60	0.01	0.8521	0.0056	0.1424	1.214	145.0	2.582	3.991	1.546	1.024	1.125	−1667.	969.	13.5	7.471
0.60	0.02	0.8472	0.0111	0.1416	1.210	145.1	2.589	3.988	1.541	1.024	1.125	−1672.	968.	13.5	7.459
0.60	0.03	0.8424	0.0167	0.1409	1.206	145.1	2.595	3.985	1.536	1.024	1.124	−1676.	967.	13.4	7.446
0.60	0.04	0.8375	0.0224	0.1402	1.202	145.2	2.602	3.983	1.531	1.024	1.124	−1680.	966.	13.4	7.434

THERMODYNAMIC AND THERMOPHYSICAL PROPERTICS—TABLE 9. Liquid–Vapor Equilibrium Data for the Argon-Nitrogen-Oxygen System* (Continued)

Liquid mole fraction		Vapor mole fraction			Temperature, °R	Relative volatility			Pressure activity coefficient			Enthalpy, Btu/(lb·mol)		Heat capacity, Btu/(lb·mol·°R)	
N₂/N₂ + O₂	Ar	N₂	Ar	O₂		N₂/Ar	N₂/O₂	Ar/O₂	N₂	Ar	O₂	Liquid	Vapor	Liquid	Vapor
0.60	0.05	0.8326	0.0280	0.1395	145.3	2.609	3.980	1.526	1.023	1.198	1.123	-1684.	965.	13.4	7.421
0.60	0.07	0.8227	0.0394	0.1380	145.5	2.622	3.975	1.516	1.023	1.190	1.122	-1692.	963.	13.3	7.396
0.60	0.10	0.8076	0.0566	0.1357	145.7	2.642	3.966	1.501	1.023	1.179	1.121	-1704.	960.	13.2	7.357
0.60	0.20	0.7557	0.1163	0.1280	146.5	2.707	3.937	1.454	1.025	1.145	1.120	-1744.	948.	12.8	7.224
0.60	0.40	0.6391	0.2507	0.1102	148.4	2.833	3.867	1.365	1.036	1.090	1.126	-1824.	923.	12.0	6.926
0.60	0.60	0.4938	0.4191	0.0871	150.6	2.945	3.777	1.283	1.056	1.049	1.143	-1903.	888.	11.3	6.556
0.60	0.80	0.2961	0.6500	0.0539	153.5	3.037	3.662	1.206	1.083	1.021	1.169	-1978.	837.	10.7	6.055
0.60	0.90	0.1647	0.8047	0.0306	155.2	3.071	3.592	1.170	1.098	1.010	1.185	-2014.	801.	10.4	5.723
0.80	0.	0.9384	0.	0.0616	142.0	2.501	3.811	1.524	1.013	1.273	1.214	-1570.	949.	14.0	7.514
0.80	0.01	0.9340	0.0047	0.0613	142.0	2.509	3.811	1.519	1.013	1.268	1.212	-1575.	948.	13.9	7.503
0.80	0.02	0.9296	0.0094	0.0610	142.1	2.517	3.812	1.515	1.013	1.263	1.210	-1580.	947.	13.9	7.492
0.80	0.03	0.9252	0.0142	0.0607	142.2	2.525	3.812	1.510	1.013	1.257	1.209	-1585.	947.	13.8	7.481
0.80	0.04	0.9207	0.0189	0.0604	142.3	2.533	3.813	1.506	1.013	1.258	1.207	-1590.	946.	13.8	7.470
0.80	0.05	0.9162	0.0237	0.0601	142.4	2.540	3.813	1.501	1.013	1.247	1.205	-1595.	945.	13.7	7.459
0.80	0.07	0.9071	0.0334	0.0595	142.6	2.556	3.814	1.492	1.012	1.238	1.202	-1606.	944.	13.6	7.437
0.80	0.10	0.8934	0.0481	0.0586	142.8	2.580	3.814	1.478	1.013	1.224	1.197	-1621.	942.	13.5	7.403
0.80	0.20	0.8452	0.0994	0.0554	143.8	2.658	3.814	1.435	1.015	1.181	1.185	-1671.	934.	13.0	7.284
0.80	0.40	0.7339	0.2177	0.0483	146.0	2.809	3.797	1.352	1.028	1.112	1.173	-1772.	916.	12.2	7.013
0.80	0.60	0.5869	0.3740	0.0391	148.7	2.943	3.751	1.275	1.051	1.062	1.173	-1870.	889.	11.4	6.662
0.80	0.80	0.3690	0.6059	0.0252	152.2	3.045	3.662	1.202	1.082	1.026	1.184	-1964.	843.	10.7	6.153
0.80	0.90	0.2117	0.7736	0.0147	154.5	3.078	3.595	1.168	1.098	1.013	1.193	-2007.	806.	10.3	5.790
0.90	0.	0.9709	0.	0.0291	140.6	2.459	3.710	1.509	1.015	1.311	1.271	-1522.	939.	14.2	7.530
0.90	0.01	0.9667	0.0044	0.0289	140.7	2.468	3.712	1.504	1.014	1.305	1.268	-1527.	938.	14.2	7.519
0.90	0.02	0.9624	0.0088	0.0288	140.8	2.476	3.714	1.500	1.014	1.299	1.265	-1533.	938.	14.1	7.509

0.90	0.03	0.9581	0.0133	0.0286	140.9	2.485	3.716	1.496	1.013	1.293	1.263	−1538.	937.	14.1	7.498
0.90	0.04	0.9538	0.0177	0.0285	141.0	2.493	3.718	1.491	1.013	1.287	1.260	−1544.	937.	14.0	7.488
0.90	0.05	0.9494	0.0222	0.0284	141.1	2.502	3.720	1.487	1.013	1.281	1.257	−1550.	936.	14.0	7.477
0.90	0.07	0.9407	0.0312	0.0281	141.3	2.519	3.724	1.478	1.013	1.270	1.252	−1561.	935.	13.8	7.456
0.90	0.10	0.9274	0.0450	0.0276	141.6	2.545	3.729	1.465	1.012	1.254	1.254	−1578.	934.	13.7	7.423
0.90	0.20	0.8808	0.0931	0.0261	142.6	2.629	3.743	1.424	1.014	1.204	1.226	−1633.	928.	13.2	7.310
0.90	0.40	0.7723	0.2048	0.0229	144.9	2.793	3.755	1.344	1.026	1.126	1.200	−1745.	912.	12.3	7.050
0.90	0.60	0.6262	0.3552	0.0186	147.8	2.938	3.733	1.270	1.050	1.069	1.190	−1853.	889.	11.4	6.708
0.90	0.80	0.4018	0.5860	0.0122	151.7	3.048	3.660	1.201	1.081	1.029	1.192	−1956.	846.	10.7	6.197
0.90	0.90	0.2339	0.7589	0.0072	154.2	3.082	3.597	1.167	1.098	1.014	1.197	−2004.	809.	10.3	5.822
0.97	0.	0.9916	0.	0.0084	139.8	2.429	3.638	1.498	1.018	1.342	1.316	−1488.	933.	14.4	7.541
0.97	0.01	0.9874	0.0042	0.0084	139.9	2.438	3.641	1.494	1.018	1.335	1.313	−1494.	932.	14.4	7.531
0.97	0.02	0.9832	0.0085	0.0083	140.0	2.447	3.644	1.489	1.017	1.329	1.309	−1500.	932.	14.3	7.520
0.97	0.03	0.9790	0.0127	0.0083	140.1	2.456	3.647	1.485	1.017	1.322	1.306	−1505.	931.	14.2	7.510
0.97	0.04	0.9748	0.0170	0.0083	140.2	2.465	3.650	1.481	1.016	1.315	1.302	−1511.	931.	14.2	7.500
0.97	0.05	0.9705	0.0213	0.0082	140.3	2.474	3.653	1.477	1.016	1.309	1.299	−1517.	930.	14.1	7.489
0.97	0.07	0.9619	0.0300	0.0081	140.5	2.492	3.658	1.468	1.015	1.296	1.293	−1529.	929.	14.0	7.469
0.97	0.10	0.9488	0.0432	0.0080	140.8	2.519	3.667	1.456	1.014	1.278	1.284	−1547.	928.	13.9	7.437
0.97	0.20	0.9032	0.0893	0.0076	141.8	2.608	3.691	1.415	1.014	1.224	1.257	−1606.	923.	13.3	7.327
0.97	0.40	0.7965	0.1969	0.0066	144.2	2.780	3.722	1.339	1.025	1.137	1.220	−1725.	910.	12.3	7.073
0.97	0.60	0.6513	0.3433	0.0054	147.2	2.934	3.718	1.267	1.049	1.075	1.202	−1841.	889.	11.4	6.737
0.97	0.80	0.4236	0.5728	0.0036	151.3	3.050	3.658	1.199	1.081	1.031	1.198	−1951.	847.	10.7	6.226
0.97	0.90	0.2489	0.7489	0.0021	153.9	3.084	3.597	1.167	1.098	1.015	1.200	−2002.	810.	10.3	5.844
1.00	0.	1.0000	0.	0.	139.4	2.416	3.607	1.493	1.021	1.357	1.338	−1473.	930.	14.5	7.546
1.00	0.01	0.9959	0.0041	0.0000	139.5	2.425	3.611	1.489	1.020	1.350	1.334	−1479.	929.	14.4	7.535
1.00	0.02	0.9917	0.0083	0.0000	139.6	2.434	3.614	1.485	1.019	1.343	1.330	−1485.	929.	14.4	7.525
1.00	0.03	0.9875	0.0125	0.0000	139.7	2.443	3.617	1.480	1.019	1.336	1.326	−1491.	929.	14.3	7.515
1.00	0.04	0.9833	0.0167	0.0000	139.8	2.452	3.621	1.476	1.018	1.329	1.322	−1497.	928.	14.3	7.505
1.00	0.05	0.9791	0.0209	0.0000	139.9	2.462	3.624	1.472	1.018	1.322	1.318	−1503.	928.	14.2	7.495
1.00	0.07	0.9705	0.0295	0.0000	140.1	2.480	3.630	1.464	1.017	1.309	1.311	−1516.	927.	14.1	7.474
1.00	0.10	0.9576	0.0424	0.0000	140.4	2.507	3.640	1.452	1.016	1.290	1.301	−1534.	926.	13.9	7.443
1.00	0.20	0.9122	0.0878	0.0000	141.5	2.598	3.668	1.412	1.015	1.232	1.271	−1595.	921.	13.4	7.333
1.00	0.40	0.8063	0.1937	0.0000	143.9	2.774	3.708	1.337	1.025	1.142	1.230	−1716.	909.	12.4	7.082
1.00	0.60	0.6616	0.3384	0.	147.0	2.932	3.712	1.266	1.048	1.077	1.208	−1836.	888.	11.4	6.749
1.00	0.80	0.4327	0.5674	0.	151.1	3.050	3.657	1.199	1.080	1.032	1.201	−1949.	848.	10.7	6.838
1.00	0.90	0.2553	0.7447	0.	153.8	3.085	3.598	1.166	1.098	1.016	1.201	−2001.	811.	10.3	5.853

THERMODYNAMIC AND THERMOPHYSICAL PROPERTICS—TABLE 9. Liquid–Vapor Equilibrium Data for the Argon-Nitrogen-Oxygen System* (Continued)

Liquid mole fraction		Vapor mole fraction			Temperature, °R	Relative volatility			Pressure activity coefficient			Enthalpy, Btu/(lb·mol)		Heat capacity, Btu/(lb·mol·°R)	
$N_2/N_2 + O_2$	Ar	N_2	Ar	O_2		N_2/Ar	N_2/O_2	Ar/O_2	N_2	Ar	O_2	Liquid	Vapor	Liquid	Vapor
					Pressure, 4 atm										
0.	0.	0.	0.	1.0000	190.6	2.020	2.776	1.375	0.987	1.111	1.000	-1466.	1224.	13.4	8.122
0.	0.01	0.	0.0137	0.9863	190.5	2.023	2.775	1.372	0.986	1.109	1.001	-1470.	1218.	13.4	8.094
0.	0.02	0.	0.0272	0.9728	190.4	2.026	2.773	1.369	0.985	1.106	1.001	-1473.	1212.	13.4	8.065
0.	0.03	0.	0.0405	0.9595	190.3	2.030	2.772	1.366	0.985	1.104	1.002	-1477.	1207.	13.4	8.037
0.	0.04	0.	0.0537	0.9463	190.2	2.033	2.770	1.362	0.984	1.102	1.002	-1480.	1201.	13.4	8.010
0.	0.05	0.	0.0668	0.9332	190.1	2.037	2.768	1.359	0.983	1.100	1.003	-1484.	1196.	13.3	7.982
0.	0.07	0.	0.0924	0.9076	189.9	2.043	2.765	1.353	0.982	1.096	1.004	-1490.	1185.	13.3	7.929
0.	0.10	0.	0.1299	0.8701	189.6	2.053	2.759	1.344	0.980	1.089	1.005	-1500.	1169.	13.3	7.850
0.	0.20	0.	0.2471	0.7529	188.8	2.086	2.738	1.313	0.974	1.070	1.013	-1533.	1121.	13.1	7.605
0.	0.40	0.	0.4548	0.5452	187.5	2.148	2.688	1.251	0.967	1.038	1.034	-1593.	1037.	12.7	7.166
0.	0.60	0.	0.6410	0.3590	186.5	2.207	2.627	1.190	0.967	1.016	1.066	-1647.	964.	12.3	6.771
0.	0.80	0.	0.8190	0.1811	185.8	2.264	2.560	1.131	0.976	1.003	1.110	-1698.	896.	11.9	6.389
0.	0.90	0.	0.9084	0.0916	185.5	2.292	2.524	1.101	0.983	1.000	1.137	-1723.	862.	11.8	6.196
0.10	0.	0.2393	0.	0.7607	186.2	2.063	2.831	1.372	0.994	1.120	1.020	-1452.	1193.	13.7	8.150
0.10	0.01	0.2363	0.0116	0.7521	186.1	2.066	2.828	1.369	0.994	1.117	1.021	-1455.	1188.	13.6	8.126
0.10	0.02	0.2334	0.0230	0.7436	186.1	2.069	2.825	1.365	0.993	1.115	1.021	-1459.	1184.	13.6	8.102
0.10	0.03	0.2305	0.0344	0.7351	186.0	2.072	2.822	1.362	0.992	1.113	1.021	-1462.	1180.	13.6	8.078
0.10	0.04	0.2277	0.0457	0.7267	186.0	2.075	2.820	1.359	0.991	1.110	1.022	-1465.	1175.	13.6	8.054
0.10	0.05	0.2248	0.0569	0.7183	186.0	2.078	2.817	1.356	0.990	1.108	1.022	-1469.	1171.	13.5	8.031
0.10	0.07	0.2192	0.0792	0.7017	185.9	2.083	2.811	1.349	0.988	1.104	1.023	-1475.	1163.	13.5	7.984
0.10	0.10	0.2109	0.1120	0.6772	185.8	2.092	2.802	1.340	0.986	1.097	1.024	-1485.	1150.	13.4	7.915
0.10	0.20	0.1843	0.2174	0.5983	185.5	2.119	2.772	1.308	0.979	1.077	1.030	-1517.	1110.	13.2	7.691

0.10	0.40	0.1353	0.4150	0.4497	185.2	2.173	2.707	1.246	0.971	1.044	1.049	-1578.	1037.	12.8	7.269
0.10	0.60	0.0896	0.6045	0.3058	185.0	2.224	2.638	1.186	0.971	1.020	1.078	-1637.	968.	12.4	6.860
0.10	0.80	0.0452	0.7961	0.1588	185.1	2.273	2.564	1.128	0.978	1.006	1.117	-1693.	900.	12.0	6.444
0.10	0.90	0.0229	0.8957	0.0814	185.2	2.296	2.526	1.100	0.985	1.002	1.141	-1720.	865.	11.8	6.226
0.20	0.	0.4170	0.	0.5831	182.4	2.094	2.861	1.366	0.997	1.128	1.041	-1431.	1166.	13.8	8.189
0.20	0.01	0.4126	0.0099	0.5776	182.4	2.097	2.857	1.363	0.996	1.125	1.041	-1434.	1162.	13.8	8.167
0.20	0.02	0.4082	0.0198	0.5721	182.4	2.100	2.854	1.359	0.995	1.123	1.042	-1438.	1159.	13.8	8.146
0.20	0.03	0.4038	0.0297	0.5666	182.4	2.102	2.851	1.356	0.994	1.120	1.042	-1441.	1155.	13.8	8.125
0.20	0.04	0.3994	0.0395	0.5611	182.4	2.105	2.847	1.353	0.993	1.118	1.042	-1445.	1152.	13.7	8.104
0.20	0.05	0.3951	0.0493	0.5557	182.4	2.107	2.844	1.349	0.992	1.116	1.042	-1448.	1149.	13.7	8.083
0.20	0.07	0.3864	0.0688	0.5448	182.4	2.112	2.837	1.343	0.990	1.111	1.043	-1455.	1142.	13.7	8.041
0.20	0.10	0.3735	0.0979	0.5286	182.5	2.120	2.827	1.333	0.988	1.104	1.044	-1465.	1132.	13.6	7.978
0.20	0.20	0.3316	0.1932	0.4752	182.6	2.145	2.792	1.301	0.982	1.083	1.049	-1498.	1099.	13.3	7.771
0.20	0.40	0.2506	0.3808	0.3686	183.0	2.194	2.720	1.240	0.974	1.049	1.065	-1562.	1035.	12.9	7.363
0.20	0.60	0.1706	0.5714	0.2580	183.6	2.239	2.645	1.181	0.974	1.024	1.090	-1625.	971.	12.4	6.944
0.20	0.80	0.0883	0.7742	0.1376	184.3	2.281	2.568	1.126	0.980	1.008	1.124	-1687.	904.	12.0	6.497
0.20	0.90	0.0452	0.8834	0.0715	184.8	2.301	2.528	1.099	0.986	1.003	1.145	-1717.	867.	11.8	6.255
0.40	0.	0.6560	0.	0.3441	176.3	2.124	2.859	1.346	0.992	1.146	1.090	-1372.	1121.	14.1	8.277
0.40	0.01	0.6506	0.0077	0.3417	176.4	2.127	2.856	1.343	0.991	1.143	1.090	-1376.	1119.	14.0	8.260
0.40	0.02	0.6453	0.0155	0.3393	176.4	2.129	2.853	1.340	0.991	1.140	1.090	-1380.	1117.	14.0	8.242
0.40	0.03	0.6400	0.0232	0.3369	176.5	2.132	2.850	1.337	0.990	1.138	1.090	-1384.	1115.	14.0	8.224
0.40	0.04	0.6347	0.0310	0.3345	176.6	2.135	2.846	1.333	0.990	1.135	1.090	-1387.	1112.	13.9	8.206
0.40	0.05	0.6293	0.0387	0.3320	176.6	2.137	2.843	1.330	2.989	1.133	1.090	-1391.	1110.	13.9	8.188
0.40	0.07	0.6186	0.0543	0.3271	176.8	2.143	2.837	1.324	0.988	1.128	1.090	-1399.	1106.	13.9	8.153
0.40	0.10	0.6025	0.0778	0.3197	177.0	2.151	2.827	1.315	0.986	1.120	1.090	-1410.	1099.	13.8	8.098
0.40	0.20	0.5482	0.1575	0.2943	177.7	2.176	2.794	1.284	0.982	1.098	1.091	-1449.	1077.	13.5	7.915
0.40	0.40	0.4348	0.3259	0.2393	179.2	2.223	2.725	1.226	0.978	1.061	1.100	-1525.	1029.	13.0	7.527
0.40	0.60	0.3104	0.5141	0.1756	180.9	2.264	2.652	1.171	0.979	1.033	1.116	-1600.	975.	12.5	7.095
0.40	0.80	0.1684	0.7334	0.0982	182.9	2.297	2.573	1.120	0.985	1.013	1.139	-1674.	910.	12.0	6.597
0.40	0.90	0.0882	0.8595	0.0523	184.1	2.309	2.531	1.096	0.989	1.006	1.153	-1710.	872.	11.8	6.312
0.60	0.	0.8076	0.	0.1924	171.7	2.120	2.798	1.320	0.986	1.173	1.154	-1296.	1086.	14.2	8.369
0.60	0.01	0.8023	0.0064	0.1913	171.8	2.123	2.796	1.317	0.986	1.170	1.153	-1301.	1084.	14.2	8.354
0.60	0.02	0.7971	0.0127	0.1902	171.9	2.127	2.794	1.314	0.985	1.167	1.152	-1305.	1083.	14.1	8.338
0.60	0.03	0.7918	0.0192	0.1891	172.0	2.130	2.792	1.311	0.985	1.164	1.152	-1310.	1082.	14.1	8.322

THERMODYNAMIC AND THERMOPHYSICAL PROPERTIES—TABLE 9. Liquid–Vapor Equilibrium Data for the Argon-Nitrogen-Oxygen System* (Continued)

Liquid mole fraction		Vapor mole fraction			Temperature, °R	Relative volatility			Pressure activity coefficient			Enthalpy, Btu/(lb·mol)		Heat capacity, Btu/(lb·mol·°R)	
$N_2/N_2 + O_2$	Ar	N_2	Ar	O_2		N_2/Ar	N_2/O_2	Ar/O_2	N_2	Ar	O_2	Liquid	Vapor	Liquid	Vapor
0.60	0.04	0.7865	0.0256	0.1879	172.1	2.133	2.790	1.308	0.985	1.161	1.151	−1315.	1080.	14.1	8.306
0.60	0.05	0.7812	0.0321	0.1868	172.2	2.137	2.788	1.305	0.984	1.159	1.150	−1319.	1079.	14.0	8.289
0.60	0.07	0.7704	0.0451	0.1845	172.4	2.143	2.784	1.299	0.984	1.153	1.149	−1329.	1076.	14.0	8.257
0.60	0.10	0.7542	0.0649	0.1810	172.6	2.153	2.778	1.290	0.983	1.144	1.148	−1343.	1072.	13.9	8.208
0.60	0.20	0.6980	0.1332	0.1688	173.7	2.184	2.757	1.262	0.981	1.119	1.143	−1389.	1056.	13.6	8.038
0.60	0.40	0.5739	0.2848	0.1413	175.9	2.239	2.708	1.209	0.980	1.076	1.141	−1482.	1021.	13.0	7.666
0.60	0.60	0.4261	0.4667	0.1073	178.5	2.283	2.648	1.160	0.984	1.043	1.145	−1573.	976.	12.5	7.228
0.60	0.80	0.2413	0.6963	0.0625	181.6	2.311	2.576	1.115	0.989	1.019	1.155	−1661.	916.	12.0	6.690
0.60	0.90	0.1293	0.8367	0.0340	183.4	2.317	2.533	1.093	0.992	1.009	1.161	−1704.	876.	11.8	6.367
0.80	0.	0.9152	0.	0.0848	168.0	2.095	2.699	1.288	0.986	1.216	1.239	−1209.	1057.	14.3	8.462
0.80	0.01	0.9102	0.0055	0.0843	168.1	2.099	2.699	1.286	0.986	1.212	1.237	−1215.	1056.	14.2	8.447
0.80	0.02	0.9052	0.0110	0.0839	168.2	2.103	2.699	1.283	0.986	1.209	1.235	−1220.	1055.	14.2	8.432
0.80	0.03	0.9001	0.0165	0.0834	168.3	2.107	2.698	1.280	0.985	1.205	1.234	−1226.	1054.	14.2	8.417
0.80	0.04	0.8950	0.0221	0.0829	168.4	2.112	2.698	1.278	0.985	1.201	1.232	−1232.	1053.	14.1	8.402
0.80	0.05	0.8899	0.0277	0.0825	168.5	2.116	2.698	1.275	0.984	1.198	1.230	−1237.	1052.	14.1	8.387
0.80	0.07	0.8795	0.0390	0.0815	168.7	2.124	2.698	1.270	0.984	1.190	1.227	−1249.	1050.	14.0	8.356
0.80	0.10	0.8638	0.0562	0.0801	169.1	2.136	2.697	1.263	0.983	1.180	1.222	−1266.	1047.	14.0	8.309
0.80	0.20	0.8087	0.1162	0.0751	170.3	2.175	2.693	1.238	0.981	1.148	1.208	−1322.	1037.	13.7	8.148
0.80	0.40	0.6826	0.2535	0.0638	173.1	2.244	2.674	1.192	0.983	1.095	1.188	−1434.	1012.	13.1	7.786
0.80	0.60	0.5231	0.4274	0.0496	176.4	2.295	2.636	1.149	0.988	1.055	1.176	−1543.	976.	12.5	7.347
0.80	0.80	0.3077	0.6624	0.0299	180.3	2.323	2.576	1.109	0.993	1.024	1.171	−1647.	920.	12.0	6.777
0.80	0.90	0.1684	0.8150	0.0166	182.7	2.325	2.535	1.090	0.994	1.012	1.169	−1698.	880.	11.8	6.420
0.90	0.	0.9596	0.	0.0404	166.3	2.077	2.641	1.271	0.990	1.245	1.292	−1163.	1044.	14.3	8.509
0.90	0.01	0.9547	0.0051	0.0402	166.4	2.081	2.641	1.269	0.990	1.241	1.289	−1169.	1043.	14.3	8.494

0.90	0.02	0.9497	0.0103	0.0399	166.5	2.086	2.642	1.267	0.989	1.236	1.287	−1175.	1042.	14.2	8.479
0.90	0.03	0.9448	0.0155	0.0397	166.7	2.091	2.643	1.264	0.989	1.232	1.284	−1181.	1042.	14.2	8.464
0.90	0.04	0.9397	0.0208	0.0395	166.8	2.095	2.644	1.262	0.988	1.228	1.281	−1188.	1041.	14.2	8.449
0.90	0.05	0.9347	0.0260	0.0393	166.9	2.100	2.645	1.260	0.988	1.224	1.279	−1194.	1040.	14.1	8.434
0.90	0.07	0.9245	0.0367	0.0388	167.2	2.109	2.646	1.255	0.987	1.215	1.274	−1206.	1039.	14.1	8.404
0.90	0.10	0.9090	0.0529	0.0381	167.5	2.122	2.649	1.248	0.986	1.203	1.267	−1225.	1036.	14.0	8.358
0.90	0.20	0.8546	0.1096	0.0358	168.8	2.166	2.653	1.225	0.984	1.167	1.246	−1287.	1028.	13.7	8.198
0.90	0.40	0.7287	0.2407	0.0305	171.8	2.242	2.651	1.182	0.985	1.107	1.215	−1409.	1007.	13.1	7.840
0.90	0.60	0.5659	0.4102	0.0239	175.3	2.299	2.627	1.143	0.990	1.062	1.193	−1527.	975.	12.5	7.400
0.90	0.80	0.3387	0.6467	0.0146	179.7	2.328	2.575	1.106	0.995	1.027	1.179	−1640.	922.	12.0	6.818
0.90	0.90	0.1873	0.8045	0.0082	182.4	2.329	2.536	1.089	0.995	1.013	1.173	−1694.	882.	11.8	6.446
0.97	0.	0.9882	0.	0.0118	165.2	2.062	2.598	1.260	0.995	1.269	1.334	−1130.	1035.	14.3	8.541
0.97	0.01	0.9834	0.0050	0.0117	165.3	2.067	2.599	1.257	0.994	1.264	1.330	−1136.	1035.	14.3	8.527
0.97	0.02	0.9784	0.0099	0.0116	165.5	2.072	2.601	1.255	0.994	1.259	1.327	−1143.	1034.	14.2	8.512
0.97	0.03	0.9735	0.0149	0.0116	165.6	2.077	2.602	1.253	0.993	1.254	1.324	−1149.	1033.	14.2	8.497
0.97	0.04	0.9685	0.0200	0.0115	165.7	2.082	2.604	1.251	0.992	1.250	1.321	−1156.	1033.	14.2	8.482
0.97	0.05	0.9635	0.0250	0.0114	165.8	2.087	2.605	1.248	0.992	1.245	1.317	−1163.	1032.	14.2	8.467
0.97	0.07	0.9534	0.0353	0.0113	166.1	2.097	2.608	1.244	0.991	1.236	1.311	−1176.	1031.	14.1	8.437
0.97	0.10	0.9380	0.0509	0.0111	166.5	2.111	2.612	1.237	0.989	1.222	1.302	−1195.	1029.	14.0	8.391
0.97	0.10	0.9380	0.0509	0.0111	166.5	2.111	2.612	1.237	0.989	1.222	1.302	−1195.	1029.	14.0	8.391
0.97	0.20	0.8840	0.1056	0.0104	167.9	2.158	2.624	1.216	0.987	1.182	1.276	−1261.	1022.	13.7	8.233
0.97	0.40	0.7584	0.2327	0.0089	170.9	2.240	2.633	1.176	0.986	1.116	1.235	−1390.	1003.	13.1	7.877
0.97	0.60	0.5940	0.3990	0.0070	174.6	2.302	2.620	1.138	0.991	1.067	1.206	−1516.	974.	12.5	7.436
0.97	0.80	0.3597	0.6361	0.0043	179.3	2.332	2.574	1.104	0.996	1.030	1.185	−1635.	923.	12.0	6.846
0.97	0.90	0.2003	0.7972	0.0024	182.1	2.331	2.537	1.088	0.996	1.014	1.176	−1692.	883.	11.8	6.464
1.00	0.	1.000	0.	0.	164.7	2.057	2.580	1.254	0.997	1.280	1.352	−1115.	1032.	14.3	8.555
1.00	0.01	0.9951	0.0049	0.0000	164.9	2.061	2.581	1.252	0.997	1.275	1.349	−1122.	1031.	14.3	8.541
1.00	0.02	0.9902	0.0098	0.0000	165.0	2.066	2.583	1.250	0.996	1.270	1.346	−1129.	1031.	14.3	8.526

Liquid-Vapor Equilibrium Data for the Argon-Nitrogen-Oxygen System* (Continued)

Liquid mole fraction		Vapor mole fraction			Temperature, °R	Relative volatility			Pressure activity coefficient			Enthalpy, Btu/ (lb·mol)		Heat capacity, Btu/ (lb·mol·°R)	
N_2/N_2+O_2	Ar	N_2	Ar	O_2		N_2/Ar	N_2/O_2	Ar/O_2	N_2	Ar	O_2	Liquid	Vapor	Liquid	Vapor
1.00	0.03	0.9853	0.0147	0.0000	165.1	2.071	2.584	1.248	0.995	1.265	1.342	−1136.	1030.	14.2	8.511
1.00	0.04	0.9803	0.0197	0.0000	165.3	2.076	2.586	1.246	0.995	1.260	1.338	−1142.	1029.	14.2	8.496
1.00	0.05	0.9753	0.0247	0.0000	165.4	2.081	2.588	1.243	0.994	1.255	1.335	−1149.	1029.	14.2	8.481
1.00	0.07	0.9653	0.0347	0.0000	165.7	2.091	2.591	1.239	0.993	1.245	1.328	−1163.	1028.	14.1	8.451
1.00	0.10	0.9499	0.0501	0.0000	166.1	2.106	2.596	1.233	0.991	1.231	1.319	−1183.	1026.	14.0	8.405
1.00	0.20	0.8960	0.1040	0.0000	167.4	2.154	2.610	1.212	0.988	1.189	1.289	−1250.	1019.	13.7	8.247
1.00	0.40	0.7706	0.2294	0.0000	170.6	2.239	2.626	1.173	0.987	1.120	1.244	−1382.	1002.	13.1	7.892
1.00	0.60	0.6055	0.3945	0.	174.4	2.303	2.617	1.136	0.992	1.069	1.211	−1511.	973.	12.5	7.451
1.00	0.80	0.3684	0.6316	0.	179.1	2.333	2.574	1.103	0.996	1.030	1.188	−1633.	923.	12.0	6.858
1.00	0.90	0.2058	0.7942	0.	182.0	2.332	2.537	1.088	0.996	1.015	1.177	−1691.	884.	11.8	6.471

* Calculated values, from Wilson, Silverberg, and Zellner, USAF Aero Propulsion Laboratory. Rep. APL TDR 64-64 (AD 603 151), 1964. Relative volatility = α_{i-j} = $(y_i/x_i)(x_j/y_j)$, where x = liquid composition, y = vapor composition. Pressure activity coefficient = y_iP/x_ip^0, where p^0 = vapor pressure. These data were confirmed by the analyses of Bender, Cryogenics, 13, 11 (1973); and by Elshayal and Lu. J. Chem. Eng. Data, 16, 31 (1971). See also Armstrong, G. T., J. M. Goldstein, et al., J. Res. N.B.S., 55, 5 (1955): 265–277; Bender, E., Cryogenics, 13, 1 (1973): 11–18; Elshayal, I. M. and B. C-y Lu. J. Chem. Eng. Data, 16, 1 (1971): 31–37; Funada, I., S. Yoshimura, et al., Advan. Cryog. Engng., 7 (1982): 893–901; and Hwang, S-C., Fluid Phase Equila., 37 (1987): 153–167.

THERMODYNAMIC AND THERMOPHYSICAL PROPERTIES — TABLE 10. International Standard Atmosphere

Z, m	T, K	P, bar	p, kg/m³	g, m/s²	M	a, m/s	μ, Pa·s	k, W/(m·K)	λ, m	H, m
0	288.15	1.0135	1.2250	9.80665	28.964	340.29	1.79,-5	2.54,-5	6.63,-8	0
1,000	281.65	0.89876	1.1117	9.8036	28.964	336.43	1.76,-5	2.49,-5	7.31,-8	1,000
2,000	275.15	0.79501	1.0066	9.8005	28.964	332.53	1.73,-5	2.43,-5	8.07,-8	2,999
3,000	268.66	0.70121	0.90925	9.7974	28.964	328.58	1.69,-5	2.38,-5	8.94,-8	2,999
4,000	262.17	0.61660	0.81935	9.7943	28.964	324.59	1.66,-5	2.33,-5	9.92,-8	3,997
5,000	255.68	0.54048	0.73643	9.7912	28.964	320.55	1.63,-5	2.28,-5	1.10,-7	4,996
6,000	249.19	0.47217	0.66011	9.7882	28.964	316.45	1.59,-5	2.22,-5	1.23,-7	5,994
7,000	242.70	0.41105	0.59002	9.7851	28.964	312.31	1.56,-5	2.17,-5	1.38,-7	6,992
8,000	236.22	0.35651	0.52579	9.7820	28.964	308.11	1.53,-5	2.12,-5	1.55,-7	7,990
9,000	229.73	0.30800	0.46706	9.7789	28.964	303.85	1.49,-5	2.06,-5	1.74,-7	8,987
10,000	223.25	0.26499	0.41351	9.7759	28.964	299.53	1.46,-5	2.01,-5	1.97,-7	9,984
15,000	216.65	0.12111	0.19476	9.7605	28.964	295.07	1.42,-5	1.95,-5	4.17,-7	14,965
20,000	216.65	0.05529	0.08891	9.7452	28.964	295.07	1.42,-5	1.95,-5	9.14,-7	19,937
25,000	221.55	0.02549	0.04008	9.7300	28.964	298.39	1.45,-5	1.99,-5	2.03,-6	24,902
30,000	226.51	0.01197	0.01841	9.7147	28.964	301.71	1.48,-5	2.04,-5	4.42,-6	29,859
40,000	250.35	2.87,-3	4.00,-3	9.6844	28.964	317.19	1.60,-5	2.23,-5	2.03,-5	39,750
50,000	270.65	8.00,-4	1.03,-3	9.6542	28.964	329.80	1.70,-5	2.40,-5	7.91,-5	49,610
60,000	247.02	2.20,-4	3.10,-4	9.6241	28.964	315.07	1.58,-5	2.21,-5	2.62,-4	59,439
70,000	219.59	5.22,-5	8.28,-5	9.5942	28.964	297.06	1.44,-5	1.98,-5	9.81,-4	69,238
80,000	198.64	1.05,-5	1.85,-5	9.5644	28.964	282.54	1.32,-5	1.80,-5	4.40,-3	79,006
90,000	186.87	1.84,-6	3.43,-6	9.5348	28.95				2.37,-2	88,744
100,000	195.08	3.20,-7	5.60,-7	9.5052	28.40				0.142	98,451
150,000	634.39	4.54,-9	2.08,-9	9.3597	24.10				33	146,542
200,000	854.56	8.47,-10	2.54,-10	9.2175	21.30				240	193,899
250,000	941.33	2.48,-10	6.07,-11	9.0785	19.19				890	240,540
300,000	976.01	8.77,-11	1.92,-11	8.9427	17.73				2600	286,480
400,000	995.83	1.45,-11	2.80,-12	8.6799	15.98				1.6,+4	376,320
500,000	999.24	3.02,-12	5.22,-13	8.4286	14.33				7.7,+4	463,540
600,000	999.85	8.21,-13	1.14,-13	8.1880	11.51				2.8,+5	548,252
800,000	999.99	1.70,-13	1.14,-14	7.7368	5.54				1.4,+6	710,574
1,000,000	1,000.00	7.51,-14	3.56,-15	7.3218	3.94				3.1,+6	864,071

THERMODYNAMIC AND THERMOPHYSICAL PROPERTIES—TABLE 11. Saturated Benzene

T, K	p, bar	v_f, m³/kg	v_g, m³/kg	h_f, kJ/kg	h_g, kJ/kg	s_f, kJ/(kg·K)	s_g, kJ/(kg·K)	c_{pf}, kJ/(kg·K)	μ_f, 10^{-4} Pa·s	k_f, W/(m·K)
290	0.0860	1.133.–10⁻³	3.569.–10	371.1	810.3	2.172	3.686	1.719	6.75	0.147
300	0.1382	1.147.–10⁻³	2.292.–10	388.3	820.4	2.229	3.670	1.746	5.80	0.144
310	0.2139	1.162.–10⁻³	1.525.–10	405.9	830.8	2.286	3.657	1.774	5.14	0.141
320	0.3206	1.176.–10⁻³	1.046.–10	423.8	841.5	2.344	3.650	1.804	4.52	0.138
330	0.4665	1.192.–10⁻³	7.379.–10⁻¹	442.1	852.4	2.400	3.643	1.836	3.95	0.135
340	0.6615	1.207.–10⁻³	5.332.–10⁻¹	460.8	863.6	2.455	3.641	1.868	3.55	0.132
350	0.9162	1.224.–10⁻³	3.938.–10⁻¹	479.6	875.0	2.510	3.641	1.890	3.23	0.129
360	1.2419	1.241.–10⁻³	2.965.–10⁻¹	498.7	886.7	2.564	3.642	1.920	2.99	0.126
370	1.6517	1.259.–10⁻³	2.233.–10⁻¹	518.1	898.6	2.617	3.646	1.950	2.72	0.123
380	2.1588	1.277.–10⁻³	1.767.–10⁻¹	537.7	910.6	2.669	3.651	1.989	2.46	0.120
390	2.7774	1.297.–10⁻³	1.393.–10⁻¹	557.6	922.9	2.592	3.657	2.030	2.24	0.117
400	3.5228	1.318.–10⁻³	1.112.–10⁻¹	577.9	935.2	2.644	3.665	2.070	2.05	0.114
410	4.4091	1.340.–10⁻³	8.972.–10⁻²	598.6	947.8	2.823	3.674	2.110	1.88	0.111
420	5.4540	1.363.–10⁻³	7.309.–10⁻²	619.7	960.4	2.873	3.684	2.160	1.73	0.107
430	6.6739	1.388.–10⁻³	6.003.–10⁻²	641.3	973.0	2.924	3.695	2.210	1.60	0.104
440	8.0861	1.415.–10⁻³	4.965.–10⁻²	663.5	985.6	2.974	3.706	2.260	1.48	0.101
450	9.7088	1.444.–10⁻³	4.131.–10⁻²	686.3	998.2	3.025	3.718	2.320	1.37	0.098
460	11.451	1.475.–10⁻³	3.455.–10⁻²	709.7	1010.7	3.075	3.730	2.380	1.28	0.095
470	13.660	1.510.–10⁻³	2.901.–10⁻²	733.8	1022.9	3.126	3.742	2.450	1.10	0.092
480	16.028	1.548.–10⁻³	2.441.–10⁻²	758.6	1034.9	3.179	3.753	2.519	1.12	0.089
490	18.685	1.591.–10⁻³	2.059.–10⁻²	784.3	1046.4	3.230	3.765	2.590	1.05	0.086
500	21.651	1.640.–10⁻³	1.736.–10⁻²	810.9	1057.3	3.284	3.777	2.670	0.98	0.083
510	24.952	1.697.–10⁻³	1.462.–10⁻²	838.5	1067.5	3.336	3.785	2.750	0.91	
520	28.613	1.765.–10⁻³	1.226.–10⁻²	867.2	1076.6	3.391	3.794	2.839	0.84	
530	32.669	1.849.–10⁻³	1.020.–10⁻²	897.2	1084.3	3.446	3.800	2.941	0.77	
540	37.161	2.126.–10⁻³	8.349.–10⁻³	928.8	1089.5	3.504	3.802		0.70	
550	42.144	2.258.–10⁻³	6.616.–10⁻³	963.2	1090.4	3.565	3.797		0.65	
560	47.696	2.512.–10⁻³	4.696.–10⁻³	1007.3	1077.6	3.642	3.769		0.60	
562.2	48.979	3.290.–10⁻³	3.290.–10⁻³	1043.0	1043.0	3.706	3.706			

THERMODYNAMIC AND THERMOPHYSICAL PROPERTIES—TABLE 12. Saturated Bromine

T, K	P, bar	v_f, m³/kg	v_g, m³/kg	h_f, kJ/kg	h_g, kJ/kg	s_f, kJ/(kg·K)	s_g, kJ/(kg·K)	c_{pf}, kJ/(kg·K)	μ_f, 10^{-4} Pa·s	k_f, W/(m·K)
260	0.042	3.106.–4	3.195	−147.2	51.8	0.903	1.669	0.486	13.4	0.131
280	0.124	3.168.–4	1.169	−138.9	56.2	0.933	1.629	0.479	11.5	0.127
300	0.310	3.232.–4	0.5002	−131.6	60.6	0.956	1.597	0.475	9.3	0.122
320	0.680	3.311.–4	0.2425	−124.2	64.8	0.978	1.570	0.473	7.8	0.118
340	1.330	3.385.–4	0.1309	−112.3	71.1	1.004	1.539	0.471	6.7	0.114
360	2.384	3.464.–4	0.0767	−108.6	73.1	1.026	1.531	0.470	5.7	0.109
380	4.010	3.550.–4	0.0477	−100.6	76.9	1.048	1.515	0.471	5.0	0.104
400	6.390	3.647.–4	0.0311	−93.4	80.6	1.063	1.501	0.475	4.5	0.099
420	9.730	3.752.–4	0.0211	−85.8	84.0	1.084	1.488	0.480	4.0	0.094
440	14.25	3.885.–4	0.0148	−77.7	87.1	1.103	1.477	0.489	3.7	0.089
460	20.17	4.023.–4	0.0107	−69.0	89.9	1.122	1.467	0.503	3.3	0.084
480	27.75	4.179.–4	0.00786	−59.7	92.2	1.142	1.457	0.527	3.1	0.079
500	37.21	4.378.–4	0.00589	−49.3	94.0	1.161	1.448	0.595	2.8	0.073
520	48.81	4.623.–4	0.00445	−37.7	95.0	1.183	1.438	0.710	2.6	0.066
540	62.80	4.938.–4	0.00337	−24.0	94.8	1.207	1.428	0.860	2.5	0.059
560	79.41	5.368.–4	0.00251	−7.1	92.5	1.237	1.414	1.063	2.3	0.050
580	98.90	6.250.–4	0.00167	18.8	82.5	1.280	1.390	2.31	2.2	0.035
584.2	103.4	8.475.–4	0.00085	64.8	64.8	1.356	1.356	∞	2.1	∞

THERMODYNAMIC AND THERMOPHYSICAL PROPERTIES—TABLE 13. Saturated Normal Butane (R600)

T, K	P, bar	v_f, m³/kg	v_g, m³/kg	h_f, kJ/kg	h_g, kJ/kg	s_f, kJ/(kg·K)	s_g, kJ/(kg·K)	c_{pf}, kJ/(kg·K)	μ_f, 10^{-4} Pa·s	k_f, W/(m·K)
134.9	6.7.–6	1.360.–3	28630	0.00	494.21	2.3056	5.9702	1.946	15.8	0.81
140	1.7.–5	1.369.–3	11635	9.95	499.96	2.3778	5.8779	1.953	14.4	0.179
150	8.7.–5	1.387.–3	2470	29.44	511.39	2.5121	5.7251	1.970	12.0	0.175
160	3.5.–4	1.405.–3	654	49.10	523.13	2.6389	5.6016	1.985	9.94	0.171
170	1.17.–3	1.424.–3	207	68.94	535.16	2.7592	5.5017	2.001	8.26	0.167
180	3.37.–3	1.443.–3	76.4	88.97	547.48	2.8738	5.4211	2.018	6.87	0.163
190	8.53.–3	1.463.–3	31.8	109.22	560.07	2.9835	5.3564	2.035	5.71	0.160
200	1.94.–2	1.484.–3	14.7	129.71	572.93	3.0887	5.3048	2.055	4.83	0.156
210	4.05.–2	1.505.–3	7.39	150.45	586.06	3.1900	5.2643	2.077	4.15	0.152
220	7.81.–2	1.528.–3	4.00	171.49	599.42	3.2879	5.2331	2.101	3.61	0.148
230	0.1411	1.551.–3	2.31	192.83	613.02	3.3828	5.2097	2.128	3.18	0.144
240	0.2408	1.575.–3	1.40	214.50	626.83	3.4749	5.1929	2.158	2.83	0.140
250	0.3915	1.601.–3	0.893	236.52	640.82	3.5647	5.1818	2.192	2.55	0.136
260	0.6100	1.628.–3	0.592	258.92	654.97	3.6523	5.1755	2.231	2.31	0.132
270	0.9155	1.656.–3	0.406	281.72	669.24	3.7380	5.1732	2.274	2.10	0.128
280	1.3297	1.686.–3	0.286	309.94	683.60	3.8220	5.1744	2.323	1.93	0.124
290	1.8765	1.718.–3	0.207	328.62	697.99	3.9046	5.1783	2.377	1.77	0.120
300	2.5811	1.752.–3	0.1533	352.77	712.36	3.9860	5.1846	2.437	1.62	0.116
310	3.4706	1.790.–3	0.1156	377.46	726.67	4.0663	5.1928	2.503	1.47	0.113
320	4.5731	1.830.–3	0.0885	402.71	740.84	4.1458	5.2025	2.577	1.34	0.109
330	5.9179	1.874.–3	0.0687	428.61	754.80	4.2248	5.2132	2.657	1.21	0.105
340	7.5354	1.923.–3	0.0539	455.25	768.49	4.3035	5.2248	2.746	1.08	0.101
350	9.4573	1.978.–3	0.0427	482.74	781.79	4.3822	5.2367	2.842	0.97	0.097
360	11.72	2.041.–3	0.0340	511.22	794.60	4.4613	5.2485	2.947	0.87	0.093
370	14.35	2.114.–3	0.0272	540.88	806.72	4.5412	5.2597	3.062	0.78	0.089
380	17.40	2.200.–3	0.0218	571.94	817.86	4.6225	5.2696	3.20	0.69	0.085
390	20.90	2.307.–3	0.0174	604.76	827.56	4.7058	5.2771	3.34	0.62	0.081
400	24.92	2.447.–3	0.0138	639.85	834.95	4.7922	5.2800	3.50	0.55	0.077
410	29.54	2.652.–3	0.0106	678.30	838.10	4.8842	5.2740	3.69	0.49	0.074
420	34.86	3.048.–3	0.0075	723.89	830.34	4.9903	5.2437	3.84	0.44	0.072
425.2	37.96	4.405.–3	0.0044	783.50	783.50	5.1290	5.1290	∞		∞

THERMODYNAMIC AND THERMOPHYSICAL PROPERTIES—TABLE 14. Superheated Normal Butane

		Temperature, K									
P, bar		150	200	250	300	350	400	450	500	600	700
1.013	v	0.00139	0.00148	0.00160	0.4106	0.4847	0.5575	0.6297	0.7013	0.8440	0.9861
	h	29.6	129.8	236.6	718.9	810.7	913.1	1026.0	1149.0	1423	1730
	s	2.512	3.088	3.564	5.334	5.616	5.889	6.155	6.414	6.913	7.386
5	v	0.00139	0.00148	0.00160	0.00175	0.0909	0.1078	0.1238	0.1393	0.1693	0.1988
	h	30.0	130.2	237.0	352.9	798.5	904.3	1019.3	1143.7	1420	1728
	s	2.511	3.088	3.563	3.985	5.363	5.645	5.916	6.178	6.680	7.155
10	v	0.00139	0.00148	0.00160	0.00175	0.00198	0.0502	0.0593	0.0677	0.0835	0.0987
	h	30.6	130.8	237.4	353.3	482.7	891.9	1010.3	1136.8	1415	1725
	s	2.510	3.087	3.562	3.983	4.382	5.524	5.803	6.069	6.575	7.052
20	v	0.00138	0.00148	0.00160	0.00174	0.00196	0.0205	0.0268	0.0318	0.0406	0.0487
	h	31.7	131.8	238.4	354.0	482.6	860.0	990.1	1122.0	1406	1718
	s	2.509	3.085	3.560	3.980	4.376	5.364	5.670	5.948	6.464	6.945
30	v	0.00138	0.00148	0.00159	0.00174	0.00195	0.00240	0.0156	0.0198	0.0263	0.0320
	h	32.8	132.9	239.3	354.7	482.6	637.3	965.5	1105.9	1396	1711
	s	2.507	3.082	3.557	3.976	4.370	4.783	5.570	5.866	6.394	6.880
40	v	0.00138	0.00148	0.00159	0.00173	0.00194	0.00234	0.0097	0.0137	0.0192	0.0237
	h	33.9	134.0	240.3	355.4	482.7	633.6	932.2	1088.1	1387	1705
	s	2.505	3.080	3.555	3.973	4.365	4.768	5.468	5.797	6.341	6.832
50	v	0.00138	0.00148	0.00159	0.00173	0.00193	0.00229	0.00549	0.0101	0.0149	0.0188
	h	35.0	135.0	241.3	356.2	428.8	631.0	877.0	1068.2	1377	1699
	s	2.503	3.078	3.552	3.970	4.360	4.755	5.329	5.734	6.297	6.792
60	v	0.00138	0.00148	0.00159	0.00172	0.00192	0.00255	0.00352	0.00764	0.0121	0.0155
	h	36.2	136.1	242.3	356.9	483.1	629.1	825.1	1046.4	1367	1692
	s	2.501	3.076	3.550	3.967	4.355	4.745	5.204	5.673	6.258	6.759
80	v	0.00138	0.00147	0.00158	0.00172	0.00190	0.00219	0.00286	0.00482	0.00868	0.0114
	h	38.4	138.3	244.2	358.5	483.7	626.5	798.1	1001.5	1347	1680
	s	2.498	3.072	3.545	3.960	4.346	4.727	5.130	5.559	6.191	6.704
100	v	0.00138	0.00147	0.00158	0.00171	0.00188	0.00214	0.00264	0.00368	0.00669	0.00901
	h	40.6	140.4	246.2	360.1	484.5	624.9	787.9	971.3	1329	1668
	s	2.495	3.069	3.540	3.954	4.337	4.712	5.095	5.310	6.134	6.658
200	v	0.00137	0.00146	0.00156	0.00167	0.00178	0.00200	0.00225	0.00258	0.00349	0.00460
	h	51.9	151.3	257.9	368.8	490.3	624.4	773.3	933.7	1270	1623
	s	2.478	3.049	3.518	3.927	4.301	4.660	5.010	5.348	5.960	6.849
300	v	0.00136	0.00145	0.00154	0.00164	0.00176	0.00191	0.00209	0.00231	0.00284	0.00345
	h	63.2	162.2	266.7	378.3	498.0	629.2	773.4	928.4	1255	1603
	s	2.462	3.032	3.498	3.903	4.273	4.623	4.962	5.288	5.884	6.419
400	v	0.00136	0.00144	0.00152	0.00162	0.00173	0.00185	0.00200	0.00217	0.00255	0.00298
	h	74.5	173.3	277.4	388.2	506.8	636.2	778.0	930.2	1253	1600
	s	2.447	3.015	3.479	3.882	4.248	4.593	4.927	5.247	5.836	6.366
500	v	0.00136	0.00143	0.00151	0.00160	0.00170	0.00181	0.00193	0.00207	0.00240	0.00272
	h	85.8	184.4	288.1	398.4	516.3	644.5	784.8	935.3	1256	1599
	s	2.432	2.999	3.461	3.863	4.226	4.569	4.898	5.215	5.799	6.328

THERMODYNAMIC AND THERMOPHYSICAL PROPERTIES—TABLE 15. Saturated Carbon Dioxide

T, K	P, bar	v_f, m³/kg	v_g, m³/kg	h_f, kJ/kg	h_g, kJ/kg	s_f, kJ/(kg·K)	s_g, kJ/(kg·K)	c_{pf}, kJ/(kg·K)	μ_f, 10⁻⁴ Pa·s	k_f, W/(m·K)
216.6	5.180	8.484.−4	0.0712	386.3	731.5	2.656	4.250	1.707		0.182
220	5.996	8.574.−4	0.0624	392.6	733.1	2.684	4.232	1.761		0.178
225	7.357	8.710.−4	0.0515	401.8	735.1	2.723	4.204			0.171
230	8.935	8.856.−4	0.0428	411.1	736.7	2.763	4.178	1.879	1.64	0.164
235	10.75	9.011.−4	0.0357	420.5	737.9	2.802	4.152			0.160
240	12.83	9.178.−4	0.0300	430.2	738.9	2.842	4.128	1.933	1.45	0.156
245	15.19	9.358.−4	0.0253	440.1	739.4	2.882	4.103			0.148
250	17.86	9.554.−4	0.0214	450.3	739.6	2.923	4.079	1.992	1.28	0.140
255	20.85	9.768.−4	0.0182	460.8	739.4	2.964	4.056			0.134
260	24.19	1.000.−3	0.0155	471.6	738.7	3.005	4.032	2.125	1.14	0.128
270	32.03	1.056.−3	0.0113	494.4	735.6	3.089	3.981	2.410	1.02	0.116
275	36.59	1.091.−3	0.0097	506.5	732.8	3.132	3.954	2.887	0.91	0.109
280	41.60	1.130.−3	0.0082	519.2	729.1	3.176	3.925	3.724	0.79	0.102
290	53.15	1.241.−3	0.0058	547.6	716.9	3.271	3.854		0.60	0.088
300	67.10	1.470.−3	0.0037	585.4	690.2	3.393	3.742			0.074
304.2ᶜ	73.83	2.145.−3	0.0021	636.6	636.6	3.558	3.558	∞	0.31	∞

* c = critical point. The notation 8.484.−4 signifies 8.484×10^{-4}.

424

THERMODYNAMIC AND THERMOPHYSICAL PROPERTIES—TABLE 16. Superheated Carbon Dioxide

					Temperature, K					
P, bar	300	350	400	450	500	600	700	800	900	1000
v	0.5639	0.6595	0.7543	0.8494	0.9439	1.1333	1.3324	1.5115	1.7005	1.8894
1 h	809.3	853.1	899.1	947.1	997.0	1102	1212	1327	1445	1567
s	4.860	4.996	5.118	5.231	5.337	5.527	5.697	5.850	5.990	6.120
v	0.1106	0.1304	0.1498	0.1691	0.1882	0.2264	0.2645	0.3024	0.3403	0.3782
5 h	805.5	850.3	897.0	945.5	995.8	1101	1211	1326	1445	1567
s	4.548	4.686	4.810	4.925	50.31	5.222	5.392	5.546	5.685	5.814
v	0.0539	0.0642	0.0742	0.0841	0.0938	0.1131	0.1322	0.1513	0.1703	0.1893
10 h	800.7	846.9	894.4	943.5	994.1	1100	1211	1326	1445	1567
s	4.405	4.548	6.674	4.790	4.897	5.089	5.260	5.414	5.555	5.683
v	0.0255	0.0311	0.0364	0.0416	0.0466	0.0564	0.0661	0.0757	0.0853	0.0948
20 h	790.2	839.8	889.3	939.4	990.8	1098	1209	1325	1444	1567
s	4.249	4.402	4.534	4.653	4.762	4.955	5.127	5.282	5.423	5.551
v	0.0159	0.0201	0.0238	0.0274	0.0309	0.0375	0.0441	0.0505	0.0570	0.0633
30 h	778.5	832.4	883.8	935.2	987.3	1096	1208	1324	1444	1566
s	4.144	4.341	4.447	4.569	4.679	4.876	5.049	5.204	5.346	5.474
v	0.0110	0.0146	0.0175	0.0203	0.0230	0.0281	0.0331	0.0379	0.0428	0.0476
40 h	764.9	824.6	878.3	931.1	984.3	1094	1205	1323	1443	1566
s	4.055	4.239	4.380	4.507	4.619	4.818	4.993	5.148	5.291	5.419
v	0.0080	0.0112	0.0138	0.0161	0.0183	0.0224	0.0265	0.0304	0.0343	0.0382
50 h	748.2	816.3	872.6	926.9	981.1	1091	1205	1322	1443	1566
s	3.968	4.179	4.330	4.457	4.572	4.773	4.948	5.104	5.247	5.377
v	0.0058	0.0090	0.0113	0.0133	0.0151	0.0187	0.0221	0.0254	0.0286	0.0318
60 h	726.9	807.7	866.9	922.7	977.8	1089	1204	1321	1442	1565
s	3.878	4.126	4.314	4.416	4.532	4.736	4.912	5.069	5.212	5.341
v		0.0062	0.0081	0.0097	0.0112	0.0140	0.0166	0.0191	0.0216	0.0240
80 h		788.4	855.1	914.2	971.3	1085	1201	1320	1441	1565
s		4.029	4.208	4.347	4.468	4.675	4.854	5.011	5.155	5.286
v		0.0045	0.0062	0.0076	0.0089	0.0111	0.0133	0.0153	0.0173	0.0193
100 h		766.2	843.0	905.7	964.9	1081	1198	1318	1440	1564
s		3.936	4.144	4.290	4.417	4.627	4.808	4.967	5.111	5.241
v		0.0023	0.0038	0.0049	0.0058	0.0074	0.0089	0.0103	0.0117	0.0130
150 h		704.5	811.9	884.8	949.4	1072	1192	1314	1437	1562
s		3.716	4.005	4.177	4.313	4.536	4.722	4.884	5.030	5.162
v		0.0017	0.0027	0.0035	0.0043	0.0056	0.0067	0.0078	0.0088	0.0099
200 h		670.0	783.2	865.2	934.9	1063	1186	1310	1435	1561
s		3.591	3.894	4.088	4.234	4.468	4.668	4.824	4.970	5.104
v			0.0017	0.0023	0.0029	0.0038	0.0046	0.0053	0.0060	0.0067
300 h			745.3	834.0	910.6	1047	1176	1303	1431	1559
s			3.747	3.956	4.118	4.367	4.573	4.743	4.886	5.021
v			0.0015	0.0018	0.0022	0.0029	0.0035	0.0041	0.0047	0.0052
400 h			728.1	814.6	893.3	1035	1168	1298	1428	1558
s			3.663	3.867	4.033	4.292	4.497	4.671	4.824	4.960
v				0.0016	0.0018	0.0024	0.0029	0.0034	0.0038	0.0043
500 h				803.5	881.9	1027	1162	1294	1426	1557
s				3.805	3.970	4.234	4.443	4.620	4.774	4.913

THERMODYNAMIC AND THERMOPHYSICAL PROPERTIES—TABLE 17.   Saturated Carbon Monoxide

T, K	P, bar	v_f, m³/kg	v_g, m³/kg	h_f, kJ/kg	h_g, kJ/kg	s_f, kJ/(kg·K)	s_g, kJ/(kg·K)
81.62	1.01	1.268.−3	0.0666	150.25	365.30	3.005	5.640
83.36	1.52	1.295.−3	0.0631	158.56	368.07	3.104	5.559
88.25	2.03	1.317.−3	0.0606	165.00	370.00	3.178	5.501
96.06	4.05	1.385.−3	0.0547	182.76	374.21	3.368	5.359
101.51	6.08	1.440.−3	0.0513	195.0	375.98	3.489	5.271
105.69	8.12	1.489.−3	0.0318	204.8	376.6	3.580	5.206
109.17	10.13	1.535.−3	0.0253	213.2	376.6	3.656	5.152
116.08	15.20	1.651.−3	0.0163	231.0	374.5	3.807	5.043
121.48	20.27	1.778.−3	0.0116	246.3	370.2	3.918	4.948
125.97	25.33	1.936.−3	0.0085	261.2	363.6	4.041	4.854
129.84	30.40	2.168.−3	0.0063	277.6	313.15	4.161	4.747
132.91	34.96	3.337.−3	0.0033				

THERMODYNAMIC AND THERMOPHYSICAL PROPERTIES—TABLE 18. Saturated Carbon Tetrachloride

T, K	P, bar	v_f, m³/kg	v_g, m³/kg	h_f, kJ/kg	h_g, kJ/kg	s_f, kJ/(kg·K)	s_g, kJ/(kg·K)	c_{pf}, kJ/(kg·K)	μ_f, 10^{-6} Pa·s	k_f, W/(m·K)	Pr
280	0.064	0.000619	2.414	205.5	420.7	1.018	1.787	0.835	1042	0.1043	8.34
290	0.105	0.000625	1.495	212.9	425.7	1.042	1.775	0.844	892	0.1020	7.38
300	0.165	0.000633	0.971	220.9	430.9	1.068	1.768	0.853	774	0.0998	6.62
310	0.251	0.000641	0.669	228.8	436.1	1.095	1.764	0.863	679	0.0975	6.01
320	0.370	0.000649	0.463	236.9	441.3	1.121	1.760	0.874	603	0.0952	5.54
330	0.531	0.000657	0.3306	246.0	446.4	1.149	1.756	0.885	539	0.0930	5.13
340	0.743	0.000666	0.2407	254.5	451.5	1.174	1.754	0.897	486	0.0907	4.81
350	1.017	0.000674	0.1802	263.1	456.6	1.199	1.752	0.910	441	0.0884	4.54
360	1.361	0.000684	0.1370	271.8	461.7	1.224	1.751	0.924	402	0.0861	4.31
370	1.795	0.000694	0.1053	280.8	466.6	1.248	1.751	0.939	368	0.0839	4.12
380	2.327	0.000704	0.0820	289.7	471.5	1.272	1.750	0.954	338	0.0816	3.95
390	2.970	0.000715	0.0651	298.1	475.8	1.295	1.751	0.970	311	0.0794	3.80
400	3.735	0.000727	0.0525	307.9	481.2	1.319	1.752	0.987	287	0.0771	3.67
410	4.642	0.000739	0.0426	317.1	485.8	1.341	1.753	1.010	265	0.0749	3.57
420	5.700	0.000753	0.0350	326.0	490.4	1.363	1.754	1.034	246	0.0726	3.50
430	6.927	0.000766	0.02899	335.2	494.9	1.384	1.756	1.060	227	0.0704	3.42
440	8.342	0.000780	0.02413	344.3	499.2	1.405	1.757	1.094	211	0.0682	3.38
450	9.958	0.000796	0.02020	353.6	503.4	1.426	1.759	1.141	195	0.0660	3.37
460	11.792	0.000801	0.01692	363.1	507.3	1.446	1.760	1.207	180	0.0638	3.36
470	13.869	0.000834	0.01425	372.8	511.1	1.467	1.761	1.240	167	0.0666	3.36
480	16.21	0.000856	0.01205	382.6	514.6	1.487	1.762	1.278	156	0.0594	3.36
490	18.83	0.000880	0.01011	392.0	517.5	1.507	1.763	1.320	145	0.0511	3.35
500	21.77	0.000858	0.00858	402.5	520.2	1.526	1.762	1.375	133	0.0549	3.35
510	25.02	0.000945	0.00722	412.9	522.6	1.546	1.761	1.44			
520	28.68	0.000987	0.00607	424.3	524.2	1.568	1.760	1.52			
530	32.71	0.001041	0.00500	436.4	524.5	1.590	1.756				
540	37.18	0.001121	0.00400	448.3	522.7	1.614	1.749				
550	44.12	0.001248	0.00309	463.4	518.2	1.638	1.738				
556.4^c	45.60	0.001792	0.00179	494.4	494.4	1.692	1.692				

c = critical point. Base points: h_f = 200 at 273.15 K = 0°C = h_A − 300 kJ/kg; s_f = 1.000 at 273.15 K = 0°C = s_A − 4.000 kJ/(kg·K).

THERMODYNAMIC AND THERMOPHYSICAL PROPERTIES—TABLE 19. Saturated Carbon Tetrafluoride (R14)

T, K	P, bar	v_f, m³/kg	v_g, m³/kg	h_f, kJ/kg	h_g, kJ/kg	s_f, kJ/(kg·K)	s_g, kJ/(kg·K)	c_{pf}, kJ/(kg·K)	μ_f, 10^{-4} Pa·s	k_f, W/(m·K)
100	0.0089	5.370.–4	10.77	495.8	648.4	5.487	7.003	0.887		0.136
110	0.0286	5.515.–4	3.648	502.7	652.9	5.556	6.919	0.887		0.128
120	0.0924	5.668.–4	1.228	510.4	657.1	5.624	6.847	0.890		0.119
130	0.2986	5.834.–4	0.4051	518.8	661.1	5.691	6.786	0.896		0.111
140	0.6901	6.018.–4	0.1855	527.7	664.8	5.757	6.736	0.904	3.56	0.104
150	1.4074	6.225.–4	0.0951	537.2	668.3	5.822	6.696	0.922	3.28	0.097
160	2.598	6.460.–4	0.0532	549.4	671.4	5.885	6.662	0.975	3.03	0.089
170	4.426	6.733.–4	0.0318	557.6	674.0	5.947	6.629	1.031	2.80	0.081
180	7.067	7.055.–4	0.0200	568.2	676.1	6.007	6.607	1.104	2.59	0.072
190	10.702	7.449.–4	0.0131	579.3	677.4	6.066	6.583	1.203	2.39	0.064
200	15.531	7.957.–4	0.0087	591.0	677.8	6.124	6.558	1.334	2.19	0.057
210	21.794	8.674.–4	0.0058	603.5	676.4	6.182	6.536	1.506	2.01	0.049
220	29.269	9.931.–4	0.0036	618.5	671.4	6.233	6.490	1.73	1.85	0.042
227.5	37.45	1.598.–3	0.0016	646.9	646.9	6.371	6.371	∞		∞

THERMODYNAMIC AND THERMOPHYSICAL PROPERTIES—TABLE 20. Saturated Cesium

T, K	P, bar	v_f, m³/kg	v_g, m³/kg	h_f, kJ/kg	h_g, kJ/kg	s_f, kJ/(kg·K)	s_g, kJ/(kg·K)	c_{pf}, kJ/(kg·K)
301.6	2.66.−9	5.444.−4	7.01.+7	74.6	637.6	0.696	2.563	0.245
400	3.83.−6	5.615.−4	6.54.+4	98.5	651.9	0.765	2.148	0.240
500	3.11.−4	5.800.−4	1001	122.0	666.1	0.817	1.905	0.232
600	5.65.−3	5.999.−4	65.63	144.9	678.4	0.859	1.748	0.224
700	0.0440	6.215.−4	9.671	167.0	688.9	0.893	1.638	0.219
800	0.2029	6.443.−4	2.353	188.7	698.3	0.922	1.559	0.217
900	0.6620	6.689.−4	0.796	210.6	707.3	0.975	1.500	0.222
1000	1.693	6.954.−4	0.335	233.2	716.4	0.972	1.455	0.231
1200	6.790	7.628.−4	0.097	281.1	736.1	1.015	1.394	0.248
1500	27.6	8.84.−4	0.029	358.8	772.2	1.072	1.345	0.275

THERMODYNAMIC AND THERMOPHYSICAL PROPERTIES—TABLE 21. Saturated Chlorine

T, °C	P, bar	v_f, m³/kg	v_g, m³/kg	h_f, kJ/kg	h_g, kJ/kg	s_f, kJ/(kg·K)	s_g, kJ/(kg·K)
−50	0.475	0.000623	0.5448	221.5	518.2	1.7650	3.0946
−40	0.773	0.000634	0.3481	231.0	522.2	1.8074	3.0562
−30	1.203	0.000645	0.2314	240.6	526.1	1.8480	3.0223
−20	1.802	0.000656	0.1593	250.3	529.9	1.8869	2.9921
−10	2.608	0.000668	0.1134	260.0	533.9	1.9243	2.9649
0	3.664	0.000681	0.0829	269.7	537.4	1.9604	2.9402
10	5.014	0.000695	0.0619	279.4	540.5	1.9953	2.9177
20	6.702	0.000710	0.0471	289.2	543.3	2.0291	2.8924
30	8.774	0.000726	0.0364	299.0	545.7	2.0622	2.8777
40	11.27	0.000744	0.0286	308.8	548.0	2.0946	2.8593
50	14.25	0.000763	0.02276	318.6	549.8	2.1264	2.8417
60	17.76	0.000784	0.01827	329.1	551.2	2.1578	2.8245
70	21.85	0.000808	0.01481	340.0	552.1	2.1892	2.8074
80	26.65	0.000834	0.01202	351.4	552.5	2.2207	2.7900
90	32.17	0.000865	0.00972	364.1	552.4	2.2528	2.7714
100	38.44	0.000901	0.00789	377.8	551.0	2.2860	2.7502
110	45.54	0.000956	0.00639	391.3	548.8	2.3207	2.7317
120	53.57	0.001016	0.00508	407.1	543.7	2.3590	2.7064
130	62.68	0.001121	0.00392	426.1	535.0	2.4032	2.6733
140	72.84	0.001335	0.00282	451.1	517.3	2.4595	2.6198
144^c	77.10	0.00177	0.00177	483.1	483.1	2.5365	2.5365

c = critical point.

c_{pf}, kJ/(kg·K)	c_{pg}, kJ/(kg·K)	μ_f, 10^{-6} Pa·s	μ_g, 10^{-6} Pa·s	k_f, W/(m·K)	k_g, W/(m·K)	Pr_f	Pr_g
0.9454	0.476	565	10.3	0.1684	0.0061	3.17	0.809
0.9474	0.484	520	10.8	0.1650	0.0065	2.99	0.815
0.9496	0.497	483	11.4	0.1613	0.0069	2.85	0.820
0.9520	0.513	452	11.9	0.1573	0.0074	2.74	0.826
0.9547	0.532	422	12.4	0.1527	0.0078	2.64	0.841
0.9579	0.554	393	13.0	0.1478	0.0083	2.55	0.864
0.9618	0.579	368	13.5	0.1427	0.0088	2.48	0.888
0.9667	0.607	348	14.1	0.1378	0.0093	2.45	0.918
0.9728	0.638	333	14.7	0.1327	0.0099	2.44	0.950
0.9816	0.674	318	15.2	0.1282	0.0104	2.43	0.985
0.9968	0.720	304	15.8	0.1230	0.0110	2.46	1.034
1.022	0.786	290	16.4	0.1171	0.0117	2.53	1.107
1.054	0.885	278	17.1	0.1122	0.0126	2.61	1.201
1.124	1.017	267	17.9	0.1050	0.0137	2.85	1.331
1.253	1.205	256	18.7	0.0986	0.0149	3.26	1.510
1.418	1.434	247	19.5	0.0916	0.0163	3.82	1.700
1.632	1.696	238	20.6	0.0850	0.0178	4.57	1.96
1.891	1.960	230	22.2	0.0775	0.0195	5.61	2.23

THERMODYNAMIC AND THERMOPHYSICAL PROPERTIES—TABLE 22. Saturated Chloroform (R20)

T, K	P, bar	v_f, m³/kg	v_g, m³/kg	h_f, kJ/kg	h_g, kJ/kg	s_f, kJ/(kg·K)	s_g, kJ/(kg·K)	c_{pf}, kJ/(kg·K)	μ_f, 10⁻⁶ Pa·s	k_f, W/(m·K)	Pr_f
280	0.115	0.000660	1.689	−46.0	219.5	−0.165	0.798		748	0.120	
300	0.293	0.000678	0.714	−32.6	230.6	−0.105	0.773		587	0.114	
320	0.620	0.000695	0.358	−13.4	241.1	−0.041	0.754		468	0.109	
340	1.224	0.000715	0.190	5.2	252.1	0.015	0.741		381	0.103	
360	2.255	0.000739	0.107	23.3	263.0	0.065	0.731		319	0.095	
380	3.830	0.000765	0.0653	41.7	273.7	0.114	0.725	1.03	273	0.0921	3.35
400	6.039	0.000795	0.0425	61.4	284.2	0.165	0.722	1.07	237	0.0863	3.17
420	9.058	0.000822	0.0288	82.8	294.2	0.217	0.721	1.11	206	0.0808	3.04
440	13.39	0.000871	0.0195	106.1	303.6	0.270	0.719	1.15	177	0.0750	2.93
460	18.80	0.000921	0.0137	131.6	311.2	0.325	0.716	1.21	155	0.0694	2.86
480	26.00	0.000980	0.00962	157.4	316.5	0.380	0.711	1.32	129.6	0.0641	2.95
500	34.66	0.001059	0.00673	186.2	320.8	0.436	0.706	1.43	105.5	0.0584	2.89
520	44.68	0.001193	0.00467	219.6	321.3	0.499	0.694	1.59	81.2	0.0518	2.87
530	50.44	0.001328	0.00359	242.7	315.7	0.540	0.678		67.7	0.0461	
536.6^c	54.72	0.00200	0.00200	284.1	284.1	0.602	0.602				

c = critical point, $h_f = s_f = 0$ at n.b.p., 334.5 K.

THERMODYNAMIC AND THERMOPHYSICAL PROPERTIES—TABLE 23. Saturated Decane

T, K	P, bar	v_f, m³/kg	v_g, m³/kg	h_f, kJ/kg	h_g, kJ/kg	s_f, kJ/(kg·K)	s_g, kJ/(kg·K)	c_{pf}, kJ/(kg·K)	μ_f, 10^{-4} Pa·s	k_f, W/(m·K)
243.5	0.00001	1.319.–3	20,750	418.1	812.5	2.561	4.092	2.119	25.0	0.149
260	0.00006	1.334.–3	3,300.	452.7	836.3	2.699	4.120	2.109	16.6	0.144
280	0.00042	1.356.–3	443.	495.3	866.9	2.856	4.158	2.155	11.3	0.139
300	0.00197	1.381.–3	88.74	539.0	899.2	3.007	4.200	2.217	8.2	0.134
320	0.00720	1.410.–3	22.73	584.0	933.2	3.153	4.246	2.286	6.5	0.129
340	0.02155	1.442.–3	8.883	631.1	968.9	3.303	4.296		5.2	0.124
360	0.05522	1.478.–3	3.763	680.1	1,006.2	3.443	4.350		4.16	0.119
380	0.1248	1.515.–3	1.750	730.7	1,045.0	3.581	4.408		3.52	0.116
400	0.2549	1.552.–3	0.892	782.0	1,085.0	3.712	4.469		2.98	0.110
420	0.4789	1.591.–3	0.490	835.6	1,126.2	3.842	4.534		2.54	
440	0.8387	1.632.–3	0.290	889.6	1,168.4	3.968	4.602		2.23	
447.3	1.0133	1.650.–3	0.243	909.4	1,184.0	4.014	4.627		2.09	
460	1.3852	1.682.–3	0.178	944.5	1,211.4	4.089	4.670			
480	2.1745	1.735.–3	0.115	1,002.6	1,255.2	4.213	4.739			
500	3.2690	1.797.–3	0.0759	1,062.7	1,299.4	4.335	4.808			
520	4.733	1.868.–3	0.0525	1,124.5	1,344.4	4.456	4.879			
540	6.633	1.952.–3	0.0369	1,190.1	1,389.5	4.573	4.949			
560	9.062	2.067.–3	0.0248	1,256.1	1,432.2	4.698	5.011			
580	12.16	2.255.–3	0.0154	1,318.5	1,468.1	4.802	5.060			
600	16.12	2.588.–3	0.0093	1,384.5	1,495.6	4.913	5.098			
617.5	20.97	4.238.–3	0.0042	1,483.2	1,483.2	5.073	5.073			

THERMODYNAMIC AND THERMOPHYSICAL PROPERTIES—TABLE 24.   Saturated
Normal Deuterium

T, K	P, bar	v_f, m³/kg	v_g, m³/kg	h_f, kJ/kg	h_g, kJ/kg	s_f, kJ/(kg·K)	s_g, kJ/(kg·K)
18.71	0.1709	0.005752	2.232	−161.1	158.6	4.54	21.62
19	0.1944	0.005771	1.988	−160.0	159.1	4.68	21.48
20	0.2944	0.005840	1.365	−152.8	163.9	4.97	20.81
21	0.4297	0.005914	0.968	−145.9	167.6	5.30	20.23
22	0.6072	0.005993	0.705	−138.7	170.6	5.63	19.69
23	0.8344	0.00608	0.5256	−131.4	173.0	5.95	19.18
24	1.1192	0.00617	0.3995	−123.8	174.6	6.26	18.70
25	1.4694	0.00627	0.3088	−116.1	175.5	6.57	18.23
26	1.8932	0.00638	0.2421	−108.2	175.7	6.87	17.79
27	2.3989	0.00650	0.1921	−100.2	175.1	7.16	17.36
28	2.995	0.00663	0.1540	−92.0	173.8	7.44	16.94
29	3.690	0.00678	0.1246	−83.6	171.7	7.72	16.52
30	4.493	0.00694	0.1015	−74.9	168.7	8.00	16.12
31	5.412	0.00713	0.0831	−65.9	165.0	8.27	15.72
32	6.457	0.00735	0.0683	−56.5	160.3	8.54	15.32
33	7.455	0.00761	0.0563	−46.4	154.7	8.83	14.92
34	8.962	0.00793	0.0465	−35.5	148.0	9.12	14.52
35	10.44	0.00834	0.0382	−23.2	140.0	9.45	14.11
36	12.09	0.00890	0.0311	−8.6	130.1	9.82	13.67
37	13.91	0.00976	0.0249	10.0	117.1	10.28	13.17
38	15.92	0.01158	0.0185	39.7	95.0	11.01	12.47
38.34	16.65	0.01433	0.0143	69.2	69.2	11.76	11.76

THERMODYNAMIC AND THERMOPHYSICAL PROPERTIES—TABLE 25. Saturated Deuterium Oxide

T, K	P, bar	v_f, m³/kg	v_g, m³/kg	h_f, kJ/kg	h_g, kJ/kg	s_f, kJ/(kg·K)	s_g, kJ/(kg·K)
277.0	0.00668	9.047.–4	172.2	0.0	2320.9	0.000	8.380
278.2	0.00720	9.045.–4	160.4	5.0	2322.5	0.0188	8.351
283.2	0.01030	9.042.–4	114.1	25.9	2330.9	0.0920	8.233
288.2	0.01449	9.043.–4	82.48	46.9	2339.3	0.166	8.122
293.2	0.02011	9.047.–4	60.45	67.8	2347.6	0.239	8.016
298.2	0.02758	9.054.–4	44.88	88.7	2356.0	0.311	7.915
303.2	0.03730	9.063.–4	33.71	109.6	2364.0	0.382	7.818
308.2	0.04990	9.075.–4	25.59	130.5	2372.3	0.450	7.725
313.2	0.06598	9.091.–4	19.66	151.5	2380.7	0.518	7.637
318.2	0.08638	9.108.–4	15.24	172.4	2388.6	0.585	7.550
323.2	0.1120	9.127.–4	11.93	193.3	2396.6	0.650	7.468
333.2	0.1831	9.170.–4	7.52	234.7	2413.3	0.776	7.315
353.2	0.4439	9.274.–4	3.27	318.4	2445.1	1.020	7.042
373.2	0.9646	9.403.–4	1.58	402.0	2474.8	1.253	6.807
398.2	2.2427	9.599.–4	0.72	507.5	2509.6	1.527	6.555
423.2	4.653	9.835.–4	0.362	612.5	2541.8	1.781	6.341
448.2	8.806	1.012.–3	0.198	718.8	2569.4	2.020	6.149
473.2	15.46	1.044.–3	0.115	826.8	2585.7	2.256	5.973
498.2	25.52	1.082.–3	0.0704	938.5	2597.0	2.483	5.812
523.2	39.99	1.133.–3	0.0447	1055.2	2598.7	2.707	5.658
548.2	60.04	1.200.–3	0.0290	1177.4	2587.0	2.930	5.501
573.2	86.97	1.276.–3	0.0191	1306.7	2555.6	3.153	5.332
598.2	122.4	1.392.–3	0.0124	1445.6	2492.4	3.356	5.132
623.2	168.3	1.596.–3	0.0075	1607.1	2366.5	3.631	4.850
644.7	218.4	2.950.–3	0.0030				

THERMODYNAMIC AND THERMOPHYSICAL PROPERTIES—TABLE 26. Deuterium Oxide Gas at 1-kg/cm³ Pressure

T, K	400	450	500	550	600	650	700	750
v, m³/kg	1.676	1.895	2.112	2.322	2.535	2.747	2.960	3.172
h, kJ/kg	2525	2619	2712	2807	2904	3002	3102	3205
s, kJ/(kg·K)	6.931	7.151	7.349	7.529	7.697	7.855	8.003	8.153

THERMODYNAMIC AND THERMOPHYSICAL PROPERTIES—TABLE 27. Saturated Diphenyl

T, K	P, bar	v_f, m³/kg	v_g, m³/kg	h_f, kJ/kg	h_g, kJ/kg	s_f, kJ/(kg·K)	s_g, kJ/(kg·K)	c_{pf}, kJ/(kg·K)	μ_f, 10^{-4} Pa·s	k_f, W/(m·K)
343	0.0010	1.010,–3	252.5	0.0	444.2	0.000	1.298	1.760	15.0	0.139
350	0.0016	1.014,–3	156.1	13.0	444.2	0.036	1.266	1.782	13.5	0.138
360	0.0029	1.021,–3	85.0	30.0	446.7	0.084	1.236	1.813	11.7	0.136
370	0.0049	1.030,–3	49.9	47.2	449.7	0.130	1.213	1.844	10.3	0.135
380	0.0064	1.037,–3	29.9	65.0	454.5	0.178	1.200	1.875	9.1	0.133
390	0.0129	1.046,–3	18.3	82.7	462.7	0.224	1.194	1.906	8.1	0.132
400	0.0200	1.054,–3	11.7	99.3	461.2	0.273	1.202	1.936	7.3	0.130
420	0.0432	1.072,–3	5.84	139.9	499.0	0.358	1.228	1.998	6.0	0.127
440	0.0879	1.092,–3	3.021	180.3	532.4	0.451	1.267	2.060	5.0	0.125
460	0.1694	1.112,–3	1.652	222.7	569.7	0.545	1.378	2.122	4.3	0.122
480	0.3112	1.132,–3	0.9594	267.6	611.6	0.652	1.367	2.184	3.7	0.119
500	0.5218	1.154,–3	0.4452	314.9	651.8	0.746	1.424	2.246	3.3	0.116
520	0.8375	1.177,–3	0.3652	361.5	687.8	0.824	1.477	2.308	2.7	0.113
540	1.290	1.204,–3	0.2261	404.5	723.8	0.915	1.529	2.370	2.4	0.110
560	1.941	1.230,–3	0.1447	457.2	762.7	1.032	1.582	2.432	2.2	0.107
580	2.818	1.258,–3	0.0977	522.3	801.7	1.125	1.635	2.494	1.90	0.105
600	3.926	1.291,–3	0.0685	563.7	842.4	1.223	1.688	2.556	1.71	0.102
620	5.408	1.326,–3	0.0504	630.4	886.4	1.316	1.740	2.618	1.54	0.099
640	7.328	1.366,–3	0.0381	689.1	930.9	1.375	1.748	2.680	1.39	0.096
660	9.572	1.412,–3	0.0301	745.9	977.1	1.457	1.791	2.741	1.24	0.093
680	12.05	1.465,–3	0.0236	802.8	1024.9	1.585	1.856	2.803	1.10	0.090
700	15.21	1.529,–3	0.0186	860.1	1073.1	1.663	1.951	2.865	0.97	0.087
720	19.14	1.56,–3	0.0147	917.5	1116.7	1.746	2.003	2.93		
740	23.93	1.70,–3	0.0113	975.2	1152.8	1.822	2.058	3.00		
760	28.71	1.95,–3	0.0085	1033.1	1182.5	1.901	2.099			
780	34.83	2.16,–3	0.0058	1091.2	1163.0	1.977	2.107			
800	42.46	3.18,–3	0.0032	1148.4	1148.4	2.047	2.047			

THERMODYNAMIC AND THERMOPHYSICAL PROPERTIES – TABLE 28. Saturated Ethane (R170)

T, K	P, bar	v_f, m³/kg	v_g, m³/kg	h_f, kJ/kg	h_g, kJ/kg	s_f, kJ/(kg·K)	s_g, kJ/(kg·K)	c_{pf}, kJ/(kg·K)	μ_f, 10⁻⁴ Pa·s	k_f, W/(m·K)
90.4†	1.131.–5	1.534.–3	21945	176.8	769.4	2.560	9.113	2.260	14.19	0.215
100	1.110.–4	1.546	2484.5	198.7	782.4	2.790	8.627	2.274	9.37	0.208
110	7.467.–3	1.573	407.0	221.5	795.0	3.008	8.222	2.284	6.57	0.201
120	3.545.–3	1.615	93.61	244.4	807.2	3.207	7.897	2.292	4.89	0.194
130	1.291.–2	1.644	27.83	267.4	819.3	3.391	7.637	2.302	3.81	0.187
140	3.831.–2	1.675	10.08	290.5	831.4	3.562	7.426	2.316	3.07	0.180
150	9.672.–2	1.708	4.263	313.7	843.5	3.722	7.254	2.333	2.55	0.174
160	0.2146	1.743	2.039	337.2	855.6	3.873	7.113	2.355	2.17	0.167
170	0.4290	1.780	1.075	360.9	867.6	4.017	6.998	2.383	1.88	0.160
180	0.7874	1.819	0.6139	384.9	879.4	4.154	6.901	2.417	1.65	0.153
190	1.347	1.862	0.3738	409.3	890.8	4.285	6.819	2.458	1.47	0.147
200	2.174	1.908	0.2395	434.2	901.7	4.412	6.750	2.508	1.33	0.140
210	3.340	1.958	0.1602	459.7	911.9	4.535	6.689	2.568	1.21	0.133
220	4.922	2.014	0.1109	485.9	921.4	4.655	6.635	2.640	1.11	0.126
230	7.004	2.076	0.0789	512.8	929.6	4.773	6.585	2.730	1.03	0.119
240	9.670	2.148	0.0573	540.8	936.6	4.890	6.539	2.843	0.96	0.112
250	13.01	2.231	0.0423	569.9	941.9	5.006	6.493	2.991	0.82	0.106
260	17.12	2.330	0.0316	600.7	945.4	5.123	6.449	3.214	0.73	0.099
270	22.10	2.452	0.0237	633.6	946.4	5.233	6.392	3.511	0.64	0.092
280	28.06	2.613	0.0177	669.3	943.6	5.370	6.350	4.011	0.55	0.085
290	35.14	2.847	0.0129	709.8	934.7	5.502	6.278	5.089	0.44	0.078
300	43.54	3.295	0.0087	761.6	910.8	5.669	6.166	9.919	0.31	0.067
305.3	48.71	4.891	0.0048	841.2	841.2	5.919	5.919	∞		

THERMODYNAMIC AND THERMOPHYSICAL PROPERTIES—TABLE 29. Superheated Ethane

P, bar		100	150	200	250	300	350	400	450	500	600	700
							Temperature, K					
1.013	v	0.00156	0.00171	0.5310	0.6725	0.8118	0.9500	1.0877	1.2250	1.3622	1.6360	1.9096
	h	198.9	313.8	909.3	984.7	1068.3	1161.5	1265.3	1379.8	1504.6	1783	2097
	s	2.790	3.722	6.993	7.330	7.634	7.921	8.198	8.467	8.730	9.237	9.720
5	v	0.00156	0.00171	0.00191	0.1288	0.1595	0.1890	0.2178	0.2464	0.2747	0.3308	0.3867
	h	199.4	314.3	434.5	973.3	1060.3	1155.6	1260.7	1376.1	1501.5	1781	2096
	s	2.789	3.720	4.411	6.858	7.175	7.468	7.748	8.020	8.284	8.793	9.227
10	v	0.00156	0.00171	0.00190	0.0590	0.0765	0.0923	0.1073	0.1220	0.1365	0.1650	0.1933
	h	200.0	314.9	435.0	956.5	1050.0	1148.2	1255.0	1371.5	1497.9	1777	2094
	s	2.788	3.719	4.408	6.618	6.959	7.262	7.547	7.821	8.087	8.598	9.083
20	v	0.00156	0.00170	0.00190	0.00222	0.0346	0.0438	0.0521	0.0599	0.0674	0.0822	0.0966
	h	201.3	316.1	435.9	569.8	1026.1	1132.3	1243.3	1362.4	1490.5	1774	2090
	s	2.785	3.715	4.404	4.999	6.710	7.038	7.334	7.614	7.884	8.399	8.886
40	v	0.00155	0.00170	0.00189	0.00219	0.0118	0.0193	0.0244	0.0288	0.0329	0.0407	0.0482
	h	203.9	318.5	437.9	569.9	947.9	1096.2	1218.6	1343.8	1475.9	1764	2083
	s	2.780	3.709	4.394	4.982	6.309	6.770	7.097	7.391	7.670	8.194	8.686
60	v	0.00155	0.00170	0.00188	0.00217	0.00290	0.0109	0.0132	0.0185	0.0215	0.0270	0.0321
	h	206.5	321.0	439.8	570.3	738.1	1050.9	1192.0	1324.8	1461.2	1754	2077
	s	2.775	3.702	4.385	4.966	5.574	6.557	6.934	7.247	7.535	8.068	8.564

80	v	0.00155	0.00169	0.00188	0.00215	0.00273	0.00667	0.0106	0.0134	0.0158	0.0201	0.0459
	h	209.1	323.4	441.9	570.9	728.1	993.8	1163.6	1305.5	1446.7	1745	2070
	s	2.769	3.696	4.377	4.951	5.522	6.345	6.800	7.135	7.432	7.975	8.476
100	v	0.00155	0.00169	0.00187	0.00213	0.00263	0.00465	0.00791	0.0104	0.0124	0.0160	0.0193
	h	211.7	325.8	443.9	571.8	722.7	924.4	1134.7	1286.3	1432.4	1736	2064
	s	2.764	3.690	4.368	4.938	5.486	6.166	6.682	7.040	7.348	7.900	8.406
150	v	0.00155	0.00168	0.00185	0.00209	0.00247	0.00328	0.00488	0.00655	0.00805	0.0107	0.0130
	h	218.1	332.0	449.2	574.6	716.4	887.4	1075.2	1242.3	1399.3	1715	2050
	s	2.752	3.674	4.348	4.907	5.423	5.955	6.457	6.851	7.182	7.758	8.274
200	v	0.00154	0.00167	0.00184	0.00205	0.00237	0.00291	0.00383	0.00495	0.00605	0.00806	0.00986
	h	224.6	338.2	454.7	578.2	714.8	870.5	1041.7	1210.2	1327.3	1697	2038
	s	2.738	3.660	4.329	4.880	5.377	5.863	6.320	6.717	7.059	7.651	8.176
300	v	0.00153	0.00166	0.00181	0.00200	0.00225	0.00259	0.00307	0.00367	0.00433	0.00563	0.00686
	h	237.6	350.6	465.9	586.8	715.9	860.9	1014.9	1175.5	1338.7	1671	2019
	s	2.715	3.632	4.294	4.833	5.309	5.757	6.168	6.547	6.891	7.496	8.032
400	v	0.00153	0.00165	0.00179	0.00195	0.00216	0.00244	0.00276	0.00316	0.00361	0.00454	0.00545
	h	250.6	363.2	477.6	596.6	723.7	861.6	1008.3	62.5	1322.7	1657	2008
	s	2.692	3.605	4.262	4.793	5.257	5.688	6.080	6.443	6.780	7.388	7.930
500	v	0.00152	0.00163	0.00176	0.00192	0.00210	0.00232	0.00258	0.00288	0.00322	0.00392	0.00465
	h	263.5	375.8	489.3	607.1	732.0	866.5	1009.3	1159.3	1316.9	1650	2003
	s	2.670	3.580	4.234	4.758	5.213	5.634	6.015	6.369	6.00	7.306	7.851

THERMODYNAMIC AND THERMOPHYSICAL PROPERTIES—TABLE 30. Saturated Ethanol

T, K	P, bar	v_f, m³/kg	v_g, m³/kg	h_f, kJ/kg	h_g, kJ/kg	s_f, kJ/(kg·K)	s_g, kJ/(kg·K)	c_{pf}, kJ/(kg·K)	μ_f, 10^{-6} Pa·s	k_f, W/(m·K)	Pr_f
250	0.0027	0.001184						2.113	295	0.177	35.2
260	0.0059	0.001196						2.167	229	0.175	28.4
270	0.0128	0.001208						2.227	193	0.173	24.8
280	0.025	0.001220						2.294	156	0.171	20.9
290	0.048	0.001233						2.369	127	0.170	17.7
300	0.088	0.001246						2.45	104	0.168	15.2
310	0.151	0.001260						2.54	86	0.165	13.2
320	0.253	0.001274						2.64	72	0.162	11.7
330	0.406	0.001288						2.75	61	0.159	10.6
340	0.632	0.001304						2.86	52	0.157	9.5
350	0.956	0.001318	0.7656	199.9	1161.9			2.99	45.0	0.155	8.7
360	1.409	0.001337	0.5052	230.1	1178.4			3.12	39.0	0.153	8.0
370	2.023	0.001357	0.3555	262.2	1193.9			3.27	34.2	0.151	7.4
380	2.837	0.001379	0.2556	295.1	1208.4			3.42	30.0	0.149	6.9
390	3.897	0.001403	0.1873	329.1	1221.5			3.58	26.1	0.147	6.3
400	5.251	0.001430	0.1398	364.2	1233.6			3.74	22.7	0.145	5.9
410	6.954	0.001461	0.1058	400.8	1244.2			3.99	20.0	0.144	5.5
420	9.063	0.001495	0.0812	435.7	1254.2			4.26	17.6	0.142	5.3
430	11.64	0.001532	0.0631	472.2	1262.3			4.55	15.3	0.140	5.0
440	14.72	0.001574	0.0493	512.7	1269.2			4.88	13.9	0.139	4.9
450	18.33	0.001623	0.0389	557.2	1274.2			5.23	12.5	0.137	4.8
460	22.61	0.001682	0.0308	605.0	1275.5						
470	27.66	0.001752	0.0243	653.7	1271.1						
480	33.55	0.001832	0.0193	704.5	1262.3						
490	40.39	0.001950	0.0148	757.7	1250.2						
500	48.28	0.002091	0.0110	818.9	1232.7						
510	57.32										
516.3ᶜ	63.90										

c = critical point.

440

THERMODYNAMIC AND THERMOPHYSICAL PROPERTIES—TABLE 31. Saturated Ethylene (Ethene—R1150)

Temperature, K	Pressure, bar	v_f, m³/kg	v_g, m³/kg	h_f, kJ/kg	h_g, kJ/kg	s_f, kJ/(kg·K)	s_g, kJ/(kg·K)	c_{pf}, kJ/(kg·K)
104.0^t	0.00123	0.001527	251.36	-323.81	244.36	-1.9901	3.4730	2.497
110	0.00334	0.001545	97.57	-309.54	251.47	-1.8571	3.2431	2.500
120	0.01380	0.001576	25.75	-284.17	263.23	-1.6362	2.9255	2.539
130	0.04456	0.001609	8.62	-259.13	274.87	-1.4358	2.6717	2.465
140	0.1191	0.001644	3.46	-234.80	286.28	-1.2554	2.4663	2.405
150	0.2747	0.001681	1.5977	-210.90	297.37	-1.0908	2.2977	2.377
160	0.5636	0.001721	0.8232	-187.12	308.00	-0.9378	2.1566	2.377
170	1.0526	0.001763	0.4625	-163.23	318.04	-0.7935	2.0375	2.395
180	1.8207	0.001810	0.2784	-139.05	327.35	-0.6559	1.9352	2.427
190	2.9574	0.001861	0.1770	-114.46	335.79	-0.5244	1.7812	2.472
200	4.560	0.001918	0.1177	-89.33	343.21	-0.3967	1.7659	2.531
210	6.730	0.001981	0.0810	-63.52	349.41	-0.2730	1.6932	2.608
220	9.575	0.002054	0.0573	-36.84	354.18	-0.1515	1.6258	2.711
230	13.206	0.002139	0.0413	-9.04	357.17	-0.0314	1.5609	2.852
240	17.742	0.002241	0.0302	20.23	357.90	0.0088	1.4957	3.055
250	23.307	0.002369	0.02222	51.55	355.37	0.2114	1.4276	3.372
260	30.046	0.002541	0.01624	85.91	348.68	0.3397	1.3503	3.945
270	38.132	0.002804	0.01152	125.79	333.71	0.4819	1.3054	5.40
280	47.834	0.003442	0.00720	183.40	292.83	0.6803	1.0711	20.0
282.3^c	50.403	0.004669	0.00467	234.55	234.55	0.8585	0.8585	

t = triple point; c = critical point. $h_f = s_f = 0$ at 233.15 K = -40°C.

441

THERMODYNAMIC AND THERMOPHYSICAL PROPERTIES—TABLE 32. Compressed Ethylene

Pressure, bar		Temperature, K							
		110	125	150	175	200	225	250	275
1	v (m³/kg)	0.001545	0.001592	0.001681	0.5036	0.5814	0.6580	0.7337	0.8091
	h (kJ/kg)	−309.4	−271.4	−210.8	324.8	357.0	389.9	424.0	459.7
	s (kJ/kg·K)	−1.858	−1.534	−1.091	2.091	2.264	2.419	2.562	2.698
5	v (m³/kg)	0.001544	0.001591	0.001680	0.001785	0.001917	0.1240	0.1407	0.1569
	h (kJ/kg)	−308.9	−271.0	−210.4	−150.8	−89.3	378.4	415.0	452.3
	s (kJ/kg·K)	−1.859	−1.535	−1.093	−0.726	−0.397	1.907	2.061	2.203
10	v (m³/kg)	0.001543	0.001591	0.001679	0.001783	0.001914	0.05643	0.06672	0.07525
	h (kJ/kg)	−308.3	−270.4	−209.8	−150.3	−89.0	361.2	402.4	442.3
	s (kJ/kg·K)	−1.860	−1.537	−1.095	−0.728	−0.400	1.646	1.820	1.973
20	v (m³/kg)	0.001542	0.001589	0.001676	0.001780	0.001908	0.002084	0.02810	0.03405
	h (kJ/kg)	−307.1	−269.2	−208.7	−149.4	−88.2	−23.0	370.3	419.7
	s (kJ/kg·K)	−1.863	−1.540	−1.098	−0.733	−0.406	−0.099	1.520	1.708
30	v (m³/kg)	0.001541	0.001588	0.001674	0.001776	0.001903	0.002072	0.002347	0.01978
	h (kJ/kg)	−305.9	−268.0	−207.6	−148.4	−87.5	−22.8	50.5	390.7
	s (kJ/kg·K)	−1.866	−1.543	−1.102	−0.737	−0.412	−0.107	0.201	1.508
40	v (m³/kg)	0.001540	0.001587	0.001672	0.001773	0.001897	0.002062	0.002318	0.01163
	h (kJ/kg)	−304.7	−266.8	−206.5	−147.4	−86.7	−22.5	49.1	344.7
	s (kJ/kg·K)	−1.869	−1.546	−1.106	−0.741	−0.418	−0.115	0.186	1.284
50	v (m³/kg)	0.001539	0.001585	0.001670	0.001770	0.001892	0.002052	0.002293	0.002846
	h (kJ/kg)	−303.5	−265.7	−205.4	−146.4	−85.9	−22.2	48.1	139.8
	s (kJ/kg·K)	−1.872	−1.550	−1.110	−0.746	−0.423	−0.123	0.173	0.521
	v (m³/kg)	0.001538	0.001584	0.001668	0.001767	0.001887	0.002043	0.002270	0.002723

Property								
60 h (kJ/kg)	-302.3	-264.5	-204.2	-145.4	-85.1	-21.8	47.4	132.3
s (kJ/kg·K)	-1.875	-1.553	-1.113	-0.750	-0.428	-0.130	0.161	0.484
v (m^3/kg)	0.001535	0.001581	0.001664	0.001761	0.001877	0.002025	0.002232	0.002585
80 h (kJ/kg)	-299.8	-262.1	-202.0	-143.4	-83.5	-20.9	46.5	124.1
s (kJ/kg·K)	-1.881	-1.559	-1.120	-0.759	-0.439	-0.145	0.139	0.434
v (m^3/kg)	0.001533	0.001579	0.001660	0.001754	0.001867	0.002009	0.002199	0.002495
100 h (kJ/kg)	-297.4	-259.7	-199.7	-141.2	-81.8	-19.9	46.1	119.6
s (kJ/kg·K)	-1.887	-1.565	-1.127	-0.767	-0.449	-0.158	0.120	0.400
v (m^3/kg)	0.001528	0.001571	0.001650	0.001740	0.001846	0.001973	0.002136	0.002356
150 h (kJ/kg)	-291.3	-253.7	-194.0	-136.0	-77.3	-16.7	46.6	114.4
s (kJ/kg·K)	-1.901	-1.580	-1.145	-0.787	-0.473	-0.188	0.079	0.337
v (m^3/kg)	0.001522	0.001565	0.001641	0.001727	0.001826	0.001943	0.002086	0.002268
200 h (kJ/kg)	-285.3	-247.7	-188.3	-130.7	-72.5	-13.0	48.6	113.2
s (kJ/kg·K)	-1.914	-1.595	-1.161	-0.806	-0.495	-0.215	0.045	0.291
v (m^3/kg)	0.001517	0.001559	0.001633	0.001715	0.001809	0.001918	0.002046	0.002203
250 h (kJ/kg)	-279.2	-241.7	-182.5	-125.2	-67.6	-8.9	51.4	113.9
s (kJ/kg·K)	-1.928	-1.610	-1.177	-0.824	-0.516	-0.240	0.015	0.253
v (m^3/kg)	0.001512	0.001552	0.001625	0.001704	0.001793	0.001895	0.002012	0.002151
300 h (kJ/kg)	-273.0	-235.7	-174.5	-119.6	-62.5	-4.4	54.9	115.9
s (kJ/kg·K)	-1.942	-1.623	-1.192	-0.841	-0.536	-0.262	-0.012	0.220
v (m^3/kg)	0.001503	0.001542	0.001609	0.001683	0.001765	0.001855	0.001957	0.002072
400 h (kJ/kg)	-260.8	-223.6	-164.8	-108.3	-51.9	5.1	63.0	122.1
s (kJ/kg·K)	-1.968	-1.650	-1.221	-0.873	-0.572	-0.303	-0.059	0.166
v (m^3/kg)	0.001499	0.001531	0.001596	0.001665	0.001740	0.001823	0.001913	0.00201
500 h (kJ/kg)	-246.9	-211.4	-152.9	-96.8	-40.9	15.3	72.3	130.1
s (kJ/kg·K)	-1.978	-1.676	-1.249	-0.906	-0.605	-0.339	-0.099	0.121

THERMODYNAMIC AND THERMOPHYSICAL PROPERTIES—TABLE 32.   Compressed Ethylene

Pressure, bar	Temperature, K						
	300	325	350	375	400	425	450
v (m³/kg)	0.8842	0.9591	1.0339	1.1084	1.1830	1.2575	1.3319
1 h (kJ/kg)	497.3	536.9	578.8	622.8	668.9	717.2	767.9
s (kJ/kg·K)	2.829	2.956	3.079	3.201	3.320	3.437	3.553
v (m³/kg)	0.1728	0.1884	0.2039	0.2193	0.2346	0.2499	0.2650
5 h (kJ/kg)	491.0	531.5	574.1	618.6	665.2	713.9	764.9
s (kJ/kg·K)	2.338	2.467	2.593	2.716	2.836	2.954	3.071
v (m³/kg)	0.08380	0.09207	0.1002	0.1081	0.1160	0.1238	0.1316
10 h (kJ/kg)	482.8	542.5	568.0	613.3	660.5	709.7	761.1
s (kJ/kg·K)	2.113	2.247	2.375	2.500	2.622	2.742	2.859
v (m³/kg)	0.03914	0.04379	0.04823	0.05257	0.05675	0.06088	0.06491
20 h (kJ/kg)	465.0	509.8	555.5	602.4	650.9	701.2	753.5
s (kJ/kg·K)	1.866	2.009	2.144	2.274	2.399	2.521	2.640
v (m³/kg)	0.02404	0.02763	0.03090	0.03400	0.03700	0.03990	0.04270
30 h (kJ/kg)	444.7	493.8	542.3	591.2	641.2	692.6	745.9
s (kJ/kg·K)	1.696	1.853	1.996	2.131	2.261	2.387	2.508
v (m³/kg)	0.01630	0.01947	0.02220	0.02473	0.02710	0.02938	0.03160
40 h (kJ/kg)	420.6	476.3	528.3	579.3	631.2	688.9	738.2
s (kJ/kg·K)	1.550	1.728	1.882	2.023	2.157	2.286	2.409
v (m³/kg)	0.01140	0.01451	0.01697	0.01916	0.02119	0.02311	0.02495
50 h (kJ/kg)	390.4	456.9	513.4	567.5	621.1	675.1	730.5
s (kJ/kg·K)	1.404	1.617	1.784	1.933	2.072	2.207	2.330
v (m³/kg)	0.007757	0.01116	0.01347	0.01546	0.01725	0.01892	0.02052
60 h (kJ/kg)	347.8	435.1	497.7	555.2	610.9	666.3	722.9
s (kJ/kg·K)	1.230	1.510	1.696	1.854	1.999	2.135	2.263
v (m³/kg)	0.003672	0.006864	0.009136	0.01085	0.01237	0.01374	0.01502
80 h (kJ/kg)	238.7	382.8	463.5	529.4	590.1	648.5	707.6
s (kJ/kg·K)	0.832	1.295	1.534	1.717	1.874	2.016	2.151
v (m³/kg)	0.003094	0.004698	0.006596	0.008163	0.009492	0.01068	0.01177
100 h (kJ/kg)	210.2	330.2	427.5	503.1	569.3	630.9	692.7
s (kJ/kg·K)	0.715	1.098	1.387	1.596	1.768	1.918	2.059
v (m³/kg)	0.002684	0.003223	0.004040	0.004983	0.005914	0.006765	0.007578
150 h (kJ/kg)	188.8	272.7	361.4	445.5	521.6	592.1	658.1
s (kJ/kg·K)	0.596	0.864	1.126	1.359	1.556	1.722	1.878
v (m³/kg)	0.002508	0.002840	0.003292	0.003838	0.004445	0.005058	0.005664
200 h (kJ/kg)	181.7	255.0	332.4	410.6	487.0	560.1	629.5
s (kJ/kg·K)	0.529	0.763	0.992	1.208	1.406	1.580	1.742
v (m³/kg)	0.002397	0.002644	0.003024	0.003327	0.003743	0.004190	0.004648
250 h (kJ/kg)	179.2	247.7	319.1	392.1	465.5	538.2	608.5
s (kJ/kg·K)	0.480	0.698	0.910	1.111	1.301	1.476	1.639
v (m³/kg)	0.002317	0.002517	0.002578	0.003037	0.003351	0.003690	0.004042
300 h (kJ/kg)	179.1	244.6	312.6	382.2	452.9	524.1	593.9
s (kJ/kg·K)	0.440	0.670	0.850	1.043	1.226	1.398	1.558
v (m³/kg)	0.002203	0.002352	0.002522	0.002711	0.002919	0.003413	0.003382
400 h (kJ/kg)	182.7	244.9	308.8	374.3	441.3	509.8	578.3
s (kJ/kg·K)	0.377	0.576	0.764	0.946	1.119	1.285	1.442
v (m³/kg)	0.002122	0.002245	0.002379	0.002524	0.002678	0.002847	0.003022
500 h (kJ/kg)	189.1	250.7	312.8	376.1	440.4	505.8	572.0
s (kJ/kg·K)	0.326	0.523	0.707	0.882	1.048	1.206	1.358

THERMODYNAMIC AND THERMOPHYSICAL PROPERTIES—TABLE 33. Saturated Fluorine

T, K	P, bar	v_f, m³/kg	v_g, m³/kg	h_f, kJ/kg	h_g, kJ/kg	s_f, kJ/(kg·K)	s_g, kJ/(kg·K)	c_{pf}, kJ/(kg·K)	μ_f, 10^{-4} Pa·s	k_f, W/(m·K)
53.5	0.0025	5.866.−4	46.2	−158.6	40.9	1.602	5.314	1.446	8.8	0.186
55	0.0041	5.898.−4	17.1	−153.5	42.0	1.642	5.235	1.442	8.0	0.184
60	0.0155	6.005.−4	8.46	−149.1	45.8	1.768	5.004	1.437	6.0	0.177
65	0.0477	6.119.−4	2.93	−141.8	49.6	1.885	4.816	1.442	4.7	0.170
70	0.1230	6.240.−4	1.24	−134.4	53.2	1.995	4.666	1.450	3.8	0.162
75	0.276	6.369.−4	0.583	−127.0	56.8	2.097	4.540	1.460	3.19	0.154
80	0.555	6.508.−4	0.309	−119.5	60.1	2.194	4.433	1.474	2.71	0.146
85	1.019	6.657.−4	0.176	−111.9	63.3	2.285	4.342	1.498	2.33	0.137
90	1.740	6.819.−4	0.108	−104.3	66.1	2.372	4.262	1.535	2.00	0.129
95	2.802	6.997.−4	0.069	−96.5	68.6	2.455	4.191	1.555	1.76	0.120
100	4.280	7.193.−4	0.0466	−88.6	70.7	2.535	4.127	1.585	1.53	0.112
105	6.280	7.412.−4	0.0323	−80.5	72.4	2.612	4.068	1.630	1.36	0.103
110	8.885	7.659.−4	0.0231	−72.2	73.6	2.688	4.012	1.692	1.21	0.095
115	12.20	7.948.−4	0.0168	−63.6	74.1	2.763	3.959	1.782	1.08	0.087
120	16.33	8.283.−4	0.0125	−54.5	73.9	2.837	3.906	1.888	0.96	0.080
125	21.37	8.696.−4	0.0093	−44.9	72.7	2.912	3.864	2.05	0.86	0.073
130	27.48	9.223.−4	0.0069	−34.5	70.2	2.989	3.795	2.33	0.74	0.066
135	34.72	9.963.−4	0.0051	−22.7	65.6	3.073	3.727	2.90	0.63	0.070
140	43.47	1.119.−3	0.0036	−8.4	56.9	3.170	3.636	3.64	0.49	0.105
144.3	52.15	1.743.−3	0.0017	23.9	23.9	3.388	3.388	∞		∞

THERMODYNAMIC AND THERMOPHYSICAL PROPERTIES—TABLE 34. Fluorine Gas at Atmospheric Pressure

T, K	84.95	90	100	120	140	160	180	200	220	240	260	280	300
v, m³/kg	0.1776	0.1892	0.2118	0.2562	0.3002	0.3439	0.3874	0.4309	0.4744	0.5176	0.5610	0.6043	0.6476
h, kJ/kg	63.22	67.30	75.27	90.96	106.53	122.06	137.62	153.2	169.0	184.9	201.0	217.2	233.7
s, kJ/(kg·K)	4.342	4.390	4.474	4.616	4.737	4.840	4.932	5.014	5.090	5.158	5.221	5.282	5.340

THERMODYNAMIC AND THERMOPHYSICAL PROPERTIES—TABLE 35. Saturated Helium

T, K	P, bar	v_f, m^3/kg	v_g, m^3/kg	h_f, kJ/kg	h_g, kJ/kg	s_f, kJ/(kg·K)	s_g, kJ/(kg·K)
1.0	0.0122	0.01222	1.72	−5.69	6.75	2.28	14.72
1.1	0.0182	0.01224	1.33	−5.49	7.40	2.65	14.34
1.2	0.0274	0.01227	1.02	−5.26	8.03	2.95	14.02
1.3	0.0370	0.01231	0.805	−5.01	8.65	3.20	13.70
1.4	0.0517	0.01236	0.649	−4.75	9.27	3.40	13.41
1.5	0.0659	0.01241	0.526	−4.47	9.88	3.60	13.13
1.6	0.0871	0.01247	0.437	−4.17	10.46	3.80	12.88
1.7	0.107	0.01254	0.363	−3.84	11.04	3.91	12.53
1.8	0.137	0.01262	0.308	−3.47	11.60	4.01	12.38
1.9	0.163	0.01271	0.260	−3.07	12.15	4.13	12.14
2.0	0.202	0.01282	0.222	−2.64	12.68	4.26	11.91
2.1	0.237	0.01294	0.189	−2.17	13.19	4.40	11.69
2.2	0.284	0.01308	0.164	−1.55	13.67	4.55	11.47
2.3	0.326	0.01324	0.142	−0.99	14.13	4.71	11.25
2.4	0.385	0.01343	0.124	−0.34	14.57	4.87	11.04
2.5	0.438	0.01365	0.109	0.36	14.98	5.03	10.84
2.6	0.508	0.01390	0.096	1.16	15.37	5.20	10.64
2.7	0.576	0.01419	0.085	2.01	15.89	5.38	10.41
2.8	0.653	0.01456	0.074	2.96	16.40	5.57	10.17
2.9	0.732	0.01497	0.064	4.01	16.37	5.77	9.92
3.0	0.803	0.01549	0.055	5.28	16.32	6.00	9.66
3.1	0.907	0.01614	0.047	6.70	16.20	6.24	9.34
3.2	1.023	0.01720	0.039	8.44	15.98	6.54	8.90
3.3	1.128	0.01902	0.028	10.66	14.50	6.96	8.35
3.32	1.165	0.02394	0.024	13.25	13.25	7.50	7.50

THERMODYNAMIC AND THERMOPHYSICAL PROPERTIES—TABLE 36. Saturated Helium

Temperature, K	Pressure, bar	v_f, m³/kg	v_g, m³/kg	h_f, kJ/kg	h_g, kJ/kg	s_f, kJ/(kg·K)	s_g, kJ/(kg·K)	c_{pf}, kJ/(kg·K)	c_{pg}, kJ/(kg·K)
0.8	1.475.−5	0.00689	1125.9	0.0019	19.42	0.0047	23.94	0.022	5.210
0.9	5.379.−5	0.00689	347.1	0.0054	19.94	0.0087	21.86	0.050	5.230
1.0	1.557.−4	0.00689	133.0	0.0127	20.44	0.0163	20.44	0.100	5.262
1.1	3.800.−4	0.00689	59.8	0.0268	20.95	0.0296	18.82	0.185	5.305
1.2	8.148.−4	0.00689	30.4	0.0518	21.44	0.0510	17.67	0.318	5.360
1.3	0.00158	0.00689	16.93	0.0932	21.92	0.0836	16.70	0.511	5.424
1.4	0.00282	0.00689	10.17	0.1579	22.40	0.1308	15.86	0.780	5.496
1.5	0.00472	0.00689	6.49	0.2543	22.87	0.1962	15.13	1.138	5.574
1.6	0.00746	0.00689	4.35	0.3923	23.32	0.2839	14.49	1.602	5.654
1.7	0.01128	0.00688	3.04	0.5836	23.77	0.3981	13.93	2.193	5.736
1.8	0.01638	0.00688	2.1993	0.8422	24.20	0.5437	13.43	2.938	5.818
1.9	0.02299	0.00687	1.6420	1.186	24.63	0.7270	12.98	3.893	5.898
2.0	0.03129	0.00686	1.2601	1.642	25.04	0.9578	12.58	5.187	5.975
2.1	0.04141	0.00685	0.9921	2.261	25.45	1.256	12.23	7.244	6.046
2.2	0.05335	0.00684	0.7994	3.090	25.85	1.638	11.92	4.222	6.111
2.3	0.06730	0.00685	0.6566	3.418	26.24	1.780	11.65	2.685	6.170
2.4	0.08354	0.00687	0.5470	3.678	26.63	1.886	11.40	2.375	6.228
2.5	0.01023	0.00690	0.4608	3.922	27.00	1.980	11.17	2.284	6.285
2.6	0.1237	0.00693	0.3923	4.161	27.37	2.068	10.96	2.320	6.344
2.7	0.1481	0.00695	0.3367	4.408	27.72	2.155	10.76	2.351	6.406
2.8	0.1755	0.00699	0.2913	4.662	28.06	2.240	10.57	2.403	6.470
2.9	0.2063	0.00703	0.2537	4.923	28.38	2.324	10.39	2.486	6.540
3.0	0.2405	0.00707	0.2223	5.195	28.69	2.408	10.22	2.597	6.616
3.1	0.2784	0.00713	0.1958	5.483	28.98	2.494	10.05	2.740	6.700
3.2	0.3201	0.00717	0.1728	5.787	29.26	2.581	9.90	2.896	6.792
3.3	0.3659	0.00723	0.1542	6.108	29.52	2.670	9.747	3.061	6.897
3.4	0.4159	0.00728	0.1376	6.448	29.76	2.780	9.600	3.273	7.015

THERMODYNAMIC AND THERMOPHYSICAL PROPERTIES—TABLE 36. Saturated Helium (*Continued*)

Temperature, K	Pressure, bar	v_f, m³/kg	v_g, m³/kg	h_f, kJ/kg	h_g, kJ/kg	s_f, kJ/(kg·K)	s_g, kJ/(kg·K)	c_{pf}, kJ/(kg·K)	c_{pg}, kJ/(kg·K)
3.5	0.4704	0.00735	0.1232	6.806	29.97	2.852	9.458	3.413	7.150
3.6	0.5296	0.00742	0.1107	7.184	30.17	2.946	9.318	3.601	7.305
3.7	0.5935	0.00749	0.0997	7.581	30.34	3.042	9.181	3.801	7.484
3.8	0.6625	0.00758	0.0900	7.998	30.48	3.140	9.046	4.017	7.694
3.9	0.7366	0.00766	0.0814	8.437	30.60	3.239	8.911	4.254	7.942
4.0	0.8162	0.00776	0.0738	8.899	30.68	3.341	8.776	4.519	8.238
4.1	0.9014	0.00786	0.0669	9.387	30.73	3.444	8.641	4.820	8.641
4.2	0.9923	0.00797	0.0606	9.901	30.74	3.551	8.504	5.170	9.033
4.3	1.089	0.00810	0.0550	10.45	30.71	3.661	8.363	5.587	9.58
4.4	1.193	0.00824	0.0499	11.02	30.62	3.775	8.218	6.097	10.29
4.5	1.303	0.00841	0.0452	11.64	30.47	3.893	8.067	6.742	11.22
4.6	1.419	0.00860	0.0408	12.31	30.24	4.018	7.906	7.590	12.50
4.7	1.543	0.00881	0.0367	13.04	29.91	4.151	7.732	8.763	14.37
4.8	1.674	0.00907	0.0329	13.85	29.45	4.296	7.539	10.51	17.32
4.9	1.813	0.00941	0.0291	14.76	28.80	4.458	7.327	13.38	22.64
5.0	1.960	0.00986	0.0252	15.85	27.83	4.649	7.041	19.02	34.93
5.1	2.116	0.01056	0.0207	17.26	26.08	4.898	6.624	34.60	95.84
5.195^c	2.275	0.01436	0.0145						

c = critical pt.

THERMODYNAMIC AND THERMOPHYSICAL PROPERTIES—TABLE 37. Superheated Helium

		Temp., °C								
P, bars		0	100	200	300	400	500	600	800	1000
1	v	5.677	7.754	9.831	11.908	13.985	16.063	18.140	22.294	26.448
	h	0.327	519.6	1039	1558	2078	2597	3116	4155	5193
	s	0.0116	1.620	2.853	3.849	4.684	5.403	6.035	7.106	7.993
5	v	1.138	1.553	1.968	2.384	2.799	3.215	3.630	4.461	5.291
	h	1.636	520.9	1040	1560	2079	2598	3117	4156	5194
	s	-3.343	-1.723	-0.490	0.506	1.341	2.060	2.692	3.763	4.650
10	v	0.5704	0.780	0.986	1.193	1.401	1.609	1.816	2.232	2.647
	h	3.272	522.5	1042	1561	2080	2600	3119	4157	5196
	s	-4.782	-3.162	-1.929	-0.934	-0.098	0.621	1.252	2.323	3.211
20	v	0.2867	0.3904	0.4942	0.5979	0.7017	0.8055	0.9093	1.1169	1.3245
	h	6.544	525.8	1045	1564	2083	2603	3122	4160	5199
	s	-6.221	-4.601	-3.368	-2.373	-1.537	-0.818	-0.187	0.884	1.771
50	v	0.1164	0.1579	0.1993	0.2408	0.2822	0.3257	0.3652	0.4481	0.5311
	h	16.360	535.5	1055	1574	2093	2612	3131	4169	5207
	s	-8.121	-6.501	-5.268	-4.273	-3.438	-2.719	-2.088	-1.017	-0.130
100	v	0.0597	0.0803	0.1010	0.1217	0.1424	0.1631	0.1838	0.2252	0.2666
	h	37.720	551.7	1071	1590	2108	2627	3146	4184	5222
	s	-9.555	-7.936	-6.703	-5.709	-4.874	-4.155	-3.524	-2.454	-1.567
150	v	0.0407	0.0545	0.0682	0.0820	0.0958	0.1095	0.1233	0.1509	0.1785
	h	49.080	567.9	1087	1605	2124	2643	3161	4199	5236
	s	-10.391	-8.773	-7.541	-6.546	-5.712	-4.994	-4.363	-3.293	-2.407
200	v	0.0312	0.0416	0.0518	0.0622	0.0725	0.0828	0.0931	0.1137	0.1344
	h	65.440	584.1	1103	1621	2140	2658	3176	4213	5250
	s	-10.983	-9.635	-8.134	-7.139	-6.306	-5.588	-4.957	-3.888	-3.002

THERMODYNAMIC AND THERMOPHYSICAL PROPERTIES—TABLE 38. Helium Gas at Atmospheric Pressure

T, K	4.224	5	10	20	30	40	50	75	100	200	300	400	500	600	800	1000
v, m³/kg	0.0591	0.0834	0.1612	0.4094	0.6161	0.8218	1.0273	1.5403	2.053	4.102	6.154	8.191	10.24	12.31	16.40	20.50
h, kJ/kg	30.30	36.18	64.91	117.95	170.24	222.4	274.4	404.4	534.2	1054	1573	2092	2612	3131	4170	5208
s, kJ/(kg·K)	8.327	9.614	13.369	17.321	19.442	20.94	22.10	24.21	25.71	29.30	31.41	32.90	34.06	35.01	36.50	37.66

Saturated n-Heptane

T, K	P, bar	v_f, m³/kg	v_g, m³/kg	h_f, kJ/kg	h_g, kJ/kg	s_f, kJ/(kg·K)	s_g, kJ/(kg·K)	c_{pf}, kJ/(kg·K)	μ_f, 10⁻⁴Pa·s	k_f, W/(m·K)
182.6		1.292.–3		284.1		2.260		2.025	39.4	0.150
200	0.00002	1.316.–3		319.4	722.6	2.441	4.457	2.011	21.0	0.148
220	0.00019	1.344.–3		359.7	757.1	2.636	4.442	2.026	12.6	0.145
240	0.00133	1.374.–3		400.5	791.4	2.814	4.443	2.063	8.52	0.142
250	0.00303	1.389.–3		421.3	808.3	2.899	4.447	2.088	7.23	0.140
260	0.00635	1.405.–3		442.3	824.9	2.981	4.453	2.117	6.52	0.137
270	0.01316	1.422.–3		463.6	841.2	3.061	4.460	2.147	5.46	0.135
280	0.02347	1.440.–3		485.2	857.8	3.140	4.471	2.180	4.83	0.132
290	0.03997	1.457.–3		507.2	874.8	3.217	4.485	2.216	4.29	0.129
300	0.06674	1.475.–3	3.744	529.6	891.9	3.293	4.501	2.252	3.85	0.126
310	0.1070	1.494.–3	2.412	552.3	908.9	3.367	4.517	2.291	3.48	0.123
320	0.1656	1.514.–3	1.596	575.4	926.0	3.441	4.537	2.329	3.17	0.121
330	0.2461	1.534.–3	1.101	598.8	943.3	3.513	4.557	2.370	2.89	0.119
340	0.3614	1.555.–3	0.7650	622.8	961.2	3.584	4.579	2.412	2.66	0.116
350	0.5130	1.578.–3	0.5510	647.0	979.1	3.655	4.604	2.454	2.45	0.114
360	0.712	1.601.–3	0.4058	671.9	997.5	3.725	4.629	2.500	2.24	0.111
370	0.967	1.625.–3	0.3036	697.1	1016.1	3.794	4.656	2.548	2.04	0.109
371.6	1.013	1.629.–3	0.2904	701.9	1019.8	3.805	4.660	2.556	2.01	0.108
380	1.289	1.651.–3	0.2308	723.9	1035.4	3.864	4.684	2.60	1.86	0.107
390	1.689	1.678.–3	0.1781	750.4	1054.2	3.932	4.711	2.65	1.71	0.105
400	2.180	1.708.–3	0.1388	777.2	1073.2	4.000	4.740	2.70	1.58	0.103
420	3.471	1.775.–3	0.0734					2.81	1.35	0.099
440	5.268	1.853.–3	0.0576					2.93	1.15	0.095
460	7.691	1.954.–3	0.0389					3.05	0.97	0.091
480	10.92	2.065.–3	0.0265					3.19	0.82	0.087
500	15.10	2.235.–3	0.0178					3.38	0.67	0.080
520	20.43	2.52.–3						3.7		
540.1	27.35	4.3.–3	0.0043							

THERMODYNAMIC AND THERMOPHYSICAL PROPERTIES—TABLE 40. Saturated Hydrazine

Temperature, K	Pressure, bar	v_f, m³/kg	v_g, m³/kg	h_f, kJ/kg	h_g, kJ/kg	s_f, kJ/(kg·K)	s_g, kJ/(kg·K)
386.6	1.013	0.001 053	0.9833	−105.9	65.4	0.5994	1.0426
390	1.135	0.001 060	0.8850	−104.8	66.0	0.6029	1.0409
400	1.560	0.001 081	0.6579	−101.4	68.2	0.6120	1.0360
410	2.102	0.001 104	0.4994	−97.6	70.6	0.6211	1.0314
420	2.786	0.001 127	0.3850	−93.9	73.0	0.6300	1.0275
440	4.732	0.001 178	0.2355	−86.1	77.6	0.6492	1.0212
460	7.610	0.001 235	0.1500	−76.9	82.1	0.6707	1.0163
480	11.76	0.001 299	0.1005	−67.1	86.6	0.6916	1.0118
500	17.42	0.001 374	0.0690	−57.3	90.8	0.7124	1.0086
520	29.59	0.001 460	0.0407	−47.8	94.6	0.7320	1.0058
540	34.75	0.001 563	0.0353	−36.0	97.7	0.7566	1.0042
560	47.09	0.001 681	0.0263	−25.2	101.2	0.7762	1.0020
580	62.44	0.001 835	0.0196	−12.4	103.6	0.8002	1.0002
600	81.17	0.002 045	0.0142	5.2	104.2	0.8335	0.9988
620	102.7	0.002 320	0.0106	23.2	103.6	0.8671	0.9967
640	128.1	0.002 86	0.0074	45.9	98.1	0.9035	0.9906
653	146.9	0.004 33	0.0043	83.7	83.7	0.9715	0.9715

THERMODYNAMIC AND THERMOPHYSICAL PROPERTIES—TABLE 41. Saturated Hydrogen

T, K	P, bar	v_f, m³/kg	v_g, m³/kg	h_f, kJ/kg	h_g, kJ/kg	s_f, kJ/(kg·K)	s_g, kJ/(kg·K)	c_{pf}, kJ/(kg·K)	μ_f, 10⁻⁴ Pa·s	k_f, W/(m·K)
13.95	0.072	0.01298	7.974	218.3	667.4	14.079	46.635	6.36	0.255	0.073
14	0.074	0.01301	7.205	219.6	669.3	14.173	46.301	6.47	0.248	0.075
15	0.127	0.01316	4.488	226.4	678.2	14.640	44.763	6.91	0.218	0.083
16	0.204	0.01332	2.954	233.8	686.7	15.104	43.418	7.36	0.194	0.089
17	0.314	0.01348	2.032	241.6	694.7	15.568	42.227	7.88	0.175	0.093
18	0.461	0.01366	1.449	249.9	702.1	16.032	41.158	8.42	0.159	0.095
19	0.654	0.01387	1.064	258.8	708.8	16.498	40.188	8.93	0.146	0.097
20	0.901	0.01407	0.8017	268.3	714.8	16.966	39.299	9.45	0.135	0.098
21	1.208	0.01430	0.6177	278.4	720.2	17.440	38.485	10.13	0.125	0.100
22	1.585	0.01455	0.4828	289.2	724.4	17.919	37.710	10.82	0.116	0.101
23	2.039	0.01483	0.3829	300.8	727.6	18.405	36.973	11.69	0.108	0.101
24	2.579	0.01515	0.3072	313.3	729.8	18.901	36.266	12.52	0.101	0.101
25	3.213	0.01551	0.2489	326.7	730.7	19.408	35.579	13.44	0.094	0.100
26	3.950	0.01592	0.2032	341.2	730.2	19.929	34.900	14.80	0.088	0.098
27	4.800	0.01639	0.1667	357.0	728.0	20.473	34.221	16.17	0.082	0.096
28	5.770	0.01696	0.1370	374.3	723.7	21.041	33.524	18.48	0.076	0.094
29	6.872	0.01765	0.1125	393.6	716.6	21.650	32.795	22.05	0.070	0.091
30	8.116	0.01854	0.0919	415.4	705.9	22.315	32.002	26.59	0.065	0.087
31	9.510	0.01977	0.0738	441.3	689.7	23.075	31.091	36.55	0.058	0.086
32	11.07	0.02174	0.0571	474.7	663.2	24.032	29.926	65.37	0.051	0.092
33.18	13.13	0.03182	0.0318	565.4	565.4	26.680	26.680	∞		∞

THERMODYNAMIC AND THERMOPHYSICAL PROPERTIES—TABLE 42. Compressed Hydrogen

Temperature, K

Pressure, bar		15	20	30	40	50	60	80	100	150	200
0.1	v	6.076	8.176	12.333	16.473	20.606	24.736	32.991	41.244	61.870	82.495
	h	679.2	731.6	835.5	938.9	1,042.3	1,146	1,356	1,575	2,172	2,826
	s	46.02	49.04	53.25	56.23	58.53	60.43	63.45	65.89	70.68	74.46
1	v	0.0131	0.0141	1.196	1.625	2.046	2.463	3.295	4.123	6.190	8.254
	h	227.3	268.3	826.0	932.7	1,037.9	1,143	1,354	1,574	2,172	2,826
	s	14.62	16.96	43.56	46.63	48.98	50.89	53.93	56.38	61.17	64.96
5	v	0.0131	0.0140	0.2006	0.3039	0.3958	0.4839	0.6553	0.8238	1.241	1.655
	h	231.7	272.1	775.0	903.4	1,017.6	1,128	1,345	1,568	2,170	2,826
	s	14.57	16.88	35.80	39.52	42.07	44.07	47.20	49.68	54.66	58.31
10	v	0.0130	0.0138	0.0181	0.1376	0.1895	0.2366	0.3255	0.4116	0.6221	0.8303
	h	237.2	277.0	412.1	861.8	991.1	1,109	1,334	1,560	2,167	2,826
	s	14.50	16.77	22.09	35.95	38.85	40.99	44.23	46.75	51.63	55.44
20	v	0.0129	0.0136	0.0167	0.0521	0.0866	0.1135	0.1611	0.2057	0.3129	0.4179
	h	248.2	286.9	406.5	752.0	934.7	1,070	1,312	1,546	2,163	2,826
	s	14.37	16.58	21.33	31.07	35.19	37.67	41.15	43.76	48.71	52.55
40	v		0.0133	0.0155	0.0216	0.0376	0.0533	0.0796	0.1033	0.1586	0.2119
	h		307.3	413.5	589.3	823.5	997	1,271	1,521	2,155	2,826
	s		16.26	20.50	25.49	30.73	33.91	37.87	40.65	45.75	49.64
60	v		0.0130	0.0147	0.0182	0.0254	0.0351	0.0532	0.0697	0.1073	0.1433
	h		328.0	427.2	570.1	757.0	940	1,237	1,499	2,149	2,828
	s		15.98	19.95	24.03	28.19	31.54	35.82	38.76	43.99	47.92
80	v		0.0127	0.0142	0.0167	0.0211	0.0273	0.0406	0.0531	0.0818	0.1090
	h		348.9	443.5	572.3	732.8	905	1,210	1,482	2,146	2,831
	s		15.74	19.53	23.21	26.78	29.93	34.34	37.37	42.72	46.69

THERMODYNAMIC AND THERMOPHYSICAL PROPERTIES—TABLE 42. Compressed Hydrogen (*Continued*)

Pressure, bar		15	20	30	40	50	60	80	100	150	200
						Temperature, K					
100	v		0.0125	0.0138	0.0158	0.0190	0.0233	0.0335	0.0434	0.0666	0.0885
	h		369.8	461.1	581.5	727.4	888	1,192	1,469	2,144	2,835
	s		15.53	19.19	22.63	25.88	28.80	33.19	36.28	41.73	45.73
200	v		0.0117	0.0125	0.0136	0.0150	0.0167	0.0207	0.0253	0.0368	0.0480
	h		474.4	556.1	658.7	776.9	908	1,182	1,458	2,156	2,869
	s		14.71	17.99	20.93	23.56	25.94	29.88	32.97	38.59	42.72
400	v			0.0113	0.0119	0.0126	0.0134	0.0151	0.0171	0.0225	0.0279
	h			751.0	841.9	945.4	1,059	1,303	1,560	2,249	2,973
	s			16.59	19.20	21.50	23.58	27.07	29.94	35.48	39.67
600	v			0.0106	0.0110	0.0115	0.0120	0.0131	0.0144	0.0178	0.0214
	h			941.5	1,027	1,124	1,231	1,463	1,709	2,385	3,107
	s			15.68	18.14	20.29	22.24	25.57	28.31	33.74	37.92
800	v				0.0104	0.0107	0.0111	0.0120	0.0130	0.0155	0.0181
	h				1,209	1,302	1,405	1,628	1,870	2,535	3,255
	s				17.35	19.43	21.30	24.50	27.20	32.54	36.70
1,000	v				0.0099	0.0102	0.0106	0.0112	0.0120	0.0140	0.0160
	h				1,387	1,478	1,578	1,796	2,032	2,692	3,403
	s				16.72	18.75	20.58	23.70	26.33	31.63	35.75

THERMODYNAMIC AND THERMOPHYSICAL PROPERTIES—TABLE 42. Compressed Hydrogen (*Continued*)

Pressure, bar		Temperature, K										
		250	300	350	400	450	500	600	700	800	900	1,000
0.1	v	103.12	123.23	144.35	164.97	185.60	206.22	247.46	288.70	329.94	371.18	412.43
	h	3,517	4,227	4,945	5,668	6,393	7,118	8,571	10,028	11,493	12,969	14,458
	s	77.53	80.13	82.34	84.27	85.98	87.51	90.15	92.40	94.36	96.10	97.66
1	v	10.32	12.38	14.44	16.50	18.57	20.63	24.75	28.88	33.00	37.13	41.25
	h	3,517	4,227	4,946	5,669	6,393	7,118	8,571	10,029	11,494	12,969	14,459
	s	68.03	70.63	72.85	74.78	76.48	78.01	80.66	82.91	84.86	86.60	88.17
5	v	2.069	2.482	2.895	3.307	3.720	4.132	4.957	5.782	6.607	7.432	8.257
	h	3,518	4,229	4,948	5,671	6,396	7,121	8,574	10,032	11,497	12,973	14,462
	s	61.39	63.99	66.21	68.14	69.84	71.37	74.02	76.27	78.23	79.96	81.53
10	v	1.038	1.245	1.451	1.658	1.864	2.070	2.483	2.896	3.308	3.720	4.133
	h	3,519	4,231	4,951	5,674	6,399	7,125	8,578	10,036	11,501	12,977	14,467
	s	58.52	61.12	63.34	65.28	66.98	68.51	71.16	73.41	75.37	77.10	78.67
20	v	0.522	0.6259	0.7294	0.8328	0.9361	1.040	1.246	1.452	1.658	1.865	2.071
	h	3,522	4,235	4,956	5,680	6,406	7,132	8,586	10,044	11,509	12,985	14,475
	s	55.65	58.26	60.48	62.41	64.12	65.65	68.30	70.55	72.51	74.24	75.81
40	v	0.2644	0.3166	0.3685	0.4204	0.4721	0.5238	0.6271	0.7303	0.8335	0.9366	1.040
	h	3,527	4,244	4,967	5,692	6,419	7,146	8,601	10,059	11,525	13,002	14,492
	s	52.76	55.38	57.61	59.55	61.26	62.79	65.44	67.69	69.65	71.39	72.95
60	v	0.1786	0.2136	0.2483	0.2829	0.3174	0.3519	0.4209	0.4897	0.5585	0.6273	0.6961
	h	3,533	4,253	4,978	5,705	6,432	7,160	8,616	10,075	11,542	13,018	14,508
	s	51.02	53.69	55.92	57.86	59.58	61.11	63.76	66.02	67.97	70.51	71.28
80	v	0.1357	0.1621	0.1882	0.2142	0.2401	0.2660	0.3177	0.3694	0.4120	0.4726	0.5242
	h	3,540	4,263	4,989	5,718	6,446	7,174	8,631	10,091	11,558	13,035	14,525
	s	49.84	52.49	54.73	56.67	58.39	59.92	62.57	64.83	66.79	68.52	70.09

THERMODYNAMIC AND THERMOPHYSICAL PROPERTIES—TABLE 42. Compressed Hydrogen (*Continued*)

Pressure, bar		250	300	350	400	450	500	600	700	800	900	1,000
						Temperature, K						
100	v	0.1099	0.1312	0.1521	0.1730	0.1937	0.2145	0.2559	0.2972	0.3385	0.3798	0.4211
	h	3,547	4,273	5,001	5,731	6,460	7,189	8,647	10,107	11,574	13,051	14,542
	s	48.89	51.55	53.79	55.74	57.46	59.00	61.65	63.90	65.87	67.60	69.17
200	v	0.0588	0.0695	0.0801	0.0905	0.1001	0.1114	0.1321	0.1528	0.1734	0.1941	0.2147
	h	3,594	4,329	5,064	5,798	6,531	7,263	8,724	10,187	11,656	13,134	14,625
	s	45.94	48.62	50.89	52.85	54.58	56.12	58.78	61.04	63.00	64.74	66.31
400	v	0.0334	0.0388	0.0441	0.0493	0.0545	0.0597	0.0701	0.0804	0.0908	0.1011	0.1114
	h	3,716	4,458	5,202	5,943	6,681	7,416	8,883	10,349	11,820	13,300	14,792
	s	042.98	45.68	47.97	49.95	51.69	53.24	55.91	58.17	60.14	61.88	63.45
600	v	.0249	0.0285	0.0321	0.0355	0.0390	0.0425	0.0494	0.0562	0.0631	0.0700	0.0768
	h	3,854	4,600	5,349	6,095	6,836	7,574	9,045	10,513	11,985	13,466	14,958
	s	41.24	43.95	46.26	48.26	50.00	51.56	54.24	56.50	58.47	60.21	61.78
800	v	0.0207	0.0234	0.0260	0.0286	0.0312	0.0338	0.0390	0.0441	0.0492	0.0543	0.0594
	h	4,003	4,748	5,501	6,249	6,993	7,734	9,207	10,677	12,150	13,631	15,124
	s	40.03	42.73	45.05	47.05	48.81	50.37	53.05	55.32	57.29	59.03	60.60
1,000	v	0.0181	0.0202	0.0223	0.0244	0.0265	0.0286	0.0327	0.0367	0.0408	0.0449	0.0490
	h	4,156	4,898	5,654	6,405	7,151	7,893	9,370	10,842	12,316	13,797	15,289
	s	39.10	41.79	44.12	46.12	47.88	49.45	52.14	54.41	56.38	58.12	59.69

THERMODYNAMIC AND THERMOPHYSICAL PROPERTIES—TABLE 43. Saturated Para-Hydrogen

T, K	P, bar	v_f, m³/kg	v_g, m³/kg	h_f, kJ/kg	h_g, kJ/kg	s_f, kJ/(kg·K)	s_g, kJ/(kg·K)	c_{pf}, kJ/(kg·K)	μ_f, 10⁻⁴ Pa·s	k_f, W/(m·K)
13.8	0.070	0.0130	7.97	−308.9	140.3	4.97	37.52	6.37	0.255	0.073
14	0.079	0.0130	7.20	−307.6	142.1	5.06	37.19	6.47	0.248	0.075
15	0.134	0.0132	4.49	−300.9	151.1	5.53	36.65	6.91	0.218	0.082
16	0.216	0.0133	2.96	−293.4	159.6	5.99	34.31	7.36	0.194	0.089
17	0.329	0.0135	2.03	−285.6	167.6	6.45	33.11	7.88	0.175	0.092
18	0.482	0.0137	1.45	−277.3	175.0	6.92	32.05	8.42	0.159	0.095
19	0.682	0.0139	1.07	−268.4	181.7	7.38	31.08	8.93	0.146	0.097
20	0.935	0.0141	0.802	−258.9	187.7	7.85	30.19	9.45	0.135	0.098
21	1.250	0.0143	0.618	−248.8	193.0	8.32	29.37	10.13	0.125	0.100
22	1.634	0.0146	0.483	−237.9	197.3	8.80	28.60	10.82	0.116	0.101
23	2.096	0.0148	0.383	−226.3	200.5	9.29	27.86	11.69	0.108	0.101
24	2.645	0.0152	0.307	−213.9	202.7	9.78	27.15	12.52	0.101	0.100
25	3.288	0.0155	0.249	−200.4	203.6	10.29	26.46	13.44	0.094	0.099
26	4.035	0.0159	0.203	−185.9	203.1	10.81	25.79	14.81	0.088	0.098
27	4.892	0.0164	0.167	−170.2	200.9	11.36	25.11	16.18	0.082	0.096
28	5.88	0.0170	0.137	−152.9	196.5	11.93	24.41	18.5	0.076	0.094
29	6.98	0.0177	0.113	−133.6	189.5	12.54	23.68	22.1	0.070	0.091
30	8.23	0.0185	0.092	−111.7	178.8	13.20	22.89	26.6	0.065	0.087
31	9.63	0.0198	0.074	−85.8	162.6	13.96	21.98	36.6	0.058	0.088
32	11.20	0.0217	0.057	−52.4	136.1	14.92	20.81	65.4	0.051	0.092
33	12.93	0.0318	0.032	38.3	38.3	17.56	17.56	∞		∞

THERMODYNAMIC AND THERMOPHYSICAL PROPERTIES — TABLE 44. Saturated Hydrogen Peroxide

T, K	P, bar	v_f, m³/kg	v_g, m³/kg	h_f, kJ/kg	h_g, kJ/kg	s_f, kJ/(kg·K)	s_g, kJ/(kg·K)	c_{pf}, kJ/(kg·K)	μ_f, 10^{-4} Pa·s	k_f, W/(m·K)
273	0.0004	0.00068	1672	−5577	−4027	2.990	8.662	1.45	18.0	0.483
300	0.0031	0.00069	235	−5510	−3995	3.224	8.269	1.48	11.3	0.481
350	0.0564	0.00072	15.1	−5376	−3933	3.631	7.758	1.54	4.3	0.474
400	0.4521	0.00076	2.12	−5238	−3878	4.032	7.440	1.61	2.2	0.464
450	2.143	0.00081	0.487	−5091	−3820	4.346	7.172	1.68	1.3	0.453
500	7.126	0.00088	0.155	−4945	−3777	4.656	6.992	1.75	0.89	0.443
550	18.56	0.00095	0.0605	−4794	−3745	4.941	6.846	1.82	0.65	0.431
600	40.75	0.00107	0.0268	−4635	−3731	5.209	6.720	1.90	0.50	0.416
650	79.27	0.00125	0.0125	−4463	−3746	5.485	6.582			
700	141.7	0.00171	0.0048	−4195	−3860	5.682	6.339			
708.5	155.3	0.00284	0.0028	−4012	−4012	5.732	5.732			

THERMODYNAMIC AND THERMOPHYSICAL PROPERTIES—TABLE 45. Saturated Isobutane (R600a)

T, K	P, bar	v_f, m³/kg	v_g, m³/kg	h_f, kJ/kg	h_g, kJ/kg	s_f, kJ/(kg·K)	s_g, kJ/(kg·K)	c_{pf}, kJ/(kg·K)	μ_f, 10^{-4} Pa·s	k_f, W/(m·K)
113.6	1.9.–7	1.349.–3	8.60.+6	0.0	485.3	1.863	6.136			
120	9.3.–7	1.360.–3	1.84.+6	11.0	491.1	1.957	5.957	1.78		
140	4.8.–5	1.396.–3	4210	46.0	510.1	2.226	5.541	1.87		0.163
160	8.2.–4	1.435.–3	278.2	82.1	530.8	2.467	5.272	1.93		0.158
180	0.0070	1.476.–3	36.66	119.5	533.0	2.688	5.097	1.99	9.46	0.149
200	0.0369	1.520.–3	7.723	158.5	576.7	2.893	4.984	2.05	6.06	0.142
220	0.1374	1.568.–3	2.265	199.0	601.5	3.086	4.916	2.12	4.21	0.134
240	0.3989	1.621.–3	0.8432	241.4	627.4	3.270	4.878	2.19	3.11	0.127
260	0.9600	1.680.–3	0.3738	285.8	654.2	3.446	4.863	2.28	2.40	0.120
270	1.4081	1.712.–3	0.2617	308.8	667.7	3.532	4.861	2.33	2.14	0.117
280	2.0020	1.746.–3	0.1882	332.3	681.3	3.617	4.863	2.39	1.93	0.113
290	2.7686	1.784.–3	0.1385	356.4	694.9	3.700	4.867	2.46	1.75	0.110
300	3.7365	1.824.–3	0.1040	381.1	708.4	3.783	4.874	2.53	1.59	0.106
310	4.934	1.868.–3	0.0794	406.4	721.7	3.865	4.882	2.61	1.46	0.102
320	6.392	1.916.–3	0.0614	432.4	734.8	3.946	4.891	2.70	1.35	0.099
330	8.140	1.971.–3	0.0481	459.2	747.7	4.028	4.902	2.81	1.25	0.095
340	10.21	2.032.–3	0.0380	486.9	760.0	4.109	4.912	2.92	1.15	0.092
350	12.64	2.103.–3	0.0301	515.7	771.8	4.191	4.923	3.04	1.05	0.088
360	15.46	2.187.–3	0.0240	545.6	782.7	4.273	4.932	3.17	0.95	0.083
370	18.72	2.289.–3	0.0190	577.1	792.3	4.357	4.939	3.31	0.85	0.080
380	22.48	2.420.–3	0.0150	610.6	799.8	4.444	4.942	3.45	0.75	0.076
390	26.82	2.604.–3	0.0115	647.1	803.7	4.536	4.937	3.62	0.63	0.071
400	31.86	2.920.–3	0.0083	689.6	799.6	4.639	4.915	3.85	0.51	0.065
408	36.55	4.464.–3	0.0045	752.5	752.5	4.791	4.791	∞		∞

THERMODYNAMIC AND THERMOPHYSICAL PROPERTIES—TABLE 46. Saturated Krypton

T, K	P, bar	v_f, m³/kg	v_g, m³/kg	h_f, kJ/kg	h_g, kJ/kg	s_f, kJ/(kg·K)	s_g, kJ/(kg·K)	c_{pf}, kJ/(kg·K)	μ_f, 10^{-4} Pa·s	k_f, W/(m·K)
10		3.235.−4		0.22		0.0256		0.070		
20		3.246.−4		1.59		0.1141		0.188		
30		3.265.−4		3.84		0.2034		0.247		
40		3.288.−4		6.49		0.2791		0.276		
50		3.313.−4		9.37		0.3431		0.295		
60		3.341.−4		12.40		0.3982		0.311		
70		3.372.−4		15.57		0.4471		0.327		
80		3.407.−4		18.97		0.4925		0.345		
90		3.446.−4		22.58		0.5353		0.366		
100		3.492.−4		26.42		0.5765		0.389		
110		3.544.−4		30.52		0.6165		0.414		
115.76		3.579.−4		33.18		0.6390		0.427		
115.76	0.732	4.090.−4	0.1529	52.78	161.8	0.8095	1.751	0.547		
119.76	1.013	4.143.−4	0.1136	54.99	162.6	0.8279	1.726	0.545		
120	1.032	4.146.−4	0.1116	55.09	162.6	0.8291	1.724	0.544	3.72	0.0900
130	2.112	4.284.−4	0.0578	60.55	164.1	0.8724	1.669	0.542	3.16	0.828
140	3.878	4.440.−4	0.0330	66.02	165.3	0.9124	1.662	0.546	2.64	0.0756
150	6.552	4.619.−4	0.0201	71.58	166.1	0.9499	1.580	0.559	2.20	0.0688
160	10.37	4.831.−4	0.0130	77.34	166.4	0.9859	1.543	0.587	1.87	0.0625
170	15.57	5.091.−4	0.0086	83.48	166.0	1.022	1.507	0.641	1.54	0.0558
180	22.41	5.423.−4	0.0059	90.26	164.6	1.058	1.472	0.734	1.28	0.0494
190	31.20	5.882.−4	0.0040	98.19	161.8	1.098	1.433	0.905	1.05	0.0433
200	42.23	6.641.−4	0.0026	108.40	156.0	1.147	1.386	1.515	0.80	0.0348
209.39	54.96	1.098.−3	0.0011	133.90	133.9	1.262	1.262	∞		∞

THERMODYNAMIC AND THERMOPHYSICAL PROPERTIES—TABLE 47. Compressed Krypton

Pressure, bar

Temperature, K		1	10	20	40	60	80	100	200	400	600	800	1000
100	v	3.49.-4	3.49.-4	3.49.-4	3.48.-4	3.48.-4	3.47.-4	3.47.-4	3.45.-4	3.42.-4	3.39.-4	3.36.-4	3.33.-4
	h	26.42	26.69	26.99	27.59	28.18	28.78	29.38	32.38	38.42	44.47	50.52	56.57
	s	0.5765	0.5760	0.5755	0.5745	0.5735	0.5724	0.5714	0.5667	0.5580	0.5503	0.5432	0.5366
200	v	0.1971	0.0184	8.39.-3	3.00.-3	6.19.-4	5.94.-4	5.76.-4	5.27.-4	4.83.-4	4.58.-4	4.41.-4	4.28.-4
	h	183.1	179.3	174.5	159.4	105.6	104.4	103.6	102.7	105.3	109.7	114.9	120.3
	s	1.859	1.618	1.533	1.405	1.129	1.116	1.106	1.073	1.037	1.013	0.993	0.977
300	v	0.2971	0.0292	0.0143	6.84.-3	4.37.-3	3.14.-3	2.41.-3	1.09.-3	6.92.-4	5.94.-4	5.44.-4	5.13.-4
	h	208.1	206.3	204.2	200.0	195.7	191.2	186.6	166.8	155.1	155.2	158.2	162.3
	s	1.961	1.728	1.654	1.575	1.525	1.485	1.451	1.333	1.239	1.196	1.169	1.149
400	v	0.3966	0.0394	0.0196	9.67.-3	6.37.-3	4.73.-3	3.75.-3	1.85.-3	1.10.-3	7.79.-4	6.76.-4	6.14.-4
	h	233.0	231.9	230.7	228.3	225.9	223.6	221.3	211.4	199.7	196.8	197.9	200.8
	s	2.032	1.802	1.730	1.657	1.612	1.579	1.552	1.463	1.368	1.317	1.284	1.259
500	v	0.4960	0.0495	0.0247	0.0123	8.20.-3	6.15.-3	4.91.-3	2.49.-3	1.33.-3	9.81.-4	8.22.-4	7.29.-4
	h	257.8	257.1	256.3	254.9	253.3	251.9	250.5	244.5	236.9	234.2	234.7	237.4
	s	2.088	1.858	1.788	1.716	1.673	1.642	1.617	1.537	1.451	1.400	1.365	1.340
600	v	0.5953	0.0596	0.0298	0.0149	9.96.-3	7.49.-3	6.01.-3	3.07.-3	1.64.-3	1.18.-3	9.67.-4	8.44.-4
	h	282.7	282.2	281.7	280.7	279.7	278.8	277.9	274.2	269.6	268.1	269.1	271.7
	s	2.133	1.904	1.834	1.763	1.721	1.691	1.667	1.591	1.511	1.462	1.428	1.403
700	v	0.6946	0.0696	0.0348	0.0175	0.0117	8.80.-3	7.07.-3	3.62.-3	1.93.-3	1.38.-3	1.11.-3	9.56.-4
	h	307.5	307.2	306.9	306.2	305.6	305.1	304.5	302.2	299.8	299.6	301.1	304.0
	s	2.171	1.942	1.873	1.803	1.761	1.732	1.708	1.634	1.557	1.511	1.478	1.453
800	v	0.7939	0.0795	0.0399	0.0200	0.0134	0.0101	8.11.-3	4.16.-3	2.21.-3	1.57.-3	1.25.-3	1.07.-3
	h	332.3	332.2	331.9	331.6	331.2	330.9	330.5	329.3	328.6	329.4	331.5	334.6
	s	2.204	1.975	1.906	1.837	1.795	1.766	1.743	1.671	1.596	1.551	1.518	1.494
900	v	0.8931	0.0895	0.0448	0.0225	0.0151	0.0114	9.13.-3	4.68.-3	2.48.-3	1.75.-3	1.39.-3	1.18.-3
	h	357.1	357.0	356.9	356.8	356.6	356.4	356.3	355.8	356.3	358.0	360.7	364.2
	s	2.233	2.005	1.936	1.866	1.825	1.796	1.773	1.702	1.628	1.584	1.553	1.528
1000	v	0.9924	0.0994	0.0498	0.0250	0.0168	0.0126	0.0102	5.20.-3	2.74.-3	1.93.-3	1.53.-3	1.29.-3
	h	381.9	381.9	381.9	381.8	381.8	381.8	381.8	381.9	383.4	385.9	389.0	392.8
	s	2.260	2.031	1.962	1.893	1.852	1.823	1.800	1.729	1.657	1.614	1.583	1.559

THERMODYNAMIC AND THERMOPHYSICAL PROPERTIES—TABLE 48. Saturated
Lithium

T, K	P, bar	v_f, m³/kg	v_g, m³/kg	h_f, kJ/kg	h_g, kJ/kg	s_f, kJ/(kg·K)	s_g, kJ/(kg·K)	c_{pf}, kJ/(kg·K)
453.7	1.78.–13	1.912.–3		1703	24259	6.776	56.492	4.30
500	8.21.–12	1.946.–3		1905	24390	7.199	52.169	4.34
600	4.18.–9	1.988.–3		2334	24674	7.983	45.216	4.23
700	3.51.–7	2.028.–3	2.40.+7	2697	24869	8.633	40.307	4.19
800	9.57.–6	2.070.–3	9.94.+5	3174	25162	9.192	36.678	4.17
900	1.24.–4	2.114.–3	8.55.+4	3590	25341	9.682	33.850	4.16
1000	9.60.–4	2.160.–3	1.22.+4	4006	25477	10.120	31.591	4.16
1200	0.0204	2.262.–3	669.3	4835	25654	10.876	28.225	4.14
1400	0.1794	2.370.–3	86.06	5668	25778	11.518	25.882	4.19
1500	0.4269	2.433.–3	38.17	6088	25845	11.808	24.979	4.20

THERMODYNAMIC AND THERMOPHYSICAL PROPERTIES—TABLE 49. Saturated Mercury

T, K	P, bar	$v_f \times 10^5$, m^3/kg	v_g, m^3/kg	h_f, kJ/kg	h_g, kJ/kg	h_{fg}, kJ/kg	s_f, kJ/(kg·K)	s_g, kJ/(kg·K)
203.15	$2.298 \cdot 10^{-11}$	7.26239	$3.665 \cdot 10^8$	33.131	342.637	309.506	0.32434	1.84787
213.15	$1.288 \cdot 10^{-10}$	7.27570	$6.862 \cdot 10^8$	34.567	343.674	309.107	0.33124	1.78142
223.15	$6.169 \cdot 10^{-10}$	7.28900	$1.499 \cdot 10^8$	35.997	344.710	308.713	0.33780	1.72123
233.15	$2.580 \cdot 10^{-9}$	7.30231	$3.746 \cdot 10^7$	37.422	345.746	308.324	0.34404	1.66647
243.15	$9.573 \cdot 10^{-9}$	7.31563	$1.053 \cdot 10^7$	38.842	346.782	307.940	0.35001	1.61647
253.15	$3.198 \cdot 10^{-8}$	7.32896	$3.281 \cdot 10^6$	40.258	347.819	307.561	0.35571	1.57065
263.15	$9.736 \cdot 10^{-8}$	7.34229	$1.120 \cdot 10^6$	41.668	348.855	307.187	0.36118	1.52852
273.15	$2.728 \cdot 10^{-7}$	7.35563	$4.150 \cdot 10^5$	43.074	349.891	306.817	0.36642	1.48967
283.15	$7.101 \cdot 10^{-7}$	7.36898	$1.653 \cdot 10^5$	44.476	350.927	306.451	0.37146	1.45375
293.15	$1.729 \cdot 10^{-6}$	7.38234	$7.026 \cdot 10^4$	45.874	351.964	306.090	0.37631	1.42045
303.15	$3.968 \cdot 10^{-6}$	7.39572	$3.167 \cdot 10^4$	47.268	353.000	305.732	0.38099	1.38951
313.15	$8.626 \cdot 10^{-6}$	7.40911	$1.505 \cdot 10^4$	48.659	354.036	305.377	0.38550	1.36068
323.15	$1.786 \cdot 10^{-5}$	7.42252	$7.501 \cdot 10^3$	50.046	355.072	305.026	0.38986	1.33378
333.15	$3.356 \cdot 10^{-5}$	7.43594	$3.905 \cdot 10^3$	51.430	356.108	304.678	0.39408	1.30862
343.15	$6.724 \cdot 10^{-5}$	7.44938	$2.115 \cdot 10^3$	52.810	357.145	304.335	0.39816	1.28505
353.15	$1.232 \cdot 10^{-4}$	7.46285	$1.188 \cdot 10^{-3}$	54.188	358.181	303.993	0.40212	1.26292
363.15	$2.182 \cdot 10^{-4}$	7.47633	$6.899 \cdot 10^{-2}$	55.563	359.217	303.654	0.40596	1.24213
373.15	$3.745 \cdot 10^{-4}$	7.48984	413.0	56.936	360.253	303.317	0.40969	1.22255
383.15	$6.247 \cdot 10^{-4}$	7.50337	254.2	58.306	361.289	302.983	0.41331	1.20408
393.15	$1.015 \cdot 10^{-3}$	7.51693	153.6	59.674	362.326	302.652	0.41684	1.18665
403.15	$1.608 \cdot 10^{-3}$	7.53052	103.9	61.039	363.362	302.323	0.42027	1.17017
413.15	$2.491 \cdot 10^{-3}$	7.55415	68.75	62.403	364.397	301.994	0.42361	1.15456
423.15	$3.778 \cdot 10^{-3}$	7.55780	46.43	63.765	365.433	301.668	0.42687	1.13978
433.15	$5.618 \cdot 10^{-3}$	7.57148	31.96	65.125	366.469	301.344	0.43004	1.12575
443.15	$8.204 \cdot 10^{-3}$	7.58520	22.39	66.484	367.504	301.020	0.43314	1.11242
453.15	$1.178 \cdot 10^{-2}$	7.59897	15.95	67.842	368.539	300.697	0.43617	1.09975
463.15	$1.664 \cdot 10^{-2}$	7.61277	11.54	69.198	369.574	300.376	0.43913	1.08768
473.15	$2.315 \cdot 10^{-2}$	7.62662	8.469	70.553	370.609	300.056	0.44203	1.07619
483.15	$3.177 \cdot 10^{-2}$	7.64051	6.301	71.908	371.642	299.734	0.44486	1.06524

THERMODYNAMIC AND THERMOPHYSICAL PROPERTIES—TABLE 49. Saturated Mercury (Continued)

T, K	P, bar	$v_f \times 10^5$, m³/kg	v_g, m³/kg	h_f, kJ/kg	h_g, kJ/kg	h_{fg}, kJ/kg	s_f, kJ/(kg·K)	s_g, kJ/(kg·K)
493.15	$4.304 \cdot 10^{-2}$	7.65444	4.746	73.261	372.676	299.415	0.44763	1.05478
503.15	$5.758 \cdot 10^{-2}$	7.66843	3.621	74.614	373.708	299.094	0.45035	1.04479
513.15	$7.614 \cdot 10^{-2}$	7.68247	2.793	75.967	374.740	298.773	0.45301	1.03525
523.15	$9.959 \cdot 10^{-2}$	7.69656	2.176	77.319	375.771	298.452	0.45562	1.02611
533.15	0.12892	7.71071	1.7132	78.671	376.800	298.129	0.45818	1.01737
543.15	0.16527	7.72491	1.3613	80.023	377.829	297.806	0.46069	1.00899
553.15	0.20993	7.73918	1.0912	81.375	378.855	297.480	0.46316	1.00095
563.15	0.26435	7.75351	0.88213	82.728	379.880	297.152	0.46558	0.99324
573.15	0.33015	7.7679	0.71874	84.080	380.904	296.824	0.46796	0.98584
583.15	0.40910	7.7823	0.59002	85.434	381.925	296.491	0.47030	0.97893
593.15	0.50320	7.7969	0.48779	86.788	382.944	296.156	0.47260	0.97190
603.15	0.61460	7.8115	0.40600	88.143	383.960	295.817	0.47487	0.96532
613.15	0.74567	7.8262	0.34008	89.499	384.973	295.474	0.47709	0.95899
623.15	0.89896	7.8409	0.28660	90.856	385.984	295.128	0.47929	0.95289
633.15	1.0772	7.8558	0.24291	92.215	386.991	294.776	0.48145	0.94702
643.15	1.2834	7.8707	0.20702	93.575	387.994	294.419	0.48358	0.94135
653.15	1.5207	7.8858	0.17735	94.937	388.994	294.057	0.48568	0.93589
663.15	1.9725	7.9008	0.15269	96.300	389.989	293.689	0.48774	0.93061
673.15	2.1024	7.9160	0.13207	97.666	390.980	293.314	0.48978	0.92552
683.15	2.454	7.9313	0.11476	99.033	391.966	292.933	0.49180	0.92059
693.15	2.852	7.9467	0.10014	100.403	392.947	292.544	0.49378	0.91583
703.15	3.299	7.9622	0.08775	101.775	393.923	292.148	0.49574	0.91123
713.15	3.801	7.9778	0.07719	103.150	394.893	291.743	0.49768	0.90677
723.15	4.362	7.9935	0.06815	104.528	395.858	291.330	0.49959	0.90245
733.15	4.986	8.0094	0.06039	105.908	396.816	290.908	0.50148	0.89827
743.15	5.679	8.0252	0.05369	107.292	397.767	290.475	0.50335	0.89422
753.15	6.446	8.0413	0.04789	108.679	398.711	290.032	0.50519	0.89029
763.15	7.292	8.0574	0.04285	110.069	399.649	289.580	0.50702	0.88647
773.15	8.222	8.074	0.03846	111.463	400.579	289.116	0.50882	0.88277
783.15	9.242	8.090	0.03462	112.861	401.501	288.640	0.51061	0.87917

793.15	10.358	8.106	0.03124	114.262	402.415	288.153	0.51238	0.87568
803.15	11.576	8.123	0.02827	115.668	403.321	287.653	0.51412	0.87228
813.15	12.901	8.140	0.02565	117.078	404.218	287.140	0.51586	0.86898
823.15	14.340	8.157	0.02333	118.492	405.106	286.614	0.51757	0.86576
833.15	15.899	8.174	0.02126	119.911	405.985	286.074	0.51927	0.86263
843.15	17.584	8.191	0.019426	121.335	406.855	285.520	0.52095	0.85959
853.15	19.403	8.209	0.017785	122.763	407.715	284.952	0.52262	0.85662
863.15	21.36	8.226	0.016317	124.197	408.565	284.368	0.52427	0.85372
873.15	23.46	8.244	0.015000	125.636	409.405	283.769	0.52591	0.85090
883.15	25.72	8.262	0.013815	127.080	410.235	283.155	0.52753	0.84815
893.15	28.14	8.280	0.012748	128.530	411.054	282.524	0.52914	0.84546
903.15	30.72	8.298	0.011784	129.986	411.861	281.875	0.53074	0.84284
913.15	33.47	8.316	0.010911	131.448	412.658	281.210	0.53232	0.84028
923.15	36.41	8.335	0.010120	132.915	413.444	280.529	0.53389	0.83777
933.15	39.53	8.353	0.009401	134.389	414.218	279.829	0.53545	0.83533
943.15	42.85	8.372	0.008746	135.869	414.980	279.111	0.53700	0.83294
953.15	46.36	8.391	0.008150	137.356	415.731	278.375	0.53854	0.83060
963.15	50.09	8.410	0.007604	138.850	416.469	277.619	0.54006	0.82831
973.15	54.03	8.430	0.007105	140.350	417.195	276.845	0.54158	0.82606
983.15	58.20	8.450	0.006648	141.858	417.909	276.051	0.54308	0.82387
993.15	62.59	8.468	0.006228	143.372	418.610	275.238	0.54458	0.82172
1003.15	67.22	8.488	0.005842	144.894	419.298	274.404	0.54607	0.81691
1013.15	72.10	8.508	0.005487	146.424	419.974	273.550	0.54754	0.81754
1023.15	77.22	8.529	0.005159	147.961	420.636	272.675	0.54901	0.81552
1033.15	82.60	8.550	0.004856	149.506	421.286	271.780	0.55047	0.81353
1043.15	88.25	8.570	0.004576	151.059	421.923	270.864	0.55192	0.81158
1053.15	94.17	8.590	0.004317	152.619	422.546	269.927	0.55336	0.80966
1063.15	100.37	8.612	0.004077	154.188	423.156	268.968	0.55479	0.80778
1073.15	106.85	8.632	0.003854	155.766	423.752	267.986	0.55621	0.80593

THERMODYNAMIC AND THERMOPHYSICAL PROPERTIES—TABLE 50. Saturated Methane

T, K	P, bar	v_f, m³/kg	v_g, m³/kg	h_f, kJ/kg	h_g, kJ/kg	s_f, kJ/(kg·K)	s_g, kJ/(kg·K)	c_{pf}, kJ/(kg·K)	μ_f, 10⁻⁴ Pa·s	k_f, W/(m·K)
90.7	0.117	2.215.–3	3.976	216.4	759.9	4.231	10.225	3.288	2.02	0.225
95	0.198	2.244.–3	2.463	232.5	769.0	4.406	10.034	3.318	1.71	0.215
100	0.345	2.278.–3	1.479	246.3	776.9	4.556	9.862	3.369	1.56	0.206
105	0.565	2.316.–3	0.940	263.2	785.7	4.719	9.710	3.425	1.33	0.197
110	0.884	2.353.–3	0.625	280.1	794.5	4.882	9.558	3.478	1.22	0.189
115	1.325	2.396.–3	0.430	297.7	802.5	5.035	9.436	3.525	1.09	0.181
120	1.919	2.438.–3	0.306	315.3	810.4	5.188	9.314	3.570	0.98	0.173
125	2.693	2.487.–3	0.223	333.5	817.3	5.332	9.062	3.620	0.89	0.165
130	3.681	2.536.–3	0.167	351.7	824.1	5.476	8.810	3.679	0.81	0.158
135	4.912	2.594.–3	0.127	370.6	829.5	5.614	8.871	3.755	0.73	0.150
140	6.422	2.652.–3	0.098	389.5	834.8	5.751	8.932	3.849	0.66	0.143
145	8.246	2.722.–3	0.077	409.5	844.4	5.885	8.891	3.965	0.61	0.136
150	10.41	2.792.–3	0.061	429.4	853.9	6.019	8.849	4.101	0.56	0.129
155	12.97	2.882.–3	0.049	450.8	848.5	6.151	8.725	4.27	0.51	0.122
160	15.94	2.971.–3	0.039	472.1	843.0	6.283	8.601	4.47	0.46	0.115
165	19.39	3.095.–3	0.032	495.4	840.0	6.417	8.513	4.75	0.42	0.108
170	23.81	3.218.–3	0.026	518.6	837.0	6.551	8.424	5.16	0.38	0.101
175	27.81	3.419.–3	0.020	545.8	827.6	6.697	8.315	5.89	0.34	0.094
180	32.86	3.619.–3	0.016	572.9	818.1	6.843	8.205	7.27	0.30	0.088
185	38.59	3.979.–3	0.012	605.4	797.7	7.017	8.049	11.1	0.25	0.085
190	45.20	4.900.–3	0.008	661.6	750.7	7.293	7.762	70	0.19	0.090
190.6	45.99	6.233.–3	0.006	704.4	704.4	7.516	7.516	∞	0.17	∞

THERMODYNAMIC AND THERMOPHYSICAL PROPERTIES—TABLE 51. Superheated Methane

P, bar		100	150	200	250	300	350	400	450	500
1	v	0.00228	0.7661	1.0299	1.2915	1.5521	1.8122	2.0719	2.3669	2.5911
	h	246.4	879.0	984.3	1090.4	1199.8	1314.8	1437.4	1568.8	1708.9
	s	4.555	10.152	10.757	11.230	11.629	11.983	12.310	12.618	12.914
5	v	0.00228	0.1434	0.2006	0.2549	0.3083	0.3611	0.4136	0.4657	0.5181
	h	247.0	865.0	976.1	1084.7	1195.5	1311.5	1434.7	1566.6	1706.9
	s	4.553	9.256	9.896	10.381	10.785	11.142	11.471	11.781	12.066
10	v	0.00227	0.0643	0.0968	0.1254	0.1528	0.1798	0.2063	0.2327	0.2590
	h	247.8	843.6	965.5	1077.9	1190.6	1307.9	1432.0	1564.1	1705.3
	s	4.549	8.797	9.501	10.002	10.414	10.775	11.106	11.417	11.715
20	v	0.00227	0.00277	0.0446	0.0606	0.0751	0.0891	0.1027	0.1162	0.1295
	h	249.4	429.8	941.9	1063.6	1180.7	1300.6	1426.5	1560.3	1702.1
	s	4.542	6.003	9.059	9.603	10.030	10.400	10.736	11.050	11.349
40	v	0.00226	0.00274	0.0176	0.0281	0.0363	0.0438	0.0510	0.0579	0.0648
	h	252.5	430.8	879.3	1032.9	1160.5	1286.0	1415.7	1552.1	1696.0
	s	4.528	5.973	8.465	9.155	9.621	10.008	10.354	10.674	10.978
60	v	0.00226	0.00271	0.00615	0.0173	0.0234	0.0287	0.0338	0.0386	0.0432
	h	255.7	432.2	734.0	999.8	1140.0	1271.7	1405.1	1544.2	1690.0
	s	4.515	5.946	7.623	8.847	9.359	9.765	10.121	10.440	10.756
80	v	0.00225	0.00268	0.00411	0.0119	0.0171	0.0213	0.0252	0.0289	0.0324
	h	258.9	433.8	660.5	964.4	1119.7	1257.7	1394.9	1536.6	1684.4
	s	4.502	5.920	7.209	8.590	9.158	9.584	9.951	10.283	10.595
100	v	0.00224	0.00266	0.00375	0.00888	0.0133	0.0169	0.0201	0.0231	0.0260
	h	262.1	435.5	644.5	928.5	1099.6	1244.2	1385.2	1529.4	1679.0
	s	4.489	5.897	7.090	8.364	8.991	9.437	9.814	10.153	10.469
150	v	0.00223	0.00261	0.00337	0.00555	0.00852	0.0111	0.0134	0.0155	0.0175
	h	270.2	440.7	630.2	860.0	1054.1	1213.1	1362.8	1513.0	1667.0
	s	4.458	5.843	6.930	7.953	8.664	9.155	9.555	9.907	10.233
200	v	0.00221	0.00256	0.00318	0.00447	0.00644	0.00837	0.0101	0.0118	0.0133
	h	278.3	446.5	626.5	825.0	1019.8	1187.2	1343.8	1498.9	1656.9
	s	4.429	5.796	6.829	7.719	8.426	8.944	9.362	9.727	10.060
300	v	0.00218	0.00249	0.00296	0.00369	0.00474	0.00593	0.00708	0.00818	0.00924
	h	294.7	459.6	629.2	804.4	982.9	1153.6	1316.8	1478.5	1642.2
	s	4.373	5.714	6.690	7.471	8.122	8.649	9.085	9.465	9.811
400	v		0.00244	0.00282	0.00336	0.00406	0.00486	0.00569	0.00560	0.00729
	h		473.8	637.7	802.4	970.1	1137.8	1303.0	1467.7	1634.7
	s		5.645	6.588	7.323	7.935	8.451	8.893	9.280	9.633
500	v		0.00239	0.00272	0.00315	0.00368	0.00428	0.00492	0.00555	0.00616
	h		488.8	648.9	807.7	969.0	1132.8	1297.8	1464.2	1633.2
	s		5.584	6.507	7.215	7.802	8.307	8.748	9.139	9.496

Temperature, K

THERMODYNAMIC AND THERMOPHYSICAL PROPERTIES – TABLE 52. Thermophysical Properties of Saturated Methanol

Pressure, bar	Temp., K	v_f, m³/kg	v_g, m³/kg	h_f, kJ/kg	h_g, kJ/kg	s_f, kJ/(kg·K)	s_g, kJ/(kg·K)	c_{pf}, kJ/(kg·K)	μ_f, 10^{-6}Pa·s	k_f, W/(m·K)	Pr_f
4×10^{-6t}	175.6	0.001057	1700000	0.0	1303.1	2.8114	10.2328				
0.1	288.4	0.001257	7.309	261.0	1440.3	3.9383	8.0281	2.531	625	0.204	7.75
0.2	301.7	0.001276	3.801	293.9	1455.4	4.0493	7.9032	2.554	525	0.196	6.84
0.5	320.7	0.001307	1.599	345.0	1476.2	4.2117	7.7386	2.669	401	0.193	5.55
1.013	337.7	0.001336	0.819	391.7	1492.1	4.3516	7.6104	2.777	329	0.189	4.83
1.5	348.0	0.001356	0.5632	421.0	1500.3	4.4361	7.5379	2.845	288	0.186	4.41
2.0	356.0	0.001371	0.4276	444.2	1505.8	4.5014	7.4836	2.894	268	0.184	4.22
2.5	362.5	0.001385	0.3443	463.6	1509.8	4.5536	7.4398	2.946	242	0.182	3.92
3.0	368.0	0.001396	0.2893	479.8	1512.4	4.5992	7.4051	2.984	227	0.181	3.74
4.0	377.1	0.001417	0.2188	507.8	1515.9	4.6728	7.3474	3.050	204	0.179	3.48
5	384.5	0.001434	0.17569	529.7	1517.4	4.7307	7.2992	3.117	187	0.178	3.27
6	390.8	0.001450	0.14683	549.6	1518.4	4.7836	7.2624	3.176	174	0.177	3.12
8	401.3	0.001479	0.11015	582.7	1518.0	4.8678	7.1988	3.265	156	0.175	2.91
10	409.8	0.001504	0.08783	610.3	1516.1	4.9366	7.1471	3.349	141	0.173	2.73
15	426.3	0.001560	0.05761	665.8	1507.9	5.0708	7.0461	3.540	117	0.171	2.42
20	438.9	0.001611	0.04224	710.5	1553.8	5.1744	6.9677	3.72	102	0.169	2.25
25	449.3	0.001666	0.03290	749.0	1486.4	5.2605	6.9017	3.91	92	0.167	2.15
30	458.2	0.001710	0.02661	783.8	1474.7	5.3355	6.8435	4.12	84	0.165	2.10
40	472.9	0.001814	0.01863	846.7	1450.1	5.4650	6.7409	4.67	72	0.160	2.10
50	484.9	0.001934	0.01373	905.2	1423.2	5.5793	6.6475	5.55	63	0.154	2.27
60	495.1	0.002086	0.01032	963.3	1391.8	5.6889	6.5543				
80	508.1	0.002507	0.00642	1065.3	1318.7	5.8803	6.3791				
80.95^c	512.6	0.003715	0.00372	1186.8	1186.8	6.0979	6.0979				

t = triple point; c = critical point.

THERMODYNAMIC AND THERMOPHYSICAL PROPERTIES—TABLE 53. Thermodynamic Properties of Compressed Methanol

Pressure, bar		200	250	300	350	400	450	500	550	600
						Temperature, K				
0.1	v (m³/kg)	0.001137	0.001203	7.630	8.942	10.23	11.56	12.84		15.45
	h (kJ/kg)	57.5	169.8	1456.7	1529.5	1607.5	1691.5	1781.7	1878.1	1980.2
	s (kJ/kg·K)	3.096	3.597	8.081	8.305	8.514	8.711	8.901	9.085	9.263
0.5	v (m³/kg)	0.001137	0.001202	0.001274	1.764	2.033	2.296	2.558	2.818	3.078
	h (kJ/kg)	57.5	169.9	290.5	1522.7	1603.0	1687.9	1778.9	1875.3	1977.4
	s (kJ/kg·K)	3.096	3.597	4.038	7.877	8.091	8.291	8.482	8.666	8.844
1.013	v (m³/kg)	0.001137	0.001202	0.001274	0.8560	0.9958	1.1283	1.2843	1.3870	1.5157
	h (kJ/kg)	57.6	169.9	290.5	1514.0	1598.7	1685.1	1795.4	1873.5	1975.8
	s (kJ/kg·K)	3.096	3.597	4.038	7.675	7.902	8.105	8.117	8.482	8.660
10	v (m³/kg)	0.001136	0.001201	0.001272	0.001357	0.001474	0.1068	0.1236	0.1381	0.1519
	h (kJ/kg)	58.4	170.7	291.2	427.4	578.8	1638.1	1751.5	1858.0	1965.2
	s (kJ/kg·K)	3.095	3.596	4.036	4.451	4.857	7.427	7.667	7.870	8.056
15	v (m³/kg)	0.001136	0.001201	0.001272	0.001356	0.001472	0.0673	0.0806	0.0911	0.1007
	h (kJ/kg)	58.8	171.1	291.6	427.7	578.9	1601.9	1735.6	1849.4	1960.2
	s (kJ/kg·K)	3.094	3.595	4.035	4.450	4.856	7.253	7.536	7.752	7.946
20	v (m³/kg)	0.001135	0.001200	0.001271	0.001355	0.001469	0.0466	0.0589	0.0675	0.0751
	h (kJ/kg)	59.2	171.6	292.0	428.1	579.0	1565.3	1717.7	1840.0	1954.8
	s (kJ/kg·K)	3.094	3.595	4.035	4.449	4.854	7.087	7.431	7.664	7.864
30	v (m³/kg)	0.001134	0.001199	0.001269	0.001355	0.001465	0.001659	0.0367	0.0436	0.0492
	h (kJ/kg)	60.1	172.4	292.9	428.8	579.4	751.3	1675.4	1818.7	1942.7
	s (kJ/kg·K)	3.092	3.593	4.036	4.447	4.851	5.264	7.253	7.526	7.743
40	v (m³/kg)	0.001133	0.001198	0.001268	0.001350	0.001461	0.001649	0.0251	0.0314	0.0361
	h (kJ/kg)	61.0	173.3	293.7	429.5	579.8	750.4	1623.0	1794.2	1928.8
	s (kJ/kg·K)	3.091	3.592	4.032	4.445	4.849	5.258	7.088	7.414	7.650

THERMODYNAMIC AND THERMOPHYSICAL PROPERTIES—TABLE 53. Thermodynamic Properties of Compressed Methanol (*Continued*)

Pressure, bar		Temperature, K								
		200	250	300	350	400	450	500	550	600
50	v (m³/kg)	0.001133	0.001197	0.001266	0.001348	0.001457	0.001637	0.0176	0.0239	0.0282
	h (kJ/kg)	61.9	174.2	294.5	430.2	580.2	749.7	1556.4	1766.7	1913.4
	s (kJ/kg·K)	3.090	3.591	4.030	4.443	4.846	5.579	6.912	7.314	7.570
60	v (m³/kg)	0.001131	0.001196	0.001265	0.001346	0.001453	0.001628	0.0120	0.0188	0.0228
	h (kJ/kg)	62.8	175.0	295.3	430.9	580.6	749.1	1461.8	1736.1	1896.6
	s (kJ/kg·K)	3.089	3.589	4.029	4.442	4.843	5.248	6.692	7.220	7.500
75	v (m³/kg)	0.001130	0.001194	0.001263	0.001343	0.001448	0.001614	0.002084	0.01359	0.0174
	h (kJ/kg)	64.1	176.3	296.6	431.9	581.2	748.3	982.1	1683.9	1869.1
	s (kJ/kg·K)	3.087	3.586	4.027	4.439	4.839	5.241	5.718	7.081	7.405
100	v (m³/kg)	0.001128	0.001191	0.001259	0.001337	0.001439	0.001595	0.001952	0.	0.01188
	h (kJ/kg)	66.3	178.5	298.6	433.8	582.4	747.5	964.8	1572.9	1818.8
	s (kJ/kg·K)	3.084	3.584	4.023	4.435	4.833	5.230	5.673	6.829	7.261
150	v (m³/kg)	0.001125	0.001186	0.001252	0.001328	0.001423	0.001562	0.001825		0.006513
	h (kJ/kg)	70.7	182.8	302.8	437.4	584.9	746.8	948.4	1248.8	1704.3
	s (kJ/kg·K)	3.078	3.578	4.016	4.426	4.822	5.211	5.622	6.302	6.997
200	v (m³/kg)	0.001121	0.001182	0.001246	0.001317	0.001408	0.001535	0.001751	0.002314	0.004091
	h (kJ/kg)	75.1	187.2	307.0	441.2	587.8	747.0	939.9	1223.5	1583.5
	s (kJ/kg·K)	3.071	3.571	4.009	4.418	4.811	5.194	5.587	6.125	6.752
300	v (m³/kg)	0.001113	0.001172	0.001234	0.001302	0.001384	0.001492	0.001656	0.001957	0.002600
	h (kJ/kg)	83.9	195.9	315.4	448.9	593.8	749.4	932.0	1173.4	1443.5
	s (kJ/kg·K)	3.060	3.559	3.996	4.403	4.791	5.166	5.537	5.996	6.466
400	v (m³/kg)	0.001107	0.001164	0.001223	0.001288	0.001363	0.001459	0.001593	0.001808	0.002182
	h (kJ/kg)	92.7	204.7	324.0	456.9	600.5	753.4	929.6	1154.4	1388.1
	s (kJ/kg·K)	3.048	3.548	3.983	4.388	4.774	5.142	5.500	5.926	6.335
500	v (m³/kg)	0.001101	0.001156	0.001213	0.001274	0.001345	0.001431	0.001546	0.001716	0.001980
	h (kJ/kg)	101.5	213.4	332.6	465.1	607.7	758.4	930.4	1145.6	1360.6
	s (kJ/kg·K)	3.037	3.536	3.971	4.375	4.757	5.121	5.470	5.880	6.254

470

THERMODYNAMIC AND THERMOPHYSICAL PROPERTIES — TABLE 54. Saturated Methyl Chloride

T, K	P, bar	v_f, m³/kg	v_g, m³/kg	h_f, kJ/kg	h_g, kJ/kg	s_f, kJ/(kg·K)	s_g, kJ/(kg·K)	c_{pf}, kJ/(kg·K)	μ_f, 10^{-4} Pa·s	k_f, W/(m·K)
175	0.0117	8.84.–4	27.90	274.5	764.3	3.529	6.328	1.469		
180	0.0165	8.91.–4	19.85	280.9	767.7	3.570	6.274	1.472		
185	0.0233	8.97.–4	14.12	287.5	771.0	3.603	6.222	1.475		
190	0.0327	9.04.–4	10.12	294.5	774.3	3.647	6.172	1.477		
195	0.0462	9.10.–4	7.208	301.7	777.5	3.684	6.124	1.480		
200	0.0653	9.17.–4	5.137	309.0	780.7	3.722	6.080	1.483	4.44	0.241
205	0.0919	9.25.–4	3.835	316.3	783.9	3.756	6.038	1.486	4.27	0.236
210	0.1315	9.33.–4	2.656	323.7	787.0	3.791	5.998	1.489	4.11	0.232
215	0.181	9.40.–4	1.975	331.0	790.1	3.825	5.961	1.492	3.96	0.228
220	0.243	9.48.–4	1.505	338.4	793.2	3.859	5.928	1.496	3.82	0.224
225	0.319	9.56.–4	1.168	345.7	796.3	3.892	5.896	1.500	3.69	0.219
230	0.417	9.65.–4	0.911	353.1	799.3	3.925	5.866	1.504	3.57	0.215
235	0.539	9.73.–4	0.718	360.5	802.3	3.957	5.845	1.508	3.46	0.211
240	0.688	9.81.–4	0.572	368.0	805.3	3.988	5.822	1.513	3.35	0.207
245	0.866	9.89.–4	0.462	375.6	808.2	4.019	5.786	1.518	3.25	0.202
250	1.076	9.98.–4	0.377	383.2	811.1	4.050	5.762	1.523	3.16	0.198
255	1.328	10.08.–4	0.311	390.7	814.0	4.080	5.740	1.528	3.08	0.194
260	1.627	10.18.–4	0.257	398.3	816.8	4.110	5.720	1.533	3.00	0.190
265	1.970	10.27.–4	0.215	406.0	819.4	4.139	5.699	1.539	2.92	0.186
270	2.364	10.36.–4	0.1807	413.7	822.0	4.168	5.680	1.546	2.85	0.182

THERMODYNAMIC AND THERMOPHYSICAL PROPERTIES — TABLE 54.   Saturated Methyl Chloride (*Continued*)

T, K	P, bar	v_f, m³/kg	v_g, m³/kg	h_f, kJ/kg	h_g, kJ/kg	s_f, kJ/(kg·K)	s_g, kJ/(kg·K)	c_{pf}, kJ/(kg·K)	μ_f, 10⁻⁴ Pa·s	k_f, W/(m·K)
275	2.830	10.46.–4	0.1524	421.5	824.4	4.197	5.662	1.554	2.78	0.177
280	3.347	10.57.–4	0.1301	429.4	826.8	4.225	5.644	1.565	2.72	0.173
285	3.936	10.68.–4	0.1115	437.3	829.0	4.253	5.628	1.574	2.66	0.169
290	4.612	10.79.–4	0.0960	445.2	831.2	4.280	5.612	1.583	2.61	0.165
295	5.361	10.91.–4	0.0830	453.2	833.2	4.308	5.597	1.594	2.56	0.160
300	6.189	11.03.–4	0.0723	461.2	835.2	4.334	5.581	1.605	2.51	0.156
305	7.110	11.15.–4	0.0632	469.3	837.0	4.361	5.567	1.617	2.46	0.152
310	8.111	11.27.–4	0.0556	477.4	838.8	4.388	5.553	1.631	2.42	0.148
315	9.243	11.40.–4	0.0489	485.6	840.5	4.414	5.540	1.644	2.37	0.143
320	10.47	11.55.–4	0.0433	493.8	841.9	4.440	5.527	1.658	2.33	0.139
325	11.78	11.70.–4	0.0386	502.1	843.3	4.465	5.516		2.30	0.135
330	13.27	11.86.–4	0.0343	510.4	844.5	4.491	5.504		2.27	0.131
340	16.52	12.17.–4	0.0282	518.8	846.4	4.542	5.481		2.12	0.124
350	20.53	12.54.–4	0.0228	538.3	847.5	4.592	5.457		1.99	0.117
360	25.29	12.97.–4	0.0186	562.9	847.6	4.643	5.434		1.87	0.110
370	30.74	13.47.–4	0.0151	581.6	845.9	4.694	5.398		1.77	0.103
380	36.99	14.11.–4	0.0117	602.8	842.6	4.747	5.382		1.67	0.095
390	44.05	14.67.–4	0.0096	622.9	837.4	4.805	5.358		1.59	0.086
400	52.29	15.66.–4	0.0075	643.6	826.4	4.870	5.323		1.51	0.075
405	56.6	16.48.–4	0.0063	663.2	819.1	4.904	5.289			
410	61.5	17.97.–4	0.0052	677.3	807.1	4.954	5.256			
415	67.4	21.10.–4	0.0038	714.1	778.6	5.025	5.200			
416	69.0	27.40.–4	0.0027	749.3	749.3	5.116	5.116			

THERMODYNAMIC AND THERMOPHYSICAL PROPERTIES—TABLE 55. Saturated Neon

T, K	P, bar	v_f, m³/kg	v_g, m³/kg	h_f, kJ/kg	h_g, kJ/kg	s_f, kJ/(kg·K)	s_g, kJ/(kg·K)	c_{pf}, kJ/(kg·K)	μ_f, 10⁻⁴ Pa·s	k_f, W/(m·K)
10		6.654,–4		0.75		0.0992		0.278		
20		6.823,–4		6.78		0.4906		0.945		
24.6		6.696,–4		11.96		0.7257		1.345		
24.6	0.434	8.012,–4	0.2266	28.22	117.0	1.388	5.006	1.802	1.57	0.146
26	0.718	8.172,–4	0.1429	30.90	118.1	1.494	4.846	1.868	1.37	0.132
28	1.321	8.413,–4	0.0817	34.75	119.3	1.634	4.653	1.955	1.16	0.124
30	2.238	8.687,–4	0.0501	38.80	120.1	1.771	4.483	2.052	1.00	0.115
32	3.552	9.001,–4	0.0323	43.06	120.6	1.905	4.329	2.163	0.84	0.106
34	5.352	9.370,–4	0.0217	47.57	120.6	2.036	4.184	2.302	0.71	0.097
36	7.728	9.820,–4	0.0149	52.34	119.9	2.166	4.043	2.506	0.59	0.088
38	10.78	1.039,–3	0.0104	57.52	118.4	2.297	3.900	2.825	0.48	0.078
40	14.62	1.116,–3	0.0073	63.33	115.8	2.435	3.749	3.436	0.38	0.069
42	19.39	1.232,–4	0.0050	69.82	111.8	2.582	3.582	5.26	0.31	0.059
44	25.22	1.538,–3	0.0031	80.83	103.0	2.812	3.316	25.000	0.25	
44.4	26.53	2.070,–3	0.0021	92.50	92.5	3.062	3.062	∞		∞

Temperature, K		Pressure, bar											
		1	10	20	40	60	80	100	200	400	600	800	1000
100	v	0.4117	0.0410	0.0204	0.0102	6.76.−3	5.08.−3	4.09.−3	2.22.−3	1.42.−3	1.18.−3	1.06.−3	9.74.−4
	h	195.4	194.0	192.4	189.4	186.6	184.0	181.6	174.1	173.3	180.0	189.2	199.2
	s	6.129	5.168	4.869	4.556	4.363	4.221	4.106	3.739	3.386	3.197	3.066	2.964
200	v	0.8243	0.0828	0.0416	0.0210	0.0142	0.0107	8.69.−3	4.61.−3	2.61.−3	1.95.−3	1.63.−3	1.43.−3
	h	298.5	298.4	298.4	298.2	298.2	298.2	298.3	299.4	304.6	312.4	321.8	332.1
	s	6.844	5.893	5.605	5.315	5.143	5.020	4.924	4.620	4.308	4.124	3.994	3.893
300	v	1.236	0.1241	0.0624	0.0315	0.0212	0.0160	0.0129	6.77.−3	3.71.−3	2.69.−3	2.18.−3	1.87.−3
	h	401.6	401.8	402.2	402.8	403.5	404.1	404.9	408.8	417.8	427.8	438.5	449.7
	s	7.262	6.312	6.026	5.739	5.570	5.450	5.357	5.065	4.769	4.593	4.469	4.372
400	v	1.648	0.1654	0.0830	0.0418	0.0281	0.0212	0.0171	8.88.−3	4.77.−3	3.40.−3	2.72.−3	2.30.−3
	h	504.6	505.0	505.4	506.4	507.4	508.3	509.3	514.4	525.3	536.7	548.4	560.2
	s	7.558	6.609	6.323	6.037	5.896	5.750	5.657	5.369	5.078	4.907	4.785	4.690
500	v	2.060	0.2066	0.1036	0.0521	0.0350	0.0264	0.0213	0.0110	5.82.−3	4.10.−3	3.24.−3	2.73.−3
	h	607.6	608.1	608.6	609.7	610.8	611.9	613.0	618.8	630.7	642.9	655.2	667.5
	s	7.788	6.839	6.553	6.267	6.100	5.981	5.889	5.601	5.313	5.144	5.023	4.929
600	v	2.472	0.2478	0.1242	0.0625	0.0419	0.0316	0.0254	0.0130	6.85.−3	4.80.−3	3.77.−3	3.15.−3
	h	710.6	711.1	711.7	712.9	714.1	715.3	716.5	722.5	735.0	747.8	760.5	773.2
	s	7.975	7.027	6.741	6.455	6.288	6.169	6.077	5.791	5.504	5.335	5.215	5.122
700	v	2.884	0.2890	0.1449	0.0728	0.0487	0.0367	0.0295	0.0151	7.89.−3	5.49.−3	4.29.−3	3.57.−3
	h	813.5	814.1	814.7	816.0	817.2	818.5	819.7	826.0	838.9	851.9	856.0	878.1
	s	8.134	7.186	6.900	6.614	6.447	6.328	6.236	5.950	5.664	5.496	5.376	5.284
800	v	3.296	0.3302	0.1655	0.0831	0.0556	0.0419	0.0336	0.0172	8.92.−3	6.18.−3	4.81.−3	3.98.−3
	h	916.5	917.1	917.7	919.0	920.3	921.6	922.9	929.3	942.4	955.7	969.0	982.2
	s	8.272	7.323	7.038	6.752	6.585	6.466	6.374	6.088	5.802	5.634	5.515	5.423
900	v	3.708	0.3714	0.1861	0.0934	0.0625	0.0470	0.0378	0.0192	9.96.−3	6.87.−3	5.32.−3	4.40.−3
	h	1020	1020	1021	1022	1023	1025	1026	1033	1046	1059	1073	1086
	s	8.393	7.444	7.159	6.873	6.706	6.588	6.496	6.210	5.924	5.756	5.637	5.545
1000	v	4.120	0.4126	0.2067	0.1037	0.0693	0.0522	0.0419	0.0213	0.0110	7.56.−3	5.84.−3	4.81.−3
	h	1123	1123	1124	1125	1126	1128	1129	1136	1149	1163	1176	1190
	s	8.502	7.553	7.267	6.982	6.815	6.696	6.604	6.318	6.032	5.856	5.746	5.654

THERMODYNAMIC AND THERMOPHYSICAL PROPERTIES—TABLE 57. Saturated Nitrogen (R728)

T, K	P, bar	v_f, 10^{-3} m³/kg	v_g, m³/kg	h_f, kJ/kg	h_g, kJ/kg	s_f, kJ/(kg·K)	s_g, kJ/(kg·K)	c_{pf}, kJ/(kg·K)	μ_f, 10^{-4} Pa·s	k_f, W/(m·K)
63.15	0.1253	1.155	1477	−148.5	64.1	2.459	5.826	1.928		0.170
65	0.1743	1.165	1091	−144.9	65.8	2.516	5.757	1.930	2.74	0.160
70	0.3859	1.193	525.6	−135.2	70.5	2.657	5.595	1.937	2.17	0.151
75	0.7609	1.224	281.8	−125.4	74.9	2.789	5.460	1.948	1.77	0.141
77.35	1.0133	1.239	216.9	−120.8	76.8	2.849	5.404	1.955	1.60	0.136
80	1.369	1.258	164.0	−115.6	78.9	2.913	5.345	1.964	1.48	0.132
85	2.287	1.297	101.7	−105.7	82.3	3.032	5.244	1.989	1.27	0.123
90	3.600	1.340	66.28	−95.6	85.0	3.147	5.152	2.028	1.10	0.114
95	5.398	1.390	44.87	−85.2	86.8	3.256	5.067	2.086	0.97	0.105
100	7.775	1.447	31.26	−74.5	87.7	3.363	4.985	2.176	0.87	0.097
105	10.83	1.514	22.23	−63.8	87.4	3.469	4.904	2.319	0.79	0.088
110	14.67	1.597	15.98	−51.4	85.6	3.575	4.820	2.566	0.71	0.800
115	19.40	1.714	11.47	−38.1	81.8	3.687	4.729	3.063	0.60	0.071
120	25.15	1.892	8.031	−21.4	74.3	3.821	4.619		0.48	0.063
125	32.05	2.324	5.016	5.1	57.2	4.024	4.444		0.32	0.052
126.25	33.96	3.289	3.289	34.8	34.8	4.252	4.252	∞		∞

THERMODYNAMIC AND THERMOPHYSICAL PROPERTIES — TABLE 58. Nitrogen (R728) at Atmospheric Pressure

T (K)	77.4[b]	80	100	120	140	160	180	200	220	240
v (m³/kg)	0.2164	0.2252	0.2871	0.3474	0.4071	0.4664	0.5255	0.5845	0.6434	0.7023
h (kJ/kg)	76.7	80.0	101.9	123.1	144.2	165.2	186.1	207.0	227.8	248.7
s (kJ/kg·K)	5.403	5.446	5.690	5.884	6.046	6.186	6.309	6.419	6.519	6.609
c_p (kJ/kg·K)	1.341	1.196	1.067	1.056	1.050	1.047	1.045	1.043	1.043	1.042
Z	0.9545	0.9610	0.9801	0.9883	0.9927	0.9952	0.9967	0.9977	0.9984	0.9990
$\bar{v}_s$ (m/s)	172	177	202	222	240	257	273	288	302	316
η (10⁻⁶ Pa·s)	5.0	5.2	6.7	8.0	9.3	10.6	11.8	12.9	14.0	15.0
k (W/m·K)	0.0074	0.0077	0.0098	0.0117	0.0136	0.0154	0.0171	0.0187	0.0203	0.0218
N_{pr}	0.913	0.811	0.728	0.727	0.723	0.721	0.720	0.719	0.718	0.717

T (K)	260	280	300	320	340	360	380	400	420	440
v (m³/kg)	0.7611	0.8199	0.8786	0.9371	0.9960	1.0546	1.1134	1.1719	1.2305	1.2892
h (kJ/kg)	269.5	290.3	311.2	332.0	352.8	373.7	394.5	415.4	436.3	457.3
s (kJ/kg·K)	6.693	6.770	6.842	6.909	6.972	7.032	7.088	7.142	7.193	7.242
c_p (kJ/kg·K)	1.042	1.041	1.041	1.042	1.042	1.043	1.044	1.045	1.047	1.048
Z	0.9994	0.9997	0.9998	0.9999	1.0000	1.0001	1.0002	1.0002	1.0002	1.0003
$\bar{v}_s$ (m/s)	329	341	359	365	376	387	397	408	417	427
η (10⁻⁶ Pa·s)	16.0	17.0	17.9	18.8	19.7	20.5	21.4	22.2	23.0	23.8
k (W/m·K)	0.0232	0.0247	0.0260	0.0273	0.0286	0.0299	0.0311	0.0324	0.0336	0.0347
N_{pr}	0.717	0.716	0.716	0.717	0.717	0.717	0.717	0.717	0.717	0.717

T (K)	460	480	500	600	700	800	900	1000	1500	2000
v (m³/kg)	1.3481	1.4065	1.4654	1.758	2.052	2.344	2.636	2.931	4.396	5.862
h (kJ/kg)	478.3	499.3	520.4	626.9	735.6	846.6	960.0	1075.7	1680.5	2313.5
s (kJ/kg·K)	7.288	7.333	7.376	7.570	7.738	7.886	8.019	8.141	8.630	8.995
c_p (kJ/kg·K)	1.051	1.053	1.056	1.075	1.098	1.122	1.146	1.167	1.244	1.284
Z	1.0003	1.0004	1.0004	1.000	1.000	1.000	1.000	1.001	1.001	1.001
$\bar{v}_s$ (m/s)	437	446	455	496	534	568	601	631	765	879
η (10⁻⁶ Pa·s)	24.5	25.3	26.0	29.5	32.8	35.9	38.8	41.6		
k (W/m·K)	0.0359	0.0371	0.0383	0.0440	0.0496	0.0551	0.0606	0.0658		
N_{pr}	0.718	0.718	0.718	0.722	0.726	0.730	0.734	0.737		

b = normal boiling point.

THERMODYNAMIC AND THERMOPHYSICAL PROPERTIES—TABLE 59.   Saturated Nitrogen Tetroxide

Pressure, bar	Temperature, K	v_f, m³/kg	v_g, m³/kg	M_f	M_g
1.0133	299.32	0.000694	0.2996	91.857	79.157
2	309.57	0.000711	0.1630	91.886	76.503
4	326.66	0.000733	0.0876	91.766	73.538
6	337.43	0.000749	0.0608	91.625	71.748
8	345.45	0.000762	0.0469	91.488	70.480
10	351.88	0.000774	0.0382	91.346	69.483
15	364.09	0.000800	0.0262	90.979	67.742
20	373.17	0.000822	0.0199	90.601	66.547
30	386.57	0.000863	0.0133	89.823	64.997
40	396.52	0.000903	0.0098	89.018	64.099
50	404.50	0.000945	0.00761	88.191	63.532
60	411.20	0.000993	0.00607	87.344	63.181
80	422.07	0.001129	0.00394	85.602	62.959
100	430.76	0.001577	0.00209	83.817	63.366

THERMODYNAMIC AND THERMOPHYSICAL PROPERTIES—TABLE 60.   Saturated Nitrous Oxide

Temp., °F	Pressure, psia	v_f, ft³/lb$_m$	v_g, ft³/lb$_m$	h_f, Btu/lb$_m$	h_g, Btu/lb$_m$	s_f, Btu/lb$_m$°R	s_g, Btu/lb$_m$°R
−127.2	14.70	0.01310	5.069	0.0	161.7	0.0000	0.4864
−100	33.68	0.01358	2.374	11.7	165.9	0.0304	0.4591
−80	56.79	0.01398	1.463	20.8	168.8	0.0534	0.4433
−60	90.29	0.01444	0.939	30.2	171.5	0.0782	0.4315
−40	136.68	0.01495	0.648	40.3	173.7	0.1044	0.4222
−20	198.62	0.01555	0.450	50.6	175.3	0.1296	0.4133
0	278.97	0.01625	0.316	60.5	176.2	0.1518	0.4036
20	380.88	0.01711	0.227	70.2	176.2	0.1718	0.3928
40	507.51	0.01819	0.164	80.3	175.0	0.1920	0.3815
60	662.69	0.01968	0.117	91.9	172.1	0.2145	0.3687
80	851.5	0.0222	0.0792	105.7	165.0	0.2382	0.3480
90	961.0	0.0247	0.0611	114.7	157.5	0.2523	0.3302
97.6	1052.2	0.0354	0.0354	136.4	136.4	0.2890	0.2890

THERMODYNAMIC AND THERMOPHYSICAL PROPERTIES—TABLE 61. Nonane

T, K	P, bar	v_f, m³/kg	v_g, m³/kg	h_f, kJ/kg	h_g, kJ/kg	s_f, kJ/(kg·K)	s_g, kJ/(kg·K)	c_{pf}, kJ/(kg·K)	μ_f, 10^{-4} Pa·s	k_f, W/(m·K)
219.7	2.6.–6			358.4		2.424		2.07	33.5	0.150
220	2.7.–6			359.2		2.427		2.07	33.0	0.150
240	3.74.–5			400.6		2.607		2.08	17.9	0.145
260	2.97.–4			442.2	828.7	2.774	4.210	2.10	12.1	0.140
280	1.61.–3			484.8	859.4	2.932	4.243	2.16	8.7	0.134
300	6.40.–3	1.404.–3	30.35	528.6	891.7	3.083	4.282	2.22	6.53	0.129
320	0.0203	1.436.–3	10.19	573.8	925.6	3.229	4.324	2.30	5.13	0.123
340	0.0547	1.471.–3	4.00	622.0	961.1	3.370	4.368		4.16	0.118
360	0.1279	1.508.–3	1.80	671.3	998.2	3.511	4.419		3.44	0.112
380	0.2678	1.548.–3	0.894	722.5	1036.5	3.650	4.476		2.91	0.107
400	0.513	1.591.–3	0.485	776.7	1076.0	3.788	4.536		2.50	0.101
420	0.911	1.637.–3	0.286	833.3	1116.6	3.927	4.601		2.18	0.096
440	1.521	1.690.–3	0.161	890.2	1157.1	4.053	4.660			0.092
460	2.401	1.748.–3	0.104	950.3	1199.2	4.186	4.727			0.089
480	3.639	1.815.–3	0.069	1012.1	1241.3	4.316	4.794			0.085
500	5.309	1.895.–3	0.045	1076.2	1282.9	4.444	4.857			0.082
520	7.437	2.00.–3	0.030	1141.3	1324.5	4.569	4.921			
540	10.20	2.13.–3	0.021	1207.7	1363.8	4.691	3.980			
560	13.76	2.35.–3	0.013	1275.4	1338.7	4.811	5.029			
580	18.02	2.78.–3	0.008	1342.9	1318.1	4.927	5.056			
594.6	22.90	4.23.–3	0.004	1305.2	1305.2	5.032	5.032			

THERMODYNAMIC AND THERMOPHYSICAL PROPERTIES—TABLE 62.   Octane

T, K	P, bar	v_f, 10^{-3} m^3/kg	v_g, 10^{-3} m^3/kg	h_f, kJ/kg	h_g, kJ/kg	s_f, kJ/(kg·K)	s_g, kJ/(kg·K)	c_{pf}, kJ/(kg·K)	μ_f, 10^{-3} Pa·s	k_f, W/(m·K)
216.4	1.49,–5			365.9		2.487		2.033	2.25	0.149
220	2.41,–5			373.2		2.520		2.035	2.01	0.148
240	2.18,–4	1.353,–3	700	414.1	811.4	2.698	4.207	2.059	1.24	0.143
260	0.0014	1.368,–3	125	455.8	842.1	2.865	4.259	2.105	0.87	0.138
280	0.0061	1.384,–3	31.9	498.4	873.5	3.023	4.312	2.165	0.65	0.133
300	0.0207	1.420,–3	10.7	542.4	906.2	3.175	4.366	2.231	0.504	0.128
320	0.0575	1.457,–3	4.01	589.8	939.8	3.325	4.419		0.405	0.123
340	0.1384	1.495,–3	1.752	637.9	974.6	3.471	4.461		0.334	0.118
360	0.3000	1.536,–3	0.844	687.1	1010.4	3.611	4.509		0.282	0.112
380	0.5856	1.582,–3	0.448	737.7	1047.3	3.747	4.562		0.244	0.107
400	1.0507	1.632,–3	0.252	790.1	1084.8	3.881	4.617		0.200	0.102
420	1.758	1.685,–3	0.155	843.1	1123.6	4.010	4.677		0.167	0.099
440	2.797	1.747,–3	0.100	897.5	1162.5	4.137	4.740		0.143	0.095
460	4.246	1.818,–3	0.066	954.8	1202.0	4.264	4.802		0.121	0.091
480	6.201	1.904,–3	0.045	1013.5	1241.8	4.388	4.864		0.103	0.087
500	8.785	2.013,–3	0.031	1072.8	1281.2	4.508	4.924		0.086	0.083
520	12.15	2.16,–3	0.021	1136.0	1318.6	4.629	4.980		0.072	
540	16.46	2.37,–3	0.014	1201.5	1352.4	4.749	5.028		0.058	
560	21.98	2.81,–3	0.008	1276.7	1370.4	4.880	5.048		0.044	
568.8	24.97	4.26,–3	0.004	1331.7	1331.7	4.977	4.977			

THERMODYNAMIC AND THERMOPHYSICAL PROPERTIES—TABLE 63. Saturated Oxygen (R732)

T, K	P, bar	v_f, 10^{-3}m^3/kg	v_g, 10^{-3}m^3/kg	h_f, kJ/kg	h_g, kJ/kg	s_f, kJ/(kg·K)	s_g, kJ/(kg·K)	c_{pf}, kJ/(kg·K)	μ_f, 10^{-4}Pa·s	k_f, W/(m·K)
54.35	0.0015	0.776	93980	−189.8	48.9	2.156	6.548			
55	0.0018	0.778	77920	−188.9	49.5	2.172	6.507			
60	0.0073	0.790	21240	−181.1	53.8	2.308	6.223			
65	0.0233	0.802	7200	−173.3	58.1	2.432	5.992			
70	0.0624	0.816	2894	−165.5	62.4	2.545	5.801			
75	0.1448	0.827	1330	−159.2	66.6	2.631	5.642	1.570	3.04	0.170
80	0.3003	0.845	680.7	−149.7	70.8	2.754	5.510	1.589	2.54	0.164
85	0.5677	0.862	379.7	−141.7	74.9	2.849	5.397	1.607	2.16	0.157
90	0.9943	0.880	227.1	−133.7	78.8	2.940	5.301	1.625	1.88	0.151
90.18	1.0133	0.881	223.2	−133.4	78.9	2.943	5.297	1.626	1.87	0.151
95	1.634	0.899	143.9	−125.4	82.4	3.045	5.216	1.645	1.66	0.144
100	2.547	0.920	95.46	−117.1	85.7	3.113	5.141	1.672	1.51	0.138
105	3.794	0.944	65.81	−108.6	88.5	3.196	5.073	1.706	1.34	0.131
110	5.443	0.970	46.81	−99.9	90.8	3.276	5.009	1.752	1.20	0.125
115	7.559	0.998	34.15	−90.0	92.6	3.354	4.950	1.814	1.07	0.118
120	10.21	1.031	25.42	−81.6	93.6	3.432	4.892	1.896	0.97	0.111
125	13.48	1.070	19.21	−71.8	93.9	3.510	4.836	2.004	0.86	0.103
130	17.44	1.116	14.67	−61.5	93.3	3.588	4.779	2.148	0.78	0.096
135	22.19	1.170	11.25	−50.6	91.6	3.667	4.720	2.341	0.70	0.088
140	27.82	1.237	8.612	−38.9	88.4	3.748	4.657	2.629	0.60	0.080
145	34.45	1.332	6.499	−25.9	82.9	3.833	4.583	3.141	0.52	0.072
150	42.23	1.487	4.705	−10.8	73.1	3.928	4.487	3.935		
154.77	50.87	2.464	2.464	35.2	35.2	4.219	4.219	∞		∞

480

THERMODYNAMIC AND THERMOPHYSICAL PROPERTIES — TABLE 64. Saturated Potassium

T, K	P, bar	v_f, m³/kg	v_g, m³/kg	h_f, kJ/kg	h_g, kJ/kg	s_f, kJ/(kg·K)	s_g, kJ/(kg·K)	c_{pf}, kJ/(kg·K)
336.4	1.37.–9	0.001208		93.8	2327	1.928	8.567	0.822
400	1.84.–7	0.001229	4.64.+6	145.5	2342	2.068	7.559	0.805
500	3.13.–5	0.001266	3.39.+4	225.1	2390	2.246	6.576	0.785
600	9.26.–4	0.001304	3164	302.7	2433	2.388	5.937	0.771
700	0.01022	0.001346	142.3	379.4	2468	2.506	5.490	0.762
800	0.06116	0.001389	26.75	455.5	2498	2.608	5.161	0.761
1000	0.7322	0.001488	2.691	609.7	2552	2.780	4.722	0.792
1200	3.913	0.001605	0.584	773.5	2610	2.929	4.459	0.846
1400	12.44	0.001742	0.207	948.0	2679	3.063	4.299	0.899
1500	20.0	0.001816	0.132	1040.0	2718	3.123	4.209	0.924

THERMODYNAMIC AND THERMOPHYSICAL PROPERTIES—TABLE 65. Saturated Propane (R290)

T, K	P, bar	v_f, m³/kg	v_g, m³/kg	h_f, kJ/kg	h_g, kJ/kg	s_f, kJ/(kg·K)	s_g, kJ/(kg·K)	c_{pf}, kJ/(kg·K)	μ_f, 10^{-4} Pa·s	k_f, W/(m·K)
85.5	3.0,–9	1.364,–3	5.37,+7	124.92	690.02	1.8738	8.3548	1.92		
90	1.5,–8	1.373,–3	1.12,+7	133.56	693.58	1.9723	8.0953	1.92		
100	3.2,–7	1.392,–3	5.85,+5	152.74	702.23	2.1743	7.6163	1.93		
110	3.9,–6	1.412,–3	53,275	172.03	711.71	2.3581	7.2377	1.94		
120	3.1,–5	1.432,–3	7,350	191.46	721.78	2.5271	6.9343	1.95		
130	1.8,–4	1.453,–3	1,400	211.03	732.27	2.6838	6.6885	1.96		
140	7.7,–4	1.475,–3	344	230.77	743.07	2.8300	6.4881	1.98		
150	2.74,–3	1.497,–3	103	250.67	754.12	2.9674	6.3237	2.00	6.61	0.191
160	8.22,–3	1.521,–3	36.8	270.78	765.37	3.0971	6.1886	2.02	5.54	0.183
170	0.0214	1.545,–3	15.0	291.10	776.80	3.2202	6.0775	2.04	4.67	0.175
180	0.0495	1.570,–3	6.84	311.66	788.40	3.3377	5.9862	2.07	3.97	0.166
190	0.1035	1.597,–3	3.43	332.48	800.15	3.4503	5.9114	2.10	3.27	0.158
200	0.1993	1.625,–3	1.868	353.61	812.03	3.5586	5.8502	2.13	2.98	0.150
210	0.3574	1.654,–3	1.087	375.07	824.01	3.6631	5.8005	2.16	2.65	0.143
220	0.6031	1.686,–3	0.669	396.90	836.04	3.7645	5.7603	2.20	2.36	0.136
230	0.9661	1.719,–3	0.432	419.16	848.08	3.8631	5.7280	2.25	2.07	0.129
240	1.4800	1.754,–3	0.290	442.07	860.07	3.9605	5.7022	2.29	1.86	0.123
250	2.1819	1.792,–3	0.2020	465.58	871.94	4.0563	5.6817	2.34	1.69	0.117
260	3.1118	1.833,–3	0.1445	489.70	883.62	4.1505	5.6656	2.41	1.53	0.111
270	4.3120	1.878,–3	0.1059	514.45	895.02	4.2433	5.6528	2.48	1.40	0.106
280	5.8278	1.927,–3	0.0791	539.88	906.03	4.3349	5.6426	2.56	1.29	0.100
290	7.7063	1.982,–3	0.0600	566.06	916.54	4.4257	5.6343	2.65	1.19	0.096
300	9.9973	2.044,–3	0.0461	593.11	926.41	4.5160	5.6270	2.76	1.10	0.091
310	12.75	2.115,–3	0.0357	621.18	935.45	4.6062	5.6200	2.89	0.93	0.086
320	16.03	2.200,–3	0.0279	650.49	943.38	4.6971	5.6124	3.06	0.82	0.082
330	19.88	2.301,–3	0.0218	681.37	949.79	4.7896	5.6030	3.28	0.72	0.078
340	24.36	2.430,–3	0.0170	714.38	953.92	4.8850	5.5896	3.62	0.62	0.073
350	29.56	2.607,–3	0.0130	750.52	954.23	4.9861	5.5681	4.23	0.52	0.069
360	35.55	2.896,–3	0.0095	792.50	946.56	5.0997	5.5277	5.98	0.40	0.066
369.8	42.42	4.566,–3	0.0046	879.20	879.20	5.3300	5.3300	∞	0.29	∞

THERMODYNAMIC AND THERMOPHYSICAL PROPERTIES—TABLE 66. Saturated Propylene (Propene, R1270)

T, K	P, bar	v_f, m³/kg	v_g, m³/kg	h_f, kJ/kg	h_g, kJ/kg	s_f, kJ/(kg·K)	s_g, kJ/(kg·K)	c_{pf}, kJ/(kg·K)	μ_f, 10^{-6} Pa·s	k_f, W/(m·K)	Pr
87.9ᵗ	9.54,–9	0.001301	1.82,+7	–290.1	279.2	–1.923	4.554				
90	2.05,–8	0.001305	8.66,+6	–285.1	281.1	–1.867	4.424	1.695	2017	0.214	15.98
100	4.81,–7	0.001325	411165	–265.4	290.2	–1.659	3.897	1.760	1526	0.209	12.85
110	6.08,–6	0.001346	35753	–247.7	299.6	–1.490	3.488	1.820	1185	0.204	10.57
120	4.88,–5	0.001367	4856	–229.8	309.3	–1.335	3.158	1.875	941	0.198	8.91
130	2.77,–4	0.001389	927.0	–211.4	319.3	–1.187	2.895	1.923	735	0.193	7.32
140	1.20,–3	0.001411	230.91	–192.4	329.4	–1.046	2.681	1.964	587	0.188	6.13
150	4.17,–3	0.001434	71.043	–172.9	339.8	–0.912	2.506	1.996	478	0.183	5.21
160	0.0122	0.001458	25.903	–153.1	350.4	–0.784	2.363	2.020	397	0.178	4.50
170	0.0309	0.001483	10.842	–133.1	361.2	–0.663	2.245	2.044	334.5	0.173	3.95
180	0.0697	0.001508	5.080	–112.7	372.1	–0.547	2.147	2.067	286.1	0.168	3.52
190	0.1425	0.001535	2.613	–92.2	383.1	–0.436	2.066	2.094	244.9	0.162	3.17
200	0.2686	0.001563	1.452	–71.4	394.2	–0.329	1.999	2.128	212.7	0.157	2.88
210	0.4727	0.001593	0.860	–50.3	405.3	–0.226	1.943	2.162	187.0	0.152	2.66
220	0.7849	0.001624	0.538	–28.8	416.3	–0.127	1.896	2.182	175.0	0.149	2.56
225.5ᵇ	1.0133	0.001642	0.4241	–16.9	422.2	–0.073	1.874	2.199	166.2	0.147	2.49
230	1.2401	0.001657	0.3515	–7.0	427.1	–0.030	1.857	2.243	149.2	0.142	2.36
240	1.8775	0.001693	0.2388	15.3	437.8	0.064	1.825	2.298	135.0	0.137	2.26
250	2.7401	0.001732	0.1674	38.0	448.2	0.157	1.797	2.369	123.0	0.131	2.22
260	3.8737	0.001774	0.1206	61.3	458.2	0.247	1.774	2.418	112.9	0.126	2.17
270	5.3269	0.001820	0.0888	85.2	467.8	0.336	1.753	2.494	106.6	0.121	2.20
280	7.1499	0.001872	0.0666	109.9	476.9	0.425	1.735	2.584	100.0	0.116	2.23
290	9.3954	0.001929	0.0507	135.3	485.3	0.512	1.719	2.693	93.0	0.112	2.24
300	12.12	0.001995	0.0390	161.6	492.8	0.600	1.704	2.842	85.6	0.109	2.23
310	15.38	0.002071	0.0303	189.0	499.3	0.688	1.688	3.007	77.8	0.104	2.25
320	19.23	0.002162	0.0236	217.7	504.3	0.776	1.672	3.335	69.6	0.097	2.39
330	23.75	0.002273	0.0184	248.2	507.4	0.867	1.652	3.723	61.0	0.090	2.52
340	29.01	0.002418	0.0142	280.9	507.6	0.961	1.627			0.082	
350	35.12	0.002628	0.0107	317.6	502.8	1.062	1.592	4.669			
360	42.20	0.003038	0.0075	364.1	486.0	1.188	1.527				
365.6ᶜ	46.65	0.004476	0.0045	433.3	433.3	1.374	1.374				

t = triple point; b = normal boiling point; c = critical point. The notation 9.54,–9 signifies 9.54×10^{-9}. $h_f = s_f = 0$ at 233.15 K = –40°C.

THERMODYNAMIC AND THERMOPHYSICAL PROPERTIES – TABLE 67. Compressed Propylene (Propene, R1270)

Pressure, bar		Temperature, K									
		225	250	275	300	325	350	375	400	425	450
1	v (m³/kg)	0.00164	0.4817	0.5334	0.5846	0.6354	0.6858	0.7361	0.7861	0.8361	0.8859
	h (kJ/kg)	−17.9	455.5	491.2	529.1	569.1	611.4	656.0	702.8	751.9	803.2
	s (kJ/kg·K)	−0.0779	2.0169	2.1530	2.2847	2.4128	2.5380	2.6610	2.7821	2.9010	3.0183
10	v (m³/kg)	0.00164	0.00173	0.00184	0.04986	0.05670	0.06295	0.06889	0.07460	0.08014	0.08557
	h (kJ/kg)	−17.2	38.5	97.6	501.2	547.0	593.2	640.8	689.7	740.6	793.3
	s (kJ/kg·K)	−0.0810	0.1535	0.3788	1.7631	1.9653	2.0466	2.1774	2.3042	2.4273	2.5480
20	v (m³/kg)	0.00163	0.00172	0.00184	0.00198	0.02324	0.02773	0.03149	0.03489	0.03803	0.04104
	h (kJ/kg)	−16.3	39.3	98.1	161.5	512.7	568.2	621.5	673.5	726.5	781.5
	s (kJ/kg·K)	−0.0841	0.1497	0.3736	0.5941	1.6920	1.8567	2.0024	2.1381	2.2674	2.3923
40	v (m³/kg)	0.00163	0.00172	0.00182	0.00196	0.00216	0.00256		0.01465	0.01682	0.01872
	h (kJ/kg)	−14.4	40.8	99.0	161.3	230.0	313.3		633.6	695.4	755.6
	s (kJ/kg·K)	−0.0908	0.1419	0.3638	0.5806	0.8001	1.0466		1.9256	2.0758	2.2131
60	v (m³/kg)	0.00162	0.00171	0.00181	0.00194	0.00211	0.00240		0.00743	0.00944	0.01126
	h (kJ/kg)	−12.6	42.3	100.1	161.5	228.2	303.8		575.4	656.7	726.1
	s (kJ/kg·K)	−0.0970	0.1345	0.3546	0.5684	0.7816	1.0055		1.7272	1.9250	2.0832
80	v (m³/kg)	0.00162	0.00170	0.00180	0.00192	0.00208	0.00231		0.00402	0.00605	0.00757
	h (kJ/kg)	−10.7	44.0	101.3	162.0	227.2	299.0		499.7	607.7	693.3
	s (kJ/kg·K)	−0.1031	0.1274	0.3458	0.5570	0.7657	0.9781		1.5107	1.7795	1.9693

100 v (m³/kg)	0.00161	0.00169	0.00179	0.00190	0.00204	0.00224	0.00256	0.00316	0.00426	0.00551
h (kJ/kg)	−8.8	45.6	102.6	162.7	226.7	296.1	373.5	464.9	567.8	660.1
s (kJ/kg·K)	−0.1091	0.1202	0.3374	0.5466	0.7514	0.9570	1.1704	1.4061	1.6456	1.8669
150 v (m³/kg)	0.00160	0.00167	0.00176	0.00186	0.00198	0.00214	0.00234	0.00262	0.00300	0.00354
h (kJ/kg)	−4.1	49.9	106.1	165.0	227.0	292.8	362.9	438.5	519.5	604.6
s (kJ/kg·K)	−0.1236	0.1038	0.3180	0.5228	0.7214	0.9163	1.1100	1.3049	1.5021	1.6958
200 v (m³/kg)	0.00159	0.00166	0.00174	0.00183	0.00194	0.00207	0.00222	0.00242	0.00266	0.00296
h (kJ/kg)	0.8	54.4	110.0	167.9	228.7	292.3	359.2	429.9	504.4	581.4
s (kJ/kg·K)	−0.1371	0.0884	0.3004	0.5021	0.6963	0.8852	1.0701	1.2521	1.4319	1.6086
300 v (m³/kg)	0.00157	0.00163	0.00170	0.00178	0.00187	0.00197	0.00208	0.00221	0.00236	0.00253
h (kJ/kg)	10.9	63.8	118.5	175.2	234.2	295.5	359.3	425.7	494.4	565.6
s (kJ/kg·K)	−0.1625	0.0601	0.2688	0.4660	0.6549	0.8367	1.0127	1.1839	1.3507	1.5133
400 v (m³/kg)	0.00155	0.00161	0.00167	0.00174	0.00182	0.00190	0.00199	0.00209	0.00220	0.00232
h (kJ/kg)	21.2	73.6	127.7	183.6	241.5	301.5	363.7	428.0	494.4	562.9
s (kJ/kg·K)	−0.1863	0.0347	0.2407	0.4351	0.6207	0.7985	0.9705	1.1362	1.2972	1.4536
500 v (m³/kg)	0.00153	0.00159	0.00165	0.00171	0.00178	0.00185	0.00193	0.00201	0.00210	0.00220
h (kJ/kg)	31.6	83.7	137.3	192.6	249.8	309.0	370.1	433.3	498.7	565.4
s (kJ/kg·K)	−0.2082	0.0112	0.2155	0.4080	0.5910	0.7664	0.9351	1.0981	1.2559	1.4092

THERMODYNAMIC AND THERMOPHYSICAL PROPERTIES—TABLE 68. Saturated Refrigerant 11

T, K	P, bar	v_f, m³/kg	v_g, m³/kg	h_f, kJ/kg	h_g, kJ/kg	s_f, kJ/(kg·K)	s_g, kJ/(kg·K)	c_{pf}, kJ/(kg·K)	μ_f, 10^{-4} Pa·s	k_f, W/(m·K)
200	0.0043	5.901.–4	28.06	–14.37	186.30	–0.0651	0.9431	0.815	1.674	0.115
220	0.0417	6.061.–4	6.272	–8.20	195.89	–0.0361	0.8925	0.828	1.142	0.110
240	0.0768	6.225.–4	1.882	4.97	205.85	0.0210	0.8581	0.842	0.831	0.104
260	0.2215	6.398.–4	0.703	21.01	216.06	0.0851	0.8353	0.856	0.635	0.098
270	0.3514	6.491.–4	0.458	29.53	221.23	0.1172	0.8272	0.863	0.563	0.095
280	0.5364	6.587.–4	0.309	38.25	226.40	0.1489	0.8209	0.870	0.504	0.093
290	0.7917	6.688.–4	0.216	47.10	231.58	0.1799	0.8160	0.878	0.454	0.090
300	1.1341	6.794.–4	0.154	56.06	236.73	0.2102	0.8124	0.887	0.413	0.087
310	1.5821	6.908.–4	0.113	65.10	241.83	0.2397	0.8099	0.897	0.377	0.084
320	2.1556	7.027.–4	0.0847	74.22	246.88	0.2686	0.8081	0.907	0.346	0.081
330	2.876	7.156.–4	0.0645	83.42	251.84	0.2967	0.8071	0.917	0.320	0.079
340	3.764	7.293.–4	0.0500	92.72	256.69	0.3243	0.8065	0.928	0.297	0.076
350	4.845	7.442.–4	0.0392	102.12	261.40	0.3513	0.8064	0.939	0.276	0.073
360	6.142	7.603.–4	0.0311	111.64	265.95	0.3778	0.8065	0.950	0.259	0.070
380	9.487	7.974.–4	0.0201	131.12	274.40	0.4298	0.8069	0.975	0.229	0.065
400	14.02	8.435.–4	0.0134	151.38	281.69	0.4808	0.8066	1.004	0.203	0.059
420	19.98	9.042.–4	0.0090	172.76	287.20	0.5317	0.8041	1.04	0.169	0.053
440	27.65	9.930.–4	0.0059	196.01	289.72	0.5840	0.7970	1.09	0.131	0.048
460	37.36	1.167.–3	0.0036	223.85	285.36	0.6435	0.7773	1.19	0.084	0.037
471.2	44.09	1.799.–3	0.0018	258.70	258.70	0.7162	0.7162	∞	0.033	∞

THERMODYNAMIC AND THERMOPHYSICAL PROPERTIES — TABLE 69. Saturated Refrigerant 12

T, K	P, bar	v_f, m³/kg	v_g, m³/kg	h_f, kJ/kg	h_g, kJ/kg	s_f, kJ/(kg·K)	s_g, kJ/(kg·K)	c_{pf}, kJ/(kg·K)	μ_f, 10^{-4} Pa·s	k_f, W/(m·K)
150	0.00091	5.767.−4	179.12	294.6	496.0	3.492	4.835	0.808	18.9	0.123
160	0.00305	5.849.−4	36.05	302.3	500.2	3.543	4.780	0.817	15.1	0.119
170	0.00871	5.926.−4	13.40	310.3	504.5	3.591	4.734	0.827	12.1	0.116
180	0.02178	6.024.−4	5.666	318.3	508.9	3.637	4.696	0.836	9.69	0.113
190	0.04877	6.118.−4	2.665	326.5	513.5	3.681	4.665	0.845	7.94	0.109
200	0.0996	6.217.−4	1.370	334.8	518.1	3.724	4.640	0.855	6.64	0.105
210	0.1879	6.139.−4	0.7589	343.2	522.7	3.765	4.620	0.864	5.65	0.102
220	0.3317	6.431.−4	0.4476	351.8	527.4	3.805	4.603	0.873	4.88	0.098
230	0.5531	6.549.−4	0.2784	360.6	531.1	3.844	4.590	0.882	4.26	0.094
240	0.8781	6.675.−4	0.1811	369.5	536.8	3.881	4.579	0.891	3.77	0.090
250	1.3359	6.810.−4	0.1225	378.0	541.5	3.918	4.570	0.902	3.37	0.087
260	1.959	6.970.−4	0.08559	387.7	546.1	3.954	4.563	0.913	3.03	0.083
270	2.784	7.112.−4	0.06147	397.0	550.7	3.989	4.558	0.926	2.75	0.080
280	3.825	7.282.−4	0.04543	406.5	555.1	4.023	4.554	0.942	2.52	0.076
290	5.184	7.470.−4	0.03888	416.1	559.4	4.057	4.551	0.959	2.31	0.072
300	6.840	7.678.−4	0.02582	426.0	563.5	4.090	4.548	0.979	2.14	0.069
310	8.860	7.912.−4	0.01992	436.0	567.3	4.122	4.546	1.005	2.00	0.065
320	11.29	8.173.−4	0.01553	446.2	570.9	4.154	4.543	1.041	1.86	0.061
330	14.17	8.478.−4	0.01218	456.8	574.0	4.186	4.541	1.093	1.74	0.058
340	17.58	8.840.−4	0.00957	467.8	576.5	4.218	4.538	1.166	1.60	0.054
350	21.57	9.286.−4	0.00750	479.4	578.2	4.250	4.533	1.264	1.45	0.050
360	26.19	9.868.−4	0.00582	492.1	578.7	4.285	4.525	1.39	1.28	0.046
370	31.56	1.072.−3	0.00439	506.4	577.2	4.322	4.514	1.55	1.06	0.041
380	37.76	1.237.−3	0.00305	524.7	571.2	4.369	4.900		0.75	
385	41.31	1.876.−3	0.00188	551.1	551.1	4.437	4.437	∞	0.31	∞

THERMODYNAMIC AND THERMOPHYSICAL PROPERTIES—TABLE 70.   Saturated Refrigerant 13

T, K	P, bar	v_f, m³/kg	v_g, m³/kg	h_f, kJ/kg	h_g, kJ/kg	s_f, kJ/(kg·K)	s_g, kJ/(kg·K)	c_{pf}, kJ/(kg·K)	μ_f, 10^{-4} Pa·s	k_f, W/(m·K)
91	3.817.−6	5.367.−4	19,557	238.1	424.9	3.080	5.133			
100	3.418.−5	5.448.−4	2,392	243.8	429.0	3.140	4.990			
110	2.563.−4	5.538.−4	347.0	251.1	433.1	3.205	4.860			
120	0.00137	5.635.−4	70.25	258.2	437.2	3.267	4.759			
130	0.00571	5.739.−4	18.15	265.8	441.3	3.327	4.677			
140	0.01895	5.850.−4	5.865	273.7	455.3	3.385	4.610			
150	0.05250	5.969.−4	2.2617	281.7	449.3	3.441	4.558	0.826	6.83	0.114
160	0.1258	6.095.−4	1.0019	290.0	453.5	3.494	4.516	0.845	5.60	0.109
170	0.2680	6.231.−4	0.4962	298.4	457.6	3.545	4.482	0.865	4.59	0.104
180	0.5186	6.380.−4	0.2689	307.1	461.8	3.594	4.454	0.884	3.83	0.099
190	0.9269	6.536.−4	0.1567	315.9	465.9	3.642	4.431	0.898	3.26	0.093
200	1.5507	6.709.−4	9.69.−2	325.0	469.9	3.688	4.413	0.910	2.82	0.088
210	2.456	6.899.−4	6.28.−2	334.3	473.8	3.732	4.397	0.924	2.48	0.083
220	3.712	7.110.−4	4.24.−2	343.8	477.5	3.777	4.385	0.943	2.20	0.078
230	5.396	7.346.−4	2.95.−2	353.6	481.0	3.820	4.374	0.972	1.97	0.072
240	7.589	7.615.−4	2.11.−2	363.5	484.1	3.862	4.364	1.014	1.79	0.067
250	10.37	7.928.−4	1.53.−2	373.9	486.1	3.903	4.355	1.072	1.63	0.062
260	13.85	8.302.−4	1.13.−2	384.7	489.1	3.944	4.346	1.151	1.50	0.057
270	18.13	8.769.−4	8.28.−3	396.2	490.5	3.986	4.336	1.255	1.34	0.051
280	23.32	9.320.−4	6.10.−3	408.8	490.6	4.029	4.323	1.386	1.14	0.045
290	29.57	1.035.−3	4.34.−3	423.6	488.3	4.080	4.303	1.549	0.87	0.038
300	37.05	1.284.−3	2.60.−3	445.3	477.5	4.151	4.257	1.75	0.52	
302.0	38.70	1.808.−3	1.81.−3	463.1	463.1	4.209	4.209	∞	0.29	∞

THERMODYNAMIC AND THERMOPHYSICAL PROPERTIES—TABLE 71. Saturated Refrigerant 13B1

T, K	P, bar	v_f, m³/kg	v_g, m³/kg	h_f, kJ/kg	h_g, kJ/kg	s_f, kJ/(kg·K)	s_g, kJ/(kg·K)	c_{pf}, kJ/(kg·K)	μ_f, 10^{-4} Pa·s	k_f, W/(m·K)
170	0.059	4.594.–4	1.6015	−40.90	90.95	−0.2033	0.5723	0.597	9.54	0.101
180	0.127	4.677.–4	0.7840	−34.75	94.37	−0.1682	0.5491	0.618	7.60	0.096
190	0.250	4.765.–4	0.4190	−28.51	97.83	−0.1345	0.5305	0.634	6.20	0.091
200	0.455	4.860.–4	0.2407	−22.17	101.32	−0.1020	0.5154	0.648	5.13	0.086
210	0.777	4.961.–4	0.1467	−15.68	104.82	−0.0704	0.5033	0.663	4.33	0.082
215.4	1.013	5.020.–4	0.1147	−12.09	106.70	−0.0536	0.4978	0.670	3.97	0.079
220	1.254	5.071.–4	0.0940	−9.02	108.28	−0.0396	0.4936	0.676	3.71	0.077
230	1.933	5.190.–4	0.0628	−2.19	111.68	−0.0094	0.4857	0.690	3.22	0.073
240	2.863	5.321.–4	0.0433	4.83	114.99	0.0202	0.4793	0.703	2.83	0.068
250	4.096	5.466.–4	0.0308	12.03	118.16	0.0494	0.4739	0.721	2.51	0.063
260	5.690	5.627.–4	0.0224	19.44	121.16	0.0781	0.4693	0.742	2.25	0.059
270	7.703	5.809.–4	0.0166	27.06	123.93	0.1064	0.4652	0.767	2.04	0.054
280	10.20	6.018.–4	0.0124	34.94	126.41	0.1345	0.4612	0.800	1.84	0.049
290	13.25	6.264.–4	0.0094	43.11	128.51	0.1625	0.4570	0.842	1.69	0.045
300	16.91	6.562.–4	0.0072	51.68	130.09	0.1908	0.4522	0.891	1.57	0.040
310	21.28	6.940.–4	0.0055	60.81	130.97	0.2197	0.4460	0.951	1.45	0.035
320	26.44	7.458.–4	0.0041	70.80	130.76	0.2503	0.4376	1.09	1.26	0.030
330	32.48	8.295.–4	0.0030	82.42	128.59	0.2845	0.4245	1.29	0.99	0.026
340.2	39.64	1.344.–3	0.0013	108.70	108.70	0.3605	0.3605	∞	0.35	∞

THERMODYNAMIC AND THERMOPHYSICAL PROPERTIES—TABLE 72. Saturated Refrigerant 21

Temperature, K	Pressure, bar	v_f, m³/kg	v_g, m³/kg	h_f, kJ/kg	h_g, kJ/kg	s_f, kJ/(kg·K)	s_g, kJ·(kg·K)
250	0.2415	0.000 677	0.8292	16.6	274.8	0.0687	1.1015
260	0.3953	0.000 687	0.5247	26.5	279.9	0.1076	1.0820
270	0.6200	0.000 698	0.3455	36.6	284.9	0.1454	1.0653
280	0.9364	0.000 709	0.2355	46.7	290.0	0.1824	1.0511
290	1.3682	0.000 722	0.1654	57.1	295.0	0.2186	1.0389
300	1.9417	0.000 735	0.1192	67.7	300.0	0.2543	1.0286
310	2.6849	0.000 748	0.0879	78.4	304.8	0.2894	1.0196
320	3.6279	0.000 763	0.0661	89.5	309.5	0.3242	1.0119
330	4.8022	0.000 778	0.0505	100.7	314.1	0.3586	1.0051
340	6.2409	0.000 794	0.0391	112.3	318.4	0.3927	0.9989
350	7.978	0.000 812	0.0307	124.1	322.4	0.4266	0.9932
360	10.049	0.000 830	0.0243	136.2	326.1	0.4602	0.9877
370	12.489	0.000 850	0.0194	148.6	329.3	0.4935	0.9820
380	15.337	0.000 870	0.0155	161.2	331.9	0.5264	0.9758
390	18.630	0.000 893	0.0125	173.9	333.8	0.5587	0.9688
400	22.41	0.000 918	0.01011	186.4	334.8	0.5896	0.9605
410	26.72	0.000 944	0.00820	198.3	334.7	0.6180	0.9506
420	31.60	0.000 972	0.00672	208.7	333.7	0.6418	0.9394
430	37.10	0.001 002	0.00564	216.4	332.4	0.6587	0.9286
440	43.26	0.001 034	0.00491	221.1	332.3	0.6682	0.9208

THERMODYNAMIC AND THERMOPHYSICAL PROPERTIES—TABLE 73. Saturated Refrigerant 22

T, K	P, bar	v_f, m³/kg	v_g, m³/kg	h_f, kJ/kg	h_g, kJ/kg	s_f, kJ/(kg·K)	s_g, kJ/(kg·K)	c_{pf}, kJ/(kg·K)	μ_f, 10^{-4} Pa·s	k_f, W/(m·K)
150	0.0017	6.209.–4	83.40	268.2	547.3	3.355	5.215	1.059		0.161
160	0.0054	6.293.–4	28.20	278.2	552.1	3.430	5.141	1.058		0.156
170	0.0150	6.381.–4	10.85	288.3	557.0	3.494	5.075	1.057	0.770	0.151
180	0.0369	6.474.–4	4.673	298.7	561.9	3.551	5.013	1.058	0.647	0.146
190	0.0821	6.573.–4	2.225	308.6	566.8	3.605	4.963	1.060	0.554	0.141
200	0.1662	6.680.–4	1.145	318.8	571.6	3.657	4.921	1.065	0.481	0.136
210	0.3116	6.794.–4	0.6370	329.1	576.5	3.707	4.885	1.071	0.424	0.131
220	0.5470	6.917.–4	0.3772	339.7	581.2	3.756	4.854	1.080	0.378	0.126
230	0.9076	7.050.–4	0.2352	350.6	585.9	3.804	4.828	1.091	0.340	0.121
240	1.4346	7.195.–4	0.1532	361.7	590.5	3.852	4.805	1.105	0.309	0.117
250	2.174	7.351.–4	0.1037	373.0	594.9	3.898	4.785	1.122	0.282	0.112
260	3.177	7.523.–4	0.07237	384.5	599.0	3.942	4.768	1.143	0.260	0.107
270	4.497	7.733.–4	0.05187	396.3	603.0	3.986	4.752	1.169	0.241	0.102
280	6.192	7.923.–4	0.03803	408.2	606.6	4.029	4.738	1.193	0.225	0.097
290	8.324	8.158.–4	0.02838	420.4	610.0	4.071	4.725	1.220	0.211	0.092
300	10.956	8.426.–4	0.02148	432.7	612.8	4.113	4.713	1.257	0.198	0.087
310	14.17	8.734.–4	0.01643	445.5	615.1	4.153	4.701	1.305	0.186	0.082
320	18.02	9.096.–4	0.01265	458.6	616.7	4.194	4.688	1.372	0.176	0.077
330	22.61	9.535.–4	9.753.–3	472.4	617.3	4.235	4.674	1.460	0.167	0.072
340	28.03	1.010.–3	7.479.–3	487.2	616.5	4.278	4.658	1.573	0.151	0.067
350	34.41	1.086.–3	5.613.–3	503.7	613.3	4.324	4.637	1.718	0.130	0.062
360	41.86	1.212.–3	4.036.–3	523.7	605.5	4.378	4.605	1.897	0.106	
369.3	49.89	2.015.–3	2.015.–3	570.0	570.0	4.501	4.501	∞		

THERMODYNAMIC AND THERMOPHYSICAL PROPERTIES—TABLE 74. Compressed R22

Pressure, bar	Property	Temperature, K							
		275	300	325	350	375	400	425	450
1	c_p (kJ/kg·K)	0.639	0.653	0.689	0.714	0.739	0.758	0.781	0.806
	μ (10^{-6}Pa·s)	11.8	12.8	13.9	14.9	15.8	16.7	17.7	18.7
	k (W/m·K)	0.0091	0.0106	0.0121	0.0136	0.0151	0.0166	0.0181	0.0196
	Pr	0.829	0.793	0.787	0.782	0.773	0.762	0.765	0.769
5	c_p (kJ/kg·K)	0.725	0.728	0.744	0.759	0.766	0.775	0.791	0.816
	μ (10^{-6}Pa·s)	11.8	12.8	13.8	15.0	16.2	17.0	18.0	18.8
	k (W/m·K)	0.0096	0.0107	0.0123	0.0138	0.0153	0.0170	0.0184	0.0199
	Pr	0.887	0.871	0.852	0.839	0.803	0.775	0.773	0.771
10	c_p (kJ/kg·K)	1.166	0.847	0.810	0.799	0.797	0.803	0.814	0.828
	μ (10^{-6}Pa·s)	211	13.7	14.4	15.1	16.1	17.1	18.1	19.0
	k (W/m·K)	0.0954	0.0121	0.0128	0.0144	0.0160	0.0175	0.0190	0.0205
	Pr	2.58	0.959	0.901	0.838	0.802	0.785	0.775	0.767
20	c_p (kJ/kg·K)	0.164	1.237		0.949	0.889	0.865	0.858	0.859
	μ (10^{-6}Pa·s)	211	159		16.5	17.3	18.0	18.8	19.6
	k (W/m·K)	0.0963	0.0849		0.0157	0.0172	0.0184	0.0199	0.0214
	Pr	2.55	2.32		0.997	0.894	0.846	0.811	0.787

40	c_p (kJ/kg·K)	1.152	1.217	1.359		1.373	1.089	0.996	0.956
	μ (10^{-6} Pa·s)	218	164	123		20.7	20.5	20.7	21.2
	k (W/m·K)	0.0980	0.0872	0.0767		0.0219	0.0210	0.0217	0.0233
	Pr	2.56	2.29	2.18		1.30	1.063	0.950	0.870
60	c_p (kJ/kg·K)	1.142	1.191	1.311	1.460		1.767	1.221	1.089
	μ (10^{-6} Pa·s)	221	170	128	94.6		24.7	24.2	23.9
	k (W/m·K)	0.0993	0.0889	0.0786			0.0305	0.0287	0.0268
	Pr	2.54	2.28	2.14			1.431	1.030	0.971
80	c_p (kJ/kg·K)	1.132	1.177	1.277	1.444	1.861		1.396	1.262
	μ (10^{-6} Pa·s)	226	175	133	101	73.6		29.9	27.7
	k (W/m·K)	0.1003	0.0904	0.0803	0.0690	0.0523		0.0374	0.0337
	Pr	2.55	2.28	2.12	2.12	2.62		1.12	1.04
100	c_p (kJ/kg·K)	1.122	1.154	1.247	1.361	1.564	2.073	1.923	1.471
	μ (10^{-6} Pa·s)	230	179	138	108	83.2	55.5	37.6	32.4
	k (W/m·K)	0.1013	0.0916	0.0817	0.0716	0.0607	0.0504	0.0421	0.0378
	Pr	2.55	2.26	2.11	2.04	2.14	2.28	1.72	1.26

THERMODYNAMIC AND THERMOPHYSICAL PROPERTIES—TABLE 75. Saturated Refrigerant 23

Temp. K	Pressure, bar	v_f, m³/kg	v_g, m³/kg	h_f, kJ/kg	h_g, kJ/kg	s_f, kJ/(kg·K)	s_g, kJ/(kg·K)	c_{pf}, kJ/(kg·K)	μ_f, 10⁻⁶ Pa·s	k_f, W/(m·K)	Pr
180	0.510	0.000678	0.4088	-66.0	181.1	-0.3179	1.0549				
190	0.950	0.000693	0.2279	-54.4	185.3	-0.2554	1.0062				
191.1[b]	1.013	0.000695	0.2139	-53.1	185.7	-0.2485	1.0011				
200	1.652	0.000710	0.1353	-42.6	189.1	-0.1948	0.9635				
210	2.709	0.000729	0.0845	-30.3	192.4	-0.1353	0.9254				
220	4.298	0.000751	0.0551	-17.5	195.4	-0.0764	0.8913				
230	6.312	0.000777	0.0372	-4.3	197.8	-0.0182	0.8602				
240	9.091	0.000807	0.0259	9.4	199.6	0.0392	0.8314	0.710	170.1	0.105	1.15
250	12.69	0.000844	0.0183	23.6	200.7	0.0957	0.8042	1.043	150.4	0.098	1.56
260	17.25	0.000889	0.0132	38.1	200.9	0.1512	0.7773	1.289	131.2	0.091	1.85
270	22.94	0.000948	0.0095	53.5	199.8	0.2071	0.7493	1.497	113.0	0.084	2.00
280	29.98	0.001031	0.0068	70.5	196.4	0.2665	0.7162				
290	38.68	0.001169	0.0046	92.0	188.1	0.3387	0.6698				
299.1[c]	48.36	0.001905	0.0019	143.0	143.0	0.5062	0.5062				

b = normal boiling point; c = critical point. $h_f = s_f = 0$ at 233.15 K = -40°C.

THERMODYNAMIC AND THERMOPHYSICAL PROPERTIES—TABLE 76.   Saturated Difluoromethane (R32)

Temp. K	Pressure, bar	v_f, m³/kg	v_g, m³/kg	h_f, kJ/kg	h_g, kJ/kg	s_f, kJ/(kg·K)	s_g, kJ/(kg·K)	c_{pf}, kJ/(kg·K)	c_{pg}, kJ/(kg·K)	μ_f, 10^{-6} Pa·s	μ_g, 10^{-6} Pa·s	k_f, W/(m·K)	k_g, W/(m·K)	Pr_f	Pr_g
200	0.2960	7.845.−4	1.0580	−52.340	351.160	−0.2418	1.7757								
210	0.5440	8.025.−4	0.5990	−37.750	356.880	−0.1652	1.7098								
220	0.9384	8.208.−4	0.3593	−20.690	361.970	−0.0909	1.6485	1.557	0.799	283.8	10.30				
230	1.5345	8.402.−4	0.2261	−4.984	366.773	−0.0215	1.5948	1.580	0.839	247.8	10.37				
240	2.3963	8.611.−4	0.1483	10.947	371.129	0.0459	1.5468	1.613	0.894	220.4	10.46				
250	3.5966	8.842.−4	0.1005	27.1778	374.971	0.1117	1.5030	1.642	0.963	198.1	10.66	0.1646	0.0097	1.98	1.06
260	5.2160	9.096.−4	0.07020	43.786	378.224	0.1763	1.4624	1.682	1.043	177.9	10.95	0.1562	0.0106	1.92	1.08
270	7.3423	9.376.−4	0.05009	60.849	380.786	0.2397	1.4247	1.730	1.138	159.1	11.31	0.1487	0.0115	1.85	1.12
280	10.070	9.696.−4	0.03643	78.456	382.525	0.3029	1.3886	1.786	1.244	141.6	11.70	0.1403	0.0125	1.81	1.16
290	13.502	1.006.−3	0.02687	96.713	383.262	0.3654	1.3534	1.863	1.375	126.1	12.21	0.1308	0.0136	1.80	1.23
300	17.749	1.049.−3	0.02001	115.754	382.737	0.4283	1.3182	1.955	1.560	112.0	12.82	0.1228	0.0149	1.78	1.34
310	22.931	1.100.−3	0.01497	135.801	380.576	0.4919	1.2815	2.084	1.810	98.8	13.71	0.1155	0.0165	1.78	1.50
320	29.186	1.166.−3	0.01117	157.212	376.163	0.5574	1.2415	2.282	2.16	86.1	14.4	0.1073	0.0184	1.83	1.69
330	36.675	1.243.−3	0.00822	180.724	368.357	0.6264	1.1950	2.620	2.62	75.1	15.3	0.0990	0.0205	1.99	1.96
340	45.603	1.394.−3	0.00581	208.262	354.460	0.7047	1.1347	3.560	4.21	65.4	17.3		0.0236		3.10
350	56.336		0.00317	274.640	337.933	0.8927	1.0735								
351.4ᶜ	57.927	0.00237	0.00237	286.675	286.675	0.9269	0.9269								

c = critical point. The notation 7.845.−4 signifies 7.845×10^{-4}.

THERMODYNAMIC AND THERMOPHYSICAL PROPERTIES—TABLE 77. Specific Heat at Constant Pressure, Thermal Conductivity, Viscosity, and Prandtl Number of R32 Gas

Temp., K	Property	P, bar			
		1	5	10	15
250	c_p (kJ/kg·K)	0.805			
	μ (10^{-6} Pa·s)	10.55			
	k (W/m·K)	0.0094			
	Pr	0.908			
260	c_p (kJ/kg·K)	0.810	1.025		
	μ (10^{-6} Pa·s)	11.00	10.96		
	k (W/m·K)	0.0100	0.0104		
	Pr	0.890	1.080		
270	c_p (kJ/kg·K)	0.818	0.991		
	μ (10^{-6} Pa·s)	11.42	11.37		
	k (W/m·K)	0.0107	0.0111		
	Pr	0.873	1.015		
280	c_p (kJ/kg·K)	0.825	0.969	1.238	
	μ (10^{-6} Pa·s)	11.82	11.77	11.68	
	k (W/m·K)	0.0116	0.0118	0.0125	
	Pr	0.860	0.967	1.157	
290	c_p (kJ/kg·K)	0.837	0.959	1.161	
	μ (10^{-6} Pa·s)	12.28	12.22	12.17	
	k (W/m·K)	0.0121	0.0125	0.0131	
	Pr	0.849	0.938	1.079	
300	c_p (kJ/kg·K)	0.849	0.951	1.118	1.370
	μ (10^{-6} Pa·s)	12.70	12.69	12.66	12.62
	k (W/m·K)	0.0128	0.0132	0.0138	0.0144
	Pr	0.842	0.914	1.026	1.201

Temp., K	Property	P, bar							
		1	5	10	15	20	25	30	40
310	c_p (kJ/kg·K)	0.861	0.945	1.084	1.279	1.560			
	μ (10^{-6} Pa·s)	13.12	13.10	13.08	13.07	13.09			
	k (W/m·K)	0.0135	0.0139	0.0144	0.0150	0.0159			
	Pr	0.878	0.891	0.985	1.114	1.284			
320	c_p (kJ/kg·K)	0.873	0.944	1.059	1.207	1.400	1.704		
	μ (10^{-6} Pa·s)	13.54	13.54	13.55	13.56	13.60	13.75		
	k (W/m·K)	0.0142	0.0146	0.0150	0.0156	0.0164	0.0173		
	Pr	0.836	0.875	0.957	1.049	1.161	1.354		
330	c_p (kJ/kg·K)	0.885	0.942	1.038	1.158	1.301	1.508	1.837	
	μ (10^{-6} Pa·s)	13.96	13.96	13.98	14.01	14.15	14.28	14.52	
	k (W/m·K)	0.0148	0.0152	0.0156	0.0162	0.0169	0.0177	0.0187	
	Pr	0.834	0.865	0.930	1.001	1.089	1.217	1.426	
340	c_p (kJ/kg·K)	0.897	0.937	1.020	1.135	1.242	1.388	1.612	2.488
	μ (10^{-6} Pa·s)	14.38	14.40	14.43	14.47	14.53	14.65	14.85	16.00
	k (W/m·K)	0.0155	0.0159	0.0163	0.0168	0.0175	0.0182	0.0190	0.0217
	Pr	0.832	0.849	0.903	0.978	1.031	1.117	1.260	1.834
350	c_p (kJ/kg·K)	0.910	0.934	1.004	1.118	1.200	1.308	1.440	1.914
	μ (10^{-6} Pa·s)	14.80	14.82	14.84	14.87	14.92	15.06	15.21	16.16
	k (W/m·K)	0.0162	0.0165	0.0169	0.0174	0.0180	0.0186	0.0194	0.0216
	Pr	0.831	0.839	0.882	0.955	0.995	1.060	1.130	1.432

THERMODYNAMIC AND THERMOPHYSICAL PROPERTIES—TABLE 78. Saturated SUVA MP 39

Temp., °C	P_f, bar	P_g, bar	v_f, m³/kg	v_g, m³/kg	h_f, kJ/kg	h_g, kJ/kg	s_f, kJ/(kg·K)	s_μ, kJ/(kg·K)	$c_{\mu f}$, kJ/(kg·K)	μ_f, 10^{-6} Pa·s	k_f, W/(m·K)	Pr_f
-40	0.733	0.533	0.000712	0.3778	154.0	385.0	0.8188	1.8244	1.078	351	0.1209	3.13
-30	1.155	0.871	0.000728	0.2391	164.9	390.6	0.8647	1.8059	1.109	323	0.1154	3.06
-20	1.748	1.361	0.000744	0.1576	176.2	396.3	0.9099	1.7907	1.137	291	0.1107	2.99
-10	2.553	2.043	0.000762	0.1075	188.6	401.8	0.9577	1.7781	1.165	266	0.1057	2.93
0	3.615	2.965	0.000781	0.0755	200.0	407.3	1.0000	1.7675	1.197	241	0.1012	2.85
10	4.984	4.177	0.000803	0.0544	212.7	412.6	1.0454	1.7587	1.233	221	0.0967	2.82
20	6.712	5.733	0.000826	0.0399	225.3	417.6	1.0884	1.7510	1.277	202	0.0922	2.80
30	8.857	7.697	0.000851	0.0298	238.3	422.2	1.1316	1.7439	1.329	186	0.0877	2.83
40	11.475	10.133	0.000878	0.0225	252.0	426.5	1.1752	1.7372	1.392	170	0.0830	2.85
50	14.628	13.112	0.000909	0.0172	266.4	430.1	1.2194	1.7304	1.468	157	0.0781	2.95
60	18.378	16.711	0.000944	0.01313	281.6	433.0	1.2647	1.7228	1.564	143	0.0737	3.04
70	22.79	21.01	0.000988	0.01005	297.9	434.9	1.3118	1.7138	1.652	131	0.0684	3.16
80	27.92	26.12	0.001028	0.00764	315.9	435.4	1.3616	1.7022	1.802	122	0.0631	3.48
90	33.83	32.13	0.001084	0.00570	336.2	433.5	1.4163	1.6858	1.958	115	0.0577	3.90
100	40.53	39.22	0.001140	0.00403	361.4	426.9	1.4820	1.6582	2.16	110	0.0533	4.46
108.0^c	46.04	46.04	0.00196	0.00196	397	397						

c = critical point. SUVA MP 39 = R401A = CHClF₂ (R22) 53% wt + CH₃CHF₂ (R152a) 13% wt + CHClFCF₃ (R124) 34% wt, near-azeotropic blend.

THERMODYNAMIC AND THERMOPHYSICAL PROPERTIES—TABLE 79. SUVA MP 39 at Atmospheric Pressure

Temp., °C	-27.01	-20	0	20	40	60	80	100	120	140
v (m²/kg)	0.2102	0.2167	0.2351	0.2534	0.2715	0.2896	0.3076	0.3256	0.3435	0.3613
h (kJ/kg)	351.7	396.9	410.4	424.5	439.2	454.4	470.3	486.6	503.5	521.2
g (kJ/kg·K)	1.8009	1.8193	1.8706	1.9204	1.9689	2.0161	2.0623	2.1073	2.1513	2.1943
c_p (kJ/kg·K)	0.648	0.669	0.698	0.727	0.757	0.787	0.811	0.836	0.859	0.883
μ (10^{-6} Pa·s)	10.17	10.13	11.18	11.93	12.68	13.42	14.17	14.89	15.61	16.32
k (W/m·K)	0.00878	0.00921	0.01041	0.01161	0.01282	0.01404	0.01536	0.01668	0.01796	0.01929
Pr	0.750	0.756	0.750	0.749	0.749	0.748	0.748	0.748	0.747	0.747
Z	0.9829	0.9852	0.9906	0.9949	0.9979	1.0005	1.0025	1.0043	1.0056	1.0060

THERMODYNAMIC AND THERMOPHYSICAL PROPERTIES — TABLE 80.   Saturated KLEA 60

Pressure, bar	T_f, K	T_g, K	v_f, m³/kg	v_g, m³/kg	h_f, kJ/kg	h_g, kJ/kg	s_f, kJ/(kg·K)	s_g, kJ/(kg·K)
1	227.3	234.0	0.0007118	0.2097	−7.80	229.64		0.9965
1.5	236.1	242.5	0.0007263	0.1433	3.92	235.02		0.9833
2	242.8	249.1	0.0007381	0.1093	12.89	239.07		0.9744
2.5	248.3	254.5	0.0007483	0.0885	20.27	242.35		0.9679
3	253.0	259.1	0.0007573	0.0744	26.57	245.08		0.9629
4	260.7	266.8	0.0007735	0.0564	37.23	249.54		0.9552
5	267.3	273.1	0.0007880	0.0442	46.12	253.07		0.9496
6	272.9	278.5	0.0008012	0.0384	53.84	255.24		0.9450
8	282.1	287.5	0.0008254	0.0286	67.02	260.70		0.9378
10	289.8	295.0	0.0008480	0.0228	78.23	263.86		0.9318
12.5	297.9	302.8	0.0008750	0.01802	90.50	266.95		0.9257
15	304.8	309.5	0.0009017	0.01481	101.51	269.12		0.9190
17.5	311.0	315.4	0.0009290	0.01247	111.64	270.58		0.9128
20	316.5	320.7	0.0009613	0.01069	121.18	271.46		0.9065
22.5	321.4	325.5	0.0009884	0.00928	130.31	271.79		0.8999
25	326.1	329.8	0.001023	0.00828	139.17	271.63		0.8927
27.5	330.4	333.9	0.001063	0.00717	147.89	270.97		0.8850
30	334.5	337.6	0.001115	0.00635	156.58	269.81		0.8765

$h_f = s_f = 0$ at 233.15 K = −40°C.

THERMODYNAMIC AND THERMOPHYSICAL PROPERTIES — TABLE 81.   Saturated KLEA 61

Pressure, bar	T_f, K	T_g, K	v_f, m³/kg	v_g, m³/kg	h_f, kJ/kg	h_g, kJ/kg	s_f, kJ/(kg·K)	s_g, kJ/(kg·K)
1	225.6	230.0	0.0006852	0.1800	−9.45	191.64		0.8433
1.5	234.3	238.5	0.0006994	0.1230	2.52	196.90		0.8341
2	241.8	245.0	0.0007110	0.0937	9.72	200.88		0.8282
2.5	246.4	250.4	0.0007211	0.0758	16.59	204.10		0.8245
3	251.1	254.9	0.0007301	0.0637	22.47	206.80		0.8215
4	258.9	262.6	0.0007463	0.04831	32.43	211.22		0.8172
5	265.4	269.0	0.0007607	0.03888	40.76	214.74		0.8141
6	270.9	274.4	0.0007740	0.03249	48.00	217.65		0.8123
8	280.2	283.4	0.0007985	0.02435	59.82	222.21		0.8080
10	287.8	290.9	0.0008214	0.01936	70.98	225.63		0.8048
12.5	295.8	298.7	0.0008491	0.01528	82.59	228.80		0.8010
15	302.8	305.5	0.0008768	0.01251	93.02	231.08		0.7971
17.5	308.8	311.4	0.0009053	0.01049	102.67	232.64		0.7929
20	314.3	316.7	0.0009353	0.00896	111.79	233.60		0.7882
22.5	319.3	321.5	0.0009680	0.00774	120.55	233.99		0.7829
25	323.9	325.9	0.001005	0.00674	129.11	233.85		0.7769
27.5	328.1	330.0	0.001048	0.00590	137.62	233.16		0.7700
30	332.1	333.7	0.001102	0.00518	146.21	231.84		0.7619

THERMODYNAMIC AND THERMOPHYSICAL PROPERTIES—TABLE 82. Saturated SUVA HP 62

Temp., °C	P_f, bar	P_g, bar	v_f, m³/kg	v_g, m³/kg	h_f, kJ/kg	h_g, kJ/kg	s_f, kJ/(kg·K)	s_g, kJ/(kg·K)	c_{pf}, kJ/(kg·K)	μ_f, 10⁻⁶ Pa·s	k_f, W/(m·K)	Pr_f
−50	0.852	0.821	0.000761	0.2244	133.1	337.3	0.7318	1.6487		370	0.0970	
−40	1.367	1.325	0.000779	0.1434	145.6	343.8	0.7862	1.6380		318		
−30	2.095	2.041	0.000799	0.0953	159.9	350.3	0.8460	1.6301	1.220	276	0.0868	3.88
−20	3.087	3.018	0.000820	0.0656	172.8	356.5	0.8975	1.6245	1.260	238	0.0834	3.60
−10	4.404	4.321	0.000843	0.0463	186.1	362.6	0.9487	1.6202	1.302	207	0.0801	3.37
0	6.111	6.013	0.000868	0.03338	200.0	368.3	1.0000	1.6188	1.351	181	0.0767	3.19
10	8.278	8.165	0.000898	0.02444	214.5	373.6	1.0515	1.6138	1.412	158	0.0733	3.04
20	10.977	10.851	0.000933	0.01809	229.9	378.3	1.1038	1.6106	1.489	138	0.0698	2.94
30	14.287	14.150	0.000977	0.01348	246.2	382.2	1.1574	1.6065	1.592	122	0.0663	2.93
40	18.292	18.148	0.001037	0.01003	263.8	385.0	1.2130	1.6005	1.753	106	0.0624	2.98
50	23.08	22.94	0.001122	0.00739	283.2	386.1	1.2723	1.5910	2.09	91	0.0583	3.26
60	28.75	28.63	0.001261	0.00527	305.8	384.2	1.3389	1.5742		76	0.0535	
70	35.58			0.00285	339.8	375.9				61		
72.1	37.32	37.32	0.00206	0.00206	361	361						

THERMODYNAMIC AND THERMOPHYSICAL PROPERTIES—TABLE 83. SUVA HP 62 at Atmospheric Pressure

Temp., °C	−45.63	−40	−20	0	20	40	60	80	100	120
v (m³/kg)	0.1866	0.1921	0.2100	0.2278	0.2455	0.2630	0.2805	0.2980	0.3153	0.3325
h (kJ/kg)	336.0	344.4	359.9	376.2	393.1	410.9	429.3	448.4	468.2	488.7
s (kJ/kg·K)	1.6599	1.6636	1.7274	1.7891	1.8491	1.9076	1.9646	2.0203	2.0747	2.1278
c_p (kJ/kg·K)	0.732	0.738	0.781	0.821	0.860	0.897	0.933	0.967	1.000	1.032
μ (10⁻⁶ Pa·s)	9.47	9.68	10.45	11.22	11.99	12.76	13.53	14.30	15.07	15.84
k (W/m·K)	0.00860	0.00932	0.01059	0.01186	0.01313	0.01440	0.01568	0.01695	0.01827	0.01949
P_r	0.806	0.767	0.771	0.777	0.785	0.795	0.805	0.816	0.827	0.839
Z	0.9755	0.9800	0.9867	0.9919	0.9961	0.9989	1.0014	1.0037	1.0050	1.0060

THERMODYNAMIC AND THERMOPHYSICAL PROPERTIES—TABLE 84. Saturated KLEA 66

Pressure, bar	T_f, K	T_g, K	v_f, m³/kg	v_g, m³/kg	h_f, kJ/kg	h_g, kJ/kg	s_f, kJ/(kg·K)	s_g, kJ/(kg·K)
0.69	221.46	228.77	0.000 7122	0.31325	−16.16	241.25		1.0729
1	228.89	236.05	0.000 7237	0.22131	−5.89	245.91		1.0580
1.5	237.69	244.69	0.000 7382	0.15140	6.22	251.38		1.0430
2	244.45	251.33	0.000 7501	0.11537	15.51	255.52		1.0330
2.5	249.99	256.76	0.000 7600	0.09104	23.12	258.95		1.0258
3	254.36	261.39	0.000 7695	0.07855	29.62	261.63		1.0201
4	262.60	269.14	0.000 7857	0.05964	40.60	266.16		1.0114
5	269.12	275.51	0.000 8001	0.04806	49.74	269.74		1.0055
6	274.70	280.98	0.000 8133	0.04021	57.69	272.68		0.9993
8	284.03	290.08	0.000 8375	0.03022	71.22	277.25		0.9913
10	291.74	297.56	0.000 8599	0.02410	82.73	280.64		0.9834
12.5	299.87	305.44	0.000 8867	0.01910	95.32	283.74		0.9770
15	306.87	312.18	0.000 9131	0.01571	106.59	285.93		0.9701
17.5	313.05	318.10	0.000 9400	0.01324	116.97	287.28		0.9633
20	318.60	323.40	0.000 9680	0.01137	126.73	288.26		0.9564
22.5	323.7	328.2	0.000 9981	0.00988	136.0	288.6		0.9493
25	328.3	332.5	0.001 032	0.00883	145.1	288.4		0.9418
27.5	332.7	336.6	0.001 072	0.00766	153.9	287.8		0.9338
30	336.7	340.4	0.001 125	0.00703	162.7	286.6		0.9251

THERMODYNAMIC AND THERMOPHYSICAL PROPERTIES — TABLE 85. Saturated SUVA MP 66

Temp, °C	P_f, bar	P_g, bar	v_f, m³/kg	v_g, m³/kg	h_f, kJ/kg	h_g, kJ/kg	s_f, kJ/(kg·K)	s_g, kJ/(kg·K)	c_{pf}, kJ/(kg·K)	μ_f, 10⁻⁶ Pa·s	k_f, W/(m·K)	Pr_f
−40	0.788	0.585	0.000710	0.3498	153.8	386.0	0.8184	1.8291	1.078	349	0.1209	3.11
−30	1.239	0.952	0.000725	0.2224	164.8	391.6	0.8643	1.8100	1.109	313	0.1154	3.01
−20	1.872	1.479	0.000740	0.1471	176.0	397.1	0.9095	1.7940	1.137	282	0.1106	2.90
−10	2.726	2.212	0.000758	0.1008	188.6	402.6	0.9577	1.7807	1.165	257	0.1057	2.83
0	3.850	3.198	0.000778	0.0710	200.0	407.8	1.0000	1.7694	1.197	236	0.1012	2.79
10	5.297	4.491	0.000801	0.05124	212.6	412.9	1.0450	1.7598	1.233	217	0.0967	2.77
20	7.120	6.146	0.000827	0.03771	225.1	417.7	1.0879	1.7512	1.277	198	0.0922	2.74
30	9.379	8.229	0.000858	0.02818	238.2	422.1	1.1311	1.7433	1.329	181	0.0877	2.74
40	12.133	10.808	0.000895	0.02131	251.9	426.1	1.1747	1.7357	1.392	168	0.0830	2.82
50	15.444	13.955	0.000939	0.01625	266.3	429.4	1.2190	1.7278	1.468	151	0.0781	2.84
60	19.378	17.750	0.000994	0.01244	281.6	431.9	1.2645	1.7191	1.564	139	0.0737	2.95
70	24.00	22.28	0.001066	0.00951	298.1	433.4	1.3120	1.7088	1.652	127	0.0684	3.07
80	29.37	27.64	0.001164	0.00721	316.3	433.2	1.3625	1.6956	1.802	116	0.0631	3.31
90	35.55	33.96	0.001313	0.00534	337.2	430.4	1.4187	1.6768			0.0577	
100	42.30										0.0533	
106.1c	46.82	46.82	0.00195	0.00195	389	389						

c = critical point SUVA MP 66 = R401 = CHClF₂ (R22) 61% wt + CH₃CHF₂ (R152a) 11% wt + CHClFCF₃ (R124) 28% wt, near-azeotropic blend.

THERMODYNAMIC AND THERMOPHYSICAL PROPERTIES — TABLE 86. SUVA MP 66 at Atmospheric Pressure

Temp., °C	−28.63b	−20	0	20	40	60	80	100	120	140
v (m³/kg)	0.2086	0.2177	0.2362	0.2545	0.2727	0.2908	0.3089	0.3269	0.3449	0.3629
h (kJ/kg)	392.2	397.9	411.2	425.1	439.6	454.6	470.1	486.2	502.7	519.4
s (kJ/kg·K)	1.8081	1.8299	1.8804	1.9295	1.9772	2.0237	2.0690	2.1132	2.1564	2.1986
c_p (kJ/kg·K)	0.641	0.652	0.688	0.716	0.744	0.771	0.796	0.822	0.844	0.866
μ (10⁻⁶ Pa·s)	9.78	10.43	11.18	11.93	12.68	13.42	14.17	14.89	15.61	16.32
k (W/m·K)	0.00817	0.00921	0.01041	0.01161	0.01282	0.01404	0.01536	0.01668	0.01796	0.01929
P_r	0.767	0.738	0.737	0.736	0.735	0.735	0.734	0.734	0.733	0.733
Z	0.9652	0.9730	0.9783	0.9822	0.9852	0.9876	0.9896	0.9912	0.9925	0.9937

b = normal boiling point.

THERMODYNAMIC AND THERMOPHYSICAL PROPERTIES—TABLE 87. Saturated SUVA HP 80

Temp., °C	P_f, bar	P_g, bar	v_f, m³/kg	v_g, m³/kg	h_f, kJ/kg	h_g, kJ/kg	s_f, kJ/(kg·K)	s_g, kJ/(kg·K)	c_{pf}, kJ/(kg·K)	μ_f, 10^{-6}Pa·s	k_f, W/(m·K)	Pr_f
−50	0.962	0.872	0.000679	0.2033	139.6	334.1	0.7578	1.6327		377	0.0970	
−40	1.520	1.403	0.000695	0.1303	150.8	339.9	0.8070	1.6206		317		
−30	2.305	2.156	0.000713	0.0869	163.1	345.6	0.8584	1.6110	1.193	283	0.0880	3.84
−20	3.370	3.188	0.000733	0.0598	174.9	351.1	0.9053	1.6034	1.217	247	0.0849	3.54
−10	4.776	4.560	0.000757	0.0423	187.6	356.4	0.9541	1.5972	1.236	215	0.0813	3.27
0	6.588	6.336	0.000785	0.03060	200.0	361.3	1.0000	1.5919	1.253	188	0.0778	3.03
10	8.877	8.592	0.000819	0.02248	213.0	365.9	1.0461	1.5870	1.286	165	0.0743	2.86
20	11.720	11.404	0.000860	0.01671	226.7	369.8	1.0927	1.5820	1.340	146	0.0708	2.76
30	15.195	14.855	0.000911	0.01250	241.2	373.1	1.1403	1.5762	1.412	128	0.0672	2.69
40	19.388	19.034	0.000977	0.00936	256.8	375.4	1.1897	1.5690	1.512	113	0.0634	2.70
50	24.39	24.04	0.001070	0.00696	273.9	376.2	1.2420	1.5589	1.64	98	0.0593	2.71
60	30.30	29.97	0.001212	0.00505	293.6	374.6	1.2998	1.5433	1.81	83	0.0551	2.79
70										68		
75.5^c	41.35	41.35	0.001850	0.00185	340	340						

c = critical point. SUVA HP 80 = R402 = CHF$_2$CF$_3$ (R125) 60% wt + CH$_3$CH$_2$CH$_3$ (R290) 2% wt + CHClF$_2$ (R22) 38% wt, near-azeotropic blend.

THERMODYNAMIC AND THERMOPHYSICAL PROPERTIES—TABLE 88. SUVA HP 80 at Atmospheric Pressure

Temp., °C	−46.95^b	−40	−20	0	20	40	60	80	100	120
v (m³/kg)	0.1768	0.1827	0.1996	0.2164	0.2331	0.2497	0.2663	0.2828	0.2992	0.3155
h (kJ/kg)	335.9	340.5	354.3	368.6	383.5	398.7	414.9	431.4	448.5	466.1
s (kJ/kg·K)	1.6286	1.6490	1.7055	1.7599	1.8124	1.8633	1.9128	1.9610	2.0081	2.0541
c_p (kJ/kg·K)	0.648	0.654	0.687	0.721	0.749	0.779	0.807	0.836	0.863	0.890
μ (10^{-6}Pa·s)	9.42	9.69	10.45	11.22	11.99	12.75	13.52	14.29	15.06	15.82
k (W/m·K)	0.00888	0.00932	0.01059	0.01186	0.01313	0.01440	0.01568	0.01695	0.01822	0.01949
P_r	0.687	0.680	0.678	0.681	0.685	0.690	0.696	0.703	0.713	0.722
Z	0.9673	0.9697	0.9758	0.9804	0.9840	0.9868	0.9892	0.9910	0.9923	0.9932

b = normal boiling pt.

THERMODYNAMIC AND THERMOPHYSICAL PROPERTIES—TABLE 89. Saturated SUVA HP 81

Temp., °C	P_f, bar	P_g, bar	v_f, m³/kg	v_g, m³/kg	h_f, kJ/kg	h_g, kJ/kg	s_f kJ/(kg·K)	s_g, kJ/(kg·K)	c_{pf}, kJ/(kg·K)	μ_f, 10^{-6} Pa·s	k_f, W/(m·K)	Pr_f
-50	0.883	0.787	0.000687	0.2425	140.3	351.7	0.7606	1.7122		383	0.1031	
-40	1.403	1.273	0.000702	0.1548	151.4	357.2	0.8092	1.6957		333	0.0983	
-30	2.135	1.967	0.000719	0.1028	163.3	362.7	0.8589	1.6820	1.178	290	0.0941	3.63
-20	3.132	2.923	0.000739	0.0707	174.9	368.0	0.9054	1.6706	1.191	253	0.0900	3.35
-10	4.451	4.198	0.000761	0.0499	187.8	373.0	0.9550	1.6611	1.204	223	0.0863	3.11
0	6.153	5.852	0.000787	0.03610	200.0	377.8	1.0000	1.6528	1.221	195	0.0818	2.91
10	8.307	7.959	0.000817	0.02656	212.7	382.2	1.0450	1.6451	1.288	173	0.0790	2.82
20	10.984	10.591	0.000854	0.01980	226.0	386.0	1.0905	1.6376	1.313	151	0.0753	2.63
30	14.261	13.827	0.000899	0.01490	240.1	389.3	1.1367	1.6299	1.37	137	0.0715	2.49
40	18.216	17.750	0.000955	0.01125	255.1	391.5	1.1842	1.6211	1.75	122	0.0676	3.16
50	22.93	22.45	0.001030	0.00848	271.4	392.8	1.2339	1.6104	2.07	106	0.0633	3.47
60	28.50	28.03	0.001136	0.00632	289.5	392.2	1.2873	1.5961		91	0.0586	
70	35.01	34.60	0.001307	0.00456	299.6	390.9	1.3164	1.5866		75	0.0544	
80												
82.6^c	44.45	44.45	0.00188	0.00188	351	351	351	351				

c = critical point. SUVA HP 81 = R402 (38/2/60) = CHF_2CF_3 (R125) 38% wt + $CH_3CH_2CH_3$ (R290) 2% wt + $CHClF_2$ (R22) 60% wt, near-azeotropic blend.

THERMODYNAMIC AND THERMOPHYSICAL PROPERTIES—TABLE 90. SUVA HP 81 at Atmospheric Pressure

Temp., °C	-44.87^b	-40	-20	0	20	40	60	80	100	120
v (m³/kg)	0.1903	0.1960	0.2142	0.2322	0.2500	0.2678	0.2856	0.3032	0.3209	0.3386
h (kJ/kg)	354.7	357.7	370.8	384.6	398.8	413.6	428.9	444.7	461.0	477.7
s (kJ/kg·K)	1.7032	1.7169	1.7711	1.8232	1.8735	1.9222	1.9696	2.0158	2.0607	2.1047
c_p (kJ/kg·K)	1.187	1.177	1.169	1.159	1.149	1.143	1.134	1.128	1.124	1.120
μ (10^{-6}Pa·s)	10.16	10.33	11.10	11.86	12.62	13.39	14.15	14.78	15.54	16.30
k (W/m·K)	0.00739	0.00768	0.00902	0.01036	0.01170	0.01304	0.01438	0.01572	0.01706	0.01840
P_r	1.632	1.583	1.439	1.327	1.239	1.174	1.124	1.061	1.024	0.992
Z	0.9622	0.9703	0.9766	0.9811	0.9843	0.9870	0.9894	0.9909	0.9926	0.9940

b = normal boiling point.

THERMODYNAMIC AND THERMOPHYSICAL PROPERTIES—TABLE 91. Saturated Refrigerant 113

T, K	P, bar	v_f, m³/kg	v_g, m³/kg	h_f, kJ/kg	h_g, kJ/kg	s_f, kJ/(kg·K)	s_g, kJ/(kg·K)	c_{pf}, kJ/(kg·K)	μ_f, 10^{-4} Pa·s	k_f, W/(m·K)
240	0.0233	5.908.−4	4.548	5.70	171.97	0.0241	0.7169	0.845	17.9	0.087
250	0.0435	5.986.−4	2.537	14.19	178.06	0.0587	0.7142	0.877	14.8	0.084
260	0.0767	6.066.−4	1.492	22.83	184.22	0.0926	0.7134	0.895	12.3	0.083
270	0.1290	6.150.−4	0.9189	31.65	190.46	0.1259	0.7141	0.916	10.4	0.081
280	0.2076	6.237.−4	0.5893	40.63	196.75	0.1585	0.7161	0.933	8.9	0.079
290	0.3217	6.328.−4	0.3917	49.77	203.08	0.1906	0.7192	0.946	7.6	0.077
300	0.4817	6.422.−4	0.2687	59.07	209.44	0.2221	0.7233	0.958	6.6	0.075
310	0.6999	6.522.−4	0.1895	68.51	215.80	0.2530	0.7281	0.971	5.9	0.073
320	0.9897	6.626.−4	0.1370	78.09	222.17	0.2833	0.7336	0.983	5.2	0.071
330	1.3657	6.737.−4	0.1012	87.80	228.53	0.3131	0.7396	0.992	4.7	0.069
340	1.8347	6.854.−4	0.0762	97.64	234.86	0.3424	0.7460	1.000	4.2	0.066
350	2.4406	6.979.−4	0.0584	107.58	241.16	0.3711	0.7528	1.013	3.8	0.065
360	3.174	7.112.−4	0.0454	117.65	247.41	0.3993	0.7598	1.029	3.4	0.062
370	4.062	7.255.−4	0.0357	127.82	253.59	0.4270	0.7669	1.042	3.2	0.060
380	5.123	7.411.−4	0.0284	138.11	259.70	0.4542	0.7742	1.059	2.9	0.058
390	6.379	7.580.−4	0.0229	148.52	265.71	0.4810	0.7815	1.084	2.7	0.056
400	7.849	7.767.−4	0.0185	159.07	271.59	0.5075	0.7888	1.109	2.46	0.054
410	9.556	7.975.−4	0.0151	169.78	277.31	0.5336	0.7958	1.14	2.28	0.052
420	11.52	8.211.−4	0.0124	180.69	282.83	0.5595	0.8027	1.18	2.10	0.050
430	13.87	8.483.−4	0.0102	191.85	288.09	0.5853	0.8091	1.22	1.93	0.047
440	16.35	8.806.−4	0.0083	203.35	292.98	0.6112	0.8149	1.27	1.75	0.045
450	19.26	9.201.−4	0.0068	215.31	297.38	0.6375	0.8198	1.32	1.58	0.042
460	22.56	9.713.−4	0.0055	227.97	301.03	0.6645	0.8234	1.38	1.33	0.039
470	26.29	1.044.−3	0.0044	241.79	303.41	0.6933	0.8244	1.45	1.07	0.035
480	30.52	1.174.−3	0.0032	258.16	303.00	0.7264	0.8198	1.54	0.77	0.031
487.5	34.11	1.754.−3	0.0018	288.10	288.10	0.7828	0.7828	∞	0.30	∞

THERMODYNAMIC AND THERMOPHYSICAL PROPERTIES—TABLE 92. Saturated Refrigerant 114

T, K	P, bar	v_f, m³/kg	v_g, m³/kg	h_f, kJ/kg	h_g, kJ/kg	s_f, kJ/(kg·K)	s_g, kJ/(kg·K)	c_{pf}, kJ/(kg·K)	μ_f, 10^{-4} Pa·s	k_f, W/(m·K)
190	0.0058	6.326.–4	15.823	−42.58	125.78	−0.2091	0.6794	0.765	23.9	0.093
200	0.0137	6.344.–4	7.094	−31.87	131.01	−0.1542	0.6648	0.787	18.2	0.090
210	0.020	6.366.–4	3.465	−21.48	136.41	−0.1035	0.6541	0.810	14.3	0.088
220	0.059	6.391.–4	1.822	−11.37	141.95	−0.0565	0.6466	0.831	11.5	0.085
230	0.109	6.421.–4	1.021	−1.50	147.61	−0.0126	0.6419	0.854	9.4	0.082
240	0.190	6.457.–4	0.604	8.18	153.36	0.0286	0.6393	0.877	7.9	0.080
250	0.317	6.500.–4	0.375	17.74	159.18	0.0676	0.6387	0.900	6.61	0.077
260	0.505	6.554.–4	0.2431	27.22	165.05	0.1047	0.6396	0.923	5.66	0.075
270	0.773	6.619.–4	0.1633	36.71	170.95	0.1405	0.6418	0.946	4.96	0.072
280	1.143	6.700.–4	0.1132	46.27	176.85	0.1751	0.6452	0.967	4.30	0.069
290	1.636	6.799.–4	0.0807	55.95	182.75	0.2090	0.6494	0.991	3.80	0.067
300	2.279	6.918.–4	0.0590	65.79	188.61	0.2422	0.6543	1.015	3.35	0.064
310	3.096	7.060.–4	0.0440	75.79	194.44	0.2748	0.6598	1.038	3.02	0.061
320	4.116	7.224.–4	0.0334	85.92	200.19	0.3067	0.6657	1.062	2.69	0.059
330	5.366	7.412.–4	0.0257	96.16	205.84	0.3379	0.6719	1.087	2.48	0.056
340	6.877	7.624.–4	0.0201	106.49	211.37	0.3685	0.6781	1.111	2.27	0.054
350	8.683	7.863.–4	0.0158	116.96	216.71	0.3984	0.6843	1.136	2.07	0.051
360	10.82	8.135.–4	0.0125	127.63	221.82	0.4280	0.6903	1.160	1.91	0.048
370	13.32	8.453.–4	0.0099	138.60	226.57	0.4575	0.6957	1.185	1.76	0.045
380	16.24	8.836.–4	0.0079	149.99	230.84	0.4872	0.7002	1.210	1.59	0.042
390	19.62	9.324.–4	0.0062	162.01	234.36	0.5176	0.7032	1.236	1.39	0.038
400	23.52	1.001.–3	0.0048	175.03	236.61	0.5496	0.7036	1.261	1.17	0.034
410	28.00	1.118.–3	0.0035	190.13	236.20	0.5857	0.6980	1.5	0.87	0.030
419.0	32.61	1.795.–3	0.0018	219.90	219.90	0.6559	0.6559	∞	0.34	∞

THERMODYNAMIC AND THERMOPHYSICAL PROPERTIES—TABLE 93.   Saturated
Refrigerant 115

Temp., °F	Pressure, lb/in² abs.	Volume, ft³/lb		Enthalpy, Btu/lb		Entropy, Btu/(lb) (°F)	
		Liquid	Vapor	Liquid	Vapor	Liquid	Vapor
−100	2.327	0.00966	10.57	−13.07	45.83	−0.0335	0.1302
−80	4.573	0.00986	5.624	−8.78	48.39	−0.0219	0.1286
−60	8.306	0.01009	3.218	−4.43	50.96	−0.0108	0.1278
−40	14.13	0.01033	1.953	0.00	53.53	0.0000	0.1275
−20	22.74	0.01060	1.245	4.50	56.07	0.0104	0.1277
0	34.94	0.01090	0.8257	9.09	58.56	0.0206	0.1282
20	51.59	0.01123	0.5657	13.76	61.00	0.0305	0.1290
40	73.65	0.01161	0.3979	18.54	63.35	0.0401	0.1298
60	102.1	0.01204	0.2857	23.45	65.60	0.0496	0.1308
80	138.1	0.01255	0.2081	28.54	67.71	0.0591	0.1317
100	182.7	0.01316	0.1530	33.85	69.63	0.0686	0.1325
120	237.3	0.01393	0.1125	39.50	71.24	0.0782	0.1330
140	303.2	0.01496	0.0817	45.67	72.36	0.0884	0.1329
160	382.0	0.01664	0.0567	52.76	72.42	0.0996	0.1314
170	427.0	0.01838	0.0444	56.56	71.33	0.1055	0.1290
175.89	457.6	0.0261	0.0261	64.30	64.30	0.1175	0.1175

THERMODYNAMIC AND THERMOPHYSICAL PROPERTIES—TABLE 94. Refrigerant 123

Pressure, bar	Temp., K	v_f, m³/kg	v_g, m³/kg	h_f, kJ/kg	h_g, kJ/kg	s_f, kJ/(kg·K)	s_g, kJ/(kg·K)	c_{pf}, kJ/(kg·K)	k_f, W/(m·K)	μ_f, 10⁻⁶ Pa·s	Pr_f
0.1	249.49	0.0006315	1.3430	13.25	198.51	0.0548	0.7977	0.849	0.0908	798.7	7.46
0.5	282.87	0.0006664	0.2993	41.72	218.53	0.1610	0.7863	0.923	0.0811	503.7	5.73
1.0	300.62	0.0006862	0.1567	58.62	229.20	0.2195	0.7869	1.000	0.0759	409.8	5.40
1.013	300.99	0.0006868	0.1546	58.99	229.43	0.2208	0.7870	1.001	0.0758	408.1	5.39
1.5	312.25	0.0007008	0.1070	70.51	236.52	0.2582	0.7892	1.038	0.0726	361.5	5.17
2.0	321.18	0.0007126	0.08139	79.90	241.76	0.2877	0.7917	1.063	0.0696	329.6	5.03
2.5	328.50	0.0007230	0.06546	87.76	246.20	0.3118	0.7942	1.079	0.0678	306.1	4.87
3.0	334.79	0.0007323	0.05525	94.59	249.96	0.3323	0.7965	1.091	0.0660	287.4	4.75
4.0	345.29	0.0007490	0.03836	106.16	256.17	0.3661	0.8006	1.108	0.0630	259.7	4.57
5.0	353.95	0.0007640	0.03358	115.83	261.17	0.3935	0.8042	1.120	0.0605	239.1	4.43
6	361.41	0.0007779	0.02799	124.23	265.34	0.4168	0.8073	1.130			
8	373.92	0.0008038	0.02090	138.48	272.04	0.4551	0.8124	1.148			
10	384.19	0.0008280	0.01675	150.35	277.18	0.4860	0.8162	1.168			
15	404.54	0.0008874	0.01062	174.49	286.01	0.5462	0.8218	1.234			
20	420.30	0.0009512	0.00751	194.19	291.01	0.5928	0.8232	1.345			
25	433.33	0.001030	0.00549	212.00	293.05	0.6334	0.8203	1.559			
30	444.10	0.001136	0.00408	228.26	291.27	0.6692	0.8112	2.005			
36.68	456.83	0.001818	0.00182	264.54	264.54	0.7393	0.7393				

$h_f = s_f = 0$ at $-40°C = 233.15$ K, s_f, s_g, c_{pf} units: kJ/kg·K.

507

THERMODYNAMIC AND THERMOPHYSICAL PROPERTIES—TABLE 95.   Saturated Refrigerant 124

Temp., °C	Pressure, bar	v_f, m³/kg	v_g, m³/kg	h_f, kJ/kg	h_g, kJ/kg	s_f, kJ/(kg·K)	s_g, kJ/(kg·K)
−40	0.2680	0.000644	0.5173	159.1	334.9	0.8384	1.5927
−30	0.4499	0.000655	0.3185	169.3	340.6	0.8813	1.5856
−20	0.7197	0.000668	0.2049	179.5	346.2	0.9222	1.5808
−10	1.1044	0.000681	0.1369	189.7	351.8	0.9616	1.5777
0	1.6348	0.000696	0.0945	200.0	357.4	1.0000	1.5762
10	2.3447	0.000711	0.06703	210.5	363.0	1.0376	1.5760
20	3.2710	0.000728	0.04867	221.3	368.5	1.0747	1.5768
30	4.4529	0.000747	0.03604	282.3	373.9	1.1115	1.5785
40	5.9320	0.000768	0.02713	243.7	379.2	1.1480	1.5808
50	7.7521	0.000791	0.02069	255.4	384.4	1.1843	1.5836
60	9.9599	0.000818	0.01594	267.5	389.3	1.2207	1.5864
70	12.605	0.000849	0.01236	280.1	393.9	1.2572	1.5890
80	15.742	0.000887	0.00961	293.2	398.0	1.2942	1.5909
90	19.432	0.000935	0.00744	307.0	401.3	1.3318	1.5915
100	23.749	0.000999	0.00569	321.9	403.4	1.3710	1.5894
110	28.787	0.001098	0.00420	338.4	403.0	1.4133	1.5820
120	34.702	0.001338	0.00269	360.6	394.9	1.4685	1.5558
122.5^c	36.340	0.001810	0.00181	378.5	378.5		

c = critical point.

THERMODYNAMIC AND THERMOPHYSICAL PROPERTIES—TABLE 96. Saturated Refrigerant 125

Temp., K	Pressure, bar	v_f, m³/kg	v_g, m³/kg	h_f, kJ/kg	h_g, kJ/kg	s_f, kJ/(kg·K)	s_g, kJ/(kg·K)	c_{pf}, kJ/(kg·K)	μ_f, 10⁻⁶ Pa·s
172.5*	0.035	0.000591	3.48						
180	0.064	0.000599	1.958						
190	0.133	0.000611	0.986						
200	0.257	0.000624	0.5312	-30.5	140.8	-0.1386	0.7183		644
210	0.465	0.000638	0.3057	-23.3	146.7	-0.1024	0.7067		531
220	0.794	0.000653	0.1854	-13.9	152.6	-0.0604	0.6981		445.8
224.9†	1.013	0.000660	0.1475	-8.8	155.5	-0.0386	0.6948		411.2
230	1.290	0.000669	0.1175	-3.4	158.4	-0.0147	0.6919	1.077	379.6
240	2.005	0.000686	0.0775	7.7	164.2	0.0324	0.6875	1.139	326.6
250	3.000	0.000705	0.0527	19.3	170.0	0.0800	0.6847	1.184	282.8
260	4.336	0.000725	0.0369	31.3	175.3	0.1274	0.6831	1.221	245.8
270	6.078	0.000749	0.0264	43.7	180.5	0.1743	0.6822	1.257	213.6
280	8.298	0.000776	0.0193	56.5	185.4	0.2206	0.6819	1.299	185.3
290	11.068	0.000809	0.0143	69.7	189.9	0.2666	0.6815	1.356	159.7
300	14.476	0.000848	0.0106	83.5	194.2	0.3126	0.6805	1.437	136.3
310	18.62	0.000898	0.0079	98.1	196.9	0.3597	0.6774	1.57	115.0
320	23.63	0.000969	0.0059	113.9	198.5	0.4079	0.6726	1.82	95.4
330	29.65	0.001088	0.0041	132.2	197.4	0.4621	0.6639		
339.4‡	35.95	0.00175	0.0018	169.0	169.0	0.5699	0.5699		

* = triple point; † = normal boiling point; ‡ = critical point. Converted, extrapolated, and interpolated from 1993 ASHRAE *Handbook—Fundamentals* (SI ed.) $h_f = s_f$ = 0 at 233.15 K = −40°C. This source also contains an enthalpy-log-pressure diagram from 0.3 to 100 bar, −65 to 175°C. An apparently identical diagram but a different saturation table is contained in Durate-Garza, H.A., Hwang, C.A. et al., ASHRAE Trans., 99, 2 (1993): 649–664. R124: The 1993 ASHRAE *Handbook—Fundamentals* (SI ed.) contains a saturation table from −60 to 122.47°C.

THERMODYNAMIC AND THERMOPHYSICAL PROPERTIES — TABLE 97. Refrigerant 134a

Pressure, bar	Temp., K	v_f, m³/kg	v_g, m³/kg	h_f, kJ/kg	h_g, kJ/kg	s_f, kJ/(kg·K)	s_g, kJ/(kg·K)	$c_{\mu f}$, kJ/(kg·K)	μ_f, 10⁻⁶ Pa·s	k_f, W/(m·K)	Pr_f
0.0039^t	169.85	0.0006285	35.263	−76.68	186.50	−0.3830	1.1665	1.147	2187		
0.5	232.69	0.0007062	0.3692	−0.57	225.27	−0.0025	0.9669	1.242	506	0.1121	5.61
0.6	236.22	0.0007113	0.3015	3.85	227.52	0.0161	0.9636	1.248	480	0.1105	5.42
0.8	242.04	0.007199	0.2375	11.15	231.19	0.0467	0.9560	1.258	438	0.1078	5.12
1.0	246.80	0.0007272	0.1924	17.14	234.15	0.0713	0.9507	1.267	408	0.1056	4.90
1.013	247.03	0.0007276	0.1902	17.50	234.33	0.0728	0.9503	1.268	406	0.1054	4.89
1.5	256.03	0.0007421	0.1312	28.96	239.86	0.1181	0.9419	1.288	358.7	0.1013	4.56
2.0	263.09	0.0007543	0.0999	38.13	244.14	0.1533	0.9364	1.306	326.6	0.0980	4.35
2.5	268.88	0.0007648	0.0806	45.75	247.60	0.1819	0.9326	1.322	303.2	0.0954	4.20
3.0	273.82	0.0007743	0.0677	52.33	250.50	0.2059	0.9297	1.337	285.1	0.0931	4.09
4.0	282.08	0.0007912	0.0512	63.50	255.22	0.2458	0.9256	1.363	257.7	0.0893	3.93
5	288.89	0.0008063	0.04116	72.87	258.99	0.2784	0.9232	1.387	237.5	0.0861	3.83
6	294.72	0.0008203	0.03434	81.04	262.09	0.3062	0.9208	1.410	221.6	0.0835	3.74
8	304.47	0.0008460	0.02565	95.00	267.01	0.3522	0.9171	1.454	197.6	0.0790	3.64
10	312.53	0.0008703	0.02035	106.86	270.74	0.3901	0.9144	1.497	179.5	0.0753	3.57
12	319.47	0.0008938	0.01675	117.34	273.65	0.4227	0.9120	1.541	165.1	0.0721	3.53
14	325.57	0.0009170	0.01414	126.80	275.92	0.4515	0.9095	1.589	153.0	0.0693	3.51
16	330.11	0.0009362	0.01247	134.00	277.40	0.4729	0.9073	1.631	144.3	0.0672	3.50
18	336.04	0.0009555	0.01059	143.68	279.01	0.5013	0.9041	1.698	133.2	0.0645	3.51
20	340.63	0.0009894	0.00931	151.39	279.95	0.5236	0.9010	1.764	124.8	0.0623	3.53
25	350.73	0.0010585	0.00695	169.30	280.64	0.5738	0.8913	1.987	106.6	0.0577	3.67
30	359.37	0.001144	0.00528	185.05	278.32	0.6212	0.8807	2.418	90.4	0.0538	4.06
35	366.89	0.001270	0.00399	203.19	273.52	0.6657	0.8574				
40	373.50	0.001606	0.00255	229.24	257.12	0.7292	0.8038				
40.56^c	374.18	0.001948	0.00195	241.22	241.22	0.7620	0.7620				

t = triple point. c = critical point. $h_f = s_f = 0$ at −40°C = 233.15 K.

THERMODYNAMIC AND THERMOPHYSICAL PROPERTIES—TABLE 98. Compressed Gaseous Refrigerant 134a

Temp., K		Pressure, bar						
		0	1	2	3	4	5	6
230	c_p (kJ/kg·K)	—	—	—	—	—	—	—
	μ (10^{-6} Pa·s)	—	—	—	—	—	—	—
	k (W/m·K)	—	—	—	—	—	—	—
	P_r	—	—	—	—	—	—	—
240	c_p (kJ/kg·K)	—	—	—	—	—	—	—
	μ (10^{-6} Pa·s)	—	—	—	—	—	—	—
	k (W/m·K)	—	—	—	—	—	—	—
	P_r	—	—	—	—	—	—	—
250	c_p (kJ/kg·K)	0.7437	0.7953	—	—	—	—	—
	μ (10^{-6} Pa·s)	10.11	10.15	—	—	—	—	—
	k (W/m·K)	0.0096	0.0097	—	—	—	—	—
	P_r	0.783	0.797	—	—	—	—	—
260	c_p (kJ/kg·K)	0.7627	0.8048	—	—	—	—	—
	μ (10^{-6} Pa·s)	10.47	10.51	—	—	—	—	—
	k (W/m·K)	0.0105	0.0107	—	—	—	—	—
	P_r	0.761	0.790	—	—	—	—	—
270	c_p (kJ/kg)	0.7831	0.8158	0.8557	—	—	—	—
	μ (10^{-6} Pa·s)	10.84	10.88	10.94	—	—	—	—
	k (W/m·K)	0.0117	0.0118	0.0118	—	—	—	—
	P_r	0.724	0.761	0.793	—	—	—	—
280	c_p (kJ/kg·K)	0.7996	0.8283	0.8604	—	—	—	—
	μ (10^{-6} Pa·s)	11.22	11.26	11.29	—	—	—	—
	k (W/m·K)	0.0122	0.0123	0.0123	—	—	—	—
	P_r	0.735	0.757	0.790	—	—	—	—
290	c_p (kJ/kg·K)	0.8176	0.8412	0.8673	0.8938	9.9335		—
	μ (10^{-6} Pa·s)	11.62	11.65	11.68	11.71	11.74		—
	k (W/m·K)	0.0130	0.0131	0.0131	0.0133			—
	P_r	0.731	0.748	0.773	0.787			—
300	c_p (kJ/kg·K)	0.8354	0.8556	0.8771	0.8972	0.9277	0.9606	0.9976
	μ (10^{-6} Pa·s)	12.05	12.06	12.08	12.10	12.15		
	k (W/m·K)	0.0139	0.0139	0.0140	0.0141	0.0142	0.0143	
	P_r	0.730	0.742	0.757	0.770	0.792	0.816	
310	c_p (kJ/kg·K)	0.8530	0.8703	0.8875	0.9046	0.9292	0.9546	0.9827
	μ (10^{-6} Pa·s)	12.44	12.45	12.47	12.49	12.52	12.54	
	k (W/m·K)	0.0145	0.0146	0.0147	0.0148	0.0149	0.0150	0.0152
	P_r	0.730	0.742	0.753	0.763	0.781	0.798	
320	c_p (kJ/kg·K)	0.8703	0.8843	0.8993	0.9163	0.9356	0.9548	0.9750
	μ (10^{-6} Pa·s)	12.83	12.84	12.86	12.88	12.90	12.93	12.97
	k (W/m·K)	0.0153	0.0153	0.0154	0.0155	0.0156	0.0157	0.0158
	P_r	0.730	0.740	0.751	0.761	0.774	0.786	0.800
330	c_p (kJ/kg·K)	0.8874	0.8996	0.9114	0.9268	0.9398	0.9569	0.9750
	μ (10^{-6} Pa·s)	13.22	13.23	13.25	13.27	13.29	13.32	13.35
	k (W/m·K)	0.0160	0.0161	0.0161	0.0162	0.0163	0.0164	0.0165
	P_r	0.729	0.739	0.750	0.759	0.766	0.777	0.789

THERMODYNAMIC AND THERMOPHYSICAL PROPERTIES—TABLE 98. Compressed Gaseous Refrigerant 134a (*Continued*)

	Pressure, bar						
Temp., K	0	1	2	3	4	5	6
c_p (kJ/kg·K)	0.9042	0.9152	0.9262	0.9372	0.9502	0.9632	0.9770
340 μ (10^{-6} Pa·s)	13.61	13.62	13.64	13.66	13.68	13.70	13.73
k (W/m·K)	0.0169	0.0169	0.0169	0.0170	0.0170	0.0171	0.071
P_r	0.728	0.738	0.748	0.755	0.765	0.772	0.780
c_p (kJ/kg·K)	0.9208	0.9307	0.9406	0.9505	0.9607	0.9695	0.9830
350 μ (10^{-6} Pa·s)	13.98	13.99	14.01	14.03	14.05	14.07	14.10
k (W/m·K)	0.0175	0.0176	0.0176	0.0177	0.0177	0.0178	0.0179
P_r	0.730	0.740	0.749	0.754	0.763	0.767	0.774

	Pressure, bar						
Temp., K	8	10	12.5	15	17.5	20	22.5
c_p (kJ/kg·K)	—	—	—	—	—	—	—
300 μ (10^{-6} Pa·s)	—	—	—	—	—	—	—
k (W/m·K)	—	—	—	—	—	—	—
P_r	—	—	—	—	—	—	—
c_p (kJ/kg·K)	1.053	—	—	—	—	—	—
310 μ (10^{-6} Pa·s)	—	—	—	—	—	—	—
k (W/m·K)	0.0155	—	—	—	—	—	—
P_r	—	—	—	—	—	—	—
c_p (kJ/kg·K)	1.028	1.097	—	—	—	—	—
320 μ (10^{-6} Pa·s)	13.05	13.13	—	—	—	—	—
k (W/m·K)	0.0161	—	—	—	—	—	—
P_r	0.833	–	—	—	—	—	—
c_p (kJ/kg·K)	1.015	1.065	1.151	1.276	—	—	—
330 μ (10^{-6} Pa·s)	13.41	13.49	13.64	13.86	—	—	—
k (W/m·K)	0.0168	0.0171	0.0177	0.0184	—	—	—
P_r	0.810	0.840	0.887	0.961	—	—	—
c_p (kJ/kg·K)	1.008	1.049	1.107	1.187	1.319	—	—
340 μ (10^{-6} Pa·s)	13.79	13.86	13.98	14.17		—	—
k (W/m·K)	0.0174	0.0177	0.181	0.0187		—	—
P_r	0.799	0.821	0.855	0.899		—	—
c_p (kJ/kg·K)	1.008	1.040	1.086	1.148	1.225	1.340	1.525
350 μ (10^{-6} Pa·s)	14.15	14.22	14.34	14.49		14.97	
k (W/m·K)	0.0181	0.0183	0.0186	0.0192	0.0198	0.0205	0.0215
Pr	0.788	0.828	0.837	0.866			

THERMODYNAMIC AND THERMOPHYSICAL PROPERTIES—TABLE 99. Refrigerant 142b

T, K	P, bar	v_f, m³/kg	v_g, m³/kg	h_f, kJ/kg	h_g, kJ/kg	s_f, kJ/(kg·K)	s_g, kJ/(kg·K)	cp_f, kJ/(kg·K)	μ_f, 10^{-4} Pa·s	k_f, W/(m·K)
200	0.0380	7.505.–4	4.337	−24.36	200.49	−0.1123	1.0119			0.123
210	0.0728	7.626.–4	2.374	−17.48	206.82	−0.0788	0.9893	1.15		0.118
220	0.1314	7.751.–4	1.373	−10.21	213.28	−0.0450	0.9708	1.17		0.114
230	0.2252	7.883.–4	0.833	−2.52	219.82	−0.0109	0.9558	1.18		0.111
240	0.3691	8.019.–4	0.527	9.92	229.74	0.0414	0.9387	1.19	0.517	0.109
250	0.5815	8.164.–4	0.346	14.32	233.05	0.0592	0.9341	1.21	0.466	0.103
260	0.8846	8.317.–4	0.234	23.54	239.66	0.0952	0.9264	1.22	0.422	0.099
270	1.3046	8.480.–4	0.162	33.32	246.18	0.1320	0.9204	1.24	0.385	0.095
280	1.8714	8.653.–4	0.115	43.68	252.57	0.1695	0.9155	1.26	0.355	0.091
290	2.6184	8.843.–4	0.0838	54.60	258.77	0.2076	0.9116	1.28	0.329	0.088
300	3.583	9.047.–4	0.0619	66.07	264.69	0.2462	0.9082	1.30	0.305	0.084
310	4.803	9.273.–4	0.0464	78.07	270.26	0.2851	0.9051	1.32	0.285	0.080
320	6.324	9.525.–4	0.0353	90.55	275.40	0.3243	0.9020	1.34	0.267	0.075
330	8.187	9.810.–4	0.0271	103.45	280.01	0.3634	0.8985		0.241	0.072
340	10.44	1.014.–3	0.0210	116.71	283.99	0.4024	0.8943		0.216	0.068
350	13.13	1.052.–3	0.0164	130.30	287.23	0.4409	0.8893		0.192	0.064
360	16.30	1.099.–3	0.0129	144.18	289.61	0.4791	0.8831			0.060
370	20.01	1.157.–3	0.0102	158.45	291.01	0.5170	0.8753			0.056
380	24.29	1.235.–3	0.0080	173.45	291.22	0.5557	0.8656			0.052
390	29.20	1.348.–3	0.0062	190.16	289.77	0.5974	0.8528			0.048
400	34.78	1.541.–3	0.0046	212.57	284.04	0.6521	0.8307			0.044
410	41.5	2.300.–3	0.0023	255.00	255.00					∞

THERMODYNAMIC AND THERMOPHYSICAL PROPERTIES—TABLE 100. Saturated Refrigerant R143a

Temp., K	Pressure, bar	v_f, m³/kg	v_g, m³/kg	h_f, kJ/kg	h_g, kJ/kg	s_f, kJ/(kg·K)	s_g, kJ/(kg·K)	c_{pf}, kJ/(kg·K)	μ_f, 10⁻⁴ Pa·s	k_f, W/(m·K)
161.82	0.01124	0.000752	14.22	53.2	320.5	0.3181	1.970	1.188	6.011	0.1416
170	0.02497	0.000764	6.709	63.0	325.5	0.3774	1.922	1.215	5.366	0.1403
180	0.05914	0.000778	2.991	75.3	331.8	0.4474	1.872	1.235	4.692	0.1372
190	0.126	0.000793	1.474	87.7	338.1	0.5147	1.832	1.252	4.121	0.1331
200	0.2458	0.000809	0.7898	100.3	344.5	0.5792	1.800	1.268	3.636	0.1281
210	0.4455	0.000826	0.4532	113.1	350.8	0.6415	1.774	1.287	3.221	0.1226
220	0.7586	0.000845	0.2754	126.1	357.1	0.7018	1.752	1.308	2.864	0.1168
225.92	1.01325	0.000856	0.2098	133.9	360.8	0.7367	1.741	1.323	2.676	0.1133
230	1.225	0.000865	0.1755	139.3	363.3	0.7604	1.734	1.333	2.556	0.1108
240	1.89	0.000886	0.1164	152.8	369.4	0.8176	1.720	1.362	2.288	0.1047
250	2.806	0.000910	0.07975	166.6	375.3	0.8736	1.708	1.394	2.055	0.09862
260	4.027	0.000936	0.05617	180.8	380.9	0.9287	1.698	1.431	1.852	0.09272
270	5.613	0.000966	0.04045	195.3	386.2	0.983	1.690	1.475	1.673	0.08682
280	7.629	0.000999	0.02964	210.3	391.1	1.037	1.682	1.527	1.496	0.08098
290	10.14	0.001038	0.02200	225.9	395.4	1.091	1.675	1.593	1.334	0.07521
300	13.23	0.001084	0.01646	242.1	399.0	1.144	1.668	1.679	1.192	0.06951
310	16.98	0.001140	0.01234	259.2	401.6	1.199	1.659	1.804	1.067	0.06381
320	21.48	0.001214	0.009182	277.4	402.7	1.255	1.647	2.006	0.9569	0.05803
330	26.85	0.001321	0.006678	297.5	401.0	1.315	1.629	2.421	0.8321	0.05202
340	33.25	0.001514	0.004520	321.8	393.4	1.385	1.595	4.021	0.6560	0.04480
346.75	38.32	0.002311	0.002311	360.6	360.6	1.471	1.471	—	—	—

THERMODYNAMIC AND THERMOPHYSICAL PROPERTIES—TABLE 101. Saturated Refrigerant R152a

Temp., K	Pressure, bar	v_f, m³/kg	v_g, m³/kg	h_f, kJ/kg	h_g, kJ/kg	s_f, kJ/(kg·K)	s_g, kJ/(kg·K)	c_{pf}, kJ/(kg·K)	μ_f, 10^{-4} Pa·s	k_f, W/(m·K)
154.56	0.000641	0.000839	303.6	14.0	419.8	0.1130	2.738	1.492	10.85	0.1932
160	0.001297	0.000846	155.2	22.2	423.5	0.1647	2.673	1.500	9.614	0.1894
170	0.004145	0.000859	51.59	37.2	430.6	0.2560	2.570	1.510	8.058	0.1822
180	0.01141	0.000873	19.82	52.4	437.9	0.3425	2.484	1.517	6.940	0.1753
190	0.02775	0.000887	8.588	67.6	445.3	0.4247	2.413	1.525	6.012	0.1685
200	0.06088	0.000902	4.110	82.9	452.8	0.5032	2.353	1.535	5.236	0.1618
210	0.1224	0.000918	2.138	98.3	460.4	0.5784	2.303	1.547	4.582	0.1552
220	0.2284	0.000935	1.193	113.8	468.0	0.6507	2.261	1.562	4.028	0.1487
230	0.4004	0.000952	0.7064	129.5	475.6	0.7205	2.225	1.580	3.556	0.1423
240	0.6647	0.000970	0.4397	145.5	483.1	0.7881	2.195	1.600	3.153	0.1361
249.12	1.01325	0.000989	0.2961	160.2	489.8	0.8481	2.171	1.622	2.834	0.1304
250	1.053	0.000991	0.2855	161.6	490.5	0.8538	2.169	1.624	2.805	0.1299
260	1.603	0.001012	0.1922	178.0	497.7	0.9178	2.147	1.651	2.500	0.1239
270	2.354	0.001035	0.1334	194.7	504.7	0.9805	2.129	1.681	2.231	0.1179
280	3.354	0.001060	0.09500	211.7	511.4	1.042	2.112	1.716	1.995	0.1121
290	4.650	0.001087	0.06916	229.1	517.8	1.103	2.098	1.756	1.789	0.1064
300	6.297	0.001118	0.05126	246.9	523.8	1.162	2.085	1.803	1.607	0.1008
310	8.351	0.001152	0.03857	265.2	529.3	1.222	2.073	1.859	1.447	0.09526
320	10.87	0.001190	0.02935	284.1	534.1	1.281	2.062	1.928	1.305	0.08986
330	13.92	0.001235	0.02251	303.7	538.2	1.340	2.051	2.015	1.180	0.08457
340	17.57	0.001289	0.01735	324.2	541.3	1.400	2.038	2.131	1.069	0.07939
350	21.90	0.001355	0.01335	345.7	542.9	1.460	2.024	2.299	1.005	0.07336
360	27.00	0.001440	0.01020	368.8	542.5	1.523	2.006	2.573	0.9225	0.06666
370	32.97	0.001563	0.007603	394.3	538.5	1.591	1.980	3.143	0.8191	0.06032
380	39.97	0.001785	0.005274	425.4	526.2	1.671	1.936	5.407	0.6638	0.05176
386.41	45.17	0.002717	0.002717	477.3	477.3	1.778	1.778	—	—	—

515

THERMODYNAMIC AND THERMOPHYSICAL PROPERTIES—TABLE 102. Saturated Refrigerant 216

Temp., °F	Pressure, lb/in² abs.	Volume, ft³/lb		Enthalpy, Btu/lb		Entropy, Btu/(lb)(°F)	
		Liquid	Vapor	Liquid	Vapor	Liquid	Vapor
−40	0.339	0.00927	59.957	0.000	62.415	0.0000	0.1487
−20	0.713	0.00942	29.749	4.778	65.276	0.0111	0.1487
0	1.382	0.00958	15.986	9.541	68.208	0.0217	0.1493
20	2.497	0.00974	9.184	14.298	71.199	0.0318	0.1504
40	4.247	0.00992	5.582	19.056	74.239	0.0415	0.1520
60	6.862	0.01010	3.558	23.821	77.319	0.0509	0.1538
80	10.612	0.01030	2.361	28.598	80.429	0.0509	0.1559
100	15.797	0.01050	1.6215	33.391	83.559	0.0686	0.1582
120	22.753	0.01073	1.1462	38.205	86.701	0.0770	0.1607
140	31.845	0.01097	0.8304	43.049	89.845	0.0852	0.1632
160	43.468	0.01124	0.6142	47.930	92.981	0.0931	0.1658
180	58.046	0.01153	0.4623	52.861	96.099	0.1009	0.1685
200	76.033	0.01186	0.3529	57.857	99.186	0.1085	0.1712
220	97.913	0.01223	0.2725	62.939	102.225	0.1161	0.1739
240	124.21	0.01266	0.2121	68.132	105.196	0.1235	0.1765
260	155.50	0.01317	0.1660	73.474	108.066	0.1309	0.1790
280	192.40	0.01378	0.1300	79.015	110.789	0.1384	0.1813
300	235.63	0.01458	0.1013	84.835	113.282	0.1460	0.1834
320	286.03	0.01570	0.0776	91.089	115.373	0.1539	0.1851
340	344.81	0.01764	0.0565	98.234	116.538	0.1628	0.1856
355.98	399.45	0.02771	0.0277	110.248	110.248	0.1773	0.1773

THERMODYNAMIC AND THERMOPHYSICAL PROPERTIES — TABLE 103.   Saturated
Refrigerant 245

T, K	P, bar	v_f, m³/kg	v_g, m³/kg	h_f, kJ/kg	h_g, kJ/kg	s_f, kJ/(kg·K)	s_g, kJ/(kg·K)
172	0.0034	6.46.–4	31.49	–63.4	133.8	–0.3131	0.8327
180	0.0076	6.57.–4	14.63	–55.9	138.7	–0.2707	0.8099
190	0.0190	6.70.–4	6.20	–46.2	145.1	–0.2182	0.7885
200	0.0425	6.83.–4	2.91	–36.0	151.7	–0.1666	0.7725
210	0.0870	6.97.–4	1.48	–25.7	158.5	–0.1157	0.7612
220	0.1654	7.11.–4	0.822	–14.8	165.4	–0.0654	0.7539
230	0.2946	7.25.–4	0.475	–3.6	172.5	–0.0156	0.7500
240	0.4958	7.40.–4	0.292	8.0	179.6	0.0337	0.7487
250	0.7946	7.55.–4	0.192	19.9	186.8	0.0824	0.7497
260	1.2204	7.72.–4	0.125	32.3	194.0	0.1305	0.7525
270	1.806	7.89.–4	0.0862	44.9	201.1	0.1781	0.7567
280	2.584	8.08.–4	0.0611	57.9	208.3	0.2249	0.7621
290	3.600	8.30.–4	0.0443	71.1	215.3	0.2711	0.7683
300	4.888	8.53.–4	0.0327	84.6	222.2	0.3161	0.7751
310	6.491	8.80.–4	0.0246	98.4	228.9	0.3614	0.7822
320	8.456	9.11.–4	0.0186	112.6	235.3	0.4057	0.7893
330	10.83	9.48.–4	0.0143	127.1	241.4	0.4497	0.7960
340	13.67	9.93.–4	0.0111	142.1	246.9	0.4937	0.8018
350	17.04	0.00105	0.0084	157.2	251.5	0.5382	0.8060
360	21.02	0.00113	0.0063	174.7	254.8	0.5844	0.8071
370	25.71	0.00125	0.0045	193.6	255.2	0.6349	0.8013
375	28.46	0.00137	0.0036	205.2	252.5	0.6649	0.7953
380.1	31.37	0.00204	0.0020	231.8	231.8	0.7341	0.7341

THERMODYNAMIC AND THERMOPHYSICAL PROPERTIES—TABLE 104.   Refrigerant C318

T, K	P, bar	v_f, m³/kg	v_g, m³/kg	h_f, kJ/kg	h_g, kJ/kg	s_f, kJ/(kg·K)	s_g, kJ/(kg·K)	c_{pf}, kJ/(kg·K)	μ_f, 10⁻⁴ Pa·s	k_f, W/(m·K)
200	0.0216	5.507.–4	3.810	353.5	498.0	3.909	4.560			
210	0.0449	5.593.–4	1.931	361.0	500.1	3.947	4.564			
220	0.0875	5.683.–4	1.038	369.2	502.2	3.984	4.569			
230	0.1608	5.778.–4	0.588	377.6	504.4	4.022	4.574	0.98	11.7	0.088
240	0.2810	5.879.–4	0.349	386.4	510.9	4.060	4.578	1.00	9.55	0.085
250	0.466	5.988.–4	0.2166	395.6	517.4	4.097	4.584	1.02	7.90	0.082
260	0.741	6.106.–4	0.1401	405.2	524.0	4.133	4.592	1.03	6.63	0.078
270	1.133	6.234.–4	0.0938	415.1	530.7	4.172	4.599	1.05	5.64	0.075
280	1.672	6.375.–4	0.0647	425.8	537.3	4.210	4.609	1.07	4.85	0.071
290	2.392	6.529.–4	0.0458	436.2	543.9	4.247	4.618	1.09	4.22	0.068
300	3.325	6.694.–4	0.0332	447.3	550.4	4.284	4.626	1.12	3.70	0.065
310	4.522	6.893.–4	0.0245	458.7	556.9	4.322	4.638	1.15	3.20	0.061
320	6.007	7.115.–4	0.0184	470.5	563.3	4.359	4.648	1.18	2.94	0.058
330	7.826	7.365.–4	0.0139	482.7	569.4	4.396	4.659	1.23	2.66	0.054
340	10.018	7.666.–4	0.0106	495.2	575.4	4.433	4.669	1.27	2.33	0.051
350	12.632	8.034.–4	0.0082	508.1	581.0	4.469	4.678	1.32	2.00	0.048
360	15.71	8.508.–4	0.0062	521.5	585.8	4.507	4.685	1.39		
370	19.33	9.172.–4	0.0047	535.6	589.9	4.544	4.691			
380	23.59	1.031.–3	0.0033	551.4	591.5	4.585	4.691			
388.5	27.83	1.613.–3	0.0016	577.2	577.2	4.651	4.651			

THERMODYNAMIC AND THERMOPHYSICAL PROPERTIES—TABLE 105. Saturated Refrigerant 500

T, K	P, bar	v_f, m^3/kg	v_g, m^3/kg	h_f, kJ/kg	h_g, kJ/kg	s_f, kJ/(kg·K)	s_g, kJ/(kg·K)	c_{pf}, kJ/(kg·K)	μ_f, 10^{-4} Pa·s	k_f, W/(m·K)
200	0.1219	6.966.−4	1.360	−29.56	185.87	−0.1363	0.9408	1.044	6.11	0.113
210	0.2258	7.090.−4	0.766	−21.03	191.25	−0.0948	0.9161	1.018	5.15	0.109
220	0.3936	7.222.−4	0.457	−12.17	196.63	−0.0536	0.8955	0.997	4.42	0.106
230	0.6511	7.361.−4	0.286	−2.97	201.96	−0.0130	0.8782	0.987	3.85	0.102
240	1.0291	7.509.−4	0.187	6.58	207.23	0.0277	0.8638	0.987	3.42	0.098
250	1.5632	7.668.−4	0.1261	16.50	212.40	0.0680	0.8517	0.997	3.04	0.094
260	2.2932	7.839.−4	0.0879	26.78	217.45	0.1082	0.8415	1.017	2.74	0.090
270	3.2624	8.024.−4	0.0628	37.44	222.35	0.1481	0.8329	1.048	2.48	0.086
280	4.5172	8.226.−4	0.0459	48.48	227.06	0.1878	0.8257	1.089	2.26	0.082
290	6.1064	8.450.−4	0.0342	59.91	231.56	0.2275	0.8194	1.140	2.08	0.078
300	8.0809	8.699.−4	0.0259	71.76	235.79	0.2671	0.8139	1.201	1.92	0.074
310	10.49	8.981.−4	0.0198	84.05	239.69	0.3067	0.8088	1.273	1.77	0.070
320	13.40	9.306.−4	0.0154	96.83	243.19	0.3464	0.8038	1.355	1.63	0.066
330	16.86	9.690.−4	0.0119	110.17	246.14	0.3864	0.7985	1.447	1.48	0.062
340	20.93	1.016.−3	0.0093	124.20	248.36	0.4271	0.7922	1.550	1.34	0.058
350	25.70	1.077.−3	0.0072	139.18	249.47	0.4689	0.7841	1.663		
360	31.25	1.162.−4	0.0055	155.66	248.71	0.5135	0.7721	1.919		
370	37.72	1.307.−4	0.0040	175.59	244.26	0.5650	0.7509	2.07		
378.6	44.26	2.012.−3	0.0020	219.50	219.50	0.6729	0.6729	∞		

THERMODYNAMIC AND THERMOPHYSICAL PROPERTIES—TABLE 106. Saturated Refrigerant 502

T, K	P, bar	v_f, m³/kg	v_g, m³/kg	h_f, kJ/kg	h_g, kJ/kg	s_f, kJ/(kg·K)	s_g, kJ/(kg·K)	c_{pf}, kJ/(kg·K)	μ_f, 10⁻⁴ Pa·s	k_f, W/(m·K)
200	0.2274	6.381.–4	0.648	–29.04	153.34	–0.1337	0.7782	1.018	5.72	0.103
210	0.4098	6.507.–4	0.374	–20.83	158.42	–0.0937	0.7599	1.036	4.88	0.099
220	0.6965	6.640.–4	0.228	–12.15	163.49	–0.0534	0.7449	1.055	4.23	0.095
230	1.1251	6.783.–4	0.146	–2.99	168.50	–0.0128	0.7328	1.075	3.71	0.091
240	1.7392	6.938.–4	0.0969	6.66	173.42	0.0280	0.7228	1.097	3.28	0.087
250	2.5867	7.105.–4	0.0665	16.78	178.20	0.0691	0.7148	1.120	2.94	0.083
260	3.7188	7.289.–4	0.0470	27.36	182.81	0.1102	0.7082	1.144	2.65	0.079
270	5.1893	7.492.–4	0.0340	38.36	187.21	0.1514	0.7027	1.170	2.41	0.075
280	7.0530	7.720.–4	0.0251	49.77	191.35	0.1923	0.6980	1.197	2.18	0.072
290	9.3660	7.979.–4	0.0188	61.55	195.16	0.2330	0.6937	1.225	1.99	0.068
300	12.19	8.280.–4	0.0143	73.68	198.56	0.2734	0.6896	1.254	1.79	0.064
310	15.57	8.637.–4	0.0109	86.17	201.43	0.3134	0.6852	1.285	1.59	0.060
320	19.60	9.081.–4	0.0084	99.06	203.57	0.3532	0.6798	1.317	1.40	0.056
330	24.35	9.666.–4	0.0064	112.53	204.62	0.3933	0.6723	1.351	1.23	0.052
340	29.95	1.053.–3	0.0048	127.13	203.71	0.4351	0.6604	1.386	1.07	0.048
350	36.62	1.220.–3	0.0033	145.44	197.82	0.4859	0.6355	1.422	0.93	0.044
355.3	40.75	1.786.–3	0.0018	174.00	174.00	0.5634	0.5634			

THERMODYNAMIC AND THERMOPHYSICAL PROPERTIES—TABLE 107. Saturated Refrigerant 503

T, K	P, bar	v_f, m³/kg	v_g, m³/kg	h_f, kJ/kg	h_g, kJ/kg	s_f, kJ/(kg·K)	s_g, kJ/(kg·K)	c_{pf}, kJ/(kg·K)	μ_f, 10⁻⁴Pa·s	k_f, W/(m·K)
150	0.0750	6.384.–4	1.894	−89.60	111.02	−0.4694	0.8681	0.482	6.12	0.128
160	0.1798	6.478.–4	0.837	−79.73	115.40	−0.4057	0.8139	0.554	5.05	0.123
170	0.3828	6.585.–4	0.414	−69.55	119.70	−0.3441	0.7691	0.620	4.16	0.116
180	0.7395	6.700.–4	0.224	−59.08	123.84	−0.2844	0.7318	0.682	3.43	0.111
190	1.3187	6.850.–4	0.130	−48.36	127.77	−0.2267	0.7003	0.747	2.94	0.105
200	2.1999	7.014.–4	0.0803	−37.45	131.45	−0.1710	0.6735	0.817	2.56	0.099
210	3.4713	7.204.–4	0.0520	−26.36	134.84	−0.1173	0.6503	0.896	2.25	0.094
220	5.2281	7.426.–4	0.0350	−15.10	137.87	−0.0656	0.6298	0.988	1.98	0.088
230	7.5713	7.687.–4	0.0242	−3.65	140.49	−0.0155	0.6112	1.017	1.73	0.082
240	10.61	8.001.–4	0.0172	8.07	142.58	0.0334	0.5939	1.227	1.52	0.076
250	14.46	8.386.–4	0.0124	20.22	143.98	0.0817	0.5767	1.382	1.33	0.070
260	19.25	8.874.–4	0.0090	33.10	144.38	0.1305	0.5585	1.57	1.17	0.065
270	25.13	9.526.–4	0.0064	47.22	143.23	0.1816	0.5373	1.79	1.03	0.059
280	32.27	1.050.–3	0.0045	63.64	139.25	0.2384	0.5085	2.03	0.91	0.054
290	40.87	1.264.–3	0.0028	86.41	127.51	0.3131	0.4548	2.35		
292.6	43.57	1.773.–3	0.0018	110.20	110.20	0.3864	0.3864	∞	∞	

THERMODYNAMIC AND THERMOPHYSICAL PROPERTIES—TABLE 108.   Saturated Refrigerant 504

Temp., °F	Pressure, lb/in² abs.	Volume, ft³/lb		Enthalpy, Btu/lb		Entropy, Btu/(lb)(°F)	
		Liquid	Vapor	Liquid	Vapor	Liquid	Vapor
−120	2.964	0.01095	15.31	−21.48	86.69	−0.0565	0.2609
−100	6.042	0.01119	7.874	−16.39	89.31	−0.0420	0.2519
−80	11.34	0.01146	4.372	−11.12	91.84	−0.0277	0.2435
−60	19.85	0.01175	2.585	−5.65	94.25	−0.0137	0.2362
−40	32.76	0.01206	1.609	0.00	96.50	0.0000	0.2299
−20	51.44	0.01242	1.045	5.85	98.58	0.0135	0.2244
0	77.41	0.01282	0.7029	11.91	100.45	0.0269	0.2195
20	112.3	0.01328	0.4859	18.22	102.09	0.0401	0.2150
40	158.0	0.01379	0.3431	24.81	103.44	0.0533	0.2107
60	216.2	0.01443	0.2458	31.78	104.41	0.0667	0.2065
80	289.2	0.01522	0.1773	39.25	104.85	0.0804	0.2020
100	379.1	0.01629	0.1274	47.43	104.49	0.0948	0.1968
120	488.3	0.01783	0.0893	56.78	102.72	0.1107	0.1899
140	618.1	0.02083	0.0578	69.97	97.70	0.1322	0.1784
150	692.2	0.02597	0.0394	76.96	89.76	0.1432	0.1642

THERMODYNAMIC AND THERMOPHYSICAL PROPERTIES—TABLE 109.   Refrigerant 507*

Temp., K	Pressure, bar	v_f, m³/kg	v_g, m³/kg	h_f, kJ/kg	h_g, kJ/kg	s_f, kJ/(kg·K)	s_g, kJ/(kg·K)
230.5	1.013	0.000574	0.1280	−3.1	143.3	−0.015	0.620
240	1.59	0.000602	0.0826	10.3	150.2	0.042	0.623
250	2.42	0.000627	0.0546	22.6	154.5	0.092	0.619
260	3.54	0.000658	0.0377	37.6	159.0	0.149	0.617
270	4.95	0.000695	0.0270	51.6	163.8	0.202	0.618
280	6.70	0.000738	0.0198	64.7	169.0	0.250	0.620
290	8.85	0.000787	0.0148	77.2	174.6	0.295	0.634
300	11.52	0.000839	0.0112	89.4	180.3	0.336	0.640
310	14.74	0.000903	0.0084	101.6	185.4	0.378	0.648
320	18.76	0.001006	0.0062	115.7	188.6	0.422	0.649
330	23.65	0.001221	0.0042	135.5	189.3	0.481	0.641
340	29.57	0.001618	0.0025	161.7	179.9	0.557	0.611
341.5	32.67	0.00197	0.0020	172.7	172.7	0.590	0.590

* Azeotropic mixture of R152a and R218, $h_f = s_f = 0$ at 233.15 K = −40°C.

THERMODYNAMIC AND THERMOPHYSICAL PROPERTIES — TABLE 110.   Saturated Rubidium

T, K	P, bar	v_f, m³/kg	v_g, m³/kg	h_f, kJ/kg	h_g, kJ/kg	s_f, kJ/(kg·K)	s_g, kJ/(kg·K)	c_{pf}, kJ/(kg·K)
312.7	2.46.−9	6.75.−4		118.7	1036	0.998	3.932	0.379
400	1.69.−6	6.98.−4	2.3.+5	151.6	1057	1.091	3.355	0.375
500	1.73.−4	7.22.−4	2790	188.8	1078	1.174	2.953	0.369
600	0.0037	7.46.−4	156.6	225.4	1096	1.241	2.692	0.362
700	0.0317	7.73.−4	20.75	261.3	1111	1.296	2.511	0.357
800	0.1584	8.10.−4	4.662	296.8	1124	1.343	2.378	0.353
1000	1.467	8.65.−4	0.605	367.6	1150	1.422	2.205	0.360
1200	6.466	9.40.−4	0.159	440.1	1179	1.490	2.104	0.385
1400	18.6	1.03.−3						
1500	28.5	1.08.−3						

THERMODYNAMIC AND THERMOPHYSICAL PROPERTIES — TABLE 111.   Saturated Seawater

Temp., °C	Pressure, bar	v, (m³/kg)10³	c_p, kJ/(kg·K)	μ, Ns/m²	k, W/(m·K)	N_{Pr}	$10^5\,\kappa$, 1/bar
0	0.005993	1.000158	4.000	0.001884	0.560	13.46	5.06
1	0.006438	1.000099	4.000	0.001827	0.563	12.98	5.02
2	0.006916	1.000057	4.000	0.001772	0.565	12.55	4.98
3	0.007427	1.000033	4.000	0.001720	0.567	12.13	4.95
4	0.007970	1.000025	4.001	0.001669	0.569	11.74	4.92
5	0.008548	1.000033	4.001	0.001620	0.571	11.35	4.89
6	0.009163	1.000057	4.001	0.001574	0.574	10.97	4.86
7	0.009816	1.000096	4.002	0.001529	0.576	10.62	4.83
8	0.010511	1.000149	4.002	0.001486	0.578	10.29	4.80
9	0.011248	1.000261	4.002	0.001445	0.580	9.97	4.78
10	0.01203	1.000298	4.003	0.001405	0.582	9.70	4.76
11	0.01286	1.000392	4.003	0.001367	0.584	9.37	4.74
12	0.01374	1.000500	4.003	0.001330	0.586	9.09	4.72
13	0.01467	1.000620	4.004	0.001294	0.588	8.81	4.70
14	0.01566	1.000727	4.004	0.001259	0.590	8.54	4.68
15	0.01671	1.000899	4.005	0.001226	0.592	8.29	4.66
16	0.01781	1.001055	4.005	0.001195	0.594	8.06	4.65
17	0.01898	1.001224	4.006	0.001165	0.595	7.82	4.63
18	0.02022	1.001404	4.006	0.001136	0.597	7.62	4.62
19	0.02153	1.001595	4.007	0.001107	0.599	7.41	4.60
20	0.02291	1.001796	4.007	0.001080	0.600	7.21	4.59
21	0.02437	1.002009	4.007	0.001054	0.602	7.02	4.57
22	0.02591	1.002232	4.008	0.001029	0.604	6.82	4.56
23	0.02753	1.002465	4.008	0.001005	0.605	6.66	4.55
24	0.02924	1.002708	4.009	0.000981	0.607	6.48	4.54
25	0.03104	1.002961	4.009	0.000958	0.608	6.31	4.53
26	0.03294	1.003224	4.009	0.000936	0.609	6.16	4.52
27	0.03494	1.003496	4.010	0.000915	0.611	6.01	4.51
28	0.03705	1.003778	4.010	0.000895	0.612	5.86	4.50
29	0.03926	1.004069	4.011	0.000875	0.614	5.72	4.49
30	0.04159	1.004369	4.011	0.000855	0.615	5.58	4.48

THERMODYNAMIC AND THERMOPHYSICAL PROPERTIES—TABLE 112.   Saturated
Sodium

Temp., K	Pressure, bar	v_f, m³/kg	v_g, m³/kg	h_f, kJ/kg	h_g, kJ/kg	s_f, kJ/(kg·K)	s_g, kJ/(kg·K)
371	1.59.−10	0.001 078	8.54.+9	207	4,739	2.259	14.475
400	1.80.−9	0.001 088	8.08.+8	247	4,757	2.920	14.195
500	8.99.−7	0.001 115	1.99.+6	382	4,817	3.222	12.092
600	5.57.−5	0.001 144	38,022	514	4,872	3.462	10.745
700	0.00105	0.001 174	2,320	642	4,921	3.661	10.631
800	0.00941	0.001 208	291.5	769	4,966	3.830	9.076
900	0.05147	0.001 242	58.8	895	5,007	3.978	8.547
1,000	0.1995	0.001 280	16.6	1,020	5,044	4.110	8.134
1,100	0.6016	0.001 323	5.95	1,146	5,079	4.230	7.805
1,154.7	1.013	0.001 347	3.89	1,215	5,097	4.290	7.652
1,200	1.50	0.001 366	2.54	1,273	5,111	4.340	7.358
1,300	3.26	0.001 416	1.24	1,402	5,140	4.444	7.319
1,400	6.30	0.001 471	0.676	1,534	5,168	4.542	7.138
1,500	11.13	0.001 531	0.400	1,671	5,193	4.636	6.984
1,600	18.28	0.001 597	0.253	1,812	5,217	4.727	6.855
1,700	28.28	0.001 675	0.168	1,959	5,238	4.816	6.745
1,800	41.61	0.001 761	0.117	2,113	5,256	4.904	6.650
1,900	58.70	0.001 862	0.084	2,274	5,268	4.992	6.568
2,000	79.91	0.001 984	0.063	2,444	5,273	5.079	6.494
2,100	105.5	0.002 174	0.0472	2,625	5,265		
2,200	135.7	0.002 320	0.0361	2,822	5,241		
2,300	170.6	0.002 584	0.0275	3,047	5,188		
2,400	210.3	0.002 985	0.0203	3,331	5,078		
2,500	254.7	0.004 19	0.0098	3,965	4,617		
2,503.7^c	256.4	0.004 57	0.0046	4,294	4,294		

c = critical point.

c_{pf}, kJ/(kg·K)	c_{pg}, kJ/(kg·K)	μ_f, 10^{-6} Pa·s	μ_g, 10^{-6} Pa·s	k_f, W/(m·K)	k_g, W/(m·K)	Pr_f	Pr_g
1.383		688		89.4		0.0106	
1.372	0.86	599		87.2		0.0094	
1.334	1.25	415		80.1		0.069	
1.301	1.80	321		73.7		0.0057	
1.277	2.28	264		68.0		0.0050	
1.260	2.59	227	19.6	62.9	0.0343	0.0045	1.48
1.252	2.72	201	20.6	58.3	0.0406	0.0043	1.38
1.252	2.70	181	23.0	54.2	0.0455	0.0042	1.36
1.261	2.62	166	25.3	50.5	0.0492	0.0042	1.35
1.271	2.56	159	26.5	48.7	0.0522	0.0041	1.30
1.279	2.51	153	27.5	47.2	0.0547	0.0041	1.26
1.305	2.43	143	29.9	44.0	0.0570	0.0042	1.27
1.340	2.39	135	32.2	41.1	0.0592	0.0044	1.30
1.384	2.36	128	34.6	38.2		0.0046	
1.437	2.34	122	37.1	34.5		0.0050	
1.500	2.41	117		32.6		0.0054	
1.574	2.46	112		29.7		0.0059	
1.661	2.53	108		26.6		0.0067	
1.764	2.66	104		23.2		0.0079	
1.926	2.91						
2.190	3.40						
2.690	4.47						
4.012	8.03						
39.3	417.						

THERMODYNAMIC AND THERMOPHYSICAL PROPERTIES — TABLE 113. Saturated Sulfur Dioxide

T, K	P, bar	v_f, m³/kg	v_g, m³/kg	h_f, kJ/kg	h_g, kJ/kg	s_f, kJ/(kg·K)	s_g, kJ/(kg·K)	c_{pf}, kJ/(kg·K)	μ_f, 10⁻⁴ Pa·s	k_f, W/(m·K)
200	0.02056	6.189.−4	12.602	7.4	433.3	0.033	2.212	1.280	12.3	
210	0.04569	6.284.−4	5.946	9.1	446.1	0.041	2.159	1.284	10.6	
220	0.09997	6.384.−4	2.876	28.6	453.8	0.123	2.075	1.288	8.37	
230	0.1844	6.488.−4	1.605	43.5	459.5	0.198	2.001	1.293	7.03	
240	0.3202	6.596.−4	0.9602	56.5	464.5	0.254	1.952	1.299	5.97	
250	0.5430	6.707.−4	0.5864	70.0	469.7	0.308	1.906	1.308	5.11	0.262
260	0.8778	6.819.−4	0.3745	85.1	474.5	0.363	1.865	1.317	4.39	0.243
270	1.3634	6.938.−4	0.2479	99.8	479.3	0.425	1.827	1.328	3.78	0.224
280	2.0402	7.057.−4	0.1699	114.8	484.3	0.473	1.793	1.343	3.30	0.206
290	2.9574	7.184.−4	0.1197	129.2	488.5	0.523	1.763	1.363	2.87	0.190
300	4.1675	7.312.−4	0.08647	143.1	492.5	0.568	1.732	1.389	2.51	0.174
310	5.7372	7.447.−4	0.06366	157.1	496.3	0.612	1.706	1.422	2.19	0.162
320	7.8226	7.590.−4	0.04707	170.1	498.9	0.649	1.678	1.459	1.91	0.151
330	10.301	7.847.−4	0.03572	183.0	501.2	0.690	1.654	1.499	1.67	0.139
340	13.229	8.066.−4	0.02792	196.0	502.5	0.731	1.633	1.546	1.46	0.128
350	16.759	8.303.−4	0.02209	211.2	502.9	0.781	1.614	1.603	1.27	0.117
360	21.01	8.571.−4	0.01755	223.7	503.1	0.817	1.593	1.68	1.11	0.108
370	26.01	8.877.−4	0.01399	239.9	502.9	0.862	1.573	1.75	0.96	0.098
380	31.92	9.236.−4	0.01110	257.9	502.7	0.910	1.555	1.84	0.84	0.089
390	38.76	9.671.−4	0.00877	277.7	500.7	0.962	1.534	1.97	0.73	0.081
400	46.67	1.023.−3	0.00685	300.2	496.7	1.020	1.511	2.12	0.63	0.072
410	55.80	1.098.−3	0.00559	326.2	489.5	1.083	1.481		0.53	0.064
420	66.19	1.235.−3	0.00387	355.6	474.1	1.155	1.436		0.44	0.055
425.1	78.81	1.906.−3	0.00191	423.6	423.6	1.304	1.304			

THERMODYNAMIC AND THERMOPHYSICAL PROPERTIES—TABLE 114. Saturated Sulfur Hexafluoride (SF_6)

Temp., K	Pressure, bar	v_f, m³/kg	v_g, m³/kg	h_f, kJ/kg	h_g, kJ/kg	s_f, kJ/(kg·K)	s_g, kJ/(kg·K)	c_{pf}, kJ/(kg·K)	c_{pg}, kJ/(kg·K)
222.4	2.200	0.0005389	0.05428	−57.55	59.08	−0.2310	0.2935		0.579
225	2.470	0.0005429	0.04861	−54.41	60.49	−0.2171	0.2936		0.583
230	3.045	0.0005507	0.03978	−48.51	63.12	−0.1913	0.2940		0.592
235	3.710	0.0005588	0.03286	−42.67	65.68	−0.1663	0.2947		0.602
240	4.475	0.0005675	0.02737	−36.87	68.18	−0.1421	0.2956		0.613
245	5.346	0.0005768	0.02296	−31.13	70.61	−0.1186	0.2966		0.626
250	6.332	0.0005866	0.01939	−25.47	72.96	−0.0960	0.2977		0.640
255	7.442	0.0005971	0.01647	−19.87	75.22	−0.0741	0.2988		0.656
260	8.684	0.0006085	0.01406	−14.33	77.39	−0.0528	0.2999		0.674
265	10.07	0.0006207	0.01205	−8.85	79.44	−0.0323	0.3009		0.695
270	11.60	0.0006341	0.01035	−3.41	81.38	−0.0123	0.3017		0.720
275	13.30	0.0006488	0.00892	2.00	83.19	0.0071	0.3024		0.748
280	15.18	0.0006652	0.00769	7.42	84.84	0.0262	0.3027		0.783
285	17.25	0.0006836	0.00663	12.88	86.30	0.0451	0.3027		0.827
290	19.52	0.0007047	0.00571	18.45	87.52	0.0639	0.3021	0.409	0.882
295	22.01	0.0007295	0.00490	24.20	88.45	0.0829	0.3008	0.631	0.941
300	24.75	0.0007594	0.00418	30.22	89.00	0.1025	0.2984	0.870	1.070
305	27.76	0.000798	0.00352	36.75	88.97	0.1233	0.2945	1.17	1.26
310	31.05	0.000851	0.00291	44.05	88.06	0.1462	0.2881	163	1.63
315	34.67	0.000949	0.00228	53.98	85.22	0.1769	0.2761	2.48	2.40
318.7	37.79	0.001372	0.00137	71.74	71.74	0.2317	0.2317	∞	∞

THERMODYNAMIC AND THERMOPHYSICAL PROPERTIES—TABLE 115. Saturated Toluene

T, K	P, bar	v_f, m³/kg	v_g, m³/kg	h_f, kJ/kg	h_g, kJ/kg	s_f, kJ/(kg·K)	s_g, kJ/(kg·K)	c_{pf}, kJ/(kg·K)	μ_f, 10^{-4} Pa·s	k_f, W/(m·K)
270	0.0076	1.127.–3	34.9	316.7	745.7	2.236	3.825	1.64	8.02	0.141
280	0.0139	1.138.–3	19.1	333.0	756.1	2.295	3.806	1.66	6.96	0.138
290	0.0246	1.150.–3	10.6	349.6	766.8	2.353	3.792	1.68	6.10	0.136
300	0.0418	1.162.–3	6.46	366.5	777.8	2.410	3.782	1.71	5.41	0.133
310	0.0682	1.175.–3	4.08	383.7	789.2	2.467	3.776	1.74	4.83	0.131
320	0.1072	1.188.–3	2.67	401.3	800.9	2.522	3.771	1.78	4.34	0.128
330	0.1633	1.201.–3	1.80	419.6	812.9	2.577	3.771	1.81	3.93	0.126
340	0.2416	1.215.–3	1.25	437.4	825.2	2.632	3.772	1.84	3.58	0.124
350	0.3480	1.230.–3	0.891	456.0	837.8	2.686	3.777	1.88	3.28	0.121
360	0.4894	1.245.–3	0.698	475.1	850.7	2.739	3.783	1.92	3.01	0.119
370	0.6736	1.261.–3	0.481	494.6	863.8	2.792	3.791	1.96	2.78	0.117
380	0.9090	1.277.–3	0.364	514.4	877.2	2.846	3.801	2.01	2.56	0.114
390	1.2049	1.294.–3	0.279	534.7	890.9	2.898	3.811	2.05	2.37	0.112
400	1.5713	1.312.–3	0.218	555.4	904.8	2.950	3.824	2.09	2.19	0.110
420	2.5589	1.350.–3	0.137	598.1	933.1	3.054	3.852	2.17	1.89	0.105
440	3.965	1.393.–3	9.00.–2	642.3	962.0	3.156	3.883	2.24	1.64	0.101
460	5.892	1.443.–3	6.11.–2	688.1	991.3	3.258	3.917	2.31		0.096
480	8.451	1.499.–3	4.26.–2	735.5	1021.1	3.358	3.953	2.38		0.091
500	11.76	1.567.–3	3.03.–2	784.4	1051.3	3.457	3.989	2.45		0.086
520	15.96	1.651.–3	2.19.–2	834.9	1081.4	3.554	4.027	2.53		0.082
540	21.99	1.761.–3	1.58.–2	887.3	1109.6	3.651	4.062	2.65		0.078
560	27.65	1.919.–3	1.13.–2	942.8	1132.1	3.750	4.088	2.82		0.074
580	35.56	2.213.–3	7.59.–3	1005.6	1142.3	3.857	4.093			
590	40.16	2.650.–3	5.28.–3	1050.2	1128.1	3.932	4.063			
591.8	41.04	3.432.–3	3.43.–3	1084.9	1084.9	3.989	3.989			

THERMODYNAMIC AND THERMOPHYSICAL PROPERTIES—TABLE 116. Saturated Solid/Vapor Water

Temp., °F	Pressure lb/in² abs.	Volume, ft³/lb		Enthalpy, Btu/lb		Entropy, Btu/(lb)(°F)	
		Solid	Vapor	Solid	Vapor	Solid	Vapor
−160	4.949.−8	0.01722	3.607.+9	−222.05	990.38	−0.4907	3.5579
−150	1.620.−7	0.01723	1.139.+9	−218.82	994.80	−0.4801	3.4387
−140	4.928.−7	0.01724	3.864.+8	−215.49	999.21	−0.4695	3.3301
−130	1.403.−6	0.01725	1.400.+8	−212.08	1003.63	−0.4590	3.2284
−120	3.757.−6	0.01726	5.386.+7	−208.58	1008.05	−0.4485	3.1330
−110	9.517.−6	0.01728	2.189.+7	−204.98	1012.47	−0.4381	3.0434
−100	2.291.−5	0.01729	9.352.+6	−201.28	1016.89	−0.4277	2.9591
−90	5.260.−5	0.01730	4.186.+6	−197.49	1021.31	−0.4173	2.8796
−80	1.157.−4	0.01731	1.955.+6	−193.60	1025.73	−0.4069	2.8045
−70	2.443.−4	0.01732	9.501.+5	−189.61	1030.15	−0.3965	2.7336
−60	4.972.−4	0.01734	4.788.+5	−185.52	1034.58	−0.3862	2.6664
−50	9.776.−4	0.01735	2.496.+5	−181.34	1039.00	−0.3758	2.6028
−45	1.354.−3	0.01736	1.824.+5	−179.21	1041.21	−0.3707	2.5723
−40	1.861.−3	0.01737	1.343.+5	−177.06	1043.42	−0.3655	2.5425
−35	2.540.−3	0.01737	9.961.+4	−174.88	1045.63	−0.3604	2.5135
−30	3.440.−3	0.01738	7.441.+4	−172.68	1047.84	−0.3552	2.4853
−25	4.627.−3	0.01739	5.596.+4	−170.46	1050.05	−0.3501	2.4577
−20	6.181.−3	0.01739	4.237.+4	−168.21	1052.26	−0.3449	2.4308
−15	8.204.−3	0.01740	3.228.+4	−165.94	1054.47	−0.3398	2.4046
−10	1.082.−2	0.01741	2.475.+4	−163.65	1056.67	−0.3347	2.3791
−5	1.419.−2	0.01741	1.909.+4	−161.33	1058.88	−0.3295	2.3541
0	1.849.−2	0.01742	1.481.+4	−158.98	1061.09	−0.3244	2.3297
5	2.396.−2	0.01743	1.155.+4	−156.61	1063.29	−0.3193	2.3039
10	3.087.−2	0.01744	9.060.+3	−154.22	1065.50	−0.3142	2.2827
15	3.957.−2	0.01744	7.144.+3	−151.80	1067.70	−0.3090	2.2600
16	4.156.−2	0.01745	6.817.+3	−151.32	1068.14	−0.3080	2.2555
18	4.581.−2	0.01745	6.210.+3	−150.34	1069.02	−0.3060	2.2466
20	5.045.−2	0.01745	5.662.+3	−149.36	1069.90	−0.3039	2.2378
22	5.552.−2	0.01746	5.166.+3	−148.38	1070.38	−0.3019	2.2291
24	6.105.−2	0.01746	4.717.+3	−147.39	1071.66	−0.2998	2.2205
26	6.708.−2	0.01746	4.311.+3	−146.40	1072.53	−0.2978	2.2119
28	7.365.−2	0.01746	3.943.+3	−145.40	1073.41	−0.2957	2.2034
30	8.080.−2	0.01747	3.608.+3	−144.40	1074.29	−0.2937	2.1950
31	8.461.−2	0.01747	3.453.+3	−143.90	1074.73	−0.2927	2.1908
32	8.858.−2	0.01747	3.305.+3	−143.40	1075.16	−0.2916	2.1867

THERMODYNAMIC AND THERMOPHYSICAL PROPERTIES—TABLE 117. Saturated Water Substance—Temperature, fps units

Temp., °F	Pressure, lb/in² abs.	Volume, ft³/lb		Enthalpy, Btu/lb		Entropy, Btu/(lb)(°F)	
		Liquid	Vapor	Liquid	Vapor	Liquid	Vapor
32.018	0.08865	0.016022	3302.4	0.000	1075.5	0.0000	2.1872
35	0.09991	0.016020	2948.1	3.002	1076.8	0.0061	2.1767
40	0.12163	0.016019	2445.8	8.027	1079.0	0.0162	2.1594
45	0.14744	0.016020	2037.8	13.044	1081.2	0.0262	2.1426
50	0.17796	0.016023	1704.8	18.054	1083.4	0.0361	2.1262
55	0.21392	0.016027	1432.0	23.059	1085.6	0.0458	2.1102
60	0.25611	0.016033	1207.6	28.060	1087.7	0.0555	2.0946
65	0.30545	0.016041	1022.1	33.057	1089.9	0.0651	2.0794
70	0.36292	0.016050	868.4	38.052	1092.1	0.0745	2.0645
75	0.42964	0.016060	740.3	43.045	1094.3	0.0839	2.0500
80	0.50683	0.016072	633.3	48.037	1096.4	0.0932	2.0359
85	0.59583	0.016085	543.6	53.027	1098.6	0.1024	2.0221
90	0.69813	0.016099	468.1	58.018	1100.8	0.1115	2.0086
95	0.81534	0.016114	404.4	63.008	1102.9	0.1206	1.9954
100	0.94294	0.016130	350.4	67.999	1105.1	0.1295	1.9825
110	1.2750	0.016165	265.39	77.98	1109.3	0.1472	1.9577
120	1.6927	0.016204	203.26	87.97	1113.6	0.1646	1.9339
130	2.2230	0.016247	157.33	97.96	1117.8	0.1817	1.9112
140	2.8892	0.016293	122.98	107.89	1122.0	0.1985	1.8895
150	3.7184	0.016343	97.07	117.95	1126.1	0.2150	1.8686
160	4.7414	0.016395	77.27	127.96	1130.2	0.2313	1.8487
170	5.9926	0.016451	62.06	137.97	1134.2	0.2473	1.8295
180	7.5110	0.016510	50.255	148.00	1138.2	0.2631	1.8111
190	9.340	0.016572	40.957	158.04	1142.1	0.2787	1.7934
200	11.526	0.016637	33.639	168.09	1146.0	0.2940	1.7764
210	14.123	0.016705	27.816	178.15	1149.7	0.3091	1.7600
212	14.696	0.016719	26.799	180.17	1150.5	0.3121	1.7568
220	17.186	0.016775	23.148	188.23	1153.4	0.3241	1.7442
230	20.779	0.016849	19.381	198.33	1157.1	0.3388	1.7290
240	24.968	0.016926	16.321	208.45	1160.6	0.3533	1.7142
250	29.825	0.017066	13.819	218.59	1164.0	0.3677	1.7000
260	35.427	0.017089	11.762	228.76	1167.4	0.3819	1.6862
270	41.856	0.017175	10.060	238.95	1170.6	0.3960	1.6729
280	49.200	0.017264	8.644	249.17	1173.8	0.4098	1.6599
290	57.550	0.01736	7.4603	259.4	1167.8	0.4236	1.6473
300	67.005	0.01745	6.4658	269.7	1179.7	0.4372	1.6351
320	89.643	0.01766	4.9138	290.4	1185.2	0.4640	1.6116
340	117.992	0.01787	3.7878	311.3	1190.1	0.4902	1.5892
360	153.01	0.01811	2.9573	332.3	1194.4	0.5161	1.5678
380	195.73	0.01836	2.3353	353.6	1198.0	0.5416	1.5473

THERMODYNAMIC AND THERMOPHYSICAL PROPERTIES—TABLE 117.   Saturated
Water Substance—Temperature, fps units (*Continued*)

Temp., °F	Pressure, lb/in² abs.	Volume, ft³/lb		Enthalpy, Btu/lb		Entropy, Btu/(lb)(°F)	
		Liquid	Vapor	Liquid	Vapor	Liquid	Vapor
400	247.26	0.01864	1.8630	375.1	1201.0	0.5667	1.5274
420	308.78	0.01894	1.4997	396.9	1203.1	0.5915	1.5080
440	381.54	0.01926	1.2169	419.0	1204.4	0.6161	1.4890
460	466.87	0.01961	0.99424	441.5	1204.8	0.6405	1.4704
480	566.15	0.02000	0.81717	464.5	1204.1	0.6648	1.4518
500	680.86	0.02043	0.67492	487.9	1202.2	0.6890	1.4333
520	812.53	0.02091	0.55956	521.0	1199.0	0.7133	1.4146
540	962.79	0.02146	0.46513	536.8	1194.3	0.7378	1.3954
560	1133.38	0.02207	0.38714	562.4	1187.7	0.7625	1.3757
580	1326.17	0.02279	0.32216	589.1	1179.0	0.7876	1.3550
600	1543.2	0.02364	0.26747	617.1	1167.7	0.8134	1.3330
620	1786.9	0.02466	0.22081	646.9	1153.2	0.8403	1.3092
640	2059.9	0.02595	0.18021	679.1	1133.7	0.8686	1.2821
660	2365.7	0.02768	0.14431	714.9	1107.0	0.8995	1.2498
680	2708.6	0.03037	0.11117	758.5	1068.5	0.9365	1.2086
700	3094.3	0.03662	0.07519	825.2	991.7	0.9924	1.1359
702	3135.5	0.03824	0.06997	835.0	979.7	1.0006	1.1210
704	3177.2	0.04108	0.06300	854.2	956.2	1.0169	1.1046
705.47	3208.2	0.05078	0.05078	906.0	906.0	1.0612	1.0612

THERMODYNAMIC AND THERMOPHYSICAL PROPERTIES—TABLE 118.   Saturated Water Substance—Temperature, SI units

Temp., K	Pressure, bar*	Volume, m³/kg		Enthalpy, kJ/kg		Entropy, kJ/kg·K)	
		Condensed†	Vapor	Condensed†	Vapor	Condensed†	Vapor
150	6.30.–11	1.073.–3	9.55.+9	–539.6	2273	–2.187	16.54
160	7.72.–10	1.074.–3	9.62.+8	–525.7	2291	–2.106	15.49
170	7.29.–9	1.076.–3	1.08.+8	–511.7	2310	–2.026	14.57
180	5.38.–8	1.077.–3	1.55.+7	–497.8	2328	–1.947	13.76
190	3.23.–7	1.078.–3	2.72.+6	–483.8	2347	–1.868	13.03
200	1.62.–6	1.079.–3	5.69.+5	–467.5	2366	–1.789	12.38
210	7.01.–6	1.081.–3	1.39.+5	–451.2	2384	–1.711	11.79
220	2.65.–5	1.082.–3	3.83.+4	–435.0	2403	–1.633	11.20
230	8.91.–5	1.084.–3	1.18.+4	–416.3	2421	–1.555	10.79
240	3.72.–4	1.085.–3	4.07.+3	–400.1	2440	–1.478	10.35
250	7.59.–4	1.087.–3	1.52.+3	–381.5	2459	–1.400	9.954
255	1.23.–3	1.087.–3	956.4	–369.8	2468	–1.361	9.768
260	1.96.–3	1.088.–3	612.2	–360.5	2477	–1.323	9.590
265	3.06.–3	1.089.–3	400.4	–351.2	2486	–1.281	9.461
270	4.69.–3	1.090.–3	265.4	–339.6	2496	–1.296	9.255
273.15	6.11.–3	1.091.–3	206.3	–333.5	2502	–1.221	9.158
273.15	0.00611	1.000.–3	206.3	0.0	2502	0.000	9.158
275	0.00697	1.000.–3	181.7	7.8	2505	0.028	9.109
280	0.00990	1.000.–3	130.4	28.8	2514	0.104	8.980
285	0.01387	1.000.–3	99.4	49.8	2523	0.178	8.857
290	0.01917	1.001.–3	69.7	70.7	2532	0.251	8.740
295	0.02617	1.002.–3	51.94	91.6	2541	0.323	8.627
300	0.03531	1.003.–3	39.13	112.5	2550	0.393	8.520
305	0.04721	1.005.–3	27.90	133.4	2559	0.462	8.417
310	0.06221	1.007.–3	22.93	154.3	2568	0.530	8.318
315	0.08132	1.009.–3	17.82	175.2	2577	0.597	8.224
320	0.1053	1.011.–3	13.98	196.1	2586	0.649	8.151
325	0.1351	1.013.–3	11.06	217.0	2595	0.727	8.046
330	0.1719	1.016.–3	8.82	237.9	2604	0.791	7.962
335	0.2167	1.018.–3	7.09	258.8	2613	0.854	7.881
340	0.2713	1.021.–3	5.74	279.8	2622	0.916	7.804
345	0.3372	1.024.–3	4.683	300.7	2630	0.977	7.729
350	0.4163	1.027.–3	3.846	321.7	2639	1.038	7.657
355	0.5100	1.030.–3	3.180	342.7	2647	1.097	7.588
360	0.6209	1.034.–3	2.645	363.7	2655	1.156	7.521
365	0.7514	1.038.–3	2.212	384.7	2663	1.214	7.456
370	0.9040	1.041.–3	1.861	405.8	2671	1.271	7.394
373.15	1.0133	1.044.–3	1.679	419.1	2676	1.307	7.356
375	1.0815	1.045.–3	1.574	426.8	2679	1.328	7.333
380	1.2869	1.049.–3	1.337	448.0	2687	1.384	7.275
385	1.5233	1.053.–3	1.142	496.2	2694	1.439	7.218
390	1.794	1.085.–3	0.980	490.4	2702	1.494	7.163
400	2.455	1.067.–3	0.731	532.9	2716	1.605	7.058
410	3.302	1.077.–3	0.553	575.6	2729	1.708	6.959
420	4.370	1.088.–3	0.425	618.6	2742	1.810	6.865
430	5.699	1.099.–3	0.331	661.8	2753	1.911	6.775
440	7.333	1.110.–3	0.261	705.3	2764	2.011	6.689
450	9.319	1.123.–3	0.208	749.2	2773	2.109	6.607
460	11.71	1.137.–3	0.167	793.5	2782	2.205	6.528
470	14.55	1.152.–3	0.136	838.2	2789	2.301	6.451
480	17.90	1.167.–3	0.111	883.4	2795	2.395	6.377
490	21.83	1.184.–3	0.0922	929.1	2799	2.479	6.312
500	26.40	1.203.–3	0.0766	975.6	2801	2.581	6.233
510	31.66	1.222.–3	0.0631	1023	2802	2.673	6.163
520	37.70	1.244.–3	0.0525	1071	2801	2.765	6.093

Specific heat, C_p, kJ/(kg·K)		Viscosity, Ns/m²		Thermal conductivity, W/(m·K)		Prandtl no.		Surface tension, N/m	Temp., K
Condensed†	Vapor	Condensed†	Vapor	Condensed†	Vapor	Condensed†	Vapor	Condensed†	
1.155				3.73					150
1.233				3.52					160
1.311				3.34					170
1.389				3.18					180
1.467				3.04					190
1.545				2.91					200
1.623				2.79					210
1.701				2.69					220
1.779				2.59					230
1.857				2.50					240
1.935				2.42					250
1.974				2.38					255
2.013				2.35					260
2.052				2.31					265
2.091				2.27					270
2.116				2.26					273.15
4.217	1.854	1750.–6	8.02.–6	0.569	0.0182	12.99	0.815	0.0755	273.15
4.211	1.855	1652.–6	8.09.–6	0.574	0.0183	12.22	0.817	0.0753	275
4.198	1.858	1422.–6	8.29.–6	0.582	0.0186	10.26	0.825	0.0748	280
4.189	1.861	1225.–6	8.49.–6	0.590	0.0189	8.81	0.833	0.0743	285
4.184	1.864	1080.–6	8.69.–6	0.598	0.0193	7.56	0.841	0.0737	290
4.181	1.868	959.–6	8.89.–6	0.606	0.0195	6.62	0.849	0.0727	295
4.179	1.872	855.–6	9.09.–6	0.613	0.0196	5.83	0.857	0.0717	300
4.178	1.877	769.–6	9.29.–6	0.620	0.0201	5.20	0.865	0.0709	305
4.178	1.882	695.–6	9.49.–6	0.628	0.0204	4.62	0.873	0.0700	310
4.179	1.888	631.–6	9.69.–6	0.634	0.0207	4.16	0.883	0.0692	315
4.180	1.895	577.–6	9.89.–6	0.640	0.0210	3.77	0.894	0.0683	320
4.182	1.903	528.–6	10.09.–6	0.645	0.0213	3.42	0.901	0.0675	325
4.184	1.911	489.–6	10.29.–6	0.650	0.0217	3.15	0.908	0.0666	330
4.186	1.920	453.–6	10.49.–6	0.655	0.0220	2.88	0.916	0.0658	335
4.188	1.930	420.–6	10.69.–6	0.660	0.0223	2.66	0.925	0.0649	340
4.191	1.941	389.–6	10.89.–6	0.665	0.0226	2.45	0.933	0.0641	345
4.195	1.954	365.–6	11.09.–6	0.668	0.0230	2.29	0.942	0.0632	350
4.199	1.968	343.–6	11.29.–6	0.671	0.0233	2.14	0.951	0.0623	355
4.203	1.983	324.–6	11.49.–6	0.674	0.0237	2.02	0.960	0.0614	360
4.209	1.999	306.–6	11.69.–6	0.677	0.0241	1.91	0.969	0.0605	365
4.214	2.017	289.–6	11.89.–6	0.679	0.0245	1.80	0.978	0.0595	370
4.217	2.029	279.–6	12.02.–6	0.680	0.0248	1.76	0.984	0.0589	373.15
4.220	2.036	274.–6	12.09.–6	0.681	0.0249	1.70	0.987	0.0586	375
4.226	2.057	260.–6	12.29.–6	0.683	0.0254	1.61	0.995	0.0576	380
4.232	2.080	248.–6	12.49.–6	0.685	0.0258	1.53	1.004	0.0566	385
4.239	2.104	237.–6	12.69.–6	0.686	0.0263	1.47	1.013	0.0556	390
4.256	2.158	217.–6	13.05.–6	0.688	0.0272	1.34	1.033	0.0536	400
4.278	2.221	200.–6	13.42.–6	0.688	0.0282	1.24	1.054	0.0515	410
4.302	2.291	185.–6	13.79.–6	0.688	0.0293	1.16	1.075	0.0494	420
4.331	2.369	173.–6	14.14.–6	0.685	0.0304	1.09	1.10	0.0472	430
4.36	2.46	162.–6	14.50.–6	0.682	0.0317	1.04	1.12	0.0451	440
4.40	2.56	152.–6	14.85.–6	0.687	0.0331	0.99	1.14	0.0429	450
4.44	2.68	143.–6	15.19.–6	0.673	0.0346	0.95	1.17	0.0407	460
4.48	2.79	136.–6	15.54.–6	0.667	0.0363	0.92	1.20	0.0385	470
4.53	2.94	129.–6	15.88.–6	0.660	0.0381	0.89	1.23	0.0362	480
4.59	3.10	124.–6	16.23.–6	0.651	0.0401	0.87	1.25	0.0339	490
4.66	3.27	118.–6	16.59.–6	0.642	0.0423	0.86	1.28	0.0316	500
4.47	3.47	113.–6	16.95.–6	0.631	0.0447	0.85	1.31	0.0293	510
4.84	3.70	108.–6	17.33.–6	0.621	0.0475	0.84	1.35	0.0269	520

Temp., K	Pressure, bar*	Volume, m³/kg		Enthalpy, kJ/kg		Entropy, kJ/kg·K)	
		Condensed†	Vapor	Condensed†	Vapor	Condensed†	Vapor
530	44.58	1.268.–3	0.0445	1119	2798	2.856	6.023
540	52.38	1.294.–3	0.0375	1170	2792	2.948	5.953
550	61.19	1.323.–3	0.0317	1220	2784	3.039	5.882
560	71.08	1.355.–3	0.0269	1273	2772	3.132	5.808
570	82.16	1.392.–3	0.0228	1328	2757	3.225	5.733
580	94.51	1.433.–3	0.0193	1384	2737	3.321	5.654
590	108.3	1.482.–3	0.0163	1443	2717	3.419	5.569
600	123.5	1.541.–3	0.0137	1506	2682	3.520	5.480
610	137.3	1.612.–3	0.0115	1573	2641	3.627	5.318
620	159.1	1.705.–3	0.0094	1647	2588	3.741	5.259
625	169.1	1.778.–3	0.0085	1697	2555	3.805	5.191
630	179.7	1.856.–3	0.0075	1734	2515	3.875	5.115
635	190.9	1.935.–3	0.0066	1783	2466	3.950	5.025
640	202.7	2.075.–3	0.0057	1841	2401	4.037	4.912
645	215.2	2.351.–3	0.0045	1931	2292	4.223	4.732
647.3‡	221.2	3.170.–3	0.0032	2107	2107	4.443	4.443

* 1 bar = 10⁵ N/m².
† Above the solid line, the condensed phase is solid, below it, liquid.
‡ Critical temperature.

Temperature, °C	$10^4 \beta$	10^4 k_t/bar	10^4 k_s/bar	v_k, m/s	μ_f, 10^{-6} Pa·s	c_p, kJ/kg·K	k, W/m·K	Pr, bar	σ, N/m
0	−0.681	0.50885	0.50855	1402.4	1.793	4.2176	0.567	13.32	0.07565
1	−0.501	0.50509	0.50493	1407.4	1.732	4.2140	0.569	12.83	0.07551
2	−0.327	0.50151	0.50143	1412.2	1.675	4.2107	0.570	12.37	0.07537
3	−0.160	0.49808	0.49806	1417.0	1.621	4.2077	0.572	11.93	0.07522
4	0.003	0.49481	0.49481	1421.6	1.569	4.2048	0.573	11.51	0.07508
5	0.160	0.49169	0.49167	1426.2	1.520	4.2022	0.575	11.11	0.07494
6	0.312	0.48871	0.48865	1430.6	1.474	4.1999	0.577	10.73	0.07480
7	0.460	0.48587	0.48573	1434.9	1.429	4.1977	0.578	10.38	0.07465
8	0.604	0.48315	0.48291	1439.1	1.387	4.1956	0.580	10.04	0.07451
9	0.744	0.48056	0.48019	1443.3	1.346	4.1938	0.581	9.72	0.07436
10	0.880	0.47809	0.47757	1447.3	1.308	4.1921	0.5828	9.41	0.07422
11	1.012	0.47573	0.47504	1451.2	1.271	4.1906	0.5844	9.11	0.07407
12	1.141	0.47347	0.47260	1455.0	1.236	4.1892	0.5859	8.84	0.07393
13	1.267	0.47133	0.47024	1458.7	1.202	4.1879	0.5875	8.57	0.07378
14	1.389	0.46928	0.46797	1462.4	1.170	4.1867	0.5891	8.32	0.07364
15	1.509	0.46733	0.46578	1465.9	1.139	4.1856	0.5906	8.07	0.07349
16	1.626	0.46548	0.46366	1469.4	1.110	4.1847	0.5922	7.84	0.07334
17	1.740	0.46371	0.46162	1472.7	1.081	4.1838	0.5937	7.62	0.07319
18	1.852	0.46203	0.45966	1476.0	1.054	4.1830	0.5953	7.41	0.07304
19	1.961	0.46043	0.45776	1479.2	1.028	4.1823	0.5968	7.20	0.07289
20	2.068	0.45892	0.45593	1482.3	1.003	4.1817	0.5983	7.01	0.07274
21	2.173	0.45748	0.45417	1485.3	0.979	4.1812	0.5999	6.82	0.07259
22	2.275	0.45612	0.45248	1488.3	0.955	4.1807	0.6014	6.64	0.07244
23	2.376	0.45484	0.45084	1491.2	0.933	4.1802	0.6029	6.47	0.07228
24	2.475	0.45362	0.44927	1493.9	0.911	4.1798	0.6044	6.30	0.07213
25	2.572	0.45247	0.44776	1496.7	0.891	4.1795	0.6059	6.15	0.07198

Specific heat, C_p, kJ/(kg·K)		Viscosity, Ns/m²		Thermal conductivity, W/(m·K)		Prandtl no.		Surface tension, N/m	Temp., K
Condensed[†]	Vapor	Condensed[†]	Vapor	Condensed[†]	Vapor	Condensed[†]	Vapor	Condensed[†]	
4.95	3.96	104.–6	17.72.–6	0.608	0.0506	0.85	1.39	0.0245	530
5.08	4.27	101.–6	18.1.–6	0.594	0.0540	0.86	1.43	0.0221	540
5.24	4.64	97.–6	18.6.–6	0.580	0.0583	0.87	1.47	0.0197	550
5.43	5.09	94.–6	19.1.–6	0.563	0.0637	0.90	1.52	0.0173	560
5.68	5.67	91.–6	19.7.–6	0.548	0.0698	0.94	1.59	0.0150	570
6.00	6.40	88.–6	20.4.–6	0.528	0.0767	0.99	1.68	0.0128	580
6.41	7.35	84.–6	21.5.–6	0.513	0.0841	1.05	1.84	0.0105	590
7.00	8.75	81.–6	22.7.–6	0.497	0.0929	1.14	2.15	0.0084	600
7.85	11.1	77.–6	24.1.–6	0.467	0.103	1.30	2.60	0.0063	610
9.35	15.4	72.–6	25.9.–6	0.444	0.114	1.52	3.46	0.0045	620
10.6	18.3	70.–6	27.0.–6	0.430	0.121	1.65	4.20	0.0035	625
12.6	22.1	67.–6	28.0.–6	0.412	0.130	2.0	4.8	0.0026	630
16.4	27.6	64.–6	30.0.–6	0.392	0.141	2.7	6.0	0.0015	635
26	42	59.–6	32.0.–6	0.367	0.155	4.2	9.6	0.0008	640
90		54.–6	37.0.–6	0.331	0.178	12	26	0.0001	645
∞	∞	45.–6	45.0.–6	0.238	0.238	∞	∞	0.0000	647.3[‡]

THERMODYNAMIC AND THERMOPHYSICAL PROPERTIES—TABLE 119.
Saturated Liquid Water—Miscellaneous Properties (*Continued*)

Temperature, °C	$10^4 \beta$	10^4 k_T/bar	10^4 k_s/bar	v_k, m/s	μ_f, 10^{-6} Pa·s	c_p, kJ/kg·K	k, W/m·K	Pr, bar	σ, N/m
26	2.667	0.45139	0.44630	1499.3	0.871	4.1792	0.6074	5.99	0.07182
27	2.761	0.45038	0.44490	1501.9	0.852	4.1790	0.6089	5.85	0.07167
28	2.852	0.44943	0.44355	1504.3	0.833	4.1788	0.6104	5.70	0.07151
30	3.032	0.44771	0.44102	1509.1	0.798	4.1785	0.6133	5.44	0.07120
32	3.206	0.44622	0.43869	1513.6	0.765	4.1783	0.6162	5.19	0.07089
34	3.375	0.44496	0.43655	1517.8	0.734	4.1782	0.6190	4.95	0.07058
36	3.539	0.44390	0.43459	1521.7	0.705	4.1783	0.6218	4.74	0.07025
38	3.698	0.44305	0.43280	1525.4	0.679	4.1784	0.6246	4.54	0.06992
40	3.853	0.44239	0.43118	1528.9	0.653	4.1786	0.6273	4.35	0.06960
42	4.004	0.44192	0.42972	1532.1	0.629	4.1789	0.6299	4.17	0.06927
44	4.152	0.44162	0.42842	1535.0	0.607	4.1792	0.6315	4.02	0.06894
46	4.296	0.44149	0.42726	1537.7	0.586	4.1797	0.6351	3.86	0.06861
48	4.438	0.44153	0.42624	1540.3	0.566	4.1801	0.6375	3.71	0.06828
50	4.576	0.44173	0.42535	1542.6	0.547	4.1807	0.6400	3.75	0.06795
55	4.910	0.44290	0.42370	1547.4	0.5043	4.1824	0.6457	3.267	0.06710
60	5.231	0.44496	0.42281	1551.0	0.4668	4.1844	0.6511	3.000	0.06624
65	5.539	0.44788	0.42262	1553.4	0.4338	4.1869	0.6561	2.768	0.06537
70	5.837	0.45162	0.42309	1554.8	0.4045	4.1897	0.6607	2.565	0.06449
75	6.128	0.45614	0.42418	1555.1	0.3784	4.1929	0.6649	2.386	0.06359
80	6.411	0.46143	0.42587	1554.4	0.3550	4.1965	0.6686	2.228	0.06268
85	6.689	0.46748	0.42812	1552.9	0.3340	4.2005	0.6721	2.088	0.06176
90	6.962	0.47429	0.43093	1550.5	0.3150	4.2050	0.6753	1.962	0.06083
95	7.233	0.48185	0.43429	1547.2	0.2979	4.2102	0.6779	1.850	0.05988
100	7.501	0.49019	0.43819	1543.1	0.2823	4.2164	0.6800	1.756	0.05892

THERMODYNAMIC AND THERMOPHYSICAL PROPERTIES—TABLE 120.   Compressed Steam

Temperature, K		Pressure, bar									
		0.1	0.5	1	5	10	20	40	60	80	100
350	v	16.12	1.027.–3	1.027.–3	1.027.–3	1.027.–3	1.026.–3	1.025.–3	1.204.–3	1.023.–3	1.023.–3
	h	2,644	321.7	231.8	322.1	322.5	323.3	324.9	326.4	328.1	329.7
	s	8.327	1.037	1.037	1.037	1.037	1.036	1.035	1.034	1.032	1.031
400	v	18.44	3.67	1.827	1.067.–3	1.067.–3	1.066.–3	1.065.–3	1.064.–3	1.063.–3	1.061.–3
	h	2,739	2,735	2,730	533.1	533.4	534.1	535.4	536.8	538.2	539.6
	s	8.581	7.831	7.502	1.601	1.600	1.599	1.597	1.595	1.593	1.592
450	v	20.75	4.14	2.063	0.410	1.124.–3	1.123.–3	1.121.–3	1.119.–3	1.118.–3	1.116.–3
	h	2,835	2,833	2,830	2,804	749.0	749.8	750.8	751.9	753.0	754.1
	s	8.811	8.061	7.736	6.949	2.110	2.107	2.105	2.102	2.099	2.097
500	v	23.07	4.61	2.298	0.452	0.221	0.104	1.201.–3	1.198.–3	1.196.–3	1.193.–3
	h	2,932	2,931	2,929	2,912.4	2,891.2	2,839.4	975.9	976.3	976.8	977.3
	s	9.012	8.261	7.944	7.177	6.823	6.422	2.578	2.575	2.571	2.567
600	v	27.7	5.53	2.76	0.548	0.271	0.133	0.0630	0.0396	0.0276	0.0201
	h	3,131	3,130	3,129	3,120	3,109	3,087	3,036	2,976	2,906	2,820
	s	9.374	8.630	8.309	7.560	7.223	6.875	6.590	6.224	5.997	5.775
700	v	32.3	6.46	3.23	0.643	0.319	0.158	0.0769	0.0500	0.0346	0.0283
	h	3,335	3,335	3,334	3,328	3,322	3,307	3,278	3,247	3,214	3,179
	s	9.692	8.946	8.625	7.877	7.550	7.215	6.864	6.644	6.431	6.334
800	v	36.9	7.38	3.69	0.736	0.367	0.182	0.0889	0.0589	0.0436	0.0343
	h	3,547	3,546	3,546	3,542	3,537	3,526	3,506	3,485	3,464	3,442
	s	9.971	9.228	8.908	8.161	7.837	7.507	7.151	6.965	6.809	6.685
900	v	41.5	8.31	4.15	0.829	0.414	0.206	0.102	0.0674	0.0501	0.0398
	h	3,765	3,765	3,764	3,761	3,757	3,750	3,737	3,719	3,704	3,688
	s	10.228	9.485	9.165	8.420	8.097	7.770	7.462	7.237	7.092	6.975
1000	v	46.2	9.23	4.615	0.921	0.460	0.229	0.114	0.0758	0.0564	0.0449
	h	3,990	3,990	3,990	3,987	3,984	3,978	3,967	3,955	3,944	3,935
	s	10.466	9.723	9.402	8.659	8.336	8.011	7.682	7.486	7.345	7.233
1500	v	69.2	13.9	6.92	1.385	0.692	0.341	0.1730	0.1153	0.0865	0.0692
	h	5,231	5,228	5,227	5,225	5,224	5,221	5,217	5,212	5,207	5,203
	s	11.47	10.77	10.40	9.66	9.34	9.015	8.693	8.503	8.368	8.262
2000	v	93.0	18.6	9.26	1.850	0.925	0.462	0.231	0.1543	0.1157	0.0926
	h	6,832	6,734	6,706	6,662	6,649	6,639	6,629	6,623	6,619	6,616
	s	12.38	11.58	11.25	10.48	10.15	9.828	9.503	9.313	9.178	9.073
2500	v	123.7	24.0	11.90	2.35	1.171	0.583	0.291	0.1942	0.1457	0.1166
	h	10,417	9,330	9,046	8,621	8,504	8,413	8,342	8,307	8,285	8,269
	s	13.95	12.73	12.28	11.35	10.80	10.62	10.26	10.06	9.920	9.810.

THERMODYNAMIC AND THERMOPHYSICAL PROPERTIES—TABLE 120. Compressed
Steam (*Continued*)

Temperature, K		Pressure, bar										
		150	200	250	300	400	500	600	700	800	900	1000
350	v	1.020.–3	1.018.–3	1.016.–3	1.014.–3	1.009.–3	1.005.–3	1.002.–3	9.977.–4	9.937.–4	9.900.–4	9.865.–4
	h	333.7	337.7	341.7	344.7	353.8	361.8	369.7	377.7	385.7	393.7	401.7
	s	1.028	1.025	1.022	1.019	1.013	1.007	1.001	0.996	0.991	0.985	0.979
400	v	1.059.–3	1.056.–3	1.053.–3	1050.–3	1.045.–3	1.041.–3	1.035.–3	1.031.–3	1.027.–3	1.022.–3	1.018.–3
	h	543.1	546.5	550.1	553.5	560.6	567.8	574.9	582.1	589.3	596.5	603.8
	s	1.587	1.583	1.578	1.574	1.565	1.557	1.549	1.541	1.533	1.526	1.518
450	v	1.112.–3	1.108.–3	1.105.–3	1.101.–3	1.094.–3	1.088.–3	1.082.–3	1.076.–3	1.070.–3	1.065.–3	1.059.–3
	h	756.8	759.5	762.3	765.2	771.0	776.9	783.0	789.6	795.3	801.6	807.9
	s	2.088	2.082	2.076	2.070	2.060	2.049	2.039	2.029	2.019	2.010	2.002
500	v	1.187.–3	1.181.–3	1.175.–3	1.170.–3	1.160.–3	1.151.–3	1.142.–3	1.134.–3	1.126.–3	1.119.–3	1.112.–3
	h	978.8	980.3	981.9	983.7	987.4	991.5	995.9	1,000.5	1,005.3	1,010.3	1,015.4
	s	2.558	2.549	2.541	2.533	2.517	2.502	2.488	2.474	2.461	2.449	2.437
600	v	1.519.–3	1.483.–3	1.454.–3	1.428.–3	1.392.–3	1.362.–3	1.337.–3	1.315.–3	1.296.–3	1.280.–3	1.265.–3
	h	1,499	1,489	1,479	1,472	1,462	1,456	1,452	1,449	1,447	1,447	1,447
	s	3.501	3.469	3.443	3.419	3.379	3.346	3.316	3.290	3.266	3.244	3.223
700	v	1.724.–2	1.157.–2	7.986.–3	5.416.–3	2.630.–3	2.038.–3	1.831.–3	1.716.–3	1.639.–3	1.589.–3	1.536.–3
	h	3,082	2,965	2,821	2,635	2,233	2,084	1,986	1,962	1,946	1,931
	s	6.037	5.770	5.494	5.179	4.554	4.308	4.192	4.116	4.058	4.012	3.972
800	v	2.195.–2	1.575.–2	1.201.–2	9.512.–3	6.391.–3	4.576.–3	3.496.–3	2.866.–3	2.484.–3	2.239.–3	2.072.–3
	h	3,386	3,325	3,261	3,193	3,047	2,895	2,734	2,648	2,567	2,508	2,465
	s	6.444	6.252	6.086	5.934	5.654	5.397	5.175	4.998	4.864	4.761	4.701
900	v	2.590.–2	1.899.–2	1.483.–2	1.207.–2	8.619.–3	6.581.–3	5.257.–3	4.348.–3	3.704.–3	3.454.–3	2.907.–3
	h	3,649	3,609	3,568	3,526	3,440	3,354	3,269	3,188	3,113	3,049	2,995
	s	6.755	6.587	6.449	6.327	6.119	5.940	5.780	5.637	5.510	5.399	5.305
1000	v	2.954.–2	2.186.–2	1.726.–2	1.420.–2	1.038.–2	8.102.–3	6.605.–3	5.557.–3	4.792.–3	4.212.–3	3.763.–3
	h	3,904	3,874	3,845	3,816	3,756	3,697	3,640	3,584	3,532	3,482	3,435
	s	7.023	6.867	6.741	6.633	6.453	6.302	6.172	6.055	5.951	5.856	5.727
1500	v	0.0461	0.0346	0.0277	0.0231	0.0173	0.0139	0.0116	0.00993	0.00871	0.00776	0.00700
	h	5,202	5,198	5,186	5,180	5,171	5,157	5,144	5,133	5,120	5,108	5,095
	s	8.074	7.936	7.827	7.738	7.597	7.484	7.391	7.310	7.239	7.176	7.118
2000	v	0.0619	0.0465	0.0372	0.0311	0.0234	0.0188	0.0157	0.0135	0.0119	0.0106	0.0096
	h	6,613	6,610	6,608	6,605	6,599	6,595	6,590	6,585	6,581	6,577	6,574
	s	8.883	8.748	8.642	8.555	8.418	8.310	8.222	8.147	8.082	8.024	7.971
2500	v	0.0778	0.0584	0.0468	0.0391	0.0294	0.0236	0.0197	0.0170	0.0149	0.0133	0.0120
	h	8,269	8,269	8,269	8,268	8,267	8,265	8,261	2,856	8,250	8,244	8,240
	s	9.610	9.468	9.358	9.270	9.129	9.020	8.930	8.854	8.788	8.730	8.677

THERMODYNAMIC AND THERMOPHYSICAL PROPERTIES—TABLE 121.   Density, Specific Heats at Constant Pressure and at Constant Volume and Velocity of Sound for Compressed Water, 1–1000 bar, 0–150°C

P, bar	0°C (ITS-90) Density, kg/m³	C_p, kJ/(kg·K)	C_v, kJ/(kg·K)	w, m/s	10°C (ITS-90) Density, kg/m³	C_p, kJ/(kg·K)	C_v, kJ/(kg·K)	w, m/s	20°C (ITS-90) Density, kg/m³	C_p, kJ/(kg·K)	C_v, kJ/(kg·K)	w, m/s	30°C (ITS-90) Density, kg/m³	C_p, kJ/(kg·K)	C_v, kJ/(kg·K)	w, m/s
1	999.702	4.1923	4.1877	1447.3	998.207	4.1812	4.1538	1482.3	995.650	4.1774	4.1148	1509.1	992.217	4.1775	4.0715	1528.9
50	1002.03	4.174	4.168	1455	1000.44	4.166	4.137	1491	997.82	4.164	4.099	1517	994.36	4.166	4.058	1537
100	1004.38	4.156	4.149	1464	1002.69	4.151	4.119	1499	1000.02	4.151	4.084	1526	996.52	4.154	4.044	1546
150	1006.71	4.139	4.130	1472	1004.93	4.137	4.103	1507	1002.19	4.139	4.069	1534	998.66	4.142	4.031	1554
200	1009.01	4.123	4.112	1480	1007.13	4.124	4.087	1516	1004.34	4.127	4.055	1543	1000.77	4.131	4.018	1563
250	1011.28	4.108	4.095	1489	1009.32	4.110	4.071	1524	1006.47	4.115	4.041	1551	1002.87	4.121	4.005	1571
300	1013.53	4.093	4.078	1497	1011.48	4.098	4.056	1532	1008.57	4.104	4.027	1559	1004.94	4.110	3.993	1579
400	1017.97	4.065	4.046	1513	1015.74	4.074	4.027	1548	1012.72	4.083	4.001	1576	1009.03	4.091	3.969	1596
500	1022.31	4.040	4.016	1529	1019.92	4.052	3.999	1565	1016.79	4.063	3.976	1592	1013.03	4.072	3.946	1612
600	1026.57	4.018	3.988	1545	1024.02	4.032	3.974	1581	1020.79	4.044	3.952	1608	1016.97	4.055	3.924	1628
800	1034.85	3.979	3.937	1577	1031.99	3.996	3.926	1613	1028.56	4.011	3.908	1640	1024.62	4.023	3.884	1660
1000	1042.83	3.948	3.892	1609	1039.68	3.967	3.884	1644	1036.06	3.982	3.869	1671	1032.00	3.995	3.847	1692

P, bar	40°C (ITS-90) Density, kg/m³	C_p, kJ/(kg·K)	C_v, kJ/(kg·K)	w, m/s	50°C (ITS-90) Density, kg/m³	C_p, kJ/(kg·K)	C_v, kJ/(kg·K)	w, m/s	60°C (ITS-90) Density, kg/m³	C_p, kJ/(kg·K)	C_v, kJ/(kg·K)	w, m/s	70°C (ITS-90) Density, kg/m³	C_p, kJ/(kg·K)	C_v, kJ/(kg·K)	w, m/s
1	988.036	4.1799	4.0248	1542.6	983.197	4.1840	3.9755	1551.0	977.766	4.1896	3.9246	1554.8	971.791	4.1967	3.8727	1554.5
50	990.16	4.169	4.012	1551	985.33	4.173	3.964	1560	979.92	4.179	3.915	1564	973.98	4.186	3.864	1564
100	992.31	4.158	4.000	1560	987.48	4.163	3.953	1568	982.09	4.169	3.905	1573	976.18	4.176	3.855	1573
150	994.44	4.147	3.988	1568	989.61	4.152	3.943	1577	984.23	4.158	3.895	1582	978.35	4.165	3.846	1582
200	996.54	4.137	3.976	1577	991.71	4.142	3.932	1586	986.36	4.148	3.885	1590	980.51	4.155	3.838	1591
250	998.62	4.126	3.965	1585	993.80	4.132	3.922	1594	988.46	4.139	3.876	1599	982.63	4.146	3.829	1600
300	1000.68	4.117	3.954	1594	995.86	4.123	3.911	1603	990.53	4.129	3.867	1608	984.74	4.136	3.821	1609
400	1004.74	4.098	3.932	1610	999.92	4.105	3.892	1620	994.62	4.111	3.849	1625	988.87	4.118	3.805	1627
500	1008.72	4.080	3.911	1627	1003.90	4.087	3.873	1637	998.62	4.094	3.832	1642	992.92	4.101	3.789	1644
600	1012.62	4.063	3.892	1643	1007.80	4.071	3.855	1653	1002.54	4.078	3.815	1659	996.88	4.085	3.774	1662
800	1020.21	4.033	3.854	1676	1015.38	4.041	3.821	1686	1010.15	4.048	3.784	1693	1004.56	4.054	3.745	1696
1000	1027.53	4.005	3.820	1707	1022.69	4.013	3.789	1718	1017.48	4.020	3.754	1726	1011.94	4.027	3.717	1730

THERMODYNAMIC AND THERMOPHYSICAL PROPERTIES—TABLE 122. Density, Specific Heats at Constant Pressure and at Constant Volume and Velocity of Sound for Compressed Water, 1–1000 bar, 0–150°C (*Concluded*)

80°C (ITS-90)

P, bar	Density, kg/m³	C_p, kJ/(kg·K)	C_v, kJ/(kg·K)	w, m/s
1	965.309	4.2056	3.8206	1550.5
50	967.54	4.195	3.813	1560
100	969.79	4.184	3.805	1569
150	972.00	4.174	3.797	1579
200	974.20	4.164	3.789	1588
250	976.36	4.154	3.782	1597
300	978.50	4.144	3.774	1607
400	982.71	4.126	3.759	1625
500	986.82	4.108	3.745	1643
600	990.83	4.092	3.731	1661
800	998.62	4.061	3.704	1696
1000	1006.06	4.033	3.678	1731

90°C (ITS-90)

P, bar	Density, kg/m³	C_p, kJ/(kg·K)	C_v, kJ/(kg·K)	w, m/s
1	958.348	4.2164	3.7689	1543.1
50	960.64	4.205	3.762	1553
100	962.94	4.194	3.755	1563
150	965.21	4.184	3.748	1572
200	967.45	4.173	3.741	1582
250	969.67	4.163	3.734	1592
300	971.85	4.153	3.727	1601
400	976.15	4.135	3.714	1620
500	980.34	4.117	3.701	1639
600	984.43	4.100	3.688	1658
800	992.34	4.068	3.663	1694
1000	999.92	4.039	3.639	1730

100°C (ITS-90)

P, bar	Density, kg/m³	C_p, kJ/(kg·K)	C_v, kJ/(kg·K)	w, m/s
1	950.927	4.2296	3.7181	1532.5
50	953.28	4.218	3.712	1543
100	955.65	4.206	3.706	1553
150	957.99	4.195	3.699	1563
200	960.30	4.184	3.693	1573
250	962.57	4.174	3.687	1583
300	964.82	4.164	3.681	1593
400	969.21	4.144	3.669	1613
500	973.50	4.126	3.657	1632
600	977.69	4.108	3.645	1652
800	985.76	4.076	3.621	1690
1000	993.47	4.046	3.598	1727

110°C (ITS-90)

P, bar	Density, kg/m³	C_p, kJ/(kg·K)	C_v, kJ/(kg·K)	w, m/s
1	943.059	4.2453	3.6684	1519.0
50	945.50	4.233	3.663	1530
100	947.95	4.221	3.657	1540
150	950.36	4.209	3.652	1551
200	952.74	4.198	3.646	1561
250	955.09	4.187	3.641	1572
300	957.40	4.176	3.635	1582
400	961.92	4.155	3.624	1603
500	966.32	4.136	3.613	1623
600	970.61	4.118	3.602	1644
800	978.87	4.084	3.579	1683
1000	986.73	4.053	3.556	1723

120°C (ITS-90)

P, bar	Density, kg/m³	C_p, kJ/(kg·K)	C_v, kJ/(kg·K)	w, m/s
1	934.749	4.2639	3.6201	1502.8
50	937.28	4.251	3.615	1514
100	939.83	4.238	3.610	1525
150	942.33	4.225	3.605	1536
200	944.79	4.213	3.600	1547
250	947.22	4.201	3.595	1558
300	949.61	4.190	3.590	1569
400	954.27	4.168	3.580	1591
500	958.81	4.148	3.569	1612
600	963.22	4.128	3.558	1634
800	971.68	4.092	3.536	1676
1000	979.72	4.059	3.512	1717

130°C (ITS-90)

P, bar	Density, kg/m³	C_p, kJ/(kg·K)	C_v, kJ/(kg·K)	w, m/s
1	925.997	4.2859	3.5733	1484.1
50	928.64	4.271	3.569	1496
100	931.29	4.257	3.564	1508
150	933.90	4.244	3.560	1519
200	936.46	4.231	3.555	1531
250	938.97	4.218	3.550	1543
300	941.45	4.206	3.545	1554
400	946.28	4.183	3.535	1577
500	950.96	4.161	3.525	1600
600	955.51	4.140	3.514	1623
800	964.20	4.101	3.491	1667
1000	972.44	4.066	3.466	1710

140°C (ITS-90)

P, bar	Density, kg/m³	C_p, kJ/(kg·K)	C_v, kJ/(kg·K)	w, m/s
1	916.797	4.3114	3.5279	1463.0
50	919.57	4.296	3.524	1475
100	922.34	4.280	3.520	1488
150	925.06	4.266	3.516	1501
200	927.73	4.251	3.511	1513
250	930.35	4.238	3.506	1525
300	932.92	4.224	3.501	1538
400	937.94	4.199	3.491	1562
500	942.79	4.176	3.480	1586
600	947.48	4.153	3.469	1610
800	956.43	4.112	3.444	1658
1000	964.88	4.073	3.417	1704

150°C (ITS-90)

P, bar	Density, kg/m³	C_p, kJ/(kg·K)	C_v, kJ/(kg·K)	w, m/s
1	907.143	4.3408	3.4848	1439.8
50	910.06	4.324	3.481	1453
100	912.97	4.307	3.477	1467
150	915.82	4.291	3.473	1480
200	918.61	4.276	3.469	1493
250	921.34	4.261	3.464	1507
300	924.03	4.246	3.459	1520
400	929.24	4.219	3.448	1546
500	934.27	4.194	3.436	1572
600	939.13	4.169	3.423	1598
800	948.36	4.124	3.395	1648
1000	957.04	4.081	3.364	1698

THERMODYNAMIC AND THERMOPHYSICAL PROPERTIES—TABLE 123. Specific Heat and Other Thermophysical Properties of Water Substance*

Pressure, bar		300	350	400	450	500	600	700	800	900	1000	1200	1400	1600	1800	2000
							Temperature, K									
1	μ	8.57.-4	3.70.-4	1.32.-5	1.52.-5	1.73.-5	2.15.-5	2.57.-5	2.98.-5	3.39.-5	3.78.-5	4.48.-5	5.06.-5	5.65.-5	6.19.-5	6.70.-5
	c_p	4.18	4.19	1.99	1.97	1.98	2.02	2.09	2.15	2.22	2.29	2.43	2.58	2.73	3.02	3.79
	k	0.614	0.668	0.0268	0.0311	0.0358	0.0464	0.0581	0.0710	0.0843	0.0981	0.13	0.16	0.21	0.33	0.57
	Pr	5.18	2.32	0.980	0.967	0.955	0.936	0.920	0.906	0.891	0.881	0.83	0.80.	0.75	0.57	0.45
5	μ	8.57.-4	3.70.-4	2.17.-4	1.49.-5	1.72.-5	2.15.-5	2.57.-5	2.98.-5	3.39.-5	3.78.-5	4.45.-5	5.06.-5	5.65.-5	6.19.-5	6.70.-5
	c_p	4.18	4.19	4.26	2.21	2.10	2.07	2.11	2.16	2.23	2.29	2.43	2.58	2.73	2.98	3.40
	k	0.614	0.668	0.689	0.0335	0.0369	0.0469	0.0585	0.0713	0.0846	0.0984	0.13	0.16	0.20	0.28	0.43
	Pr	5.82	2.32	1.34	0.983	0.973	0.947	0.925	0.907	0.892	0.881	0.83	0.81	0.77	0.65	0.53
10	μ	8.57.-4	3.70.-4	2.17.-4	1.51.-4	1.71.-5	2.15.-5	2.58.-5	2.99.-5	3.39.-5	3.78.-5	4.45.-5	5.06.-5	5.65.-5	6.19.-5	6.70.-5
	c_p	4.18	4.19	4.25	4.39	2.29	2.13	2.13	2.18	2.24	2.30	2.44	2.58	2.73	2.95	3.29
	k	0.615	0.668	0.689	0.677	0.0380	0.0474	0.0590	0.0717	0.0851	0.0988	0.13	0.16	0.20	0.26	0.39
	Pr	5.82	2.32	1.34	0.981	1.028	0.963	0.931	0.908	0.892	0.881	0.84	0.82	0.78	0.70	0.57
20	μ	8.56.-4	3.71.-4	2.18.-4	1.51.-4	1.68.-5	2.15.-5	2.59.-5	3.00.-5	3.40.-5	3.79.-5	4.46.-5	5.06.-5	5.65.-5	6.19.-5	6.70.-5
	c_p	4.17	4.19	4.25	4.39	2.84	2.26	2.19	2.21	2.26	2.32	2.45	2.59	2.73	2.92	3.21
	k	0.616	0.669	0.689	0.679	0.0402	0.0485	0.0599	0.0726	0.0859	0.0996	0.13	0.16	0.20	0.25	0.36
	Pr	5.80	2.32	1.34	0.979	1.19	0.999	0.946	0.912	0.893	0.881	0.84	0.82	0.79	0.72	0.60
40	μ	8.55.-4	3.71.-4	2.18.-4	1.52.-4	1.19.-4	2.15.-5	2.61.-5	3.02.-5	3.42.-5	3.80.-5	4.47.-5	5.07.-5	5.65.-5	6.19.-5	6.70.-5
	c_p	4.17	4.19	4.25	4.38	4.65	2.60	2.32	2.28	2.30	2.34	2.46	2.59	2.73	2.90	3.14
	k	0.617	0.671	0.690	0.680	0.644	0.516	0.0620	0.0744	0.0877	0.101	0.13	0.16	0.19	0.24	0.33
	Pr	5.78	2.31	1.34	0.977	0.862	1.08	0.975	0.924	0.895	0.881	0.84	0.82	0.80	0.73	0.63
60	μ	8.54.-4	3.72.-4	2.19.-4	1.53.-4	1.20.-4	2.14.-5	2.63.-5	3.04.-5	3.43.-5	3.82.-5	4.48.-5	5.07.-5	5.66.-5	6.19.-5	6.70.-5
	c_p	4.16	4.18	4.24	4.37	4.63	3.11	2.47	2.35	2.34	2.37	2.48	2.60	2.73	2.89	3.11
	k	0.619	0.672	0.692	0.682	0.646	0.0561	0.0645	0.0764	0.0895	0.103	0.13	0.16	0.19	0.24	0.32
	Pr	5.74	2.31	1.34	0.976	0.859	1.19	1.008	0.934	0.899	0.879	0.84	0.82	0.81	0.74	0.65
	μ	8.53.-4	3.72.-4	2.19.-4	1.53.-4	1.20.-4	2.14.-5	2.66.-5	3.06.-5	3.45.-5	3.83.-5	4.48.-5	5.08.-5	5.66.-5	6.19.-5	6.70.-5

Temp																
80	c_p	4.16	4.18	4.24	4.36	4.62	3.88	2.65	2.43	2.39	2.40	2.49	2.61	2.73	2.88	3.09
	k	0.620	0.674	0.693	0.684	0.648	0.0628	0.0672	0.0785	0.0914	0.105	0.13	0.16	0.10	0.24	0.31
	Pr	5.72	2.31	1.34	0.975	0.856	1.33	1.046	0.946	0.902	0.877	0.84	0.83	0.81	0.74	0.66
100	μ	8.52.-4	3.73.-4	2.20.-4	1.53.-4	1.21.-4	2.14.-5	2.69.-5	3.08.-5	3.47.-5	3.85.-5	4.49.-5	5.08.-5	5.66.-5	6.19.-5	6.70.-5
	c_p	4.15	4.17	4.23	4.35	4.60	5.22	2.85	2.52	2.44	2.44	2.50	2.62	2.73	2.88	3.08
	k	0.622	0.675	0.694	0.685	0.651	0.0730	0.0704	0.0807	0.0934	0.107	0.13	0.16	0.19	0.24	0.31
	Pr	5.69	2.31	1.34	0.975	0.853	1.74	1.088	0.960	0.905	0.876	0.84	0.83	0.81	0.74	0.67
150	μ	8.51.-4	3.74.-4	2.22.-4	1.56.-4	1.22.-4	8.22.-5	2.72.-5	3.12.-5	3.51.-5	3.89.-5	4.52.-5	5.09.-5	5.67.-5	6.19.-5	6.70.-5
	c_p	4.14	4.16	4.22	4.34	4.54		3.55	2.74	2.57	2.53	2.54	2.65	2.75	2.88	3.06
	k	0.624	0.678	0.699	0.693	0.657	0.520	0.079	0.086	0.098	0.110	0.14	0.16	0.19	0.23	0.31
	Pr	5.64	2.30	1.34	0.974	0.842		1.22	0.994	0.916	0.891	0.84	0.83	0.82	0.76	0.67
200	μ	8.50.-4	3.75.-4	2.24.-4	1.57.-4	1.23.-4	8.32.-5	2.80.-5	3.17.-5	3.54.-5	3.93.-5	4.54.-5	5.11.-5	5.67.-5		
	c_p	4.12	4.15	4.21	4.32	4.51		4.67	3.04	2.71	2.62	2.57	2.67	2.76	2.88	3.05
	k	0.626	0.681	0.702	0.697	0.661	0.525	0.095	0.095	0.104	0.113	0.14	0.16	0.19		
	Pr	5.59	2.29	1.34	0.974	0.833		1.38	1.014	0.925	0.903	0.84	0.83	0.82		
250	μ	8.49.-4	3.76.-4	2.26.-4	1.59.-4	1.23.-4	8.41.-5	2.89.-5	3.24.-5	3.59.-5	3.98.-5	4.56.-5	5.12.-5	5.68.-5		
	c_p	4.12	4.14	4.20	4.30	4.49	5.90	6.16	3.40	2.86	2.71	2.61	2.69	2.77	2.89	3.04
	k	0.627	0.683	0.705	0.701	0.672	0.537	0.112	0.103	0.110	0.119	0.136	0.16			
	Pr	5.57	2.28	1.34	0.974	0.826	0.924	1.590	1.070	0.940	0.910	0.85	0.84			
300	μ	8.49.-4	3.77.-4	2.28.-4	1.60.-4	1.24.-4	8.50.-5	3.7.-5	3.4.-5	3.64.-5	4.02.-5	4.59.-5	5.14.-5	5.68.-5		
	c_p	4.10	4.13	4.19	4.29	4.44	5.60	10.20	3.82	3.03	2.81	2.65	2.72	2.78	2.90	3.04
	k	0.629	0.685	0.708	0.704	0.675	0.548	0.173	0.113	0.113	0.123	0.14				
	Pr	5.53	2.27	1.34	0.973	0.820	0.859	2.18	1.149	0.976	0.917	0.87				
400	μ	8.49.-4	3.80.-4	2.30.-4	1.62.-4	1.26.-4	8.64.-5	5.3.-5	3.6.-5	3.8.-5	4.1.-5	4.6.-5	5.17.-5			
	c_p	4.08	4.12	4.16	4.26	4.42	5.31	13.20	4.86	3.39	3.01	2.70	2.77	2.81	2.91	3.04
	k	0.631	0.689	0.714	0.710	0.676	0.567	0.327	0.145	0.129	0.134	0.15				
	Pr	5.49	2.26	1.34	0.971	0.817	0.799	2.14	1.207	0.999	0.926					
	μ	8.50.-4	3.82.-4	2.31.-4	1.64.-4	1.28.-4	8.83.-5	5.8.-5	4.0.-5	4.0.-5	4.2.-5	4.7.-5				

Specific Heat and Other Thermophysical Properties of Water Substance *(Continued)*

Temperature, K

Pressure, bar		300	350	400	450	500	600	700	800	900	1000	1200	1400	1600	1800	2000
500	c_p	4.06	4.10	4.15	4.23	4.38	5.08	8.44	5.70	3.90	3.21	2.77	2.81	2.84	2.92	3.04
	k	0.634	0.695	0.719	0.717	0.693	0.583	0.378	0.186	0.147	0.145					
	Pr	5.44	2.25	1.33	0.971	0.814	0.773	1.30	1.225	1.061	0.932					
600	μ	8.51.–4		2.32.–4	1.66.–4	1.30.–4	9.17.–5	6.5.–5	4.4.–5	4.2.–5	4.4.–5					
	c_p	4.04	4.08	4.13	4.20	4.33	4.92	6.93	6.83	4.19	3.38	2.87	2.86	2.86	2.92	3.04
	k	0.639	0.699	0.725	0.725	0.700	0.597	0.420	0.239	0.170	0.159					
	Pr	5.38	2.24	1.32	0.970	0.812	0.755	1.073	1.175	1.035	0.935					
700	μ	8.52.–4		2.33.–4	1.69.–4	1.33.–4	9.50.–5	6.9.–5	4.9.–5	4.5.–5	4.6.–5					
	c_p	4.01	4.07	4.12	4.17	4.29	4.78	6.12	6.26	4.62	3.59	2.94	2.91	2.88	2.93	3.05
	k	0.644	0.706	0.730	0.732	0.707	0.614	0.442	0.279	0.198	0.177					
	Pr	5.33	2.23	1.32	0.970	0.810	0.739	1.047	1.098	1.010	0.935					
800	μ	8.53.–4		2.34.–4	1.72.–4	1.36.–4	9.82.–5	7.3.–5	5.4.–5	4.8.–5	4.8.–5					
	c_p	3.99	4.05	4.10	4.15	4.26	4.67	5.60	6.09	4.77	3.75	3.01	2.96	2.91	2.95	3.05
	k	0.648	0.709	0.735	0.736	0.714	0.625	0.478	0.320	0.228	0.193					
	Pr	5.28	2.23	1.31	0.970	0.808	0.725	0.855	1.028	1.003	0.933					
900	μ	8.54.–4		2.35.–4	1.74.–4	1.38.–4	1.00.–4	7.6.–5	5.8.–5	5.1.–5	5.0.–5					
	c_p	3.98	4.03	4.08	4.13	4.23	4.57	5.29	5.86	4.85	3.86	3.08	3.00	2.94	2.97	3.06
	k	0.651	0.713	0.738	0.742	0.724	0.636	0.496	0.351	0.260	0.210					
	Pr	5.23	2.22	1.30	0.969	0.806	0.712	0.810	0.968	0.950	0.919					
1000	μ	8.56.–4		2.36.–4	1.76.–4	1.40.–4	1.02.–4	7.9.–5	6.2.–5	5.4.–5	5.1.–5					
	c_p	3.97	4.02	4.06	4.11	4.20	4.47	5.08	5.51	4.88	3.96	3.16	3.05	2.97	2.98	3.07
	k	0.653	0.717	0.743	0.747	0.731	0.650	0.516	0.372	0.288	0.228					
	Pr	5.19	2.22	1.30	0.968	0.804	0.701	0.778	0.918	0.900	0.886					

*μ = viscosity, Ns/m²; c_p = specific heat at constant pressure, kJ/(kg·K); k = thermal conductivity, W/(m·K); Pr = Prandtl number.

THERMODYNAMIC AND THERMOPHYSICAL PROPERTIES—TABLE 124. Water Substance along the Melting Line

P, bar	T, °C	$10^3\, v_f$, m³/kg	h_f, kJ/kg	s_f, kJ/kg·K	c_{pf}, kJ/kg·K	c_{melt}, kJ/kg·K	$10^6\, \alpha_f$, K⁻¹	$10^5 \mathbf{K}_{f,T}$ bar⁻¹
6.117,-5	0.0100	1.00021	0	0	4.219	3.969	-67.42	50.90
1.01325	0.0026	1.00016	0.0719	-0.0001	4.218	3.970	-67.17	50.88
50	-0.3618	0.99770	3.5140	-0.0054	4.196	3.997	-54.92	50.30
100	-0.7410	0.99523	6.9794	-0.0110	4.174	4.023	-42.52	49.73
150	-1.1249	0.99278	10.3964	-0.0167	4.152	4.047	-30.24	49.17
200	-1.5166	0.99037	13.7648	-0.0225	4.132	4.070	-18.05	48.63
250	-1.9151	0.98798	17.0843	-0.0285	4.112	4.092	-5.93	48.11
300	-2.3206	0.98562	20.3547	-0.0347	4.092	4.113	6.12	47.59
400	-3.1532	0.98098	26.7472	-0.0474	4.056	4.150	30.09	46.61
500	-4.0156	0.97643	32.9403	-0.0607	4.022	4.184	53.97	45.68
600	-4.909	0.97196	38.932	-0.0747	3.992	4.215	77.87	44.80
800	-6.790	0.96326	50.300	-0.1046	3.937	4.270	126.18	43.19
1000	-8.803	0.95493	60.836	-0.1371	3.893	4.320	175.98	41.74

THERMODYNAMIC AND THERMOPHYSICAL PROPERTIES—TABLE 125. Saturated Xenon

T, K	P, bar	v_f, m³/kg	v_g, m³/kg	h_f, kJ/kg	h_g, kJ/kg	s_f, kJ/(kg·K)	s_g, kJ/(kg·K)	c_{pf}, kJ/(kg·K)	μ_f, 10^{-4} Pa·s	k_f, W/(m·K)
10		2.642.−4		0.19		0.0236		0.058		
20		2.650.−4		1.21		0.0901		0.133		
30		2.661.−4		2.74		0.1510		0.164		
40		2.675.−4		4.47		0.2003		0.178		
50		2.689.−4		6.31		0.2410		0.186		
60		2.704.−4		8.21		0.2755		0.191		
80		2.737.−4		12.14		0.3319		0.202		
100		2.776.−4		16.30		0.3783		0.214		
120		2.820.−4		20.81		0.4197		0.231		
140		2.874.−4		25.67		0.4581		0.251		
160		2.941.−4		30.94		0.4946		0.270		
161.4		2.946.−4		31.30		0.4969		0.271		
161.4	0.816	3.372.−4	0.1219	48.98	145.5	0.6072	1.206	0.350		
170	1.336	3.439.−4	0.0776	52.01	146.5	0.6253	1.181	0.349	4.50	0.0707
180	2.218	3.523.−4	0.0487	55.52	147.5	0.6452	1.156	0.349	3.99	0.0663
190	3.480	3.615.−4	0.0321	59.04	148.3	0.6641	1.134	0.352	3.51	0.0622
200	5.212	3.715.−4	0.0220	62.61	148.9	0.6820	1.113	0.357	3.09	0.0582
210	7.504	3.828.−4	0.0156	66.25	149.2	0.6994	1.095	0.365	2.71	0.0542
220	10.45	3.955.−4	0.0113	70.00	149.4	0.7163	1.077	0.379	2.39	0.0506
230	14.16	4.100.−4	0.0084	73.91	149.2	0.7330	1.060	0.400	2.09	0.0468
240	18.72	4.271.−4	0.0063	78.05	148.5	0.7498	1.044	0.432	1.83	0.0429
250	24.25	4.476.−4	0.0047	82.54	147.5	0.7671	1.027	0.482	1.60	0.0393
260	30.87	4.730.−4	0.0036	87.52	145.7	0.7855	1.009	0.560	1.38	0.0355
270	38.69	5.079.−4	0.0027	93.30	142.8	0.8058	0.989	0.685	1.18	0.0313
280	47.86	5.689.−4	0.0019	100.6	138.0	0.8308	0.964	0.995	0.95	0.0275
289.7	58.21	9.091.−4	0.0009	120.0	120.0	0.8962	0.896	∞		∞

THERMODYNAMIC AND THERMOPHYSICAL PROPERTIES—TABLE 126. Compressed Xenon

T, K		1	100	200	300	400	500	600	700	800	900	1000
						Pressure, bar						
100	v	2.776.-4	2.764.-4	2.752.-4	2.742.-4	2.731.-4	2.721.-4	2.711.-4	2.702.-4	2.693.-4	2.684.-4	2.675.-4
	h	16.30	18.84	21.40	23.95	26.50	29.05	31.59	34.13	36.67	39.21	41.74
	s	0.3783	0.3762	0.3742	0.3723	0.3704	0.3686	0.3669	0.3652	0.3636	0.3621	0.3802
200	v	0.1245	3.623.-4	3.547.-4	3.484.-4	3.430.-4	3.383.-4	3.342.-4	3.304.-4	3.270.-4	3.240.-4	3.211.-4
	h	151.8	64.22	66.14	68.19	70.34	72.56	74.83	77.13	79.46	81.81	84.18
	s	1.228	0.6727	0.6643	0.6570	0.6505	0.6446	0.6391	0.6340	0.6292	0.6247	0.6204
300	v	0.1890	5.729.-4	4.769.-4	4.431.-4	4.220.-4	4.068.-4	3.955.-4	3.862.-4	3.783.-4	3.716.-4	3.657.-4
	h	168.0	106.4	101.6	101.3	102.0	103.3	104.9	106.7	108.5	110.6	112.8
	s	1.294	0.8401	0.8073	0.7908	0.7789	0.7691	0.7608	0.7540	0.7477	0.7424	0.7370
400	v	0.2527	1.998.-3	8.759.-4	6.425.-4	5.604.-4	5.141.-4	4.839.-4	4.622.-4	4.457.-4	4.325.-4	4.217.-4
	h	183.9	164.2	145.4	137.4	134.7	134.1	134.5	135.5	136.8	138.3	140.0
	s	1.340	1.012	0.9330	0.8945	0.8730	0.8581	0.8467	0.8373	0.8292	0.8220	0.8162
500	v	0.3163	2.899.-3	1.389.-3	9.449.-4	7.577.-4	6.593.-4	5.986.-4	5.570.-4	5.268.-4	5.038.-4	4.859.-4
	h	199.8	187.8	177.1	169.4	165.1	163.0	162.3	162.4	163.1	164.3	165.7
	s	1.375	1.065	1.004	0.9664	0.9409	0.9228	0.9088	0.8975	0.8881	0.8801	0.8731
600	v	0.3798	3.673.-3	1.823.-3	1.240.-3	9.699.-4	8.206.-4	7.273.-4	6.636.-4	6.172.-4	5.820.-4	5.545.-4
	h	215.7	207.4	200.3	194.8	191.1	188.9	187.9	187.6	188.0	188.8	189.9
	s	1.404	1.101	1.047	1.013	0.9885	0.9700	0.9555	0.9435	0.9334	0.9247	0.9172
700	v	0.4432	4.397.-3	2.217.-3	1.513.-3	1.175.-3	9.815.-4	8.583.-4	7.734.-4	7.115.-4	6.642.-4	6.268.-4
	h	231.5	225.6	220.6	216.7	213.8	212.2	211.3	211.1	211.3	212.0	213.1
	s	1.428	1.129	1.078	1.047	1.023	1.006	0.9916	0.9797	0.9695	0.9606	0.9528
800	v	0.5066	5.093.-3	2.587.-3	1.769.-3	1.370.-3	1.137.-3	9.870.-4	8.824.-4	8.057.-4	7.469.-4	7.005.-4
	h	247.4	243.0	239.5	236.7	234.8	233.6	233.0	232.9	233.3	234.0	235.0
	s	1.450	1.152	1.103	1.073	1.052	1.035	1.021	1.009	0.9988	0.9901	0.9823
900	v	0.5700	5.773.-3	2.944.-3	2.014.-3	1.557.-3	1.288.-3	1.112.-3	9.893.-4	8.989.-4	8.289.-4	7.737.-4
	h	263.2	260.1	257.5	255.7	254.4	253.6	253.4	253.7	254.2	254.9	256.1
	s	1.468	1.172	1.125	1.096	1.075	1.058	1.045	1.033	1.023	1.015	1.007
1000	v	0.6333	6.441.-3	3.291.-3	2.252.-3	1.738.-3	1.435.-3	1.235.-3	1.094.-3	9.899.-4	9.097.-4	8.461.-4
	h	279.1	276.8	275.1	273.9	273.2	272.9	273.0	273.4	274.1	275.1	276.2
	s	1.485	1.190	1.143	1.115	1.095	1.079	1.065	1.054	1.044	1.036	1.028

THERMODYNAMIC AND THERMOPHYSICAL PROPERTIES—TABLE 127. Surface Tension (N/m) of Saturated Liquid Refrigerants

R no.					Temperature, °C				
	−50	−25	0	25	50	75	100	125	150
11	0.0279	0.0244	0.0210	0.0178	0.0146	0.0116	0.0087	0.0060	0.0036
12	0.0188	0.0152	0.0118	0.0085	0.0055	0.0029	0.0007	—	—
13	0.0092	0.0056	0.0025	0.0002	—	—	—	—	—
22	0.0197	0.0156	0.0117	0.0081	0.0047	0.0018	—	—	—
23	0.0115	0.0065	0.0025						
32				0.0069	0.0032	0.0002	—	—	—
113	—	0.0231	0.0201	0.0172	0.0144	0.0118	0.0092	0.0067	0.0045
114			0.0138	0.0109	0.0082	0.0056	0.0033	0.0012	—
115			0.0075	0.0047	0.0022	—	—	—	—
134	0.0192	0.0154	0.0117	0.0082	0.0050	0.0021	0.0000	—	—
142	0.0213	0.0178	0.0145	0.0113	0.0083	0.0055	0.0029		—
152	0.0201	0.0166	0.0132	0.0100	0.0068	0.0038	0.0011	—	—
170	0.0100	0.0051	0.0032	0.0005	—	—	—	—	—
290			0.0101	0.0082	0.0041	0.0016	—	—	—
C318	—	0.0143	0.0113	0.0085	0.0048	0.0033	0.0011	—	—
502	0.0159	0.0121	0.0086	0.0054	0.0026	—	—	—	—
503	0.0094	0.0053	0.0018	—	—	—	—	—	—
600		0.0180	0.0150	0.0122	0.0094	0.0068	0.0043	0.0020	0.0001
600			0.0132	0.0101	0.0073	0.0047	0.0024	0.0005	—
718	—	—	0.0755	0.0720	0.0680	0.0636	0.0590	0.0540	0.0488
744		0.0096	0.0044	0.0005	—	—	—	—	—
1150	0.0100	0.0055	0.0013						
1270	0.0171	0.0136	0.0102	0.0070	0.0041	0.0014			

THERMODYNAMIC AND THERMOPHYSICAL PROPERTIES—TABLE 128. Velocity of Sound (m/s) in Gaseous Refrigerants at Atmospheric Pressure

R. no.					Temperature, °C				
	−50	−25	0	25	50	75	100	125	150
11	—	—	—	141	147	153	158	163	168
12	—	136	143	150	156	162	168	173	179
13	142	150	157	164	170	176	182	188	193
14	158	166	173	180	187	194	200	206	212
22	—	166	174	182	189	196	202	208	215
23	179	188	197	205	212	220	227	234	240
32									

HERMODYNAMIC AND THERMOPHYSICAL PROPERTIES—TABLE 128. Velocity of Sound (m/s) in Gaseous Refrigerants at Atmospheric Pressure (*Continued*)

R. no.	Temperature, °C								
	−50	−25	0	25	50	75	100	125	150
113	—	—	—	—	121	126	131	135	140
114	—	—	—	120	126	131	136	141	146
134		146	154	162	169	175	180	186	192
170	272	286	299	311	323	334	344	355	364
290	—	227	238	249	258	268	277	286	294
600	—	—	200	210	220	228	237	245	252
600	—	—	201	211	221	229	237	246	253
718	—	—	—	—	—	—	473	490	505
744		248	258	269	279	288	297	307	316
1150	290	305	318	330	341	352	363	373	384
1270	—	235	246	257	267	277	286	295	303

THERMODYNAMIC AND THERMOPHYSICAL PROPERTIES—TABLE 129. Velocity of Sound (m/s) in Saturated Liquid Refrigerants

R. no.	Temperature, °C								
	−50	−25	0	25	50	75	100	125	150
11	933	843	772	705	639	569	493	408	323
12	829	695	564	434				—	—
13	602	444	302		—	—	—	—	—
14	182	—	—	—	—	—	—	—	—
22	899	790	682	571	446	319	—	—	—
23		538	348	191	—	—	—	—	—
32							—	—	—
113	—	871	786	700	633				
114	853	726	623	540	453	371	284	183	—
115			454	346	255		—	—	—
134	858	743	626	517	387	262	105	—	—
290	1210	982	884	719	551	367	—	—	—
600	1290	1163	1031	896	759	609	477	325	142
600	1205	1078	947	812	661	528	378	208	—
718	—	—	1402	1495	1542	1554	1543	1514	1468
744		751	525	272	—	—	—	—	—
1150	874	644	372						
1270	1184	1022	859	694	524	335			

THERMODYNAMIC AND THERMOPHYSICAL PROPERTIES—TABLE 130. Miscellaneous Saturated Liquids

Substance	Property	-50	-40	-30	-20	-10	0	10	20	30	40	50	60	70	80	90	100
Acetaldehyde	ρ (kg/m³)	863	852	840	828	816	804	794	783								
	c_p (kJ/kg·K)	2.05	2.08	2.11	2.14	2.17	2.20	2.24	2.28								
	μ (10⁻⁶ Pa·s)	460	404	358	321	290	263	241	222								
	k (W/m·K)	0.211	0.206	0.200	0.195	0.189	0.184	0.182	0.180								
	Pr	4.47	4.08	3.78	3.52	3.33	3.14	2.97	2.81								
Acetic acid	ρ (kg/m³)								1,049	1,039	1,028	1,018	1,006	995	984	972	960
	c_p (kJ/kg·K)								2.031								
	μ (10⁻⁶ Pa·s)								1,210	1,102	1,010	795	600				
	k (W/m·K)								0.173	0.170	0.168	0.167	0.165	0.163	0.161		
	Pr								14.2								
Aniline	ρ (kg/m³)	—	—	—	—	—	1,039	1,030	1,022	1,013	1,005	996	987	978	969	960	951
	c_p (kJ/kg·K)	—	—	—	—	—	2.024	2.047	2.071	2.093	2.113	2.132	2.17	2.20	2.23	2.27	2.32
	μ (10⁻⁶ Pa·s)	—	—	—	—	—	10,200	6,500	4,400	3,160	2,370	1,850	1,510	1,270	1,090	935	825
	k (W/m·K)	—	—	—	—	—	0.186	0.184	0.182	0.180	0.177	0.174	0.171	0.169	0.168	0.167	0.167
	Pr	—	—	—	—	—	111	72	50	36.7	28.3	22.7	19.2	16.5	14.5	12.7	11.5
Butanol	ρ (kg/m³)	845	841	837	833	829	825	817	810	803	797	791	784	776	768	760	753
	c_p (kJ/kg·K)	1.947	1.996	2.046	2.100	2.153	2.202	2.262	2.345	2.437	2.524	2.621					
	μ (10⁻⁶ Pa·s)	34,700	22,400	14,700	10,300	7,400	5,190	3,870	2,950	2,300	1,780	1,410	1,140	930	760	630	535
	k (W/m·K)	0.175	0.174	0.173	0.172	0.171	0.170	0.168	0.167	0.166	0.165	0.164	0.163	0.162	0.161	0.160	0.159
	Pr	3,860	2,570	1,740	1,260	930	670	120	41	33.8	27.2	22.5					
Carbon disulfide	ρ (kg/m³)	1,362	1,348	1,334	1,320	1,306	1,292	1,278	1,263								
	c_p (kJ/kg·K)	0.988	0.989	0.990	0.991	0.993	0.996	1.004	1.017								
	μ (10⁻⁶ Pa·s)	630	580	535	496	463	435	405	375	350	330						
	k (W/m·K)	0.194	0.190	0.186	0.182	0.178	0.174	0.170	0.166	0.161	0.158	0.156	0.154	0.152	0.150		
	Pr	3.21	3.02	2.85	2.70	2.58	2.49	2.39	2.30								
Cyclohexane	ρ (kg/m³)	—	—	—	—	—	—	789	779	769	759	750	740	731	721		
	c_p (kJ/kg·K)	—	—	—	—	—	—	2.068	2.081	2.094	2.106	2.119					
	μ (10⁻⁶ Pa·s)	—	—	—	—	—	—	1,175	980	820	710	605	540				
	k (W/m·K)	—	—	—	—	—	—	0.122	0.120	0.119	0.118	0.117	0.116	0.114	0.112		
	Pr	—	—	—	—	—	—	19.9	17.0	14.4	12.7	11.0					

Properties of saturated liquids (continued). Values are listed from the warm end to the cold end, as read across the table.

Substance	Property	Values (warm → cold)
Ethanol	ρ (kg/m³)	716, 725, 735, 745, 754, 763, 776, 781, 789, 798, 806
	c_p (kJ/kg·K)	3.30, 3.19, 3.03, 2.93, 2.83, 2.73, 2.62, 2.52, 2.43, 2.35, 2.27, 2.19, 2.13, 2.08, 2.04, 2.01
	μ (10^{-6} Pa·s)	314, 370, 435, 500, 590, 700, 835, 1,000, 1,200, 1,470, 1,770, 2,220, 2,825, 3,650, 4,790, 6,400
	k (W/m·K)	0.151, 0.153, 0.156, 0.159, 0.162, 0.165, 0.168, 0.171, 0.173, 0.175, 0.177, 0.179, 0.181, 0.184, 0.186, 0.188
	Pr	6.9, 7.7, 8.4, 9.2, 10.3, 11.6, 13.0, 14.7, 16.9, 19.7, 22.7, 27.2, 33.2, 41.3, 52.5, 68.4
Ethyl acetate	ρ (kg/m³)	797, 811, 825, 838, 851, 863, 876, 888, 901, 912, 924, 935, 947, 1,090
	c_p (kJ/kg·K)	2.01
	μ (10^{-6} Pa·s)	220, 230, 250, 280, 310, 345, 370, 400, 455, 510, 580
	k (W/m·K)	0.119, 0.123, 0.127, 0.130, 0.133, 0.136, 0.139, 0.142, 0.145
	Pr	6.3
Ethylamine	ρ (kg/m³)	607, 620, 633, 646, 658, 671, 683, 695, 707, 718, 729, 739, 750, 761
	c_p (kJ/kg·K)	3.03, 3.01, 3.00, 2.98, 2.97, 2.95
	μ (10^{-6} Pa·s)	320, 350, 390, 435, 500, 580
	k (W/m·K)	0.191, 0.194, 0.196, 0.199, 0.201, 0.204
	Pr	5.08, 5.43, 5.97, 6.51, 7.39, 8.39
Ethyl ether	ρ (kg/m³)	640, 653, 666, 676, 689, 702, 714, 725, 736, 747, 758, 769, 780, 790
	c_p (kJ/kg·K)	2.51, 2.47, 2.43, 2.39, 2.36, 2.332, 2.299, 2.265, 2.233, 2.205, 2.179, 2.156, 2.135
	μ (10^{-6} Pa·s)	153, 166, 181, 197, 214, 233, 265, 290, 330, 365, 410, 470, 550
	k (W/m·K)	0.112, 0.116, 0.120, 0.125, 0.129, 0.134, 0.139, 0.140, 0.144, 0.147, 0.151, 0.155, 0.159
	Pr	3.43, 3.54, 3.67, 3.77, 3.92, 4.05, 4.38, 4.69, 5.12, 5.48, 5.92, 6.54, 7.39
Ethyl iodide	c_p (kJ/kg·K)	0.724, 0.718, 0.712, 0.705, 0.698, 0.691, 0.684, 0.677, 0.670, 0.663, 0.656
	μ (10^{-6} Pa·s)	390, 420, 455, 495, 539, 590, 655, 730
	k (W/m·K)	0.080, 0.081, 0.083, 0.085, 0.086, 0.088, 0.090, 0.092
	Pr	3.53, 3.72, 3.90, 4.11, 4.30, 4.63, 4.98, 5.37
Ethylene glycol	ρ (kg/m³)	1,056, 1,063, 1,070, 1,077, 1,085, 1,092, 1,099, 1,106, 1,113, 1,120, 1,127
	c_p (kJ/kg·K)	2.779, 2.734, 2.685, 2.636, 2.586, 2.536, 2.484, 2.431, 2.381, 2.327, 2.272
	μ (10^{-6} Pa·s)	2,000, 2,440, 3,000, 3,450, 4,000, 7,070, 9,400, 13,100, 20,200, 33,300, 57,000
	k (W/m·K)	0.260, 0.259, 0.258, 0.256, 0.255, 0.254
	Pr	69.0, 87.3, 126, 190, 305, 510
Formic acid	ρ (kg/m³)	1,108, 1,124, 1,140, 1,156, 1,170, 1,184, 1,196, 1,209, 1,220, 1,231, 1,241
	μ (10^{-6} Pa·s)	550, 615, 680, 780, 890, 1,030, 1,220, 1,470, 1,800, 2,260
	k (W/m·K)	0.232, 0.236, 0.240, 0.243, 0.246, 0.250, 0.253, 0.257, 0.257, 0.261, 0.265

THERMODYNAMIC AND THERMOPHYSICAL PROPERTIES—TABLE 130. Miscellaneous Saturated Liquids (Continued)

Substance	Property	Temperature, °C															
		-50	-40	-30	-20	-10	0	10	20	30	40	50	60	70	80	90	100
Gasoline	ρ (kg/m³)				784	775	767	759	751	743	735	721	717	708	699	690	681
	c_p (kJ/kg·K)				1.88	1.92	1.97	2.02	2.06	2.11	2.15	2.20	2.25	2.30	2.35	2.41	2.46
	μ (10^{-6} Pa·s)	1,710	1,400	1,170	990	850	735	645	530	464	410	367	330	298	270	246	225
	k (W/m·K)	0.131	0.128	0.125	0.123	0.121	0.120	0.118	0.116	0.114	0.112	0.110	0.108	0.106	0.104	0.102	0.100
	Pr				15.1	13.5	12.1	11.0	9.41	8.59	7.87	7.34	6.88	6.47	6.10	5.81	5.54
Glycerol	ρ (kg/m³)	—	—	—	—	—	1,276	1,270	1,260	1,254	1,248	1,242					
	c_p (kJ/kg·K)								2.393	2.406	2.457	2.504	2.548	2.588	2.625	2.657	2.686
	μ (10^{-6} Pa·s)						1.2+7	4.0+6	1.5+6								
	k (W/m·K)								0.284	0.285	0.287	0.288	0.289	0.291	0.293	0.294	0.295
	Pr								12,650								
Kerosene	ρ (kg/m³)	1,150					781	774	767	760	754	748	742				
	c_p (kJ/kg·K)						1.91	1.96	2.02	2.07	2.13	2.18	2.23	2.28	2.32	2.35	2.38
	μ (10^{-6} Pa·s)		725	500	360	275	215	173	149	126	108	95	83	73	66	60	55
	k (W/m·K)						0.140	0.139	0.139	0.138	0.138	0.137	0.137				
	Pr						2.93	2.44	2.17	1.89	1.67	1.51	1.35				
Methanol	ρ (kg/m³)									783	774	766	756	746	736	725	711
	c_p (kJ/kg·K)	2.30	2.32	2.35	2.37	2.40	2.42	2.45	2.47	2.49	2.52	2.55	2.65	2.78	2.94	3.13	3.30
	μ (10^{-6} Pa·s)	2,305	1,800	1,410	1,170	975	820	692	590	510	455	400	355	315	271	240	218
	k (W/m·K)	0.225	0.222	0.219	0.216	0.212	0.209	0.206	0.203	0.199	0.195	0.192	0.189	0.187	0.184	0.182	0.180
	Pr	23.6	18.8	15.1	12.9	11.0	9.53	8.23	7.18	6.38	5.88	5.31	4.98	4.68	4.34	4.13	3.99
Methyl formate	ρ (kg/m³)	1,069	1,056	1,043	1,030	1,017	1,003	989	975	960	944	929	913	897	880	863	845
	c_p (kJ/kg·K)	1.84	1.86	1.88	1.90	1.92	1.95	1.99	2.03	2.08							
	μ (10^{-6} Pa·s)	830	711	618	544	481	430	380	345	315							
	k (W/m·K)	0.217	0.213	0.209	0.205	0.200	0.195	0.191	0.186	0.180							
	Pr	7.04	6.21	5.56	5.04	4.62	4.30	3.96	3.77	3.64							

Substance	Property	Values (in order of increasing temperature)
Oil, castor	ρ (kg/m^3)	
	c_p (kJ/kg·K)	
	μ (10^{-6} Pa·s)	43,000 74,000 125,000 231,000 451,000 986,000 2,420,000
	k (W/m·K)	0.17 0.174 0.175 0.176 0.177 0.178 0.179 0.180 0.181 0.182
	Pr	
Oil, olive	ρ (kg/m^3)	914
	c_p (kJ/kg·K)	1.633
	μ (10^{-6} Pa·s)	12,400 17,000 24,500 36,300 52,000 84,000 138,000
	k (W/m·K)	0.164 0.164 0.165 0.165 0.166 0.167 0.168 0.169 0.170
	Pr	810
Pentane	ρ (kg/m^3)	538 550 562 574 585 596 606 616 626 636 646 656 665 674 684 693
	c_p (kJ/kg·K)	2.273 2.239 2.206 2.167 2.137 2.110 2.084 2.060
	μ (10^{-6} Pa·s)	113 124 137 148 161 175 190 209 234 254 279 307 339 379 428 489
	k (W/m·K)	0.091 0.095 0.098 0.101 0.105 0.108 0.112 0.115 0.119 0.122 0.125 0.128 0.132 0.136 0.139 0.142
	Pr	4.47 4.66 4.92 5.20 5.49 5.88 6.42 7.14
Propanol	ρ (kg/m^3)	743 747 752 760 770 779 788 796 811 814 819 849
	c_p (kJ/kg·K)	2.219 2.239 2.245 1.955
	μ (10^{-6} Pa·s)	447 508 630 760 921 1,130 1,400 1,720 2,900 3,900 5,110 6,900 9,500 13,500 20,200
	k (W/m·K)	0.162 0.163 0.164 0.165 0.166 0.167 0.168 0.169 0.171
	Pr	236
Sulfuric acid	ρ (kg/m^3)	1,834
	c_p (kJ/kg·K)	1.382
	μ (10^{-6} Pa·s)	5,190 6,090 7,220 8,820 11,500 15,700 25,400 35,200 48,400
	k (W/m·K)	0.314
	Pr	
Toluene	ρ (kg/m^3)	790 800 810 820 829 839 848 858 867 876 886 895 904 913 923 932
	c_p (kJ/kg·K)	1.97 1.92 1.87 1.83 1.80 1.76 1.73 1.701 1.675 1.652 1.633 1.602 1.579 1.556 1.535 1.514
	μ (10^{-5} Pa·s)	270 295 325 355 380 420 470 520 590 670 770 915 1,100 1,345 1,670 2,120
	k (W/m·K)	0.114 0.117 0.119 0.122 0.124 0.126 0.129 0.132 0.134 0.137 0.139 0.142 0.144 0.147 0.149 0.152
	Pr	4.7 4.8 5.1 5.3 5.5 5.9 6.3 6.7 7.4 8.1 9.0 10.3 12.1 14.2 17.8 21.1
Turpentine	ρ (kg/m^3)	675 730 820 925 1,070 1,270 1,490 1,780 2,250
	c_p (kJ/kg·K)	1.93 1.80 1.76 1.72
	μ (10^{-6} Pa·s)	
	k (W/m·K)	0.125 0.126 0.127 0.128 0.129 0.130
	Pr	14.3 16.1 18.4 20.9 24.3 29.8

THERMODYNAMIC AND THERMOPHYSICAL PROPERTIES—TABLE 131.   Selected Nonmetallic Solid Substances

Material	Density, kg/m^3	Emissivity	Specific heat, kJ/(kg·K)	Thermal conductivity, W/(m·K)	Thermal diffusivity, m^2/s × 10^6
Alumina	3975		0.765	36	11.9
Asphalt	2110		0.920	0.06	0.03
Bakelite	1300		1.465	1.4	0.74
Beryllia	3000	0.82	1.030	270	88
Brick	1925	0.93	0.835	0.72	0.45
Brick, fireclay	2640	0.93	0.960	1.0	0.39
Carbon, amorphous	1950	0.86	0.724	1.6	1.13
Clay	1460	0.91	0.880	1.3	1.01
Coal	1350	0.80	1.26	0.26	0.15
Cotton	80		1.30	0.06	0.58
Diamond	3500		0.509	2300	1290
Granite	2630		0.775	2.79	1.37
Hardboard	1000		1.38	0.15	0.11
Magnesite	3025	0.38	1.13	4.0	1.2
Magnesia	3635	0.72	0.943	48	14
Oak	770	0.90	2.38	0.18	0.10
Paper	930	0.83	1.34	0.011	0.01
Pine	525	0.84	2.75	0.12	0.54
Plaster board	800	0.91		0.17	
Plywood	540		1.22	0.12	0.18
Pyrex	2250	0.92	0.835	1.4	0.74
Rubber	1150	0.92	2.00	0.2	0.09
Rubber, foam	70	0.90		0.03	
Salt		0.34	0.854	7.1	
Sandstone	2150	0.59	0.745	2.9	1.8
Silica		0.79	0.743	1.3	
Sapphire	3975	0.48	0.765	46	15
Silicon carbide	3160	0.86	0.675	490	230
Soil	2050	0.38	1.84	0.52	0.14
Teflon	2200	0.92	0.35	0.26	0.34
Thoria	4160	0.28	0.71	14	4.7
Urethane foam	70		1.05	0.03	0.36
Vermiculite	120		0.84	0.06	0.60

ULTRAVIOLET SPECTROSCOPY

Ultraviolet spectroscopy is the study of the ultraviolet radiation that is characteristically absorbed by molecular substances.

ULTRAVIOLET SPECTROSCOPY—TABLE 1. Ultraviolet Cutoff Wavelengths for Various Solvents

Solvent	UV cutoff (nm)
Acetone	330
Acetonitrile	190
n-Butyl acetate	254
n-Butyl alcohol	215
iso-Butyl alcohol	220
n-Butyl chloride	220
Chlorobenzene	287
Chloroform	245
Cyclohexane	200
Cyclopentane	198
o-Dichlorobenzene	295
Dichloromethane	233
Diethyl ether	215
Dimethyl acetamide	268
N,N-Dimethylformamide	268
Dimethyl sulfoxide	268
1,4-Dioxane	215
Ethyl acetate	256
Ethyl alcohol (ethanol)	210
Ethylene dichloride	228
Ethylene glycol monomethyl ether (glyme)	220
n-Heptane	200
n-Hexane	195
2-Methoxyethanol	210
Methyl alcohol (methanol)	205
Methyl iso-amyl ketone	330
Methyl t-butyl ether	210
Methyl iso-butyl ketone	334
Methyl ethyl ketone	329
Methyl n-propyl ketone	331
N-Methylpyrrolidone	285
iso-Octane	215
Pentane	190
n-Propyl alcohol	210
iso-Propyl alcohol	205
Propylene carbonate	220
Tetrahydrofuran	212
Toluene	284
1,2,4-Trichlorobenzene	308
1,1,2-Trichlorotrifluoroethane	231
Trifluoroacetic acid	210
o-Xylene	288

VAPOR PRESSURE

Vapor pressure is the pressure exerted by a pure component at equilibrium at any temperature when both liquid and vapor phases exist and thus extends from a minimum at the triple-point temperature to a maximum at the critical temperature. The critical pressure is the most important of the basic thermodynamic properties affecting liquids and vapors.

Except at very high total pressures (above about 10 MPa), total pressure has no effect on vapor pressure. If such an effect is present, a correction can be applied. The pressure exerted above a solid–vapor mixture may also be called vapor pressure but is normally available only as experimental data for common compounds that sublime.

Correlation Methods

Vapor pressure is correlated as a function of temperature by numerous methods that are mainly derived from the Clapeyron equation. The classic simple equation used for correlation of low to moderate vapor pressures is:

$$\ln P^{\text{sut}} = A + \frac{B}{T+C} \tag{141}$$

This equation does not work accurately with data much above the normal boiling point. Thus, as regression by computer is now standard, more accurate expressions applicable to the critical point have become usable. Thus:

$$\ln P^{\text{sut}} = A + \frac{B}{T} + C\ln T + DT^E \tag{142}$$

A, B, C, and D are regression constants and E is an exponent equal to 1, 2, or 6, depending on which regression gives the most accurate fit of the data.

Another equation (Equation 143) fits the hydrocarbon data over the entire pressure range:

$$\ln P^{\text{sut}} = A + \frac{B}{T} + C\ln T + DT^2 + \frac{E}{T^2} \tag{143}$$

Both equations (i.e., Equations 142 and 143) can be extrapolated above the critical temperature where necessary for thermodynamic calculations.

Yet another equation (Equation 144) has the advantage that it will match critical data exactly, although it cannot be extrapolated above the critical point.

$$\ln P_r = aX_1 + bX_2 + cX_3 + dX_4 \tag{144}$$

where $\quad X_1 = \dfrac{1-T_r}{T_r}, X_2 = \dfrac{(1-T_r)^{1.5}}{T_r}, X_3 = \dfrac{(1-T_r)^{2.6}}{T_r}, X_4 = \dfrac{(1-T_r)^5}{T_r}$

$\qquad P_r = P^{\text{sat}}/P_c \qquad T_r = T^{\text{sat}}/T_c$

Prediction Methods

Two methods have gained almost universal acceptance for prediction of the vapor pressure of pure hydrocarbons. One is the preferred method if the critical temperature and the critical pressure of the hydrocarbon are known or can be reasonably predicted by the methods just provided. The corresponding states method (Equation 145) with the simple fluid and correction terms can be calculated (from Equations 146 and 147, respectively) for any T_r.

$$\ln P_r^{sat} = \left(\ln P_r^{sat}\right)^{(0)} + \omega \left(\ln P_r^{sat}\right)^{(1)} \tag{145}$$

$$\left(\ln P_r^{sat}\right)^{(0)} = 5.92714 - 6.09648/T_r - 1.28862 \ln T_r + 0.169347 T_r^6 \tag{146}$$

$$\left(\ln P_r^{sat}\right)^{(1)} = 15.2518 - 15.6875/T_r - 13.4721 \ln T_r + 0.43577 T_r^6 \tag{147}$$

This method is applicable at reduced temperatures above 0.30 or the freezing point (whichever is higher) and below the critical point. The method is most reliable when $0.5 < T_r < 0.95$, where errors in prediction average 3.5% when experimental critical properties are known. Errors are higher for predicted critical properties.

VAPOR PRESSURE—TABLE 1. Vapor Pressures of Various Elements at Different Temperatures

Element	Atomic number	Atomic symbol	Boiling point, °C	Vapor Pressure Temperature, °C								
				E-08	E-07	E-06	E-05	E-04	E-03	E-02	E-01	1
Aluminum	13	Al	2467	685	742	812	887	972	1082	1217	1367	1557
Antimony	52	Sb	1750	279	309	345	383	425	475	533	612	757
Arsenic	33	As	613	104	127	150	174	204	237	277	317	372
Barium	56	Ba	1140	272	310	354	402	462	527	610	711	852
Beryllium	4	Be	2970	707	762	832	907	997	1097	1227	1377	1557
Bismuth	83	Bi	1560	347	367	409	459	517	587	672	777	897
Boron	5	B	2550	1282	1367	1467	1582	1707	1867	2027	2247	2507
Cadmium	48	Cd	765	74	95	119	146	177	217	265	320	392
Calcium	20	Ca	1484	282	317	357	405	459	522	597	689	802
Carbon	6	C	4827	1657	1757	1867	1987	2137	2287	2457	2657	2897
Cobalt	27	Co	2870	922	992	1067	1157	1257	1382	1517	1687	1907
Chromium	24	Cr	2672	837	902	977	1062	1157	1267	1397	1552	1737
Copper	29	Cu	2567	722	787	852	937	1027	1132	1257	1417	1617
Dysprosium	66	Dy	2562	625	682	747	817	897	997	1117	1262	1437
Erbium	68	Er	2510	649	708	777	852	947	1052	1177	1332	1527
Europium	63	Eu	1597	283	319	361	409	466	532	611	708	827
Gallium	31	Ga	2403	619	677	742	817	907	1007	1132	1282	1472
Germanium	32	Ge	2830	812	877	947	1037	1137	1257	1397	1557	1777
Gold	79	Au	2807	807	877	947	1032	1132	1252	1397	1567	1767
Indium	77	In	2000	488	539	597	664	742	837	947	1082	1247
Iron	26	Fe	2750	892	957	1032	1127	1227	1342	1477	1647	1857
Lanthanum	57	La	3469	1022	1102	1192	1297	1422	1562	1727	1927	2177
Lead	82	Pb	1740	342	383	429	485	547	625	715	832	977
Lithium	49	Li	1347	235	268	306	350	404	467	537	627	747
Magnesium	12	Mg	1107	185	214	246	282	327	377	439	509	605

Element												
Manganese	25	Mn	1962	505	554	611	675	747	837	937	1082	1217
Mercury	80	Hg	357	−72	−59	−44	−27	7	16	46	80	125
Molybdenum	42	Mo	4612	1592	1702	1822	1957	2117	2307	2527	2787	3117
Nickel	28	Ni	2732	927	997	1072	1157	1262	1382	1527	1697	1907
Niobium	41	Nb	4927	1762	1867	1987	2127	2277	2447	2657	2897	3177
Palladium	46	Pd	2927	842	912	992	1082	1192	1317	1462	1647	1877
Phosphorus	15	P	2804	54	69	88	108	129	157	185	222	261
Platinum	78	Pt	3827	1292	1382	1492	1612	1747	1907	2097	2317	2587
Potassium	19	K	774	21	42	65	91	123	161	208	267	345
Praseodymium	59	Pr	3127	797	867	947	1042	1147	1277	1427	1617	1847
Rhenium	75	Re	5627	1947	2077	2217	2387	2587	2807	3067	3407	3807
Rhodium	45	Rh	3727	1277	767	1472	1582	1707	1857	2037	2247	2507
Scandium	21	Sc	2832	772	837	917	1007	1107	1232	1377	1567	1797
Selenium	34	Se	685	63	83	107	133	164	199	243	297	363
Silicon	14	Si	4827	992	1067	1147	1237	1337	1472	1632	1817	2057
Silver	47	Ag	2212	574	626	685	752	832	922	1027	1162	1322
Sodium	11	Na	553	74	97	123	155	193	235	289	357	441
Strontium	38	Sr	1384	241	273	309	353	394	465	537	627	732
Sulfur	16	S	45	−10	3	17	37	55	80	109	147	189
Tantalum	73	Ta	5425	1957	2097	2237	2407	2587	2807	3057	3357	3707
Tellurium	52	Te	990	155	181	209	242	280	323	374	433	518
Thallium	81	Tl	1457	283	319	359	407	463	530	609	706	827
Tin	50	Sn	2270	682	747	807	897	997	1107	1247	1412	1612
Titanium	22	Ti	3287	1062	1137	1227	1327	1442	1577	1737	1937	2177
Tungsten	74	W	5660	2117	2247	2407	2567	2757	2977	3227	3537	3917
Ytterbium	70	Yb	1466	247	279	317	365	417	482	557	647	787
Yttrium	39	Y	3337	957	1032	1117	1217	1332	1467	1632	1832	2082
Zinc	30	Zn	907	123	147	177	209	247	292	344	408	487

VAPOR PRESSURE—TABLE 2. Vapor Pressures of Inorganic Compounds, up to 1 atm

Compound Name	Formula	Pressure, mm Hg Temperature, °C										Melting point, °C
		1	5	10	20	40	60	100	200	400	760	
Aluminum	Al	1284				1635	1684	1749	1844	1947	2056	660
borohydride	Al(BH₄)₃		-52.2	-42.9	-32.5	-20.9	-13.4	-3.9	+11.2	28.1	45.9	-64
bromide	AlBr₃	81.3	103.8	118.0	134.0	150.6	161.7	176.1	199.8	227.0	256.3	97
chloride	Al₂Cl₆	100.0	116.4	123.8	131.8	139.9	145.4	152.0	161.8	171.6	180.2	192.4
fluoride	AlF₃	1238	1298	1324	1350	1378	1398	1422	1457	1496	1537	1040
iodide	AlI₃	178.0	207.7	225.8	244.2	265.0	277.8	294.5	322.0	354.0	385.5	
oxide	Al₂O₃	2148	2306	2385	2465	2549	2599	2665	2766	2874	2977	2050
Ammonia	NH₃	-109.1	-97.5	-91.9	-85.8	-79.2	-74.3	-68.4	-57.0	-45.4	-33.6	-77.7
heavy	ND₃						-74.0	-67.4	-57.0	-45.4	-33.4	-74.0
Ammonium bromide	NH₄Br	198.3	234.5	252.0	270.6	290.0	303.8	320.0	345.3	370.8	396.0	
carbamate	N₂H₆CO₂	-26.1	-10.4	-2.9	+5.3	14.0	19.6	26.7	37.2	48.0	58.3	520
chloride	NH₄Cl	160.4	193.8	209.8	226.1	245.0	256.2	271.5	293.2	316.5	337.8	
cyanide	NH₄CN	-50.6	-35.7	-28.6	-20.9	-12.6	-7.4	-0.5	+9.6	20.5	31.7	36
hydrogen sulfide	NH₄HS	-51.1	-36.0	-28.7	-20.8	-12.3	-7.0	0.0	+10.5	21.8	33.3	
iodide	NH₄I	210.9	247.0	263.5	282.8	302.8	316.0	331.8	355.8	381.0	404.9	
Antimony	Sb	886	984	1033	1084	1141	1176	1223	1288	1364	1440	630.5
tribromide	SbBr₃	93.9	126.0	142.7	158.3	177.4	188.1	203.5	225.7	250.2	275.0	96.6
trichloride	SbCl₃	49.2	71.4	85.2	100.6	117.8	128.3	143.3	165.9	192.2	219.0	73.4
pentachloride	SbCl₅	22.7	48.6	61.8	75.8	91.0	101.0	114.1				2.8
triiodide	SbI₃	163.6	203.8	223.5	244.8	267.8	282.5	303.5	333.8	368.5	401.0	167
trioxide	Sb₄O₆	574	626	666	729	812	873	957	1085	1242	1425	656
Argon	A	-218.2	-213.9	-210.9	-207.9	-204.9	-202.9	-200.5	-195.6	-190.6	-185.6	-189.2
Arsenic	As	372	416	437	459	483	498	518	548	579	610	814
Arsenic tribromide	AsBr₃	41.8	70.6	85.2	101.3	118.7	130.0	145.2	167.7	193.6	220.0	
trichloride	AsCl₃	-11.4	+11.7	+23.5	36.0	50.0	58.7	70.9	89.2	109.7	130.4	-18
trifluoride	AsF₃					-2.5	+4.2	13.2	26.7	41.4	56.3	-5.9
pentafluoride	AsF₅	-117.9	-108.0	-103.1	-98.0	-92.4	-88.5	-84.3	-75.5	-64.0	-52.8	-79.8
trioxide	As₂O₃	212.5	242.6	259.7	279.2	299.2	310.3	332.5	370.0	412.2	457.2	312.8

Name	Formula											
Arsine	AsH₃	-142.6	-130.8	-124.7	-117.7	-110.2	-104.8	-98.0	-87.2	-75.2	-62.1	-116.3
Barium	Ba		984	1049	1120	1195	1240	1301	1403	1518	1638	850
Beryllium borohydride	Be(BH₄)₂	+1.0	19.8	28.1	36.8	46.2	51.7	58.6	69.0	79.7	90.0	123
bromide	BeBr₂	289	325	342	361	379	390	405	427	451	474	490
chloride	BeCl₂	291	328	346	365	384	395	411	435	461	487	405
iodide	BeI₂	283	322	341	361	382	394	411	435	461	487	488
Bismuth	Bi	1021	1099	1136	1177	1217	1240	1271	1319	1370	1420	271
tribromide	BiBr₃	261	282	305	316	327	340	360	392	425	461	218
trichloride	BiCl₃	242	264	287	299	311	324	343	372	405	441	230
Diborane hydrobromide	B₂H₅Br	-93.3	-75.3	-66.3	-56.4	-45.4	-38.2	-29.0	-15.4	0.0	+16.3	-104.2
Borine carbonyl	BH₃CO	-139.2	-127.3	-121.1	-114.1	-106.6	-101.9	-95.3	-85.5	-74.8	-64.0	-137.0
triamine	B₃N₃H₆	-63.0	-45.0	-35.3	-25.0	-13.2	-5.8	+4.0	18.5	34.3	50.6	-58.2
Boron hydrides												
dihydrodecaborane	B₁₀H₁₄	60.0	80.8	90.2	100.0	117.4	127.8	142.3	163.8			99.6
dihydrodiborane	B₂H₆	-159.7	-149.5	-144.3	-138.5	-131.6	-127.2	-120.9	-111.2	-99.8	-86.5	-169
dihydropentaborane	B₅H₉	-50.2	-40.4	-30.7	-20.0	-8.0	-0.4	+9.6	24.6	40.8	58.1	-47.0
tetrahydropentaborane	B₅H₁₁	-90.9	-29.9	-19.9	-9.2	+2.7	10.2	20.1	34.8	51.2	67.0	
tetrahydrotetraborane	B₄H₁₀	-41.4	-20.4	-10.1	+1.5	14.0	22.1	33.5	50.3	70.0	91.7	-119.9
Boron tribromide	BBr₃	-91.5	-75.2	-66.9	-57.9	-47.8	-41.2	-32.4	-18.9	-3.6	+12.7	-45
trichloride	BCl₃	-154.6	-145.4	-141.3	-136.4	-131.0	-127.6	-123.0	-115.9	-108.3	-100.7	-107
trifluoride	BF₃	-48.7	-32.8	-25.0	-16.8	-8.0	-0.6	+9.3	24.3	41.0	58.2	-126.8
Bromine	Br₂											-7.3
pentafluoride	BrF₅	-69.3	-51.0	-41.9	-32.0	-21.0	-14.0	-4.5	+9.9	25.7	40.4	-61.4
Cadmium	Cd	394	455	484	516	553	578	611	658	711	765	320.9
chloride	CdCl₂		618	656	695	736	762	797	847	908	967	568
fluoride	CdF₂	1112	1231	1286	1344	1400	1436	1486	1561	1651	1751	520
iodide	CdI₂	416	481	512	546	584	608	640	688	742	796	385
oxide	CdO	1000	1100	1149	1200	1257	1295	1341	1409	1484	1559	
Calcium	Ca		926	983	1046	1111	1152	1207	1288			851
Carbon (graphite)	C	3586	3828	3946	4069	4196	4273	4373	4516	4660	4827	
dioxide	CO₂	-134.3	-124.4	-119.5	-114.4	-108.6	-104.8	-100.2	-93.0	-85.7	-78.2	-57.5
disulfide	CS₂	-73.8	-54.3	-44.7	-34.3	-22.5	-15.3	-5.1	+10.4	28.0	46.5	-110.8
monoxide	CO	-222.0	-217.2	-215.0	-212.8	-210.0	-208.1	-205.7	-201.3	-196.3	-191.3	-205.0
oxyselenide	COSe	-117.1	-102.3	-95.0	-86.3	-76.4	-70.2	-61.7	-49.8	-35.6	-21.9	
oxysulfide	COS	-132.4	-119.8	-113.3	-106.0	-98.3	-93.0	-85.9	-75.0	-62.7	-49.9	-138.8

VAPOR PRESSURE—TABLE 2. Vapor Pressures of Inorganic Compounds, up to 1 atm (*Continued*)

Compound Name	Formula	Pressure, mm Hg — Temperature, °C										Melting point, °C
		1	5	10	20	40	60	100	200	400	760	
selenosulfide	$CSeS$	-47.3	-26.5	-16.0	-4.4	+8.6	17.0	28.3	45.7	65.2	85.6	-75.2
subsulfide	C_3S_2	14.0	41.2	54.9	69.3	85.6	96.0	109.9	130.8			+0.4
tetrabromide	CBr_4					96.3	106.3	119.7	139.7	163.5	189.5	90.1
tetrachloride	CCl_4	-50.0	-30.0	-19.6	-8.2	+4.3	12.3	23.0	38.3	57.8	76.7	-22.6
tetrafluoride	CF_4	-184.6	-174.1	-169.3	-164.3	-158.8	-155.4	-150.7	-143.6	-135.5	-127.7	-183.7
Cesium	Cs	279	341	375	409	449	474	509	561	624	690	28.5
bromide	$CsBr$	748	838	887	938	993	1026	1072	1140	1221	1300	636
chloride	$CsCl$	744	837	884	934	989	1023	1069	1139	1217	1300	646
fluoride	CsF	712	798	844	893	947	980	1025	1092	1170	1251	683
iodide	CsI	738	828	873	923	976	1009	1055	1124	1200	1280	621
Chlorine	Cl_2	-118.0	-106.7	-101.6	-93.3	-84.5	-79.0	-71.7	-60.2	-47.3	-33.8	-100.7
fluoride	ClF		-143.4	-139.0	-134.3	-128.8	-125.3	-120.8	-114.4	-107.0	-100.5	-145
trifluoride	ClF_3		-80.4	-71.8	-62.3	-51.3	-44.1	-34.7	-20.7	-4.9	+11.5	-83
monoxide	Cl_2O	-98.5	-81.6	-73.1	-64.3	-54.3	-48.0	-39.4	-26.5	-12.5	+2.2	-116
dioxide	ClO_2			-59.0	-51.2	-42.8	-37.2	-29.4	-17.8	-4.0	+11.1	-59
heptoxide	Cl_2O_7	-45.3	-23.8	-13.2	-2.1	+10.3	+18.2	29.1	44.6	62.2	78.8	-91
Chlorosulfonic acid	HSO_3Cl	32.0	53.5	64.0	75.3	87.6	95.2	105.3	120.0	136.1	151.0	-80
Chromium	Cr	1616	1768	1845	1928	2013	2067	2139	2243	2361	2482	1615
carbonyl	$Cr(CO)_6$	36.0	58.0	68.3	79.5	91.2	98.3	108.0	121.8	137.2	151.0	
oxychloride	CrO_2Cl_2	-18.4	+3.2	13.8	25.7	38.5	46.7	58.0	75.2	95.2	117.1	
Cobalt chloride	$CoCl_2$					770	801	843	904	974	1050	735
nitrosyl tricarbonyl	$Co(CO)_3NO$				-1.3	+11.0	18.5	29.0	44.4	62.0	80.0	-11
Columbium fluoride	CbF_3			86.3	103.0	121.5	133.2	148.5	172.2	198.0	225.0	75.5
Copper	Cu	1628	1795	1879	1970	2067	2127	2207	2325	2465	2595	1083
Cuprous bromide	Cu_2Br_2	572	666	718	777	844	887	951	1052	1189	1355	504
chloride	Cu_2Cl_2	546	645	702	766	838	886	960	1077	1249	1490	422
iodide	Cu_2I_2		610	656	716	786	836	907	1018	1158	1336	605

Compound	Formula											m.p.
Cyanogen	C_2N_2	-95.8	-83.2	-76.8	-70.1	-62.7	-57.9	-51.8	-42.6	-33.0	-21.0	-34.4
bromide	CNBr	-35.7	-18.3	-10.0	-1.0	+8.6	14.7	22.6	33.8	46.0	61.5	58
chloride	CNCl	-76.7	-61.4	-53.8	-46.1	-37.5	-32.1	-24.9	-14.1	-2.3	+13.1	-6.5
fluoride	CNF	-134.4	-123.8	-118.5	-112.8	-106.4	-102.3	-97.0	-89.2	-80.5	-72.6	-12
Deuterium cyanide	DCN	-68.9	-54.0	-46.7	-38.8	-30.1	-24.7	-17.5	-5.4	+10.0	26.2	
Fluorine	F_2	-223.0	-216.9	-214.1	-211.0	-207.7	-205.6	-202.7	-198.3	-193.2	-187.9	-223
oxide	F_2O	-196.1	-186.6	-182.3	-177.8	-173.0	-170.0	-165.8	-159.0	-151.9	-144.6	-223.9
Germanium bromide	$GeBr_4$		43.3	56.8	71.8	88.1	98.8	113.2	135.4	161.6	189.0	26.1
chloride	$GeCl_4$	-45.0	-24.9	-15.0	-4.1	+8.0	16.2	27.5	44.4	63.8	84.0	-49.5
hydride	GeH_4	-163.0	-151.0	-145.3	-139.2	-131.6	-126.7	-120.3	-111.2	-100.2	-88.9	-165
Trichlorogermane	$GeHCl_3$	-41.3	-22.3	-13.0	-3.0	+8.8	16.2	26.5	41.6	58.3	75.0	-71.1
Tetramethylgermane	$Ge(CH_3)_4$	-73.2	-54.6	-45.2	-35.0	-23.4	-16.2	-6.3	+8.8	26.0	44.0	-88
Digermane	Ge_2H_6	-88.7	-69.8	-60.1	-49.9	-38.2	-30.7	-20.3	-4.7	+3.3	31.5	-109
Trigermane	Ge_3H_6	-36.9	-12.8	-0.9	+11.8	26.3	35.5	47.9	67.0	88.6	110.8	-105.6
Gold	Au	1869	2059	2154	2256	2363	2431	2521	2657	2807	2966	1063
Helium	He	-271.7	-271.5	-271.3	-271.1	-270.7	-270.6	-270.3	-269.8	-269.3	-268.6	
para-Hydrogen	H_2	-263.3	-261.9	-261.3	-260.4	-259.6	-258.9	-257.9	-256.3	-254.5	-252.5	-259.1
Hydrogen bromide	HBr	-138.8	-127.4	-121.8	-115.4	-108.3	-103.8	-97.7	-88.1	-78.0	-66.5	-87.0
chloride	HCl	-150.8	-140.7	-135.6	-130.0	-123.8	-119.6	-114.0	-105.2	-95.3	-84.8	-114.3
cyanide	HCN	-71.0	-55.3	-47.7	-39.7	-30.9	-25.1	-17.8	-5.3	+10.2	25.9	-13.2
fluoride	H_2F_2		-74.7	-65.8	-56.0	-45.0	-37.9	-28.2	-13.2	+2.5	19.7	-83.7
iodide	HI	-123.3	-109.6	-102.3	-94.5	-85.6	-79.8	-72.1	-60.3	-48.3	-35.1	-50.9
oxide (water)	H_2O	-17.3	+1.2	11.2	22.1	34.0	41.5	51.6	66.5	83.0	100.0	0.0
sulfide	H_2S	-134.3	-122.4	-116.3	-109.7	-102.3	-97.9	-91.6	-82.3	-71.8	-60.4	-85.5
disulfide	HSSH	-43.2	-24.4	-15.2	-5.1	+6.0	12.8	22.0	35.3	49.6	64.0	-89.7
selenide	H_2Se	-115.3	-103.4	-97.9	-91.8	-84.7	-80.2	-74.2	-65.2	-53.6	-41.1	-64
telluride	H_2Te	-96.4	-82.4	-75.4	-67.8	-59.1	-53.7	-45.7	-32.4	-17.2	-2.0	-49.0
Iodine	I_2	38.7	62.2	73.2	84.7	97.5	105.4	116.5	137.3	159.8	183.0	112.9
heptafluoride	IF	-87.0	-70.7	-63.0	-54.5	-45.3	-39.4	-31.9	-20.7	-8.3	+4.0	5.5
Iron	Fe	1787	1957	2039	2128	2224	2283	2360	2475	2605	2735	1535
pentacarbonyl	$Fe(CO)_5$		-6.5	+4.6	16.7	30.3	39.1	50.3	68.0	86.1	105.0	-21
Ferric chloride	Fe_2Cl_6	194.0	221.8	235.5	246.0	256.8	263.7	272.5	285.0	298.0	319.0	304
Ferrous chloride	$FeCl_2$			700	737	779	805	842	897	961	1026	
Krypton	Kr	-199.3	-191.3	-187.2	-182.9	-178.4	-175.7	-171.8	-165.9	-159.0	-152.0	-156.7

VAPOR PRESSURE—TABLE 2. Vapor Pressures of Inorganic Compounds, up to 1 atm (*Continued*)

Compound Name	Formula	Pressure, mm Hg										Melting point, °C
		Temperature, °C										
		1	5	10	20	40	60	100	200	400	760	
Lead	Pb	973	1099	1162	1234	1309	1358	1421	1519	1630	1744	327.5
bromide	PbBr$_2$	513	578	610	646	686	711	745	796	856	914	373
chloride	PbCl$_2$	547	615	648	684	725	750	784	833	893	954	501
fluoride	PbF$_2$		861	904	950	1003	1036	1080	1144	1219	1293	855
iodide	PbI$_2$	479	540	571	605	644	668	701	750	807	872	402
oxide	PbO	943	1039	1085	1134	1189	1222	1265	1330	1402	1472	890
sulfide	PbS	852	928	975	1005	1048	1074	1108	1160	1221	1281	1114
Lithium	Li	723	838	881	940	1003	1042	1097	1178	1273	1372	186
bromide	LiBr	748	840	888	939	994	1028	1076	1147	1226	1310	547
chloride	LiCl	783	880	932	987	1045	1081	1129	1203	1290	1382	614
fluoride	LiF	1047	1156	1211	1270	1333	1372	1425	1503	1591	1681	870
iodide	LiI	723	802	841	883	927	955	993	1049	1110	1171	446
Magnesium	Mg	621	702	743	789	838	868	909	967	1034	1107	651
chloride	MgCl$_2$	778	877	930	968	1050	1088	1142	1223	1316	1418	712
Manganese	Mn	1292	1434	1505	1583	1666	1720	1792	1900	2029	2151	1260
chloride	MnCl$_2$		736	778	825	879	913	960	1028	1108	1190	650
Mercury	Hg	126.2	164.8	184.0	204.6	228.8	242.0	261.7	290.7	323.0	357.0	−38.9
Mercuric bromide	HgBr$_2$	136.5	165.3	179.8	194.3	211.5	221.0	237.8	262.7	290.0	319.0	237
chloride	HgCl$_2$	136.2	166.0	180.2	195.8	212.5	222.2	237.0	256.5	275.5	304.0	277
iodide	HgI$_2$	157.5	189.2	204.5	220.0	238.2	249.0	261.8	291.0	324.2	354.0	259
Molybdenum	Mo	3102	3393	3535	3690	3859	3964	4109	4322	4553	4804	2622
hexafluoride	MoF$_6$	−65.5	−49.0	−40.8	−32.0	−22.1	−16.2	−8.0	+4.1	17.2	36.0	17
oxide	MoO$_3$	734	785	814	851	892	917	955	1014	1082	1151	795
Neon	Ne	−257.3	−255.5	−254.6	−253.7	−252.6	−251.9	−251.0	−249.7	−248.1	−246.0	−248.7
Nickel	Ni	1810	1979	2057	2143	2234	2289	2364	2473	2603	2732	1452
carbonyl	Ni(CO)$_4$					−23.0	−15.9	−6.0	+8.8	25.8	42.5	−25
chloride	NiCl$_2$	671	731	759	789	821	840	866	904	945	987	1001

Name	Formula											m.p.
Nitrogen	N_2	-226.1	-221.3	-219.1	-216.8	-214.0	-212.3	-209.7	-205.6	-200.9	-195.8	-210.0
Nitric oxide	NO	-184.5	-180.6	-178.2	-175.3	-171.7	-168.9	-166.0	-162.3	-156.8	-151.7	-161
Nitrogen dioxide	NO_2	-55.6	-42.7	-36.7	-30.4	-23.9	-19.9	-14.7	-5.0	+8.0	21.0	-9.3
Nitrogen pentoxide	N_2O_5	-36.8	-23.0	-16.7	-10.0	-2.9	+1.8	7.4	15.6	24.4	32.4	30
Nitrous oxide	N_2O	-143.4	-133.4	-128.7	-124.0	-118.3	-114.9	-110.3	-103.6	-96.2	-85.5	-90.9
Nitrosyl chloride	$NOCl$					-60.2	-54.2	-46.3	-34.0	-20.3	-6.4	-64.5
fluoride	NOF	-132.0	-120.3	-114.3	-107.8	-100.3	-95.7	-88.8	-79.2	-68.2	-56.0	-134
Osmium tetroxide (yellow)	OsO_4	3.2	22.0	31.3	41.0	51.7	59.4	71.5	89.5	109.3	130.0	56
(white)	OsO_4	-5.6	+15.6	26.0	37.4	50.5	59.4	71.5	89.5	109.3	130.0	42
Oxygen	O_2	-219.1	-213.4	-210.6	-207.5	-204.1	-201.9	-198.8	-194.0	-188.8	-183.1	-218.7
Ozone	O_3	-180.4	-168.6	-163.2	-157.2	-150.7	-146.7	-141.0	-132.6	-122.5	-111.1	-251
Phosgene	$COCl_2$	-92.9	-77.0	-69.3	-60.3	-50.3	-44.0	-35.6	-22.3	-7.6	+8.3	-104
Phosphorus (yellow)	P	76.6	111.2	128.0	146.2	166.7	179.8	197.3	222.7	251.0	280.0	44.1
(violet)	P	237	271	287	306	323	334	349	370	391	417	590
tribromide	PBr_3	7.8	34.4	47.8	62.4	79.0	89.8	103.6	125.2	149.7	175.3	-40
trichloride	PCl_3	-51.6	-31.5	-21.3	-10.2	+2.3	10.2	21.0	37.6	56.9	74.2	-111.8
pentachloride	PCl_5	55.5	74.0	83.2	92.5	102.5	108.3	117.0	131.3	147.2	162.0	
Phosphine	PH_3					-129.4	-125.0	-118.8	-109.4	-98.3	-87.5	-132.5
Phosphonium bromide	PH_4Br	-43.7	-28.5	-21.2	-13.3	-5.0	+0.3	7.4	17.6	28.0	38.3	
chloride	PH_4Cl	-91.0	-79.6	-74.0	-68.0	-61.5	-57.3	-52.0	-44.0	-35.4	-27.0	-28.5
iodide	PH_4I	-25.2	-9.0	-1.1	+7.3	16.1	21.9	29.3	39.9	51.6	62.3	
Phosphorus trioxide	P_4O_6		39.7	53.0	67.8	84.0	94.2	108.3	129.0	150.3	173.1	22.5
pentoxide	P_4O_{10}	384	424	442	462	481	493	510	532	556	591	569
oxychloride	$POCl_3$		2.0	2.0	13.6	27.3	35.8	47.4	65.0	84.3	105.1	2
thiobromide	$PSBr_3$	50.0	72.4	83.6	95.5	108.0	116.0	126.3	141.8	157.8	175.0	38
thiochloride	$PSCl_3$	-18.3	+4.6	16.1	29.0	42.7	51.8	63.8	82.0	102.3	124.0	-36.2
Platinum	Pt	2730	3007	3146	3302	3469	3574	3714	3923	4169	4407	1755
Potassium	K	341	408	443	483	524	550	586	643	708	774	62.3
bromide	KBr	795	892	940	994	1050	1087	1137	1212	1297	1383	730
chloride	KCl	821	919	968	1020	1078	1115	1164	1239	1322	1407	790
fluoride	KF	885	988	1039	1096	1156	1193	1245	1323	1411	1502	880
hydroxide	KOH	719	814	863	918	976	1013	1064	1142	1233	1327	380
iodide	KI	745	840	887	938	995	1030	1080	1152	1238	1324	723
Radon	Rn	-144.2	-132.4	-126.3	-119.2	-111.3	-106.2	-99.0	-87.7	-75.0	-61.8	-71
Rhenium heptoxide	Re_2O_7	212.5	237.5	248.0	261.0	272.0	280.0	289.0	307.0	336.0	362.4	296
Rubidium	Rb	297	358	389	422	459	482	514	563	620	679	38.5

Compound Name	Formula	Pressure, mm Hg — Temperature, °C										Melting point, °C
		1	5	10	20	40	60	100	200	400	760	
bromide	RbBr	781	876	923	975	1031	1066	1114	1186	1267	1352	682
chloride	RbCl	792	887	937	990	1047	1084	1133	1207	1294	1381	715
fluoride	RbF	921	982	1016	1052	1096	1123	1168	1239	1322	1408	760
iodide	RbI	748	839	884	935	991	1026	1072	1141	1223	1304	642
Selenium	Se	356	413	442	473	506	527	554	594	637	680	217
dioxide	SeO$_2$	157.0	187.7	202.5	217.5	234.1	244.6	258.0	277.0	297.7	317.0	340
hexafluoride	SeF$_6$	-118.6	-105.2	-98.9	-92.3	-84.7	-80.0	-73.9	-64.8	-55.2	-45.8	-34.7
oxychloride	SeOCl$_2$	34.8	59.8	71.9	84.2	98.0	106.5	118.0	134.6	151.7	168.0	8.5
tetrachloride	SeCl$_4$	74.0	96.3	107.4	118.1	130.1	137.8	147.5	161.0	176.4	191.5	
Silicon	Si	1724	1835	1888	1942	2000	2036	2083	2151	2220	2287	1420
dioxide	SiO$_2$			1732	1798	1867	1911	1969	2053	2141	2227	1710
tetrachloride	SiCl$_4$	-63.4	-44.1	-34.4	-24.0	-12.1	-4.8	+5.4	21.0	38.4	56.8	-68.8
tetrafluoride	SiF$_4$	-144.0	-134.8	-130.4	-125.9	-120.8	-117.5	-113.3	-107.2	-100.7	-94.8	-90
Trichlorofluorosilane	SiFCl$_3$	-92.6	-76.4	-68.3	-59.0	-48.8	-42.2	-33.2	-19.3	-4.0	+12.2	-120.8
Iodosilane	SiH$_3$I		-53.0	-47.7	-33.4	-21.8	-14.3	-4.4	+10.7	27.9	45.4	-57.0
Diiodosilane	SiH$_2$I$_2$		3.8	18.0	34.1	52.6	64.0	79.4	101.8	125.5	149.5	-1.0
Disiloxan	(SiH$_3$)$_2$O	-112.5	-95.8	-88.2	-79.8	-70.4	-64.2	-55.9	-43.5	-29.3	-15.4	-144.2
Trisilane	Si$_3$H$_8$	-68.9	-49.7	-40.0	-29.0	-16.9	-9.0	+1.6	17.8	35.5	53.1	-117.2
Trisilazane	(SiH$_3$)$_3$N	-68.7	-49.9	-40.4	-30.0	-18.5	-11.0	-1.1	+14.0	31.0	48.7	-105.7
Tetrasilane	Si$_4$H$_{10}$	-27.7	-6.2	+4.3	15.8	28.4	36.6	47.4	63.6	81.7	100.0	-93.6
Octachlorotrisilane	Si$_3$Cl$_8$	46.3	74.7	89.3	104.2	121.5	132.0	146.0	166.2	189.5	211.4	
Hexachlorodisiloxane	(SiCl$_3$)$_2$O	-5.0	17.8	29.4	41.5	55.2	63.8	75.4	92.5	113.6	135.6	-33.2
Hexachlorodisilane	Si$_2$Cl$_6$	+4.0	27.4	38.8	51.5	65.3	73.9	85.4	102.2	120.6	139.0	-1.2
Tribromosilane	SiHBr$_3$	-30.5	-8.0	+3.4	16.0	30.0	39.2	51.6	70.2	90.2	111.8	-73.5
Trichlorosilane	SiHCl$_3$	-80.7	-62.6	-53.4	-43.8	-32.9	-25.8	-16.4	-1.8	+14.5	31.8	-126.6
Trifluorosilane	SiHF$_3$	-152.0	-142.7	-138.2	-132.9	-127.3	-123.7	-118.7	-111.3	-102.8	-95.0	-131.4
Dibromosilane	SiH$_2$Br$_2$	-60.9	-40.0	-29.4	-18.0	-5.2	+3.2	14.1	31.6	50.7	70.5	-70.2
Difluorosilane	SiH$_2$F$_2$	-146.7	-136.0	-130.4	-124.3	-117.6	-113.3	-107.3	-98.3	-87.6	-77.8	

Name	Formula											
Monobromosilane	SiH_3Br	-117.8	-85.7	-77.3	-68.3	-57.8	-51.1	-42.3	-28.6	-13.3	+2.4	-93.9
Monochlorosilane	SiH_3Cl	-153.0	-104.3	-97.7	-90.1	-81.8	-76.0	-68.5	-57.0	-44.5	-30.4	
Monofluorosilane	SiH_3F	-46.1	-145.5	-141.2	-136.3	-130.8	-127.2	-122.4	-115.2	-106.8	-98.0	
Tribromofluorosilane	$SiFBr_3$	-124.7	-25.4	-15.1	-3.7	+9.2	17.4	28.6	45.7	64.6	83.8	-82.5
Dichlorodifluorosilane	SiF_2Cl_2	-144.0	-110.5	-102.9	-94.5	-85.0	-78.6	-70.3	-58.0	-45.0	-31.8	-139.7
Trifluorobromosilane	SiF_3Br	-81.0	-133.0	-127.0	-120.5	-112.8	-108.2	-101.7	-69.8	-55.9	-41.7	-70.5
Trifluorochlorosilane	SiF_3Cl	-86.5	-68.8	-63.1	-57.0	-50.6	-46.7	-41.7	-91.7	-81.0	-70.0	-142
Hexafluorodisilane	Si_2F_6	-65.2	-68.4	-59.0	-48.8	-37.0	-29.0	-19.5	-34.2	-26.4	-18.9	-18.6
Dichlorofluorobromosilane	$SiFCl_2Br$	-179.3	-45.5	-35.6	-24.5	-12.0	-4.7	+6.3	-3.2	+15.4	35.4	-112.3
Dibromochlorofluorosilane	$SiFClBr_2$	-114.8	-168.6	-163.0	-156.9	-150.3	-146.3	-140.5	23.0	43.0	59.5	-99.3
Silane	SiH_4								-131.6	-122.0	-111.5	-185
Disilane	Si_2H_6		-99.3	-91.4	-82.7	-72.8	-66.4	-57.5	-44.6	-29.0	-14.3	-132.6
Silver	Ag	1357	1500	1575	1658	1743	1795	1865	1971	2090	2212	960.5
chloride	AgCl	912	1019	1074	1134	1200	1242	1297	1379	1467	1564	455
iodide	AgI	820	927	983	1045	1111	1152	1210	1297	1400	1506	552
Sodium	Na	439	511	549	589	633	662	701	758	823	892	97.5
bromide	NaBr	806	903	952	1005	1063	1099	1148	1220	1304	1392	755
chloride	NaCl	865	967	1017	1072	1131	1169	1220	1296	1379	1465	800
cyanide	NaCN	817	928	983	1046	1115	1156	1214	1302	1401	1497	564
fluoride	NaF	1077	1186	1240	1300	1363	1403	1455	1531	1617	1704	992
hydroxide	NaOH	739	843	897	953	1017	1057	1111	1192	1286	1378	318
iodide	NaI	767	857	903	952	1005	1039	1083	1150	1225	1304	651
Strontium	Sr		847	898	953	1018	1057	1111	1192	1285	1384	800
Strontium oxide	SrO	2068	2198	2262	2333	2410						2430
Sulfur	S	183.8	223.0	243.8	264.7	288.3	305.5	327.2	359.7	399.6	444.6	112.8
monochloride	S_2Cl_2	-7.4	+15.7	27.5	40.0	54.1	63.2	75.3	93.5	115.4	138.0	-80
hexafluoride	SF_5	-132.7	-120.6	-114.7	-108.4	-101.5	-96.8	-90.9	-82.3	-72.6	-63.5	-50.2
Sulfuryl chloride	SO_2Cl_2		-35.1	-24.8	-13.4	-1.0	+7.2	17.8	33.7	51.3	69.2	-54.1
Sulfur dioxide	SO_2	-95.5	-83.0	-76.8	-69.7	-60.5	-54.6	-46.9	-35.4	-23.0	-10.0	-73.2
trioxide (α)	SO_3	-39.0	-23.7	-16.5	-9.1	-1.0	+4.0	10.5	20.5	32.6	44.8	16.8
trioxide (β)	SO_3	-34.0	-19.2	-12.3	-4.9	+3.2	8.0	14.3	23.7	32.6	44.8	32.3
trioxide (γ)	SO_3	-15.3	-2.0	+4.3	11.1	17.9	21.4	28.0	35.8	44.0	51.6	62.1
Tellurium	Te	520	605	650	697	753	789	838	910	997	1087	452
chloride	$TeCl_4$			233	253	273	287	304	330	360	392	224
fluoride	TeF_5	-111.3	-98.8	-92.4	-86.0	-78.4	-73.8	-67.9	-57.3	-48.2	-38.6	-37.8
Thallium	Tl	825	931	983	1040	1103	1143	1196	1274	1364	1457	3035

VAPOR PRESSURE—TABLE 2. Vapor Pressures of Inorganic Compounds, up to 1 atm (*Continued*)

Compound Name	Formula	Pressure, mmHg										Melting point, °C
		Temperature, °C										
		1	5	10	20	40	60	100	200	400	760	
Thallous bromide	TlBr		490	522	559	598	621	653	703	759	819	460
chloride	TlCl		487	517	550	589	612	645	694	748	807	430
iodide	TlI	440	502	531	567	607	631	663	712	763	823	440
Thionyl bromide	$SOBr_2$	-6.7	+18.4	31.0	44.1	58.8	68.3	80.6	99.0	119.2	139.5	-52.2
Thionyl chloride	$SOCl_2$	-52.9	-32.4	-21.9	-10.5	+2.2	10.4	21.4	37.9	56.5	75.4	-104.5
Tin	Sn	1492	1634	1703	1777	1855	1903	1968	2063	2169	2270	231.9
Stannic bromide	$SnBr_4$		58.3	72.7	88.1	105.5	116.2	131.0	152.8	177.7	204.7	31.0
Stannous chloride	$SnCl_2$	316	366	391	420	450	467	493	533	577	623	246.8
Stannic chloride	$SnCl_4$	-22.7	-1.0	+10.0	22.0	35.2	43.5	54.7	72.0	92.1	113.0	-30.2
iodide	SnI_4		156.0	175.8	196.2	218.8	234.2	254.2	283.5	315.5	348.0	144.5
hydride	SnH_4	-140.0	-125.8	-118.5	-111.2	-102.3	-96.6	-89.2	-78.0	-65.2	-52.3	-149.9
Tin tetramethyl	$Sn(CH_3)_4$	-51.3	-31.0	-20.6	-9.3	+3.5	11.7	22.8	39.8	58.5	78.0	
trimethyl-ethyl	$Sn(CH_3)_3 \cdot C_2H_5$	-30.0	-7.6	+3.8	16.1	30.0	38.4	50.0	67.3	87.6	108.8	
trimethyl-propyl	$Sn(CH_3)_3 \cdot C_3H_7$	-12.0	+10.7	21.8	34.0	48.5	57.5	69.8	88.0	109.6	131.7	
Titanium chloride	$TiCl_4$	-13.9	+9.4	21.3	34.2	48.4	58.0	71.0	90.5	112.7	136.0	-30
Tungsten	W	3990	4337	4507	4690	4886	5007	5168	5403	5666	5927	3370
Tungsten hexafluoride	WF_6	-71.4	-56.5	-49.2	-41.5	-33.0	-27.5	-20.3	-10.0	+1.2	17.3	-0.5
Uranium hexafluoride	UF_6	-38.8	-22.0	-13.8	-5.2	+4.4	10.4	18.2	30.0	42.7	55.7	69.2
Vanadyl trichloride	$VOCl_3$	-23.2	+0.2	12.2	26.6	40.0	49.8	62.5	82.0	103.5	127.2	-28
xenon	Xe	-168.5	-158.2	-152.8	-147.1	-141.2	-137.7	-132.8	-125.4	-117.1	-108.0	-111.6
Zinc	Zn	487	558	593	632	673	700	736	788	844	907	419.4
chloride	$ZnCl_2$	428	481	508	536	566	584	610	648	689	732	365
fluoride	ZnF_2	970	1055	1086	1129	1175	1207	1254	1329	1417	1497	872
diethyl	$Zn(C_2H_5)_2$	-22.4	0.0	+11.7	24.2	38.0	47.2	59.1	77.0	97.3	118.0	-28
Zirconium bromide	$ZrBr_4$	207	237	250	266	281	289	301	318	337	357	450
chloride	$ZrCl_4$	190	217	230	243	259	268	279	295	312	331	437
iodide	ZrI_4	264	297	311	329	344	355	369	389	409	431	499

VAPOR PRESSURE—TABLE 3. Vapor Pressure of Inorganic and Organic Liquids

Cmpd. no.	Name	Formula	CAS no.	C1	C2	C3	C4	C5	T_{min}, K	P_s at T_{min}	T_{max}, K	P_s at T_{max}
1	Methane	CH_4	74,828	39.205	−1,324.4	−3.4366	3.1019E−05	2	90.69	1.1687E+04	190.56	4.5897E+06
2	Ethane	C_2H_6	74,840	51.857	−2,598.7	−5.1283	1.4913E−05	2	90.35	1.1273E+00	305.32	4.8522E+06
3	Propane	C_3H_8	74,986	59.078	−3,492.6	−6.0669	1.0919E−05	2	85.47	1.6788E−04	369.83	4.2135E+06
4	n-Butane	C_4H_{10}	106,978	66.343	−4,363.2	−7.046	9.4509E−06	2	134.86	6.7441E−01	425.12	3.7699E+06
5	n-Pentane	C_5H_{12}	109,660	78.741	−5,420.3	−8.8253	9.6171E−06	2	143.42	6.8642E−02	469.7	3.3642E+06
6	n-Hexane	C_6H_{14}	110,543	104.65	−6,995.5	−12.702	1.2381E−05	2	177.83	9.0169E−01	507.6	3.0449E+06
7	n-Heptane	C_7H_{16}	142,825	87.829	−6,996.4	−9.8802	7.2099E−06	2	182.57	1.8269E−01	540.2	2.7192E+06
8	n-Octane	C_8H_{18}	111,659	96.084	−7,900.2	−11.003	7.1802E−06	2	216.38	2.1083E−00	568.7	2.4673E+06
9	n-Nonane	C_9H_{20}	111,842	109.35	−9,030.4	−12.882	7.8544E−06	2	219.66	4.3058E−01	594.6	2.3054E+06
10	n-Decane	$C_{10}H_{22}$	124,185	112.73	−9,749.6	−13.245	7.1266E−06	2	243.51	1.3930E−00	617.7	2.0908E+06
11	n-Undecane	$C_{11}H_{24}$	1,120,214	131	−11,143	−15.855	8.1871E−06	2	247.57	4.0836E−01	639	1.9493E+06
12	n-Dodecane	$C_{12}H_{26}$	112,403	137.47	−11,976	−16.698	8.0906E−06	2	263.57	6.1534E−01	658	1.8223E+06
13	n-Tridecane	$C_{13}H_{28}$	629,505	137.45	−12,549	−16.543	7.1275E−06	2	267.76	2.5096E−01	675	1.6786E+06
14	n-Tetradecane	$C_{14}H_{30}$	629,594	140.47	−13,231	−16.859	6.5877E−06	2	279.01	2.5268E−01	693	1.5693E+06
15	n-Pentadecane	$C_{15}H_{32}$	629,629	135.57	−13,478	−16.022	5.6136E−06	2	283.07	1.2884E−01	708	1.4743E+06
16	n-Hexadecane	$C_{16}H_{34}$	544,763	156.06	−15,015	−18.941	6.8172E−06	2	291.31	9.2265E−02	723	1.4106E+06
17	n-Heptadecane	$C_{17}H_{36}$	629,787	156.95	−15,557	−18.966	6.4559E−06	2	295.13	4.6534E−02	736	1.3438E+06
18	n-Octadecane	$C_{18}H_{38}$	593,453	157.68	−16,093	−18.954	5.9272E−06	2	301.31	3.3909E−02	747	1.2555E+06
19	n-Nonadecane	$C_{19}H_{40}$	629,925	182.54	−17,897	−22.498	7.4008E−06	2	305.04	1.5909E−02	758	1.2078E+06
20	n-Eicosane	$C_{20}H_{42}$	112,958	203.66	−19,441	−25.525	8.8382E−06	2	309.58	9.2574E−03	768	1.1746E+06
21	2-Methylpropane	C_4H_{10}	75,285	100.18	−4,841.9	−13.541	2.0063E−02	1	113.54	1.4051E−02	408.14	3.6199E+06
22	2-Methylbutane	C_5H_{12}	78,784	72.35	−5,010.9	−7.883	8.9795E−06	2	113.25	1.1569E−04	460.43	3.3709E+06
23	2,3-Dimethylbutane	C_6H_{14}	79,298	77.235	−5,695.9	−8.5109	8.0163E−06	2	145.19	1.5081E−02	499.98	3.1255E+06
24	2-Methylpentane	C_6H_{14}	107,835	77.36	−5,791.7	−8.4912	7.7939E−06	2	119.55	9.2204E−06	497.5	3.0192E+06
25	2,3-Dimethylpentane	C_7H_{16}	565,593	78.282	−6,347	−8.502	6.4169E−06	2	160	1.2631E−02	537.35	2.8823E+06
26	2,3,3-Trimethylpentane	C_8H_{18}	560,214	83.105	−6,903.7	−9.1858	6.4703E−06	2	172.22	1.6820E−02	573.5	2.8116E+06
27	2,2,4-Trimethylpentane	C_8H_{18}	540,841	87.868	−6,831.7	−9.9783	7.7279E−06	2	165.78	1.6187E−02	543.96	2.5630E+06
28	Ethylene	C_2H_4	74,851	74.242	−2,707.2	−9.8462	2.2457E−02	1	104	1.2361E+02	282.34	5.0296E+06

VAPOR PRESSURE—TABLE 3. Vapor Pressure of Inorganic and Organic Liquids (*Continued*)

Cmpd. no.	Name	Formula	CAS no.	C1	C2	C3	C4	C5	T_{min}, K	P_s at T_{min}	T_{max}, K	P_s at T_{max}
29	Propylene	C_3H_6	115,071	57.263	−3,382.4	−5.7707	1.0431E−05	2	87.89	9.3867E−04	365.57	4.6346E+06
30	1-Butene	C_4H_8	106,989	68.49	−4,350.2	−7.4124	1.0503E−05	2	87.8	7.1809E−07	419.95	4.0391E+06
31	cis-2-Butene	C_4H_8	590,181	102.62	−5,260.3	−13.764	1.9183E−02	1	134.26	2.4051E−01	435.58	4.2388E+06
32	trans-2-Butene	C_4H_8	624,646	70.589	−4,530.4	−7.7229	1.0928E−05	2	167.62	7.4729E+01	428.63	4.0811E+06
33	1-Pentene	C_5H_{10}	109,671	120.15	−6,192.4	−16.597	2.1922E−02	1	107.93	3.5210E−06	464.78	3.5557E+06
34	1-Hexene	C_6H_{12}	592,416	85.3	−6,171.7	−9.702	8.9604E−06	2	133.39	2.5272E−04	504.03	3.1397E+06
35	1-Heptene	C_7H_{14}	592,767	92.68	−7,055.2	−10.679	8.4459E−06	2	154.27	1.2810E−03	537.29	2.8225E+06
36	1-Octene	C_8H_{16}	111,660	97.57	−7,836	−11.272	7.7267E−06	2	171.45	2.7570E−02	566.65	2.5735E+06
37	1-Nonene	C_9H_{18}	124,118	144.45	−9,676.2	−19.446	1.8031E−02	1	191.78	8.5514E−03	593.25	2.3308E+06
38	1-Decene	$C_{10}H_{20}$	872,059	78.808	−8,367.9	−7.9553	8.7442E−18	6	206.89	1.7308E−02	616.4	2.2092E+06
39	2-Methylpropene	C_4H_8	115,117	102.5	−5,021.8	−13.88	2.0296E−02	1	132.81	6.2213E−01	417.9	3.9760E+06
40	2-Methyl-1-butene	C_5H_{10}	563,462	97.33	−5,631.8	−12.589	1.5395E−02	1	135.58	1.9687E−02	465	3.4544E+06
41	2-Methyl-2-butene	C_5H_{10}	513,359	82.605	−5,606.6	−9.4236	1.0512E−05	2	139.39	1.9447E−02	471	3.3769E+06
42	1,2-Butadiene	C_4H_6	590,192	39.714	−3,769.9	−2.6407	6.9379E−18	6	136.95	4.4720E−01	452	4.3613E+06
43	1,3-Butadiene	C_4H_6	106,990	73.522	−4,564.3	−8.1958	1.1580E−05	2	164.25	6.9110E−01	425.17	4.3041E+06
44	2-Methyl-1,3-butadiene	C_5H_8	78,795	79.656	−5,239.6	−9.4314	9.5850E−03	1	127.27	2.4768E−03	484	3.8509E+06
45	Acetylene	C_2H_2	74,862	172.06	−5,318.5	−27.223	5.4619E−02	1	192.4	1.2603E+05	308.32	6.1467E+06
46	Methylacetylene	C_3H_4	74,997	119.42	−5,364.5	−16.81	2.5523E−02	1	170.45	3.7264E+02	402.39	5.6206E+06
47	Dimethylacetylene	C_4H_6	503,173	66.592	−4,999.8	−6.8387	6.6793E−06	2	240.91	6.1212E+03	473.2	4.8699E+06
48	3-Methyl-1-butyne	C_5H_8	598,232	69.459	−5,250	−7.1125	7.9289E−17	6	183.45	4.3551E+01	463.2	4.1986E+06
49	1-Pentyne	C_5H_8	627,190	82.805	−5,683.8	−9.4301	1.0767E−05	2	167.45	2.3990E+00	481.2	4.1701E+06
50	2-Pentyne	C_5H_8	627,214	137.29	−7,447.1	−19.01	2.1415E−02	1	163.83	2.0462E−01	519	4.0198E+06
51	1-Hexyne	C_6H_{10}	693,027	133.2	−7,492.9	−18.405	2.2062E−02	1	141.25	3.9157E−04	516.2	3.6352E+06
52	2-Hexyne	C_6H_{10}	764,352	123.71	−7,639	−16.451	1.6495E−02	1	183.65	5.4026E−01	549	3.5301E+06
53	3-Hexyne	C_6H_{10}	928,494	47.091	−5,104	−3.6371	5.1621E−04	1	170.05	2.1950E−01	544	3.5397E+06
54	1-Heptyne	C_7H_{12}	628,717	66.447	−6,395.6	−6.3848	1.1250E−17	6	192.22	6.7026E−01	559	3.1343E+06
55	1-Octyne	C_8H_{14}	629,050	82.353	−7,240.6	−9.1843	5.8038E−03	1	193.55	1.0092E−01	585	2.8202E+06
56	Vinylacetylene*	C_4H_4	689,974	55.682	−4,439.3	−5.0136	1.9650E−17	6	173.15	6.6899E+01	454	4.8874E+06

57	Cyclopentane	C$_5$H$_{10}$	287,923	51.434	-4,770.6	-4.3515	1.9605E-17	6	179.28	9.4420E+00	511.76	4.5028E+06
58	Methylcyclopentane	C$_6$H$_{12}$	96,377	79.673	-6,086.6	-8.7933	7.4046E-06	2	130.73	6.7059E-05	532.79	3.7808E+06
59	Ethylcyclopentane	C$_7$H$_{14}$	1,640,897	88.622	-7,011	-10.038	7.4481E-06	2	134.71	3.7061E-06	569.52	3.3970E+06
60	Cyclohexane	C$_6$H$_{12}$	110,827	116.51	-7,103.3	-15.49	1.6959E-02	1	279.69	5.3802E-03	553.58	4.0958E+06
61	Methylcyclohexane	C$_7$H$_{14}$	108,872	92.611	-7,077.8	-10.684	8.1239E-06	2	146.58	1.5256E-04	572.19	3.4828E+06
62	1,1-Dimethylcyclohexane	C$_8$H$_{16}$	590,669	81.184	-6,927	-8.8498	5.4580E-06	2	239.66	6.0584E+01	591.15	2.9387E+06
63	Ethylcyclohexane	C$_8$H$_{16}$	1,678,917	80.208	-7,203.2	-8.6023	4.5901E-06	2	161.84	3.5747E-04	609.15	3.0411E+06
64	Cyclopentene	C$_5$H$_8$	142,290	49.88	-4,649.7	-4.1191	1.9564E-17	6	138.13	1.6884E-02	507	4.8062E+06
65	1-Methylcyclopentene	C$_6$H$_{10}$	693,890	52.732	-5,286.9	-4.4509	1.0883E-17	6	146.62	3.9787E-03	542	4.1303E+06
66	Cyclohexene	C$_6$H$_{10}$	110,838	88.184	-6,624.9	-10.059	8.2566E-06	2	169.67	1.0377E-01	560.4	4.3922E+06
67	Benzene	C$_6$H$_6$	71,432	83.918	-6,517.7	-9.3453	7.1182E-06	2	278.68	4.7620E+03	562.16	4.8819E+06
68	Toluene	C$_7$H$_8$	108,883	80.877	-6,902.4	-8.7761	5.8034E-06	2	178.18	4.2348E-02	591.8	4.1012E+06
69	o-Xylene	C$_8$H$_{10}$	95,476	90.356	-7,948.7	-10.081	5.9756E-06	2	247.98	2.1968E+01	630.33	3.7424E+06
70	m-Xylene	C$_8$H$_{10}$	108,383	84.782	-7,598.3	-9.2612	5.5445E-06	2	225.3	3.2099E+00	617.05	3.5286E+06
71	p-Xylene	C$_8$H$_{10}$	106,423	85.475	-7,595.8	-9.378	5.6875E-06	2	286.41	5.8144E+02	616.23	3.4984E+06
72	Ethylbenzene	C$_8$H$_{10}$	100,414	88.09	-7,688.3	-9.7708	5.8844E-06	2	178.15	4.0140E-03	617.2	3.5968E+06
73	Propylbenzene	C$_9$H$_{12}$	103,651	136.83	-9,544.8	-18.190	1.6590E-02	1	324.18	2.0014E+03	638.32	3.2001E+06
74	1,2,4-Trimethylbenzene	C$_9$H$_{12}$	95,636	60.658	-7,260.4	-5.3772	4.5816E-18	6	229.33	7.9735E-01	649.13	3.2533E+06
75	Isopropylbenzene	C$_9$H$_{12}$	98,828	143.62	-9,687.7	-19.305	1.7703E-02	1	177.14	3.8034E-04	631.1	3.1837E+06
76	1,3,5-Trimethylbenzene	C$_9$H$_{12}$	108,678	48.603	-6,545.2	-3.6412	1.9307E-18	6	228.42	1.1889E+00	637.36	3.1119E+06
77	p-Isopropyltoluene	C$_{10}$H$_{14}$	99,876	107.71	-9,402.7	-12.545	6.6661E-06	6	205.25	9.9261E-03	653.15	2.7957E+06
78	Naphthalene	C$_{10}$H$_8$	91,203	62.447	-8,109	-5.5571	2.0800E-18	6	353.43	9.9229E+02	748.35	3.9941E+06
79	Biphenyl	C$_{12}$H$_{10}$	92,524	76.811	-9,878.5	-7.4384	2.0436E-18	6	342.2	9.3752E+01	789.26	3.8615E+06
80	Styrene	C$_8$H$_8$	100,425	105.93	-8,685.9	-12.42	7.5583E-06	6	242.54	1.0613E+01	636	3.8234E+06
81	m-Terphenyl	C$_{18}$H$_{14}$	92,068	88.044	-13,367	-8.6482	8.7874E-19	6	360	1.0112E+00	924.85	3.5297E+06
82	Methanol	CH$_4$O	67,561	81.768	-6,876	-8.7078	7.1926E-06	2	175.47	1.1147E-01	512.64	8.1402E+06
83	Ethanol	C$_2$H$_6$O	64,175	74.475	-7,164.3	-7.327	3.1340E-06	2	159.05	4.8459E-04	513.92	6.1171E+06
84	1-Propanol	C$_3$H$_8$O	71,238	88.134	-8,498.6	-9.0766	8.3303E-18	6	146.95	3.0828E-07	536.78	5.1214E+06
85	1-Butanol	C$_4$H$_{10}$O	71,363	93.173	-9,185.9	-9.7464	4.7796E-18	6	184.51	5.7220E-04	563.05	4.3392E+06
86	2-Butanol	C$_4$H$_{10}$O	78,922	152.54	-11,111	-19.025	1.0426E-05	6	158.45	1.1323E-06	536.05	4.2014E+06
87	2-Propanol	C$_3$H$_8$O	67,630	76.964	-7,623.8	-7.4924	5.9436E-18	6	185.28	3.6606E-02	508.3	4.7908E+06
88	2-Methyl-2-propanol	C$_4$H$_{10}$O	75,650	172.31	-11,590	-22.118	1.3709E-05	2	298.97	5.9356E-03	506.21	3.9910E+06
89	1-Pentanol	C$_5$H$_{12}$O	71,410	168.96	-12,659	-21.366	1.1591E-05	2	195.56	3.1816E-04	586.15	3.8657E+06

Cmpd. no.	Name	Formula	CAS no.	C1	C2	C3	C4	C5	T_{min}, K	P_s at T_{min}	T_{max}, K	P_s at T_{max}
90	2-Methyl-1-butanol	$C_5H_{12}O$	137,326	410.44	−20,262	−62.366	6.3353E−02	1	203	3.7992E−04	565	3.8749E+06
91	3-Methyl-1-butanol	$C_5H_{12}O$	123,513	107.02	−10,237	−11.695	6.8003E−18	6	155.95	2.1036E−08	577.2	3.9013E+06
92	1-Hexanol	$C_6H_{14}O$	111,273	117.31	−11,239	−13.149	9.3676E−18	6	228.55	3.7401E−02	611.35	3.4557E+06
93	1-Heptanol	$C_7H_{16}O$	111,706	160.08	−14,095	−19.211	1.7043E−17	6	239.15	1.6990E−02	631.9	3.1810E+06
94	Cyclohexanol	$C_6H_{12}O$	108,930	135.01	−12,238	−15.702	1.0349E−17	6	296.6	7.9382E+01	650	4.2456E+06
95	Ethylene glycol	$C_2H_6O_2$	107,211	79.276	−10,105	−7.521	7.3408E−19	6	260.15	2.4834E−01	719.7	7.7100E+06
96	1,2-Propylene glycol	$C_3H_8O_2$	57,556	212.8	−15,420	−28.109	2.1564E−05	2	213.15	9.2894E−05	626	6.0413E+06
97	Phenol	C_6H_6O	108,952	95.444	−10,113	−10.09	6.7603E−18	6	314.06	1.8798E−02	694.25	6.0585E+06
98	o-Cresol	C_7H_8O	95,487	210.88	−13,928	−29.483	2.5182E−02	1	304.19	6.5326E+01	697.55	5.0583E+06
99	m-Cresol	C_7H_8O	108,394	95.403	−10,581	−10.004	4.3032E−18	6	285.39	5.8624E+00	705.85	4.5221E+06
100	p-Cresol	C_7H_8O	106,445	118.53	−11,957	−13.293	8.6988E−18	6	307.93	3.4466E+01	704.65	5.1507E+06
101	Dimethyl ether	C_2H_6O	115,106	44.704	−3,525.6	−3.4444	5.4574E−17	6	131.65	3.0496E+00	400.1	5.2735E+06
102	Methyl ethyl ether	C_3H_8O	540,670	205.79	−9,834.5	−28.739	3.5317E−05	2	160	5.3423E−01	437.8	4.4658E+06
103	Methyl n-propyl ether	$C_4H_{10}O$	557,175	50.83	−4,781.7	−4.1773	9.4076E−18	6	133.97	4.8875E−03	476.3	3.7721E+06
104	Methyl isopropyl ether	$C_4H_{10}O$	598,538	55.096	−4,793.2	−4.8689	2.9518E−17	6	127.93	2.4971E−03	464.5	3.8892E+06
105	Methyl n-butyl ether	$C_5H_{12}O$	628,284	102.04	−6,954.9	−12.278	1.2131E−05	2	157.48	1.9430E−02	510	3.3089E+06
106	Methyl isobutyl ether	$C_5H_{12}O$	625,445	58.165	−5,362.1	−5.2568	2.0194E−17	6	150	1.9801E−02	497	3.4130E+06
107	Methyl tert-butyl ether	$C_5H_{12}O$	1,634,044	55.875	−5,131.6	−4.9604	1.9123E−17	6	164.55	5.3566E−01	497.1	3.4106E+06
108	Diethyl ether	$C_4H_{10}O$	60,297	136.9	−6,954.3	−19.254	2.4508E−02	1	156.85	3.9545E−01	466.7	3.6412E+06
109	Ethyl propyl ether	$C_5H_{12}O$	628,320	143.11	−8,353.7	−18.751	2.0620E−05	2	145.65	7.3931E−04	500.23	3.3729E+06
110	Ethyl isopropyl ether	$C_5H_{12}O$	625,547	57.723	−5,236.9	−5.2136	2.2998E−17	6	140	4.3092E−03	489	3.4145E+06
111	Methyl phenyl ether	C_7H_8O	100,663	128.06	−9,307.7	−16.693	1.4919E−02	1	235.65	2.4466E+00	645.6	4.2731E+06
112	Diphenyl ether	$C_{12}H_{10}O$	101,848	59.969	−8,585.5	−5.1538	1.9983E−18	6	300.03	7.0874E+00	766.8	3.0971E+06
113	Formaldehyde	CH_2O	50,000	101.51	−4,917.2	−13.765	2.2031E−02	1	181.15	8.8700E+02	408	3.0977E+06
114	Acetaldehyde	C_2H_4O	75,070	193.69	−8,036.7	−29.502	4.3678E−02	1	150.15	3.2320E−01	466	5.5652E+06
115	1-Propanal	C_3H_6O	123,386	80.581	−5,896.1	−8.9301	8.2236E−06	2	170	1.3133E+00	504.4	4.9189E+06
116	1-Butanal	C_4H_8O	123,728	99.33	−7,083.6	−11.733	1.0027E−05	2	176.75	3.1699E−01	537.2	4.3232E+06
117	1-Pentanal	$C_5H_{10}O$	110,623	149.58	−8,890	−20.697	2.2101E−02	1	182	5.2282E−02	566.1	3.9685E+06

118	1-Hexanal	C$_6$H$_{12}$O	66,251	81.507	−7,776.8	−8.4516	1.5143E−17	6	217.15	12.473E+00	591	3.4607E+06
119	1-Heptanal	C$_7$H$_{14}$O	111,717	107.17	−9,070.3	−12.503	7.4446E−06	2	229.8	1.1177E+00	617	3.1829E+06
120	1-Octanal	C$_8$H$_{16}$O	124,130	250.25	−16,162	−33.927	2.2349E−05	2	246	4.1640E−01	638.1	2.9704E+06
121	1-Nonanal	C$_9$H$_{18}$O	124,196	337.71	−18,506	−50.224	4.7345E−02	1	255.15	3.4172E−01	658	2.7430E+06
122	1-Decanal	C$_{10}$H$_{20}$O	112,312	201.64	−15,133	−26.264	1.4625E−05	2	267.15	4.8648E−01	674.2	2.5989E+06
123	Acetone	C$_3$H$_6$O	67,641	69.006	−5,599.6	−7.0985	6.2237E−06	2	178.45	2.7851E+00	508.2	4.7091E+06
124	Methyl ethyl ketone	C$_4$H$_8$O	78,933	72.698	−6,143.6	−7.5779	5.6476E−06	2	186.48	1.3904E+00	535.5	4.1201E+06
125	2-Pentanone	C$_5$H$_{10}$O	107,879	84.635	−7,078.4	−9.3	6.2702E−06	2	196.29	7.5235E−01	561.08	3.7062E+06
126	Methyl isopropyl ketone	C$_5$H$_{10}$O	563,804	308.74	−13,693	−47.557	5.7002E−02	1	181.15	2.2648E−02	553	3.8413E+06
127	2-Hexanone	C$_6$H$_{12}$O	591,786	65.841	−7,042	−6.1376	7.2196E−18	6	217.35	1.5111E+00	587.05	3.3120E+06
128	Methyl isobutyl ketone	C$_6$H$_{12}$O	108,101	153.23	−10,055	−19.848	1.6426E−05	2	189.15	3.3536E−02	571.4	3.2659E+06
129	3-Methyl-2-pentanone	C$_6$H$_{12}$O	565,617	64.641	−6,457.4	−6.218	3.4543E−06	2	167.15	3.2662E−03	573	3.3212E+06
130	3-Pentanone	C$_5$H$_{10}$O	96,220	44.286	−5,415.1	−3.0913	1.8580E−18	6	234.18	7.3422E+01	560.95	3.6993E+06
131	Ethyl isopropyl ketone	C$_6$H$_{12}$O	565,695	206.77	−12,537	−27.894	2.2462E−05	2	200	6.0339E−02	567	3.3424E+06
132	Diisopropyl ketone	C$_7$H$_{14}$O	565,800	96.919	8,014.2	−11.093	7.3452E−06	2	204.81	3.9036E−01	576	3.0606E+06
133	Cyclohexanone	C$_6$H$_{10}$O	108,941	95.118	−8,300.4	−10.796	6.5037E−06	2	242	6.9667E+00	653	4.0126E+06
134	Methyl phenyl ketone	C$_8$H$_8$O	98,862	62.688	−8,088.8	−5.5434	2.0774E−18	6	292.81	3.5899E+01	709.5	3.8451E+06
135	Formic acid	CH$_2$O$_2$	64,186	50.323	−5,378.2	−4.203	3.4697E−06	2	281.45	2.4024E+03	588	5.8074E+06
136	Acetic acid	C$_2$H$_4$O$_2$	64,197	53.27	−6,304.5	−4.2985	8.8865E−18	6	289.81	1.2769E−03	591.95	5.7390E+06
137	Propionic acid	C$_3$H$_6$O$_2$	79,094	54.552	−7,194.4	−4.2769	1.1843E−18	6	252.45	1.3142E+01	600.81	4.6080E+06
138	n-Butyric acid	C$_4$H$_8$O$_2$	107,926	93.815	−9,942.2	−9.8019	9.3124E−18	6	267.95	6.7754E+00	615.7	4.0705E+06
139	Isobutyric acid	C$_4$H$_8$O$_2$	79,312	110.38	−10,540	−12.262	1.4310E−17	6	227.15	7.8244E−02	605	3.6834E+06
140	Benzoic aicd	C$_7$H$_6$O$_2$	65,850	88.513	−11,829	−8.6826	2.3248E−19	6	395.45	7.9550E−02	751	4.4691E+06
141	Acetic anhydride	C$_4$H$_6$O$_3$	108,247	100.95	−8,873.2	−11.451	6.1316E−06	2	200.15	2.1999E−02	606	3.9702E+06
142	Methyl formate	C$_2$H$_4$O$_2$	107,313	77.184	−5,606.1	−8.392	7.8468E−06	2	174.15	6.8808E+00	487.2	5.9829E+06
143	Methyl acetate	C$_3$H$_6$O$_2$	79,209	61.267	−5,618.6	−5.6473	2.1080E−17	6	175.15	1.0170E+00	506.55	4.6948E+06
144	Methyl propionate	C$_4$H$_8$O$_2$	554,121	70.717	−6,439.7	−6.9845	2.0129E−17	6	185.65	6.3409E−01	530.6	4.0278E+06
145	Methyl n-butyrate	C$_5$H$_{10}$O$_2$	623,427	71.87	−6,885.7	−7.0944	1.4903E−17	6	187.35	1.3435E−01	554.5	3.4797E+06
146	Ethyl formate	C$_3$H$_6$O$_2$	109,944	73.833	−5,817	−7.809	6.3200E−06	2	193.55	1.8119E−01	508.4	4.7080E+06
147	Ethyl acetate	C$_4$H$_8$O$_2$	141,786	66.824	−6,227.6	−6.41	1.7914E−17	6	189.6	1.4318E+00	523.3	3.8502E+06
148	Ethyl propinate	C$_5$H$_{10}$O$_2$	105,373	105.64	−8,007	−12.477	9.0000E−06	2	199.25	7.7988E−01	546	3.3365E+06
149	Ethyl n-butyrate	C$_6$H$_{12}$O$_2$	105,544	57.661	−6,346.5	−5.032	8.2534E−18	6	175.15	1.0390E−02	571	2.9352E+06

VAPOR PRESSURE—TABLE 3. Vapor Pressure of Inorganic and Organic Liquids (*Continued*)

Cmpd. no.	Name	Formula	CAS no.	C1	C2	C3	C4	C5	T_{min}, K	P_s at T_{min}	T_{max}, K	P_s at T_{max}
150	n-Propyl formate	$C_4H_8O_2$	110,747	104.08	−7,535.9	−12.348	9.6020E−06	2	180.25	2.1101E−01	538	4.0310E+06
151	n-Propyl acetate	$C_5H_{10}O_2$	109,604	115.16	−8,433.9	−13.934	1.0346E−05	2	178.15	1.7113E−02	549.73	3.3657E+06
152	n-Butyl acetate	$C_6H_{12}O_2$	123,864	71.34	−7,285.8	−6.9459	9.9895E−18	6	199.65	1.4347E−01	579.15	3.1097E+06
153	Methyl benzoate	$C_8H_8O_2$	93,583	82.976	−9,226.1	−8.4427	5.9115E−18	6	260.75	1.8653E+00	693	3.5896E+06
154	Ethyl benzoate	$C_9H_{10}O_2$	93,890	53.024	−7,676.8	−4.1593	1.6850E−18	6	238.45	1.4385E−01	698	3.2190E+06
155	Vinyl acetate	$C_4H_6O_2$	108,054	57.406	−5,702.8	−5.0307	1.1042E−17	6	180.35	7.0586E−01	519.13	3.9298E+06
156	Methylamine	CH_5N	74,895	75.206	−5,082.8	−8.0919	8.1130E−06	2	179.69	1.7671E+02	430.05	7.4139E+06
157	Dimethylamine	C_2H_7N	124,403	71.738	−5,302	−7.3324	6.4200E−17	6	180.96	7.5575E+01	437.2	5.2583E+06
158	Trimethylamine	C_3H_9N	75,503	134.68	−6,055.8	−19.415	2.8619E−02	1	156.08	9.9206E+00	433.25	4.1020E+06
159	Ethylamine	C_2H_7N	75,047	81.56	−5,596.9	−9.0779	8.7920E−06	2	192.15	1.5183E+02	456.15	5.5937E+06
160	Diethylamine	$C_4H_{11}N$	109,897	49.314	−4,949	−3.9256	9.1978E−18	6	223.35	3.7411E+02	496.6	3.6744E+06
161	Triethylamine	$C_6H_{15}N$	121,448	56.55	−5,681.9	−4.9815	1.2363E−17	6	158.45	1.0646E−02	535.15	3.0373E+06
162	n-Propylamine	C_3H_9N	107,108	58.398	−5,312.7	−5.2876	1.9913E−06	2	188.36	1.3004E+01	496.95	4.7381E+06
163	di-n-Propylamine	$C_6H_{15}N$	142,847	54	−6,018.5	−4.4981	9.9684E−18	6	210.15	3.6942E+00	550	3.1113E+06
164	Isopropylamine	C_3H_9N	75,310	136.66	−7,201.5	−18.934	2.2255E−02	1	177.95	7.7251E+00	471.85	4.5404E+06
165	Diisopropylamine	$C_6H_{15}N$	108,189	462.84	−18,227	−73.734	9.2794E−02	1	176.85	4.4724E−03	523.1	3.1987E+06
166	Aniline	C_6H_7N	62,533	66.287	−8,207.1	−6.0132	2.8414E−18	6	267.13	7.1322E+00	699	5.3514E+06
167	N-Methylaniline	C_7H_9N	100,618	70.843	−8,517.5	−6.7007	5.6411E−18	6	216.15	1.0207E−02	701.55	5.1935E+06
168	N,N-Dimethylaniline	$C_8H_{11}N$	121,697	51.352	−7,160	−4.0127	8.1481E−07	2	275.5	1.7940E+00	687.15	3.6262E+06
169	Ethylene oxide	C_2H_4O	75,218	91.944	−5,293.4	−11.682	1.4902E−02	1	160.65	7.7879E+00	469.15	7.2553E+06
170	Furan	C_4H_4O	110,009	74.738	−5,417	−8.0636	7.4700E−06	2	187.55	5.0026E+01	490.15	5.5497E+06
171	Thiophene	C_4H_4S	110,021	89.171	−6,860.3	−10.104	7.4769E−06	2	234.94	1.8538E+02	579.35	5.7145E+06
172	Pyridine	C_5H_5N	110,861	82.154	−7,211.3	−8.8646	5.2528E−06	2	231.51	2.0535E+01	619.95	5.6356E+06
173	Formamide	CH_3NO	75,127	100.3	−10,763	−10.946	3.8503E−06	2	275.6	1.0350E+00	771	7.7514E+06
174	N,N-Dimethylformamide	C_3H_7NO	68,122	82.762	−7,955.5	−8.8038	4.2431E−06	2	212.72	1.9532E−01	649.6	4.3653E+06
175	Acetamide	C_2H_5NO	60,355	125.81	−12,376	−14.589	5.0824E−06	2	353.33	3.3637E+02	761	6.5688E+06
176	N-Methylacetamide	C_3H_7NO	79,163	79.128	−9,523.9	−7.7355	3.1616E−18	6	301.15	2.8618E+01	718	4.9973E+06

177	Acetonitrile	C_2H_3N	75,058	58.302	−5,385.6	−5.4954	5.3634E−06	2	229.32	1.8694E+02	545.5	4.8517E+06
178	Propionitrile	C_3H_5N	107,120	82.699	−6,703.5	−9.1506	7.5424E−06	2	180.26	1.6936E−01	564.4	4.1906E+06
179	*n*-Butyronitrile	C_4H_7N	109,740	66.32	−6,714.9	−6.3087	1.3516E−17	6	161.25	6.1777E−04	582.25	3.7870E+06
180	Benzonitrile	C_7H_5N	100,470	55.463	−7,430.8	−4.548	1.7501E−18	6	260.4	5.1063E+00	699.35	4.2075E+06
181	Methyl mercaptan	CH_4S	74,931	54.15	−4,337.7	−4.8127	4.5000E−17	6	150.18	3.1479E+00	469.95	7.2309E+06
182	Ethyl mercaptan	C_2H_6S	75,081	65.551	−5,027.4	−6.6853	6.3208E−06	2	125.26	1.1384E−03	499.15	5.4918E+06
183	*n*-Propyl mercaptan	C_3H_8S	107,039	62.165	−5,624	−5.8595	2.0597E−17	6	159.95	6.5102E−02	536.6	4.6272E+06
184	*n*-Butyl mercaptan	$C_4H_{10}S$	109,795	65.382	−6,262.4	−6.2585	1.4943E−17	6	157.46	2.3532E−03	570.1	3.9730E+06
185	Isobutyl mercaptan	$C_4H_{10}S$	513,440	61.736	−5,909.2	−5.7554	1.5119E−17	6	128.31	4.7502E−06	559	4.0603E+06
186	*sec*-Butyl mercaptan	$C_4H_{10}S$	513,531	60.649	−5,785.9	−5.6113	1.5877E−17	6	133.02	3.3990E−05	554	4.0598E+06
187	Dimethyl sulfide	C_2H_6S	75,183	83.485	−5,711.7	−9.4999	9.8449E−06	2	174.88	7.9009E+00	503.04	5.5324E+06
188	Methyl ethyl sulfide	C_3H_8S	624,895	79.07	−6,114.1	−8.631	6.5333E−06	2	167.23	2.2456E−01	533	4.2610E+06
189	Diethyl sulfide	$C_4H_{10}S$	352,932	60.867	−5,969.6	−5.5979	1.4530E−17	6	169.2	4.3401E−02	557.15	3.9629E+06
190	Fluoromethane	CH_3F	593,533	59.123	−3,043.7	−6.1845	1.6637E−05	2	131.35	4.3287E+02	317.42	5.8754E+06
191	Chloromethane	CH_3Cl	74,873	64.697	−4,048.1	−6.8066	1.0371E−05	2	175.43	8.7091E+02	416.25	6.6905E+06
192	Trichloromethane	$CHCl_3$	67,663	146.43	−7,792.3	−20.614	2.4578E−02	1	207.15	5.2512E+01	536.4	5.5543E+06
193	Tetrachloromethane	CCl_4	56,235	78.441	−6,228.1	−8.5766	6.8465E−06	2	250.33	1.1225E+03	556.35	4.5436E+06
194	Bromomethane	CH_3Br	74,839	72.586	−4,698.6	−7.9966	1.1553E−05	2	179.47	1.9544E+02	467	7.9972E+06
195	Fluoroethane	C_2H_5F	353,366	56.639	−3,576.5	−5.5801	9.8969E−06	2	129.95	8.3714E+00	375.31	5.0060E+06
196	Chloroethane	C_2H_5Cl	75,003	70.159	−4,786.7	−7.5387	9.3370E−06	2	134.8	1.1658E−01	460.35	5.4578E+06
197	Bromoethane	C_2H_5Br	74,964	62.217	−5,113.3	−5.9761	4.7174E−17	6	154.55	3.7155E−01	503.8	6.2903E+06
198	1-Chloropropane	C_3H_7Cl	540,545	79.24	−5,718.8	−8.789	8.4486E−06	2	150.35	6.9630E−02	503.15	4.5812E+06
199	2-Chloropropane	C_3H_7Cl	75,296	46.854	−4,445.5	−3.6533	1.3260E−17	6	155.97	9.0844E−01	489	4.5097E+06
200	1,1-Dichloropropane	$C_3H_6Cl_2$	78,999	83.495	−6,661.4	−9.2386	6.7652E−06	2	200	4.5248E+00	560	4.2394E+06
201	1,2-Dichloropropane	$C_3H_6Cl_2$	78,875	65.955	−6,015.6	−6.5509	4.3172E−06	2	172.71	8.2532E−02	572	4.2319E+06
202	Vinyl chloride	C_2H_3Cl	75,014	91.432	−5,141.7	−10.981	1.4318E−05	2	119.36	1.9178E−02	432	5.7495E+06
203	Fluorobenzene	C_6H_5F	462,066	51.915	−5,439	−4.2896	8.7527E−18	6	230.94	1.5142E+02	560.09	4.5437E+06
204	Chlorobenzene	C_6H_5Cl	108,907	54.144	−6,244.4	−4.5343	4.7030E−18	6	227.95	8.4456E+00	632.35	4.5293E+06
205	Bromobenzene	C_6H_5Br	108,861	63.749	−7,130.2	−5.879	5.2136E−18	6	242.43	7.8364E+00	670.15	4.4196E+06
206	Air†		132,259,100	21.662	−692.39	−0.39208	4.7574E−03	1	59.15	5.6421E+03	132.45	3.7934E+06
207	Hydrogen	H_2	1,333,740	12.69	−94.896	1.1125	3.2915E−04	2	13.95	7.2116E+03	33.19	1.3154E+06
208	Helium-4‡	He	7,440,597	11.533	−8.99	−0.6724	2.7430E−01	1	1.76	1.4625E+03	5.2	2.2845E+05
209	Neon	Ne	7,440,019	29.755	−271.06	−2.6081	5.2700E−04	2	24.56	4.3800E+04	44.4	2.6652E+06

VAPOR PRESSURE—TABLE 3. Vapor Pressure of Inorganic and Organic Liquids (*Continued*)

Cmpd. no.	Name	Formula	CAS no.	C1	C2	C3	C4	C5	T_{min}, K	P_s at T_{min}	T_{max}, K	P_s at T_{max}
210	Argon	Ar	7,440,371	42.127	−1,093.1	−4.1425	5.7254E−05	2	83.78	6.8721E+04	150.86	4.8963E+06
211	Fluorine	F$_2$	7,782,414	42.393	−1,103.3	−4.1203	5.7815E−05	2	53.48	2.5272E+02	144.12	5.1674E+06
212	Chlorine	Cl$_2$	7,782,505	71.334	−3,855	−8.5171	1.2378E−02	1	172.12	1.3660E+03	417.15	7.7930E+06
213	Bromine	Br$_2$	7,726,956	108.26	−6,592	−14.16	1.6043E−02	1	265.85	5.8534E+03	584.15	1.0276E+07
214	Oxygen	O$_2$	7,782,447	51.245	−1,200.2	−6.4361	2.8405E−02	1	54.36	1.4754E+02	154.58	5.0206E+06
215	Nitrogen	N$_2$	7,727,379	58.282	−1,084.1	−8.3144	4.4127E−02	1	63.15	1.2508E+04	126.2	3.3906E+06
216	Ammonia	NH$_3$	7,664,417	90.483	−4,669.7	−11.607	1.7194E−02	1	195.41	6.1111E+03	405.65	1.1301E+07
217	Hydrazine	N$_2$H$_4$	302,012	76.858	−7,245.2	−8.22	6.1557E−03	1	274.69	4.0847E+02	653.15	1.4731E+07
218	Nitrous oxide	N$_2$O	10,024,972	96.512	−4,045	−12.277	2.8860E−05	2	182.3	8.6908E+04	309.57	7.2782E+06
219	Nitric oxide	NO	10,102,439	72.974	−2,650	−8.261	9.7000E−15	6	109.5	2.1956E+04	180.15	6.5156E+06
220	Cyanogen	C$_2$N$_2$	460,195	88.589	−5,059.9	−10.483	1.5403E−05	2	245.25	7.3385E+04	400.15	5.9438E+06
221	Carbon monoxide	CO	630,080	45.698	−1,076.6	−4.8814	7.5673E−05	2	68.15	1.5430E+04	132.92	3.4940E+06
222	Carbon dioxide	CO$_2$	124,389	140.54	−4,735	−21.268	4.0909E−02	1	216.58	5.1867E+05	304.21	7.3896E+06
223	Carbon disulfide	CS$_2$	75,150	67.114	−4,820.4	−7.5303	9.1695E−03	1	161.11	1.4944E+00	552	8.0408E+06
224	Hydrogen fluoride	HF	7,664,393	59.544	−4,143.8	−6.1764	1.4161E−05	2	189.79	3.3683E+02	461.15	6.4872E+06
225	Hydrogen chloride	HCl	7,647,010	104.27	−3,731.2	−15.047	3.1340E−02	1	158.97	1.3522E+04	324.65	8.3564E+06
226	Hydrogen bromide§	HBr	10,035,106	29.315	−2,424.5	−1.1354	2.3806E−18	6	185.15	2.9501E+04	363.15	8.4627E+06
227	Hydrogen cyanide	HCN	74,908	36.75	−3,927.1	−2.1245	3.8948E−17	6	259.83	1.8687E+04	456.65	5.3527E+06
228	Hydrogen sulfide	H$_2$S	7,783,064	85.584	−3,839.9	−11.199	1.8848E−02	6	187.68	2.2873E+04	373.53	8.9988E+06
229	Sulfur dioxide	SO$_2$	7,446,095	47.365	−4,084.5	−3.6469	1.7990E−17	6	197.67	1.6743E+03	430.75	7.8596E+06
230	Sulfur trioxide	SO$_3$	7,446,119	180.99	−12,060	−22.839	7.2350E−17	6	289.95	2.0934E+04	490.85	8.1919E+06
231	Water	H$_2$O	7,732,185	73.649	−7,258.2	−7.3037	4.1653E−06	2	273.16	6.1056E+02	647.13	2.1940E+07

Temperatures are in K; vapor pressures are in Pa.

$P_s \times 9.869233\text{E}{-}06 = \text{atm}$; $P_s \times 1.450377\text{E}{-}04 = \text{psia}$; vapor pressure $= \exp[C1 + (C2/T) + C3 \times \ln(T) + C4 \times T^{C5}]$.

* Decomposes violently on heating. Forms explosive peroxides with air or oxygen. Polymerizes under pressure and heat.

† At the bubble point.

‡ Exhibits superfluid properties below 2.2 K.

§ Coefficients are hypothetical above the decomposition temperature.

VAPOR PRESSURE–TABLE 4. Vapor Pressures of Organic Compounds, up to 1 atm

| Compound | | Pressure, mm Hg | | | | | | | | | | Melting point, °C |
Name	Formula	1	5	10	20	40	60	100	200	400	760	
					Temperature, °C							
Acenaphthalene	$C_{12}H_{10}$		114.8	131.2	148.7	168.2	181.2	197.5	222.1	250.0	277.5	95
Acetal	$C_6H_{14}O_2$	−23.0	−2.3	+8.0	19.6	31.9	39.8	50.1	66.3	84.0	102.2	
Acetaldehyde	C_2H_4O	−81.5	−65.1	−56.8	−47.8	−37.8	−31.4	−22.6	−10.0	+4.9	20.2	−123.5
Acetamide	C_2H_5NO	65.0	92.0	105.0	120.0	135.8	145.8	158.0	178.3	200.0	222.0	81
Acetanilide	C_8H_9NO	114.0	146.6	162.0	180.0	199.6	211.8	227.2	250.5	277.0	303.8	113.5
Acetic acid	$C_2H_4O_2$	−17.2	+6.3	17.5	29.9	43.0	51.7	63.0	80.0	99.0	118.1	16.7
acetic acid anhydride	$C_4H_6O_3$	1.7	24.8	36.0	48.3	62.1	70.8	82.2	100.0	119.8	139.6	−73
Acetone	C_3H_6O	−59.4	−40.5	−31.1	−20.8	−9.4	−2.0	+7.7	22.7	39.5	56.5	−94.6
Acetonitrile	C_2H_3N	−47.0	−26.6	−16.3	−5.0	+7.7	15.9	27.0	43.7	62.5	81.8	−41
Acetophenone	C_8H_8O	37.1	64.0	78.0	92.4	109.4	119.8	133.6	154.2	178.0	202.4	20.5
Acetyl chloride	C_2H_3OCl	−50.0	−35.0	−27.6	−19.6	−10.4	−4.5	+3.2	16.1	32.0	50.8	−112.0
Acetylene	C_2H_2	−142.9	−133.0	−128.2	−122.8	−116.7	−112.8	−107.9	−100.3	−92.0	−84.0	−81.5
Acridine	$C_{13}H_9N$	129.4	165.8	184.0	203.5	224.2	238.7	256.0	284.0	314.3	346.0	110.5
Acrolein (2-propenal)	C_3H_4O	−64.5	−46.0	−36.7	−26.3	−15.0	−7.5	+2.5	17.5	34.5	52.5	−87.7
Acrylic acid	$C_3H_4O_2$	+3.5	27.3	39.0	52.0	66.2	75.0	86.1	103.3	122.0	141.0	14
Adipic acid	$C_6H_{10}O_4$	159.5	191.0	205.5	222.0	240.5	251.0	265.0	287.8	312.5	337.5	152
Allene (propadiene)	C_3H_4	−120.6	−108.0	−101.0	−93.4	−85.2	−78.8	−72.5	−61.3	−48.5	−35.0	−136
Allyl alcohol (propen-1-ol-3)	C_3H_6O	−20.0	+0.2	10.5	21.7	33.4	40.3	50.0	64.5	80.2	96.6	−129
chloride (3-chloropropene)	C_3H_5Cl	−70.0	−52.0	−42.9	−32.8	−21.2	−14.1	−4.5	10.4	27.5	44.6	−136.4
isopropyl ether	$C_6H_{12}O$	−43.7	−23.1	−12.9	−1.8	+10.9	18.7	29.0	44.3	61.7	79.5	
isothiocyanate	C_4H_5NS	−2.0	+25.3	38.3	52.1	67.4	76.2	89.5	108.0	129.8	150.7	−80
n-propyl ether	$C_6H_{12}O$	−39.0	−18.2	−7.9	+3.7	16.4	25.0	35.8	52.6	71.4	90.5	
4-Allylveratrole	$C_{11}H_{14}O_2$	85.0	113.9	127.0	142.8	158.3	169.6	183.7	204.0	226.2	248.0	
iso-Amyl acetate	$C_7H_{14}O_2$	0.0	+23.7	35.2	47.8	62.1	71.0	83.2	101.3	121.5	142.0	
n-Amyl alcohol	$C_5H_{12}O$	+13.6	34.7	44.9	55.8	68.0	75.5	85.8	102.0	119.8	137.8	

VAPOR PRESSURE–TABLE 4. Vapor Pressures of Organic Compounds, up to 1 atm (*Continued*)

Name	Formula	Pressure, mm Hg										Melting point, °C
		1	5	10	20	40	60	100	200	400	760	
		Temperature, °C										
iso-Amyl alcohol	$C_5H_{12}O$	+10.0	30.9	40.8	51.7	63.4	71.0	80.7	95.8	113.7	130.6	–117.2
sec-Amyl alcohol (2-pentanol)	$C_5H_{12}O$	+1.5	22.1	32.2	42.6	54.1	61.5	70.7	85.7	102.3	119.7	
tert-Amyl alcohol	$C_5H_{12}O$	–12.9	+7.2	17.2	27.9	38.8	46.0	55.3	69.7	85.7	101.7	–11.9
sec-Amylbenzene	$C_{11}H_{16}$	29.0	55.8	69.2	83.8	100.0	110.4	124.1	145.2	168.0	193.0	
iso-Amyl benzoate	$C_{12}H_{16}O_2$	72.0	104.5	121.6	139.7	158.3	171.4	186.8	210.2	235.8	262.0	
bromide (1-bromo-3-methylbutane)	$C_5H_{11}Br$	–20.4	+2.1	13.6	26.1	39.8	48.7	60.4	78.7	99.4	120.4	
n-butyrate	$C_9H_{18}O_2$	21.2	47.1	59.9	74.0	90.0	99.8	113.1	133.2	155.3	178.6	
formate	$C_6H_{12}O_2$	–17.5	+5.4	17.1	30.0	44.0	53.3	65.4	83.2	102.7	123.3	
iodide (1-iodo-3-methylbutane)	$C_5H_{11}I$	–2.5	+21.9	34.1	47.6	62.3	71.9	84.4	103.8	125.8	148.2	
isobutyrate	$C_9H_{18}O_2$	14.8	40.1	52.8	66.6	81.8	91.7	104.4	124.2	146.0	168.8	
Amyl isopropionate	$C_8H_{16}O_2$	+8.5	33.7	46.3	60.0	75.5	85.2	97.6	117.3	138.4	160.2	
iso-Amyl isovalerate	$C_{10}H_{20}O_2$	27.0	54.4	68.6	83.8	100.6	110.3	125.1	146.1	169.5	194.0	
n-Amyl levulinate	$C_{10}H_{18}O_3$	81.3	110.0	124.0	139.7	155.8	165.2	180.5	203.1	227.4	253.2	
iso-Amyl levulinate	$C_{10}H_{18}O_3$	75.6	104.0	118.8	134.4	151.7	162.6	177.0	198.1	222.7	247.9	
nitrate	$C_5H_{11}NO_3$	+5.2	28.8	40.3	53.5	67.6	76.3	88.6	106.7	126.5	147.5	
4-*tert*-Amylphenol	$C_{11}H_{16}O$	62.6	109.8	125.5	142.3	160.3	172.6	189.0	213.0	239.5	266.0	93
Anethole	$C_{10}H_{12}O$	62.6	91.6	106.0	121.8	139.3	149.8	164.2	186.1	210.5	235.3	22.5
Angelonitrile	C_5H_7N	–8.0	+15.0	28.0	41.0	55.8	65.2	77.5	96.3	117.7	140.0	
Aniline	C_6H_7N	34.8	57.9	69.4	82.0	96.7	106.0	119.9	140.1	161.9	184.4	–6.2
2-Anilinoethanol	$C_8H_{11}NO$	104.0	134.3	149.6	165.7	183.7	194.0	209.5	230.6	254.5	279.6	
Anisaldehyde	$C_8H_8O_2$	73.2	102.6	117.8	133.5	150.5	161.7	176.7	199.0	223.0	248.0	2.5
o-Anisidine (2-methoxyaniline)	C_7H_9NO	61.0	88.0	101.7	116.1	132.0	142.1	155.2	175.3	197.3	218.5	5.2
Anthracene	$C_{14}H_{10}$	145.0	173.5	187.2	201.9	217.5	231.8	250.0	279.0	310.2	342.0	217.5
Anthraquinone	$C_{14}H_8O_2$	190.0	209.4	234.2	248.3	264.3	273.3	285.0	314.6	346.2	379.9	286
Azelaic acid	$C_9H_{16}O_4$	178.3	210.4	225.5	242.4	260.0	271.8	286.5	309.6	332.8	356.5	106.5
Azelaldehyde	$C_9H_{18}O$	33.3	58.4	71.6	85.0	100.2	110.0	123.0	142.1	163.4	185.0	
Azobenzene	$C_{12}H_{10}N_2$	103.5	135.7	151.5	168.3	187.9	199.8	216.0	240.0	266.1	293.0	68
Benzal chloride (α,α-Dichlorotoluene)	$C_7H_6Cl_2$	35.4	64.0	78.7	94.3	112.1	123.4	138.3	160.7	187.0	214.0	–16.1
Benzaldehyde	C_7H_6O	26.2	50.1	62.0	75.0	90.1	99.6	112.5	131.7	154.1	179.0	–26
Benzanthrone	$C_{17}H_{10}O$	225.0	274.5	297.2	322.5	350.0	368.8	390.0	426.5			174

Temperatures (°C) at the following pressures (mm Hg):

Name	Formula	1	5	10	20	40	60	100	200	400	760	mp °C
Benzene	C_6H_6	-36.7	-19.6	-11.5	-2.6	+7.6	15.4	26.1	42.2	60.6	80.1	+5.5
Benzenesulfonylchloride	$C_6H_5ClO_2S$	65.9	96.5	112.0	129.0	147.7	158.2	174.5	198.0	224.0	251.5	14.5
Benzil	$C_{14}H_{10}O_2$	128.4	165.2	183.0	202.8	224.5	238.2	255.8	283.5	314.3	347.0	95
Benzoic acid	$C_7H_6O_2$	96.0	119.5	132.1	146.7	162.6	172.8	186.2	205.8	227.0	249.2	121.7
anhydride	$C_{14}H_{10}O_3$	143.8	180.0	198.0	218.0	239.8	252.7	270.4	299.1	328.8	360.0	42
Benzoin	$C_{14}H_{12}O_2$	135.6	170.2	188.1	207.0	227.6	241.7	258.0	284.4	313.5	343.0	132
Benzonitrile	C_7H_5N	28.2	55.3	69.2	83.4	99.6	109.8	123.5	144.1	166.7	190.6	-12.9
Benzophenone	$C_{13}H_{10}O$	108.2	141.7	157.6	175.8	195.7	208.2	224.4	249.8	276.8	305.4	48.5
Benzotrichloride (α,α,α-Trichlorotoluene)	$C_7H_5Cl_3$	45.8	73.7	87.6	102.7	119.8	130.0	144.3	165.6	189.2	213.5	-21.2
Benzotrifluoride (α,α,α-Trifluorotoluene)	$C_7H_5F_3$	-32.0	-10.3	-0.4	12.2	25.7	34.0	45.3	62.5	82.0	102.2	-29.3
Benzoyl bromide	C_7H_5BrO	47.0	75.4	89.8	105.4	122.6	133.4	147.7	169.2	193.7	218.5	0
chloride	C_7H_5ClO	32.1	59.1	73.0	87.6	103.8	114.7	128.0	149.5	172.8	197.2	-0.5
nitrile	C_8H_5NO	44.5	71.7	85.5	100.2	116.6	127.0	141.0	161.3	185.0	208.0	33.5
Benzyl acetate	$C_9H_{10}O_2$	45.0	73.4	87.6	102.3	119.6	129.8	144.0	165.5	189.0	213.5	-51.5
alcohol	C_7H_8O	58.0	80.8	92.6	105.8	119.8	129.3	141.7	160.0	183.0	204.7	-15.3
Benzylamine	C_7H_9N	29.0	54.8	67.7	81.8	97.3	107.3	120.0	140.0	161.3	184.5	-4
Benzyl bromide (α-bromotoluene)	C_7H_7Br	32.2	59.6	73.4	88.3	104.8	115.6	129.8	150.8	175.2	198.5	-39
chloride (α-chlorotoluene)	C_7H_7Cl	22.0	47.8	60.8	75.0	90.7	100.5	114.2	134.0	155.8	179.4	39
cinnamate	$C_{16}H_{14}O_2$	173.8	206.3	221.5	239.3	255.8	267.0	281.5	303.8	326.7	350.0	
Benzyldichlorosilane	$C_7H_8Cl_2Si$	45.3	70.2	83.2	96.7	111.8	121.3	133.5	152.0	173.0	194.3	
Benzyl ethyl ether	$C_9H_{12}O$	26.0	52.0	65.0	79.6	95.4	105.5	118.9	139.6	161.5	185.0	
phenyl ether	$C_{13}H_{12}O$	95.4	127.7	144.0	160.7	180.1	192.6	209.2	233.2	259.8	287.0	
isothiocyanate	C_8H_7NS	79.5	107.8	121.8	137.0	153.0	163.8	177.7	198.0	220.4	243.0	
Biphenyl	$C_{12}H_{10}$	70.6	101.8	117.0	134.2	152.5	165.2	180.7	204.2	229.4	254.9	69.5
1-Biphenyloxy-2,3-epoxypropane	$C_{15}H_{14}O_2$	135.5	169.9	187.2	205.8	226.3	239.7	255.0	280.4	309.8	340.0	29
d-Bornyl acetate	$C_{12}H_{20}O_2$	46.9	75.7	90.2	106.0	123.7	135.7	149.8	172.0	197.5	223.0	
Bornyl n-butyrate	$C_{14}H_{24}O_2$	74.0	103.4	118.0	133.8	150.7	161.8	176.4	198.0	222.2	247.0	
formate	$C_{11}H_{18}O_2$	47.0	74.8	89.3	104.0	121.2	131.7	145.8	166.4	190.2	214.0	
isobutyrate	$C_{14}H_{24}O_2$	70.0	99.8	114.0	130.0	147.2	157.6	172.2	194.2	218.2	243.0	
propionate	$C_{13}H_{22}O_2$	64.6	93.7	108.0	123.7	140.4	151.2	165.7	187.5	211.2	235.0	
Brassidic acid	$C_{22}H_{42}O_2$	209.6	241.7	256.0	272.9	290.0	301.5	316.2	336.8	359.6	382.5	61.5
Bromoacetic acid	$C_2H_3BrO_2$	54.7	81.6	94.1	108.2	124.0	133.8	146.3	165.8	186.7	208.0	49.5
4-Bromoanisole	C_7H_7BrO	48.8	77.8	91.9	107.8	125.0	136.0	150.1	172.7	197.5	223.0	12.5
Bromobenzene	C_6H_5Br	+2.9	27.8	40.0	53.8	68.6	78.1	90.8	110.1	132.3	156.2	-30.7

VAPOR PRESSURE–TABLE 4. Vapor Pressures of Organic Compounds, up to 1 atm (*Continued*)

Name	Formula	Pressure, mm Hg										Melting point, °C
		1	5	10	20	40	60	100	200	400	760	
		Temperature, °C										
4-Bromobiphenyl	$C_{12}H_9Br$	98.0	133.7	150.6	169.8	190.8	204.5	221.8	248.2	277.7	310.0	90.5
1-Bromo-2-butanol	C_4H_9BrO	23.7	45.4	55.8	67.2	79.5	87.0	97.6	112.1	128.3	145.0	
1-Bromo-2-butanone	C_4H_7BrO	+6.2	30.0	41.8	54.2	68.2	77.3	89.2	107.0	126.3	147.0	
cis-1-Bromo-1-butene	C_4H_7Br	-44.0	-23.2	-12.8	-1.4	+11.5	19.8	30.8	47.8	66.8	86.2	-100.3
trans-1-Bromo-1-butene	C_4H_7Br	-38.4	-17.0	-6.4	+5.4	18.4	27.2	38.1	55.7	75.0	94.7	-133.4
2-Bromo-1-butene	C_4H_7Br	-47.3	-27.0	-16.8	-5.3	+7.2	15.4	26.3	42.8	61.9	81.0	-111.2
cis-2-Bromo-2-butene	C_4H_7Br	-39.0	-17.9	-7.2	+4.6	17.7	26.2	37.5	54.5	74.0	93.9	-114.6
trans-2-Bromo-2-butene	C_4H_7Br	-45.0	-24.1	-13.8	-2.4	+10.5	18.7	29.9	46.5	66.0	85.5	
1,4-Bromochlorobenzene	C_6H_4BrCl	32.0	59.5	72.7	87.8	103.8	114.8	128.0	149.5	172.6	196.9	16.6
1-Bromo-1-chloroethane	C_2H_4BrCl	-36.0	-18.0	-9.4	0.0	+10.4	17.0	28.0	44.7	63.4	82.7	-16.6
1-Bromo-2-chloroethane	C_2H_4BrCl	-28.8	-7.0	+4.1	16.0	29.7	38.0	49.5	66.8	86.0	106.7	
2-Bromo-4,6-dichlorophenol	$C_6H_3BrCl_2O$	84.0	115.6	130.8	147.7	165.8	177.6	193.2	216.5	242.0	268.0	68
1-Bromo-4-ethyl)-benzene	C_8H_9Br	30.4	42.5	74.0	90.2	108.5	121.0	135.5	156.5	182.0	206.0	-45.0
(2-Bromoethyl)-benzene	C_8H_9Br	48.0	76.2	90.5	105.8	123.2	133.8	148.2	169.8	194.0	219.0	
2-Bromoethyl 2-chloroethyl ether	C_4H_8BrClO	36.5	63.2	76.3	90.8	106.6	116.4	129.8	150.0	172.3	195.8	
(2-Bromoethyl)-cyclohexane	$C_8H_{15}Br$	38.7	66.6	80.5	95.8	113.0	123.7	138.0	160.0	186.2	213.0	
1-Bromoethylene	C_2H_3Br	-95.4	-77.8	-68.8	-58.8	-48.1	-41.2	-31.9	-17.2	-1.1	+15.8	-138
Bromoform (tribromomethane)	$CHBr_3$		22.0	34.0	48.0	63.6	73.4	85.9	106.1	127.9	150.5	8.5
1-Bromonaphthalene	$C_{10}H_7Br$	84.2	117.5	133.6	150.2	170.2	183.5	198.8	224.2	252.0	281.1	5.5
2-Bromo-4-phenylphenol	$C_{12}H_9BrO$	100.0	135.4	152.3	171.8	193.8	207.0	224.5	251.0	280.2	311.0	95
3-Bromopyridine	C_5H_4BrN	16.8	42.0	55.2	69.1	84.1	94.1	107.8	127.7	150.0	173.4	
2-Bromotoluene	C_7H_7Br	24.4	49.7	62.3	76.0	91.0	100.0	112.0	133.6	157.3	181.8	-28
3-Bromotoluene	C_7H_7Br	14.8	50.8	64.0	78.1	93.9	104.1	117.8	138.0	160.0	183.7	39.8
4-Bromotoluene	C_7H_7Br	10.3	47.5	61.1	75.2	91.8	102.3	116.4	137.4	160.2	184.5	28.5
3-Bromo-2,4,6-trichlorophenol	$C_6H_2BrCl_3O$	112.4	146.2	163.2	181.8	200.5	213.0	229.3	253.0	278.0	305.0	
2-Bromo-1,4-xylene	C_8H_9Br	37.5	65.0	78.8	94.0	110.6	121.6	135.7	156.4	181.0	206.7	+9.5
1,2-Butadiene (methyl allene)	C_4H_6	-89.0	-72.7	-64.2	-54.9	-44.3	-37.5	-28.3	-14.2	+1.8	18.5	-108.9
1,3-Butadiene	C_4H_6	-102.8	-87.6	-79.7	-71.0	-61.3	-55.1	-46.8	-33.9	-19.3	-4.5	-135
n-Butane	C_4H_{10}	-101.5	-85.7	-77.8	-68.9	-59.1	-52.8	-44.2	-31.2	-16.3	-0.5	

Name	Formula											
iso-Butane (2-methylpropane)	C_4H_{10}	-109.2	-94.1	-86.4	-77.9	-68.4	-62.4	-54.1	-41.5	-27.1	-11.7	-145
1,3-Butanediol	$C_4H_{10}O_2$	22.2	67.5	85.3	100.0	117.4	127.5	141.2	161.0	183.8	206.5	77
1,2,3-Butanetriol	$C_4H_{10}O_3$	102.0	132.0	146.0	161.0	178.0	188.0	202.5	222.0	243.5	264.0	
1-Butene	C_4H_8	-104.8	-89.4	-81.6	-73.0	-63.4	-57.2	-48.9	-36.2	-21.7	-6.3	-130
cis-2-Butene	C_4H_8	-96.4	-81.1	-73.4	-64.6	-54.7	-48.4	-39.8	-26.8	-12.0	+3.7	-138.9
trans-2-Butene	C_4H_8	-99.4	-84.0	-76.3	-67.5	-57.6	-51.3	-42.7	-29.7	-14.8	+0.9	-105.4
3-Butenenitrile	C_4H_5N	-19.6	+2.9	14.1	26.6	40.0	48.8	60.2	78.0	98.0	119.0	
iso-Butyl acetate	$C_6H_{12}O_2$	-21.2	+1.4	12.8	25.5	39.2	48.0	59.7	77.6	97.5	118.0	-98.9
n-Butyl acrylate	$C_7H_{12}O_2$	-0.5	+23.5	35.5	48.6	63.4	72.6	85.1	104.0	125.2	147.2	-64.6
n-Butyl alcohol	$C_4H_{10}O$	-1.2	+20.0	30.2	41.5	53.4	60.3	70.1	84.3	100.8	117.5	-79.9
iso-Butyl alcohol	$C_4H_{10}O$	-9.0	+11.6	21.7	32.4	44.1	51.7	61.5	75.9	91.4	108.0	-108
sec-Butyl alcohol	$C_4H_{10}O$	-12.2	+7.2	16.9	27.3	38.1	45.2	54.1	67.9	83.9	99.5	-114.7
tert-Butyl alcohol	$C_4H_{10}O$	-20.4	-3.0	+5.5	14.3	24.5	31.0	39.8	52.7	68.0	82.9	25.3
iso-Butyl amine	$C_4H_{11}N$	-50.0	-31.0	-21.0	-10.3	+1.3	8.8	18.8	32.0	50.7	68.6	-85.0
n-Butylbenzene	$C_{10}H_{14}$	22.7	48.8	62.0	76.3	92.4	102.6	116.2	136.9	159.2	183.1	-88.0
iso-Butylbenzene	$C_{10}H_{14}$	14.1	40.5	53.7	67.8	83.3	93.3	107.0	127.2	149.6	172.8	-51.5
sec-Butylbenzene	$C_{10}H_{14}$	18.6	44.2	57.0	70.6	86.2	96.0	109.5	128.8	150.3	173.5	-75.5
tert-Butylbenzene	$C_{10}H_{14}$	13.0	39.0	51.7	65.6	80.8	90.6	103.8	123.7	145.8	168.5	-58
iso-Butyl benzoate	$C_{11}H_{14}O_2$	64.0	93.6	108.6	124.2	141.8	152.0	166.4	188.2	212.8	237.0	
n-Butyl bromide (1-bromobutane)	C_4H_9Br	-33.0	-11.2	-0.3	+11.6	24.8	33.4	44.7	62.0	81.7	101.6	-112.4
iso-Butyl n-butyrate	$C_8H_{16}O_2$	+4.6	30.0	42.2	56.1	71.7	81.3	94.0	113.9	135.7	156.9	
Butyl carbitol (diethylene glycol butyl ether)	$C_8H_{18}O_3$	70.0	83.7	96.4	110.1	125.3	134.6	147.2	165.7	186.0	206.5	65
n-Butyl chloride (1-chlorobutane)	C_4H_9Cl	70.0	95.7	107.8	120.5	135.5	146.0	159.8	181.2	205.0	231.2	
iso-Butyl chloride	C_4H_9Cl	-49.0	-28.9	-18.6	-7.4	+5.0	13.0	24.0	40.0	58.8	77.8	-123.1
sec-Butyl chloride (2-Chlorobutane)	C_4H_9Cl	-53.8	-34.3	-24.5	-13.8	-1.9	+5.9	16.0	32.0	50.0	68.9	-131.2
tert-Butyl chloride	C_4H_9Cl	-60.2	-39.8	-29.2	-17.7	-5.0	+3.4	14.2	31.5	50.0	68.0	-131.3
sec-Butyl chloroacetate	$C_6H_{11}ClO_2$					-19.0	-11.4	-1.0	+14.6	32.6	51.0	-26.5
2-tert-Butyl-4-cresol	$C_{11}H_{16}O$	17.0	41.8	54.6	68.2	83.6	93.0	105.5	124.1	146.0	167.8	
4-tert-Butyl-2-cresol	$C_{11}H_{16}O$	70.0	98.0	112.0	127.2	143.9	153.7	167.0	187.8	210.0	232.6	
iso-Butyl dichloroacetate	$C_6H_{10}Cl_2O_2$	74.3	103.7	118.0	134.0	150.8	161.7	176.2	197.8	221.8	247.0	
2,3-Butylene glycol (2,3-butanediol)	$C_4H_{10}O_2$	28.6	54.3	67.5	81.4	96.7	106.6	119.8	139.2	160.0	183.0	22.5
2-Butyl-2-ethylbutane-1,3-diol	$C_{10}H_{22}O_2$	44.0	68.4	80.3	93.4	107.8	116.3	127.8	145.6	164.0	182.0	
		94.1	122.6	136.8	151.2	167.8	178.0	191.9	212.0	233.5	255.0	

VAPOR PRESSURE–TABLE 4. Vapor Pressures of Organic Compounds, up to 1 atm (Continued)

Name	Formula	Pressure, mm Hg										Melting point, °C
		Temperature, °C										
		1	5	10	20	40	60	100	200	400	760	
2-tert-Butyl-4-ethylphenol	C$_{12}$H$_{15}$O	76.3	106.2	121.0	137.0	154.0	165.4	179.0	200.3	223.8	247.8	
n-Butyl formate	C$_5$H$_{10}$O$_2$	−26.4	−4.7	+6.1	18.0	31.6	39.8	51.0	67.9	86.2	106.0	
iso-Butyl formate	C$_5$H$_{10}$O$_2$	−32.7	−11.4	−0.8	+11.0	24.1	32.4	43.4	60.0	79.0	98.2	−95.3
sec-Butyl formate	C$_5$H$_{10}$O$_2$	−34.4	−13.3	−3.1	+8.4	21.3	29.6	40.2	56.8	75.2	93.6	
sec-Butyl glycolate	C$_6$H$_{12}$O$_3$	28.3	53.6	66.0	79.8	94.2	104.0	116.4	135.5	155.6	177.5	
iso-Butyl iodide (1-iodo-2-methylpropane)	C$_4$H$_9$I	−17.0	+5.8	17.0	29.8	42.8	51.8	63.5	81.0	100.3	120.4	−90.7
isobutyrate	C$_8$H$_{16}$O$_2$	+4.1	28.0	39.9	52.4	67.2	75.9	88.0	106.3	126.3	147.5	−80.7
isovalerate	C$_9$H$_{18}$O$_2$	16.0	41.2	53.8	67.7	82.7	92.4	105.2	124.8	146.4	168.7	
levulinate	C$_9$H$_{16}$O$_3$	65.0	92.1	105.9	120.2	136.2	147.0	160.2	181.8	205.5	229.9	
naphthylketone (1-isovaleronaphthone)	C$_{15}$H$_{16}$O	136.0	167.9	184.0	201.6	219.7	231.5	246.7	269.7	294.0	320.0	
2-sec-Butylphenol	C$_{10}$H$_{14}$O	57.4	86.0	100.8	116.1	133.4	143.9	157.3	179.7	203.8	228.0	
2-tert-Butylphenol	C$_{10}$H$_{14}$O	56.6	84.2	98.1	113.0	129.2	140.0	153.5	173.8	196.3	219.5	
4-iso-Butylphenol	C$_{10}$H$_{14}$O	72.1	100.9	115.5	130.3	147.2	157.0	171.2	192.1	214.7	237.0	
4-sec-Butylphenol	C$_{10}$H$_{14}$O	71.4	100.5	114.8	130.3	147.8	157.9	172.4	194.3	217.6	242.1	
4-tert-Butylphenol	C$_{10}$H$_{14}$O	70.0	99.2	114.0	129.5	146.0	156.0	170.2	191.5	214.0	238.0	99
2-(4-tert-Butylphenoxy)ethyl acetate	C$_{14}$H$_{20}$O$_3$	118.0	150.0	165.8	183.3	201.5	212.8	228.0	250.3	277.6	304.4	
4-tert-Butylphenyl dichlorophosphate	C$_{10}$H$_{13}$Cl$_2$O$_2$P	96.0	129.6	146.0	164.0	184.3	197.2	214.3	240.0	268.2	299.0	
tert-Butyl phenyl ketone (pivalophenone)	C$_{11}$H$_{14}$O	57.8	85.7	99.0	114.3	130.4	140.8	154.0	175.0	197.7	220.0	
iso-Butyl propionate	C$_7$H$_{14}$O$_2$	−2.3	+20.9	32.3	44.8	58.5	67.6	79.5	97.0	116.4	136.8	−71
4-tert-Butyl-2,5-xylenol	C$_{12}$H$_{18}$O	88.2	119.8	135.0	151.0	169.8	180.3	195.0	217.5	241.3	265.3	
4-tert-Butyl-2,6-xylenol	C$_{12}$H$_{18}$O	74.0	103.9	119.0	135.0	152.2	163.6	176.0	196.0	217.8	239.8	
6-tert-Butyl-2,4-xylenol	C$_{12}$H$_{18}$O	70.3	100.2	115.0	131.0	148.5	158.2	172.0	192.3	214.2	236.5	
6-tert-Butyl-3,4-xylenol	C$_{12}$H$_{18}$O	83.9	113.6	127.0	143.0	159.7	170.0	184.0	204.5	226.7	249.5	
Butyric acid	C$_4$H$_8$O$_2$	25.5	49.8	61.5	74.0	88.0	96.5	108.0	125.5	144.5	163.5	−74
iso-Butyric acid	C$_4$H$_8$O$_2$	14.7	39.3	51.2	64.0	77.8	86.3	98.0	115.8	134.5	154.5	−47
Butyronitrile	C$_4$H$_7$N	−20.0	+2.1	13.4	25.7	38.4	47.3	59.0	76.7	96.8	117.5	
iso-Valerophenone	C$_{11}$H$_{14}$O	58.3	87.0	101.4	116.8	133.8	144.6	158.0	180.1	204.2	228.0	

Compound	Formula										
Camphene	$C_{10}H_{16}$			47.2	60.4	75.7	85.0	97.9	117.5	138.7	160.5	50
Campholenic acid	$C_{10}H_{16}O_2$	97.6	125.7	139.8	153.9	170.0	180.0	193.7	212.7	234.0	256.0	
d-Camphor	$C_{10}H_{16}O$	41.5	68.6	82.3	97.5	114.0	124.0	138.0	157.9	182.0	209.2	178.5
Camphylamine	$C_{10}H_{19}N$	45.3	74.0	83.7	97.6	112.5	122.0	134.6	153.0	173.8	195.0	
Capraldehyde	$C_{10}H_{20}O$	51.9	78.8	92.0	106.3	122.2	132.0	145.3	164.8	186.3	208.5	
Capric acid	$C_{10}H_{20}O_2$	125.0	142.0	152.2	165.0	179.9	189.8	200.0	217.1	240.3	268.4	31.5
n-Caproic acid	$C_6H_{12}O_2$	71.4	89.5	99.5	111.8	125.0	133.3	144.0	160.8	181.0	202.0	-1.5
iso-Caproic acid	$C_6H_{12}O_2$	66.2	83.0	94.0	107.0	120.4	129.6	141.4	158.3	181.0	207.7	-35
iso-Caprolactone	$C_6H_{10}O_2$	38.3	66.4	80.3	95.7	112.3	123.2	137.2	157.8	182.1	207.0	
Capronitrile	$C_6H_{11}N$	9.2	34.6	47.5	61.7	76.9	86.8	99.8	119.7	141.0	163.7	
Capryl alcohol (2-octanol)	$C_8H_{18}O$	32.8	57.6	70.0	83.3	98.0	107.4	119.8	138.0	157.5	178.5	-38.6
Caprylaldehyde	$C_8H_{16}O$	73.4	92.0	101.2	110.2	120.0	126.0	133.9	145.4	156.5	168.5	
Caprylic acid (octanoic acid)	$C_8H_{16}O_2$	92.3	114.1	124.0	136.4	150.6	160.0	172.2	190.3	213.9	237.5	16
Caprylonitrile	$C_8H_{15}N$	43.0	67.6	80.4	94.6	110.6	121.2	134.8	155.2	179.5	204.5	
Carbazole	$C_{12}H_9N$						248.2	265.0	292.5	323.0	354.8	244.8
Carbon dioxide	CO_2	-134.3	-124.4	-119.5	-114.4	-108.6	-104.8	-100.2	-93.0	-85.7	-78.2	-57.5
disulfide	CS_2	-73.8	-54.3	-44.7	-34.3	-22.5	-15.3	-5.1	+10.4	28.0	46.5	-110.8
monoxide	CO	-222.0	-217.2	-215.0	-212.8	-210.0	-208.1	-205.7	-201.3	-196.3	-191.3	-205.0
oxyselenide (carbonyl selenide)	$COSe$	-117.1	-102.3	-95.0	-86.3	-76.4	-70.2	-61.7	-49.8	-35.6	-21.9	
oxysulfide (carbonyl sulfide)	COS	-132.4	-119.8	-113.3	-106.0	-98.3	-93.0	-85.9	-75.0	-62.7	-49.9	-138.8
tetrabromide	CBr_4					96.3	106.3	119.7	139.7	163.5	189.5	90.1
tetrachloride	CCl_4	-50.0	-30.0	-19.6	-8.2	+4.3	12.3	23.0	38.3	57.8	76.7	-22.6
tetrafluoride	CF_4	-184.6	-174.1	-169.3	-164.3	-158.8	-155.4	-150.7	-143.6	-135.5	-127.7	-183.7
Carvacrol	$C_{10}H_{14}O$	70.0	98.4	113.2	127.9	145.2	155.3	169.7	191.2	213.8	237.0	+0.5
Carvone	$C_{10}H_{14}O$	57.4	86.1	100.4	116.1	133.0	143.8	157.3	179.6	203.5	227.5	
Chavibetol	$C_{10}H_{12}O_2$	83.6	113.3	127.0	143.2	159.8	170.7	185.5	206.8	229.8	254.0	
Chloral (trichloroacetaldehyde)	C_2HCl_3O	-37.8	-16.0	-5.0	+7.2	20.2	29.1	40.2	57.8	77.5	97.7	-57
hydrate (trichloroacetaldehyde hydrate)	$C_2H_3Cl_3O_2$	-9.8	+10.0	19.5	29.2	39.7	46.2	55.0	68.0	82.1	96.2	51.7
Chloranil	$C_6Cl_4O_2$	70.7	89.3	97.8	106.4	116.1	122.0	129.5	140.3	151.3	162.6	290
Chloroacetic acid	$C_2H_3ClO_2$	43.0	68.3	81.0	94.2	109.2	118.3	130.7	149.0	169.0	189.5	61.2
anhydride	$C_4H_4Cl_2O_3$	67.2	94.1	108.0	122.4	138.2	148.0	159.8	177.8	197.0	217.0	46
2-Chloroaniline	C_6H_6ClN	46.3	72.3	84.8	99.2	115.6	125.7	139.5	160.0	183.7	208.8	0
3-Chloroaniline	C_6H_6ClN	63.5	89.8	102.0	116.7	133.6	144.1	158.0	179.5	203.5	228.5	-10.4
4-Chloroaniline	C_6H_5Cl	59.3	87.9	102.1	117.8	135.0	145.8	159.9	182.3	206.6	230.5	70.5

VAPOR PRESSURE–TABLE 4. Vapor Pressures of Organic Compounds, up to 1 atm (*Continued*)

| Name | Formula | \multicolumn{10}{Pressure, mmHg} | Melting point, °C |

Name	Formula	1	5	10	20	40	60	100	200	400	760	Melting point, °C
						Temperature, °C						
Chlorobenzene	C_6H_5Cl	−13.0	+10.6	22.2	35.3	49.7	58.3	70.7	89.4	110.0	132.2	−45.2
2-Chlorobenzotrichloride (2-α,α,α-tetrachlorotoluene)	$C_7H_4Cl_4$	69.0	101.8	117.9	135.8	155.0	167.8	185.0	208.0	233.0	262.1	28.7
2-Chlorobenzotrifluoride (2-chloro-α,α,α-trifluorotoluene)	$C_7H_4ClF_3$	0.0	24.7	37.1	50.6	65.9	75.4	88.3	108.3	130.0	152.2	−6.0
2-Chlorobiphenyl	$C_{12}H_9Cl$	89.3	109.8	134.7	151.2	169.9	182.1	197.0	219.6	243.8	267.5	34
4-Chlorobiphenyl	$C_{12}H_9Cl$	96.4	129.8	146.0	164.0	183.8	196.0	212.5	237.8	264.5	292.9	75.5
α-Chlorocrotonic acid	$C_4H_5ClO_2$	70.0	95.6	108.0	121.2	135.6	144.4	155.9	173.8	193.2	212.0	
Chlorodifluoromethane	$CHClF_2$	−122.8	−110.2	−103.7	−96.5	−88.6	−83.4	−76.4	−65.8	−53.6	−40.8	−160
Chlorodimethylphenylsilane	$C_8H_{11}ClSi$	29.8	56.7	70.0	84.7	101.2	111.5	124.7	145.5	168.6	193.5	
1-Chloro-2-ethoxybenzene	C_8H_9ClO	45.8	72.8	86.5	101.5	117.8	127.8	141.8	162.0	185.5	208.0	
2-(2-Chloroethoxy) ethanol	$C_4H_9ClO_2$	53.0	78.3	90.7	104.1	118.4	127.5	139.5	157.2	176.5	196.0	
bis-2-Chloroethyl acetacetal	$C_6H_{12}Cl_2O_2$	56.2	83.7	97.6	112.2	127.8	138.0	150.7	169.8	190.5	212.6	
1-Chloro-2-ethylbenzene	C_8H_9Cl	17.2	43.0	56.1	70.3	86.2	96.4	110.0	130.2	152.2	177.6	−80.2
1-Chloro-3-ethylbenzene	C_8H_9Cl	18.6	45.2	58.1	73.0	89.2	99.6	113.6	133.8	156.7	181.1	−53.3
1-Chloro-4-ethylbenzene	C_8H_9Cl	19.2	46.4	60.0	75.5	91.8	102.0	116.0	137.0	159.8	184.3	−62.6
2-Chloroethyl chloroacetate	$C_4H_6Cl_2O_2$	46.0	72.1	86.0	100.0	116.0	126.2	140.0	159.8	182.2	205.0	
2-Chloroethyl 2-chloroisopropyl ether	$C_5H_{10}Cl_2O$	24.7	50.1	63.0	77.2	92.4	102.2	115.8	135.7	156.5	180.0	
2-Chloroethyl 2-chloropropyl ether	$C_5H_{10}Cl_2O$	29.8	56.5	70.0	84.8	101.5	111.8	125.6	146.3	169.8	194.1	
2-Chloroethyl α-methylbenzyl ether	$C_{10}H_{13}ClO$	62.3	91.4	106.0	121.8	139.6	150.0	164.8	186.3	210.8	235.0	
Chloroform (trichloromethane)	$CHCl_3$	−58.0	−39.1	−29.7	−19.0	−7.1	+0.5	10.4	25.9	42.7	61.3	−63.5
1-Chloronaphthalene	$C_{10}H_7Cl$	80.6	104.8	118.6	134.4	153.2	165.6	180.4	204.2	230.8	259.3	−20
4-Chlorophenethyl alcohol	C_8H_9ClO	84.0	114.3	129.0	145.0	162.0	173.5	188.1	210.0	234.5	259.3	
2-Chlorophenol	C_6H_5ClO	12.1	38.2	51.2	65.9	82.0	92.0	106.0	126.4	149.8	174.5	7
3-Chlorophenol	C_6H_5ClO	44.2	72.0	86.1	101.7	118.0	129.4	143.0	164.8	188.7	214.0	32.5
4-Chlorophenol	C_6H_5ClO	49.8	78.2	92.2	108.1	125.0	136.1	150.0	172.0	196.0	220.0	42
2-Chloro-3-phenylphenol	$C_{12}H_9ClO$	118.0	152.2	169.7	186.7	207.4	219.6	237.0	261.3	289.4	317.5	+6
2-Chloro-6-phenylphenol	$C_{12}H_9ClO$	119.8	153.7	170.7	189.8	208.2	220.0	237.1	261.6	289.5	317.0	
Chloropicrin (trichloronitromethane)	CCl_3NO_2	−25.5	−3.3	+7.8	20.0	33.8	42.3	53.8	71.8	91.8	111.9	−64

Compound	Formula											
1-Chloropropene	C_3H_5Cl	−81.3	−63.4	−54.1	−44.0	−32.7	−25.1	−15.1	+1.3	18.0	37.0	−99.0
2-Chloropyridine	C_5H_4ClN	13.3	38.8	51.7	65.8	81.7	91.6	104.6	125.0	147.7	170.2	
3-Chlorostyrene	C_8H_7Cl	25.3	51.3	65.2	80.0	96.5	107.2	121.2	142.2	165.7	190.0	
4-Chlorostyrene	C_8H_7Cl	28.0	54.5	67.5	82.0	98.0	108.5	122.0	143.5	166.0	191.0	−15.0
1-Chlorotetradecane	$C_{14}H_{29}Cl$	98.5	131.8	148.2	166.2	187.0	199.8	215.5	240.3	267.5	296.0	+0.9
2-Chlorotoluene	C_7H_7Cl	+5.4	30.6	43.2	56.9	72.0	81.8	94.7	115.0	137.1	159.3	
3-Chlorotoluene	C_7H_7Cl	+4.8	30.3	43.2	57.4	73.0	83.2	96.3	116.6	139.7	162.3	
4-Chlorotoluene	C_7H_7Cl	+5.5	31.0	43.8	57.8	73.5	83.3	96.6	117.1	139.8	162.3	+7.3
Chlorotriethylsilane	$C_6H_{15}ClSi$	−4.9	+19.8	32.0	45.5	60.2	69.5	82.3	101.6	123.6	146.3	
1-Chloro-1,2,2-trifluoroethylene	C_2ClF_3	−116.0	−102.5	−95.9	−88.2	−79.7	−74.1	−66.7	−55.0	−41.7	−27.9	−157.5
Chlorotrifluoromethane	$CClF_3$	−149.5	−139.2	−134.1	−128.5	−121.9	−117.3	−111.7	−102.5	−92.7	−81.2	
Chlorotrimethylsilane	C_3H_9ClSi	−62.8	−43.6	−34.0	−23.2	−11.4	−4.0	+6.0	21.9	39.4	57.9	
trans-Cinnamic acid	$C_9H_8O_2$	127.5	157.8	173.0	189.5	207.1	217.8	232.4	253.3	276.7	300.0	133
Cinnamyl alcohol	$C_9H_{10}O$	72.6	102.5	117.8	133.7	151.0	162.0	177.8	199.8	224.6	250.0	33
Cinnamaldehyde	C_9H_8O	76.1	105.8	120.0	135.7	152.2	163.7	177.7	199.3	222.4	246.0	−7.5
Citraconic anhydride	$C_5H_4O_3$	47.1	74.8	88.9	103.8	120.3	131.3	145.4	165.8	189.8	213.5	
cis-α-Citral	$C_{10}H_{16}O$	61.7	90.0	103.9	119.4	135.9	146.3	160.0	181.8	205.0	228.0	
d-Citronellal	$C_{10}H_{18}O$	44.0	71.4	84.8	99.8	116.1	126.2	140.1	160.0	183.8	206.5	
Citronellic acid	$C_{10}H_{18}O_2$	99.5	127.3	141.4	155.6	171.9	182.1	195.4	214.5	236.6	257.0	
Citronellol	$C_{10}H_{20}O$	66.4	93.6	107.0	121.5	137.2	147.2	159.8	179.8	201.0	221.5	
Citronellyl acetate	$C_{12}H_{22}O_2$	74.7	100.2	113.0	126.0	140.5	149.7	161.0	178.8	197.8	217.0	
Coumarin	$C_9H_6O_2$	106.0	137.8	153.4	170.0	189.0	200.5	216.5	240.0	264.7	291.0	70
o-Cresol (2-cresol; 2-methylphenol)	C_7H_8O	38.2	64.0	76.7	90.5	105.8	115.5	127.4	146.7	168.4	190.8	30.8
m-Cresol (3-cresol; 3-methylphenol)	C_7H_8O	52.0	76.0	87.8	101.4	116.0	125.8	138.0	157.3	179.0	202.8	10.9
p-Cresol (4-cresol; 4-methylphenol)	C_7H_8O	53.0	76.5	88.6	102.3	117.7	127.0	140.0	157.3	179.4	201.8	35.5
cis-Crotonic acid	$C_4H_6O_2$	33.5	57.4	69.0	82.0	96.0	104.5	116.3	133.9	152.2	171.9	15.5
trans-Crotonic acid	$C_4H_6O_2$			80.0	93.0	107.8	116.7	128.0	146.0	165.5	185.0	72
cis-Crotononitrile	C_4H_5N	−29.0	−7.1	+4.0	16.4	30.0	38.5	50.1	68.0	88.0	108.0	
trans-Crotononitrile	C_4H_5N	−19.5	+3.5	15.0	27.8	41.8	50.9	62.8	81.1	101.5	122.8	
Cumene	C_9H_{12}	+2.9	26.8	38.3	51.5	66.1	75.4	88.1	107.3	129.2	152.4	−96.0
4-Cumidene	$C_9H_{13}N$	60.0	88.2	102.2	117.8	134.2	145.0	158.0	180.0	203.2	227.0	
Cuminal	$C_{10}H_{12}O$	58.0	87.3	102.0	117.9	135.2	146.0	160.0	182.8	206.7	232.0	
Cuminyl alcohol	$C_{10}H_{14}O$	74.2	103.7	118.0	133.8	150.3	161.7	176.2	197.9	221.7	246.6	
2-Cyano-2-n-butyl acetate	$C_7H_{11}NO_2$	42.0	68.7	82.0	96.2	111.8	121.5	133.8	152.2	173.4	195.2	

VAPOR PRESSURE–TABLE 4. Vapor Pressures of Organic Compounds, up to 1 atm (*Continued*)

Name	Formula	Pressure, mm Hg										Melting point, °C
		Temperature, °C										
		1	5	10	20	40	60	100	200	400	760	
Cyanogen	C_2N_2	−95.8	−83.2	−76.8	−70.1	−62.7	−57.9	−51.8	−42.6	−33.0	−21.0	−34.4
bromide	CBrN	−35.7	−13.3	−10.0	−1.0	+8.6	14.7	22.6	33.8	46.0	61.5	58
chloride	CClN	−76.7	−61.4	−53.8	−46.1	−37.5	−32.1	−24.9	−14.1	−2.3	+13.1	−6.5
iodide	CIN	25.2	47.2	57.7	68.6	80.3	88.0	97.6	111.5	126.1	141.1	
Cyclobutane	C_4H_8	−92.0	−76.0	−67.9	−58.7	−48.4	−41.8	−32.8	−18.9	−3.4	+12.9	−50
Cyclobutene	C_4H_6	−99.1	−83.4	−75.4	−66.6	−56.4	−50.0	−41.2	−27.8	−12.2	+2.4	
Cyclohexane	C_6H_{12}	−45.3	−25.4	−15.9	−5.0	+6.7	14.7	25.5	42.0	60.8	80.7	+6.6
Cyclohexaneethanol	$C_8H_{16}O$	50.4	77.2	90.0	104.0	119.8	129.8	142.7	161.7	183.5	205.4	
Cyclohexanol	$C_6H_{12}O$	21.0	44.0	56.0	68.8	83.0	91.8	103.7	121.7	141.4	161.0	23.9
Cyclohexanone	$C_6H_{10}O$	+1.4	26.4	38.7	52.5	67.8	77.5	90.4	110.3	132.5	155.6	−45.0
2-Cyclohexyl-4,6-dinitrophenol	$C_{12}H_{14}N_2O_5$	132.8	161.8	175.9	191.2	206.7	216.0	229.0	248.7	269.8	291.5	
Cyclopentane	C_5H_{10}	−68.0	−49.6	−40.4	−30.1	−18.6	−11.3	−1.3	+13.8	31.0	49.3	−93.7
Cyclopropane	C_3H_6	−116.8	−104.2	−97.5	−90.3	−82.3	−77.0	−70.0	−59.1	−46.9	−33.5	−126.6
Cymene	$C_{10}H_{14}$	17.3	43.9	57.0	71.1	87.0	97.2	110.8	131.4	153.5	177.2	−68.2
cis-Decalin	$C_{10}H_{18}$	22.5	50.1	64.2	79.8	97.2	108.0	123.2	145.4	169.9	194.6	−43.3
trans-Decalin	$C_{10}H_{18}$	−0.8	+30.6	47.2	65.3	85.7	98.4	114.6	136.2	160.1	186.7	−30.7
Decane	$C_{10}H_{22}$	16.5	42.3	55.7	69.8	85.5	95.5	108.6	128.4	150.6	174.1	−29.7
Decan-2-one	$C_{10}H_{20}O$	44.2	71.9	85.8	100.7	117.1	127.8	142.0	163.2	186.7	211.0	+3.5
1-Decene	$C_{10}H_{20}$	14.7	40.3	53.7	67.8	83.3	93.5	106.5	126.7	149.2	172.0	
Decyl alcohol	$C_{10}H_{22}O$	69.5	97.3	111.3	125.8	142.1	152.0	165.8	186.2	208.8	231.0	+7
Decyltrimethylsilane	$C_{13}H_{30}Si$	67.4	96.4	111.0	126.5	144.0	154.3	169.5	191.0	215.5	240.0	
Dehydroacetic acid	$C_8H_8O_4$	91.7	122.0	137.3	153.0	171.0	181.5	197.5	219.5	244.5	269.0	
Desoxybenzoin	$C_{14}H_{12}O$	123.3	156.2	173.5	192.0	212.0	224.5	241.3	265.2	293.0	321.0	60
Diacetamide	$C_4H_7NO_2$	70.0	95.0	108.0	122.6	138.2	148.0	160.6	180.8	202.0	223.0	78.5
Diacetylene (1,3-butadiyne)	C_4H_2	−82.5	−68.0	−61.2	−53.8	−45.9	−41.0	−34.0	−20.9	−6.1	+9.7	−34.9
Diallyldichlorosilane	$C_6H_{10}Cl_2Si$	+9.5	34.8	47.4	61.3	76.4	86.3	99.7	119.4	142.0	165.3	
Diallyl sulfide	$C_6H_{10}S$	−9.5	+14.4	26.6	39.7	54.2	63.7	75.8	94.8	116.1	138.6	−83
Diisoamyl ether	$C_{10}H_{22}O$	18.6	44.3	57.0	70.7	86.3	96.0	109.6	129.0	150.3	173.4	

Name	Formula											
oxalate	C₁₂H₂₂O₄	85.4	116.0	131.4	147.7	165.7	177.0	192.2	215.0	240.0	265.0	
sulfide	C₁₀H₂₂S	43.0	73.0	87.6	102.7	120.0	130.6	145.3	166.4	191.0	216.0	-26
Dibenzylamine	C₁₄H₁₅N	118.3	149.8	165.6	182.2	200.2	212.2	227.3	249.8	274.3	300.0	
Dibenzyl ketone (1,3-diphenyl-2-propanone)	C₁₅H₁₄O	125.5	159.8	177.6	195.7	216.6	229.4	246.6	272.3	301.7	330.5	34.5
1,4-Dibromobenzene	C₆H₄Br₂	61.0	79.3	87.7	103.6	120.8	131.6	146.5	168.5	192.5	218.6	87.5
1,2-Dibromobutane	C₄H₈Br₂	7.5	33.2	46.1	60.0	76.0	86.0	99.8	120.2	143.5	166.3	-64.5
dl-2,3-Dibromobutane	C₄H₈Br₂	+5.0	30.0	41.6	56.4	72.0	82.0	95.3	115.7	138.0	160.5	
meso-2,3-Dibromobutane	C₄H₈Br₂	+1.5	26.6	39.3	53.2	68.0	78.0	91.7	111.8	134.2	157.3	-34.5
1,2-Dibromodecane	C₁₀H₂₀Br₂	95.7	123.6	137.3	151.0	167.4	177.5	190.2	209.6	229.8	250.4	
Di(2-bromoethyl) ether	C₄H₈Br₂O	47.7	75.3	88.5	103.6	119.8	130.0	144.0	165.0	188.0	212.5	
α,β-Dibromomaleic anhydride	C₄H₂Br₂O₃	50.0	78.0	92.0	106.7	123.5	133.8	147.7	168.0	192.0	215.0	
1,2-Dibromo-2-methylpropane	C₄H₈Br₂	-28.8	-3.0	+10.5	25.7	42.3	53.7	68.8	92.1	119.8	149.0	-70.3
1,3-Dibromo-2-methylpropane	C₄H₈Br₂	14.0	40.0	53.0	67.5	83.5	93.7	107.4	117.8	150.8	174.6	
1,2-Dibromopentane	C₅H₁₀Br₂	19.8	45.4	58.0	72.0	87.4	97.4	110.1	130.2	151.8	175.0	
1,2-Dibromopropane	C₃H₆Br₂	-7.0	+17.3	29.4	42.3	57.2	66.4	78.7	97.8	118.5	141.6	-55.5
1,3-Dibromopropane	C₃H₆Br₂	+9.7	35.4	48.0	62.1	77.8	87.8	101.3	121.7	144.1	167.5	-34.4
2,3-Dibromopropene	C₃H₄Br₂	-6.0	+17.9	30.0	43.2	57.8	67.0	79.5	98.0	119.5	141.2	
2,3-Dibromo-1-propanol	C₃H₆Br₂O	57.0	84.5	98.2	113.5	129.8	140.0	153.0	173.8	196.0	219.0	
Diisobutylamine	C₈H₁₉N	-5.1	+18.4	30.6	43.7	57.8	67.0	79.2	97.6	118.0	139.5	-70
2,6-Ditert-butyl-4-cresol	C₁₅H₂₄O	85.8	116.2	131.0	147.0	164.1	175.2	190.0	212.8	237.6	262.5	
4,6-Ditert-butyl-2-cresol	C₁₅H₂₄O	86.2	117.3	132.4	149.0	167.4	179.0	194.0	217.5	243.4	269.3	
4,6-Ditert-butyl-3-cresol	C₁₅H₂₄O	103.7	135.2	150.0	167.0	185.3	196.1	211.0	233.0	257.1	282.0	
2,6-Ditert-butyl-4-ethylphenol	C₁₆H₂₆O	89.1	121.4	137.0	154.0	172.1	183.9	198.0	220.0	244.0	268.6	
4,6-Ditert-butyl-3-ethylphenol	C₁₆H₂₆O	111.5	142.6	157.4	174.0	192.3	204.4	218.0	241.7	264.6	290.0	
Diisobutyl oxalate	C₁₀H₁₈O₄	63.2	91.2	105.3	120.3	137.5	147.8	161.8	183.5	205.8	229.5	
2,4-Ditert-butylphenol	C₁₄H₂₂O	84.5	115.4	130.0	146.0	164.3	175.8	190.0	212.5	237.0	260.8	
Dibutyl phthalate	C₁₆H₂₂O₄	148.2	182.1	198.2	216.2	235.8	247.8	263.7	287.0	313.5	340.0	-79.7
sulfide	C₈H₁₈S	+21.7	51.8	66.4	80.5	96.0	105.8	118.6	138.0	159.0	182.0	
Diisobutyl d-tartrate	C₁₂H₂₂O₆	117.8	151.8	169.0	188.0	208.5	221.6	239.5	264.7	294.0	324.0	73.5
Dicarvacryl-mono-(6-chloro-2-xenyl) phosphate	C₃₂H₃₄ClO₄P	204.2	234.5	249.3	264.5	280.5	290.7	304.9	323.8	342.0	361.0	
Dicarvacryl-2-tolyl phosphate	C₂₇H₃₃O₄P	180.2	209.3	221.8	237.0	251.5	260.3	272.5	290.0	309.8	330.0	

Name	Formula	Pressure, mm Hg										Melting point, °C
		Temperature, °C										
		1	5	10	20	40	60	100	200	400	760	
Dichloroacetic acid	$C_2H_2Cl_2O_2$	44.0	69.8	82.6	96.3	111.8	121.5	134.0	152.3	173.7	194.4	9.7
1,2-Dichlorobenzene	$C_6H_4Cl_2$	20.0	46.0	59.1	73.4	89.4	99.5	112.9	133.4	155.8	179.0	-17.6
1,3-Dichlorobenzene	$C_6H_4Cl_2$	12.1	39.0	52.0	66.2	82.0	92.2	105.0	125.9	149.0	173.0	-24.2
1,4-Dichlorobenzene	$C_6H_4Cl_2$			54.8	69.2	84.8	95.2	108.4	128.3	150.2	173.9	53.0
1,2-Dichlorobutane	$C_4H_8Cl_2$	-23.6	-0.3	+11.5	24.5	37.7	47.8	60.2	79.7	100.8	123.5	
2,3-Dichlorobutane	$C_4H_8Cl_2$	-25.2	-3.0	+8.5	21.2	35.0	43.9	56.0	74.0	94.2	116.0	-80.4
1,2-Dichloro-1,2-difluoroethylene	$C_2Cl_2F_2$	-82.0	-65.6	-57.3	-48.3	-38.2	-31.8	-23.0	-10.0	+5.0	20.9	-112
Dichlorodifluoromethane	CCl_2F_2	-118.5	-104.6	-97.8	-90.1	-81.6	-76.1	-68.6	-57.0	-43.9	-29.8	
Dichlorodiphenyl silane	$C_{12}H_{10}Cl_2Si$	109.6	142.4	158.0	176.0	195.5	207.5	223.8	248.0	275.5	304.0	
Dichlorodiisopropyl ether	$C_6H_{12}Cl_2O$	29.6	55.2	68.2	82.2	97.3	106.9	119.7	139.0	159.8	182.7	
Di(2-chloroethoxy) methane	$C_5H_{10}Cl_2O_2$	53.0	80.4	94.0	109.5	125.5	135.8	149.6	170.0	192.0	215.0	
Dichloroethoxymethylsilane	$C_4H_8Cl_2OSi$	-33.8	-12.1	-1.3	+11.3	24.4	32.6	44.1	61.0	80.3	100.6	
1,2-Dichloro-3-ethylbenzene	$C_8H_8Cl_2$	46.0	75.0	90.0	105.9	123.8	135.0	149.8	172.0	197.0	222.1	-40.8
1,2-Dichloro-4-ethylbenzene	$C_8H_8Cl_2$	47.0	77.2	92.3	109.6	127.5	139.0	153.3	176.0	201.7	226.6	-76.4
1,4-Dichloro-2-ethylbenzene	$C_8H_8Cl_2$	38.5	68.0	83.2	99.8	118.0	129.0	144.0	166.2	191.5	216.3	-61.2
cis-1,2-Dichloroethylene	$C_2H_2Cl_2$	-58.4	-39.2	-29.9	-19.4	-7.9	-0.5	+9.5	24.6	41.0	59.0	-80.5
trans-1,2-Dichloro ethylene	$C_2H_2Cl_2$	-65.4	-47.2	-38.0	-28.0	-17.0	-10.0	-0.2	+14.3	30.8	47.8	-50.0
Di(2-chloroethyl) ether	$C_4H_8Cl_2O$	23.5	49.3	62.0	76.0	91.5	101.5	114.5	134.0	155.4	178.5	
Dichlorofluoromethane	$CHCl_2F$	-91.3	-75.5	-67.5	-58.6	-48.8	-42.6	-33.9	-20.9	-6.2	+8.9	-135
1,5-Dichlorohexamethyltrisiloxane	$C_6H_{18}Cl_2O_2Si_3$	26.0	52.0	65.1	79.0	94.8	105.0	118.2	138.3	160.2	184.0	-53.0
Dichloromethylphenylsilane	$C_7H_8Cl_2Si$	35.7	63.5	77.4	92.4	109.5	120.0	134.2	155.5	180.2	205.5	
1,1-Dichloro-2-methylpropane	$C_4H_8Cl_2$	-31.0	-8.4	+2.6	14.6	28.2	37.0	48.2	65.8	85.4	106.0	
1,2-Dichloro-2-methylpropane	$C_4H_8Cl_2$	-25.8	-4.2	+6.7	18.7	32.0	40.2	51.7	68.9	87.8	108.0	
1,3-Dichloro-2-methylpropane	$C_4H_8Cl_2$	-3.0	+20.6	32.0	44.8	58.6	67.5	78.8	96.1	115.4	135.0	
2,4-Dichlorophenol	$C_6H_4Cl_2O$	53.0	80.0	92.8	107.7	123.4	133.5	146.0	165.2	187.5	210.0	45.0
2,6-Dichlorophenol	$C_6H_4Cl_2O$	59.5	87.6	101.0	115.5	131.6	141.8	154.6	175.5	197.7	220.0	
α,α-Dichlorophenylacetonitrile	$C_8H_5Cl_2N$	56.0	84.0	98.1	113.8	130.0	141.0	154.5	176.2	199.5	223.5	
Dichlorophenylarsine	$C_6H_5AsCl_2$	61.8	100.0	116.0	133.1	151.0	163.2	178.9	202.8	228.8	256.5	

Compound	Formula											
1,2-Dichloropropane	$C_3H_6Cl_2$	−38.5	−17.0	−6.1	+6.0	19.4	28.0	39.4	57.0	76.0	96.8	
2,3-Dichlorostyrene	$C_8H_6Cl_2$	61.0	90.1	104.6	120.5	137.8	149.0	163.5	185.7	210.0	235.0	
2,4-Dichlorostyrene	$C_8H_6Cl_2$	53.5	82.2	97.4	111.8	129.2	140.0	153.8	176.0	200.0	225.0	
2,5-Dichlorostyrene	$C_8H_6Cl_2$	55.5	83.9	98.2	114.0	131.0	142.0	155.8	178.0	202.5	227.0	
2,6-Dichlorostyrene	$C_8H_6Cl_2$	47.8	75.7	90.0	105.5	122.4	133.3	147.6	169.0	193.5	217.0	
3,4-Dichlorostyrene	$C_8H_6Cl_2$	57.2	86.0	100.4	116.2	133.7	144.6	158.2	181.5	205.7	230.0	
3,5-Dichlorostyrene	$C_8H_6Cl_2$	53.5	82.2	97.4	111.8	129.2	140.0	153.8	176.0	200.0	225.0	
1,2-Dichlorotetraethylbenzene	$C_{14}H_{20}Cl_2$	105.6	138.7	155.0	172.5	192.2	204.8	220.7	245.6	272.8	302.0	
1,4-Dichlorotetraethylbenzene	$C_{14}H_{20}Cl_2$	91.7	126.1	143.8	162.0	183.2	195.8	212.0	238.5	265.8	296.5	
1,2-Dichloro-1,1,2,2-tetrafluoroethane	$C_2Cl_2F_4$	−95.4	−80.0	−72.3	−63.5	−53.7	−47.5	−39.1	−26.3	−12.0	+3.5	−94
Dichloro-4-tolylsilane	$C_7H_8Cl_2Si$	46.2	71.7	84.2	97.8	113.2	122.6	135.5	153.5	175.2	196.3	
3,4-Dichloro-α,α,α-trifluorotoluene	$C_7H_3Cl_2F_3$	11.0	38.3	52.2	67.3	84.0	95.0	109.2	129.0	150.5	172.8	−12.1
Dicyclopentadiene	$C_{10}H_8$		34.1	47.6	62.0	77.9	88.0	101.7	121.8	144.2	166.6	32.9
Diethoxydimethylsilane	$C_6H_{16}O_2Si$	−19.1	+2.4	13.3	25.3	38.0	46.3	57.6	74.2	93.2	113.5	
Diethoxydiphenylsilane	$C_{16}H_{20}O_2Si$	111.5	142.8	157.6	174.3	193.2	205.0	220.0	243.8	259.7	296.0	
Diethyl adipate	$C_{10}H_{18}O_4$	74.0	106.6	123.0	138.3	154.6	165.8	179.0	198.2	219.1	240.0	−21
Diethylamine	$C_4H_{11}N$			−33.0	−22.6	−11.3	−4.0	+6.0	21.0	38.0	55.5	−38.9
N-Diethylaniline	$C_{10}H_{15}N$	49.7	78.0	91.9	107.2	123.6	133.8	147.3	168.2	192.4	215.5	−34.4
Diethyl arsanilate	$C_{10}H_{16}AsNO_3$	38.0	62.6	74.8	88.0	102.6	111.8	123.8	141.9	161.0	181.0	
1,2-Diethylbenzene	$C_{10}H_{14}$	22.3	48.7	62.0	76.4	92.5	102.6	116.2	136.7	159.0	183.5	−31.4
1,3-Diethylbenzene	$C_{10}H_{14}$	20.7	46.8	59.9	74.5	90.4	100.7	114.4	134.8	156.9	181.1	−83.9
1,4-Diethylbenzene	$C_{10}H_{14}$	20.7	47.1	60.3	74.7	91.1	101.3	115.3	136.1	159.0	183.8	−43.2
Diethyl carbonate	$C_5H_{10}O_3$	−10.1	+12.3	23.8	36.0	49.5	57.9	69.7	86.5	105.8	125.8	−43
cis-Diethyl citraconate	$C_9H_{14}O_4$	59.8	88.3	103.0	118.2	135.7	146.2	160.0	182.3	206.5	230.3	
Diethyl dioxosuccinate	$C_8H_{10}O_5$	70.0	98.0	112.0	126.8	143.8	153.7	167.7	188.0	210.8	233.5	
Diethylene glycol	$C_4H_{10}O_3$	91.8	120.0	133.8	148.0	164.3	174.0	187.5	207.0	226.5	244.8	
Diethyleneglycol-bis-chloroacetate	$C_8H_{12}Cl_2O_5$	148.3	180.0	195.8	212.0	229.0	239.5	252.0	271.5	291.8	313.0	
Diethylene glycol dimethyl ether												
Di(2-methoxyethyl) ether	$C_6H_{14}O_3$	13.0	37.6	50.0	63.0	77.5	86.8	99.5	118.0	138.5	159.8	
glycol ethyl ether	$C_6H_{14}O_3$	45.3	72.0	85.8	100.3	116.7	126.8	140.3	159.0	180.3	201.9	
Diethyl ether	$C_4H_{10}O$	−74.3	−56.9	−48.1	−38.5	−27.7	−21.8	−11.5	+2.2	17.9	34.6	−116.3
ethylmalonate	$C_9H_{16}O_4$	50.8	77.8	91.6	106.0	122.4	132.4	146.0	166.0	188.7	211.5	
fumarate	$C_8H_{12}O_4$	53.2	81.2	95.3	110.2	126.7	137.7	151.1	172.2	195.5	218.5	+0.6
glutarate	$C_9H_{16}O_4$	65.6	94.7	109.7	125.4	142.8	153.2	167.8	189.5	212.8	237.0	

VAPOR PRESSURE–TABLE 4. Vapor Pressures of Organic Compounds, up to 1 atm (*Continued*)

Name	Formula	Pressure, mm Hg										Melting point, °C
		1	5	10	20	40	60	100	200	400	760	
		Temperature, °C										
Diethylhexadecylamine	$C_{20}H_{43}N$	139.8	175.8	194.0	213.5	235.0	248.5	265.5	292.8	324.6	355.0	
Diethyl itaconate	$C_9H_{14}O_4$	51.3	80.2	95.2	111.0	128.2	139.3	154.3	177.5	203.1	227.9	−42
ketone (3-pentanone)	$C_5H_{10}O$	−12.7	+7.5	17.2	27.9	39.4	46.7	56.2	70.6	86.3	102.7	
malate	$C_8H_{14}O_5$	80.7	110.4	125.3	141.2	157.8	169.0	183.9	205.3	229.5	253.4	
maleate	$C_8H_{12}O_4$	57.3	85.6	100.0	115.3	131.8	142.4	156.0	177.8	201.7	225.0	−49.8
malonate	$C_7H_{12}O_4$	40.0	67.5	81.3	95.9	113.3	123.0	136.2	155.5	176.8	198.9	
mesaconate	$C_9H_{14}O_4$	62.8	91.0	105.3	120.3	137.3	147.9	161.6	183.2	205.8	229.0	
oxalate	$C_6H_{10}O_4$	47.4	71.8	83.8	96.8	110.6	119.7	130.8	147.9	166.2	185.7	−40.6
phthalate	$C_{12}H_{14}O_4$	108.8	140.7	156.0	173.6	192.1	204.1	219.5	243.0	267.5	294.0	
sebacate	$C_{14}H_{26}O_4$	125.3	156.2	172.1	189.8	207.5	218.4	234.4	255.8	280.3	305.5	1.3
2,5-Diethylstyrene	$C_{12}H_{16}$	49.7	78.4	92.6	108.5	125.8	136.8	151.0	173.2	198.0	223.0	
Diethyl succinate	$C_8H_{14}O_4$	54.6	83.0	96.6	111.7	127.8	138.2	151.1	171.7	193.8	216.5	−20.8
isosuccinate	$C_8H_{14}O_4$	39.8	66.7	80.0	94.7	111.0	121.4	134.8	155.1	177.7	201.3	
sulfate	$C_4H_{10}O_4S$	47.0	74.0	87.7	102.1	118.0	128.6	142.5	162.5	185.5	209.5	−25.0
sulfide	$C_4H_{10}S$	−39.6	−18.6	−8.0	+3.5	16.1	24.2	35.0	51.3	69.7	88.0	−99.5
sulfite	$C_4H_{10}O_3S$	10.0	34.2	46.4	59.7	74.2	83.8	96.3	115.8	137.0	159.0	
d-Diethyl tartrate	$C_8H_{14}O_6$	102.0	133.0	148.0	164.2	182.3	194.0	208.5	230.4	254.8	280.0	17
dl-Diethyl tartrate	$C_8H_{14}O_6$	100.0	131.7	147.2	163.8	181.7	193.2	208.0	230.0	254.3	280.0	
3,5-Diethyltoluene	$C_{11}H_{16}$	34.0	61.5	75.3	90.2	107.0	117.7	131.7	152.4	176.5	200.7	
Diethylzinc	$C_4H_{10}Zn$	−22.4	0.0	+11.7	24.2	38.0	47.2	59.1	77.0	97.3	118.0	−28
1-Dihydrocarvone	$C_{10}H_{16}O$	46.6	75.5	90.0	106.0	123.7	134.7	149.7	171.8	197.0	223.0	
Dihydrocitronellol	$C_{10}H_{22}O$	68.0	91.7	103.0	115.0	127.6	136.7	145.9	160.2	176.8	193.5	
1,4-Dihydroxyanthraquinone	$C_{14}H_8O_4$	196.7	239.8	259.8	282.0	307.4	323.3	344.5	377.8	413.0	450.0	194
Dimethylacetylene (2-butyne)	C_4H_6	−73.0	−57.9	−50.5	−42.5	−33.9	−27.8	−18.8	−5.0	+10.6	27.2	−32.5
Dimethylamine	C_2H_7N	−87.7	−72.2	−64.6	−56.0	−46.7	−40.7	−32.6	−20.4	−7.1	+7.4	−96
N,N-Dimethylaniline	$C_8H_{11}N$	29.5	56.3	70.0	84.8	101.6	111.9	125.8	146.5	169.2	193.1	+2.5
Dimethyl arsanilate	$C_8H_{12}AsNO_3$	15.0	39.6	51.8	65.0	79.7	88.6	101.0	119.8	140.3	160.5	
Di(α-methylbenzyl) ether	$C_{16}H_{18}O$	96.7	128.3	144.0	160.3	179.6	191.5	206.8	229.7	254.8	281.9	
2,2-Dimethylbutane	C_6H_{14}	−69.3	−50.7	−41.5	−31.1	−19.5	−12.1	−2.0	+13.4	31.0	49.7	−99.8
2,3-Dimethylbutane	C_6H_{14}	−63.6	−44.5	−34.9	−24.1	−12.4	−4.9	+5.4	21.1	39.0	58.0	−128.2

Compound	Formula											
Dimethyl citraconate	$C_7H_{10}O_4$	50.8	78.2	91.8	106.5	122.6	132.7	145.8	165.8	188.0	210.5	
1,1-Dimethylcyclohexane	C_8H_{16}	−24.4	−1.4	+10.3	23.0	37.3	45.7	57.9	76.2	97.2	119.5	−34
cis-1,2-Dimethylcyclohexane	C_8H_{16}	−15.9	+7.3	18.4	31.1	45.3	54.4	66.8	85.6	107.0	129.7	−50.0
trans-1,2-Dimethylcyclohexane	C_8H_{16}	−21.1	+1.7	13.0	25.6	39.7	48.7	61.0	79.6	100.9	123.4	−88.0
trans-1,3-Dimethylcyclohexane	C_8H_{16}	−19.4	+3.4	14.9	27.4	41.4	50.4	62.5	81.0	102.1	124.4	−92.0
cis-1,3-Dimethylcyclohexane	C_8H_{16}	−22.7	0.0	+11.2	23.6	37.5	46.4	58.5	76.9	97.8	120.1	−76.2
cis-1,4-Dimethylcyclohexane	C_8H_{16}	−20.0	+3.2	14.5	27.1	41.1	50.1	62.3	80.8	101.9	124.3	−87.4
trans-1,4-Dimethylcyclohexane	C_8H_{16}	−24.3	−1.7	+10.1	22.6	36.5	45.4	57.6	76.0	97.0	119.3	−36.9
Dimethyl ether	C_2H_6O	−115.7	−101.1	−93.1	−85.2	−76.2	−70.4	−62.7	−50.9	−37.8	−23.7	−138.5
2,2-Dimethylhexane	C_8H_{18}	−29.7	−7.9	+3.1	15.0	28.2	36.7	48.2	65.7	85.6	106.8	
2,3-Dimethylhexane	C_8H_{18}	−23.0	−1.1	+9.9	22.1	35.6	44.2	56.0	73.8	94.1	115.6	
2,4-Dimethylhexane	C_8H_{18}	−26.9	−5.3	+5.2	17.2	30.5	39.0	50.6	68.1	88.2	109.4	
2,5-Dimethylhexane	C_8H_{18}	−26.7	−5.5	+5.3	17.2	30.4	38.9	50.5	68.0	87.9	109.1	−90.7
3,3-Dimethylhexane	C_8H_{18}	−25.8	−4.4	+6.1	18.2	31.7	40.4	52.5	70.0	90.4	112.0	
3,4-Dimethylhexane	C_8H_{18}	−22.1	+0.2	11.3	23.5	37.1	45.8	57.7	75.6	96.0	117.7	
Dimethyl itaconate	$C_7H_{10}O_4$	69.3	94.0	106.6	119.7	133.7	142.6	153.7	171.0	189.8	208.0	38
1-Dimethyl malate	$C_6H_{10}O_5$	75.4	104.0	118.3	133.8	150.1	160.4	175.1	196.3	219.5	242.6	
Dimethyl maleate	$C_6H_8O_4$	45.7	73.0	86.4	101.3	117.2	127.1	140.4	160.0	182.2	205.0	−62
malonate	$C_5H_8O_4$	35.0	59.8	72.0	85.0	100.0	109.7	121.9	140.0	159.8	180.7	
trans-Dimethyl mesaconate	$C_7H_{10}O_4$	46.8	74.0	87.8	102.1	118.0	127.8	141.5	161.0	183.5	206.0	
2,7-Dimethyloctane	$C_{10}H_{22}$	+6.3	30.5	42.3	55.8	71.2	80.8	93.9	114.0	136.0	159.7	−52.8
Dimethyl oxalate	$C_4H_6O_4$	20.0	44.0	56.0	69.4	83.6	92.8	104.8	123.3	143.3	163.3	
2,2-Dimethylpentane	C_7H_{16}	−49.0	−28.7	−18.7	−7.5	+5.0	13.0	23.9	40.3	59.2	79.2	−123.7
2,3-Dimethylpentane	C_7H_{16}	−42.0	−20.8	−10.3	+1.1	13.9	22.1	33.3	50.1	69.4	89.8	−135
2,4-Dimethylpentane	C_7H_{16}	−48.0	−27.4	−17.1	−5.9	+6.5	14.5	25.4	41.8	60.6	80.5	−119.5
3,3-Dimethylpentane	C_7H_{16}	−45.9	−25.0	−14.4	−2.9	+9.9	18.1	29.3	46.2	65.5	86.1	−135.0
2,3-Dimethylphenol (2,3-xylenol)	$C_8H_{10}O$	56.0	83.8	97.6	112.0	129.2	139.5	152.2	173.0	196.0	218.0	75
2,4-Dimethylphenol (2,4-xylenol)	$C_8H_{10}O$	51.8	78.0	91.3	105.0	121.5	131.0	143.0	161.5	184.2	211.5	25.5
2,5-Dimethylphenol (2,5-xylenol)	$C_8H_{10}O$	51.8	78.0	91.3	105.0	121.5	131.0	143.0	161.5	184.2	211.5	74.5
3,4-Dimethylphenol (3,4-xylenol)	$C_8H_{10}O$	66.2	93.8	107.7	122.0	138.0	148.0	161.0	181.5	203.6	225.2	62.5
3,5-Dimethylphenol (3,5-xylenol)	$C_8H_{10}O$	62.0	89.2	102.4	117.0	133.3	143.5	156.0	176.2	197.8	219.5	68
Dimethylphenylsilane	$C_8H_{12}Si$	+5.3	30.3	42.6	56.2	71.4	81.3	94.2	114.2	136.4	159.3	
Dimethyl phthalate	$C_{10}H_{10}O_4$	100.3	131.8	147.6	164.0	182.8	194.0	210.0	232.7	257.8	283.7	
3,5-Dimethyl-1,2-pyrone	$C_7H_8O_2$	78.6	107.6	122.0	136.4	152.7	163.8	177.5	198.0	221.0	245.0	51.5
4,6-Dimethylresorcinol	$C_8H_{10}O_2$	49.0	76.8	90.7	105.8	122.5	133.2	147.3	167.8	192.0	215.0	
Dimethyl sebacate	$C_{12}H_{22}O_4$	104.0	139.8	156.2	175.8	196.0	208.0	222.6	245.0	269.6	293.5	38
2,4-Dimethylstyrene	$C_{10}H_{12}$	34.2	61.9	75.8	90.8	107.7	118.0	132.3	153.2	177.5	202.0	

VAPOR PRESSURE–TABLE 4. Vapor Pressures of Organic Compounds, up to 1 atm (*Continued*)

Name	Formula	Pressure, mm Hg										Melting point, °C
		1	5	10	20	40	60	100	200	400	760	
		Temperature, °C										
2,5-Dimethylstyrene	C$_{10}$H$_{12}$	29.0	55.9	69.0	84.0	100.2	110.7	124.7	145.6	168.7	193.0	
α,α-Dimethylsuccinic anhydride	C$_6$H$_8$O$_3$	61.4	88.1	102.0	116.3	132.3	142.4	155.3	175.8	197.5	219.5	
Dimethyl sulfide	C$_2$H$_6$S	−75.6	−58.0	−49.2	−39.4	−28.4	−21.4	−12.0	+2.6	18.7	36.0	−83.2
d-Dimethyl tartrate	C$_6$H$_{10}$O$_6$	102.1	133.2	148.2	164.3	182.4	193.8	208.8	230.5	255.0	280.0	61.5
dl-Dimethyl tartrate	C$_6$H$_{10}$O$_6$	100.4	131.8	147.5	164.0	182.4	193.8	209.5	232.3	257.4	282.0	89
N,N-Dimethyl-2-toluidine	C$_9$H$_{13}$N	28.8	54.1	66.2	80.2	95.0	105.2	118.1	138.3	161.5	184.8	
N,N-Dimethyl-4-toluidine	C$_9$H$_{13}$N	50.1	74.3	86.7	100.0	116.3	126.4	140.3	161.6	185.4	209.5	−61
Di(nitrosomethyl) amine	C$_2$H$_5$N$_3$O$_2$	+3.2	27.8	40.0	53.7	68.2	77.7	90.3	110.0	131.3	153.0	
Diosphenol	C$_{10}$H$_{16}$O$_2$	66.7	95.4	109.0	124.0	141.2	151.3	165.6	186.2	209.5	232.0	
1,4-Dioxane	C$_4$H$_8$O$_2$	−35.8	−12.8	−1.2	+12.0	25.2	33.8	45.1	62.3	81.8	101.1	10
Dipentene	C$_{10}$H$_{16}$	14.0	40.4	53.8	68.2	84.3	94.6	108.3	128.2	150.5	174.6	
Diphenylamine	C$_{12}$H$_{11}$N	108.3	141.7	157.0	175.2	194.3	206.9	222.8	247.5	274.1	302.0	52.9
Diphenyl carbinol (benzhydrol)	C$_{13}$H$_{12}$O	110.0	145.0	162.0	180.9	200.0	212.0	227.5	250.0	275.6	301.0	68.5
Diphenyl chlorophosphate	C$_{12}$H$_{10}$ClPO$_3$	121.5	160.5	182.0	203.8	227.9	244.2	265.0	299.5	337.3	378.0	
Diphenyl disulfide	C$_{12}$H$_{10}$S$_2$	131.6	164.0	180.0	197.0	214.8	226.2	241.3	262.6	285.8	310.0	61
1,2-Diphenylethane (dibenzyl)	C$_{14}$H$_{14}$	86.8	119.8	136.0	153.7	173.7	186.0	202.8	227.8	255.0	284.0	51.5
Diphenyl ether	C$_{12}$H$_{10}$O	66.1	97.8	114.0	130.8	150.0	162.0	178.8	203.3	230.7	258.5	27
1,1-Diphenylethylene	C$_{14}$H$_{12}$	87.4	119.6	135.0	151.8	170.8	183.4	198.6	222.8	249.8	277.0	
trans-Diphenylethylene	C$_{14}$H$_{12}$	113.2	145.8	161.0	179.8	199.0	211.5	227.4	251.7	278.3	306.5	124
1,1-Diphenylhydrazine	C$_{12}$H$_{12}$N$_2$	126.0	159.3	176.1	194.0	213.5	225.9	242.5	267.2	294.0	322.2	44
Diphenylmethane	C$_{13}$H$_{12}$	76.0	107.4	122.8	139.8	157.8	170.2	186.3	210.7	237.5	264.5	26.5
Diphenyl sulfide	C$_{12}$H$_{10}$S	96.1	129.0	145.0	162.0	182.8	194.8	211.8	236.8	263.9	292.5	
Diphenyl-2-tolyl thiophosphate	C$_{18}$H$_{17}$O$_3$PS	159.7	179.6	201.6	215.5	230.6	240.4	252.5	270.3	290.0	310.0	
1,2-Dipropoxyethane	C$_8$H$_{18}$O$_2$	−38.8	−10.3	+5.0	22.3	42.3	55.8	74.2	103.8	140.0	180.0	
1,2-Diisopropylbenzene	C$_{12}$H$_{18}$	40.0	67.8	81.8	96.8	114.0	124.3	138.7	159.8	184.3	209.0	
1,3-Diisopropylbenzene	C$_{12}$H$_{18}$	34.7	62.3	76.0	91.2	107.9	118.2	132.3	153.7	177.6	202.0	−105
Dipropylene glycol	C$_6$H$_{14}$O$_3$	73.8	102.1	116.2	131.3	147.4	156.5	169.9	189.9	210.5	231.8	
Dipropyleneglycol monobutyl ether	C$_{10}$H$_{22}$O$_3$	64.7	92.0	106.0	120.4	136.3	146.3	159.8	180.0	203.8	227.0	
isopropyl ether	C$_9$H$_{20}$O$_3$	46.0	72.8	86.2	100.8	117.0	126.8	140.3	160.0	183.1	205.6	
Di-n-propyl ether	C$_6$H$_{14}$O	−43.3	−22.3	−11.8	0.0	+13.2	21.6	33.0	50.3	69.5	89.5	−122

Compound	Formula											
Diisopropyl ether	$C_6H_{14}O$	-57.0	-37.4	-27.4	-16.7	-4.5	+3.4	13.7	30.0	48.2	67.5	-60
Di-n-propyl ketone (4-heptanone)	$C_7H_{14}O$	23.0	44.4	55.0	66.2	78.1	85.8	96.0	111.2	127.3	143.7	-32.6
Di-n-propyl oxalate	$C_8H_{14}O_4$	53.4	80.2	93.9	108.6	124.6	134.8	148.1	168.0	190.3	213.5	
Diisopropyl oxalate	$C_8H_{14}O_4$	43.2	69.0	81.9	95.6	110.5	120.0	132.6	151.2	171.8	193.5	
Di-n-propyl succinate	$C_{10}H_{18}O_4$	77.5	107.6	122.2	138.0	154.8	166.0	180.3	202.5	226.5	250.8	
Di-n-propyl d-tartrate	$C_{10}H_{18}O_6$	115.6	147.7	163.5	180.4	199.7	211.7	227.0	250.1	275.6	303.0	
Diisopropyl d-tartrate	$C_{10}H_{18}O_6$	103.7	133.7	148.2	164.0	181.8	192.6	207.3	228.2	251.8	275.0	
Divinyl acetylene (1,5-hexadiene-3-yne)	C_6H_6	-45.1	-24.4	-14.0	-2.8	+10.0	18.1	29.5	46.0	64.4	84.0	-66.9
1,3-Divinylbenzene	$C_{10}H_{10}$	32.7	60.0	73.8	88.7	105.5	116.0	130.0	151.4	175.2	199.5	44.5
Docosane	$C_{22}H_{46}$	157.8	195.4	213.0	233.5	254.5	268.3	286.0	314.2	343.5	376.0	-9.6
n-Dodecane	$C_{12}H_{26}$	47.8	75.8	90.0	104.6	121.7	132.1	146.2	167.2	191.0	216.2	-31.5
1-Dodecene	$C_{12}H_{24}$	47.2	74.0	87.8	102.4	118.6	128.5	142.3	162.2	185.5	208.0	24
n-Dodecyl alcohol	$C_{12}H_{26}O$	91.0	120.2	134.7	150.0	167.2	177.8	192.0	213.0	235.7	259.0	
Dodecylamine	$C_{12}H_{27}N$	82.8	111.8	127.8	141.6	157.4	168.0	182.1	203.0	225.0	248.0	
Dodecyltrimethylsilane	$C_{15}H_{34}Si$	91.2	122.1	137.7	153.8	172.1	184.2	199.5	222.0	248.0	273.0	51.5
Elaidic acid	$C_{18}H_{34}O_2$	171.3	206.7	223.5	242.3	260.8	273.0	288.0	312.4	337.0	362.0	-25.6
Epichlorohydrin	C_3H_5ClO	-16.5	+5.6	16.6	29.0	42.0	50.6	62.0	79.3	98.0	117.9	
1,2-Epoxy-2-methylpropane	C_4H_8O	-69.0	-50.0	-40.3	-29.5	-17.3	-9.7	+1.2	17.5	36.0	55.5	33.5
Erucic acid	$C_{22}H_{42}O_2$	206.7	239.7	254.5	270.6	289.1	300.2	314.4	336.5	358.8	381.5	
Estragole (p-methoxy allyl benzene)	$C_{10}H_{12}O$	52.6	80.0	93.7	108.4	124.6	135.2	148.5	168.7	192.0	215.0	
Ethane	C_2H_6	-159.5	-148.5	-142.9	-136.7	-129.8	-125.4	-119.3	-110.2	-99.7	-88.6	-183.2
Ethoxydimethylphenylsilane	$C_{10}H_{16}OSi$	36.3	63.1	76.2	91.0	107.2	127.5	131.4	151.5	175.0	199.5	
Ethoxytrimethylsilane	$C_5H_{14}OSi$	-50.9	-31.0	-20.7	-9.8	+3.7	11.5	22.1	38.1	56.3	75.7	
Ethoxytriphenylsilane	$C_{20}H_{20}OSi$	167.0	198.2	213.5	230.0	247.0	258.3	273.5	295.0	319.5	344.0	-82.4
Ethyl acetate	$C_4H_8O_2$	-43.4	-23.5	-13.5	-3.0	+9.1	16.6	27.0	42.0	59.3	77.1	-45
acetoacetate	$C_6H_{10}O_3$	28.5	54.0	67.3	81.1	96.2	106.0	118.5	138.0	158.2	180.8	
Ethylacetylene (1-butyne)	C_4H_6	-92.5	-76.7	-68.7	-59.9	-50.5	-43.4	-34.9	-21.6	-6.9	+8.7	-130
Ethyl acrylate	$C_5H_8O_2$	-29.5	-8.7	+2.0	13.0	26.0	33.5	44.5	61.5	80.0	99.5	-71.2
α-Ethylacrylic acid	$C_5H_8O_2$	47.0	70.7	82.0	94.4	108.1	116.7	127.5	144.0	160.7	179.2	
α-Ethylacrylonitrile	C_5H_7N	-29.0	-6.4	+5.0	17.7	31.8	40.6	53.0	71.6	92.2	114.0	
Ethyl alcohol (ethanol)	C_2H_6O	-31.3	-12.0	-2.3	+8.0	19.0	26.0	34.9	48.4	63.5	78.4	-112
Ethylamine	C_2H_7N	-82.3	-66.4	-58.3	-48.6	-39.8	-33.4	-25.1	-12.3	+2.0	16.6	-80.6
4-Ethylaniline	$C_8H_{11}N$	52.0	80.0	93.8	109.0	125.7	136.0	149.8	170.6	194.2	217.4	-4
N-Ethylaniline	$C_8H_{11}N$	38.5	66.4	80.6	96.0	113.2	123.6	137.3	156.9	180.8	204.0	-63.5

VAPOR PRESSURE—TABLE 4. Vapor Pressures of Organic Compounds, up to 1 atm (*Continued*)

Name	Formula	Pressure, mm Hg										Melting point, °C
		1	5	10	20	40	60	100	200	400	760	
		Temperature, °C										
2-Ethylanisole	$C_9H_{12}O$	29.7	55.9	69.0	83.1	98.9	109.0	122.3	142.1	164.2	187.1	
3-Ethylanisole	$C_9H_{12}O$	33.7	60.3	73.9	88.5	104.8	115.5	129.2	149.7	172.8	196.5	
4-Ethylanisole	$C_9H_{12}O$	33.5	60.2	73.9	88.5	104.7	115.5	128.4	149.2	172.3	196.5	
Ethylbenzene	C_8H_{10}	-9.8	+13.9	25.9	38.6	52.8	61.8	74.1	92.7	113.8	136.2	-94.9
Ethyl benzoate	$C_9H_{10}O_2$	44.0	72.0	86.0	101.4	118.2	129.0	143.2	164.8	188.4	213.4	-34.6
benzoylacetate	$C_{11}H_{12}O_3$	107.6	136.4	150.3	166.8	181.8	191.9	205.0	223.8	244.7	265.0	
bromide	C_2H_5Br	-74.3	-56.4	-47.5	-37.8	-26.7	-19.5	-10.0	+4.5	21.0	38.4	-117.8
α-bromoisobutyrate	$C_6H_{11}BrO_2$	10.6	35.8	48.0	61.8	77.0	86.7	99.8	119.7	141.2	163.6	
n-butyrate	$C_6H_{12}O_2$	-18.4	+4.0	15.3	27.8	41.5	50.1	62.0	79.8	100.0	121.0	-93.3
isobutyrate	$C_6H_{12}O_2$	-24.3	-2.4	+8.4	20.6	33.8	42.3	53.5	71.0	90.0	110.0	-88.2
Ethylcamphoronic anhydride	$C_{11}H_{16}O_5$	118.2	149.8	165.0	181.8	199.8	211.5	226.6	248.5	272.8	298.0	
Ethyl isocaproate	$C_8H_{16}O_2$	11.0	35.8	48.0	61.7	76.3	85.8	98.4	117.8	139.2	160.4	
carbamate	$C_3H_7NO_2$		65.8	77.8	91.0	105.6	114.8	126.2	144.2	164.0	184.0	49
carbanilate	$C_9H_{11}NO_2$	107.8	131.8	143.7	155.5	168.8	177.3	187.9	203.8	220.0	237.0	52.5
Ethylcetylamine	$C_{18}H_{39}N$	133.2	168.2	186.0	205.5	226.5	239.8	256.8	283.3	313.0	342.0	
Ethyl chloride	C_2H_5Cl	-89.8	-73.9	-65.8	-56.8	-47.0	-40.6	-32.0	-18.6	-3.9	+12.3	-139
chloroacetate	$C_4H_7ClO_2$	+1.0	25.4	37.5	50.4	65.2	74.0	86.0	103.8	123.8	144.2	-26
chloroglyoxylate	$C_4H_5ClO_3$	-5.1	+18.0	29.9	42.0	56.0	65.2	76.6	94.5	114.7	135.0	
α-chloropropionate	$C_5H_9ClO_2$	+6.6	30.2	41.9	54.3	68.2	77.3	89.3	107.2	126.2	146.5	
trans-cinnamate	$C_{11}H_{12}O_2$	87.6	108.5	134.0	150.3	169.2	181.2	196.0	219.3	245.0	271.0	
3-Ethylcumene	$C_{11}H_{16}$	28.3	55.5	68.8	83.6	99.9	110.2	124.3	145.4	168.2	193.0	
4-Ethylcumene	$C_{11}H_{16}$	31.5	58.4	72.0	86.7	103.3	113.8	127.2	148.3	171.8	195.8	12
Ethyl cyanoacetate	$C_5H_7NO_2$	67.8	93.5	106.0	119.8	133.8	142.1	152.8	169.8	187.8	206.0	
Ethylcyclohexane	C_8H_{16}	-14.5	+9.2	20.6	33.4	47.6	56.7	69.0	87.8	109.1	131.8	
Ethylcyclopentane	C_7H_{14}	-32.2	-10.8	-0.1	+11.7	25.0	33.4	45.0	62.4	82.3	103.4	-111.3
Ethyl dichloroacetate	$C_4H_6Cl_2O_2$	9.6	34.0	46.3	59.5	74.0	83.6	96.1	115.2	135.9	156.5	-138.6
N,N-diethyloxamate	$C_8H_{15}NO_3$	76.0	106.3	121.7	137.7	154.4	166.0	180.3	202.8	226.5	252.0	
N-Ethyldiphenylamine	$C_{14}H_{15}N$	98.3	130.2	146.0	162.8	182.0	193.7	209.8	233.0	258.8	286.0	
Ethylene	C_2H_4	-168.3	-158.3	-153.2	-147.6	-141.3	-137.3	-131.8	-123.4	-113.9	-103.7	-169

Compound	Formula											
Ethylene-bis-(chloroacetate)	$C_6H_8Cl_2O_4$	112.0	142.4	158.0	173.5	191.0	201.8	215.0	237.3	259.5	283.5	-69
Ethylene chlorohydrin (2-chloroethanol)	C_2H_5ClO	-4.0	+19.0	30.3	42.5	56.0	64.1	75.0	91.8	110.0	128.8	8.5
diamine (1,2-ethanediamine)	$C_2H_8N_2$	-11.0	+10.5	21.5	33.0	45.8	53.8	62.5	81.0	99.0	117.2	10
dibromide (1,2-dibromethane)	$C_2H_4Br_2$	-27.0	+4.7	18.6	32.7	48.0	57.9	70.4	89.8	110.1	131.5	-35.3
dichloride (1,2-dichloroethane)	$C_2H_4Cl_2$	-44.5	-24.0	-13.6	-2.4	+10.0	18.1	29.4	45.7	64.0	82.4	-15.6
glycol (1,2-ethanediol)	$C_2H_6O_2$	53.0	79.7	92.1	105.8	120.0	129.5	141.8	158.5	178.5	197.3	
glycol diethyl ether (1,2-diethoxyethane)	$C_6H_{14}O_2$	-33.5	-10.2	+1.6	14.7	29.7	39.0	51.8	71.8	94.1	119.5	
glycol dimethyl ether (1,2-dimethoxyethane)	$C_4H_{10}O_2$	-48.0	-26.2	-15.3	-3.0	+10.7	19.7	31.8	50.0	70.8	93.0	
glycol monomethyl ether (2-methoxyethanol)	$C_3H_8O_2$	-13.5	+10.2	22.0	34.3	47.8	56.4	68.0	85.3	104.3	124.4	
oxide	C_2H_4O	-89.7	-73.8	-65.7	-56.6	-46.9	-40.7	-32.1	-19.5	-4.9	+10.7	-111.3
Ethyl α-ethylacetoacetate	$C_8H_{14}O_3$	40.5	67.3	80.2	94.6	110.3	120.6	133.8	153.2	175.6	198.0	
fluoride	C_2H_5F	-117.0	-103.8	-97.7	-90.0	-81.8	-76.4	-69.3	-58.0	-45.5	-32.0	
formate	$C_3H_6O_2$	-60.5	-42.2	-33.0	-22.7	-11.5	-4.3	-5.4	20.2	37.1	54.3	-79
2-furoate	$C_7H_8O_3$	37.6	63.8	77.1	91.5	107.5	117.5	130.4	150.1	172.5	195.0	34
glycolate	$C_4H_8O_3$	14.3	38.8	50.5	63.9	78.1	87.6	99.8	117.8	138.0	158.2	
3-Ethylhexane	C_8H_{18}	-20.0	+2.1	12.8	25.0	38.5	47.1	58.9	76.7	97.0	118.5	
2-Ethylhexyl acrylate	$C_{11}H_{20}O_2$	50.0	77.7	91.8	106.3	123.7	134.0	147.9	168.2	192.2	216.0	
Ethylidene chloride (1,1-dichloroethane)	$C_2H_4Cl_2$	-60.7	-41.9	-32.3	-21.9	-10.2	-2.9	+7.2	22.4	39.8	57.4	-96.7
fluoride (1,1-difluoroethane)	$C_2H_4F_2$	-112.5	-98.4	-91.7	-84.1	-75.8	-70.4	-63.2	-52.0	-39.5	-26.5	-117
Ethyl iodide	C_2H_5I	-54.4	-34.3	-24.3	-13.1	-0.9	+7.2	18.0	34.1	52.3	72.4	-105
Ethyl l-leucinate	$C_8H_{17}NO_2$	27.8	57.3	72.1	88.0	106.0	117.8	131.8	149.8	167.3	184.0	
Ethyl levulinate	$C_7H_{12}O_3$	47.3	74.0	87.3	101.8	117.7	127.6	141.3	160.2	183.0	206.2	
Ethyl mercaptan (ethanethiol)	C_2H_6S	-76.7	-59.1	-50.2	-40.7	-29.8	-22.4	-13.0	+1.5	17.7	35.0	-121
Ethyl methylcarbamate	$C_4H_9NO_2$	26.5	51.0	63.2	76.1	91.0	100.0	112.0	130.0	149.8	170.0	
Ethyl methyl ether	C_3H_8O	-91.0	-75.6	-67.8	-59.1	-49.4	-43.3	-34.8	-22.0	-7.8	+7.5	
1-Ethylnaphthalene	$C_{12}H_{12}$	70.0	101.4	116.8	133.8	152.0	164.1	180.0	204.6	230.8	258.1	-27
Ethyl α-naphthyl ketone (1-propionaphthone)	$C_{13}H_{12}O$	124.0	155.5	171.0	188.1	206.9	218.2	233.5	255.5	280.2	306.0	
Ethyl 3-nitrobenzoate	$C_9H_9NO_4$	108.1	140.2	155.0	173.6	192.6	205.0	220.3	244.6	270.6	298.0	47
3-Ethylpentane	C_7H_{16}	-37.8	-17.0	-6.8	+4.7	17.5	25.7	36.9	53.8	73.0	93.5	-118.6
4-Ethylphenetole	$C_{10}H_{14}O$	48.5	75.7	89.5	103.8	119.8	129.8	143.5	163.2	185.7	208.0	
2-Ethylphenol	$C_8H_{10}O$	46.2	73.4	87.0	101.5	117.9	127.9	141.8	161.6	184.5	207.5	-45

VAPOR PRESSURE–TABLE 4. Vapor Pressures of Organic Compounds, up to 1 atm (Continued)

Name	Formula	1	5	10	20	40	60	100	200	400	760	Melting point, °C
						Temperature, °C						
3-Ethylphenol	$C_8H_{10}O$	60.0	86.8	100.2	114.5	130.0	139.8	152.0	171.8	193.3	214.0	−4
4-Ethylphenol	$C_8H_{10}O$	59.3	86.5	100.2	115.0	131.3	141.7	154.2	175.0	197.4	219.0	46.5
Ethyl phenyl ether (phenetole)	$C_8H_{10}O$	18.1	43.7	56.4	70.3	86.6	95.4	108.4	127.9	149.8	172.0	−30.2
Ethyl propionate	$C_5H_{10}O_2$	−28.0	−7.2	+3.4	14.3	27.2	35.1	45.2	61.7	79.8	99.1	−72.6
Ethyl propyl ether	$C_5H_{12}O$	−64.3	−45.0	−35.0	−24.0	−12.0	−4.0	+6.8	23.3	41.6	61.7	
Ethyl salicylate	$C_9H_{10}O_3$	61.2	90.0	104.2	119.3	136.7	147.6	161.5	183.7	207.0	231.5	1.3
3-Ethylstyrene	$C_{10}H_{12}$	28.3	55.0	68.3	82.8	99.2	109.6	123.2	144.0	167.2	191.5	
4-Ethylstyrene	$C_{10}H_{12}$	26.0	52.7	66.3	80.8	97.3	107.6	121.5	142.0	165.0	189.0	
Ethylisothiocyanate	C_3H_5NS	13.2	+10.6	22.8	36.1	50.8	59.8	71.9	90.0	110.1	131.0	−5.9
2-Ethyltoluene	C_9H_{12}	9.4	34.8	47.6	61.2	76.4	86.0	99.0	119.0	141.4	165.1	
3-Ethyltoluene	C_9H_{12}	7.2	32.3	44.7	58.2	73.3	82.9	95.9	115.5	137.8	161.3	−95.5
4-Ethyltoluene	C_9H_{12}	7.6	32.7	44.9	58.5	73.6	83.2	96.3	116.1	136.4	162.0	
Ethyl trichloroacetate	$C_4H_5Cl_3O_2$	20.7	45.5	57.7	70.6	85.5	94.4	107.4	125.8	146.0	167.0	
Ethyltrimethylsilane	$C_5H_{14}Si$	−60.6	−41.4	−31.8	−21.0	−9.0	−1.2	+9.2	25.0	42.8	62.0	
Ethyltrimethyltin	$C_5H_{14}Sn$	−30.0	−7.6	+3.8	16.1	30.0	38.4	50.0	67.3	87.6	108.8	
Ethyl isovalerate	$C_7H_{14}O_2$	−6.1	+17.0	28.7	41.3	55.2	64.0	75.9	93.8	114.0	134.3	−99.3
2-Ethyl-1,4-xylene	$C_{10}H_{14}$	25.7	52.0	65.6	79.8	96.0	106.2	120.0	140.2	163.1	186.9	
4-Ethyl-1,3-xylene	$C_{10}H_{14}$	26.3	53.0	66.4	80.6	97.2	107.4	121.2	141.8	164.4	188.4	
5-Ethyl-1,3-xylene	$C_{10}H_{14}$	22.1	48.8	62.1	76.5	92.6	103.0	116.5	137.4	159.6	183.7	
Eugenol	$C_{10}H_{12}O_2$	78.4	108.1	123.0	138.7	155.8	167.3	182.2	204.7	228.3	253.5	
iso-Eugenol	$C_{10}H_{12}O_2$	86.3	117.0	132.4	149.0	167.0	178.2	194.0	217.2	242.3	267.5	−10
Eugenyl acetate	$C_{12}H_{14}O_3$	101.6	132.3	148.0	164.2	183.0	194.0	209.7	232.5	257.4	282.0	295
Fencholic acid	$C_{10}H_{16}O_2$	101.7	128.7	142.3	155.8	171.8	181.5	194.0	215.0	237.8	264.1	19
d-Fenchone	$C_{10}H_{16}O$	28.0	54.7	68.3	83.0	99.5	109.8	123.6	144.0	166.8	191.0	5
dl-Fenchyl alcohol	$C_{10}H_{18}O$	45.8	70.3	82.1	95.6	110.8	120.2	132.3	150.0	173.2	201.0	35
Fluorene	$C_{13}H_{10}$		129.3	146.0	164.2	185.2	197.8	214.7	240.3	268.6	295.0	113
Fluorobenzene	C_6H_5F	−43.4	−22.8	−12.4	−1.2	+11.5	19.6	30.4	47.2	65.7	84.7	−42.1
2-Fluorotoluene	C_7H_7F	−24.2	−2.2	+8.9	21.4	34.7	43.7	55.3	73.0	92.8	114.0	−80
3-Fluorotoluene	C_7H_7F	−22.4	−0.3	+11.0	23.4	37.0	45.8	57.5	75.4	95.4	116.0	−110.8
4-Fluorotoluene	C_7H_7F	−21.8	+0.3	11.8	24.0	37.8	46.5	58.1	76.0	96.1	117.0	

Name	Formula											mp, °C
Formaldehyde	CH_2O										−19.5	−92
Formamide	CH_3NO	70.5	96.3	109.5	122.5	137.5	147.0	157.5	175.5	193.5	210.5	
Formic acid	CH_2O_2	−20.0	−5.0	+2.1	10.3	24.0	32.4	43.8	61.4	80.3	100.6	8.2
trans-Fumaryl chloride	$C_4H_2Cl_2O_2$	+15.0	38.5	51.8	65.0	79.5	89.0	101.0	120.0	140.0	160.0	
Furfural (2-furaldehyde)	$C_5H_4O_2$	18.5	42.6	54.8	67.8	82.1	91.5	103.4	121.8	141.8	161.8	
Furfuryl alcohol	$C_5H_6O_2$	31.8	56.0	68.0	81.0	95.7	104.0	115.9	133.1	151.8	170.0	
Geraniol	$C_{10}H_{18}O$	69.2	96.8	110.0	125.6	141.8	151.5	165.3	185.6	207.8	230.0	
Geranyl acetate	$C_{12}H_{20}O_2$	73.5	102.7	117.9	133.0	150.0	160.3	175.2	196.3	219.8	243.3	
Geranyl n-butyrate	$C_{14}H_{24}O_2$	96.8	125.2	139.0	153.8	170.1	180.2	193.8	214.0	235.0	257.4	
Geranyl isobutyrate	$C_{14}H_{24}O_2$	90.9	119.6	133.0	147.9	164.0	174.0	187.7	207.6	228.5	251.0	
Geranyl formate	$C_{11}H_{18}O_2$	61.8	90.3	104.3	119.8	136.2	147.2	160.7	182.6	205.8	230.0	
Glutaric acid	$C_5H_8O_4$	155.5	183.8	196.0	210.5	226.3	235.5	247.0	265.0	283.5	303.0	97.5
Glutaric anhydride	$C_5H_6O_3$	100.8	133.3	149.5	166.0	185.5	196.2	212.5	236.5	261.0	287.0	
Glutaronitrile	$C_5H_6N_2$	91.3	123.7	140.0	156.5	176.4	189.5	205.5	230.0	257.3	286.2	
Glutaryl chloride	$C_5H_6Cl_2O_2$	56.1	84.0	97.8	112.3	128.3	139.1	151.8	172.4	195.3	217.0	
Glycerol	$C_3H_8O_3$	125.5	153.8	167.2	182.2	198.0	208.0	220.1	240.0	263.0	290.0	17.9
Glycerol dichlorohydrin (1,3-dichloro-2-propanol)	$C_3H_6Cl_2O$	28.0	52.2	64.7	78.0	93.0	102.0	114.8	133.3	153.5	174.3	
Glycol diacetate	$C_6H_{10}O_4$	38.3	64.1	77.1	90.8	106.1	115.8	128.0	147.8	168.3	190.5	−31
Glycolide (1,4-dioxane-2,6-dione)	$C_4H_4O_4$		103.0	116.6	132.0	148.6	158.2	173.2	194.0	217.0	240.0	97
Guaiacol (2-methoxyphenol)	$C_7H_8O_2$	52.4	79.1	92.0	106.0	121.6	131.0	144.0	162.7	184.1	205.0	28.3
Heneicosane	$C_{21}H_{44}$	152.6	188.0	205.4	223.2	243.4	255.3	272.0	296.5	323.8	350.5	40.4
Heptacosane	$C_{27}H_{56}$	211.7	248.6	266.8	284.6	305.7	318.0	333.5	359.4	385.0	410.6	59.5
Heptadecane	$C_{17}H_{36}$	115.0	145.2	160.0	177.7	195.8	207.3	223.0	247.8	274.5	303.0	22.5
Heptaldehyde (enanthaldehyde)	$C_7H_{14}O$	12.0	32.7	43.0	54.0	66.3	74.0	84.0	102.0	125.5	155.0	−42
n-Heptane	C_7H_{16}	−34.0	−12.7	−2.1	+9.5	22.3	30.6	41.8	58.7	78.0	98.4	−90.6
Heptanoic acid (enanthic acid)	$C_7H_{14}O_2$	78.0	101.3	113.2	125.6	139.5	148.5	160.0	179.5	199.6	221.5	−10
1-Heptanol	$C_7H_{16}O$	42.4	64.3	74.7	85.8	99.8	108.0	119.5	136.6	155.6	175.8	34.6
Heptanoyl chloride (enanthyl chloride)	$C_7H_{13}ClO$	34.2	54.6	64.6	75.0	86.4	93.5	102.7	116.3	130.7	145.0	
2-Heptene	C_7H_{14}	−35.8	−14.1	−3.5	+8.3	21.5	30.0	41.3	58.6	78.1	98.5	
Heptylbenzene	$C_{13}H_{20}$	64.0	94.6	110.0	126.0	144.0	154.8	170.2	193.3	217.8	244.0	
Heptyl cyanide (enanthonitrile)	$C_7H_{13}N$	21.0	47.8	61.6	76.3	92.6	103.0	116.8	137.7	160.0	184.6	
Hexachlorobenzene	C_6Cl_6	114.4	149.3	166.4	185.7	206.0	219.0	235.5	258.5	283.5	309.4	230
Hexachloroethane	C_2Cl_6	32.7	49.8	73.5	87.6	102.3	112.0	124.2	143.1	163.8	185.6	186.6

VAPOR PRESSURE–TABLE 4. Vapor Pressures of Organic Compounds, up to 1 atm (*Continued*)

Name	Formula	Pressure, mmHg										Melting point, °C
		1	5	10	20	40	60	100	200	400	760	
		Temperature, °C										
Hexacosane	C$_{26}$H$_{54}$	204.0	240.0	257.4	275.8	295.2	307.8	323.2	348.4	374.6	399.8	56.6
Hexadecane	C$_{16}$H$_{34}$	105.3	135.2	149.8	164.7	181.3	193.2	208.5	231.7	258.3	287.5	18.5
1-Hexadecene	C$_{16}$H$_{32}$	101.6	131.7	146.2	162.0	178.8	190.8	205.3	226.8	250.0	274.0	4
n-Hexadecyl alcohol (cetyl alcohol)	C$_{16}$H$_{34}$O	122.7	158.3	177.8	197.8	219.8	234.3	251.7	280.2	312.7	344.0	49.3
n-Hexadecylamine (cetylamine)	C$_{16}$H$_{35}$N	123.6	157.8	176.0	195.7	215.7	228.8	245.8	272.2	300.4	330.0	
Hexaethylbenzene	C$_{18}$H$_{30}$		134.3	150.3	168.0	187.7	199.7	216.0	241.7	268.5	298.3	130
n-Hexane	C$_6$H$_{14}$	−53.9	−34.5	−25.0	−14.1	−2.3	+5.4	15.8	31.6	49.6	68.7	−95.3
1-Hexanol	C$_6$H$_{14}$O	24.4	47.2	58.2	70.3	83.7	92.0	102.8	119.6	138.0	157.0	−51.6
2-Hexanol	C$_6$H$_{14}$O	14.6	34.8	45.0	55.9	67.9	76.0	87.3	103.7	121.8	139.9	
3-Hexanol	C$_6$H$_{14}$O	+2.5	25.7	36.7	49.0	62.2	70.7	81.8	98.3	117.0	135.5	
1-Hexene	C$_6$H$_{12}$	−57.5	−38.0	−28.1	−17.2	−5.0	+2.8	13.0	29.0	46.8	66.0	−98.5
n-Hexyl levulinate	C$_{11}$H$_{20}$O$_3$	90.0	120.0	134.7	150.2	167.8	179.0	193.6	215.7	241.0	266.8	
n-Hexyl phenyl ketone (enanthophenone)	C$_{13}$H$_{18}$O	100.0	130.3	145.5	161.0	178.9	189.8	204.2	225.0	248.3	271.3	
Hydrocinnamic acid	C$_9$H$_{10}$O$_2$	102.2	133.5	148.7	165.0	183.3	194.0	209.0	230.8	255.0	279.8	48.5
Hydrogen cyanide (hydrocyanic acid)	CHN	−71.0	−55.3	−47.7	−39.7	−30.9	−25.1	−17.8	−5.3	+10.2	25.9	−13.2
Hydroquinone	C$_6$H$_6$O$_2$	132.4	153.3	163.5	174.6	192.0	203.0	216.5	238.0	262.5	286.2	170.3
4-Hydroxybenzaldehyde	C$_7$H$_6$O$_2$	121.2	153.2	169.7	186.8	206.0	217.5	233.5	256.8	282.6	310.0	115.5
α-Hydroxyisobutyric acid	C$_4$H$_8$O$_3$	73.5	98.5	110.5	123.8	138.0	146.4	157.7	175.2	193.8	212.0	79
α-Hydroxybutyronitrile	C$_5$H$_9$NO	41.0	65.8	77.8	90.7	104.8	113.9	125.0	142.0	159.8	178.8	
4-Hydroxy-3-methyl-2-butanone	C$_5$H$_{10}$O$_2$	44.6	69.3	81.0	94.0	108.2	117.4	129.0	146.5	165.5	185.0	
4-Hydroxy-4-methyl-2-pentanone	C$_6$H$_{12}$O$_2$	22.0	46.7	58.8	72.0	86.7	96.0	108.2	126.8	147.5	167.9	−47
3-Hydroxypropionitrile	C$_3$H$_5$NO	58.7	87.8	102.0	117.9	134.1	144.7	157.7	178.0	200.0	221.0	
Indene	C$_9$H$_8$	16.4	44.3	58.5	73.9	90.7	100.8	114.7	135.6	157.8	181.6	−2
Iodobenzene	C$_6$H$_5$I	24.1	50.6	64.0	78.3	94.4	105.0	118.3	139.8	163.9	188.6	−28.5
Iodononane	C$_9$H$_{19}$I	70.0	96.2	109.0	123.0	138.1	147.7	159.8	179.0	199.3	219.5	
2-Iodotoluene	C$_7$H$_7$I	37.2	65.9	79.8	95.6	112.4	123.8	138.1	160.0	185.7	211.0	
α-Ionone	C$_{13}$H$_{20}$O	79.5	108.8	123.0	139.0	155.6	166.3	181.2	202.5	225.2	250.0	
Isoprene	C$_5$H$_8$	−79.8	−62.3	−53.3	−43.5	−32.6	−25.4	−16.0	−1.2	+15.4	32.6	−146.7
Lauraldehyde	C$_{12}$H$_{24}$O	77.7	108.4	123.7	140.2	157.8	168.7	184.5	207.8	231.8	257.0	44.5
Lauric acid	C$_{12}$H$_{24}$O$_2$	121.0	150.6	166.0	183.6	201.4	212.7	227.5	249.8	273.8	299.2	48

Compound	Formula											
Levulinaldehyde	C5H8O2	28.1	54.9	68.0	82.7	98.3	108.4	121.8	142.0	164.0	187.0	33.5
Levulinic acid	C5H8O3	102.0	128.1	141.8	154.1	169.5	178.0	190.2	208.3	227.4	245.8	-96.9
d-Limonene	C10H16	14.0	40.4	53.8	68.2	84.3	94.6	108.3	128.5	151.4	175.0	
Linalyl acetate	C12H20O2	55.4	82.5	96.0	111.4	127.7	138.1	151.8	173.3	196.2	220.0	
Maleic anhydride	C4H2O3	44.0	63.4	78.7	95.0	111.8	122.0	135.8	155.9	179.5	202.0	58
Menthane	C10H20	+9.7	35.7	48.3	62.7	78.3	88.6	102.1	122.7	146.0	169.5	
1-Menthol	C10H20O	56.0	83.2	96.0	110.3	126.1	136.1	149.4	168.3	190.2	212.0	42.5
Menthyl acetate	C12H22O2	57.4	85.8	100.0	115.4	132.1	143.2	156.7	178.8	202.8	227.0	
benzoate	C17H24O2	123.2	154.2	170.0	186.3	204.3	215.8	230.4	253.2	277.1	301.0	54.5
formate	C11H20O2	47.3	75.8	90.0	105.8	123.0	133.8	148.0	169.8	194.2	219.0	
Mesityl oxide	C6H10O	-8.7	+14.1	26.0	37.9	51.7	60.4	72.1	90.0	109.8	130.0	-59
Methacrylic acid	C4H6O2	25.5	48.5	60.0	72.7	86.4	95.3	106.6	123.9	142.5	161.0	15
Methacrylonitrile	C5H5N	-44.5	-23.3	-12.5	-0.6	+12.8	21.5	32.8	50.0	70.3	90.3	
Methane	CH4	-205.9	-199.0	-195.5	-191.8	-187.7	-185.1	-181.4	-175.5	-168.8	-161.5	-182.5
Methanethiol	CH4S	-90.7	-75.3	-67.5	-58.8	-49.2	-43.1	-34.8	-22.1	-7.9	+6.8	-121
Methoxyacetic acid	C3H6O3	52.5	79.3	92.0	106.5	122.0	131.8	144.5	163.5	184.2	204.0	
N-Methylacetanilide	C9H11NO		103.8	118.6	135.1	152.2	164.2	179.8	202.3	227.4	253.0	102
Methyl acetate	C3H6O2	-57.2	-38.6	-29.3	-19.1	-7.9	-0.5	+9.4	24.0	40.0	57.8	-98.7
acetylene (propyne)	C3H4	-111.0	-97.5	-90.5	-82.9	-74.3	-68.8	-61.3	-49.8	-37.2	-23.3	-102.7
acrylate	C4H6O2	-43.7	-23.6	-13.5	-2.7	+9.2	17.3	28.0	43.9	61.8	80.2	
alcohol (methanol)	CH4O	-44.0	-25.3	-16.2	-6.0	+5.0	12.1	21.2	34.8	49.9	64.7	-97.8
Methylamine	CH5N	-95.8	-81.3	-73.8	-65.9	-56.9	-51.3	-43.7	-32.4	-19.7	-6.3	-93.5
N-Methylaniline	C7H9N	36.0	62.8	76.2	90.5	106.0	115.8	129.8	149.3	172.0	195.5	-57
Methyl anthranilate	C8H9NO2	77.6	109.0	124.2	141.5	159.7	172.0	187.8	212.4	238.5	266.5	24
benzoate	C8H8O2	39.0	64.4	77.3	91.8	107.8	117.4	130.8	151.4	174.7	199.5	-12.5
2-Methylbenzothiazole	C8H7NS	70.0	97.5	111.2	125.5	141.2	150.4	163.9	183.2	204.5	225.5	15.4
α-Methylbenzyl alcohol	C8H10O	49.0	75.2	88.0	102.1	117.8	127.4	140.3	159.0	180.7	204.0	
Methyl bromide	CH3Br	-96.3	-80.6	-72.8	-64.0	-54.2	-48.0	-39.4	-26.5	-11.9	+3.6	-93
2-Methyl-1-butene	C5H10	-89.1	-72.8	-64.3	-54.8	-44.1	-37.3	-28.0	-13.8	+2.5	20.2	-135
2-Methyl-2-butene	C5H10	-75.4	-57.0	-47.9	-37.9	-26.7	-19.4	-9.9	+4.9	21.6	38.5	-133
Methyl isobutyl carbinol (2-methyl-4-pentanol)	C6H14O	-0.3	+22.1	33.3	45.4	58.2	67.0	78.0	94.9	113.5	131.7	
n-butyl ketone (2-hexanone)	C6H12O	+7.7	28.8	38.8	50.0	62.0	69.8	79.8	94.3	111.0	127.5	-56.9
isobutyl ketone (4-methyl-2-pentanone)	C6H12O	-1.4	+19.7	30.0	40.8	52.8	60.4	70.4	85.6	102.0	119.0	-84.7

VAPOR PRESSURE–TABLE 4. Vapor Pressures of Organic Compounds, up to 1 atm (*Continued*)

| Name | Formula | Pressure, mm Hg | | | | | | | | | | Melting point, °C |
| | | 1 | 5 | 10 | 20 | 40 | 60 | 100 | 200 | 400 | 760 | |
		Temperature, °C										
n-butyrate	$C_5H_{10}O_2$	-26.8	-5.5	+5.0	16.7	29.6	37.4	48.0	64.3	83.1	102.3	-84.7
isobutyrate	$C_5H_{10}O_2$	-34.1	-13.0	-2.9	+8.4	21.0	28.9	39.6	55.7	73.6	92.6	-18
caprate	$C_{11}H_{22}O_2$	63.7	93.5	108.0	123.0	139.0	148.6	161.5	181.6	202.9	224.0	
caproate	$C_7H_{14}O_2$	+5.0	30.0	42.0	55.4	70.0	79.7	91.4	109.8	129.8	150	
caprylate	$C_9H_{18}O_2$	34.2	61.7	74.9	89.0	105.3	115.3	128.0	148.1	170.0	193.0	-40
chloride	CH_3Cl		-99.5	-92.4	-84.8	-76.0	-70.4	-63.0	-51.2	-38.0	-24.0	-97.7
chloroacetate	$C_3H_5ClO_2$	-2.9	19.0	30.0	41.5	54.5	63.0	73.5	90.5	109.5	130.3	-31.9
cinnamate	$C_{10}H_{10}O_2$	77.4	108.1	123.0	140.0	157.9	170.0	185.8	209.6	235.0	263.0	33.4
α-Methylcinnamic acid	$C_{10}H_{10}O_2$	125.7	155.0	169.8	185.2	201.8	212.0	224.8	245.0	266.8	288.0	
Methylcyclohexane	C_7H_{14}	-35.9	-14.0	-3.2	+8.7	22.0	30.5	42.1	59.6	79.6	100.9	-126.4
Methylcyclopentane	C_8H_{12}	-53.7	-33.8	-23.7	-12.8	-0.6	+7.2	17.9	34.0	52.3	71.8	-142.4
Methylcyclopropane	C_4H_8	-96.0	-80.6	-72.8	-64.0	-54.2	-48.0	-39.3	-26.0	-11.3	+4.5	
Methyl *n*-decyl ketone (*n*-dodecan-2-one)	$C_{12}H_{24}O$	77.1	106.0	120.4	136.0	152.4	163.8	177.5	199.0	222.5	246.5	
dichloroacetate	$C_3H_4Cl_2O_2$	3.2	26.7	38.1	50.7	64.7	73.6	85.4	103.2	122.6	143.0	
N-Methyldiphenylamine	$C_{13}H_{13}N$	103.5	134.0	149.7	165.8	184.0	195.4	210.1	232.8	257.0	282.0	-7.6
Methyl *n*-dodecyl ketone (2-tetradecanone)	$C_{14}H_{28}O$	99.3	130.0	145.5	161.3	179.8	191.4	206.0	228.2	253.3	278.0	
Methylene bromide (dibromomethane)	CH_2Br_2	-35.1	-13.2	-2.4	+9.7	23.3	31.6	42.3	58.5	79.0	98.6	-52.8
chloride (dichloromethane)	CH_2Cl_2	-70.0	-52.1	-43.3	-33.4	-22.3	-15.7	-6.3	+8.0	24.1	40.7	-96.7
Methyl ethyl ketone (2-butanone)	C_4H_8O	-48.3	-28.0	-17.7	-6.5	+6.0	14.0	25.0	41.6	60.0	79.6	-85.9
2-Methyl-3-ethylpentane	C_8H_{18}	-24.0	-1.8	+9.5	21.7	35.2	43.9	55.7	73.6	94.0	115.6	-114.5
3-Methyl-3-ethylpentane	C_8H_{18}	-23.9	-1.4	+9.9	22.3	36.2	45.0	57.1	75.3	96.2	118.3	-90
Methyl fluoride	CH_3F	-147.3	-137.0	-131.6	-125.9	-119.1	-115.0	-109.0	-99.9	-89.5	-78.2	
formate	$C_2H_4O_2$	-74.2	-57.0	-48.6	-39.2	-28.7	-21.9	-12.9	+0.8	16.0	32.0	-99.8
α-Methylglutaric anhydride	$C_6H_8O_3$	93.8	125.4	141.8	157.7	177.5	189.9	205.0	229.1	255.5	282.5	
Methyl glycolate	$C_3H_6O_3$	+9.6	33.7	45.3	58.1	72.3	81.8	93.7	111.8	131.7	151.5	
2-Methylheptadecane	$C_{18}H_{38}$	119.8	152.0	168.7	186.0	204.8	216.3	231.5	254.5	279.8	306.5	
2-Methylheptane	C_8H_{18}	-21.0	+1.3	12.3	24.4	37.9	46.6	58.3	76.0	96.2	117.6	-109.5

Name	Formula											m.p.
3-Methylheptane	C₈H₁₈	−19.8	+2.6	13.3	25.4	38.9	47.6	59.4	77.1	97.4	118.9	−120.8
4-Methylheptane	C₈H₁₈	−20.4	+1.5	12.4	24.5	38.0	46.6	58.3	76.1	96.3	117.7	−121.1
2-Methyl-2-heptene	C₈H₁₆	−16.1	+6.7	17.8	30.4	44.0	52.8	64.6	82.3	102.2	122.5	
6-Methyl-3-hepten-2-ol	C₈H₁₆O	41.6	65.0	76.7	89.3	102.7	111.5	122.6	139.5	156.6	175.5	
6-Methyl-5-hepten-2-ol	C₈H₁₆O	41.9	66.0	77.8	90.4	104.0	112.8	123.8	140.0	156.6	174.3	−118.2
2-Methylhexane	C₇H₁₆	−40.4	−19.5	−9.1	+2.3	14.9	23.0	34.1	50.8	69.8	90.0	
3-Methylhexane	C₇H₁₆	−39.0	−18.1	−7.8	+3.6	16.4	24.5	35.6	52.4	71.6	91.9	−64.4
Methyl iodide	CH₃I		−55.0	−45.8	−35.6	−24.2	−16.9	−7.0	+8.0	25.3	42.4	5
laurate	C₁₃H₂₆O₂	87.8	117.9	133.2	149.0	166.0	176.8	190.8				
levulinate	C₆H₁₀O₃	39.8	66.4	79.7	93.7	109.5	119.3	133.0	153.4		197.7	
methacrylate	C₅H₈O₂	−30.5	−10.0	+1.0	11.0	25.5	34.5	47.0	63.0	82.0	101.0	18.5
myristate	C₁₅H₃₀O₂	115.0	145.7	160.8	177.8	195.8	207.5	222.6	245.3	269.8	295.8	
α-naphthyl ketone (1-acetonaphthone)	C₁₂H₁₀O	115.6	146.3	161.5	178.4	196.8	208.6	223.8	246.7	270.5	295.5	55.5
β-naphthyl ketone (2-acetonaphthone)	C₁₂H₁₀O	120.2	152.3	168.5	185.7	203.8	214.7	229.8	251.6	275.8	301.0	15
n-nonyl ketone (undecan-2-one)	C₁₁H₂₂O	68.2	95.5	108.9	123.1	139.0	148.6	161.0	181.2	202.3	224.0	30
palmitate	C₁₇H₃₄O₂	134.3	166.8	184.3	202.0	214.3	226.7	242.0	265.8	291.7	319.5	
n-pentadecyl ketone (2-heptdecanone)	C₁₇H₃₄O	129.6	161.6	178.0	196.4							
2-Methylpentane	C₆H₁₄	−60.9	−41.7	−32.1	−21.4	−9.7	−1.9	+8.1	24.1	41.6	60.3	−154
3-Methylpentane	C₆H₁₄	−59.0	−39.8	−30.1	−19.4	−7.3	+0.1	10.5	26.5	44.2	63.3	−118
2-Methyl-1-pentanol	C₆H₁₄O	15.4	38.0	49.6	61.6	74.7	83.4	94.2	111.3	129.8	147.9	
2-Methyl-2-pentanol	C₆H₁₄O	−4.5	+16.8	27.6	38.8	51.3	58.8	69.2	85.0	102.6	121.2	−103
Methyl n-pentyl ketone (2-heptanone)	C₇H₁₄O	19.3	43.6	55.5	67.7	81.2	89.8	100.0	116.1	133.2	150.2	
phenyl ether (anisole)	C₇H₈O	+5.4	30.0	42.2	55.8	70.7	80.1	93.0	112.3	133.8	155.5	−37.3
2-Methylpropene	C₄H₈	−105.1	−96.5	−81.9	−73.4	−63.8	−57.7	−49.3	−36.7	−22.2	−6.9	−140.3
Methyl propionate	C₄H₈O₂	−42.0	−21.5	−11.8	−1.0	+11.0	18.7	29.0	44.2	61.8	79.8	−87.5
4-Methylpropiophenone	C₁₀H₁₂O	59.6	89.3	103.8	120.2	138.0	149.3	164.2	187.4	212.7	238.5	
2-Methylpropionyl bromide	C₄H₇BrO	13.5	38.4	50.6	64.1	79.4	88.8	101.6	120.5	141.7	163.0	
Methyl propyl ether	C₄H₁₀O	−72.2	−54.3	−45.4	−35.4	−24.3	−17.4	−8.1	+6.0	22.5	39.1	
n-propyl ketone (2-pentanone)	C₅H₁₀O	−12.0	+8.0	17.9	28.5	39.8	47.3	56.8	71.0	86.8	103.3	−77.8
isopropyl ketone (3-methyl-2-butanone)	C₅H₁₀O	−19.9	−1.0	+8.3	18.3	29.6	36.2	45.5	59.0	73.8	88.9	−92
2-Methylquinoline	C₁₀H₉N	75.3	104.0	119.0	134.0	150.8	161.7	176.2	197.8	211.7	246.5	−1
Methyl salicylate	C₈H₈O₃	54.0	81.6	95.3	110.0	126.2	136.7	150.0	172.6	197.5	223.2	−8.3
α-Methyl styrene	C₉H₁₀	7.4	34.0	47.1	61.8	77.8	88.3	102.2	121.8	143.0	165.4	−23.2
4-Methyl styrene	C₉H₁₀	16.0	42.0	55.1	69.2	85.0	95.0	108.6	128.7	151.2	175.0	

VAPOR PRESSURE-TABLE 4. Vapor Pressures of Organic Compounds, up to 1 atm (*Continued*)

Name	Formula	Pressure, mm Hg Temperature, °C										Melting point, °C
		1	5	10	20	40	60	100	200	400	760	
Methyl *n*-tetradecyl ketone (2-hexadecanone)	$C_{16}H_{32}O$	109.8	151.5	167.3	184.6	203.7	215.0	230.5	254.4	279.8	307.0	
thiocyanate	C_2H_3NS	-14.0	+9.8	21.6	34.5	49.0	58.1	70.4	89.8	110.8	132.9	-51
isothiocyanate	C_2H_3NS	-34.7	-8.3	+5.4	20.4	38.2	47.5	59.3	77.5	97.8	119.0	35.5
undecyl ketone (2-tridecanone)	$C_{13}H_{26}O$	86.8	117.0	131.8	147.8	165.7	176.6	191.5	214.0	238.3	262.5	28.5
isovalerate	$C_6H_{12}O_2$	-19.2	+2.9	14.0	26.4	39.8	48.2	59.8	77.3	96.7	116.7	
Monovinylacetylene (butenyne)	C_4H_4	-93.2	-77.7	-70.0	-61.3	-51.7	-45.3	-37.1	-24.1	-10.1	+5.3	
Myrcene	$C_{10}H_{16}$	14.5	40.0	53.2	67.0	82.6	92.6	106.0	126.0	148.3	171.5	
Myristaldehyde	$C_{14}H_{28}O$	99.0	132.0	148.3	166.2	186.0	198.3	214.5	240.4	267.9	297.8	23.5
Myristic acid (tetradecanoic acid)	$C_{14}H_{28}O_2$	142.0	174.1	190.8	207.6	223.5	237.2	250.5	272.3	294.6	318.0	57.5
Naphthalene	$C_{10}H_8$	52.6	74.2	85.8	101.7	119.3	130.2	145.5	167.7	193.2	217.9	80.2
1-Naphthoic acid	$C_{11}H_8O_2$	156.0	184.0	196.8	211.2	225.0	234.5	245.8	263.5	281.4	300.0	160.5
2-Naphthoic acid	$C_{11}H_8O_2$	160.8	189.7	202.8	216.9	231.5	241.3	252.7	270.3	289.5	308.5	184
1-Naphthol	$C_{10}H_8O$	94.0	125.5	142.0	158.0	177.8	190.0	206.0	229.6	255.8	282.5	96
2-Naphthol	$C_{10}H_8O$		128.6	145.5	161.8	181.7	193.7	209.8	234.0	260.6	288.0	122.5
1-Naphthylamine	$C_{10}H_9N$	104.3	137.7	153.8	171.6	191.5	203.8	220.0	244.9	272.2	300.8	50
2-Naphthylamine	$C_{10}H_9N$	108.0	141.6	157.6	175.8	195.7	208.1	224.3	249.7	277.4	306.1	111.5
Nicotine	$C_{10}H_{14}N_2$	61.8	91.8	107.2	123.7	142.1	154.7	169.5	193.8	219.8	247.3	
2-Nitroaniline	$C_6H_6N_2O_2$	104.0	135.7	150.4	167.7	186.0	197.8	213.0	236.3	260.0	284.5	71.5
3-Nitroaniline	$C_6H_6N_2O_2$	119.3	151.5	167.8	185.5	204.2	216.5	232.1	255.3	280.2	305.7	114
4-Nitroaniline	$C_6H_6N_2O_2$	142.4	177.6	194.4	213.2	234.2	245.9	261.8	284.5	310.2	336.0	146.5
2-Nitrobenzaldehyde	$C_7H_5NO_3$	85.8	117.7	133.4	150.0	168.8	180.7	196.2	220.0	246.8	273.5	40.9
3-Nitrobenzaldehyde	$C_7H_5NO_3$	96.2	127.4	142.8	159.0	177.7	189.5	204.3	227.4	252.1	278.3	58
Nitrobenzene	$C_6H_5NO_2$	44.4	71.6	84.9	99.3	115.4	125.8	139.9	161.2	185.8	210.6	+5.7
Nitroethane	$C_2H_5NO_2$	-21.0	+1.5	12.5	24.8	38.0	46.5	57.8	74.8	94.0	114.0	-90
Nitroglycerin	$C_3H_5N_3O_9$	127	167	188	210	235	251					11
Nitromethane	CH_3NO_2	-29.0	-7.9	+2.8	14.1	27.5	35.5	46.6	63.5	82.0	101.2	-29

Compound	Formula											
2-Nitrophenol	$C_6H_5NO_3$	49.3	76.8	90.4	105.8	122.1	132.6	146.4	167.6	191.0	214.5	45
2-Nitrophenyl acetate	$C_8H_7NO_4$	100.0	128.0	142.0	155.8	172.8	181.7	194.1	213.0	233.5	253.0	
1-Nitropropane	$C_3H_7NO_2$	-9.6	+13.5	25.3	37.9	51.8	60.5	72.3	90.2	110.6	131.6	-108
2-Nitropropane	$C_3H_7NO_2$	-18.8	4.1	15.8	28.2	41.8	50.3	62.0	80.0	99.8	120.3	-93
2-Nitrotoluene	$C_7H_7NO_2$	50.0	79.1	93.8	109.6	126.3	137.6	151.5	173.7	197.7	222.3	-4.1
3-Nitrotoluene	$C_7H_7NO_2$	50.2	1.0	96.0	112.8	130.7	142.5	156.9	180.3	206.8	231.9	15.5
4-Nitrotoluene	$C_7H_7NO_2$	53.7	85.0	100.5	117.7	136.0	147.9	163.0	186.7	212.5	238.3	51.9
4-Nitro-1,3-xylene (4-nitro-m-xylene)	$C_8H_9NO_2$	65.6	95.0	109.8	125.8	143.3	153.8	168.5	191.7	217.5	244.0	+2
Nonacosane	$C_{29}H_{60}$	234.2	260.8	286.4	303.6	323.2	334.8	350.0	373.2	397.2	421.8	63.8
Nonadecane	$C_{19}H_{40}$	133.3	166.3	183.5	200.8	220.0	232.8	248.0	271.8	299.8	330.0	32
n-Nonane	C_9H_{20}	+1.4	25.8	38.0	51.2	66.0	75.5	88.1	107.5	128.2	150.8	-53.7
1-Nonanol	$C_9H_{20}O$	59.5	86.1	99.7	113.8	129.0	139.0	151.3	170.5	192.1	213.5	-5
2-Nonanone	$C_9H_{18}O$	32.1	59.0	72.3	87.2	103.4	113.8	127.4	148.2	171.2	195.0	-19
Octacosane	$C_{28}H_{58}$	226.5	260.3	277.4	295.4	314.2	326.8	341.8	364.8	388.9	412.5	61.6
Octadecane	$C_{18}H_{38}$	119.6	152.1	169.6	187.5	207.4	219.7	236.0	260.6	288.0	317.0	28
n-Octane	C_8H_{18}	-14.0	+8.3	19.2	31.5	45.1	53.8	65.7	83.6	104.0	125.6	-56.8
n-Octanol (1-octanol)	$C_8H_{18}O$	54.0	76.5	88.3	101.0	115.2	123.8	135.2	152.0	173.8	195.2	-15.4
2-Octanone	$C_8H_{18}O$	23.6	48.4	60.9	74.3	89.8	90.0	111.7	130.4	151.0	172.9	-16
n-Octyl acrylate	$C_{11}H_{20}O_2$	58.5	87.7	102.0	117.8	135.6	145.6	159.1	180.2	204.0	227.0	
iodide (1-Iodooctane)	$C_8H_{17}I$	45.8	74.8	90.0	105.9	123.8	135.4	150.0	173.3	199.3	225.5	-45.9
Oleic acid	$C_{18}H_{34}O_2$	176.5	208.5	223.0	240.0	257.2	269.8	286.0	309.8	334.7	360.0	14
Palmitaldehyde	$C_{16}H_{32}O$	121.6	154.6	171.8	190.0	210.0	222.6	239.5	264.1	292.3	321.0	34
Palmitic acid	$C_{16}H_{32}O_2$	153.6	188.1	205.8	223.8	244.4	256.0	271.5	298.7	326.0	353.8	64.0
Palmitonitrile	$C_{16}H_{31}N$	134.3	168.3	185.8	204.2	223.8	236.6	251.5	277.1	304.5	332.0	31
Pelargonic acid	$C_9H_{18}O_2$	108.2	126.0	137.4	149.8	163.7	172.3	184.4	203.1	227.5	253.5	12.5
Pentachlorobenzene	C_6HCl_5	98.6	129.7	144.3	160.0	178.5	190.1	205.5	227.0	251.6	276.0	85.5
Pentachloroethane	C_2HCl_5	+1.0	27.2	39.8	53.9	69.9	80.0	93.5	114.0	137.2	160.5	-22
Pentachloroethylbenzene	$C_6H_5Cl_5$	96.2	130.0	148.0	166.0	186.2	199.0	216.0	241.8	269.3	299.0	
Pentachlorophenol	C_6HCl_5O	192.2	211.2	223.4	239.6	261.8	285.0	290.4	298.0	303.8	309.3	188.5
Pentacosane	$C_{25}H_{52}$	194.2	230.0	248.2	266.1	285.6	298.4	314.0	339.0	365.4	390.3	53.3
Pentadecane	$C_{15}H_{32}$	91.6	121.0	135.4	150.2	167.7	178.4	194.0	216.1	242.8	270.5	10
1,3-Pentadiene	C_5H_8	-71.8	-53.8	-45.0	-34.8	-23.4	-16.5	-6.7	+8.0	24.7	42.1	
1,4-Pentadiene	C_5H_8	-83.5	-66.2	-57.1	-47.7	-37.0	-30.0	-20.6	-6.7	+8.3	26.1	
Pentaethylbenzene	$C_{16}H_{26}$	86.0	120.0	135.8	152.4	171.9	184.2	200.0	224.1	250.2	277.0	

VAPOR PRESSURE–TABLE 4. Vapor Pressures of Organic Compounds, up to 1 atm (*Continued*)

Pressure, mm Hg — Temperature, °C

Name	Formula	1	5	10	20	40	60	100	200	400	760	Melting point, °C
Pentaethylchlorobenzene	$C_{16}H_{25}Cl$	90.0	183.8	140.7	158.1	178.2	191.0	208.0	230.3	257.2	285.0	
n-Pentane	C_5H_{12}	−76.6	−62.5	−50.1	−40.2	−29.2	−22.2	−12.6	+1.9	18.5	36.1	−129.7
iso-Pentane (2-methylbutane)	C_5H_{12}	−82.9	−65.8	−57.0	−47.3	−36.5	−29.6	−20.2	−5.9	+10.5	27.8	−159.7
neo-Pentane (2,2-dimethylpropane)	C_5H_{12}	−102.0	−85.4	−76.7	−67.2	−56.1	−49.0	−39.1	−23.7	−7.1	+9.5	−16.6
2,3,4-Pentanetriol	$C_5H_{12}O_3$	155.0	159.3	204.5	220.5	239.6	249.8	263.5	284.5	307.0	327.2	
1-Pentene	C_5H_{10}	−80.4	−63.3	−54.5	−46.0	−34.1	−27.1	−17.7	−3.4	+12.8	30.1	
α-Phellandrene	$C_{10}H_{16}$	20.0	45.7	58.0	72.1	87.8	97.6	110.6	130.6	152.0	175.0	
Phenanthrene	$C_{14}H_{10}$	118.2	154.3	173.0	193.7	215.8	229.9	249.0	277.1	308.0	340.2	99.5
Phenethyl alcohol (phenyl cellosolve)	$C_8H_{10}O_2$	58.2	85.9	100.0	114.8	130.5	141.2	154.0	175.0	197.5	219.5	
2-Phenetidine	$C_8H_{11}NO$	67.0	94.7	108.6	123.7	139.9	149.8	163.5	184.0	207.0	228.0	
Phenol	C_6H_6O	40.1	62.5	73.8	86.0	100.1	108.4	121.4	139.0	160.0	181.9	40.6
2-Phenoxyethanol	$C_8H_{10}O_2$	78.0	196.6	121.2	136.0	152.2	163.2	176.5	197.6	221.0	245.3	11.6
2-Phenoxyethyl acetate	$C_{10}H_{12}O_3$	82.6	143.5	128.0	144.5	162.3	174.0	189.2	211.3	235.0	259.7	−6.7
Phenyl acetate	$C_8H_8O_2$	38.2	64.8	78.0	92.3	108.1	118.1	131.6	151.2	173.5	195.9	
Phenylacetic acid	$C_8H_8O_2$	97.0	127.0	141.3	156.0	173.6	184.5	198.2	219.5	243.0	265.5	76.5
Phenylacetonitrile	C_8H_7N	60.0	89.0	103.5	119.4	136.3	147.7	161.8	184.2	208.5	233.5	−23.8
Phenylacetyl chloride	C_8H_7ClO	48.0	75.3	89.0	103.6	119.8	129.8	143.5	163.8	186.0	210.0	
Phenyl benzoate	$C_{13}H_{10}O_2$	106.8	141.5	157.8	177.0	197.6	210.8	227.8	254.0	283.5	314.0	70.5
4-Phenyl-3-buten-2-one	$C_{10}H_{10}O$	81.7	112.2	127.4	143.8	161.3	172.6	187.8	211.0	235.4	261.0	41.5
Phenyl isocyanate	C_7H_5NO	10.6	36.0	48.5	62.5	77.7	87.7	100.6	120.8	142.7	165.6	
isocyanide	C_7H_5N	12.0	37.0	49.7	63.4	78.3	88.0	101.0	120.8	142.3	165.0	
Phenylcyclohexane	$C_{12}H_{16}$	67.5	96.5	111.3	126.4	144.0	154.2	169.3	191.3	214.6	240.0	+75
Phenyl dichlorophosphate	$C_6H_5Cl_2O_2P$	66.7	95.9	110.0	125.9	143.4	153.6	168.0	189.8	213.0	239.5	
m-Phenylene diamine (1,3-phenylenediamine)	$C_6H_8N_2$	99.8	131.2	147.0	163.8	182.5	194.0	209.9	233.0	259.0	285.5	62.8
Phenylglyoxal	$C_8H_6O_2$		75.0	87.8	100.7	115.5	124.2	136.2	153.8	173.5	193.5	73
Phenylhydrazine	$C_6H_8N_2$	75.8	101.6	115.8	131.5	148.2	158.7	173.5	195.4	218.2	243.5	19.5
N-Phenyliminodiethanol	$C_{10}H_{15}NO_2$	145.0	170.2	195.8	213.4	233.0	245.3	260.6	284.5	311.3	337.8	

Compound	Formula											fp
1-Phenyl-1,3-pentanedione	$C_{11}H_{12}O_2$	98.0	128.5	144.0	159.9	178.0	189.8	204.5	226.7	251.2	276.5	56.5
2-Phenylphenol	$C_{12}H_{10}O$	100.0	131.6	146.2	163.3	180.3	192.2	205.9	227.9	251.8	275.0	164.5
4-Phenylphenol	$C_{12}H_{10}O$			176.2	193.8	213.0	225.3	240.9	263.2	285.5	308.0	
3-Phenyl-1-propanol	$C_9H_{12}O$	74.7	102.4	116.0	131.2	147.4	156.8	170.3	191.2	212.8	235.0	
Phenyl isothiocyanate	C_7H_5NS	47.2	75.6	89.8	115.5	122.5	133.3	147.7	169.6	194.0	218.5	−21.0
Phorone	$C_9H_{14}O$	42.0	63.3	81.5	95.6	111.3	121.4	134.0	153.5	175.3	197.2	28
iso-Phorone	$C_9H_{14}O$	38.0	66.7	81.2	96.8	114.5	125.6	140.6	163.3	188.7	215.2	
Phosgene (carbonyl chloride)	CCl_2O	−92.9	−77.0	−69.3	−60.3	−50.3	−44.0	−35.6	−22.3	−7.6	+8.3	−104
Phthalic anhydride	$C_8H_4O_3$	96.5	124.3	134.0	151.7	172.0	185.3	202.3	228.0	256.8	284.5	130.8
Phthalide	$C_8H_6O_2$	95.5	127.7	144.0	161.3	181.0	193.5	210.0	234.5	261.8	290.0	73
Phthaloyl chloride	$C_8H_4Cl_2O_2$	86.3	118.3	134.2	151.0	170.0	182.2	197.8	222.0	248.3	275.8	88.5
2-Picoline	C_6H_7N	−11.1	+12.6	24.4	37.4	51.2	59.9	71.4	89.0	108.4	128.8	−70
Pimelic acid	$C_7H_{12}O_4$	163.4	196.2	212.0	229.3	247.0	258.2	272.0	294.5	318.5	342.1	103
α-Pinene	$C_{10}H_{16}$	−1.0	+24.6	37.3	51.4	66.8	76.8	90.1	110.2	132.3	155.0	−55
β-Pinene	$C_{10}H_{16}$	+4.2	30.0	42.3	58.1	71.5	81.2	94.0	114.1	136.1	158.3	−9
Piperidine	$C_5H_{11}N$	−7.0	−7.0	+3.9	15.8	29.2	37.7	49.0	66.2	85.7	106.0	37
Piperonal	$C_8H_6O_3$	87.0	117.4	132.0	148.0	165.7	177.0	191.7	214.3	238.5	263.0	−187.1
Propane	C_3H_8	−128.9	−115.4	−108.5	−100.9	−92.4	−87.0	−79.6	−68.4	−55.6	−42.1	−30.1
Propenylbenzene	C_9H_{10}	17.5	43.8	57.0	71.5	87.7	97.8	111.7	132.0	154.7	179.0	79
Propionamide	C_3H_7NO	65.0	91.0	105.0	119.0	134.8	144.3	156.0	174.2	194.0	213.0	−22
Propionic acid	$C_3H_6O_2$	4.6	28.0	39.7	52.0	65.8	74.1	85.8	102.5	122.0	141.1	−45
Propionic acid anhydride	$C_6H_{10}O_3$	20.6	45.3	57.7	70.4	85.6	94.5	107.2	127.8	146.0	167.0	−91.9
Propionitrile	C_3H_5N	−35.0	−13.6	−3.0	+8.8	22.0	30.1	41.4	58.2	77.7	97.1	21
Propiophenone	$C_9H_{10}O$	50.0	77.9	92.2	107.6	124.3	135.0	149.3	170.2	194.2	218.0	−92.5
n-Propyl acetate	$C_5H_{10}O_2$	−26.7	−5.4	+5.0	16.0	28.8	37.0	47.8	64.0	82.0	101.8	
iso-Propyl acetate	$C_5H_{10}O_2$	−38.3	−17.4	−7.2	+4.2	17.0	25.1	35.7	51.7	69.8	89.0	−127
n-Propyl alcohol (1-propanol)	C_3H_8O	−15.0	+5.0	14.7	25.3	36.4	43.5	52.8	66.8	82.0	97.8	−85.8
iso-Propyl alcohol (2-propanol)	C_3H_8O	−26.1	−7.0	+2.4	12.7	23.8	30.5	39.5	53.0	67.8	82.5	−83
n-Propylamine	C_3H_9N	−64.4	−46.3	−37.2	−27.1	−16.0	−9.0	+0.5	15.0	31.5	48.5	−99.5
Propylbenzene	C_9H_{12}	6.3	31.3	43.4	56.8	71.6	81.1	94.0	113.5	135.7	159.2	−51.6
Propyl benzoate	$C_{10}H_{12}O_2$	54.6	83.8	98.0	114.3	131.8	143.3	157.4	180.1	205.2	231.0	−109.9
n-Propyl bromide (1-bromopropane)	C_3H_7Br	−53.0	−33.4	−23.3	−12.4	−0.3	+7.5	18.0	34.0	52.0	71.0	−89.0
iso-Propyl bromide (2-bromopropane)	C_3H_7Br	−61.8	−42.5	−32.8	−22.0	−10.1	−2.5	+8.0	23.8	41.5	60.0	
n-Propyl n-butyrate	$C_7H_{14}O_2$	−1.6	+22.1	34.0	47.0	61.5	70.3	82.6	101.0	121.7	142.7	−95.2

VAPOR PRESSURE–TABLE 4. Vapor Pressures of Organic Compounds, up to 1 atm (*Continued*)

Name	Formula	Pressure, mm Hg — Temperature, °C										Melting point, °C
		1	5	10	20	40	60	100	200	400	760	
isobutyrate	$C_7H_{14}O_2$	-6.2	+16.8	28.3	40.6	54.3	63.0	73.9	91.8	112.0	133.9	
iso-Propyl isobutyrate	$C_7H_{14}O_2$	-16.3	+5.8	17.0	29.0	42.4	51.4	62.3	80.2	100.0	120.5	
Propyl carbamate	$C_4H_9NO_2$	52.4	77.6	90.0	103.2	117.7	126.5	138.3	155.8	175.8	195.0	
n-Propyl chloride (1-chloropropane)	C_3H_7Cl	-68.3	-50.0	-41.0	-31.0	-19.5	-12.1	-2.5	+12.2	29.4	46.4	-122.8
iso-Propyl chloride (2-chloropropane)	C_3H_7Cl	-78.8	-61.1	-52.0	-42.0	-31.0	-23.5	-13.7	+1.3	18.1	36.5	-117
iso-Propyl chloroacetate	$C_5H_9ClO_2$	+3.8	28.1	40.2	53.9	68.7	78.0	90.3	108.8	128.0	148.6	
Propyl chloroglyoxylate	$C_5H_7ClO_3$	9.7	32.3	43.5	55.6	68.8	77.2	88.0	104.7	123.0	150.0	
Propylene	C_3H_6	-131.9	-120.7	-112.1	-104.7	-96.5	-91.3	-84.1	-73.3	-60.9	-47.7	-185
Propylene glycol (1,2-Propanediol)	$C_3H_8O_2$	45.5	70.8	83.2	96.4	111.2	119.9	132.0	149.7	168.1	188.2	
Propylene oxide	C_3H_6O	-75.0	-57.8	-49.0	-39.3	-28.4	-21.3	-12.0	+2.1	17.8	34.5	-112.1
n-Propyl formate	$C_4H_8O_2$	-43.0	-27.7	-12.6	-1.7	+10.8	18.8	29.5	45.3	62.6	81.3	-92.9
iso-Propyl formate	$C_4H_8O_2$	-52.0	-32.7	-22.7	-12.1	-0.2	+7.5	17.8	33.6	50.5	68.3	
4,4'-iso-Propylidenebisphenol	$C_{15}H_{16}O_2$	193.0	224.2	240.8	255.5	273.0	282.9	297.0	317.5	339.0	360.5	
n-Propyl iodide (1-iodopropane)	C_3H_7I	-36.0	-13.5	-2.4	+10.0	23.6	32.1	43.8	61.8	81.8	102.5	-98.8
iso-Propyl iodide (2-iodopropane)	C_3H_7I	-43.3	-22.1	-11.7	0.0	+13.2	21.6	32.8	50.0	69.5	89.5	-90
n-Propyl levulinate	$C_8H_{14}O_3$	59.7	86.3	99.9	114.0	130.1	140.6	154.0	175.6	198.0	221.2	
iso-Propyl levulinate	$C_8H_{14}O_3$	48.0	74.5	88.0	102.4	118.1	127.8	141.8	161.6	185.2	208.2	
Propyl mercaptan (1-propanethiol)	C_3H_8S	-56.0	-36.3	-26.3	-15.4	-3.2	+4.6	15.3	31.5	49.2	67.4	-112
2-iso-Propylnaphthalene	$C_{13}H_{14}$	76.0	107.9	123.4	140.3	159.0	171.4	187.6	211.8	238.5	266.0	
iso-Propyl β-naphthyl ketone (2-isobutyronaphthone)	$C_{14}H_{14}O$	133.2	165.4	181.0	197.7	215.6	227.0	242.3	264.0	288.2	313.0	
2-iso-Propylphenol	$C_9H_{12}O$	56.6	83.8	97.0	111.7	127.5	137.7	150.3	170.1	192.6	214.5	15.5
3-iso-Propylphenol	$C_9H_{12}O$	62.0	90.3	104.1	119.8	136.2	146.6	160.2	182.0	205.0	228.0	26
4-iso-Propylphenol	$C_9H_{12}O$	67.0	94.7	108.0	123.4	139.8	149.7	163.3	184.0	206.1	228.2	61
Propyl propionate	$C_6H_{12}O_2$	-14.2	+8.0	19.4	31.6	45.0	53.8	65.2	82.7	102.0	122.4	-76
4-iso-Propylstyrene	$C_{11}H_{14}$	34.7	62.3	76.0	91.2	108.0	118.4	132.8	153.9	178.0	202.5	
Propyl isovalerate	$C_8H_{16}O_2$	+8.0	32.8	45.1	58.0	72.8	82.3	95.0	113.9	135.0	155.9	
Pulegone	$C_{10}H_{16}O$	58.3	82.5	94.0	106.8	121.7	130.2	143.1	162.5	189.8	221.0	
Pyridine	C_5H_5N	-18.9	+2.5	13.2	24.8	38.0	46.8	57.8	75.0	95.6	115.4	-42

Name	Formula											
Pyrocatechol	$C_6H_6O_2$		104.0	118.3	134.0	150.6	161.7	176.0	197.7	221.5	245.5	105
Pyrocatechol diacetate (1,2-phenylene diacetate)	$C_{10}H_{10}O_4$	98.0	129.8	145.7	161.8	179.8	191.6	206.5	228.7	253.3	278.0	
Pyrogallol	$C_6H_6O_3$		151.7	167.7	185.3	204.2	216.3	232.0	255.3	281.5	309.0	133
Pyrotartaric anhydride	$C_5H_6O_3$	69.7	99.7	114.2	130.0	147.8	158.6	173.8	196.1	221.0	247.4	
Pyruvic acid	$C_3H_4O_3$	21.4	45.8	57.9	70.8	85.3	94.1	106.5	124.7	144.7	165.0	13.6
Quinoline	C_9H_7N	59.7	89.6	103.8	119.8	136.7	148.1	163.2	186.2	212.3	237.7	-15
iso-Quinoline	C_9H_7N	63.5	92.7	107.8	123.7	141.6	152.0	167.6	190.0	214.5	240.5	24.6
Resorcinol	$C_6H_6O_2$	108.4	138.0	152.1	168.0	185.3	195.8	209.8	230.8	253.4	276.5	110.7
Safrole	$C_{10}H_{10}O_2$	63.8	93.0	107.6	123.0	140.1	150.3	165.1	186.2	210.0	233.0	11.2
Salicylaldehyde	$C_7H_6O_2$	33.0	60.1	73.8	88.7	105.2	115.7	129.4	150.0	173.7	196.5	-7
Salicylic acid	$C_7H_6O_3$	113.7	136.0	146.2	156.8	172.2	182.0	193.4	210.0	230.5	256.0	159
Sebacic acid	$C_{10}H_{18}O_4$	183.0	215.7	232.0	250.0	268.2	279.8	294.5	313.2	332.8	352.3	134.5
Selenophene	C_4H_4Se	-39.0	-16.0	-4.0	+9.1	24.1	33.8	47.0	66.7	89.8	114.3	
Skatole	C_9H_9N	95.0	124.2	139.6	154.3	171.9	183.6	197.4	218.8	242.5	266.2	95
Stearaldehyde	$C_{18}H_{36}O$	140.0	174.6	192.1	210.6	230.8	244.2	260.0	285.0	313.8	342.5	63.5
Stearic acid	$C_{18}H_{36}O_2$	173.7	209.0	225.0	243.4	263.3	275.5	291.0	316.5	343.0	370.0	69.3
Stearyl alcohol (1-octadecanol)	$C_{18}H_{36}O$	150.3	185.6	202.0	220.0	240.4	252.7	269.4	293.5	320.3	349.5	58.5
Styrene	C_8H_8	-7.0	+18.0	30.8	44.6	59.8	69.5	82.0	101.3	122.5	145.2	-30.6
Styrene dibromide [(1,2-dibromoethyl)benzene]	$C_8H_8Br_2$	86.0	115.6	129.8	145.2	161.8	172.2	186.3	207.8	230.0	245.0	
Suberic acid	$C_8H_{14}O_4$	172.8	205.5	219.5	238.2	254.6	265.4	279.8	300.5	322.8	345.5	142
Succinic anhydride	$C_4H_4O_3$	92.0	115.0	128.2	145.3	163.0	174.0	189.0	212.0	237.0	261.0	119.6
Succinimide	$C_4H_5NO_2$	115.0	143.2	157.0	174.0	192.0	203.0	217.4	240.0	263.5	287.5	125.5
Succinyl chloride	$C_4H_4Cl_2O_2$	39.0	65.0	78.0	91.8	107.5	117.2	130.0	149.3	170.0	192.5	17
α-Terpineol	$C_{10}H_{18}O$	52.8	80.4	94.3	109.8	126.0	136.3	150.1	171.2	194.3	217.5	35
Terpenoline	$C_{10}H_{16}$	32.3	58.0	70.6	84.8	100.0	109.8	122.7	142.0	163.5	185.0	
1,1,1,2-Tetrabromoethane	$C_2H_2Br_4$	58.0	83.3	95.7	108.5	123.2	132.0	144.0	161.5	181.0	200.0	
1,1,2,2-Tetrabromoethane	$C_2H_2Br_4$	65.0	95.5	110.0	126.0	144.0	155.1	170.0	192.5	217.5	243.5	
Tetraisobutylene	$C_{16}H_{32}$	63.8	93.7	108.5	124.5	142.2	152.6	167.5	190.0	214.6	240.0	
Tetracosane	$C_{24}H_{50}$	183.8	219.6	237.6	255.3	276.3	288.4	305.2	330.5	358.0	386.4	51.1
1,2,3,4-Tetrachlorobenzene	$C_6H_2Cl_4$	68.5	99.6	114.7	131.2	149.2	160.0	175.7	198.0	225.5	254.0	46.5
1,2,3,5-Tetrachlorobenzene	$C_6H_2Cl_4$	58.2	89.0	104.1	121.6	140.0	152.0	168.0	193.7	220.0	246.0	54.5
1,2,4,5-Tetrachlorobenzene	$C_6H_2Cl_4$					146.0	157.7	173.5	196.0	220.5	245.0	139

VAPOR PRESSURE–TABLE 4. Vapor Pressures of Organic Compounds, up to 1 atm (*Continued*)

Name	Formula	Pressure, mm Hg										Melting point, °C
		1	5	10	20	40	60	100	200	400	760	
		Temperature, °C										
1,1,2,2-Tetrachloro-1,2-difluoroethane	$C_2Cl_4F_2$	-37.5	-16.0	-5.0	+6.7	19.8	28.1	33.6	55.0	73.1	92.0	26.5
1,1,2-Tetrachloroethane	$C_2H_2Cl_4$	-16.3	+7.4	19.3	32.1	46.7	56.0	68.0	87.2	108.2	130.5	-68.7
1,1,2,2-Tetrachloroethane	$C_2H_2Cl_4$	-3.8	+20.7	33.0	46.2	60.8	70.0	83.2	102.2	124.0	145.9	-36
1,2,3,5-Tetrachloro-4-ethylbenzene	$C_8H_6Cl_4$	77.0	110.0	126.0	143.7	162.1	175.0	191.6	215.3	243.0	270.0	
Tetrachloroethylene	C_2Cl_4	-20.6	+2.4	13.8	26.3	40.1	49.2	61.3	79.8	100.0	120.8	-19.0
2,3,4,6-Tetrachlorophenol	$C_6H_2Cl_4O$	100.0	130.3	145.3	161.0	179.1	190.0	205.2	227.2	250.4	275.0	69.5
3,4,5,6-Tetrachloro-1,2-xylene	$C_8H_6Cl_4$	94.4	125.0	140.3	156.0	174.2	185.8	200.5	223.0	248.3	273.5	
Tetradecane	$C_{14}H_{30}$	76.4	106.0	120.7	135.6	152.7	164.0	178.5	201.8	226.8	252.5	5.5
Tetradecylamine	$C_{14}H_{31}N$	102.6	135.8	152.0	170.0	189.0	200.2	215.7	239.8	264.6	291.2	
Tetradecyltrimethylsilane	$C_{17}H_{38}Si$	120.0	150.7	166.2	183.5	201.5	213.3	227.8	250.0	275.0	300.0	
Tetraethoxysilane	$C_8H_{20}O_4Si$	16.0	40.3	52.6	65.8	81.1	90.7	103.6	123.5	146.2	168.5	
1,2,3,4-Tetraethylbenzene	$C_{14}H_{22}$	65.7	96.2	111.6	127.7	145.8	156.7	172.4	196.0	221.4	248.0	11.6
Tetraethylene glycol	$C_8H_{18}O_5$	153.9	183.7	197.1	212.3	228.0	237.8	250.0	268.4	288.0	307.8	
Tetraethylene glycol chlorohydrin	$C_8H_{17}ClO_4$	110.1	141.8	156.1	172.6	190.0	200.5	214.7	236.5	258.2	281.5	
Tetraethyllead	$C_8H_{20}Pb$	38.4	63.6	74.8	88.0	102.4	111.7	123.8	142.0	161.8	183.0	-136
Tetraethylsilane	$C_8H_{20}Si$	-1.0	+23.9	36.3	50.0	65.3	74.8	88.0	108.0	130.2	153.0	
Tetralin	$C_{10}H_{12}$	38.0	65.3	79.0	93.8	110.4	121.3	135.3	157.2	181.8	207.2	-31.0
1,2,3,4-Tetramethylbenzene	$C_{10}H_{14}$	42.6	68.7	81.8	95.8	111.5	121.8	135.7	155.7	180.0	204.4	-6.2
1,2,3,5-Tetramethylbenzene	$C_{10}H_{14}$	40.6	65.8	77.8	91.0	105.8	115.4	128.3	149.9	173.7	197.9	-24.0
1,2,4,5-Tetramethylbenzene	$C_{10}H_{14}$	45.0	65.0	74.6	88.0	104.2	114.8	128.1	149.5	172.1	195.9	79.5
2,2,3,3-Tetramethylbutane	C_8H_{18}	-17.4	+3.2	13.5	24.6	36.8	44.5	54.8	70.2	87.4	106.3	-102.2
Tetramethylene dibromide (1,4-dibromobutane)	$C_4H_8Br_2$	32.0	58.8	72.4	87.6	104.0	115.1	128.7	149.8	173.8	197.5	-20
Tetramethyllead	$C_4H_{12}Pb$	-29.0	-6.8	+4.4	16.6	30.3	39.2	50.8	68.8	89.0	110.0	-27.5
Tetramethyltin	$C_4H_{12}Sn$	-51.3	-31.0	-20.6	-9.3	+3.5	11.7	22.8	39.8	58.5	78.0	
Tetrapropylene glycol monoisopropyl ether	$C_{15}H_{32}O_5$	116.6	147.8	163.0	179.8	197.7	209.0	223.3	245.0	268.3	292.7	
Thioacetic acid (mercaptoacetic acid)	$C_2H_4O_2S$	60.0	87.7	101.5	115.8	131.8	142.0	154.0				-16.5
Thiodiglycol (2,2'-thiodiethanol)	$C_4H_{10}O_2S$	42.0	96.0	128.0	165.0	210.0	240.5	285				
Thiophene	C_4H_4S	-40.7	-20.8	-10.9	0.0	+12.5	20.1	30.5	46.5	64.7	84.4	-38.3

Compound	Formula											
Thiophenol (benzenethiol)	C6H6S	18.6	43.7	56.0	69.7	84.2	93.9	106.6	125.8	146.7	168.0	
α-Thujone	C10H16O	38.3	65.7	79.3	93.7	110.0	120.2	134.0	154.2	177.8	201.0	
Thymol	C10H14O	64.3	92.8	107.4	122.6	139.8	149.8	164.1	185.5	209.6	231.8	51.5
Tiglaldehyde	C5H8O	−25.0	−1.6	+10.0	23.2	37.0	45.8	57.7	75.4	95.5	116.4	
Tiglic acid	C5H8O2	52.0	77.8	90.2	103.8	119.0	127.8	140.5	158.0	179.2	198.5	64.5
Tiglonitrile	C5H7N	−25.5	−2.4	+9.2	22.1	36.7	46.0	58.2	77.8	99.7	122.0	
Toluene	C7H8	−26.7	−4.4	+6.4	18.4	31.8	40.3	51.9	69.5	89.5	110.6	−95.0
Toluene-2,4-diamine	C7H10N2	106.5	137.2	151.7	167.9	185.7	196.2	211.5	232.8	256.0	280.0	99
2-Toluic nitrile (2-tolunitrile)	C8H7N	36.7	64.0	77.9	93.0	110.0	120.8	135.0	156.0	180.0	205.2	−13
4-Toluic nitrile (4-tolunitrile)	C8H7N	42.5	71.3	85.8	101.7	109.5	130.0	145.2	167.3	193.0	217.6	29.5
2-Toluidine	C7H9N	44.0	69.3	81.4	95.1	110.0	119.8	133.0	153.0	176.2	199.7	−16.3
3-Toluidine	C7H9N	41.0	68.0	82.0	96.7	113.5	123.8	136.7	157.6	180.6	203.3	−31.5
4-Toluidine	C7H9N	42.0	68.2	81.8	95.8	111.5	121.5	133.7	154.0	176.9	200.4	44.5
2-Tolyl isocyanide	C8H7N	25.2	51.0	64.0	78.2	94.0	104.0	117.7	137.8	159.9	183.5	
4-Tolylhydrazine	C7H10N2	82.4	110.0	123.8	138.6	154.1	165.0	178.0	198.0	219.5	242.0	65.5
Tribromoacetaldehyde	C2HBr3O	18.5	45.0	58.0	72.1	87.8	97.5	110.2	130.0	151.6	174.0	
1,1,2-Tribromobutane	C4H7Br3	45.0	73.5	87.8	103.2	120.2	131.6	146.0	167.8	192.0	216.2	
1,2,2-Tribromobutane	C4H7Br3	41.0	69.0	83.2	98.6	116.0	127.0	141.8	163.5	188.0	213.8	
2,2,3-Tribromobutane	C4H7Br3	38.2	66.0	79.8	94.6	111.8	122.2	136.3	157.8	182.2	206.5	
1,1,2-Tribromoethane	C2H3Br3	32.6	58.0	70.6	84.2	100.0	110.0	123.5	143.5	165.4	188.4	−26
1,2,3-Tribromopropane	C3H5Br3	47.5	75.8	90.0	105.8	122.8	134.0	148.0	170.0	195.0	220.0	16.5
Triisobutylamine	C12H27N	32.3	57.4	69.8	83.0	97.8	107.3	119.7	138.0	157.8	179.0	−22
Triisobutylene	C12H24	18.0	44.0	56.5	70.0	86.7	96.7	110.0	130.2	153.0	179.0	
2,4,6-Tritertbutylphenol	C18H30O	95.2	126.1	142.0	158.0	177.4	188.0	203.0	226.2	250.6	276.3	57
Trichloroacetic acid	C2HCl3O2	51.0	76.0	88.2	101.8	116.3	125.9	137.8	155.4	175.2	195.6	
Trichloroacetic anhydride	C4Cl6O3	56.2	85.3	99.6	114.3	131.2	141.8	155.2	176.2	199.8	223.0	
Trichloroacetyl bromide	C2BrCl3O	−7.4	+16.7	29.3	42.1	57.2	66.7	79.5	98.4	120.2	143.0	
2,4,6-Trichloroaniline	C6H4Cl3N	134.0	157.8	170.0	182.6	195.8	204.5	214.6	229.8	246.4	262.0	78
1,2,3-Trichlorobenzene	C6H3Cl3	40.0	70.0	85.6	101.8	119.8	131.5	146.0	168.2	193.5	218.5	52.5
1,2,4-Trichlorobenzene	C6H3Cl3	38.4	67.3	81.7	97.2	114.8	125.7	140.0	162.0	187.7	213.0	17
1,3,5-Trichlorobenzene	C6H3Cl3		63.8	78.0	93.7	110.8	121.8	136.0	157.7	183.0	208.4	63.5
1,2,3-Trichlorobutane	C4H7Cl3	+0.5	27.2	40.0	55.0	71.5	82.0	96.2	118.0	143.0	169.0	
1,1,1-Trichloroethane	C2H3Cl3	−52.0	−32.0	−21.9	−10.8	+1.6	9.5	20.0	36.2	54.6	74.1	−30.6
1,1,2-Trichloroethane	C2H3Cl3	−24.0	−2.0	+8.3	21.6	35.2	44.0	55.7	73.3	93.0	113.9	−36.7

VAPOR PRESSURE–TABLE 4. Vapor Pressures of Organic Compounds, up to 1 atm (*Continued*)

Name	Formula	Pressure, mm Hg — Temperature, °C										Melting point, °C
		1	5	10	20	40	60	100	200	400	760	
Trichloroethylene	C_2HCl_3	-43.8	-22.8	-12.4	-1.0	+11.9	20.0	31.4	48.0	67.0	86.7	-73
Trichlorofluoromethane	CCl_3F	-84.3	-67.6	-59.0	-49.7	-39.0	-32.3	-23.0	-9.1	+6.8	23.7	
2,4,5-Trichlorophenol	$C_6H_3Cl_3O$	72.0	102.1	117.3	134.0	151.5	162.5	178.0	201.5	226.5	251.8	62
2,6-Trichlorophenol	$C_6H_3Cl_3O$	76.5	105.9	120.2	135.8	152.2	163.5	177.8	199.0	222.5	246.0	68.5
Tri-2-chlorophenylthiophosphate	$C_{18}H_{12}Cl_3O_3PS$	188.2	217.2	231.2	246.7	261.7	271.5	283.8	302.8	322.0	341.3	
1,1,1-Trichloropropane	$C_3H_5Cl_3$	-28.8	-7.0	+4.2	16.2	29.9	38.3	50.0	67.7	87.5	108.2	-77.7
1,2,3-Trichloropropane	$C_3H_5Cl_3$	+9.0	33.7	46.0	59.3	74.0	83.6	96.1	115.6	137.0	158.0	-14.7
1,1,2-Trichloro-1,2,2-trifluoroethane	$C_2Cl_3F_3$	-68.0	-49.4	-40.3	-30.0	-18.5	-11.2	-1.7	+13.5	30.2	47.6	-35
Tricosane	$C_{23}H_{48}$	170.0	206.3	223.0	242.0	261.3	273.8	289.8	313.5	339.8	366.5	47.7
Tridecane	$C_{13}H_{28}$	59.4	98.3	104.0	120.2	137.7	148.2	162.5	185.0	209.4	234.0	-6.2
Tridecanoic acid	$C_{13}H_{26}O_2$	137.8	166.3	181.0	195.8	212.4	222.0	236.0	255.2	276.5	299.0	41
Triethoxymethylsilane	$C_7H_{18}O_3Si$	-1.5	+22.8	34.6	47.2	61.7	70.4	82.7	101.0	121.8	143.5	
Triethoxyphenylsilane	$C_{12}H_{20}O_3Si$	71.0	98.8	112.6	127.2	143.5	153.2	167.5	188.0	210.5	233.5	
1,2,4-Triethylbenzene	$C_{12}H_{18}$	46.0	74.2	88.5	104.0	121.7	132.2	146.8	168.3	193.7	218.0	
1,3,4-Triethylbenzene	$C_{12}H_{18}$	47.9	76.0	90.2	105.8	122.6	133.4	147.7	168.3	193.2	217.5	
Triethylborine	$C_6H_{15}B$			-148.0	-140.6	-131.4	-125.2	-116.0	-101.0	-81.0	-56.2	
Triethyl camphoronate	$C_{15}H_{26}O_6$		150.2	166.0	183.6	201.8	213.5	228.6	250.8	276.0	301.0	135
citrate	$C_{12}H_{20}O_7$	107.0	138.7	144.0	171.1	190.4	202.5	217.8	242.2	267.5	294.0	
Triethyleneglycol	$C_6H_{14}O_4$	114.0	144.0	158.1	174.0	191.3	201.5	214.6	235.2	256.6	278.3	
Triethylheptylsilane	$C_{13}H_{30}Si$	70.0	99.8	114.6	130.3	148.0	158.2	174.0	196.0	221.0	247.0	
Triethyloctylsilane	$C_{14}H_{32}Si$	73.7	104.8	120.6	137.7	155.7	168.0	184.3	208.0	235.0	262.0	
Triethyl orthoformate	$C_7H_{16}O_3$	+5.5	29.2	40.5	53.4	67.5	76.0	88.0	106.0	125.7	146.0	
phosphate	$C_6H_{15}O_4P$	39.6	67.8	82.1	97.8	115.7	126.3	141.6	163.7	187.0	211.0	
Triethylthallium	$C_6H_{15}Tl$	+9.3	37.6	51.7	67.7	85.4	95.7	112.1	136.0	163.5	192.1	-63.0

Compound	Formula											
Trifluorophenylsilane	$C_6H_5F_3Si$	-31.0	-9.7	+0.8	12.3	25.4	33.2	44.2	60.1	78.7	98.3	
Trimethallyl phosphate	$C_{12}H_{21}PO_4$	93.7	131.0	149.8	169.8	192.0	207.0	225.7	255.0	288.5	324.0	
2,3,5-Trimethylacetophenone	$C_{11}H_{14}O$	79.0	108.0	122.3	137.5	154.2	165.7	179.7	201.3	224.3	247.5	
Trimethylamine	C_3H_9N	-97.1	-81.7	-73.8	-65.0	-55.2	-48.8	-40.3	-27.0	-12.5	+2.9	-117.1
2,4,5-Trimethylaniline	$C_9H_{13}N$	68.4	95.9	109.0	123.7	139.8	149.5	162.0	182.3	203.7	234.5	67
1,2,3-Trimethylbenzene	C_9H_{12}	16.8	42.9	55.9	69.9	85.4	95.3	108.8	129.0	152.0	176.1	-25.5
1,2,4-Trimethylbenzene	C_9H_{12}	13.6	38.3	50.7	64.5	79.8	89.5	102.8	122.7	145.4	169.2	-44.1
1,3,5-Trimethylbenzene	C_9H_{12}	9.6	34.7	47.4	61.0	76.1	85.8	98.9	118.6	141.0	164.7	-44.8
2,2,3-Trimethylbutane	C_7H_{16}			-18.8	-7.5	+5.2	13.3	24.4	41.2	60.4	80.9	-25.0
Trimethyl citrate	$C_9H_{14}O_7$	106.2	146.2	160.4	177.2	194.2	205.5	219.6	241.3	264.2	287.0	78.5
Trimethyleneglycol (1,3-propanediol)	$C_3H_8O_2$	59.4	87.2	100.6	115.5	131.0	141.1	153.4	172.8	193.8	214.2	
1,2,4-Trimethyl-5-ethylbenzene	$C_{11}H_{16}$	43.7	71.2	84.6	99.7	106.0	126.3	140.3	160.3	184.5	208.1	
1,3,5-Trimethyl-2-ethylbenzene	$C_{11}H_{16}$	38.8	67.0	80.5	96.0	113.2	123.8	137.9	158.4	183.5	208.0	
2,2,3-Trimethylpentane	C_8H_{18}	-29.0	-7.1	+3.9	16.0	29.5	38.1	49.9	67.8	88.2	109.8	-112.3
2,2,4-Trimethylpentane	C_8H_{18}	-36.5	-15.0	-4.3	+7.5	20.7	29.1	40.7	58.1	78.0	99.2	-107.3
2,3,3-Trimethylpentane	C_8H_{18}	-25.8	-3.9	+6.9	19.2	33.0	41.8	53.8	72.0	92.7	114.8	-101.5
2,3,4-Trimethylpentane	C_8H_{18}	-26.3	-4.1	+7.1	19.3	32.9	41.6	53.4	71.3	91.8	113.5	-109.2
2,2,4-Trimethyl-3-pentanone	$C_8H_{16}O$	14.7	36.0	46.4	57.6	69.8	77.3	87.6	102.2	118.4	135.0	
Trimethyl phosphate	$C_3H_9O_4P$	26.0	53.7	67.8	83.0	100.0	110.0	124.0	145.0	167.8	192.7	
2,4,5-Trimethylstyrene	$C_{11}H_{14}$	48.1	77.0	91.6	107.1	124.2	135.5	149.8	171.8	196.1	221.2	
2,4,6-Trimethylstyrene	$C_{11}H_{14}$	37.5	65.7	79.7	94.8	111.8	122.3	136.8	157.8	182.3	207.0	
Trimethylsuccinic anhydride	$C_7H_{10}O_3$	53.5	82.6	97.4	113.8	131.0	142.2	156.5	179.8	205.5	231.0	
Triphenylmethane	$C_{19}H_{16}$	169.7	188.4	197.0	206.8	215.5	221.2	228.4	239.7	249.8	259.2	93.4
Triphenylphosphate	$C_{18}H_{15}O_4P$	193.5	230.4	249.8	269.7	290.3	305.2	322.5	349.8	379.2	413.5	49.4
Tripropyleneglycol	$C_9H_{20}O_4$	96.0	125.7	140.5	155.8	173.7	184.6	199.0	220.2	244.3	267.2	
Tripropyleneglycol monobutyl ether	$C_{13}H_{28}O_4$	101.5	131.6	147.0	161.8	179.8	190.2	204.4	224.4	247.0	269.5	
Tripropyleneglycol monoisopropyl ether	$C_{12}H_{26}O_4$	82.4	112.4	127.3	143.7	161.4	173.2	187.8	209.7	232.8	256.6	
Tritolyl phosphate	$C_{21}H_{21}O_4P$	154.6	184.2	198.0	213.2	229.7	239.8	252.2	271.8	292.7	313.0	
Undecane	$C_{11}H_{24}$	32.7	59.7	73.9	85.6	104.4	115.2	128.1	149.3	171.9	195.8	-25.6
Undecanoic acid	$C_{11}H_{22}O_2$	101.4	133.1	149.0	166.0	185.6	197.2	212.5	237.8	262.8	290.0	29.5
10-Undecenoic acid	$C_{11}H_{20}O_2$	114.0	142.8	156.3	172.0	188.7	199.5	213.5	232.8	254.0	275.0	24.5
Undecan-2-ol	$C_{11}H_{24}O$	71.1	99.0	112.8	127.5	143.7	153.7	167.2	187.7	209.8	232.0	
n-Valeric acid	$C_5H_{10}O_2$	42.2	67.7	79.8	93.1	107.8	116.6	128.3	146.0	165.0	184.4	-34.5
iso-Valeric acid	$C_5H_{10}O_2$	34.5	59.6	71.3	84.0	98.0	107.3	118.9	136.2	155.2	175.1	-37.6

VAPOR PRESSURE–TABLE 4. Vapor Pressures of Organic Compounds, up to 1 atm (*Continued*)

Name	Formula	Pressure, mm Hg										Melting point, °C
		1	5	10	20	40	60	100	200	400	760	
		Temperature, °C										
γ-Valerolactone	$C_5H_8O_2$	37.5	65.8	79.8	95.2	101.9	122.4	136.5	157.7	182.3	207.5	
Valeronitrile	C_5H_9N	−6.0	+18.1	30.0	43.3	57.8	66.9	78.6	97.7	118.7	140.8	
Vanillin	$C_8H_8O_3$	107.0	138.4	154.0	170.5	188.7	199.8	214.5	237.3	260.0	285.0	81.5
Vinyl acetate	$C_4H_6O_2$	−48.0	−28.0	−18.0	−7.0	+5.3	13.0	23.3	38.4	55.5	72.5	
2-Vinylanisole	$C_9H_{10}O$	41.9	68.0	81.0	94.7	110.0	119.8	132.3	151.0	172.1	194.0	
3-Vinylanisole	$C_9H_{10}O$	43.4	69.9	83.0	97.2	112.5	122.3	135.3	154.0	175.8	197.5	
4-Vinylanisole	$C_9H_{10}O$	45.2	72.0	85.7	100.0	116.0	126.1	139.7	159.0	182.0	204.5	
Vinyl chloride (1-chloroethylene)	C_2H_3Cl	−105.6	−90.8	−83.7	−75.7	−66.8	−61.1	−53.2	−41.3	−28.0	−13.8	−153.7
cyanide (acrylonitrile)	C_3H_3N	−51.0	−30.7	−20.3	−9.0	+3.8	11.8	22.8	38.7	58.3	78.5	−82
fluoride (1-fluoroethylene)	C_2H_3F	−149.3	−138.0	−132.2	−125.4	−118.0	−113.0	−106.2	−95.4	−84.0	−72.2	−160.5
Vinylidene chloride (1,1-dichloroethene)	$C_2H_2Cl_2$	−77.2	−60.0	−51.2	−41.7	−31.1	−24.0	−15.0	−1.0	+14.8	31.7	−122.5
4-Vinylphenetole	$C_{10}H_{12}O$	64.0	91.7	105.6	120.3	136.3	146.4	159.8	180.0	202.8	225.0	
2-Xenyl dichlorophosphate	$C_{12}H_9Cl_2PO$	138.2	171.1	187.0	205.0	223.8	236.0	251.5	275.3	301.5	328.5	
2,4-Xylaldehyde	$C_9H_{10}O$	59.0	85.9	99.0	114.0	129.7	139.8	152.2	172.3	194.1	215.5	75
2-Xylene (2-xylene)	C_8H_{10}	−3.8	+20.2	32.1	45.1	59.5	68.8	81.3	100.2	121.7	144.4	−25.2
3-Xylene (3-xylene)	C_8H_{10}	−6.9	+16.8	28.3	41.1	55.3	64.4	76.8	95.5	116.7	139.1	−47.9
4-Xylene (4-xylene)	C_8H_{10}	−8.1	+15.5	27.3	40.1	54.4	63.5	75.9	94.6	115.9	138.3	+13.3
2,4-Xylidine	$C_8H_{11}N$	52.6	79.8	93.0	107.6	123.8	133.7	146.8	166.4	188.3	211.5	
2,6-Xylidine	$C_8H_{11}N$	44.0	72.6	87.0	102.7	120.2	131.5	146.0	168.0	193.7	217.9	

Compiled from D.R. Stull, *Ind. Eng. Chem.*, 1947, 39, 517.

VAPOR PRESSURE—TABLE 5. Vapor Pressure of Liquid Water from 0 to 100°C

mmHg

t, °C	0.0	0.1	0.2	0.3	0.4	0.5	0.6	0.7	0.8	0.9
0	4.579	4.613	4.647	4.681	4.715	4.750	4.785	4.820	4.855	4.890
1	4.926	4.962	4.998	5.034	5.070	5.107	5.144	5.181	5.219	5.256
2	5.294	5.332	5.370	5.408	5.447	5.486	5.525	5.565	5.605	5.645
3	5.685	5.725	5.766	5.807	5.848	5.889	5.931	5.973	6.015	6.058
4	6.101	6.144	6.187	6.230	6.274	6.318	6.363	6.408	6.453	6.498
5	6.543	6.589	6.635	6.681	6.728	6.775	6.822	6.869	6.917	6.965
6	7.013	7.062	7.111	7.160	7.209	7.259	7.309	7.360	7.411	7.462
7	7.513	7.565	7.617	7.669	7.722	7.775	7.828	7.882	7.936	7.990
8	8.045	8.100	8.155	8.211	8.267	8.323	8.380	8.437	8.494	8.551
9	8.609	8.668	8.727	8.786	8.845	8.905	8.965	9.025	9.086	9.147
10	9.209	9.271	9.333	9.395	9.458	9.521	9.585	9.649	9.714	9.779
11	9.844	9.910	9.976	10.042	10.109	10.176	10.244	10.312	10.380	10.449
12	10.518	10.588	10.658	10.728	10.799	10.870	10.941	11.013	11.085	11.158
13	11.231	11.305	11.379	11.453	11.528	11.604	11.680	11.756	11.833	11.910
14	11.987	12.065	12.144	12.223	12.302	12.382	12.462	12.543	12.624	12.706
15	12.788	12.870	12.953	13.037	13.121	13.205	13.290	13.375	13.461	13.547
16	13.634	13.721	13.809	13.898	13.987	14.076	14.166	14.256	14.347	14.438
17	14.530	14.622	14.715	14.809	14.903	14.997	15.092	15.188	15.284	15.380
18	15.477	15.575	15.673	15.772	15.871	15.971	16.071	16.171	16.272	16.374
19	16.477	16.581	16.685	16.789	16.894	16.999	17.105	17.212	17.319	17.427
20	17.535	17.644	17.753	17.863	17.974	18.085	18.197	18.309	18.422	18.536
21	18.650	18.765	18.880	18.996	19.113	19.231	19.349	19.468	19.587	19.707
22	19.827	19.948	20.070	20.193	20.316	20.440	20.565	20.690	20.815	20.941
23	21.068	21.196	21.324	21.453	21.583	21.714	21.845	21.977	22.110	22.243
24	22.377	22.512	22.648	22.785	22.922	23.060	23.198	23.337	23.476	23.616

VAPOR PRESSURE—TABLE 5. Vapor Pressure of Liquid Water from 0 to 100°C (*Continued*)

mmHg

t, °C	0.0	0.1	0.2	0.3	0.4	0.5	0.6	0.7	0.8	0.9
25	23.756	23.897	24.039	24.182	24.326	24.471	24.617	24.764	24.912	25.060
26	25.209	25.359	25.509	25.660	25.812	25.964	26.117	26.271	26.426	26.582
27	26.739	26.897	27.055	27.214	27.374	27.535	27.696	27.858	28.021	28.185
28	28.349	28.514	28.680	28.847	29.015	29.184	29.354	29.525	29.697	29.870
29	30.043	30.217	30.392	30.568	30.745	30.923	31.102	31.281	31.461	31.642
30	31.824	32.007	32.191	32.376	32.561	32.747	32.934	33.122	33.312	33.503
31	33.695	33.888	34.082	34.276	34.471	34.667	34.864	35.062	35.261	35.462
32	35.663	35.865	36.068	36.272	36.477	36.683	36.891	37.099	37.308	37.518
33	37.729	37.942	38.155	38.369	33.584	38.801	39.018	39.237	39.457	39.677
34	39.898	40.121	40.344	40.569	40.796	41.023	41.251	41.480	41.710	41.942
35	42.175	42.409	42.644	42.880	43.117	43.355	43.595	43.836	44.078	44.320
36	44.563	44.808	45.054	45.301	45.549	45.799	46.050	46.302	46.556	46.811
37	47.067	47.324	47.582	47.841	48.102	48.364	48.627	48.891	49.157	49.424
38	49.692	49.961	50.231	50.502	50.774	51.048	51.323	51.600	51.879	52.160
39	52.442	52.725	53.009	53.294	53.580	53.867	54.156	54.446	54.737	55.030
40	55.324	55.61	55.91	56.21	56.51	56.81	57.11	57.41	57.72	58.03
41	58.34	58.65	58.96	59.27	59.58	59.90	60.22	60.54	60.86	61.18
42	61.50	61.82	62.14	62.47	62.80	63.13	63.46	63.79	64.12	64.46
43	64.80	65.14	65.48	65.82	66.16	66.51	66.86	67.21	67.56	67.91
44	68.26	68.61	68.97	69.33	69.69	70.05	70.41	70.77	71.14	71.51
45	71.88	72.25	72.62	72.99	73.36	73.74	74.12	74.50	74.88	75.26
46	75.65	76.04	76.43	76.82	77.21	77.60	78.00	78.40	78.80	79.20
47	79.60	80.00	80.41	80.82	81.23	81.64	82.05	82.46	82.87	83.29
48	83.71	84.13	84.56	84.99	85.42	85.85	86.28	86.71	87.14	87.58
49	88.02	88.46	88.90	89.34	89.79	90.24	90.69	91.14	91.59	92.05

mmHg

t, °C	0	1	2	3	4	5	6	7	8	9
50	92.51	97.20	102.09	107.20	112.51	118.04	123.80	129.82	136.08	142.60
60	149.38	156.43	163.77	171.38	179.31	187.54	196.09	204.96	214.17	223.73
70	233.7	243.9	254.6	265.7	277.2	289.1	301.4	314.1	327.3	341.0
80	355.1	369.7	384.9	400.6	416.8	433.6	450.9	468.7	487.1	506.1
90	525.76	527.76	529.77	531.78	533.80	535.82	537.86	539.90	541.95	544.00
91	546.05	548.11	550.18	552.26	554.35	556.44	558.53	560.64	562.75	564.87
92	566.99	569.12	571.26	573.40	575.55	577.71	579.87	582.04	584.22	586.41
93	588.60	590.80	593.00	595.21	597.43	599.66	601.89	604.13	606.38	608.64
94	610.90	613.17	615.44	617.72	620.01	622.31	624.61	626.92	629.24	631.57
95	633.90	636.24	638.59	640.94	643.30	645.67	648.05	650.43	652.82	655.22
96	657.62	660.03	662.45	664.88	667.31	669.75	672.20	674.66	677.12	679.69
97	682.07	684.55	687.04	689.54	692.05	694.57	697.10	699.63	702.17	704.71
98	707.27	709.83	712.40	714.98	717.56	720.15	722.75	725.36	727.98	730.61
99	733.24	735.88	738.53	741.18	743.85	748.52	749.20	751.89	754.58	757.29
100	760.00	762.72	765.45	768.19	770.93	773.68	776.44	779.22	782.00	784.78
101	787.57	790.37	793.18	796.00	796.82	801.66	804.50	807.35	810.21	813.06

From the Physikalisch-technische Reichsanstalt, Holborn, Scheel, and Henning, *Wärmetabellen*, Friedrich Vieweg & Sohn, Brunswick, 1909. By permission. For data at 50(0.2)101.8°C, see *Handbook of Chemistry and Physics*, 40th ed., p. 2326, Chemical Rubber Publishing Co. For a tabulation of temperature for pressures 700(1)779 mm Hg, see Atack, *Handbook of Chemical Data*, p. 117, Reinhold, New York, 1957. For a tabulation of pressure for 105(5)200(10)370°C, see Atack, p. 134, and for 100(1)374°C, see *Handbook of Chemistry and Physics*, 40th ed., pp. 2328–2330, Chemical Rubber Publishing Co.

VAPOR PRESSURE—TABLE 6. Partial Pressures of Water over Aqueous Solutions of HCl*

$\log_{10} p_{mm} = A - B/T$, ($T$ in K), which however, agrees only approximately with the table. The table is more nearly correct.
Partial pressure of H_2O, mmHg, °C

% HCl	A	B	0°	5°	10°	15°	20°	25°	30°	35°	40°	45°	50°	60°	70°	80°	90°	100°	110°
6	8.99156	2282	4.18	6.04	8.45	11.7	15.9	21.8	29.1	39.4	50.6	66.2	86.0	139	220	333	492	715	
10	8.99864	2295	3.84	5.52	7.70	10.7	14.6	20.0	26.8	35.5	47.0	61.5	80.0	130	204	310	463	677	960
14	8.97075	2300	3.39	4.91	6.95	9.65	13.1	18.0	24.1	31.9	42.1	55.3	72.0	116	185	273	425	625	892
18	8.98014	2323	2.87	4.21	5.92	8.26	11.3	15.4	20.6	27.5	36.4	47.9	62.5	102	162	248	374	550	783
20	8.97877	2334	2.62	3.83	5.40	7.50	10.3	14.1	19.0	25.1	33.3	43.6	57.0	93.5	150	230	345	510	729
22	9.02708	2363	2.33	3.40	4.82	6.75	9.30	12.6	17.1	22.8	30.2	39.8	52.0	85.6	138	211	317	467	670
24	8.96022	2356	2.05	3.04	4.31	6.03	8.30	11.4	15.4	20.4	27.1	35.7	46.7	77.0	124	194	290	426	611
26	9.01511	2390	1.76	2.60	3.71	5.21	7.21	9.95	13.5	18.0	24.0	31.7	41.5	69.0	112	173	261	387	555
28	8.97611	2395	1.50	2.24	3.21	4.54	6.32	8.75	11.8	15.8	21.1	27.9	36.5	60.7	99.0	154	234	349	499
30	9.00117	2422	1.26	1.90	2.73	3.88	5.41	7.52	10.2	13.7	18.4	24.3	32.0	53.5	87.5	136	207	310	444
32	9.03317	2453	1.04	1.57	2.27	3.25	4.55	6.37	8.70	11.7	15.7	21.0	27.7	46.5	76.5	120	184	275	396
34	9.07143	2487	0.85	1.29	1.87	2.70	3.81	5.35	7.32	9.95	13.5	18.1	24.0	40.5	66.5	104	161	243	355
36	9.11815	2526	0.68	1.03	1.50	2.19	3.10	4.41	6.08	8.33	11.4	15.4	20.4	34.8	57.0	90.0	140	212	311
38	9.20783	2579	0.53	0.81	1.20	1.75	2.51	3.60	5.03	6.92	9.52	13.0	17.4	29.6	49.1	77.5	120	182	266
40	9.33923	2647	0.41	0.63	0.94	1.37	2.00	2.88	4.09	5.68	7.85	10.7	14.5	25.0	42.1	67.3	105	158	230
42	9.44953	2709	0.31	0.48	0.72	1.06	1.56	2.30	3.28	4.60	6.45	8.90	12.1	21.2	35.8	57.2	89.2	135	195

* Accuracy, ca. 2% for solutions of 15–30% HCl between 0 and 100°; for solutions of >30% HCl the accuracy is ca. 5% at the higher temperatures. Below 15% HCl, the accuracy is ca. 5% at the lower temperatures and ca. 15% at the higher temperatures and higher strengths to ca. 15–20% the lower strengths and perhaps 15–20% at the higher temperatures and lower strengths.

VAPOR PRESSURE—TABLE 7. Partial Pressures of HCl over Aqueous Solutions of HCl*

$\log_{10} p_{mm} = A - B/T$, ($T$ in K), which, however, agrees only approximately with the table. The table is more nearly correct. mmHg, °C

% HCl	A	B	0°	5°	10°	15°	20°	25°	30°	35°	40°	45°	50°	60°	70°	80°	90°	100°	110°
2	11.8037	4736			0.0000117	0.000023	0.000044	0.000084	0.000151	0.000275	0.00047	0.00083	0.00140	0.00380	0.0100	0.0245	0.058	0.132	0.280
4	11.6400	4471	0.000018	0.000036	.000069	.000131	.00024	.00044	.00077	.00134	.0023	.00385	.0064	.0165	.0405	.095	.21	.46	.93
6	11.2144	4202	.000066	.000125	.000234	.000425	.00076	.00131	.00225	.0038	.0062	.0102	.0163	.040	.094	.206	.44	.92	1.76
8	11.0406	4042	.000118	.000323	.000583	.00104	.00178	.0031	.00515	.0085	.0136	.022	.0344	.081	.183	.39	.82	1.64	3.10
10	10.9311	3908	.00042	.00075	.00134	.00232	.00395	.0067	.0111	.0178	.0282	.045	.069	.157	.35	.73	1.48	2.9	5.4
12	10.7900	3765	.00099	.00175	.00305	.0052	.0088	.0145	.0234	.037	.058	.091	.136	.305	.66	1.34	2.65	5.1	9.3
14	10.6954	3636	.0024	.00415	.0071	.0118	.0196	.0316	.050	.078	.121	.185	.275	.60	1.25	2.50	4.8	9.0	16.0
16	10.6261	3516	.0056	.0095	.016	.0265	.0428	.0685	.106	.163	.247	.375	.55	1.17	2.40	4.66	8.8	16.1	28
18	10.4957	3376	.0135	.0225	.037	.060	.095	.148	.228	.345	.515	.77	1.11	2.3	4.55	8.6	15.7	28	48
20	10.3833	3245	.0316	.052	.084	.132	.205	.32	.48	.72	1.06	1.55	2.21	4.4	8.5	15.6	28.1	49	83
22	10.3172	3125	.0734	.119	.187	.294	.45	.68	1.02	1.50	2.18	3.14	4.42	8.6	16.3	29.3	52	90	146
24	10.2185	2995	.175	.277	.43	.66	1.00	1.49	2.17	3.14	4.5	6.4	8.9	16.9	31.0	54.5	94	157	253
26	10.1303	2870	.41	.64	.98	1.47	2.17	3.20	4.56	6.50	9.2	12.7	17.5	32.5	58.5	100	169	276	436
28	10.0115	2732	1.0	1.52	2.27	3.36	4.90	7.05	9.90	13.8	19.1	26.4	35.7	64	112	188	309	493	760
30	9.8763	2593	2.4	3.57	5.23	7.60	10.6	15.1	21.0	28.6	39.4	53	71	124	208	340	542	845	
32	9.7523	2457	5.7	8.3	11.8	16.8	23.5	32.5	44.5	60.0	81	107	141	238	390	623	970		
34	9.6061	2316	13.1	18.8	26.4	36.8	50.5	68.5	92	122	161	211	273	450	720				
36	9.5262	2229	29.0	41.0	56.4	78	105.5	142	188	246	322	416	535	860					
38	9.4670	2094	63.0	87.0	117	158	210	277	360	465	598	758	955						
40	9.2156	1939	130	176	233	307	399	515	627	830									
42	8.9925	1800	253	332	430	560	709	900											
44	8.8621	1681	510	655	840														
46			940																

* Accuracy, ca. 2% for solutions of 15–30% HCl between 0 and 100°; for solutions of >30% HCl the accuracy is ca. 5% at the higher temperatures ca. 15% at the lower temperatures. Below 15% HCl, the accuracy is ca. 5% at the lower temperatures and higher strengths to ca. 15–20% at the lower strengths and perhaps 15–20% at the higher temperatures and lower strengths.

VAPOR PRESSURE—TABLE 8. Partial Pressures of H_2O and SO_2 over Aqueous Solutions of Sulfur Dioxide
Partial pressures of H_2O and SO_2, mmHg, °C

g SO$_2$/ 100g H$_2$O	Temperature, °C								
	0	10	20	30	40	50	60	90	120
0.01	0.02	0.04	0.07	0.12	0.19	0.29	0.43	1.21	2.28
0.05	0.38	0.66	1.07	1.68	2.53	3.69	5.24	12.9	27.0
0.10	1.15	1.91	3.03	4.62	6.80	9.71	13.5	31.7	63.9
0.15	2.10	3.44	5.37	8.07	11.7	16.5	22.7	52.2	104
0.20	3.17	5.13	7.93	11.8	17.0	23.8	32.6	73.7	145
0.25	4.34	6.93	10.6	15.7	22.5	31.4	42.8	95.8	186
0.30	5.57	8.84	13.5	19.8	28.2	39.2	53.3	118	229
0.40	8.17	12.8	19.4	28.3	40.1	55.3	74.7	164	316
0.50	10.9	17.0	25.6	37.1	52.3	72.0	96.8	211	404
1.00	25.8	39.5	58.4	83.7	117	159	212	454	856
2.00	58.6	88.5	129	183	253	342	453	955	
3.00	93.2	139	202	285	393	530	700		
4.00	129	192	277	389	535	720			
5.00	165	245	353	496	679				
6.00	202	299	430	602	824				
8.00	275	407	585	818					
10.00	351	517	741						
15.00	542	796							
20.00	735								

VAPOR PRESSURE—TABLE 9. Water Partial Pressure, bar, over Aqueous Sulfuric Acid Solutions

Weight percent, H_2SO_4

°C	10.0	20.0	30.0	40.0	50.0	60.0	70.0	75.0	80.0	85.0
0	.582E-02	.534E-02	.448E-02	.326E-02	.193E-02	.836E-03	.207E-03	.747E-04	.197E-04	.343E-05
10	.117E-01	.107E-01	.909E-02	.670E-02	.405E-02	.180E-02	.467E-03	.175E-03	.490E-04	.952E-05
20	.223E-01	.205E-01	.174E-01	.130E-01	.802E-02	.367E-02	.995E-03	.388E-03	.115E-03	.245E-04
30	.404E-01	.373E-01	.319E-01	.241E-01	.151E-01	.710E-02	.201E-02	.811E-03	.253E-03	.589E-04
40	.703E-01	.649E-01	.558E-01	.427E-01	.272E-01	.131E-01	.387E-02	.162E-02	.531E-03	.133E-03
50	.117	.109	.939E-01	.725E-01	.470E-01	.232E-01	.715E-02	.309E-02	.106E-02	.286E-03
60	.189	.175	.152	.119	.782E-01	.395E-01	.127E-01	.565E-02	.204E-02	.584E-03
70	.296	.275	.239	.188	.126	.651E-01	.217E-01	.997E-02	.376E-02	.114E-02
80	.449	.417	.365	.290	.196	.104	.360E-01	.170E-01	.668E-02	.213E-02
90	.664	.617	.542	.434	.298	.161	.578E-01	.281E-01	.115E-01	.383E-02
100	.957	.891	.786	.634	.441	.244	.905E-01	.452E-01	.192E-01	.666E-02
110	1.349	1.258	1.113	.904	.638	.360	.138	.708E-01	.312E-01	.112E-01
120	1.863	1.740	1.544	1.264	.903	.519	.206	.108	.493E-01	.183E-01
130	2.524	2.361	2.101	1.732	1.253	.734	.301	.162	.760E-01	.291E-01
140	3.361	3.149	2.810	2.333	1.708	1.020	.481	.236	.115	.451E-01
150	4.404	4.132	3.697	3.090	2.289	1.392	.605	.339	.170	.682E-01
160	5.685	5.342	4.793	4.031	3.021	1.870	.837	.478	.246	.101
170	7.236	6.810	6.127	5.185	3.930	2.475	1.138	.662	.350	.147
180	9.093	8.571	7.731	6.584	5.045	3.233	1.525	.902	.489	.208
190	11.289	10.658	9.640	8.259	6.397	4.169	2.017	1.212	.673	.291
200	13.861	13.107	11.887	10.245	8.020	5.312	2.632	1.606	.913	.401
210	16.841	15.951	14.505	12.576	9.948	6.696	3.395	2.101	1.220	.542
220	20.264	19.225	17.529	15.287	12.217	8.354	4.331	2.714	1.609	.724
230	24.160	22.960	20.992	18.414	14.864	10.322	5.466	3.467	2.096	.952
240	28.561	27.188	24.927	21.992	17.929	12.641	6.831	4.381	2.699	1.237
250	33.494	31.939	29.364	26.056	21.452	15.351	8.458	5.480	3.435	1.587
260	38.984	37.240	34.334	30.642	25.472	18.496	10.382	6.788	4.326	2.012

VAPOR PRESSURE–TABLE 9. Water Partial Pressure, bar, over Aqueous Sulfuric Acid Solutions (*Continued*)

Weight percent, H_2SO_4

°C	10.0	20.0	30.0	40.0	50.0	60.0	70.0	75.0	80.0	85.0
270	45.055	43.116	39.865	35.784	30.030	22.121	12.640	8.333	5.395	2.525
280	51.726	49.590	45.984	41.514	35.168	26.274	15.269	10.142	6.663	3.136
290	59.015	56.681	52.715	47.865	40.926	31.003	18.311	12.242	8.155	3.857
300	66.934	64.407	60.081	54.868	47.346	36.360	21.808	14.665	9.897	4.701
310	75.495	72.781	68.100	62.553	54.470	42.395	25.804	17.438	11.912	5.680
320	84.705	81.816	76.792	70.947	62.337	49.164	30.343	20.591	14.227	6.806
330	94.567	91.518	86.172	80.077	70.988	56.721	35.473	24.153	16.867	8.093
340	105.083	101.894	96.252	89.969	80.463	65.123	41.240	28.154	19.855	9.551
350	116.251	112.946	107.043	100.646	90.802	74.426	47.692	32.622	23.217	11.193

Weight percent, H_2SO_4

°C	90.0	92.0	94.0	96.0	97.0	98.0	98.5	99.0	99.5	100.0
0	.518E-06	.242E-06	.107E-06	.401E-07	.218E-07	.980E-08	.569E-08	.268E-08	.775E-09	.196E-09
10	.159E-05	.762E-06	.344E-06	.130E-06	.713E-07	.323E-07	.188E-07	.888E-08	.258E-08	.655E-09
20	.448E-05	.220E-05	.101E-05	.390E-06	.215E-06	.978E-07	.572E-07	.271E-07	.789E-08	.201E-08
30	.117E-04	.587E-05	.275E-05	.108E-05	.598E-06	.275E-06	.161E-06	.766E-07	.224E-07	.575E-08
40	.285E-04	.146E-04	.696E-05	.278E-05	.155E-05	.720E-06	.424E-06	.202E-06	.595E-07	.153E-07
50	.652E-04	.341E-04	.166E-04	.672E-05	.379E-05	.177E-05	.105E-05	.503E-06	.149E-06	.384E-07
60	.141E-03	.754E-04	.372E-04	.154E-04	.875E-05	.413E-05	.245E-05	.118E-05	.350E-06	.910E-07
70	.290E-03	.158E-03	.795E-04	.334E-04	.192E-04	.912E-05	.544E-05	.263E-05	.784E-06	.205E-06
80	.569E-03	.316E-03	.162E-03	.691E-04	.400E-04	.192E-04	.115E-04	.559E-05	.168E-05	.439E-06
90	.107E-02	.606E-03	.315E-03	.137E-03	.801E-04	.388E-04	.234E-04	.114E-04	.343E-05	.903E-06

100	.194E-02	.112E-02	.590E-03	.261E-03	.154E-03	.752E-04	.455E-04	.223E-04	.674E-05	.178E-05
110	.338E-02	.198E-02	.107E-02	.479E-03	.285E-03	.141E-03	.855E-04	.420E-04	.128E-04	.339E-05
120	.571E-02	.341E-02	.186E-02	.815E-03	.511E-03	.254E-03	.155E-03	.766E-04	.233E-04	.623E-05
130	.938E-02	.569E-02	.315E-02	.146E-02	.886E-03	.445E-03	.278E-03	.135E-03	.414E-04	.111E-04
140	.150E-01	.923E-02	.519E-02	.245E-02	.149E-02	.757E-03	.467E-03	.232E-03	.711E-04	.191E-04
150	.233E-01	.146E-01	.832E-02	.399E-02	.245E-02	.125E-02	.776E-03	.387E-03	.119E-03	.321E-04
160	.354E-01	.225E-01	.130E-01	.633E-02	.393E-02	.202E-02	.126E-02	.629E-03	.194E-03	.526E-04
170	.526E-01	.340E-01	.199E-01	.983E-02	.614E-02	.319E-02	.199E-02	.999E-03	.309E-03	.840E-04
180	.766E-01	.502E-01	.298E-01	.149E-01	.941E-02	.492E-02	.309E-02	.155E-02	.482E-03	.131E-03
190	.110	.729E-01	.438E-01	.222E-01	.141E-01	.744E-02	.469E-02	.236E-02	.735E-03	.201E-03
200	.154	.104	.631E-01	.325E-01	.208E-01	.110E-01	.698E-02	.352E-02	.110E-02	.300E-03
210	.213	.146	.894E-01	.467E-01	.300E-01	.161E-01	.102E-01	.516E-02	.161E-02	.442E-03
220	.290	.201	.125	.660E-01	.427E-01	.230E-01	.147E-01	.743E-02	.232E-02	.638E-03
230	.389	.273	.171	.918E-01	.598E-01	.325E-01	.208E-01	.105E-01	.329E-02	.906E-03
240	.514	.366	.232	.126	.825E-01	.451E-01	.290E-01	.147E-01	.460E-02	.127E-02
250	.673	.485	.310	.170	.112	.618E-01	.398E-01	.202E-01	.633E-02	.174E-02
260	.870	.635	.409	.227	.151	.835E-01	.540E-01	.274E-01	.858E-02	.237E-02
270	1.112	.822	.534	.300	.200	.111	.723E-01	.366E-01	.115E-01	.317E-02
280	1.407	1.052	.689	.391	.263	.147	.957E-01	.485E-01	.152E-01	.420E-02
290	1.763	1.335	.880	.505	.341	.192	.125	.634E-01	.199E-01	.548E-02
300	2.190	1.676	1.112	.646	.437	.248	.162	.820E-01	.257E-01	.708E-02
310	2.696	2.088	1.394	.817	.556	.316	.208	.105	.328E-01	.905E-02
320	3.292	2.578	1.732	1.025	.701	.400	.264	.133	.415E-01	.114E-01
330	3.990	3.159	2.133	1.274	.875	.502	.331	.167	.520E-01	.143E-01
340	4.801	3.843	2.608	1.571	1.083	.624	.413	.208	.646E-01	.178E-01
350	5.738	4.641	3.164	1.922	1.331	.770	.511	.256	.795E-01	.218E-01

VAPOR PRESSURE—TABLE 10. Sulfur Trioxide Partial Pressure, bar, over Aqueous Sulfuric Acid Solutions

Weight percent, H_2SO_4

°C	10.0	20.0	30.0	40.0	50.0	60.0	70.0	75.0	80.0	85.0
0	.644E−29	.103E−27	.205E−26	.688E−25	.368E−23	.341E−21	.784E−19	.174E−17	.531E−16	.229E−14
10	.149E−27	.223E−26	.395E−25	.113E−23	.522E−22	.415E−20	.796E−18	.158E−16	.417E−15	.141E−13
20	.278E−26	.394E−25	.626E−24	.156E−22	.621E−21	.426E−19	.685E−17	.121E−15	.280E−14	.767E−13
30	.426E−25	.577E−24	.832E−23	.181E−21	.630E−20	.376E−18	.509E−16	.808E−15	.164E−13	.371E−12
40	.549E−24	.714E−23	.941E−22	.181E−20	.555E−19	.288E−17	.331E−15	.473E−14	.851E−13	.162E−11
50	.602E−23	.757E−22	.921E−21	.158E−19	.429E−18	.195E−16	.191E−14	.246E−13	.395E−12	.643E−11
60	.573E−22	.699E−21	.789E−20	.122E−18	.294E−17	.118E−15	.985E−14	.116E−12	.165E−11	.234E−10
70	.477E−21	.567E−20	.599E−19	.843E−18	.181E−16	.643E−15	.461E−13	.492E−12	.634E−11	.791E−10
80	.352E−20	.410E−19	.408E−18	.524E−17	.101E−15	.319E−14	.197E−12	.192E−11	.223E−10	.249E−09
90	.233E−19	.266E−18	.250E−17	.269E−16	.516E−15	.145E−13	.775E−12	.693E−11	.731E−10	.734E−09
100	.139E−18	.157E−17	.140E−16	.153E−15	.242E−14	.606E−13	.283E−11	.232E−10	.223E−09	.204E−08
110	.756E−18	.844E−17	.719E−16	.730E−15	.105E−13	.236E−12	.961E−11	.729E−10	.641E−09	.538E−08
120	.377E−17	.418E−16	.340E−15	.323E−14	.424E−13	.858E−12	.307E−10	.215E−09	.174E−08	.135E−07
130	.174E−16	.191E−15	.150E−14	.133E−13	.160E−12	.293E−11	.922E−10	.601E−09	.446E−08	.324E−07
140	.743E−16	.815E−15	.615E−14	.517E−13	.569E−12	.943E−11	.262E−09	.159E−08	.109E−07	.745E−07
150	.297E−15	.325E−14	.237E−13	.188E−12	.191E−11	.287E−10	.710E−09	.403E−08	.256E−07	.165E−06
160	.111E−14	.122E−13	.862E−13	.649E−12	.608E−11	.833E−10	.183E−08	.974E−08	.575E−07	.351E−06
170	.393E−14	.430E−13	.296E−12	.212E−11	.184E−10	.231E−09	.453E−08	.226E−07	.125E−06	.725E−06
180	.131E−13	.144E−12	.967E−12	.622E−11	.532E−10	.610E−09	.107E−07	.505E−07	.260E−06	.145E−05
190	.415E−13	.458E−12	.301E−11	.197E−10	.147E−09	.155E−08	.246E−07	.109E−06	.527E−06	.282E−05
200	.125E−12	.139E−11	.893E−11	.561E−10	.391E−09	.379E−08	.542E−07	.228E−06	.103E−05	.534E−05
210	.362E−12	.404E−11	.254E−10	.154E−09	.100E−08	.894E−08	.116E−06	.462E−06	.198E−05	.986E−05
220	.100E−11	.112E−10	.695E−10	.405E−09	.246E−08	.204E−07	.240E−06	.911E−06	.368E−05	.178E−04
230	.265E−11	.301E−10	.183E−09	.103E−08	.587E−08	.450E−07	.482E−06	.175E−05	.668E−05	.314E−04
240	.678E−11	.777E−10	.465E−09	.253E−08	.135E−07	.965E−07	.944E−06	.328E−05	.119E−04	.543E−04
250	.167E−10	.193E−09	.114E−08	.602E−08	.303E−07	.201E−06	.180E−05	.600E−05	.206E−04	.923E−04

°C	90.0	92.0	94.0	96.0	97.0	98.0	98.5	99.0	99.5	100.0
260	.399E−10	.466E−09	.272E−08	.139E−07	.660E−07	.408E−06	.336E−05	.108E−04	.352E−04	.154E−03
270	.920E−10	.109E−08	.628E−08	.312E−07	.140E−06	.807E−06	.612E−05	.189E−04	.590E−04	.253E−03
280	.206E−09	.247E−08	.141E−07	.683E−07	.288E−06	.156E−05	.109E−04	.326E−04	.973E−04	.408E−03
290	.449E−09	.545E−08	.308E−07	.145E−06	.580E−06	.295E−05	.191E−04	.553E−04	.158E−03	.649E−03
300	.953E−09	.117E−07	.657E−07	.302E−06	.114E−05	.546E−05	.329E−04	.921E−04	.253E−03	.102E−02
310	.197E−08	.245E−07	.136E−06	.614E−06	.220E−05	.990E−05	.556E−04	.151E−03	.398E−03	.158E−02
320	.397E−08	.502E−07	.277E−06	.122E−05	.414E−05	.176E−04	.923E−04	.245E−03	.621E−03	.242E−02
330	.782E−08	.100E−06	.551E−06	.237E−05	.766E−05	.308E−04	.151E−03	.391E−03	.956E−03	.367E−02
340	.151E−07	.196E−06	.107E−05	.452E−05	.139E−04	.529E−04	.243E−03	.617E−03	.145E−02	.550E−02
350	.285E−07	.376E−06	.204E−05	.846E−05	.246E−04	.893E−04	.387E−03	.963E−03	.219E−02	.815E−02

Weight percent, H_2SO_4

°C	90.0	92.0	94.0	96.0	97.0	98.0	98.5	99.0	99.5	100.0
0	.671E−13	.216E−12	.677E−12	.240E−11	.500E−11	.124E−10	.224E−10	.502E−10	.182E−09	.755E−09
10	.345E−12	.107E−11	.326E−11	.114E−10	.234E−10	.578E−10	.104E−09	.232E−09	.839E−09	.347E−08
20	.159E−11	.475E−11	.141E−10	.482E−10	.986E−10	.241E−09	.433E−09	.961E−09	.346E−08	.142E−07
30	.664E−11	.192E−10	.557E−10	.186E−09	.376E−09	.911E−09	.163E−08	.360E−08	.129E−07	.528E−07
40	.254E−10	.709E−10	.201E−09	.655E−09	.131E−08	.315E−08	.562E−08	.123E−07	.440E−07	.179E−06
50	.897E−10	.242E−09	.669E−09	.214E−08	.424E−08	.101E−07	.179E−07	.391E−07	.139E−06	.560E−06
60	.294E−09	.771E−09	.207E−08	.647E−08	.127E−07	.299E−07	.528E−07	.115E−06	.405E−06	.163E−05
70	.904E−09	.230E−08	.602E−08	.184E−07	.357E−07	.833E−07	.146E−06	.316E−06	.111E−05	.444E−05
80	.261E−08	.643E−08	.165E−07	.492E−07	.946E−07	.218E−06	.381E−06	.820E−06	.286E−05	.114E−04
90	.712E−08	.171E−07	.426E−07	.124E−06	.237E−06	.541E−06	.940E−06	.201E−05	.698E−05	.276E−04
100	.184E−07	.430E−07	.105E−06	.300E−06	.565E−06	.127E−05	.220E−05	.470E−05	.162E−04	.638E−04
110	.456E−07	.103E−06	.247E−06	.689E−06	.128E−05	.287E−05	.494E−05	.105E−04	.359E−04	.141E−03

VAPOR PRESSURE – TABLE 10. Sulfur Trioxide Partial Pressure, bar, over Aqueous Sulfuric Acid Solutions (*Continued*)

°C	Weight percent, H_2SO_4									
	10.0	20.0	30.0	40.0	50.0	60.0	70.0	75.0	80.0	85.0
120	.108E-06	.238E-06	.555E-06	.152E-05	.280E-05	.619E-05	.106E-04	.224E-04	.764E-04	.298E-03
130	.244E-06	.526E-06	.120E-05	.321E-05	.586E-05	.128E-04	.219E-04	.459E-04	.156E-03	.606E-03
140	.533E-06	.112E-05	.250E-05	.656E-05	.118E-04	.257E-04	.435E-04	.910E-04	.308E-03	.119E-02
150	.112E-05	.230E-05	.504E-05	.129E-04	.231E-04	.497E-04	.837E-04	.174E-03	.588E-03	.226E-02
160	.229E-05	.459E-05	.983E-05	.247E-04	.438E-04	.932E-04	.156E-03	.324E-03	.109E-02	.226E-02
170	.453E-05	.886E-05	.186E-04	.459E-04	.806E-04	.170E-03	.283E-03	.586E-03	.196E-02	.746E-02
180	.870E-05	.166E-04	.343E-04	.829E-04	.144E-03	.301E-03	.499E-03	.103E-02	.343E-02	.130E-01
190	.163E-04	.304E-04	.615E-04	.146E-03	.252E-03	.520E-03	.859E-03	.177E-02	.587E-02	.222E-01
200	.297E-04	.543E-04	.108E-03	.251E-03	.429E-03	.878E-03	.144E-02	.296E-02	.981E-02	.370E-01
210	.528E-04	.946E-04	.185E-03	.422E-03	.714E-03	.145E-02	.237E-02	.486E-02	.161E-01	.603E-01
220	.919E-04	.161E-03	.309E-03	.694E-03	.117E-02	.235E-02	.383E-02	.781E-02	.258E-01	.965E-01
230	.157E-03	.269E-03	.508E-03	.112E-02	.187E-02	.373E-02	.605E-02	.123E-01	.405E-01	.152
240	.261E-03	.441E-03	.819E-03	.178E-02	.293E-02	.582E-02	.939E-02	.191E-01	.627E-01	.234
250	.428E-03	.708E-03	.130E-02	.276E-02	.453E-02	.891E-02	.143E-01	.291E-01	.955E-01	.356
260	.690E-03	.112E-02	.202E-02	.423E-02	.688E-02	.134E-01	.215E-01	.437E-01	.143	.532
270	.109E-02	.174E-02	.309E-02	.638E-02	.103E-01	.200E-01	.319E-01	.646E-01	.212	.786
280	.170E-02	.266E-02	.466E-02	.948E-02	.152E-01	.293E-01	.465E-01	.943E-01	.309	1.144
290	.261E-02	.401E-02	.694E-02	.139E-01	.221E-01	.423E-01	.670E-01	.136	.444	1.646
300	.395E-02	.595E-02	.102E-01	.201E-01	.318E-01	.604E-01	.953E-01	.193	.632	2.339
310	.589E-02	.873E-02	.148E-01	.287E-01	.451E-01	.852E-01	.134	.272	.889	3.289
320	.868E-02	.126E-01	.211E-01	.405E-01	.632E-01	.119	.186	.378	1.236	4.575
330	.126E-01	.181E-01	.299E-01	.565E-01	.877E-01	.164	.256	.520	1.703	6.303
340	.181E-01	.255E-01	.418E-01	.780E-01	.120	.224	.348	.708	2.323	8.603
350	.258E-01	.357E-01	.578E-01	.107	.164	.303	.470	.956	3.142	11.640

VAPOR PRESSURE—TABLE 11. Sulfuric Acid Partial Pressure, bar, over Aqueous Sulfuric Acid

°C	Weight Percent, H_2SO_4									
	10.0	20.0	30.0	40.0	50.0	60.0	70.0	75.0	80.0	85.0
0	.576E-21	.843E-20	.141E-18	.344E-17	.109E-15	.438E-14	.249E-12	.200E-11	.161E-10	.121E-09
10	.634E-20	.874E-19	.131E-17	.276E-16	.769E-15	.273E-13	.135E-11	.101E-10	.743E-10	.490E-09
20	.588E-19	.769E-18	.104E-16	.193E-15	.474E-14	.149E-12	.649E-11	.447E-10	.305E-09	.179E-08
30	.468E-18	.584E-17	.721E-16	.119E-14	.259E-13	.725E-12	.278E-10	.178E-09	.113E-08	.594E-08
40	.324E-17	.389E-16	.441E-15	.649E-14	.127E-12	.317E-11	.108E-09	.643E-09	.379E-08	.181E-07
50	.197E-16	.229E-15	.241E-14	.320E-13	.562E-12	.126E-10	.380E-09	.212E-08	.117E-07	.513E-07
60	.107E-15	.121E-14	.119E-13	.144E-12	.228E-11	.462E-10	.124E-08	.646E-08	.334E-07	.135E-06
70	.526E-15	.581E-14	.535E-13	.592E-12	.851E-11	.156E-09	.373E-08	.183E-07	.888E-07	.336E-06
80	.235E-14	.254E-13	.221E-12	.225E-11	.295E-10	.492E-09	.105E-07	.485E-07	.222E-06	.786E-06
90	.960E-14	.102E-12	.844E-12	.798E-11	.956E-10	.145E-08	.279E-07	.121E-06	.522E-06	.175E-05
100	.353E-13	.381E-12	.300E-11	.264E-10	.291E-09	.402E-08	.698E-07	.287E-06	.117E-05	.371E-05
110	.127E-12	.132E-11	.997E-11	.824E-10	.835E-09	.106E-07	.166E-06	.644E-06	.249E-05	.752E-05
120	.418E-12	.432E-11	.312E-10	.243E-09	.227E-08	.264E-07	.375E-06	.138E-05	.508E-05	.147E-04
130	.129E-11	.132E-10	.924E-10	.678E-09	.589E-08	.631E-07	.814E-06	.285E-05	.995E-05	.277E-04
140	.375E-11	.385E-10	.259E-09	.181E-08	.146E-07	.144E-06	.169E-05	.565E-05	.188E-04	.503E-04
150	.103E-10	.106E-09	.694E-09	.460E-08	.346E-07	.316E-06	.340E-05	.108E-04	.343E-04	.889E-04
160	.272E-10	.279E-09	.178E-08	.112E-07	.789E-07	.670E-06	.659E-05	.200E-04	.608E-04	.152E-03
170	.682E-10	.702E-09	.436E-08	.264E-07	.174E-06	.137E-05	.124E-04	.359E-04	.104E-03	.255E-03
180	.164E-09	.170E-08	.103E-07	.599E-07	.369E-06	.271E-05	.225E-04	.627E-04	.175E-03	.416E-03
190	.378E-09	.394E-08	.234E-07	.131E-06	.760E-06	.521E-05	.400E-04	.107E-03	.286E-03	.663E-03
200	.842E-09	.883E-08	.514E-07	.278E-06	.152E-05	.975E-05	.691E-04	.177E-03	.457E-03	.104E-02
210	.181E-08	.191E-07	.109E-06	.573E-06	.295E-05	.178E-04	.117E-03	.288E-03	.715E-03	.159E-02
220	.376E-08	.401E-07	.226E-06	.115E-05	.559E-05	.316E-04	.193E-03	.459E-03	.110E-02	.239E-02
230	.758E-08	.817E-07	.455E-06	.224E-05	.103E-04	.549E-04	.311E-03	.717E-03	.166E-02	.354E-02
240	.148E-07	.162E-06	.889E-06	.427E-05	.186E-04	.935E-04	.494E-03	.110E-02	.245E-02	.515E-02
250	.283E-07	.312E-06	.170E-05	.793E-05	.329E-04	.156E-03	.770E-03	.166E-02	.358E-02	.740E-02
260	.526E-07	.588E-06	.316E-05	.144E-04	.569E-04	.255E-03	.118E-02	.247E-02	.516E-02	.105E-01

VAPOR PRESSURE—TABLE 11. Sulfuric Acid Partial Pressure, bar, over Aqueous Sulfuric Acid (*Continued*)

Weight Percent, H_2SO_4

°C	10.0	20.0	30.0	40.0	50.0	60.0	70.0	75.0	80.0	85.0
270	.954E-07	.108E-05	.577E-05	.257E-04	.965E-04	.411E-03	.178E-02	.362E-02	.733E-02	.147E-01
280	.169E-06	.194E-05	.103E-04	.450E-04	.161E-03	.650E-03	.265E-02	.524E-02	.103E-01	.203E-01
290	.294E-06	.342E-05	.180E-04	.771E-04	.263E-03	.101E-02	.389E-02	.750E-02	.143E-01	.278E-01
300	.500E-06	.591E-05	.309E-04	.130E-03	.424E-03	.156E-02	.563E-02	.106E-01	.196E-01	.376E-01
310	.834E-06	.100E-04	.522E-04	.215E-03	.672E-03	.236E-02	.805E-02	.148E-01	.266E-01	.504E-01
320	.137E-05	.167E-04	.865E-04	.352E-03	.105E-02	.352E-02	.114E-01	.205E-01	.359E-01	.670E-01
330	.220E-05	.273E-04	.141E-03	.565E-03	.162E-02	.519E-02	.159E-01	.281E-01	.480E-01	.883E-01
340	.349E-05	.440E-04	.227E-03	.895E-03	.246E-02	.757E-02	.221E-01	.382E-01	.636E-01	.116
350	.544E-05	.698E-04	.360E-03	.140E-02	.369E-02	.109E-01	.303E-01	.516E-01	.836E-01	.150

Weight percent, H_2SO_4

°C	90.0	92.0	94.0	96.0	97.0	98.0	98.5	99.0	99.5	100.0
0	.534E-09	.803E-09	.112E-08	.148E-08	.167E-08	.187E-08	.196E-08	.206E-08	.217E-08	.228E-08
10	.200E-08	.296E-08	.409E-08	.540E-08	.609E-08	.679E-08	.714E-08	.750E-08	.788E-08	.827E-08
20	.677E-08	.993E-08	.136E-07	.179E-07	.201E-07	.244E-07	.236E-07	.247E-07	.260E-07	.273E-07
30	.211E-07	.306E-07	.415E-07	.543E-07	.611E-07	.680E-07	.714E-07	.749E-07	.786E-07	.824E-07
40	.607E-07	.870E-07	.117E-06	.153E-06	.171E-06	.191E-06	.200E-06	.210E-06	.220E-06	.230E-06
50	.163E-06	.231E-06	.309E-06	.400E-06	.449E-06	.498E-06	.523E-06	.548E-06	.574E-06	.600E-06
60	.411E-06	.575E-06	.765E-06	.985E-06	.110E-05	.122E-05	.128E-05	.134E-05	.140E-05	.147E-05
70	.976E-06	.135E-05	.179E-05	.229E-05	.256E-05	.283E-05	.297E-05	.310E-05	.325E-05	.339E-05
80	.220E-05	.302E-05	.396E-05	.504E-05	.562E-05	.622E-05	.652E-05	.681E-05	.712E-05	.743E-05
90	.473E-05	.642E-05	.835E-05	.106E-04	.118E-04	.130E-04	.136E-04	.143E-04	.149E-04	.155E-04

100	.973E-05	.131E-04	.169E-04	.213E-04	.237E-04	.261E-04	.274E-04	.285E-04	.298E-04	.310E-04
110	.192E-04	.256E-04	.328E-04	.412E-04	.457E-04	.503E-04	.527E-04	.549E-04	.572E-04	.595E-04
120	.366E-04	.482E-04	.614E-04	.767E-04	.849E-04	.935E-04	.977E-04	.102E-03	.106E-03	.110E-03
130	.672E-04	.879E-04	.111E-03	.138E-03	.153E-03	.168E-03	.175E-03	.182E-03	.190E-03	.197E-03
140	.120E-03	.155E-03	.195E-03	.241E-03	.266E-03	.292E-03	.304E-03	.316E-03	.329E-03	.341E-03
150	.207E-03	.266E-03	.332E-03	.408E-03	.449E-03	.493E-03	.514E-03	.534E-03	.554E-03	.574E-03
160	.348E-03	.444E-03	.550E-03	.673E-03	.740E-03	.810E-03	.844E-03	.876E-03	.909E-03	.941E-03
170	.572E-03	.723E-03	.889E-03	.108E-02	.119E-02	.130E-02	.135E-02	.140E-02	.145E-02	.150E-02
180	.917E-03	.115E-02	.140E-02	.170E-02	.186E-02	.204E-02	.212E-02	.220E-02	.227E-02	.235E-02
190	.144E-02	.179E-02	.217E-02	.262E-02	.286E-02	.312E-02	.325E-02	.336E-02	.348E-02	.359E-02
200	.221E-02	.273E-02	.329E-02	.395E-02	.431E-02	.470E-02	.488E-02	.505E-02	.522E-02	.538E-02
210	.333E-02	.408E-02	.490E-02	.585E-02	.637E-02	.693E-02	.720E-02	.744E-02	.768E-02	.791E-02
220	.494E-02	.601E-02	.715E-02	.850E-02	.924E-02	.100E-01	.104E-01	.108E-01	.111E-01	.114E-01
230	.719E-02	.869E-02	.103E-01	.122E-01	.132E-01	.143E-01	.149E-01	.153E-01	.158E-01	.162E-01
240	.103E-01	.124E-01	.146E-01	.171E-01	.186E-01	.201E-01	.209E-01	.215E-01	.221E-01	.227E-01
250	.146E-01	.174E-01	.203E-01	.238E-01	.257E-01	.278E-01	.289E-01	.297E-01	.305E-01	.314E-01
260	.203E-01	.240E-01	.279E-01	.326E-01	.352E-01	.380E-01	.394E-01	.405E-01	.416E-01	.427E-01
270	.279E-01	.329E-01	.380E-01	.441E-01	.475E-01	.513E-01	.531E-01	.545E-01	.560E-01	.574E-01
280	.380E-01	.444E-01	.510E-01	.589E-01	.633E-01	.683E-01	.706E-01	.725E-01	.744E-01	.762E-01
290	.510E-01	.592E-01	.676E-01	.778E-01	.835E-01	.900E-01	.930E-01	.954E-01	.978E-01	.100
300	.678E-01	.782E-01	.888E-01	.102	.109	.117	.121	.124	.127	.130
310	.892E-01	.102	.115	.132	.141	.151	.156	.160	.164	.167
320	.116	.132	.149	.169	.180	.193	.199	.204	.209	.213
330	.150	.170	.190	.214	.228	.245	.252	.258	.263	.269
340	.192	.216	.240	.270	.287	.307	.317	.328	.330	.386
350	.243	.272	.301	.337	.358	.383	.394	.402	.410	.417

VAPOR PRESSURE – TABLE 12. Total Pressure, bar, of Aqueous Sulfuric Acid Solutions

	Weight percent, H_2SO_4									
°C	10.0	20.0	30.0	40.0	50.0	60.0	70.0	75.0	80.0	85.0
0	.582E-02	.534E-02	.448E-02	.326E-02	.193E-02	.836E-03	.207E-03	.747E-04	.197E-04	.343E-05
10	.117E-01	.107E-01	.909E-02	.670E-02	.405E-02	.180E-02	.467E-03	.175E-03	.490E-04	.952E-05
20	.223E-01	.205E-01	.174E-01	.130E-01	.802E-02	.367E-02	.995E-03	.388E-03	.115E-03	.245E-04
30	.404E-01	.373E-01	.319E-01	.241E-01	.151E-01	.710E-02	.201E-02	.811E-03	.253E-03	.589E-04
40	.703E-01	.649E-01	.558E-01	.427E-01	.272E-01	.131E-01	.387E-02	.162E-02	.531E-03	.134E-03
50	.117	.109	.939E-01	.725E-01	.470E-01	.232E-01	.715E-02	.309E-02	.106E-02	.286E-03
60	.189	.175	.152	.119	.782E-01	.395E-01	.127E-01	.565E-02	.204E-02	.584E-03
70	.296	.275	.239	.188	.126	.651E-01	.217E-01	.997E-02	.376E-02	.114E-02
80	.449	.417	.365	.290	.196	.104	.360E-01	.170E-01	.668E-02	.213E-02
90	.664	.617	.542	.434	.298	.161	.578E-01	.281E-01	.115E-01	.383E-02
100	.957	.891	.786	.634	.441	.244	.905E-01	.452E-01	.192E-01	.666E-02
110	1.349	1.258	1.113	.904	.638	.360	.138	.708E-01	.312E-01	.112E-01
120	1.863	1.740	1.544	1.264	.903	.519	.206	.108	.493E-01	.183E-01
130	2.524	2.361	2.101	1.732	1.253	.734	.301	.162	.760E-01	.291E-01
140	3.361	3.149	2.810	2.333	1.708	1.020	.431	.236	.115	.451E-01
150	4.404	4.132	3.697	3.090	2.289	1.392	.605	.339	.170	.683E-01
160	5.685	5.342	4.793	4.031	3.021	1.870	.837	.478	.246	.101
170	7.236	6.810	6.127	5.185	3.930	2.475	1.138	.662	.350	.147
180	9.093	8.571	7.731	6.584	5.045	3.233	1.525	.902	.489	.209
190	11.289	10.658	9.640	8.259	6.397	4.169	2.017	1.212	.673	.292
200	13.861	13.107	11.887	10.245	8.020	5.312	2.633	1.606	.913	.402
210	16.841	15.951	14.505	12.576	9.948	6.696	3.396	2.101	1.221	.544
220	20.264	19.225	17.529	15.287	12.217	8.354	4.331	2.715	1.610	.726
230	24.160	22.960	20.992	18.414	14.864	10.322	5.466	3.468	2.098	.956
240	28.561	27.188	24.927	21.992	17.929	12.641	6.832	4.382	2.701	1.242
250	33.494	31.939	29.364	26.056	21.452	15.351	8.459	5.481	3.439	1.594
260	38.984	37.240	34.334	30.642	25.472	18.496	10.384	6.791	4.332	2.023

	100.0	99.5	99.0	98.5	98.0	97.0	96.0	94.0	92.0	90.0
270	2.540	5.402	8.337	12.642	22.122	30.030	35.784	39.865	43.116	45.055
280	3.157	6.673	10.147	15.272	26.275	35.168	41.514	45.984	49.590	51.726
290	3.886	8.170	12.250	18.315	31.004	40.926	47.866	52.715	56.681	59.015
300	4.740	9.916	14.675	21.814	36.361	47.347	54.869	60.081	64.407	66.934
310	5.732	11.939	17.453	25.812	42.398	54.470	62.553	68.101	72.781	75.495
320	6.876	14.264	20.611	30.355	49.168	62.338	70.947	76.792	81.816	84.705
330	8.185	16.916	24.182	35.489	56.727	70.990	80.078	86.172	91.518	94.567
340	9.672	19.920	28.193	41.262	65.130	80.466	89.970	96.252	101.894	105.083
350	11.351	23.303	32.674	47.723	74.437	90.806	100.647	107.043	112.947	116.251

Weight percent, H_2SO_4

°C	100.0	99.5	99.0	98.5	98.0	97.0	96.0	94.0	92.0	90.0
0	.323E-08	.313E-08	.479E-08	.786E-08	.117E-07	.235E-07	.416E-07	.109E-06	.243E-06	.518E-06
10	.124E-07	.113E-07	.166E-07	.261E-07	.391E-07	.774E-07	.136E-06	.348E-06	.765E-06	.159E-05
20	.435E-07	.373E-07	.528E-07	.812E-07	.121E-06	.235E-06	.407E-06	.102E-05	.221E-05	.449E-05
30	.141E-06	.114E-06	.155E-06	.234E-06	.344E-06	.659E-06	.113E-05	.279E-05	.590E-05	.117E-04
40	.425E-06	.323E-06	.425E-06	.630E-06	.914E-06	.173E-05	.293E-05	.708E-05	.147E-04	.385E-04
50	.120E-05	.861E-06	.109E-05	.159E-05	.228E-05	.425E-05	.712E-05	.169E-04	.344E-04	.653E-04
60	.319E-05	.216E-05	.264E-05	.379E-05	.538E-05	.987E-05	.164E-04	.380E-04	.759E-04	.141E-03
70	.804E-05	.514E-05	.605E-05	.856E-05	.120E-04	.218E-04	.357E-04	.813E-04	.159E-03	.291E-03
80	.193E-04	.117E-04	.132E-04	.184E-04	.257E-04	.458E-04	.742E-04	.166E-03	.319E-03	.571E-03
90	.441E-04	.253E-04	.277E-04	.390E-04	.524E-04	.921E-04	.148E-03	.324E-03	.612E-03	.107E-02
100	.966E-04	.527E-04	.555E-04	.751E-04	.103E-03	.178E-03	.283E-03	.607E-03	.113E-02	.195E-02
110	.204E-03	.106E-03	.107E-03	.143E-03	.194E-03	.332E-03	.521E-03	.110E-02	.201E-02	.340E-02
120	.414E-03	.206E-03	.201E-03	.263E-03	.354E-03	.598E-03	.929E-03	.192E-02	.346E-02	.575E-02
130	.314E-03	.387E-03	.363E-03	.470E-03	.626E-.3	.104E-02	.161E-02	.327E-02	.578E-02	.944E-02
140	.155E-02	.708E-03	.639E-03	.815E-03	.107E-02	.177E-02	.270E-02	.539E-02	.939E-02	.151E-01
150	.287E-02	.126E-02	.109E-02	.137E-02	.180E-02	.293E-02	.441E-02	.866E-02	.149E-01	.235E-01
160	.516E-02	.219E-02	.183E-02	.226E-02	.293E-02	.471E-02	.703E-02	.136E-01	.230E-01	.357E-01
170	.905E-02	.372E-02	.299E-02	.363E-02	.466E-02	.741E-02	.110E-01	.208E-01	.347E-01	.532E-01
180	.155E-01	.619E-02	.478E-02	.517E-02	.726E-02	.114E-01	.167E-01	.312E-01	.514E-01	.775E-01

Total Pressure, bar, of Aqueous Sulfuric Acid Solutions (*Continued*)

°C	Weight percent, H_2SO_4									
	10.0	20.0	30.0	40.0	50.0	60.0	70.0	75.0	80.0	85.0
190	.111	.747E-01	.460E-01	.250E-01	.172E-01	.111E-01	.880E-02	.749E-02	.101E-01	.260E-01
200	.156	.107	.665E-01	.367E-01	.255E-01	.166E-01	.133E-01	.115E-01	.161E-01	.427E-01
210	.216	.150	.944E-01	.530E-01	.371E-01	.245E-01	.198E-01	.175E-01	.253E-01	.687E-01
220	.295	.207	.132	.752E-01	.531E-01	.354E-01	.289E-01	.260E-01	.392E-01	.109
230	.396	.282	.182	.105	.749E-01	.505E-01	.417E-01	.382E-01	.596E-01	.169
240	.525	.379	.247	.145	.104	.710E-01	.592E-01	.553E-01	.895E-01	.258
250	.688	.503	.331	.197	.143	.985E-01	.830E-01	.790E-01	.132	.389
260	.881	.660	.439	.264	.193	.135	.115	.112	.193	.577
270	1.141	.856	.575	.351	.258	.153	.157	.156	.279	.846
280	1.447	1.099	.744	.460	.341	.245	.213	.215	.398	1.225
290	1.817	1.398	.954	.597	.446	.324	.285	.295	.562	1.751
300	2.261	1.761	1.211	.767	.578	.425	.379	.399	.785	2.476
310	2.791	2.199	1.524	.977	.742	.553	.498	.536	1.085	3.465
320	3.417	2.723	1.901	1.234	.944	.713	.649	.714	1.486	4.800
330	4.153	3.347	2.353	1.545	1.191	.911	.840	.944	2.018	6.586
340	5.011	4.084	2.889	1.919	1.491	1.156	1.078	1.239	2.718	8.957
350	6.006	4.949	3.523	2.366	1.852	1.456	1.374	1.614	3.631	12.079

VAPOR PRESSURE—TABLE 13. Partial Pressures of HNO_3 and H_2O over Aqueous Solutions of HNO_3

mmHg
Percentages are weight % HNO_3 in solution.

°C	20%		25%		30%		35%		40%		45%		50%	
	HNO_3	H_2O	HNO_3	H_2O	HNO_3	H_2O	HNO_3	H_2O	HNO_3	H_2O	HNO_3	H_2O	HNO_3	H_2O
0		4.1		3.8		3.6		3.3		3.0		2.6		2.1
5		5.7		5.4		5.0		4.6		4.2		3.6		3.0
10		8.0		7.6		7.1		6.5		5.8		5.0	0.12	4.2
15		10.9		10.3		9.7		8.9		8.0		6.9	.18	5.8
20		15.2		14.2		13.2		12.0		10.8	0.10	9.4	.27	7.9
25		20.6		19.2		17.8		16.2	0.12	14.6	.15	12.7	.39	10.7
30		27.6		25.7		23.8	0.09	21.7	.17	19.5	.23	16.9	.56	14.4
35		36.5		33.8	0.11	31.1	.13	28.3	.25	25.5	.33	22.3	.80	19.0
40		47.5	0.09	44	.17	41	.20	37.7	.36	33.5	.48	29.3	1.13	25.0
45		62	.13	57.5	.25	53	.28	48	.52	43	.68	38.0	1.57	32.5
50		80	.18	75	.35	69	.42	63	.75	56	.96	49.5	2.18	42.5
55	0.09	100	.28	94	.51	87	.59	79	1.04	71	1.35	62.5	2.95	54
60	.13	128	.40	121	.71	113	.85	102	1.48	90	1.83	80	4.05	70
65	.19	162	.54	151	1.00	140	1.18	127	2.05	114	2.54	100	5.46	88
70	.27	200	.77	187	1.38	174	1.63	159	2.80	143	3.47	126	7.25	110
75	.38	250	1.05	234	1.87	217	2.26	198	3.80	178	4.65	158	9.6	138
80	.53	307	1.44	287	2.53	267	3.07	243	5.10	218	6.20	195	12.5	170
85	.74	378	1.95	352	3.38	325	4.15	297	6.83	268	8.15	240	16.3	211
90	1.01	458	2.62	426	4.53	393	5.50	359	9.0	325	10.7	292	20.9	258
95	1.37	555	3.50	517	6.05	478	7.32	436	11.7	394	13.7	355	26.8	315
100	1.87	675	4.65	628	7.90	580	9.7	530	15.5	480	17.8	430	34.2	383
105	2.50	800		745		690	12.7	631	20.0	573	23.0	520	43.0	463
110							16.5	755	25.7	688	29.2	625	54.5	560
115									32.5	810	37.0	740	67	665
120											46		84	785

VAPOR PRESSURE—TABLE 13. Partial Pressures of HNO₃ and H₂O over Aqueous Solutions of HNO₃ (Continued)

mmHg
Percentages are weight % HNO₃ in solution.

°C	55% HNO₃	55% H₂O	60% HNO₃	60% H₂O	65% HNO₃	65% H₂O	70% HNO₃	70% H₂O	80% HNO₃	80% H₂O	90% HNO₃	90% H₂O	100% HNO₃
0		1.8	0.19	1.5	0.41	1.3	0.79	1.1	2				11
5	0.14	2.5	.28	2.1	.60	1.8	1.12	1.6	3		5.5		15
10	.21	3.5	.41	3.0	.86	2.6	1.58	2.2	4	1.2	8		22
15	.31	4.9	.59	4.1	1.21	3.5	2.18	3.0	6	1.7	11		30
20	.45	6.7	.84	5.6	1.68	4.9	3.00	4.1	8	2.4	15		42
25	.66	9.1	1.21	7.7	2.32	6.6	4.10	5.5	10.5	3.2	20	1	57
30	.93	12.2	1.66	10.3	3.17	8.8	5.50	7.4	14	4	27	1.3	77
35	1.30	16.1	2.28	13.6	4.26	11.6	7.30	9.8	18.5	5.5	36	1.8	102
40	1.82	21.3	3.10	18.1	5.70	15.5	9.65	12.8	24.5	7	47	2.4	133
45	2.50	28.0	4.20	23.7	7.55	20.0	12.6	16.7	32	9.5	62	3	170
50	3.41	36.3	5.68	31	10.0	26.0	16.5	21.8	41	12	80	4	215
55	4.54	46	7.45	39	12.8	33.0	21.0	27.3	52	15	103	5	262
60	6.15	60	9.9	51	16.8	43.0	27.1	35.3	67	20	127	6.5	320
65	8.18	76	13.0	64	21.7	54.5	34.5	44.5	85	25	157	8	385
70	10.7	95	16.8	81	27.5	68	43.3	56	106	31	192	10	460
75	13.9	120	21.8	102	35.0	86	54.5	70	130	38	232	13	540
80	18.0	148	27.5	126	43.5	106	67.5	86	158	48	282	16	625
85	23.0	182	34.8	156	54.5	131	83	107	192	60	338	20	720
90	29.4	223	43.7	192	67.5	160	103	130	230	73	405	24	820
95	37.3	272	55.0	233	83.5	195	125	158	278	89	480	29	
100	47	331	69.5	285	103	238	152	192	330	108	570	35	
105	58.5	400	84.5	345	124	288	183	231	392	129	675	42	
110	73	485	103	417	152	345	221	278	465	155	790		
115	90	575	126	495	181	410	262	330	545	185			
120	110	685	156	590	218	490	312	393	640	219			
125			187	700	260	580	372	469					

VAPOR PRESSURE—TABLE 14. Partial Pressures of H_2O and HBr over Aqueous Solutions of HBr at 20 to 55°C

% HBr	mmHg							
	20°C		25°C		50°C		55°C	
	HBr	H_2O	HBr	H_2O	HBr	H_2O	HBr	H_2O
32			0.0016					
34			.0022					
36			.0033					
38			.0061					
40			.011					
42			.023					
44			.048					
46			.10					
48	0.09	6.2	.13	8.2	1.3	30.2	2.0	38
50	.23	4.5	.37	6.1	3.2	24.3	4.6	31
52	.71	3.3	1.1	4.5	7.2	19.3	10.2	25
54	2.2	2.4	3.2	3.3	17	16.0	23.0	21
56	6.8	1.7	9.3	2.4	40	13.3	51	18
58	21	1.3	27	1.9	91	10.4	115	14
60							260	11.4

VAPOR PRESSURE—TABLE 15. Partial Pressures of HI over Aqueous Solutions of HI at 25°C

%HI	mmHg						
	4	46	48	50	52	54	56
p_{HI}	0.0064	0.0010	0.0022	0.0050	0.013	0.035	0.10

VAPOR PRESSURE—TABLE 16. Total Vapor Pressures of Aqueous Solutions of CH_3COOH

°C	Percentages of weight % acetic acid in the solution mmHg		
	25%	50%	75%
20	16.3	15.7	15.3
25	22.1	21.4	20.8
30	29.6	28.8	27.8
35	39.4	38.3	36.6
40	51.7	50.2	48.1
45	67.0	65.0	62.0
50	87.2	85.0	80.1
55	110	107	102
60	141	138	130
65	178	172	162
70	223	216	203
75	277	269	251
80	342	331	310
85	419	407	376
90	510	497	458
95	618	602	550
100	743	725	666

VAPOR PRESSURE—TABLE 17. Partial Pressures of H$_2$O over Aqueous Solutions of NH$_3$

Pressures are in pounds per square inch absolute

Molal concentration of ammonia in the solutions in percentages
(Weight concentration of ammonia in the solution in percentages)

t, °F	0 (0)	5 (4.74)	10 (9.50)	15 (14.29)	20 (19.10)	25 (23.94)	30 (28.81)	35 (33.71)	40 (38.64)	45 (43.59)	50 (48.57)	55 (53.58)	60 (58.62)	65 (63.69)	70 (68.79)	75 (73.91)	80 (79.07)	85 (84.26)	90 (89.47)	95 (94.72)
32	0.09	0.084	0.079	0.074	0.070	0.065	0.060	0.056	0.051	0.047	0.042	0.038	0.034	0.030	0.025	0.021	0.017	0.013	0.008	0.004
40	0.12	.115	.108	.101	.095	.089	.083	.076	.070	.064	.058	.052	.046	.040	.035	.029	.023	.015	.012	.006
50	0.18	.17	.16	.15	.14	.13	.12	.11	.10	.094	.085	.076	.068	.059	.051	.042	.034	.025	.017	.008
60	0.26	.24	.23	.21	.20	.19	.17	.16	.15	.13	.12	.11	.097	.085	.073	.061	.049	.037	.024	.012
70	0.36	.34	.32	.30	.28	.26	.25	.23	.21	.19	.17	.15	.14	.12	.10	.086	.069	.052	.034	.017
80	0.51	.48	.45	.42	.40	.37	.34	.32	.29	.27	.24	.22	.19	.17	.14	.12	.096	.072	.048	.024
90	0.70	.66	.63	.58	.55	.51	.47	.44	.40	.37	.33	.30	.26	.23	.20	.16	.13	.10	.066	.033
100	0.95	.90	.85	.79	.74	.69	.64	.59	.55	.50	.45	.41	.36	.31	.27	.22	.18	.13	.090	.045
110	1.27	1.20	1.14	1.07	1.00	.93	.86	.80	.73	.67	.60	.54	.48	.42	.36	.30	.24	.18	.120	.061
120	1.69	1.60	1.51	1.42	1.33	1.24	1.15	1.06	.97	.89	.80	.72	.64	.56	.48	.40	.32	.24	.160	.081
130	2.22	2.10	1.98	1.86	1.74	1.62	1.51	1.39	1.28	1.17	1.05	.95	.84	.74	.63	.53	.42	.32	.210	.100
140	2.89	2.73	2.57	2.42	2.26	2.11	1.96	1.81	1.66	1.52	1.37	1.23	1.10	.96	.82	.69	.55	.41	.270	.140
150	3.72	3.15	3.31	3.11	2.91	2.72	2.52	2.33	2.14	1.95	1.76	1.59	1.41	1.24	1.06	.88	.71	.53	.350	.180
160	4.74	4.48	4.22	3.97	3.71	3.46	3.22	2.97	2.73	2.49	2.25	2.02	1.80	1.58	1.35	1.12	.90	.67	.450	.220
170	5.99	5.66	5.34	5.02	4.70	4.38	4.07	3.75	3.45	3.15	2.84	2.56	2.28	1.99	1.71	1.42	1.13	.85	.570	.300
180	7.51	7.10	6.69	6.30	5.89	5.49	5.10	4.71	4.33	3.94	3.57	3.21	2.85	2.50	2.14	1.77	1.42	1.06		
190	9.34	8.83	8.32	7.82	7.32	6.83	6.34	5.86	5.38	4.91	4.44	3.99	3.55	3.10	2.65					
200	11.53	10.90	10.27	9.65	9.04	8.43	7.83	7.23	6.64	6.06	5.48	4.93	4.38	3.81						
210	14.12	13.35	12.58	11.82	11.07	10.32	9.59	8.86	8.13	7.42	6.71	6.04	5.34							
220	17.19	16.25	15.32	14.39	13.48	12.57	11.67	10.78	9.90	9.03	8.17	7.31								
230	20.78	19.64	18.51	17.40	16.29	15.19	14.11	13.03	11.97	10.91	9.87									
240	24.97	23.60	22.25	20.91	19.58	18.26	16.95	15.66	14.38	13.12	11.86									
250	29.83	28.20	26.58	25.00	23.39	21.82	20.25	18.71	17.18	15.67										

VAPOR PRESSURE—TABLE 18. Mole Percentages of H_2O over Aqueous Solutions of NH_3

Molal concentration of ammonia in the solutions in percentages
(Weight concentration of ammonia in the solutions in percentages)

t, °F	0 (0)	5 (4.74)	10 (9.50)	15 (14.29)	20 (19.10)	25 (23.94)	30 (28.81)	35 (33.71)	40 (38.64)	45 (43.59)	50 (48.57)	55 (53.58)	60 (58.62)	65 (63.69)	70 (68.79)	75 (73.91)	80 (79.07)	85 (84.26)	90 (89.47)	95 (94.72)	100 (100.00)
32	100	24.3	13.2	7.63	4.43	2.50	1.43	0.856	0.514	0.335	0.216	0.151	0.109	0.0816	0.0585	0.0457	0.0345	0.0249	0.0146	0.00689	0.00
40	100	25.3	14.1	8.15	4.73	2.74	1.59	.943	.581	.372	.248	.172	.124	.0914	.0706	.0533	.0395	.0243	.0185	.00879	
50	100	26.6	15.2	9.09	5.24	3.03	1.78	1.060	.652	.434	.290	.202	.148	.1095	.0838	.0630	.0477	.0332	.0215	.00959	
60	100	27.9	16.2	9.50	5.69	3.42	1.97	1.210	.777	.481	.331	.238	.172	.1290	.0986	.0754	.0566	.0406	.0251	.01125	
70	100	29.1	17.4	10.30	6.14	3.65	2.27	1.390	.873	.569	.383	.266	.205	.1510	.112	.0882	.0656	.0474	.0296	.0135	
80	100	31.6	18.5	11.20	6.89	4.08	2.45	1.550	.978	.659	.444	.323	.230	.1750	.130	.103	.0772	.0528	.0351	.0167	
90	100	32.7	20.0	12.00	7.40	4.47	2.73	1.730	1.100	.742	.505	.366	.267	.2020	.157	.115	.0884	.0647	.0408	.0194	
100	100	34.4	21.0	12.90	7.92	4.85	3.00	1.890	1.250	.834	.574	.420	.307	.2290	.179	.135	.104	.0714	.0473	.0226	
110	100	35.9	22.2	13.80	8.59	5.29	3.30	2.110	1.370	.932	.644	.466	.347	.2640	.208	.157	.118	.0846	.0540	.0262	
120	100	37.5	23.4	14.70	9.22	5.75	3.63	2.320	1.520	1.044	.714	.529	.395	.3020	.233	.180	.135	.0970	.0619	.0300	
130	100	39.0	24.5	15.60	9.85	6.18	3.95	2.550	1.690	1.160	.811	.596	.444	.3430	.263	.205	.154	.1117	.0703	.0339	
140	100	40.7	25.8	16.50	10.50	6.69	4.28	2.790	1.860	1.286	.906	.663	.501	.3840	.297	.232	.175	.124	.0786	.0385	
150	100	42.3	27.1	17.50	11.20	7.19	4.63	3.080	2.040	1.410	1.004	.741	.558	.4320	.334	.257	.197	.140	.0892	.0439	
160	100	44.1	28.3	18.40	11.90	7.69	5.01	3.300	2.230	1.550	1.110	.818	.617	.4800	.372	.287	.218	.154	.1005	.0499	
170	100	45.6	29.6	19.40	12.70	8.22	5.38	3.580	2.430	1.700	1.220	.904	.689	.5300	.414	.320	.242	.174	.112	.0567	
180	100	47.3	30.9	20.40	13.40	8.76	5.78	3.870	2.640	1.850	1.340	.994	.756	.5860	.456	.352	.268	.192			
190	100	48.7	32.2	21.40	14.10	9.31	6.18	4.160	2.860	2.020	1.460	1.087	.830	.6420	.501						
200	100	50.4	33.4	22.30	14.90	9.88	6.59	4.470	3.080	2.190	1.580	1.187	.907	.7010							
210	100	52.1	34.7	23.40	15.70	10.45	7.03	4.780	3.310	2.360	1.720	1.272	.983								
220	100	53.7	36.1	24.40	16.40	11.05	7.48	5.100	3.560	2.540	1.860	1.390									
230	100	55.2	37.3	25.40	17.30	11.63	7.91	5.440	3.810	2.730	2.000										
240	100	56.8	38.6	26.50	18.00	12.24	8.36	5.780	4.060	2.920	2.150										
250	100	58.4	39.8	27.50	18.80	12.88	8.82	6.120	4.340	3.120											

VAPOR PRESSURE—TABLE 19. Partial Pressures of NH₃ over Aqueous Solutions of NH₃

Pressures are in pounds per square inch absolute

Molal concentration of ammonia in the solutions in percentages
(Weight concentration of ammonia in the solutions in percentages)

t, °F	5 (4.74)	10 (9.50)	15 (14.29)	20 (19.10)	25 (23.94)	30 (28.81)	35 (33.71)	40 (38.64)	45 (43.59)	50 (48.57)	55 (53.58)	60 (58.62)	65 (63.69)	70 (68.79)	75 (73.91)	80 (79.07)	85 (84.26)	90 (89.47)	95 (94.72)
32	0.26	0.52	0.90	1.51	2.67	4.27	6.54	8.93	14.13	19.36	25.12	31.13	36.74	42.69	45.92	49.26	52.13	54.89	58.01
40	.33	.66	1.14	1.92	3.16	5.13	7.98	11.98	17.14	23.33	30.15	37.15	43.69	49.56	54.40	58.31	61.62	64.77	68.31
50	.47	.89	1.50	2.53	4.16	6.63	10.24	15.24	21.56	29.17	37.46	45.86	53.79	60.82	66.63	71.26	75.22	79.05	83.40
60	.62	1.19	2.00	3.21	5.36	8.48	13.06	19.15	26.92	36.14	46.12	56.22	65.81	73.99	80.90	86.44	91.04	95.67	100.65
70	.83	1.52	2.60	4.28	6.87	10.76	16.33	23.84	33.20	44.25	56.29	68.32	79.42	89.26	97.42	104.01	109.55	114.83	120.61
80	1.04	1.98	3.34	5.45	8.69	13.52	20.29	29.40	40.69	53.84	67.97	82.36	95.52	107.06	116.42	124.20	130.57	136.35	143.70
90	1.36	2.52	4.25	6.88	10.89	16.76	25.04	35.94	49.45	64.99	81.61	98.35	113.79	127.22	138.18	147.02	154.46	161.74	169.73
100	1.72	3.20	5.34	8.60	13.53	20.68	30.57	43.57	59.49	77.85	97.27	116.81	134.70	150.23	162.94	173.22	181.97	190.13	199.17
110	2.14	4.00	6.65	10.64	16.65	25.21	37.01	52.43	71.20	92.59	115.16	137.62	158.42	176.18	190.85	203.02	212.71	222.22	232.79
120	2.67	4.95	8.21	13.09	20.30	30.54	44.56	62.62	84.44	109.40	135.48	161.44	185.14	205.81	222.28	236.05	247.14	258.24	270.02
130	3.28	6.09	10.05	15.93	24.58	36.74	53.16	74.27	99.69	128.45	158.45	188.16	215.14	238.70	257.87	272.88	286.08	298.46	311.80
140	3.97	7.41	12.21	19.23	29.43	43.77	62.97	87.53	116.72	149.93	184.17	218.18	248.70	275.33	297.12	314.45	328.99	342.93	358.46
150	4.78	8.92	14.70	23.09	35.09	51.91	74.28	102.51	136.15	173.64	212.91	251.24	286.00	316.24	340.82	360.39	376.57	392.45	409.62
160	5.68	10.70	17.57	27.45	41.56	61.03	86.91	119.37	157.71	200.45	244.98	288.38	327.82	361.75	389.08	411.30	429.73	447.35	466.38
170	6.75	12.67	20.85	32.41	48.89	71.48	101.09	138.30	181.95	230.36	280.54	329.42	373.61	411.59	442.28	466.67	487.85	507.63	528.50
180	7.90	14.96	24.56	38.13	57.19	83.07	116.97	159.37	208.66	263.43	319.89	374.25	424.10	466.26	500.63	528.08	551.24		
190	9.23	17.55	28.78	44.49	66.49	96.22	134.89	182.72	238.39	299.86	363.11	424.15	479.40	526.15					
200	10.70	20.45	33.49	51.58	76.90	110.85	154.58	208.56	270.94	340.02	410.17	478.62	539.79						
210	12.26	23.68	38.76	59.65	88.48	126.83	176.24	236.97	307.08	383.99	462.36	537.56							
220	14.02	27.15	44.61	68.43	101.24	144.74	200.46	268.30	346.07	431.43	518.19								
230	15.95	31.09	51.06	78.14	115.45	164.17	226.67	302.53	389.29	483.53									
240	17.92	35.40	58.00	89.02	130.94	185.79	255.26	339.72	435.78	540.44									
250	20.12	40.09	65.74	100.69	147.66	209.37	286.89	380.42	486.73										

VAPOR PRESSURE—TABLE 20. Total Vapor Pressures of Aqueous Solutions of NH₃

Pressures are in pounds per square inch absolute

Molal concentration of ammonia in the solutions in percentages
(Weight concentration of ammonia in the solutions in percentages)

t, °F	0 (0)	5 (4.74)	10 (9.50)	15 (14.29)	20 (19.10)	25 (23.94)	30 (28.81)	35 (33.71)	40 (38.64)	45 (43.59)	50 (48.57)	55 (53.58)	60 (58.62)	65 (63.69)	70 (68.79)	75 (73.91)	80 (79.07)	85 (84.26)	90 (89.47)	95 (94.72)	100 (100.00)
32	0.09	0.34	0.60	0.97	1.58	2.60	4.20	6.54	9.93	14.18	19.40	25.16	31.16	36.77	42.72	45.94	49.28	52.14	54.90	58.01	62.29
40	0.12	.45	.77	1.24	2.01	3.25	5.21	8.06	12.05	17.20	23.39	30.20	37.20	43.73	49.60	54.43	58.33	61.64	64.78	68.32	73.32
50	0.18	.64	1.05	1.65	2.67	4.29	6.75	10.35	15.34	21.65	29.26	37.54	45.93	53.85	60.87	66.67	71.29	75.25	79.07	83.41	89.19
60	0.26	.86	1.42	2.21	3.51	5.55	8.65	13.22	19.30	27.05	36.26	46.23	56.32	65.90	74.06	80.96	86.49	91.08	95.69	100.66	107.6
70	0.36	1.17	1.84	2.90	4.56	7.13	11.01	16.56	24.05	33.39	44.42	56.44	68.46	79.54	89.36	97.51	104.08	109.60	114.86	120.63	128.8
80	.51	1.52	2.43	3.76	5.85	9.06	13.86	20.61	29.69	40.96	54.08	68.19	82.55	95.69	107.20	116.54	124.30	130.64	136.40	143.72	153.0
90	.70	2.02	3.15	4.83	7.43	11.40	17.23	25.48	36.34	49.82	65.32	81.91	98.61	114.02	127.42	138.34	147.15	154.56	161.81	169.76	180.6
100	.95	2.62	4.05	6.13	9.34	14.22	21.32	31.16	44.12	59.99	78.30	97.68	117.17	135.01	150.50	163.16	173.40	182.10	190.22	199.22	211.9
110	1.27	3.34	5.14	7.72	11.64	17.58	26.07	37.81	53.16	71.87	93.19	115.7	138.10	158.84	176.54	191.15	203.26	212.89	222.34	232.85	247.0
120	1.69	4.27	6.46	9.63	14.42	21.54	31.69	45.62	63.59	85.33	110.2	136.2	162.08	185.70	206.29	222.68	236.37	247.38	258.40	270.1	286.4
130	2.22	5.38	8.07	11.91	17.67	26.20	38.25	54.55	75.55	100.86	129.5	159.0	189.00	215.88	239.33	258.40	273.3	286.4	298.67	311.9	330.3
140	2.89	6.70	9.98	14.63	21.49	31.54	45.73	64.78	89.19	118.24	151.3	185.4	219.28	249.66	276.15	297.81	315.0	329.4	343.2	358.6	379.1
150	3.72	8.29	12.23	17.81	26.00	37.81	54.43	76.61	104.65	138.1	175.4	214.5	252.65	287.24	317.3	341.7	361.1	377.1	392.8	409.8	432.2
160	4.74	10.16	14.92	21.54	31.16	45.02	64.25	89.88	122.10	160.2	202.7	247.0	290.18	329.4	363.1	390.2	412.2	430.4	447.8	466.6	492.8
170	5.99	12.41	18.01	25.87	37.11	53.27	75.55	104.84	141.75	185.1	233.2	283.1	331.7	375.6	413.3	443.7	467.8	488.7	508.2	528.8	558.4
180	7.51	15.00	21.65	30.86	44.02	62.68	88.17	121.68	163.7	212.6	267.0	323.1	377.1	426.6	468.4	502.4	529.5	552.3			
190	9.34	18.06	25.87	36.60	51.81	73.32	102.56	140.75	188.1	243.3	304.3	367.1	427.7	482.5	528.8						
200	11.53	21.60	30.72	43.14	60.62	85.33	118.68	161.81	215.2	277.0	345.5	415.1	483.0	543.6							
210	14.12	25.61	36.26	50.58	70.72	98.80	136.42	185.10	245.1	314.5	390.7	468.4	542.9								
220	17.19	30.27	42.47	59.00	81.91	113.81	156.41	211.24	278.2	355.1	439.6	525.5									
230	20.78	35.59	49.60	68.46	94.43	130.64	178.28	239.70	314.5	400.2	493.4										
240	24.97	41.52	57.65	78.91	108.60	149.20	202.74	270.92	354.1	448.9	552.3										
250	29.83	48.32	66.67	90.74	124.08	169.48	229.62	305.60	397.6	502.4											

VAPOR PRESSURE—TABLE 21. Partial Pressures of H_2O over Aqueous Solutions of Sodium Carbonate, mmHg

t, °C	\%Na_2CO_3						
	0	5	10	15	20	25	30
0	4.5	4.5					
10	9.2	9.0	8.8				
20	17.5	17.2	16.8	16.3			
30	31.8	31.2	30.4	29.6	28.8	27.8	26.4
40	55.3	54.2	53.0	57.6	50.2	48.4	46.1
50	92.5	90.7	88.7	86.5	84.1	81.2	77.5
60	149.5	146.5	143.5	139.9	136.1	131.6	125.7
70	239.8	235	230.5	225	219	211.5	202.5
80	355.5	348	342	334	325	315	301
90	526.0	516	506	494	482	467	447
100	760.0	746	731	715	697	676	648

VAPOR PRESSURE—TABLE 22. Partial Pressures of H_2O and CH_3OH over Aqueous Solutions of Methyl Alcohol

Mole fraction CH_3OH	39.9°C		Mole fraction CH_3OH	59.4°C	
	P_{H_2O}, mmHg	P_{CH_3OH}, mmHg		P_{H_2O}, mmHg	P_{CH_3OH}, mmHg
0	54.7	0	0	145.4	0
14.99	39.2	66.1	22.17	106.9	210.1
17.85	38.5	75.5	27.40	102.2	240.2
21.07	37.2	85.2	33.24	96.6	272.1
27.31	35.8	100.6	39.80	91.7	301.9
31.06	34.9	108.8	47.08	84.8	335.6
40.1	32.8	127.7	55.5	76.9	373.7
47.0	31.5	141.6	69.2	57.8	439.4
55.8	27.3	158.4	78.5	43.8	486.6
68.9	20.7	186.6	85.9	30.1	526.9
86.0	10.1	225.2	100.0	0	609.3
100.0	0	260.7			

VAPOR PRESSURE—TABLE 23. Partial Pressures of H_2O over Aqueous Solutions of Sodium Hydroxide, mmHg

Conc. g NaOH/ 100g H_2O	Temperature, °C											
	0	20	40	60	80	100	120	160	200	250	300	350
0	4.6	17.5	55.3	149.5	355.5	760.0	1,489	4,633	11,647	29,771	64,200	123,600
5	4.4	16.9	53.2	143.5	341.5	730.0	1,430	4,450	11,200	28,600	61,800	118,900
10	4.2	16.0	50.6	137.0	325.5	697.0	1,365	4,260	10,750	27,500	59,300	114,100
20	3.6	13.9	44.2	120.5	288.5	621.0	1,225	3,860	9,800	25,300	54,700	105,400
30	2.9	11.3	36.6	101.0	246.0	537.0	1,070	3,460	8,950	23,300	50,800	98,000
40	2.2	8.7	28.7	81.0	202.0	450.0	920	3,090	8,150	21,500	47,200	91,600
50		6.3	20.7	62.5	160.5	368.0	770	2,690	7,400	19,900	44,100	85,800
60		4.4	15.5	47.0	124.0	294.0	635	2,340	6,750	18,400	41,200	80,700
70		3.0	10.9	34.5	94.0	231.0	515	2,030	6,100	17,100	38,700	76,000
80		2.0	7.6	24.5	70.5	179.0	415	1,740	5,500	15,800	36,300	71,900
90		1.3	5.2	17.5	53.0	138.0	330	1,490	5,000	14,700	34,200	68,100
100		0.9	3.6	12.5	38.5	105.0	262	1,300	4,500	13,650	32,200	64,600
120			1.7	6.3	20.5	61.0	164	915	3,650	11,800	28,800	58,600
140				3.0	11.0	35.5	102	765	2,980	10,300	25,900	53,400
160				1.5	6.0	20.5	63	470	2,430	8,960	23,300	49,000
180					3.5	12.0	40	340	1,980	7,830	21,200	45,100
200					2.0	7.0	25	245	1,620	6,870	19,200	41,800
250					0.5	2.0	8	110	985	5,000	15,400	35,000
300					0.1	0.5	2.7	50	610	3,690	12,500	29,800
350							0.9	23	380	2,750	10,300	25,700
400								11	240	2,080	8,600	22,400
500									100	1,210	6,100	17,500
700										440	3,300	11,500
1000											1,470	6,800
2000											150	1,760
4000												120
8000												7

WATER

WATER—TABLE 1. Boiling Points of Water

PSIA	Boiling point, °F	PSIA	Boiling point, °F	PSIA	Boiling point, °F
0.5	79.6	44	273.1	150	358.5
1	101.7	46	275.8	175	371.8
2	126.0	48	278.5	200	381.9
3	141.4	50	281.0	225	391.9
4	125.9	52	283.5	250	401.0
5	162.2	54	285.9	275	409.5
6	170.0	56	288.3	300	417.4
7	176.8	58	290.5	325	424.8
8	182.8	60	292.7	350	431.8
9	188.3	62	294.9	375	438.4
10	193.2	64	297.0	400	444.7
11	197.7	66	299.0	425	450.7
12	201.9	68	301.0	450	456.4
13	205.9	70	303.0	475	461.9
14	209.6	72	304.9	500	467.1
14.69	212.0	74	306.7	525	472.2
15	213.0	76	308.5	550	477.1
16	216.3	78	310.3	575	481.8
17	219.4	80	312.1	600	486.3
18	222.4	82	313.8	625	490.7
19	225.2	84	315.5	650	495.0
20	228.0	86	317.1	675	499.2
22	233.0	88	318.7	700	503.2
24	237.8	90	320.3	725	507.2
26	242.3	92	321.9	750	511.0
28	246.4	94	323.4	775	514.7
30	250.3	96	324.9	800	518.4
32	254.1	98	326.4	825	521.9
34	257.6	100	327.9	850	525.4
36	261.0	105	331.4	875	528.8
38	264.2	110	334.8	900	532.1
40	267.3	115	338.1	950	538.6
42	270.2	120	341.3	1000	544.8

WATER—TABLE 2. Thermodynamic Properties of Water

Temp., °F	Press., psia	Specific volume, ft³/lb			Enthalpy, Btu/lb		
		v_1	v_{1g}	v_g	h_1	h_{1g}	h_g
0	0.018502	0.01743	14,797	14,797	−158.94	1,220.00	1,061.06
1	0.019495	0.01743	14,073	14,073	−158.47	1,219.96	1,061.50
2	0.020537	0.01743	13,388	13,388	−157.99	1,219.93	1,061.94
3	0.021629	0.01743	12,740	12,740	−157.52	1,219.90	1,062.38
4	0.022774	0.01743	12,125	12,125	−157.05	1,219.87	1,062.82
5	0.023975	0.01743	11,543	11,543	−156.57	1,219.83	1,063.26
6	0.025233	0.01743	10,991	10,991	−156.09	1,219.80	1,063.70
7	0.026552	0.01744	10,468	10,468	−155.62	1,219.76	1,064.14
8	0.027933	0.01744	9,971	9,971	−155.14	1,219.72	1,064.58
9	0.029379	0.01744	9,500	9,500	−154.66	1,219.68	1,065.03
10	0.030894	0.01744	9,054	9,054	−154.18	1,219.64	1,065.47
11	0.032480	0.01744	8,630	8,630	−153.70	1,219.60	1,065.91
12	0.034140	0.01744	8,228	8,228	−153.21	1,219.56	1,066.35
13	0.035878	0.01745	7,846	7,846	−152.73	1,219.52	1,066.79
14	0.037696	0.01745	7,483	7,483	−152.24	1,219.47	1,067.23
15	0.039597	0.01745	7,139	7,139	−151.76	1,219.43	1,067.67
16	0.041586	0.01745	6,811	6,811	−151.27	1,219.38	1,068.11
17	0.043666	0.01745	6,501	6,501	−150.78	1,219.33	1,068.55
18	0.045841	0.01745	6,205	6,205	−150.30	1,219.28	1,068.99
19	0.048113	0.01745	5,924	5,924	−149.81	1,219.23	1,069.43
20	0.050489	0.01746	5,657	5,657	−149.32	1,219.18	1,069.87
21	0.052970	0.01746	5,404	5,404	−148.82	1,219.13	1,070.31
22	0.055563	0.01746	5,162	5,162	−148.33	1,219.08	1,070.75
23	0.058271	0.01746	4,932	4,932	−147.84	1,219.02	1,071.19
24	0.061099	0.01746	4,714	4,714	−147.34	1,218.97	1,071.63
25	0.064051	0.01746	4,506	4,506	−146.85	1,218.91	1,072.07
26	0.067133	0.01747	4,308	4,308	−146.35	1,218.85	1,072.50
27	0.070340	0.01747	4,119	4,119	−145.85	1,218.80	1,072.94
28	0.073706	0.01747	3,940	3,940	−145.35	1,218.74	1,073.38
29	0.077207	0.01747	3,769	3,769	−144.85	1,218.68	1,073.82
30	0.080860	0.01747	3,606	3,606	−144.35	1,218.61	1,074.26
31	0.084669	0.01747	3,450	3,450	−143.85	1,218.55	1,074.70
32	0.08865	0.01602	3,302.07	3,302.09	−0.02	1,075.15	1,075.14
33	0.09229	0.01602	3,178.15	3,178.16	0.99	1,074.59	1,075.58
34	0.09607	0.01602	3,059.47	3,059.49	2.00	1,074.02	1,076.01
35	0.09998	0.01602	2,945.66	2,945.68	3.00	1,073.45	1,076.45
36	0.10403	0.01602	2,836.60	2,836.61	4.01	1,072.88	1,076.89
37	0.10822	0.01602	2,732.13	2,732.15	5.02	1,072.32	1,077.33
38	0.11257	0.01602	2,631.88	2,631.89	6.02	1,071.75	1,077.77
39	0.11707	0.01602	2,535.86	2,535.88	7.03	1,071.18	1,078.21
40	0.12172	0.01602	2,443.67	2,443.69	8.03	1,070.62	1,078.65
41	0.12654	0.01602	2,355.22	2,355.24	9.04	1,070.05	1,079.09
42	0.13153	0.01602	2,270.42	2,270.43	10.04	1,069.48	1,079.52
43	0.13669	0.01602	2,189.02	2,189.04	11.04	1,068.92	1,079.96
44	0.14203	0.01602	2,110.92	2,110.94	12.05	1,068.35	1,080.40
45	0.14755	0.01602	2,035.91	2,035.92	13.05	1,067.79	1,080.84
46	0.15326	0.01602	1,963.85	1,963.87	14.05	1,067.22	1,081.28

WATER—TABLE 2. Thermodynamic Properties of Water (*Continued*)

Temp., °F	Press., psia	Specific volume, ft³/lb			Enthalpy, Btu/lb		
		v_1	v_{1g}	v_g	h_1	h_{1g}	h_g
47	0.15917	0.01602	1,894.71	1,894.73	15.06	1,066.66	1,081.71
48	0.16527	0.01602	1,828.28	1,828.30	16.06	1,066.09	1,082.15
49	0.17158	0.01602	1,764.44	1,764.46	17.06	1,065.53	1,082.59
50	0.17811	0.01602	1,703.18	1,703.20	18.06	1,064.96	1,083.03
51	0.18484	0.01602	1,644.25	1,644.26	19.06	1,064.40	1,083.46
52	0.19181	0.01603	1,587.64	1,587.65	20.07	1,063.83	1,083.90
53	0.19900	0.01603	1,533.22	1,533.24	21.07	1,063.27	1,084.34
54	0.20643	0.01603	1,480.89	1,480.91	22.07	1,062.71	1,084.77
55	0.21410	0.01603	1,430.61	1,430.62	23.07	1,062.14	1,085.21
56	0.22202	0.01603	1,382.19	1,382.21	24.07	1,061.58	1,085.65
57	0.23020	0.01603	1,335.65	1,335.67	25.07	1,061.01	1,086.08
58	0.23864	0.01603	1,290.85	1,290.87	26.07	1,060.45	1,086.52
59	0.24735	0.01603	1,247.76	1,247.78	27.07	1,059.89	1,086.96
60	0.25635	0.01604	1,206.30	1,206.32	28.07	1,059.32	1,087.39
61	0.26562	0.01604	1,166.38	1,166.40	29.07	1,058.76	1,087.83
62	0.27519	0.01604	1,127.93	1,127.95	30.07	1,058.19	1,088.27
63	0.28506	0.01604	1,090.94	1,090.96	31.07	1,057.63	1,088.70
64	0.29524	0.01604	1,055.32	1,055.33	32.07	1,057.07	1,089.14
65	0.30574	0.01604	1,020.98	1,021.00	33.07	1,056.50	1,089.57
66	0.31656	0.01604	987.95	987.97	34.07	1,055.94	1,090.01
67	0.32772	0.01605	956.11	956.12	35.07	1,055.37	1,090.44
68	0.33921	0.01605	925.44	925.45	36.07	1,054.81	1,090.88
69	0.35107	0.01605	895.86	895.87	37.07	1,054.24	1,091.31
70	0.36328	0.01605	867.34	867.36	38.07	1,053.68	1,091.75
71	0.37586	0.01605	839.87	839.88	39.07	1,053.11	1,092.18
72	0.38882	0.01606	813.37	813.39	40.07	1,052.55	1,092.61
73	0.40217	0.01606	787.85	787.87	41.07	1,051.98	1,093.05
74	0.41592	0.01606	763.19	763.21	42.06	1,051.42	1,093.48
75	0.43008	0.01606	739.42	739.44	43.06	1,050.85	1,093.92
76	0.44465	0.01606	716.51	716.53	44.06	1,050.29	1,094.35
77	0.45966	0.01607	694.38	694.40	45.06	1,049.72	1,049.78
78	0.47510	0.01607	673.05	673.06	46.06	1,049.16	1,095.22
79	0.49100	0.01607	652.44	652.46	47.06	1,048.59	1,095.65
80	0.50736	0.01607	632.54	632.56	48.06	1,048.03	1,096.08
81	0.52419	0.01608	613.35	613.37	49.06	1,047.46	1,096.51
82	0.54150	0.01608	594.82	594.84	50.05	1,046.89	1,096.95
83	0.55931	0.01608	576.90	576.92	51.05	1,046.33	1,097.38
84	0.57763	0.01608	559.63	559.65	52.05	1,045.76	1,097.81
85	0.59647	0.01609	542.93	542.94	53.05	1,045.19	1,098.24
86	0.61584	0.01609	526.80	526.81	54.05	1,044.63	1,098.67
87	0.63575	0.01609	511.21	511.22	55.05	1,044.06	1,099.11
88	0.65622	0.01609	496.14	496.15	56.05	1,043.49	1,099.54
89	0.67726	0.01610	481.60	481.61	57.04	1,042.92	1,099.97
90	0.69889	0.01610	467.52	467.53	58.04	1,042.36	1,100.40
91	0.72111	0.01610	453.91	453.93	59.04	1,041.79	1,100.83
92	0.74394	0.01611	440.76	440.78	60.04	1,041.22	1,101.26
93	0.76740	0.01611	428.04	428.06	61.04	1,040.65	1,101.69
94	0.79150	0.01611	415.74	415.76	62.04	1,040.08	1,102.12

WATER—TABLE 2. Thermodynamic Properties of Water (*Continued*)

Temp., °F	Press., psia	Specific volume, ft³/lb			Enthalpy, Btu/lb		
		v_1	v_{1g}	v_g	h_1	h_{1g}	h_g
95	0.81625	0.01612	403.84	403.86	63.03	1,039.51	1,102.55
96	0.84166	0.01612	392.33	392.34	64.03	1,038.95	1,102.98
97	0.86776	0.01612	381.20	381.21	65.03	1,038.38	1,103.41
98	0.89456	0.01612	370.42	370.44	66.03	1,037.81	1,103.84
99	0.92207	0.01613	359.99	360.01	67.03	1,037.24	1,104.26
100	0.95031	0.01613	349.91	349.92	68.03	1,036.67	1,104.69
101	0.97930	0.01613	340.14	340.15	69.03	1,036.10	1,105.12
102	1.00904	0.01614	330.69	330.71	70.02	1,035.53	1,105.55
103	1.03956	0.01614	321.53	321.55	71.02	1,034.95	1,105.98
104	1.07088	0.01614	312.67	312.69	72.02	1,034.38	1,106.40
105	1.10301	0.01615	304.08	304.10	73.02	1,033.81	1,106.83
106	1.13597	0.01615	295.76	295.77	74.02	1,033.24	1,107.26
107	1.16977	0.01616	287.71	287.73	75.01	1,032.67	1,107.68
108	1.20444	0.01616	279.91	279.92	76.01	1,032.10	1,108.11
109	1.23999	0.01616	272.34	272.36	77.01	1,031.52	1,108.54
110	1.27644	0.01617	265.02	265.03	78.01	1,030.95	1,108.96
111	1.31381	0.01617	257.91	257.93	79.01	1,030.38	1,109.39
112	1.35212	0.01617	251.02	251.04	80.01	1,029.80	1,109.81
113	1.39138	0.01618	244.36	244.38	81.01	1,029.23	1,110.24
114	1.43162	0.01618	237.89	237.90	82.00	1,028.66	1,110.66
115	1.47286	0.01619	231.62	231.63	83.00	1,028.08	1,111.09
116	1.51512	0.01619	225.53	225.55	84.00	1,027.51	1,111.51
117	1.55842	0.01619	219.63	219.65	85.00	1,026.93	1,111.93
118	1.60277	0.01620	213.91	213.93	86.00	1,026.36	1,112.36
119	1.64820	0.01620	208.36	208.37	87.00	1,025.78	1,112.78
120	1.69474	0.01620	202.98	202.99	88.00	1,025.20	1,113.20
121	1.74240	0.01621	197.76	197.76	89.00	1,023.62	1,113.62
122	1.79117	0.01621	192.69	192.69	90.00	1,024.05	1,114.05
123	1.84117	0.01622	187.78	187.78	90.99	1,024.47	1,114.47
124	1.89233	0.01622	182.98	182.99	91.99	1,022.90	1,114.89
125	1.94470	0.01623	178.34	178.36	92.99	1,022.32	1,115.31
126	1.99831	0.01623	173.85	173.86	93.99	1,021.74	1,115.73
127	2.05318	0.01623	169.47	169.49	94.99	1,021.16	1,116.15
128	2.10934	0.01624	165.23	165.25	95.99	1,020.58	1,116.57
129	2.16680	0.01624	161.11	161.12	96.99	1,020.00	1,116.99
130	2.22560	0.01625	157.11	157.12	97.99	1,019.42	1,117.41
131	2.28576	0.01625	153.22	153.23	98.99	1,018.84	1,117.83
132	2.34730	0.01626	149.44	149.46	99.99	1,018.26	1,118.25
133	2.41025	0.01626	145.77	145.78	100.99	1,017.68	1,118.67
134	2.47463	0.01627	142.21	142.23	101.99	1,017.10	1,119.08
135	2.54048	0.01627	138.74	138.76	102.99	1,016.52	1,119.50
136	2.60782	0.01627	135.37	135.39	103.98	1,015.93	1,119.92
137	2.67667	0.01628	132.10	132.12	104.98	1,015.35	1,120.34
138	2.74707	0.01628	128.92	128.94	105.98	1,014.77	1,120.75
139	2.81903	0.01629	125.83	125.85	106.98	1,014.18	1,121.17
140	2.89260	0.01629	122.82	122.84	107.98	1,013.60	1,121.58
141	2.96780	0.01630	119.90	119.92	108.98	1,013.01	1,122.00
142	3.04465	0.01630	117.05	117.07	109.98	1,012.43	1,122.41

WATER—TABLE 2. Thermodynamic Properties of Water (*Continued*)

Temp., °F	Press., psia	Specific volume, ft³/lb			Enthalpy, Btu/lb		
		v_1	v_{1g}	v_g	h_1	h_{1g}	h_g
143	3.12320	0.01631	114.29	114.31	110.98	1,011.84	1,122.83
144	3.20345	0.01631	111.60	111.62	111.98	1,011.26	1,123.24
145	3.28546	0.01632	108.99	109.00	112.98	1,010.67	1,123.66
146	3.36924	0.01632	106.44	106.45	113.98	1,010.09	1,124.07
147	3.45483	0.01633	103.96	103.98	114.98	1,009.50	1,124.48
148	3.54226	0.01633	101.55	101.57	115.98	1,008.91	1,124.89
149	3.63156	0.01634	99.21	99.22	116.98	1,008.32	1,125.31
150	3.72277	0.01634	96.93	96.94	117.98	1,007.73	1,125.72
151	3.81591	0.01635	94.70	94.72	118.99	1,007.14	1,126.13
152	3.91101	0.01635	92.54	92.56	119.99	1,006.55	1,126.54
153	4.00812	0.01636	90.44	90.46	120.99	1,005.96	1,126.95
154	4.10727	0.01636	88.39	88.41	121.99	1,005.37	1,127.36
155	4.20848	0.01637	86.40	86.41	122.99	1,004.78	1,127.77
156	4.31180	0.01637	84.45	84.47	123.99	1,004.19	1,128.18
157	4.41725	0.01638	82.56	82.58	124.99	1,003.60	1,128.59
158	4.52488	0.01638	80.72	80.73	125.99	1,003.00	1,128.99
159	4.63472	0.01639	78.92	78.94	126.99	1,002.41	1,129.40
160	4.7468	0.01639	77.175	77.192	127.99	1,001.82	1,129.81
161	4.8612	0.01640	75.471	75.488	128.99	1,001.22	1,130.22
162	4.9778	0.01640	73.812	73.829	130.00	1,000.63	1,130.62
163	5.0969	0.01641	72.196	72.213	131.00	1,000.03	1,131.03
164	5.2183	0.01642	70.619	70.636	132.00	999.43	1,131.43
165	5.3422	0.01642	69.084	69.101	133.00	998.84	1,131.84
166	5.4685	0.01643	67.587	67.604	134.00	998.24	1,132.24
167	5.5974	0.01643	66.130	66.146	135.00	997.64	1,132.64
168	5.7287	0.01644	64.707	64.723	136.01	997.04	1,133.05
169	5.8627	0.01644	63.320	63.336	137.01	996.44	1,133.45
170	5.9993	0.01645	61.969	61.989	138.01	995.84	1,133.85
171	6.1386	0.01646	60.649	60.666	139.01	995.24	1,134.25
172	6.2806	0.01646	59.363	59.380	140.01	994.64	1,134.66
173	6.4253	0.01647	58.112	58.128	141.02	994.04	1,135.06
174	6.5729	0.01647	56.887	56.904	142.02	993.44	1,135.46
175	6.7232	0.01648	55.694	55.711	143.02	992.83	1,135.86
176	6.8765	0.01648	54.532	54.549	144.03	992.23	1,136.26
177	7.0327	0.01649	53.397	53.414	145.03	991.63	1,136.65
178	7.1918	0.01650	52.290	52.307	146.03	991.02	1,137.05
179	7.3539	0.01650	51.210	51.226	147.03	990.42	1,137.45
180	7.5191	0.01651	50.155	50.171	148.04	989.81	1,137.85
181	7.6874	0.01651	49.126	49.143	149.04	989.20	1,138.24
182	7.8589	0.01652	48.122	48.138	150.04	988.60	1,138.64
183	8.0335	0.01653	47.142	47.158	151.05	987.99	1,139.03
184	8.2114	0.01653	46.185	46.202	152.05	987.38	1,139.43
185	8.3926	0.01654	45.251	45.267	153.05	986.77	1,139.82
186	8.5770	0.01654	44.339	44.356	154.06	986.16	1,140.22
187	8.7649	0.01655	43.448	43.465	155.06	985.55	1,140.61
188	8.9562	0.01656	42.579	42.595	156.07	984.94	1,141.00
189	9.1510	0.01656	41.730	41.746	157.07	984.32	1,141.39
190	9.3493	0.01657	40.901	40.918	158.07	983.71	1,141.78

WATER—TABLE 2. Thermodynamic Properties of Water (*Continued*)

Temp., °F	Press., psia	Specific volume, ft³/lb			Enthalpy, Btu/lb		
		v_1	v_{1g}	v_g	h_1	h_{1g}	h_g
191	9.5512	0.01658	40.092	40.108	159.08	983.10	1,142.18
192	9.7567	0.01658	39.301	39.317	160.08	982.48	1,142.57
193	9.9659	0.01659	38.528	38.544	161.09	981.87	1,142.95
194	10.1788	0.01659	37.774	37.790	162.09	981.25	1,143.34
195	10.3955	0.01660	37.035	37.052	163.10	980.63	1,143.73
196	10.6160	0.01661	36.314	36.331	164.10	980.02	1,144.12
197	10.8404	0.01661	35.611	35.628	165.11	979.40	1,144.51
198	11.0687	0.01662	34.923	34.940	166.11	978.78	1,144.89
199	11.3010	0.01663	34.251	34.268	167.12	978.16	1,145.28
200	11.5374	0.01663	33.594	33.610	168.13	977.54	1,145.66
201	11.7779	0.01664	32.951	32.968	169.13	976.92	1,146.05
202	12.0225	0.01665	32.324	32.340	170.14	976.29	1,146.43
203	12.2713	0.01665	31.710	31.726	171.14	975.67	1,146.81
204	12.5244	0.01666	31.110	31.127	172.15	975.05	1,147.20
205	12.7819	0.01667	30.523	30.540	173.16	974.42	1,147.58
206	13.0436	0.01667	29.949	29.965	174.16	973.80	1,147.96
207	13.3099	0.01668	29.388	29.404	175.17	973.17	1,148.34
208	13.5806	0.01669	28.839	28.856	176.18	972.54	1,148.72
209	13.8558	0.01669	28.303	28.319	177.18	971.92	1,149.10
210	14.1357	0.01670	27.778	27.795	178.19	971.29	1,149.48
212	14.7096	0.01671	26.763	26.780	180.20	970.03	1,150.23
214	15.3025	0.01673	25.790	25.807	182.22	968.76	1,150.98
216	15.9152	0.01674	24.861	24.878	184.24	967.50	1,151.73
218	16.5479	0.01676	23.970	23.987	186.25	966.23	1,152.48
220	17.2013	0.01677	23.118	23.134	188.27	964.95	1,153.22
222	17.8759	0.01679	22.299	22.316	190.29	963.67	1,153.96
224	18.5721	0.01680	21.516	21.533	192.31	962.39	1,154.70
226	19.2905	0.01682	20.765	20.782	194.33	961.11	1,155.43
228	20.0316	0.01683	20.045	20.062	196.35	959.82	1,156.16
230	20.7961	0.01684	19.355	19.372	198.37	958.52	1,156.89
232	21.5843	0.01686	18.692	18.709	200.39	957.22	1,157.62
234	22.3970	0.01688	18.056	18.073	202.41	955.92	1,158.34
236	23.2345	0.01689	17.466	17.463	204.44	954.62	1,159.06
238	24.0977	0.01691	16.860	16.877	206.46	953.31	1,159.77
240	24.9869	0.01692	16.298	16.314	208.49	952.00	1,160.48
242	25.9028	0.01694	15.757	15.774	210.51	950.68	1,161.19
244	26.8461	0.01695	15.238	15.255	212.54	948.35	1,161.90
246	27.8172	0.01697	14.739	14.756	214.57	948.03	1,162.60
248	28.8169	0.01698	14.259	14.276	216.60	946.70	1,163.29
250	29.8457	0.01700	13.798	13.815	218.63	945.36	1,163.99
252	30.9043	0.01702	13.355	13.372	220.66	944.02	1,164.68
254	31.9934	0.01703	12.928	12.945	222.69	942.68	1,165.37
256	33.1135	0.01705	12.526	12.147	224.73	939.99	1,166.72
258	34.2653	0.01707	12.123	12.140	226.76	939.97	1,166.73
260	35.4496	0.01708	11.742	11.759	228.79	938.61	1,167.40
262	36.6669	0.01710	11.376	11.393	230.83	937.25	1,168.08
264	37.9180	0.01712	11.024	11.041	232.87	935.88	1,168.74
266	39.2035	0.01714	10.684	10.701	234.90	934.50	1,169.41

WATER—TABLE 2. Thermodynamic Properties of Water (*Continued*)

Temp., °F	Press., psia	Specific volume, ft³/lb			Enthalpy, Btu/lb		
		v_1	v_{1g}	v_g	h_1	h_{1g}	h_g
268	40.5241	0.01715	10.357	10.374	236.94	933.12	1,170.07
270	41.8806	0.01717	10.042	10.059	238.98	931.74	1,170.72
272	43.2736	0.01719	9.737	9.755	241.03	930.35	1,171.38
274	44.7040	0.01721	9.445	9.462	243.07	928.95	1,172.02
276	46.1723	0.01722	9.162	9.179	245.11	927.55	1,172.67
278	47.6794	0.01724	8.890	8.907	247.16	926.15	1,173.31
280	49.2260	0.01726	8.627	8.644	249.20	924.74	1,173.94
282	50.8128	0.01728	8.373	8.390	251.25	923.32	1,174.57
284	52.4406	0.01730	8.128	8.146	253.30	921.90	1,175.20
286	54.1103	0.01731	7.892	7.910	255.35	920.47	1,175.82
288	55.8225	0.01733	7.664	7.681	257.40	919.03	1,176.44
290	57.5780	0.01735	7.444	7.461	259.45	917.59	1,177.05
292	59.3777	0.01737	7.231	7.248	261.51	916.15	1,177.66
294	61.2224	0.01739	7.026	7.043	263.56	914.69	1,178.26
296	63.1128	0.01741	6.827	6.844	265.62	913.24	1,178.86
298	65.0498	0.01743	6.635	6.652	267.68	911.77	1,179.45
300	67.03	0.01745	6.450	6.467	269.74	910.3	1,180.04
302	69.01	0.01747	6.275	6.292	271.79	909.0	1,180.79
304	71.09	0.01749	6.102	6.119	273.86	907.5	1,181.36
306	73.22	0.01751	5.933	5.951	275.93	906.0	1,181.93
308	75.40	0.01753	5.771	5.789	278.00	904.5	1,182.50
310	77.64	0.01755	5.614	5.632	280.06	903.0	1,183.06
312	79.92	0.01757	5.462	5.480	282.13	901.5	1,183.63
314	82.26	0.01759	5.315	5.333	284.21	899.9	1,184.11
316	84.65	0.01761	5.172	5.190	286.28	898.4	1,184.68
318	87.10	0.01763	5.034	5.052	288.36	896.9	1,185.26
320	89.60	0.01765	4.901	4.919	290.43	895.3	1,185.73
322	92.16	0.01767	4.772	4.790	292.51	893.8	1,186.31
324	94.78	0.01770	4.647	4.665	294.59	892.2	1,186.79
326	97.46	0.01772	4.525	4.543	296.67	890.7	1,187.37
328	100.20	0.01774	4.408	4.426	298.76	889.1	1,187.86
330	103.00	0.01776	4.294	4.312	300.84	887.5	1,188.34
332	105.86	0.01778	4.183	4.201	302.93	885.9	1,188.83
334	108.78	0.01780	4.076	4.094	305.02	884.3	1,189.32
336	111.76	0.01783	3.973	3.991	307.11	882.7	1,189.81
338	114.82	0.01785	3.872	3.890	309.21	881.1	1,190.31
340	117.93	0.01787	3.774	3.792	311.30	879.5	1,190.80
342	121.11	0.01789	3.680	3.698	313.39	877.9	1,191.29
344	124.36	0.01792	3.588	3.606	315.49	876.3	1,191.79
346	127.68	0.01794	3.499	3.517	317.59	874.6	1,192.19
348	131.07	0.01796	3.412	3.430	319.70	873.0	1,192.70
350	134.53	0.01799	3.328	3.346	321.80	871.3	1,193.10
352	138.06	0.01801	3.247	3.265	323.91	869.6	1,193.51
354	141.66	0.01804	3.167	3.185	326.02	868.0	1,194.02
356	145.34	0.01806	3.091	3.109	328.13	866.3	1,194.43
358	149.09	0.01808	3.286	3.304	330.24	864.6	1,194.84
360	152.92	0.01811	2.943	2.961	332.35	862.9	1,195.25
362	156.82	0.01813	2.873	2.891	334.47	861.2	1,195.67

WATER—TABLE 2. Thermodynamic Properties of Water (*Continued*)

Temp., °F	Press., psia	Specific volume, ft³/lb			Enthalpy, Btu/lb		
		v_l	v_{lg}	v_g	h_l	h_{lg}	h_g
364	160.80	0.01816	2.804	2.822	336.59	859.5	1,196.09
366	164.87	0.01818	2.738	2.756	338.71	857.7	1,196.41
368	169.01	0.01821	2.673	2.691	340.83	856.0	1,196.83
370	173.23	0.01823	5.283	2.628	342.96	854.2	1,197.16
372	177.53	0.01826	2.549	2.567	345.08	852.5	1,197.58
374	181.92	0.01828	2.325	2.508	347.21	850.7	1,197.91
376	186.39	0.01831	2.432	2.450	349.35	848.9	1,198.25
378	190.95	0.01834	2.376	2.394	351.48	847.2	1,198.68
380	195.60	0.01836	2.321	2.339	353.62	845.4	1,199.02
382	200.33	0.01839	2.268	2.286	355.76	843.6	1,199.36
384	205.15	0.01842	2.216	2.234	357.90	841.7	1,199.60
386	210.06	0.01844	2.165	2.183	360.04	839.9	1,199.94
388	215.06	0.01847	2.116	2.134	362.19	838.1	1,200.29
390	220.2	0.01850	2.069	2.087	364.34	836.2	1,200.54
392	225.3	0.01853	2.021	2.040	366.49	834.4	1,200.89
394	230.6	0.01855	1.976	1.995	368.64	832.5	1,201.14
396	236.0	0.01858	1.932	1.951	370.80	830.6	1,204.40
398	241.5	0.01861	1.889	1.908	372.96	828.7	1,201.66
400	247.1	0.01864	1.847	1.866	375.12	826.8	1,201.92
405	261.4	0.01871	1.747	1.766	308.53	822.0	1,202.53
410	276.5	0.01878	1.654	1.673	385.97	817.2	1,203.17
415	292.1	0.01886	1.566	1.585	391.42	812.2	1,203.62
420	308.5	0.01894	1.483	1.502	396.89	807.2	1,204.09
425	325.6	0.01901	1.406	1.425	402.38	802.1	1,204.48
430	343.3	0.01909	1.333	1.352	407.89	796.9	1,204.79
435	361.9	0.01918	1.265	1.284	413.42	791.7	1,205.12
440	381.2	0.01926	1.200	1.219	418.98	786.3	1,205.28
445	401.2	0.01935	1.139	1.158	424.55	780.9	1,205.45
450	422.1	0.01943	1.082	1.101	430.20	775.4	1,205.60
455	443.8	0.01952	1.027	1.047	435.80	769.8	1,205.60
460	466.3	0.01961	0.976	0.996	441.40	764.1	1,205.50
465	489.8	0.01971	0.928	0.948	447.10	758.3	1,205.40
470	514.1	0.01980	0.883	0.903	452.80	752.4	1,205.20
475	539.3	0.01990	0.840	0.8594	458.5	746.4	1,204.9
480	565.5	0.02000	0.799	0.8187	464.3	740.3	1,204.6
485	592.6	0.02011	0.760	0.7801	470.1	734.1	1,204.2
490	620.7	0.02021	0.723	0.7436	475.9	727.8	1,203.7
495	649.8	0.02032	0.689	0.7090	481.8	721.3	1,203.1
500	680.0	0.02043	0.656	0.6761	487.7	714.8	1,202.5
525	847.1	0.02104	0.514	0.5350	517.8	680.0	1,197.8
550	1,044.0	0.02175	0.406	0.4249	549.1	641.6	1,190.6
575	1,274.0	0.02259	0.315	0.3378	581.9	598.6	1,180.4
600	1,541.0	0.02363	0.244	0.2677	616.7	549.7	1,166.4

INDEX